THE MODERN NUMBER-SYSTEM

GEORGES IFRAH, now aged fifty, was the despair of his maths teachers at school – he lingered near the bottom of the class. Nevertheless he grew up to become a maths teacher himself and, in order to answer a pupil's question as to where numbers came from, he devoted some ten years to travelling the world in search of the answers, earning his keep as a night clerk, waiter, taxi-driver. Today he is a maths encyclopaedia on two legs, and his book has been translated into fourteen languages.

DAVID BELLOS is Professor of French at Princeton University and author of *Georges Perec: A Life in Words* and *Jacques Tati*. E. F. HARDING has taught at Aberdeen, Edinburgh and Cambridge and is a Director of the Statistical Advisory Unit at Manchester Institute of Science and Technology. SOPHIE WOOD is a specialist in technical translation from French and Spanish. IAN MONK, while skilled in technical translation, is better known for his translations of Georges Perec and Daniel Pennac.

THE UNIVERSAL HISTORY OF
NUMBERS

THE MODERN
NUMBER-SYSTEM

GEORGES IFRAH

Translated from the French by
David Bellos, E. F. Harding, Sophie Wood, and Ian Monk

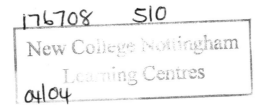

THE HARVILL PRESS
LONDON

First published in France with the title *Histoire universelle des chiffres*
by Editions Robert Laffont, Paris, in 1994

Part Two of the original edition
first published in Great Britain in 1998
and reissued as a two-volume set in 2000
by The Harvill Press,
2 Aztec Row, Berners Road
London N1 0PW

www.harvill.com

This translation has been published with the financial support
of the European Commission and of the
French Ministry of Culture and Communications

1 3 5 7 9 8 6 4 2

Copyright © 1981, 1994 by Editions Robert Laffont S.A., Paris
Translation copyright © 1998 by The Harvill Press Ltd

Georges Ifrah asserts the moral right to be
identified as the author of this work

A CIP catalogue record for this book
is available from the British Library

ISBN 1 86046 791 1

Designed and typeset in Stone Print at
Libanus Press, Marlborough, Wiltshire

Printed and bound in Italy

CONDITIONS OF SALE

SUMMARY TABLE OF CONTENTS

CHAPTER 24   PART I   Indian Civilisation: the Cradle of
                              Modern Numerals   699

CHAPTER 24   PART II   Dictionary of the Numeral Symbols
                               of Indian Civilisation   867

CHAPTER 25   Indian Numerals and Calculation in the Islamic World   1008

CHAPTER 26   The Slow Progress of Indo-Arabic Numerals in
                         Western Europe   1139

CHAPTER 27   Beyond Perfection   1168

List of Abbreviations   1183

Bibliography   1201

Index of Names and Subjects   1231

THE UNIVERSAL HISTORY OF NUMBERS

CHAPTER 24

INDIAN CIVILISATION

THE CRADLE OF MODERN NUMERALS

As G. Beaujouan (1950) has said, "the origin of the so-called 'Arabic' numerals has been written about so often that every view on the question seems plausible, and the only way of choosing between them is by personal conviction." Most of the literature (much of which is indeed of great value and has been used in the following pages) deals with one particular discipline from the many that are relevant to this tricky question of the origin of Arabic numerals. The few comprehensive works on the subject (Cajori, Datta and Singh, Guitel, Menninger, Pihan, Smith and Karpinski, or Woepcke) are now several decades old, and many discoveries have been made in more recent years. Since the beginning of the twentieth century, a wealth of reliable information has been compiled from the various specialised fields, and the findings all point to the fact that the number-system that we use today is of Indian origin. But no collective work has been produced that contained rigorous reasoning or an entirely satisfactory methodology. Moreover, the problem has been tackled in a somewhat loose manner in the past and was seen from a more limited and biased perspective than it is today. So it is well worth while going back to square one and looking at the question from a completely new angle, not only in the light of the results seen in the previous chapter and those of certain recent developments, but also, and most importantly, using a multidisciplinary process which takes into account the events of Indian civilisation.[*]

First, however, it is necessary (in order to eliminate them once and for all) to remember some of the main and rather unlikely theories which are still in circulation today on this subject.

FANCIFUL EXPLANATIONS FOR THE ORIGIN OF "ARABIC" NUMERALS

According to a popular tradition that still persists in Egypt and northern Africa, the "Arabic" numerals were the invention of a glassmaker-geometer from the Maghreb who came up with the idea of giving each of the nine

---

[*] Due to the complex nature of this civilisation, a "Dictionary of the numerical symbols of Indian Civilisation" has been compiled (see the end of the present chapter), which acts as both a thematic index and a glossary of the many notions which it is necessary to understand in order to grasp the ideas introduced in the following pages. In the present chapter, each word (whether in Sanskrit or in English) which is also found in the dictionary, is accompanied by an asterisk. (Examples: *anka, *ankakramena, *Ashvin, *Indian Astronomy, *Infinity, *Numeral, *sthâna, *Symbols, * yuga, *zero, etc.)

numerals a shape, the number of angles each one possessed being equal to
the number it denoted: *one angle* represented the number 1, *two angles* the
number 2, *three angles* the number 3, and so on (Fig. 24.1A).

1	2	3	4	5	6	7	8	9

FIG. 24.1A. *The first unlikely hypothesis on the origin of our numerals: the number of angles each
numeral possesses*

At the end of the nineteenth century, P. Voizot, a Frenchman, put forward
the same theory, apparently influenced by a Genoese author. But he also
thought that it was "equally probable" that the numerals were formed by
certain numbers of lines (Fig. 24.1B).

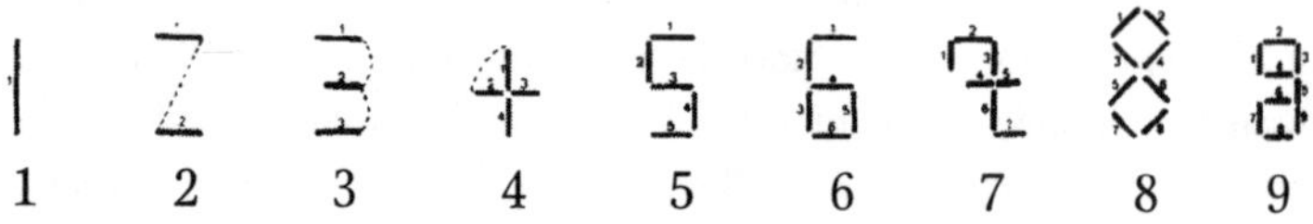

1	2	3	4	5	6	7	8	9

FIG. 24.1B. *Second unlikely hypothesis: the number of lines each numeral possesses*

Another similar hypothesis was put forward in 1642 by the Italian Jesuit
Mario Bettini, which was taken up in 1651 by the German Georg Philip
Harsdörffer. This time the idea was that the ideographical representation of
the nine units would have been based on a number of points which were
joined up to form the nine signs (Fig. 24.1C). In 1890, the Frenchman
Georges Dumesnil also adopted this theory, believing the system to be of
Greek invention: attributing the form of our present-day numerals to the
Pythagoreans, his argument stated that the joining of the points to form
the geometrical representations of the whole numbers played an important
role for the members of this group.

1	2	3	4	5	6	7	8	9

FIG. 24.1C. *Third unlikely hypothesis: the number of points*

A corresponding theory was put forward by Wiedler in 1737 which he
attributed to the tenth-century astrologer Abenragel: according to him, the
invention of numerals was the result of the division into parts of the shape
which is formed by a circle and two of its diameters. In other words,
according to Wiedler, all the figures could be made from this one geometrical shape "as if they were inside a shell": thus the vertical diameter would

have formed 1; the same diameter plus two arcs at either end formed 2; a semi-circle plus a median horizontal radius made 3, and so on until zero, which was said to be formed by the complete circle (Fig. 24.1D).

FIG. 24.1D. *Fourth unlikely hypothesis: the shapes formed by a circle and its diameters*

It is also worthwhile mentioning the theories of the Spaniard Carlos Le Maur (1778), who believed that the signs in question acquired their shape from a particular arrangement of counting stones (Fig. 24.1C) or from the number of angles that can be obtained from certain shapes formed by a rectangle, its diagonals, its medians, etc. (Fig. 24.1E).

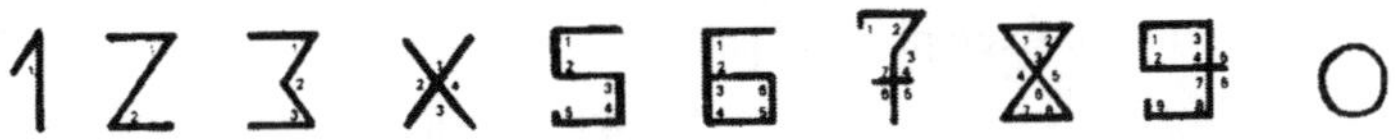

FIG. 24.1E. *Fifth unlikely hypothesis: a variation of Fig. 24.1A*

Finally, Jacob Leupold, in 1727, offered an "explanation" which goes by the name of the legend of Solomon's ring. According to this theory, the numerals were formed successively by the ring inscribing a square and its diagonals (Fig. 24.1F).

FIG. 24.1F. *Sixth unlikely hypothesis: the numerals come from a square (legend of Solomon's ring)*

If we were to believe any of these theories, it would mean that the appearance of the numerals that we use today would have to have been the fruit of one isolated individual's imagination. An individual who would have given each number a specific shape through a system based either on the use of different numbers of lines, angles or dots to add up to the amount of units the sign represents, or through the use of geometrical representations such as a triangle, rectangle, square or circle, which would mean that the signs were created according to a simple process of geometrical ordering.

These theories, then, all have one thing in common: their "explanation" for the appearance of our numerals is that these figures were the result of some kind of spontaneous generation; their shape, right from the outset, being perfectly logical. In fact, as F. Cajori (1928) explains, "the validity of any hypothesis depends upon the way in which the established facts are presented and the extent to which it opens the door to new research." In other words, a hypothesis can only acquire "scientific" value if it has the potential to broaden our knowledge of a given subject.

The hypotheses that have been mentioned so far in this chapter are basically sterile. None of them offers any explanation for the fact that the nine figures have appeared in an immense variety of shapes and forms over the centuries and in different parts of the world. Their approach is merely to consider the final product, in other words the numerals that we use today (as they appear in print), which fails to take account of the fact that these figures appear at the end of a very long story and have slowly evolved over several millennia.

These a posteriori hypotheses are flawed because they are the fruit of the pseudo-scientific imaginations of men who are fooled by appearances and who jump to conclusions which completely contradict both historical facts and the results of epigraphic and palaeographic research.*

It is still widely believed that the number-system that we use today was invented by the Arabs.

However, it definitely was not the Arabs who invented what we know as "Arabic" numerals. Historians have known for some time now that the name was coined as the result of a serious historical error. Significantly, and curiously, no trace of this belief is to be found in actual Arabic documents.

In fact, many Arabic works that concern mathematics and arithmetic reveal that Muslim Arabic authors, without the slightest hint of prejudice or complex, have always acknowledged that they were not the ones that made the discovery. But whilst it is incorrect, the name which was given to our numerals is not totally unfounded. There is always some basis for an historical error, no matter how widespread or long-standing it is, and this one is no exception, especially considering the fact that we are dealing with a broad geographical area and a duration of many centuries.

The belief that our numerals were invented by the Arabs is only found in Europe and probably originated in the late Middle Ages. This theory was only really voiced by mathematicians or arithmeticians who, in order to

---

* Moreover, this book demonstrates quite clearly that despite the significance and vast number of inventions that have punctuated the history of numerals as a whole, the findings have always been anonymous. Men would work for and in groups and gain no qualifications for their work. Certain documents made of stone, papyrus, paper and fabric immortalising the names of men who are sometimes associated with numbers mean nothing to us. Names of those who made use of and who reported numbers and counting systems are also known. But those of the inventors themselves are irretrievably lost, perhaps because their discoveries were made so long ago, or even because these brilliant inventions belonged to relatively humble men whose names were not deemed worthy of recording. It is also possible that the discoveries could be a result of the work of a team of men and so they could not really be attributed to a specific person. The "inventor" of zero, a meticulous scribe and arithmetician, whose main concern was to define a specific point in a series of numbers ruled by the place-value system, was probably never aware of the revolution that he had made possible. All of this proves the absurdity of the preceding hypotheses.

distinguish themselves from the masses, wanted to fill what they perceived as a void with random hypotheses based on preconceived ideas, and thus sacrificing historical truth to satisfy the whims of their own individual inspiration. To the uninitiated, the writings of these mathematicians would have seemed to constitute the linchpin of a doctrine that was sure to survive for many centuries. This is due to the fact that numerals and calculation have always been considered (rightly or wrongly) to be the very essence of mathematical science. The cause of the error is more easily understood now that it is known that the numerals in question arrived in the West at the end of the tenth century via the Arabs. At that time, the Arabs were relatively superior to Western civilisations in terms of both culture and science. Therefore the figures were given the name "Arabic".

This theory, however, was just one of the many explanations offered.

As the following evidence shows, European Renaissance authors offered many similar and equally unreliable theories, attributing the invention of our numerals to the Egyptians, the Phoenicians, the Chaldaeans and the Hebrews alike, all of whom are totally unconnected to this discovery.

It is interesting to note that even in the twentieth century certain authors, whilst being known for the quality of their work, have fallen into the trap of supporting unsatisfactory explanations and taking things solely at face value. At the turn of the century, historical scientists (G. R. Kaye, N. Bubnov, and B. Carra de Vaux, etc., who strongly opposed the idea that our number-system could be of Indian origin) alleged that our numerals were developed in Ancient Greece [see JPAS 8 (1907), pp. 475 ff.; N. Bubnov (1908); SC 21 (1917), pp. 273 ff.].

These men believed that the system originated in Neo-Pythagorean circles shortly before the birth of Christianity. They claimed that the system came to Rome from the port of Alexandria and soon after made its way to India via the trade route; it also travelled from Rome to Spain and the North African provinces, where it was discovered some centuries later by the Muslim Arabic conquerors. As for Middle Eastern Arabs, they picked up the system from Indian merchants. According to this view of things, European and North African numerals were formed by the "Western" transmission, and the radically different Indian and Eastern Arabic figures emerged by the "Eastern" route.

This tempting explanation is in fact an amalgam of the speculations of the early humanists, as we can see from the list of quotations that follows:

1. Köbel, *Rechenbiechlin*, first published in 1514: *Vom welchen Arabischen auch disz Kunst entsprungen ist*: "This art was also invented by the Arabs". [Köbel (1531), f° 13]

2. N. Tartaglia, *General trattato di numeri et misuri* ("General treatise of numbers and measures") first published in 1556: . . . *& que esto fu trouato di fare da gli Arabi con diece figure*: " . . . and this is what the Arabs did with ten figures [ten numerals]." [Tartaglia (1592), f° 9]

3. Robert Recorde, *The Grounde of Artes*: *In that thinge all men do agree, that the Chaldays, whiche fyrste inuented thys arte, did set these figures as thei set all their letters. For they wryte backwarde as you tearme it, and so doo they reade. And that may appeare in all Hebrew, Chaldaye and Arabike bookes . . . where as the Greekes, Latines, and all nations of Europe, do wryte and reade from the lefte hand towarde the ryghte.* [Recorde (1558), f° C, 5]

4. Peletarius, *Commentaire sur l'Arithmétique de Gemma Frisius* ("Commentary on Arithmetic by Gemma Frisius") first published in 1563: *La valeur des Figures commence au coste dextre vers le coste senestre: au rebours de notre maniere d'escrire par ce que la premiere prattique est venue des Chaldees: ou des Phéniciens, qui ont été les premiers traffiquers de marchandise*: "The figures read in ascending order from right to left which is the opposite of our way of writing. This is because the former practice comes from the Chaldaeans: or the Phoenicians, who were the first to trade their merchandise." [Peletarius, f° 77]

5. Ramus, *Arithmetic*, published in 1569: *Alii referunt ad Phoenices inventores arithmeticae, propter eandem commerciorum caussam: Alii ad Indos: Ioannes de Sacrobosco, cujus sepulchrum est Lutetiae . . . , refert ad Arabes:* "Others attribute the invention of arithmetic to the Phoenicians, for the same commercial reasons; others credit the Indians. Jean de Sacrobosco, whose tomb is in Paris . . . attributes the discovery to the Arabs." [Ramus (1569), p. 112]

6. Conrad Dasypodius, *Institutionum Mathematicarum*, published in 1593-1596: *Qui est harum Cyphrarum auctor? A quibus hae usitatae syphrarum notae sint inventae: hactenus incertum fuit: meo tamen iudicio, quod exiguum esse fateor: a Graecis librarijs (quorum olim magna fuit copia) literae Graecorum quibus veteres Graeci tamquam numerorum notis usunt usu fuerunt corruptae, vt ex his licet videre. Graecorum Literae corruptae.*

α	β	Γ	δ	ε	5	Ⴭ	N	ঽ
I	Ի	Ʋ	ᶴ	ꝺ	ᶌ	<	v	9
1	2	3	4	5	6	7	8	9

Sed qua ratione graecorum literae ita fuerunt corruptae? Finxerunt has corruptas Graecorum literarum notas: vel abiectione vt in nota binarij numeri, vel inuersione vt in septenarij numeri nota nostrae notae, quibus hodie utimur, ab his sola differunt elegantia, vt apparet: "Who invented these signs that are used as numerals? Until now no one has really known;

however, as far as I know (and I admit I know little), the letters that Ancient Greeks used to denote numbers were distorted and transformed through their use by Greek scribes (of which there were many), as one can see below. This is how the distorted letters look:

$$1\ 2\ 3\ 4\ 5\ 6\ 7\ 8\ 9$$

(as shown above).

"But how were these letters corrupted? The sign for number two has been reversed, the sign for the number seven has been inverted. The only difference between the signs that we use today and the Greek signs is that our signs are more elegant in appearance." [Dasypodius, quoted in Bayer]

7. Erpenius, *Grammatica Arabica*, published in 1613: "Arabic" numerals are "actually the figures used by Toledo's men of law", which he believes would have been transmitted to them by the Pythagoreans of Ancient Greece. But Golius, who published the book after the death of the author, realised that Erpenius had been mistaken, and suppressed that particular passage in the 1636 edition. [Erpenius]

8. Laurembergus, *The Mathematical Institution*, first published in 1636: *Supersunt volgares illi characteres Barbari, quibus hodie utitur universus fere orbis. Suntque universum novem: 1, 2, 3, 4, 5, 6, 7, 8, 9, queis additur o cyphra: seu figura nihili, Nulla, Zero Arabibus. Nonnullorum sententia est, primos harum figurarum inventores fuisse Arabes (alii Phoenices malunt; alii Indos) quae sane opinio non est a veritate aliena. Nam sicut Arabes olim totius fere orbis potiti sunt, ita credibile est, scientiarum quoque fuisse propagatores. Quicunque sit Inventor maxima sane illi debetur gratia:* "These ordinary, barbaric characters have survived the ages and are used throughout most of the modern world. There are nine altogether: 1 2 3 4 5 6 7 8 9, to which the figure 0 can be added, which denotes "nothing", the Arabic zero. Some think that it was the Arabs who originally invented these signs (whilst others believe it was the Phoenicians or even the Indians), and this is highly probable; the Arabs once dominated most of the world and it is likely that they invented the sciences. Whoever is to thank for the existence of our numerals deserves the highest recognition." [Laurembergus, p. 20, 1. 14; p. 21, 1. 2]

9. I. Vossius, *De Universae matheseos Natura et constitutione* (c.1604): "Arabic" numerals passed from "the Hindus or Persians to the Arabs, then to the Moors in Spain, then finally to the Spanish and the rest of Europe". His theory that the series was originally passed from the Greeks to the Hindus is without foundation. [Vossius (1660), pp. 39–40]

10. Nottnagelus, *The Mathematical Institution*, first published in 1645: *Computatores autem ob majorem supputandi commoditatem peculiares sibi finxerunt notas (quarum quidem inventionem nonnulli Phoenicibus adscribunt, quidam, ut Valla et Cardanus, Indis assignant, plerique vero Arabibus et Saracenis acceptam referunt) quas tamen alii ab antiqua vel potius corrupta Graecarum literarum forma, nonnulli vero aliunde derivatas autumant. Atque his posterioribus hodierni quoque utuntur Arithmetici*: "To facilitate calculation, arithmeticians invented their own unique signs (some believe it was the Phoenicians who invented them, others, such as Valla and Cardanus, believe it was the Indians; most people attribute the invention to the Arabs or Saracens); however, others claim that the numerals originated from the ancient, or rather, distorted shape of Greek letters; some even suggest another origin. The signs are still used by arithmeticians today." [Nottnagelus, p. 185]

11. Theophanes, *Chronicle*, first published in 1655: *Hinc numerorum notas et characteres, cifras vulgo dictos, Arabicum inventum aut Arabicos nulla ratione vocandos, qui haec legerit, mecum contendet . . .*: "The reader can appreciate that I can find no reason why the signs and characters that express numbers – which we vulgarly refer to as figures – are an Arabic invention . . ."

The following is an extract from a note written by Father Goar, which comments on the above passage: *Notas itaque characteresque, quibus numeros summatim exaramus, 1, 2, 3, 4, 5, 6, 7, 8, 9, ab Indis et Chaldaeis usque ad nos venisse scite magis advocat Glareanus in Arithmaticae praeludiis.* "In his Preludes to Arithmetic, Glareanus claims that the signs and characters which we use to write the numbers in an abbreviated form (1, 2, 3, etc.) actually came from the Indians and the Chaldaeans." [*Theophanis Chronographia*, p. 616, 2nd col., and p. 314]

12. P. D. Huet, Bishop of Avranches in his *Demonstratio Evangelica ad serenissimum Delphinum* claims that mediaeval European numerals were invented by the Pythagoreans. [Huet (1690)]

13. Dom Calmet (1707) upholds the theory of the Greek origin of our numerals [Calmet], as does J. F. Weidler, in *Spicilegium observationum ad historiam notarum numeralium pertinentium*. [Weidler (1755)]

14. C. Levias (1905), a contributor to the *Jewish Encyclopaedia*, states that our numerals were invented by the people of Israel and were introduced in Islamic countries around 800 CE by the Jewish scholar Mashallah. [Levias (1905), IX, p. 348]

15. Levi della Vida upholds the theory of the Greek origin of our numerals. [Levi della Vida (1933)]

16. M. Destombes says that European numerals are derived from the following letters of the Graeco-Byzantine alphabet I, Θ, H, Z, ..., Γ, B, by reversing the series of letters: B, Γ, ... Z, H, Θ, I, written in capitals and graphically adapted to the "shapes of the Visigothic letters from the third quarter of the tenth century CE". [Destombes (1962)]

The basis for all these hypotheses is, of course, invalid, because no evidence has ever been found to support the theory that the Greeks used a similar system to our own. However, rather than admit defeat in the face of solid counter-arguments firmly based in reality, the authors of these hypotheses persevered stubbornly, using all their imagination to come up with something resembling proof or confirmation of their unlikely theories.

As A. Bouché-Leclerq (1879) remarks, "it is almost tempting to admire the cunning way in which an unshakeable belief can transform into proof the very objections which threaten to destroy it, and nothing better demonstrates the psychological history of humanity than the irresistible prestige of the preconceived idea." Bouché-Leclerq is actually denouncing certain charlatans of Ancient Greece who mastered the art of exploiting trusting souls through the use of divinatory practices that were based on the interpretation of numerological dreams using the numeral letters of the Greek alphabet: "Perhaps the most embarrassing case", he explains, "was one which involved a dream which promised an elderly man a number of years that was too high to be added on to his current age and too low to represent his life-span as a whole. The charlatan, however, found a way to overcome such a dilemma. If a man of seventy heard someone say, 'You will live for fifty years,' he would live for another thirteen years. He has already lived for over fifty years and it is impossible that he will live for another fifty years, being seventy already. So the man will live for another thirteen years (according to the charlatans) because the letter *Nu* (N), whilst representing the number fifty, comes thirteenth in the Greek alphabet!"

It is likely that the same author would have also condemned the methods of historical scientists, who have been known to be somewhat economical with the truth. No doubt he would have said something similar about them if he had heard one particular historian's rather flimsy "explanation" which was soon adopted by all of his peers. When a shrewd man asked him why the Greeks had left no written trace of zero or of decimal place-value numeration, the historian in question, not to be deterred, replied: "That is because of the level of importance that they placed in oral tradition and also the great secrecy with which the Neo-Pythagoreans surrounded their knowledge"! If everyone reasoned in this way, history would amount to little more than a fairy tale.

Bearing in mind the fact that these authors were ardent admirers of Hellenistic civilisation, it is easy to understand why their theory was supported solely by claims that were unaccompanied by any shred of evidence, their main aim being to glorify the famous "Greek miracle".

The admiration that these authors display for Greek civilisation is, of course, perfectly justified. The Greeks were responsible for innovations in such varied fields as art, literature, philosophy, medicine, mathematics, astronomy, the sciences and engineering; their enormous contribution to our sciences and culture is undeniable. The paradox lies in the fact that the very men who wished to add to these achievements that are already acknowledged by the rest of the world, were unaware of the real story surrounding the scholars and mathematicians upon whom they wanted to bestow this undeserved honour. This clearly demonstrates narrow-mindedness on their part, attributing the development of our place-value notation solely to the origin of the graphical representation of the nine numerals in question.

J. F. Montucla (1798) quite rightly points out that "if the characters originate from Greek letters, they have drastically changed somewhere along the way. In fact, these letters could only resemble our numerals if they were shortened and turned about in a very odd fashion.* Moreover, the appearance of these characters is much less important than the ingenious way in which they are used; using only ten characters, it is possible to express absolutely any number. The Greeks were a highly intelligent race, and if this had been their invention, or even if they had simply got wind of it, they certainly would have made use of it."

Ancient Greece only had two systems of numerical notation: the first was the mathematical equivalent of the Roman system and the other was alphabetical, like the one used by the Hebrews. With a few exceptions towards the end of the era, neither of these systems were based on the rule of position, nor did they possess zero. Therefore the systems were not really of much practical use when it came to mathematical calculations, which were generally carried out using abacuses, upon which there were different columns for each decimal order.

Considering that the Greeks had invented such an instrument, the next logical step would have been their discovery of the place-value system and zero, through eliminating the columns of the instrument.

---

* Using such methods, it is always possible to find a way of promoting a theory: it is easy to manipulate the nine characters in order to "prove" that our nine numerals originate from them. This is precisely how certain extravagant comtemporary authors, ignoring not only the history of mathematical notation and writing, but also and above all the laws of palaeography, have come to "demonstrate" that these numerals derive from the first nine Hebrew letters, or even from the graphical representations for the twelve signs of the Zodiac. This goes to show that you can put the words of a song to any tune you like; in other words, appearances can be deceptive.

This would have provided them with the fully operational counting system that we use today.

However, the Greeks did not bother themselves with such practical concerns.

INDIA: THE TRUE BIRTHPLACE OF OUR NUMERALS

The real inventors of this fundamental discovery, which is no less important than such feats as the mastery of fire, the development of agriculture, or the invention of the wheel, writing or the steam engine, were the mathematicians and astronomers of Indian civilisation: scholars who, unlike the Greeks, were concerned with practical applications and who were motivated by a kind of passion for both numbers and numerical calculations.

There is a great deal of evidence to support this fact, and even the Arabo-Muslim scholars themselves have often voiced their agreement.

EVIDENCE FROM EUROPE WHICH SUPPORTS THE CLAIM THAT MODERN NUMERATION ORIGINATED IN INDIA

The following is a succession of historical accounts in favour of this theory, given in chronological order, beginning with the most recent.

1. P. S. Laplace (1814): "The ingenious method of expressing every possible number using a set of ten symbols (each symbol having a place value and an absolute value) emerged in India. The idea seems so simple nowadays that its significance and profound importance is no longer appreciated. Its simplicity lies in the way it facilitated calculation and placed arithmetic foremost amongst useful inventions. The importance of this invention is more readily appreciated when one considers that it was beyond the two greatest men of Antiquity, Archimedes and Apollonius." [Dantzig, p. 26]

2. J. F. Montucla (1798): "The ingenious number-system, which serves as the basis for modern arithmetic, was used by the Arabs long before it reached Europe. It would be a mistake, however, to believe that this invention is Arabic. There is a great deal of evidence, much of it provided by the Arabs themselves, that this arithmetic originated in India." [Montucla, I, p. 375]

3. John Wallis (1616–1703) referred to the nine numerals as *Indian figures* [Wallis (1695), p. 10]

4. Cataneo (1546) *le noue figure de gli Indi*, "the nine figures from

India". [Smith and Karpinski (1911), p. 3]

5. Willichius (1540) talks of *Zyphrae Indicae*, "Indian figures".
[Smith and Karpinski (1911) p. 3]

6. *The Crafte of Nombrynge* (c. 1350), the oldest known English arithmetical tract: II * fforthermore ye most vndirstonde that in this craft ben vsed teen figurys, as here bene writen for esampul 0 9 8 ∧ 6 5 4 3 2 1 . . . in the quych we vse teen figurys of Inde. Questio.* II *why ten figurys of Inde? Solucio. For as I have sayd afore thei were fonde fyrst in Inde.* [D. E. Smith (1909)]

7. Petrus of Dacia (1291) wrote a commentary on a work entitled *Algorismus* by Sacrobosco (John of Halifax, c. 1240), in which he says the following (which contains a mathematical error): *Non enim omnis numerus per quascumque figuras Indorum repraesentatur . . .*: "Not every number can be represented in Indian figures". [Curtze (1897), p. 25]

8. Around the year 1252, Byzantine monk Maximus Planudes (1260–1310) composed a work entitled *Logistike Indike* ("Indian Arithmetic") in Greek, or even *Psephophoria kata Indos* ("The Indian way of counting"), where he explains the following: "There are only nine figures. These are:

$$1\ 2\ 3\ 4\ 5\ 6\ 7\ 8\ 9$$

[figures given in their Eastern Arabic form].

"A sign known as *tziphra* can be added to these, which, according to the Indians, means 'nothing'. The nine figures themselves are Indian, and *tziphra* is written thus: 0". [B. N., Paris. *Ancien Fonds grec*, Ms 2428, fº 186 rº]

9. Around 1240, Alexandre de Ville-Dieu composed a manual in verse on written calculation (algorism). Its title was *Carmen de Algorismo*, and it began with the following two lines: *Haec algorismus ars praesens dicitur, in qua Talibus Indorum fruimur bis quinque figuris*: "*Algorism* is the art by which at present we use those Indian figures, which number two times five". [Smith and Karpinski (1911), p. 11]

10. In 1202, Leonard of Pisa (known as Fibonacci), after voyages that took him to the Near East and Northern Africa, and in particular to Bejaia (now in Algeria), wrote a tract on arithmetic entitled *Liber Abaci* ("a tract about the abacus"), in which he explains the following: *Cum genitor meus a patria publicus scriba in duana bugee pro pisanis mercatoribus ad eam confluentibus preesset, me in pueritia mea ad se uenire faciens, inspecta utilitate et commoditate futura, ibi me studio abaci per aliquot dies stare uoluit et doceri. Vbi ex mirabili magisterio in arte per nouem figuras Indorum introductus . . . Novem figurae Indorum hae sunt:*

$$9\ 8\ 7\ 6\ 5\ 4\ 3\ 2\ 1$$

cum his itaque novem figuris, et cum hoc signo o. Quod arabice zephirum appellatur, scribitur qui libet numerus: "My father was a public scribe of Bejaia, where he worked for his country in Customs, defending the interests of Pisan merchants who made their fortune there. He made me learn how to use the abacus when I was still a child because he saw how I would benefit from this in later life. In this way I learned the art of counting using the nine Indian figures . . .

The nine Indian figures are as follows:

9 8 7 6 5 4 3 2 1

[figures given in contemporary European cursive form].

"That is why, with these nine numerals, and with this sign 0, called *zephirum* in Arab, one writes all the numbers one wishes."

[Boncompagni (1857), vol.I]

11. C. 1150, Rabbi Abraham Ben Meïr Ben Ezra (1092–1167), after a long voyage to the East and a period spent in Italy, wrote a work in Hebrew entitled: *Sefer ha mispar* ("Number Book"), where he explains the basic rules of written calculation.

He uses the first nine letters of the Hebrew alphabet to represent the nine units. He represents zero by a little circle and gives it the Hebrew name of *galgal* ("wheel"), or, more frequently, *sifra* ("void") from the corresponding Arabic word.

However, all he did was adapt the Indian system to the first nine Hebrew letters (which he naturally had used since his childhood).

In the introduction, he provides some graphic variations of the figures, making it clear that they are of Indian origin, after having explained the place-value system: "That is how the learned men of India were able to represent any number using nine shapes which they fashioned themselves specifically to symbolise the nine units." [Silberberg (1895), p. 2; Smith and Ginsburg (1918); Steinschneider (1893)]

12. Around the same time, John of Seville began his *Liber algoarismi de practica arismetrice* ("Book of Algoarismi on practical arithmetic") with the following:

Numerus est unitatum collectio, quae quia in infinitum progreditur (multitudo enim crescit in infinitum), ideo a peritissimis Indis sub quibusdam regulis et certis limitibus infinita numerositas coarcatur, ut de infinitis difinita disciplina traderetur et fuga subtilium rerum sub alicuius artis certissima lege teneretur: "A number is a collection of units, and because the collection is infinite (for multiplication can continue indefinitely), the Indians ingeniously enclosed this infinite multiplicity within certain rules and limits so that infinity could be scientifically defined; these strict rules enabled them to pin down this subtle concept.

[B. N., Paris, Ms. lat. 16 202, f° 51; Boncompagni (1857), vol. I, p. 26]

13. C. 1143, Robert of Chester wrote a work entitled: *Algoritmi de numero Indorum* ("Algoritmi: Indian figures"), which is simply a translation of an Arabic work about Indian arithmetic. [Karpinski (1915); Wallis (1685), p. 12]

14. C. 1140, Bishop Raimundo of Toledo gave his patronage to a work written by the converted Jew Juan de Luna and archdeacon Domingo Gondisalvo: the *Liber Algorismi de numero Indorum* ("Book of Algorismi of Indian figures) which is simply a translation into a Spanish and Latin version of an Arabic tract on Indian arithmetic. [Boncompagni (1857), vol. I]

15. C. 1130, Adelard of Bath wrote a work entitled: *Algoritmi de numero Indorum* ("Algoritmi: of Indian figures"), which is simply a translation of an Arabic tract about Indian calculation. [Boncompagni (1857), vol. I]

16. C. 1125, The Benedictine chronicler William of Malmesbury wrote *De gestis regum Anglorum*, in which he related that the Arabs adopted the Indian figures and transported them to the countries they conquered, particularly Spain. He goes on to explain that the monk Gerbert of Aurillac, who was to become Pope Sylvester II (who died in 1003) and who was immortalised for restoring sciences in Europe, studied in either Seville or Cordoba, where he learned about Indian figures and their uses and later contributed to their circulation in the Christian countries of the West. [Malmesbury (1596), f° 36 r°; Woepcke (1857), p. 35]

17. Written in 976 in the convent of Albelda (near the town of Logroño, in the north of Spain) by a monk named Vigila, the *Codex Vigilanus* contains the nine numerals in question, but not zero. The scribe clearly indicates in the text that the figures are of Indian origin: *Item de figuris aritmetice. Scire debemus Indos subtilissimum ingenium habere et ceteras gentes eis in arithmetica et geometrica et ceteris liberalibus disciplinis concedere. Et hoc manifestum est in novem figuris, quibus designant unum quenque gradum cuiuslibet gradus. Quarum hec sunt forma:*

9 8 7 6 5 4 3 2 1.

"The same applies to arithmetical figures. It should be noted that the Indians have an extremely subtle intelligence, and when it comes to arithmetic, geometry and other such advanced disciplines, other ideas must make way for theirs. The best proof of this is the nine figures with which they represent each number no matter how high. This is how the figures look:

9 8 7 6 5 4 3 2 1."

(In the original, the figures are presented in a style very close to the North African Arabic written form.) [Bibl. San Lorenzo del Escorial, Ms. lat. d.I.2, fº 9v°; Burnam (1912), II, pl. XXIII; Ewald (1883)]

EVIDENCE FROM ARABIC SOURCES WHICH SUGGESTS THAT MODERN NUMERATION ORIGINATED IN INDIA

The following evidence proves that for over a thousand years, Arabo-Muslim authors never ceased to proclaim, in a praiseworthy spirit of openness, that the discovery of the decimal place-value system was made by the Indians.*

1. In *Khulasat al hisab* ("Essence of Calculation"), written c. 1600, Beha' ad din al 'Amuli, in reference to the figures in question, remarks that: "It was actually the Indians who invented the nine characters." [Marre (1864), p. 266]

2. C. 1470, in a commentary on an arithmetical tract, Abu'l Hasan al Qalasadi (d. 1486) wrote the following in reference to the nine figures used in Muslim Spain and Northern Africa: "Their origin is traditionally attributed to an Indian." [Woepcke (1863), p. 59]

3. In "Prolegomena" (*Muqqadimah*), written c. 1390, Abd ar Rahman ibn Khaldun (1332–1406) says that the Arabs first learned about science from the Indians along with their figures and methods of calculation in the year 156 of the Hegira (= 776 CE). [Ibn Khaldun, vol. III, p. 300]

4. In *Talkhis fi a 'mal al hisab* ("Brief guide to mathematical operations") written c. 1300, Abu'l 'abbas ahmad ibn al Banna al Marrakushi (1256–1321) makes a direct reference to the Indian origin of the figures and counting techniques. [Marre (1865); Suter (1900), p. 162]

5. C. 1230, Muwaffaq al din Abu Muhammad al Baghdadi wrote a tract entitled *Hisab al hindi* ("Indian Arithmetic"). [Suter (1900), p. 138]

6. C. 1194, Persian encyclopaedist Fakhr ad din al Razi (1149–1206) wrote a work entitled *Hada'iq al anwar*, which included a chapter called *Hisab al hindi* ("Indian Calculation"). [B. N., Anc. Fds pers., Ms. 213, fº 173r]

7. C. 1174, mathematician As Samaw'al ibn Yahya ibn 'abbas al Maghribi al andalusi, a Jew converted to Islam, wrote a work entitled *Al bahir fi 'ilm al hisab* ("The lucid book of arithmetic"), in which a direct reference is also made concerning the Indian origin of the figures and the methods of calculation. [Suter (1900), p. 124; Rashed and Ahmed (1972)]

---

* Henceforth, the scientific transcription of Arabic words will not be scrupulously adhered to. "Kh", "gh" and "sh" will be used in the place of h, g, and s to facilitate the reading of Arabic for those who are not specialists.

8. In 1172 Mahmud ibn qa'id al 'Amuni Saraf ad din al Meqi wrote a tract entitled *Fi'l handasa wa'l arqam al hindi* ("Indian geometry and figures"). [Suter (1900), p. 126]

9. C. 1048, 'Ali ibn Abi'l Rijal abu'l Hasan, alias Abenragel, in a preface to a treatise on astronomy, wrote that "the invention of arithmetic using the nine figures belongs to the Indian philosophers". [Suter (1900), p. 100]

10. C. 1030, Abu'l Hasan 'Ali ibn Ahmad an Nisawi wrote a work entitled *al muqni 'fi'l hisab al hindi* ("Complete guide to Indian arithmetic"). [Suter (1900), p. 96]

11. Between 1020 and 1030, in his autobiography, Al Husayn ibn Sina (Avicenna) tells of how, when he was very young, he heard conversations between his father and his brother which were often about Indian philosophy, geometry and calculation, and when he was ten (in the year 990), his father sent him to a merchant who was well-versed in numerical matters to learn the art of Indian calculation.

In his tract on speculative arithmetic, Ibn Sina writes the following: "As for the verification of squares using the Indian method (*fi'l tariq al hindasi*) . . . One of the properties of a cube consists of the way of verifying it using the methods of Indian calculation (*al hisab al hindasi*) . . ." [Woepcke (1863), pp. 490, 491, 502, 504; Leiden Univ. Lib., Ms. legs Warnerien, no. 84]

12. C. 1020, Abu'l Hasan Kushiyar ibn Labban al-Gili (971–1029) wrote a work which carries the Arabic title, *Fi usu'l hisab al hind* ("Elements of Indian calculation"), the opening words of which being: "This [tract] of calculations [written] in Indian [figures] is formed by . . ." [Library of Aya Sofia. Istanbul. Ms 4,857, f° 267 r; Mazaheri (1975)]

13. In roughly the same year, mathematician Abu Ali al Hasan ibn al Hasan ibn al Haytham, from Basra, wrote *Maqalat fi 'ala 'l hisab al hind* ("Principles of Indian calculation"). [Woepcke (1863), p. 489]

14. Astronomer and mathematician Muhammad ibn Ahmad Abu'l Rayhan al Biruni (973–1048), after living in India for thirty years, and having been introduced to Indian sciences, wrote a number of works between 1010 and 1030, including *Kitab al arqam* ("Book of figures"), and *Tazkira fi'l hisab wa'l mad bi'l arqam al sind wa'l hind* ("Arithmetic and counting using Sind and Indian figures").

In his work entitled *Kitab fi tahqiq i ma li'l hind* (which is one of the most important works about India to be written at that time), in which he mentions the diversity of the graphical forms of the figures used in India, and insists that the figures used by the Arabs originated in India, he makes the following remark: "Like us, the Indians use these numerical signs in their arithmetic. I have written a tract which shows, in as much detail as possible, how much more advanced the Indians are than we are in this field."

And in *Athar wu 'l baqiya* ("Vestiges of the past", or "Chronology of ancient nations"), he calls the nine figures *arqam al hind* ("Indian figures"), and demonstrates both how they differ from the sexagesimal system (which is Babylonian in origin), and their superiority over the Arab system of numeral letters. [Al-Biruni (1879) and (1910); Smith and Karpinski (1911), pp. 6–7; Datta and Singh (1938), pp. 98–9; Woepcke (1900), pp. 275–6]

15. Curiously, in his "Book of creation and history" (c. 1000), Mutahar ibn Tahir gives, in the *Nâgarî* form of the figures, the decimal positional expression of a number which the Indians believed represented the age of the planet. [Smith and Karpinski (1911), p. 7]

16. In 987, historian and biographer Ya 'qub ibn al Nadim of Baghdad wrote one of the most important works on the history of Arabic Islamic people and literature: the *Al Kitab al Fihrist al 'ulum* ("Book and index of the sciences"), in which he particularly refers to the work of the great Arabic Muslim astronomers and mathematicians of his time, and in which he constantly refers to methods of calculation as *hisab al hindi* ("Indian calculation"). [Dodge (1970); Suter (1892) and (1900); Karpinski (1915)]

17. Before 987, Sinan ibn al Fath min ahl al Harran (quoted in *Fihrist* by Ibn al Nadim) wrote a work entitled *Kitab al takht fi'l hisab al hindi* ("Tract on the wooden tablets used in Indian calculation"). [Suter (1892), pp. 37–8; Woepcke (1863), p. 490]

18. Also before 987, Ahmad Ben 'Umar al Karabisi (quoted in Ibn al Nadim's *Fihrist*) wrote *Kitab al hisab al hindi* ("A tract on Indian calculation"). [Suter (1900), p. 63; Woepcke (1863), p. 493]

19. Before 987 again, 'Ali Ben Ahmad Abu'l Qasim al Mujitabi al Antaki al Mu'aliwi (who died in 987) wrote a tract entitled *Kitab al takht al kabir fi'l hisab al hindi* ("Book of wooden tablets relating to Indian calculation"). [Suter (1900), p. 63; Woepcke (1863), p. 493]

20. Before 986, Al Sufi (who died in 986) wrote a work entitled *Kitab al hisab al hind* "Treatise on Indian calculation". [Smith and Karpinski (1911)]

21. C. 982, Abu Nasr Muhammad Ben 'Abdallah al Kalwadzani wrote *Kitab al takht fi'l hisab al hindi* ("Treatise on the tablet relative to Indian calculation"), quoted in *Fihrist* by Ibn al Nadim. [Suter (1900), p. 74; Woepcke (1863), p. 493]

22. C. 952, Abu'l Hasan Ahmad ibn Ibrahim al Uqlidisi wrote a work entitled: *Kitab al fusul fi'l hisab al hind* ("Treatise on Indian arithmetic"). [Saidan (1966)]

23. In 950, Abu Sahl ibn Tamim, a native of Kairwan (now Tunisia), wrote a commentary on *Sefer Yestsirah* (a Hebrew work concerning Cabbala) in which he explains the following: "The Indians invented the nine signs which denote units. I have already spoken about these at great length in a book which I wrote on Indian mathematics [he uses the expression *hisab al hindi*], known as *hisab al ghubar* ("calculations in the dust"). [Reinaud, p. 399; Datta and Singh (1938), p. 98]

24. C. 900, arithmetician Abu Kamil Shuja' ibn Aslam ibn Muhammad al Hasib al Misri (his last two names meaning "the Egyptian arithmetician") wrote an arithmetical work using the rule of the two false positions, which he attributed to the Indians. This work, which is only found in Latin translation, is called: "Book of enlargement and reduction, entitled 'the calculation of conjecture', after the achievements of the wise men of India and the information that Abraham[?] compiled according to the 'Indian' volume". [Suter, BM3; Folge, 3 (1902)]

25. Before 873, Abu Yusuf Ya 'qub ibn Ishaq al Kindi wrote *Kitab risalat fi isti mal 'l hisab al hindi arba' maqalatan* ("Thesis on the use of Indian calculation, in four volumes"), quoted in *Fihrist* by Ibn al Nadim. [Woepcke (1900), p. 403]

26. C. 850, the Arabic philosopher Al Jahiz (who died in 868) refers to the figures as *arqam al hind* ("figures from India") and remarks that "high numbers can be represented easily [using the Indian system]", even though the author expresses contempt for the Indian system. He asks the following question: "Who invented Indian figures . . . and calculation using the figures?" [Carra de Vaux (1917); Datta and Singh (1938), p. 97]

27. C. 820, Sanad Ben 'Ali, a Jewish mathematician who was converted to Islam, and who was one of Caliph al Ma'mun's astronomers, wrote a tract entitled: *Kitab al hisab al hindi* ("A treatise on Indian calculation") quoted by Ibn al Nadim in *Fihrist*. [Smith and Karpinski (1911), p. 10; Woepcke (1900), p. 490]

28. C. 810, Abu Ja 'far Muhammad ibn Musa al Khuwarizmi wrote: *Kitab al jam' wa'l tafriq bi hisab al hind* ("Indian technique of addition and subtraction"), of which there are Latin translations dating from the twelfth century. The tract begins thus:

" . . . we have decided to explain Indian calculating techniques using the nine characters and to show how, because of their simplicity and conciseness, these characters are capable of expressing any number."

He goes on to give a detailed explanation of the positional principle of decimal numeration, with reference to the Indian origin of the nine numerical symbols and of "the tenth figure in the shape of a circle" (zero), which he advises be used "so as not to confuse the positions". [Allard (1975); Boncompagni (1857)]; Vogel (1963); Youschkevitch (1976)]

HOW RELIABLE IS THIS EVIDENCE?

All the above evidence points to the same conclusion: the numerical symbols that are used in the modern world were created in India.

However, there still remains the task of judging how reliable this evidence is. According to E. Claparède (1937), "reliable evidence is not the rule but the exception". This idea is perhaps best expressed by Charles Péguy, through the character of Clio, Muse of History (*Oeuvres complètes*, VIII, 301–302): "Humankind lies most when giving evidence (because the testimony becomes part of history), and . . . people lie even more when giving formal evidence. In everyday life, it is important to be truthful. When giving evidence, it is necessary to be twice as truthful. It is a well-known fact, however, that people lie all the time, but people lie less when not testifying than when they are testifying."

Etymologically, "testimony" derives from the Latin *testis* ("witness"), from which we get the verbs "to attest", "to contest", etc. Thus "testimony" means "the written or verbal declaration with which a person certifies the reality of a fact of which they have had direct knowledge" (P. Foulquié, 1982).

Often, however, the fact in question is certified by an anterior declaration given by an eye-witness, as if one was testifying to a scene which a friend had seen and then recounted.

This is precisely the conditions in which nearly all the above declarations were written.

By its very nature, a testimony is never objective:

It is always marred by the subjectivity of its author, the unreliability of his memory, as well as gaps in perception and the unavoidable distortions of human memory (it is estimated that these errors increase at a rate of 0.33 per cent per day). Swiss psychologist Édouard Claparède and Belgian criminologist L. Vervaeck, using their pupils as subjects, found that correct testimonies were rare (only 5 per cent) and that the feeling of certainty increased with time . . . at the same rate as the increase in errors! [N. Sillamy (1967)].

It is because of its capital role in courtroom cases that the study of testimony plays such a major part in the applications of judicial psychology (see H. Piéron, 1979). The courtroom saying, *testis unus, testis nullus* (one sole witness is as useful as no witness at all) does not apply here because the origin of the numerals has been mentioned many times in the space of more than a thousand years. This case would in fact seem highly plausible.

But are all these accounts really completely independent of one another? If all these concurring pieces of evidence originate from one single source, then the proof might as well not exist at all.

The following example, taken from M. Bloch (1949), illustrates this point very clearly:

> Two contemporaries of Marbot – the Count of Ségur and General Pelet – gave accounts of Marbot's alleged crossing of the Danube which were analogous to Marbot's own account. Ségur's evidence came after Pelet's: he read the latter's account and did little more than copy it. It made no difference if Pelet wrote his account before Marbot; he was Marbot's friend and there is no doubt that Pelet had often heard Marbot recount his fictitious heroic deeds. This leaves Marbot as the only witness because his would-be guarantors both based their accounts on what he himself had related about the event.

In this kind of situation there is quite literally no witness at all.

However, Planudes, Fibonacci, Ibn Khaldun, Avicenna, al-Biruni, al-Khwarizmi and others, of whom many were actual eye-witnesses to the event, are neither Pelets, nor Ségurs, and certainly not Marbots. Their evidence and their accounts, as will be seen later, are firmly rooted in reality. These men are all in agreement, but this stems from neither a similar state of mind nor a phenomenon of collective psychology.

Despite the basic unreality of memory and the gaps and distortions which characterise the evidence given by any member of the human race, these accounts as a whole might still be an important item to add to the file for this investigation.

EVIDENCE FROM PRE-ISLAMIC SYRIA

The Arabs and the Europeans were not the first to offer evidence about the origin of our digits. There were others; people who were around long before and who lived far beyond the frontiers of Islam. Proof is to be found in the Middle East, at a time when Muslim religion was only just beginning to emerge, shortly after the first Ommayad caliph came to power in Damascus.

At that time there lived a Syrian bishop named Severus Sebokt. He studied philosophy, mathematics and astronomy at the monastery of Keneshre on the banks of the Euphrates: a place that was exposed to a great wealth of knowledge because of its situation at the crossroads of Greek, Mesopotamian and Indian learning.

Severus Sebokt, then, knew Greek and Babylonian sciences as well as Indian science. Irritated by the belief that Greek learning was superior to that of other civilisations, he wrote a short article in the hope of bringing the Greeks down a peg or two.

Nau, who wrote a commentary on and published this manuscript, explains the circumstances under which it was written:

> In the Greek year 973 (662 in our calendar), Severus Sebokt, clearly offended by Greek pride, reclaimed the invention of astronomy for the Syrians. He explained that the Greeks had gleaned their knowledge from the Chaldaeans and the Babylonians, who he claimed were in fact Syrians. He quite rightly concludes that science belongs to everyone and that it is accessible to any race or individual who takes the trouble to understand it; it is not the property of the Greeks [F. Nau (1910)].

It is in order to reinforce this point that Severus uses the Indians as an example:

> The Hindus, who are not even Syrians, have made subtle discoveries in the field of astronomy which are even more ingenious than those of the Greeks [sic] and the Babylonians; as for their skilful methods of calculation and their computing which belies description, they use only nine figures. If those who think they are the sole pioneers of science, simply because they speak Greek, had known of these innovations, they would have realised (albeit a little late) that there are others who speak different languages who are also knowledgeable.

This piece of evidence is indispensable. The "computing that belies description which uses only nine figures" is, to Sebokt's mind, infinitely superior to spoken numeration: it is not possible to express all numbers using the latter method (because, like most oral methods of numeration, it involves a hybrid principle, using addition and multiplication of the names of the basic numbers); the Indian system makes it possible to write any number using only the nine figures.

In other words, the Indian system, as described by Severus Sebokt, has an unlimited capacity for representation because it has positional numeration.

This numeration is decimal because it uses nine digits.

It might seem curious that Sebokt does not mention the use of zero, but this is probably because he only had an abacus upon which to carry out his mathematical operations. It is likely that his "abacus" was a board sprinkled with sand or dust upon which he would write numbers using the nine Indian symbols within various columns corresponding to the consecutive decimal denominations. Therefore, zero was not physically represented: the absence of a unit in a given column was communicated by means of an empty space.

Sebokt's evidence proves that the Indian counting system was known and esteemed outside India by the middle of the seventh century CE.

FROM THE EVIDENCE TO THE ACTUAL EVENT

The above evidence proves that all the preceding accounts are independent of each other but, however reliable these accounts are, they merely serve as confirmation of the truth. Alone, they do not constitute what is known as "historical truth". As F. de Coulanges said, "History is a science: it is a product of observation, not imagination; in order for the observation to be accurate, authentic documentation is needed."

A. Cuvillier (1954) explains that history, in the scientific sense of the word, is

> the study of human facts through time. So defined, historical facts are distinguished from those that are the subject of other sciences by their unique nature . . . Suspended in time, historical facts are, as a rule, in the past. Even when dealing with contemporary facts, the historian is still only personally privy to a very small percentage of the facts. The first task of a historian is to establish the facts through the use of documents, in other words the traces of these facts which still remain in the present.

Sociologist F. Simiand said that history is "information gleaned from left-over traces". The "traces" which are of interest here are the surviving written documents from Indian civilisation or from any culture connected to it.

Of course, it is essential to ensure that these documents are authentic. The traces in question came from an area of incredible diversity which, whilst proving the wonderful fertility of Indian civilisation, also shows an infinite complexity, with an added difficulty (to name but one): the considerable number of fakes produced by members of this same civilisation.

This, then, is the terrain the historian must embark upon; one of undeniable cultural wealth, even exuberance, yet it is crucial to remain extremely cautious when faced with documentation which is often tricky to date and which has to be closely examined in order to separate the genuine from the counterfeit, the ancient from the modern, the collective work from the individual work, a commentary from a copy of the original, etc.*

However, the vital work of historians from India and Southeast Asia must not be forgotten. For over a century, they have been separating the authentic from the fake, establishing the source and the date of a great many documents (even if this chronology is only approximate), restoring documents which had been damaged by the passage of time to their original state, studying the content and the allusions made in each work, and carrying out many other indispensable tasks.

All these results were collected in random order. To paraphrase H. Poincaré (1902), the science of history is built out of bricks; but an accumulation of historical facts is no more a science than a pile of bricks is a house.

PROOF OF THE EVENT

In the previous chapter we offered a classification of written numbering systems that are historically attested, and through it we drew out a genuine chronological logic: the guiding thread, leading through centuries and civilisations, taking the human mind from the most rudimentary systems to the most evolved. It enabled us to identify the foundation stone (and, more generally, the abstract structure) of the contemporary written numeral system, the most perfect and efficient of all time. And it is precisely this chronological logic of the mind which shows us the path to follow in order to arrive at a historical synthesis. A synthesis intended to show just how the invention of numerals actually "worked", and to place it in its overall context, in terms of period, sequence of events, influences, etc.

Using this approach, we will be able to tell the story much more rigorously and to track the invention of the Indian system very closely indeed.

Drawing on all the available evidence to prove that India really was the cradle of modern numeration, the problem will be divided into the following subsections:

---

* Indian history is a constantly shifting terrain, where "forgeries" or "modern documents presented as ancient ones" abound in great quantities. It is an area where even documents that are believed to be authentic could quite possibly have been the fruit of several successive corrections or re-workings and the result of some apparently homogenous fusion of various commentaries, even commentaries on the commentaries themselves, so that the seemingly authentic document might have absolutely nothing in common with what the author to whom the work is attributed orginally intended. It is a field where certain specialists, who have not always been as rigorous as they might have been, have confused the issue by supporting their arguments with documents that have no historical worth whatsoever. This would appear to explain why the origin of the decimal point system was such an enigma for so long.

   To untangle this apparently inextricable knot was no simple task because it involved the elimination of all unreliable sources (which are still used in a great many scientific publications) in order to include, as far as possible, nothing but trustworthy sources, from the most ancient documents on Indian civilisation.

1. To show that this civilisation discovered, and put into practice, the place-value system;

2. To prove that this same civilisation invented the concept of zero, which the Indian mathematicians knew could represent both the idea of an "empty space" and that of a "zero number";

3. To establish that the Indians formed their basic figures in the absence of any direct visual intuition;

4. To show that the early form of their symbols prefigured not only all the varieties currently in use in India and in Central and Southeast Asia, but also the respective shapes of Eastern and Western Arabic figures as well as the appearance of those figures used today and their various European predecessors of the same kind;

5. To prove that the learned men of that civilisation perfected the modern system of numeration for integers;

6. Finally, to establish once and for all that these discoveries took place in India, independent of any outside influence.

Historical reality, it can be seen, is not as simple as is generally thought: it is in any case not as simple as what an expression like "the invention of Arabic numerals", so cherished by the general public, seems to signify. For in terms of "invention" there would have to have been not only quite an exceptional combination of circumstances but also and above all an improbable conjunction of several great thoughts, created over fifteen centuries ago thanks to the genius of Indian scholars.

This would have taken exceptional powers of reflection, guided over a long period of time, not by logic or conscience, but by chance and necessity; chance discoveries and the need to remedy the problems engendered.

A. Vandel said, "A new idea is never the result of conscious or logical work. It emerges one day, fully formed, after a long gestation period which takes place within the subconscious."

It is true, as J. Duclaux says, that "the essential characteristic of scientific discoveries is that they cannot be made to order", because "the mind only makes discoveries when it is thinking of nothing".

INDIAN NUMERICAL NOTATION

With the aim of establishing the Indian origin of modern numerical figures, the following is a review of the numerical notations in common use in India before and since this colossal event, beginning with the symbols currently in use in this particular part of the world.*

* Henceforth, the references given relate to the works which write out each of the styles in question. As for the geographical location of the regions concerned, these are taken mainly from L. Frédéric's *Dictionnaire de la civilisation indienne*.

It should be made clear straight away that the modern figures 1, 2, 3, 4, 5, 6, 7, 8, 9, 0 acquired their present form in the fifteenth century in the West, modelled on specific prototypes and adopted permanently when the printing press was "invented" in Europe. Today they are used all over the world, thus constituting a kind of universal language which can be understood by East and West alike.

However, this form is not the only one which can express the decimal positional system. Particular symbols representing the same numbers still coexist with the figures that we all know in several oriental countries.

From the Near East and the Middle East to Muslim India, Indonesia and Malaysia, the following symbols are preferred:

1	2	3	4	5	6	7	8	9	0	Ref.
١	٢	٣	۴	٥	٦	٧	٨	٩	·	EIS Peignot and Adamoff Pihan Smith and Karpinski
				٤	٥					

Geographical area (see Fig. 25. 3):
Used in Libya, Egypt, Jordan, Syria, Saudi Arabia, Yemen, the Lebanon, Syria, Iraq, Iran, etc., as well as in Afghanistan, Pakistan, Muslim India, Indonesia Malaysia and formerly in Madagascar.

FIG. 24.2. *Current Eastern Arabic numerals (known as "Hindi" numerals)*

This is also the case in non-Muslim India, Central and Southeast Asia.

In these countries, symbols are still used that are graphically different from our own, and whose cursive form varies considerably from one region to another, according to the local style of writing.

Of course, this diversity dates back to ancient times, as the following pages will prove.

Nâgarî figures

In his *Kitab fi tahqiq i ma li'l hind*, (an account of what he had witnessed in India, written around 1030) al-Biruni, the Muslim astronomer of Persian origin, after having lived in India and Sind for nearly thirty years, described the great diversity of the graphical forms of figures in common use at that time in different regions of India; his commentary begins thus [see al-Biruni (1910); Woepcke (1863), pp. 275–6]:

Whilst we use letters for calculation according to their numerical value, the Indians do not use their letters at all for arithmetic.

And just as the shape of the letters [that they use for writing] is different in [different regions of] their country, so the numerical symbols [vary].

These are called *anka.

What we [the Arabs] use [for figures] is a selection of the best [and most regular] figures in India.

Their shapes are not important, however, as long as their meaning is understood.

The Kashmiris number their pages using figures which resemble ornamental drawings or letters [= characters used for writing] invented by the Chinese, which take a long time and a lot of effort to learn, but which are not used in calculation [which is carried out] in the dust (*hisab 'ala 't turab*).

Amongst the figures which were used long ago and are still used today most commonly in the various regions of India, the most regular are *Nâgarî*, which are also called *Devanâgarî*, from the name of the superb writing which they belong to (the words literally means "writing of the gods" in Sanskrit) (Fig. 24.3).

Al-Biruni (who mastered written and spoken Sanskrit), was alluding to precisely these figures when he said that the Arabs, in adopting the place-value system from India, had taken, as a means of notation for the nine units, "the best and most regular figures".

1	2	3	4	5	6	7	8	9	0	Ref.
૧	૨	૩	૪	૫	૬	૭	૮	૯	૦	Desgranges Frédéric, DCI Pihan Renou and Filliozat
૧	૨	૩	૪	૫	૬	૭	૮	૮	૦	
૧	૨	૩	૪		૬		૮	૬	૦	
૧	૨	૩	૪	૫	૬	૭	૮	૬	૦	
૧			૪	૫	૬		૮	૬	૦	
							૮	૬		

Geographical area (Fig. 24. 27 and 24. 53) :
Used in the Indian states of Madhya Pradesh (Central Province), Uttar Pradesh (Northern Province), Rajasthan, Haryana, Himachal Pradesh (the Himalayas) and Delhi.

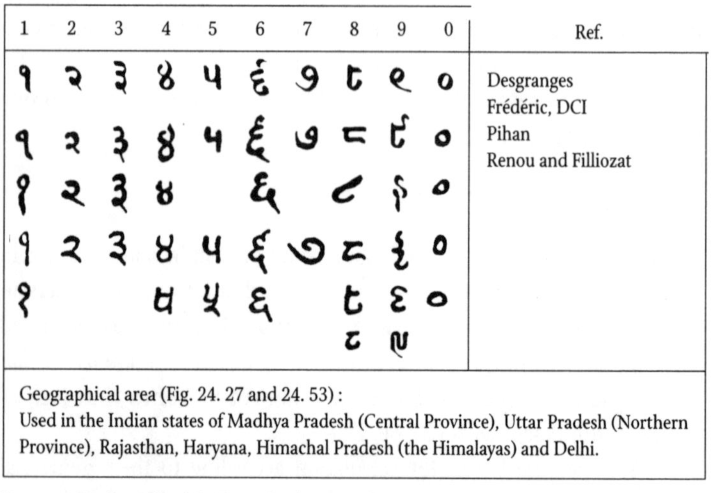

FIG. 24.3. *Modern Nâgarî (or Devanâgarî) numerals*

Indian Numerical Notation 725

This point will be confirmed later in a palaeographical study, where it will be shown how these figures, or at least their ancestors, were over the years transformed by the hands of Arabic Muslim scribes to provide:

- in the Near East, the forms of the symbols in Fig. 24.2;
- in Northwestern Africa, other graphical representations, which would gradually be transformed, this time by European scribes, into the figures that we use today.

Furthermore, a striking resemblance still persists between the first three and the last of these signs and our own numerals 1, 2, 3 and 0.

Marâthî figures

These figures are used in the west of India, in the state-province of Maharashtra (capital, Bombay). They are, as a rule, the cursive form of their corresponding *Nâgarî*, except for a slight variation in the shape of the 5 and the 6 (Fig. 24.3). There is a resemblance between these symbols for 2, 3, and 0 and our own, and the *Marâthî* nine is symmetrical to the European nine.

1	2	3	4	5	6	7	8	9	0	Ref.
৭	২	३	४	ৼ	६	७	८	९	०	Drummond Frédéric, DCI
৭	২	३	४	ৼ	६	७	८	९	०	Pihan

Geographical area (Fig. 24. 27 and 24. 53) :
Used in the area bordered in the west by the coasts of Konkan and Daman, and in the north by Gujarat and Madhya Pradesh, in the south by Karnataka and in the southeast by Andhra Pradesh.

FIG. 24.4. *Modern Marâthî numerals*

Punjabi figures

Used in the state of Punjab (capital, Chandigarh), in the northwest of India, bordering Pakistan. These are the same as the corresponding *Nâgarî* figures, except for the shape of the 7 (Fig. 24.3). There are similarities between these symbols and our figures 2, 3, 7 and 0:

1	2	3	4	5	6	7	8	9	0	Ref.
੧	੨	੩	੪	੫	੬	੭	੮	੯	੦	Pihan

Geographical area (Fig. 24. 27 and 24. 53) :
Used in the northwest of India bordering Pakistan where the Indus, the Chenab, the Jhelam, the Ravi and the Satlej rivers meet; as well as in the states of Himachal Pradesh and Haryana.

FIG. 24.5. *Modern Punjabi numerals*

Sindhî figures

These are symbols used in Sind, whose name derives from that of the river Sindh (the Indus). These signs are more or less identical to their corresponding *Nagârî*, but their shape is generally more cursive than the latter (Fig. 24.3). The figures 2, 3 and 0 are similar to our own, and the *Sindhî* 5 is rather like a symmetrical version of the European 4.

1	2	3	4	5	6	7	8	9	0	Ref.
٩	٨	౩	౮	౬	౬	౮	౬	౮	౦	Pihan Stack
٩	౨	౩	౮	౬	౬	౫	౬	౮	౦	

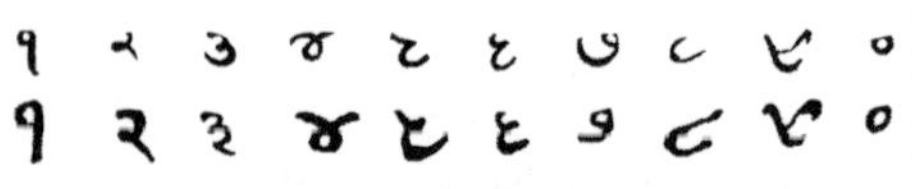

Geographical area (Fig. 24. 27 and 24. 53):
Used south of Punjab, on the lower banks of the Indus, in a region bordered in the south by the Gulf of Oman and in the west by the Thar desert.

FIG. 24.6. *Modern Sindhî numerals*

Gurûmukhî figures

In the city of Hyderabad (on the River Indus, to the east of Karachi, not to be confused with the other Hyderabad, capital of Andhra Pradesh), the merchants used to use a slight variant of the preceding figures, known as *Khudawadî*.

The traders of Shikarpur and Sukkur, on the other hand, sometimes used *Sindhî* or *Punjâbî* figures, sometimes eastern Arabic figures and sometimes *Gurûmukhî* figures, which are a mixture of *Sindhî* and *Punjâbî* styles (Fig. 24.5 and 24.6):

1	2	3	4	5	6	7	8	9	0	Ref.
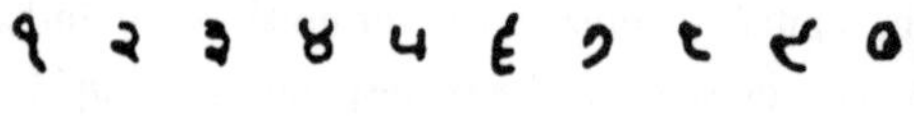 ٩	౨	౩	౮	౫	౬	౩	౬	౮	౦	Datta and Singh Stack

Geographical area (Fig. 24. 27):
Used in Sind and Punjab.

FIG. 24.7. *Gurûmukhî numerals*

Gujarâtî figures

These are used in Gujarat State (capital, Ahmadabad), on the edge of the Indian Ocean, between Bombay and the border of Pakistan. Again, these are derived from *Nâgarî* figures, but they are more cursive in form, particularly the 6 (Fig. 24.3). There are similarities between the *Gujarâtî* figures 2, 3 and 0 and our own numerals, as well as the figure 6.

1	2	3	4	5	6	7	8	9	0	Ref.
૧	૨	૩	૪	૫	૬	૭	૮	૯	૦	Drummond Forbes
?	૨	૩	૪	૪	૬	૭	૮	૪	૦	Frédéric, DCI Pihan
Geographical area (Fig. 24. 27 and 24. 53): Used in the west of India, bordering the Indian Ocean, between Bombay and the border with Pakistan, on the Gulf of Cambay.										

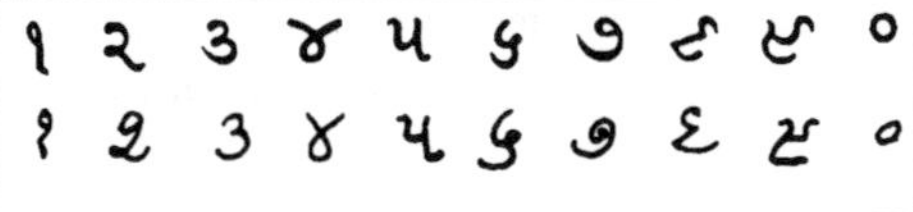

FIG. 24.8. *Modern Gujarâtî numerals*

Kaîthî figures

Used mainly in Bihar State, in Eastern India, and sometimes in Gujarat State. They evidently derive from *Nâgarî* figures and are similar in form to *Gujarâtî* figures (Fig. 24.3 and 24.8):

1	2	3	4	5	6	7	8	9	0	Ref.
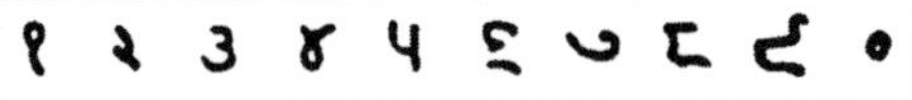 ૧	૨	૩	૪	૫	૬	૭	૮	૯	●	Datta and Singh
Geographical area (Fig. 24. 27): Used in the east of India, in the region bordered in the east by Bengal, in the north by Nepal, in the west by Uttar Pradesh and in the south by Orissa. Also sometimes used in Gujarat.										

FIG. 24.9. *Modern Kaîthî numerals*

Bengâlî figures

1	2	3	4	5	6	7	8	9	0	Ref.
১	২	৩	৪	৫	৬	৭	৮	৯	০	Frédéric, DCI Pihan
১	২	৩	৪	৩	৬	৭	৮	৯	০	Renou and Filliozat
১	২	৩	৪	৫	৬	৭	৮	৯	০	
	৬	৪	৪				৮	৯	০	
Geographical area (Fig. 24. 27 and 24. 53): Used in the regions in the northwest of the Indian sub-continent, between Bihar, Nepal, Assam, Sikkim, Bhutan, and the Bay of Bengal. Also widely used in Assam (along the Brahmaputra).										

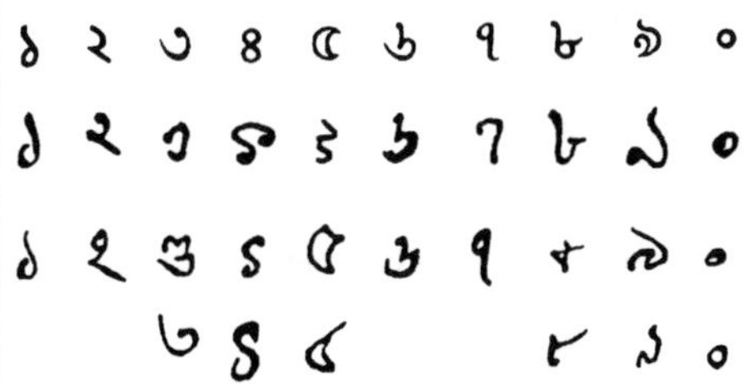

FIG. 24.10. *Modern Bengâlî numerals*

Used in the northeast of the Indian sub-continent in Bangladesh (capital, Dacca), in the Indian state of West Bengal (capital, Calcutta), and in much of central Assam (along the Brahmaputra River).

Of all the *Bengâlî* figures, there are four which resemble *Nâgarî* figures: 2, 4, 7 and 0 (Fig. 24.3). The others, however, are very different from those used in other parts of India. In one of the following variants, our figures 2, 3, 7 and 0 are recognisable; one of the variants of 8 also constitutes a sort of prefiguration of our 8.

Maithilî figures

Used mainly in the north of Bihar State, these derive mainly from *Bengâlî* figures (Fig. 24.10):

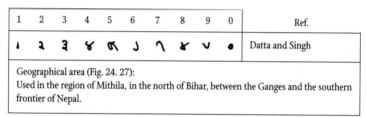

1	2	3	4	5	6	7	8	9	0	Ref.
ı	૨	૩	४	൘	Ɉ	٦	४	∨	●	Datta and Singh

Geographical area (Fig. 24. 27):
Used in the region of Mithila, in the north of Bihar, between the Ganges and the southern frontier of Nepal.

FIG. 24.11. *Modern Maithilî numerals*

Oriya figures

Used mainly in Orissa State (capital, Bhubaneswar), these are also known as *Orissî* figures. Although they derive from the same source as *Nâgarî* figures, they present significant differences (Fig. 24.3):

1	2	3	4	5	6	7	8	9	0	Ref.
୧	୨	୩	୪	୫	୬	୭	୮	୯	୦	Frédéric, DCI Pihan
୧	୨	୩	୪	୫	୬	୯	୮	୯	୦	Renou and Filliozat Sutton

Geographical area (Fig. 24. 27 and 24. 53):
Used in the region to the south of the eastern coast of Deccan, bordered in the north by Bengal and Bihar, in the west by Madhya Pradesh and in the south by Andhra Pradesh.

FIG. 24.12. *Modern Oriyâ (or Orissî) numerals*

Tâkarî figures

In everyday use in Kashmir, alongside eastern Arabic figures. They are also called *Tankrî* figures, of which a variant, *Dogrî*, is used in the Indian part of Jammu (in southwestern Kashmir):

1	2	3	4	5	6	7	8	9	0	Ref.
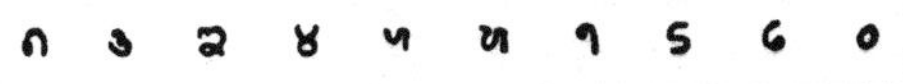 ∩	ꙅ	ꙅ	୪	ꙑ	ꙑ	ꙅ	ꙅ	ꙅ	ꙅ	Datta and Singh

Geographical area (Fig. 24. 27):
Used in the region in the extreme northwest of the Indian sub-continent, currently divided by the Indian-Pakistani border, joining the country of Jammu to the north of Himachal Pradesh, the plain of Kashmir in the high basin of the Jhelum, the valley of Zaskar in the north of the Himalayas and that of Ladakh, adjoining Tibet and China.

FIG. 24.13. *Modern Tâkarî (or Tankrî) numerals*

Shârâdâ figures

The figures that were used for many centuries in Kashmir and Punjab, from which, among others, *Dogrî* and *Tâkarî* figures derived (Fig. 24.13).

1	2	3	4	5	6	7	8	9	0	Ref.
○	3	ᴣ	ᴋ	ᴎ	⟩	ꙅ	ᴣ	⊍	·	Pihan
○	ꙅ	ꙅ	ꜰ	ᴋ	⟩	ꙅ	ꙅ	ꙅ	·	Renou and Filliozat
∩	ꙅ	ꙅ	ᴛ	ᴎ	ꙅ	ᴎ	ᴢ	⊍	●	Smith and Karpinski
ꙅ	ꙅ	ꙅ	ᴛ	ᴎ	ꙅ	ᴎ	ꙅ	ꙅ		
			ᴛ		ꙅ					

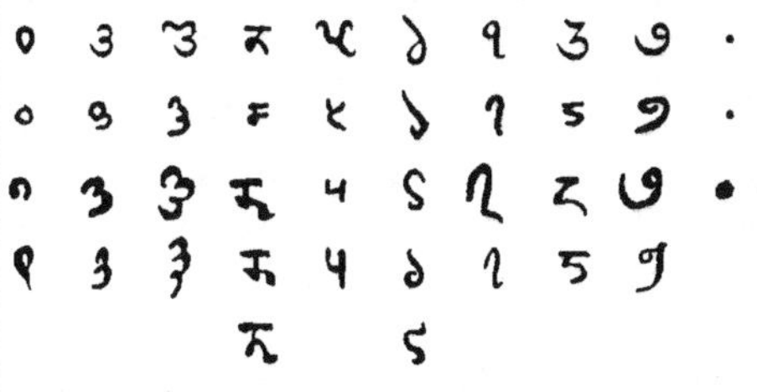

Geographical area (Fig. 24. 27 and 24. 53):
Formerly used in Kashmir and Punjab (before the sixteenth century).

FIG. 24.14. *Shârâdâ numerals (relatively recent forms)*

These figures are connected to *Shârâdâ* writing, which was used in the region at least since the ninth century, before it was replaced, relatively recently, by the Persian Arabic characters that are used for writing.

This notation (even in its most recent form) deserves special attention, because instead of representing zero with an oval or a small circle, it uses a dot, the circle being used to denote the number 1 (the shape was slightly modified according to the base).

The *Shârâdâ* 2 is like the *Nâgarî* 3, except that the lower appendage is absent from the *Shârâdâ* figure.

To the untrained eye, it should be pointed out, the figures 2 and 3 are not sufficiently distinct from one another, although the top of the 3 differs from that of the 2 because it is long and snaking.

The 6 is symmetrical to the European 6, whilst the 8 is very similar to the hand-written form of our 3.

As for the 7 and the 9, they are, respectively, almost identical to the figures 1 and 7 of *Nâgarî* notation (Fig. 24.3).

Nepâlî figures

Used mainly in the independent state of Nepal (capital, Kathmandu), these are also called *Gurkhalî* figures.

In one of the following variations, our 1, 2, 3, 4, 7 and 0 can be recognised, as well as our 8 to a certain extent (first set of figures, Fig. 24.15).

1	2	3	4	5	6	7	8	9	0	Ref.
٦	২	३	४	६	८	٩	⊢	৶	०	Datta and Singh
٦	₿	₿	४	९	६	٦	८	৹		Renou and Filliozat
﹨	३	३	६	২	६	٩	Γ	N		
৴	₿	३	८	௭	६	𝄐	Γ	৶	০	
﹨	ᘔ	₿		₷	੬	𝄐			০	

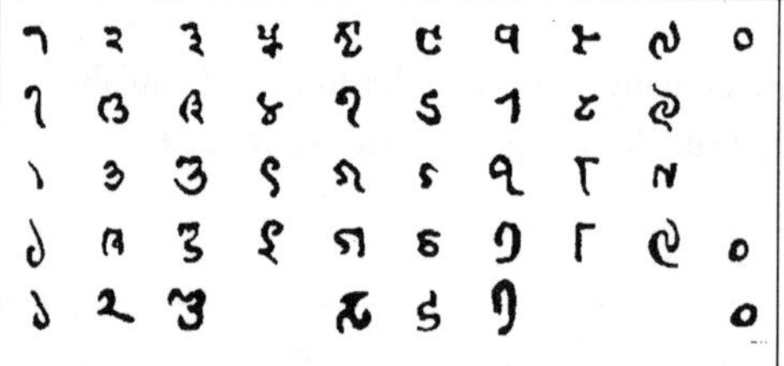

Geographical area (Fig. 24. 27 and 24. 53):
Formerly used in Kashmir and Punjab (before the sixteenth century).

FIG. 24.15. *Current Nepâlî numerals*

There is an obvious similarity between these figures and the *Nâgarî* and *Shâradâ* figures, with which they share a common source (Fig. 24.3 and 24.14).

Tibetan figures

These are the figures used in Tibet. They are similar to *Devanâgarî* figures (Tibetan writing comes from the same source as *Nâgarî*, introduced to the region in the seventh century CE at the same time as Buddhism). The 2, 3, the 9 (written backwards) and the 0 are alike.

1	2	3	4	5	6	7	8	9	0	Ref.
ʔ	۷	३	‿	⅄	↰	Ʋ	∠	(	०	Foucaux
٦	٦	₹	ﻉ	५	৬	𝑉	﹤	♪	০	Pihan
﹨	₹	₹	₠	٦	৬	∨	﹤	₽	০	Renou and Filliozat
										Smith and Karpinski

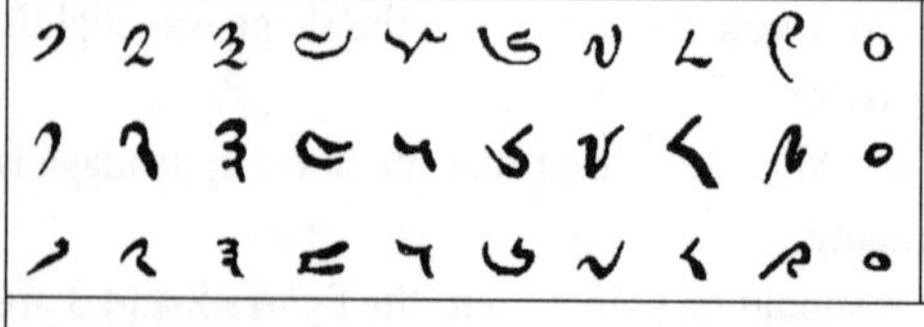

Geographical area (Fig. 24.27 and 24.53):
Used in regions of Tibet, from the border of Pakistan to the border of Burma and Bhutan.

FIG. 24.16. *Tibetan numerals*

Tamil figures

Unlike Northern and Central India, in Southern India, namely Tamil Nadu, Karnataka, Andhra Pradesh and Kerala, the Dravidian people do not speak Indo-European languages.

Tamil figures are used in Southeast India, in Tamil Nadu state (capital, Madras):

1	2	3	4	5	6	7	8	9	0	Ref.
௧	௨	௩	௪	௫	௬	௭	௮	௯		Frédéric, DCI
௧	௨	௩	௪	௫	௬	௭	௮	௯		Pihan
௧	௨	௩	௪	௫	௬	௭	௮	௯		Renou and Filliozat

Geographical area (Fig. 24. 27 and 24. 53):
Used in the region on the eastern coast of the Indian peninsula, from the north of Madras to the tip of Cape Comorin (Kanya Kumari) and bordered in the east by the Bay of Bengal, in the west by Kerala, in the northwest by Karnataka and in the north by Andhra Pradesh. Also used in the north and northwest of Sri Lanka.

FIG. 24.17. *Current Tamil numerals (or "Tamoul" numerals, according to an erroneous transcription)*

It should be noted, however, that the Tamils do not use zero in this system, which is only vaguely based on the place-value system.

Along with the signs for nine units, their system actually possesses a specific sign for 10, 100 and 1,000. To express multiples of 10, or hundreds or thousands, the sign for 10, 100 or 1,000 is proceeded by that of the corresponding units, which thus play the part of multiplier.

In other words, the *Tamil* system is based upon a principle which is at once additive and multiplicative, known as the *hybrid principle* and which has been used in many systems since early antiquity (see Fig. 23.20).

Equally, in terms of their appearance, these figures have nothing in common with the preceding notations.

For these reasons, it was believed that the Tamil figures were an original creation of the Dravidians, after they came up with the idea of using certain letters of their alphabet as signs for counting with.

It is true that there is a degree of resemblance between the first ten figures and what might constitute the corresponding letters of the Tamil alphabet, although the correspondence is not always very rigorous:

Comparison between the numeral and the letter				*Tamil name for the corresponding number*	
1	௧	க	ka, ga	ûru	1
2	௨	உ	u	irandu	2
3	௩	ஞ	ña	mûnru	3
4	௪	ச	sha	nâlu, nângu	4
5	௫	ரு	ra	aïndu, andju	5
7	௬	சா	cha	âru	6
7	௭	எ	ê	êrla, êzha	7
8	௮	அ	a	ettu	8
9	௯	கூ	kû, gû	onbadu	9

FIG. 24.18.

There is one question that cries out to be asked: if the theory is correct, why were these particular letters used to denote these numerical values? The obvious answer would be that the initials of the Tamil names for the numbers were used, but this is not the case, as the preceding table clearly demonstrates.

Then why were these letters singled out to represent numbers? Why did these people not give a numerical value to all the Tamil letters, as the Greeks and the Jews did with their respective alphabets when they created their systems of numeral letters?

This theory is rather far-fetched; it is merely a coincidence that these figures resemble the above Tamil letters. Moreover, the correspondence can only be established using the modern forms of the letters.

In fact, Tamil letters and figures are connected to all the other systems used in India: they all derive from the same source. Tamil writing, however, evolved in an entirely different manner from the others, both in terms of appearance and linguistic structure, introducing innovations which gave it its distinctive character. In particular, the characters and numerical symbols are considerably more rounded, with curves and volutes. It is not impossible that the material on which the characters were written played a role in this evolution, if it did not actually cause it.

In other words, the first nine Tamil figures are from the same family as the other corresponding Indian numerical symbols, the difference lying in their style and their adaptation to the unique shape of Tamil writing.

Malayâlam figures

These figures are used by the Dravidian people of Kerala State, on the ancient coast of Malabar, in the southwest of India. They have the same name as the form of writing used in the area.

1	2	3	4	5	6	7	8	9	0	Ref.
௧	௨	௩	௪	௫	௬	௭	௮	௯		Drummond Frédéric, DCI Peet, J. Pihan Renouand Fillozat
௧	௨	௩	௪	௫	௬	௭	௮	௯		
௧	௨	௩	௪	௫	௬	௭	௮	௯		

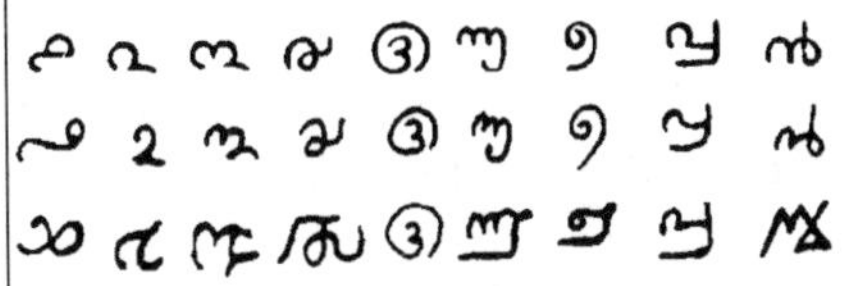

Geographical area (Fig. 24. 27 and 24. 53):
Used in the region stretching the length of the southeast coast of India, from Mangalore in the north to the southernmost point of India, and which is made up of a long coastal strip stretching from the coast of Malabar and by the Ghats encompassing the peaks of the Cardamoms.

FIG. 24.19. *Current Malayâlam numerals*

Like the Tamils, the people of Kerala did not use zero in their notation system for many centuries: *Malayâlam* figures are not based on the place-value system, and there are specific figures for 10, 100 and 1,000. It was only since the middle of the nineteenth century, under the influence of Europe, that zero was introduced and combined with the symbols for the nine units according to the positional principle.

Thus the Tamil and *Malayâlam* figures were the only ones in India that did not include zero and were not based on the positional principle until relatively recently.

However, it should be noted that Tamil figures, a few centuries ago, before they evolved into their current forms, closely resembled their *Malayâlam* cousins which have conserved a style close to the original.

The graphical link with the numerical signs of other regions of India is more easily seen through examining the original appearance of the Tamil figures than through looking at their modern form (Fig. 24.17 and 24.19):

The *Nâgarî* 1 is easily recognised, whose former shape was almost horizontal (Fig. 24.39) and which evolved in Tibet into a form constituting a sort of intermediate with the *Malayâlam* 1 (Fig. 24.16).

The *Nâgarî* 2 is also recognisable, although the "head" of the sign is very neatly rounded at the bottom.

On the other hand, the *Malayâlam* 3 is much closer to the corresponding *Oriyâ* figure (Fig. 24.12), with an extra "tail" which the *Nâgarî* 3 also has (Fig. 24.3).

The 4 is similar to its *Sindhi* equivalent except for the characteristic curve on the left (Fig. 24.6).

The 5 is very similar to one of the corresponding *Bengâli* figure (Fig. 24.10) and is reminiscent of the *Malayâlam* style.

The 6 resembles its *Sindhî* counterpart (Fig. 24.6), but it has an extra loop on the top, the whole figure being in a position which is obtained by rotating it through 90° anti-clockwise.

The 7 resembles its *Marâthî*, *Gujarâtî* and *Oriyâ* equivalents (Fig. 24.4, 24.8 and 24.12), whose prototype is found in the ancient *Nâgarî* style (Fig. 24.39).

The 8 is the symmetrical equivalent of the *Gujarâtî* 8 (Fig. 24.4).

As for the 9, it particularly resembles the *Nâgarî* style of the ninth century CE.

There can be no doubt: the Dravidian figures for the nine units have the same origin as all the others; the similarities found scattered amongst these diverse figures could not possibly be the product of chance.

The following two varieties of Dravidian figures serve as confirmation of this fact.

Telugu figures

These are the numerical symbols used by Dravidian people of the former Telingana, the Indian state of Andhra Pradesh (capital, Hyderabad). They are also called *Telinga* figures (Fig. 24.20).

1	2	3	4	5	6	7	8	9	0	Ref.
∩	﹍₀	3	४	ૠ	Ɛ	₹	ᴖ	ᖴ	ο	Burnell Campbell Datta and Singh Pihan Renou and Filliozat Smith and Karpinski
∩	﹍ₒ	?	४	ᴎ	ᒻ	2	ơ	ᖴ	ο	
⌒	♪	3	४	ᴐᴇ	Ɛ	2	ᴳ	ᖴ	ο	
∩	♪	3	४	ᴎ	e	₹	ᴗ	ᖴ	ο	

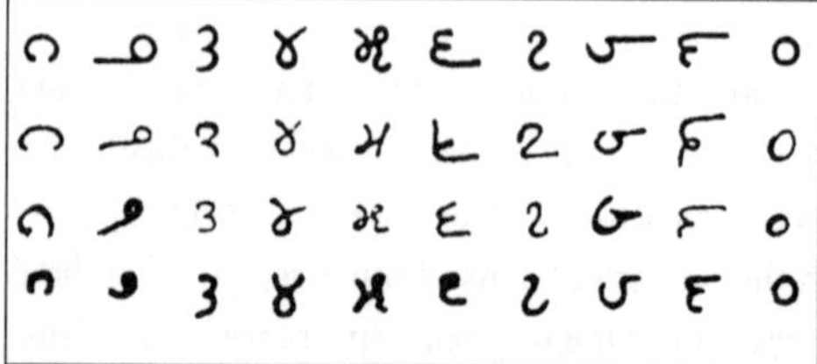

Geographical area (Fig. 24. 27 and 24, 53):
Used in the southeast of India, bordered in the southeast by the Bay of Bengal, in the north by the States of Orissa and Madhya Pradesh, in the northwest by Maharashtra, in the west by Karnataka and in the south by Tamil Nadu.

FIG. 24.20. *Modern Telugu (or Telinga) numerals*

Kannara figures

Used by the Dravidian people of central Deccan, including the state of Karnataka (capital, Bangalore) and part of Andhra Pradesh:

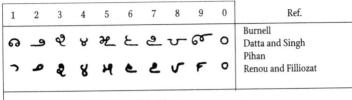

1	2	3	4	5	6	7	8	9	0	Ref.
										Burnell
										Datta and Singh
										Pihan
										Renou and Filliozat

Geographical area (Fig. 24. 27 and 24. 53):
Used mainly in the region stretching from the Mysore mountains to the eastern coast of the Indian sub-continent, between the Gulf of Oman and the Western Ghats.

FIG. 24.21. *Modern Kannara (or Kannada or Karnata) numerals*

Sinhalese figures

Used mainly in Sri Lanka and in the Maldives as well as in the islands to the north of the latter. (In the north and northwest of Sri Lanka, Tamil figures are also used due to the high number of Tamil people who live in these areas of the island.)

1	2	3	4	5	6	7	8	9	0	Ref.
										Alwis (de)
										Charter
										Frédéric, DCI
										Pihan
										Renou and Filliozat

Geographical area (Fig. 24. 27 and 24. 53):
Used in Sri Lanka, in the Maldives, as well as in the islands to the north of the Maldives.

FIG. 24.22. *Current Sinhalese (or Sinhala) numerals*

It should be noted that although Sinhalese writing is linked to Dravidian forms of writing (even though it is more stylish, striving as it does towards an ornamental effect), the language of this writing is not Dravidian. Sinhalese is an Indo-European language: "it is a language that belongs to Prakrit (dialects) of 'Middle Indian', as several inscriptions written in *Brâhmî* dating from around the second century BCE show. However, after the fifth century CE, the Sinhalese language, separated from India's Indo-European languages by the Tamil area, developed in an individual style, as did its writing. The two seem to have changed little since 1250" (L. Frédéric).

There are twenty Sinhalese figures. This number of numerical signs is due to the absence of zero and the fact that the system, which is not based upon the place-value system, uses a specific figure for every ten units, as well as special figures that represent 10, 100 and 1,000 (see Fig. 23.18).

Burmese figures

Used in Burma. Formerly used in the kingdom of Magadha, these were once known as *cha lum* figures, they are part of Burmese writing, which itself derives from the former *Pâli* alphabet, introduced to the region by Buddhists (Fig. 24.23).

1	2	3	4	5	6	7	8	9	0	Ref.
‌	‌	‌	‌	‌	‌	‌	‌	‌	‌	Carey
‌	‌	‌	‌	‌	‌	‌	‌	‌	‌	Datta and Singh
‌	‌	‌	‌	‌	‌	‌	‌	‌	‌	Latter
‌	‌	‌	‌	‌	‌	‌	‌	‌	‌	Pihan
‌	‌	‌	‌	‌	‌	‌	‌	‌	‌	

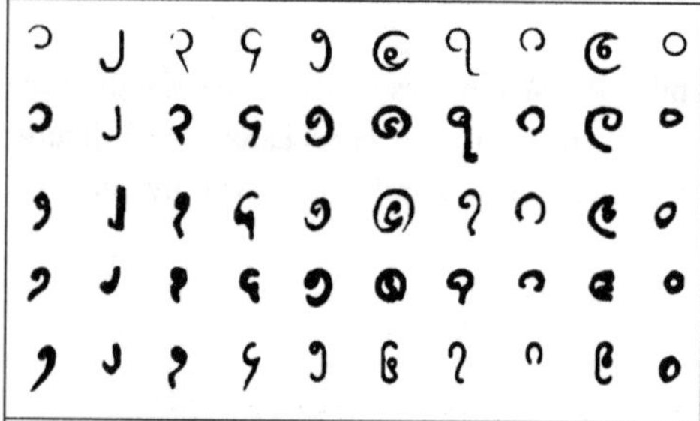

Geographical area (Fig. 24.27 and 24.53):
Used in the region stretching from Laos to the Bay of Bengal, and from Manipur to Pegu; also, in a slightly modified form, around Tenasserim and along the coast from Chittagong.

Fig. 24.23. *Modern Burmese numerals*

In modern Burmese writing, the principal element of the shape of the letters is a little circle, the value of which varies according to the breaks, juxtapositions or appendages.

The same applies to the figures, or at least to three of them, whose shapes should not be confused.

These are:

- the 1, formed by a circle, a quarter open on the left;
- the 8, which is a circle that is a quarter open at the bottom;
- and the 0 which is a whole circle.

The 3 is an open circle like the 1, with an appendage which slants towards the right, and the 4 is formed by the mirror image of the 3.

As for the 9, it is the 6 turned upside-down.

However this graphical rationalisation is relatively recent: the Indian origin (via former *Pâlî* figures) of the Burmese figures was still unknown in the seventeenth century.

Thai-Khmer figures

These are the official numerical symbols of Thailand, Laos and Cambodia. They also belong to the family of numerical signs that are of Indian origin, actually belonging to the former *Pâlî* style.

1	2	3	4	5	6	7	8	9	0	Ref.
๚	๒	๓	๔	๕	๖	๗	๘	๙	๐	Pihan Rosny
๑	๒	๓	๔	๕	๖	๗	๘	๙	๐	
๑	๒	๓	๔	๕	๖	๗	๘	๙	๐	
๑	๒	๓	๔	๕	๖	๗	๘	๙	๐	
๑	๒	๓	๔	๕	๖	๗	๘	๙	๐	
๑	๒	๓	๔	๕	๖	๗	๘	๙	๐	

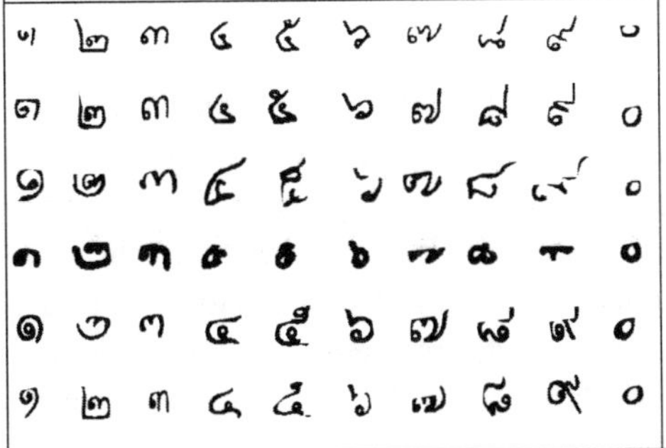

Geographical area (Fig. 24.53):
Used in Thailand, Laos, Kampuchea, in the State of Chan to the east of Burma, in some parts of Vietnam, in China in the provinces of Guangxi and Yunnan, as well as in the Nicobar islands.

Fig. 24.24. *Modern Thai-Khmer (known as "Siamese") numerals*

Some of these figures look so alike that they are easily confused. Unlike the various "true" Indian figures, the Thai-Khmer 2 is more complicated than the 3. The 5 only differs from the 4 because it has an extra loop at the top. The 8 is more or less symmetrical to the 6, and the figure 7 is easily confused with the 9.

Balinese figures

These are from Bali, and also developed from the *Pâlî* figures.

1	2	3	4	5	6	7	8	9	0	Ref.
 ᬯ	ᬯ	ᬳ	᭓	ᬕ	ᬦ	ᬯ	ᬯ	ᬯ	᭠	Renou and Filliozat

Geographical area (Fig. 24.53):
Used in Bali, Borneo and the Celebes islands.

Fig. 24.25. *Modern Balinese numerals*

Javanese figures

The final figures in this list of numerical symbols currently in use in Asia are those from the island of Java:

1	2	3	4	5	6	7	8	9	0	Ref.
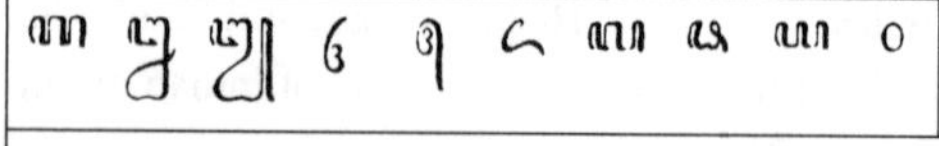										De Hollander Pihan
Geographical area (Fig. 24.53): Used in Java, Sunda, Bali, Madura and Lombok.										

FIG. 24.26. *Modern Javanese numerals*

Apart from the figures 0 and 5 (whose Indian origin is obvious), this notation actually corresponds to a relatively recent artificial innovation, the appearance of the figures curiously having been made to resemble the shape of certain letters of the current Javanese alphabet. Before this, however, the Javanese people used a notation which belonged to the *Pâlî* group of the family of Indian figures: the notation known as *Kawi* (attested since the seventh century CE), which belongs to the writing of the same name (from which the current Javanese alphabet derives).

Brâhmî, "mother" of all Indian writing

Despite the high number of graphical representations of the nine units, there is no doubt as to their common origin.

Leaving European and Arabic numerals on one side for a moment, each of the preceding styles were graphically connected to one of the various styles of writing belonging to either India, Central or Southeast Asia: it is clear from extensive palaeographical research that they all derive, directly or indirectly, from the same source.

Therefore, it is worthwhile saying a few words about the history of the styles of writing of this region.

The oldest known writing of the sub-continent of India appeared on the stamps and plaques of the civilisation of the Indus (c. 2500 – 1500 BCE), discovered mainly in the ruins of the ancient cities of Mohenjo-daro and Harappâ. However, as this writing has not yet been deciphered, the corresponding language remains unknown; therefore there is a large gulf separating these inscriptions of the first known texts in Indian writing and the language, assuming that a link exists between the two systems.

In fact, the history of Indian writing begins with the inscriptions of Asoka, third emperor of the dynasty of the Mauryas of the Magadha, who reigned in India from c. 273 to 235 BCE, whose empire stretched from

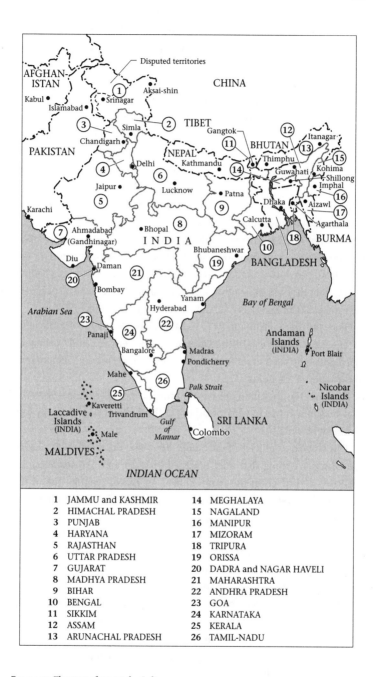

1	JAMMU and KASHMIR	14	MEGHALAYA
2	HIMACHAL PRADESH	15	NAGALAND
3	PUNJAB	16	MANIPUR
4	HARYANA	17	MIZORAM
5	RAJASTHAN	18	TRIPURA
6	UTTAR PRADESH	19	ORISSA
7	GUJARAT	20	DADRA and NAGAR HAVELI
8	MADHYA PRADESH	21	MAHARASHTRA
9	BIHAR	22	ANDHRA PRADESH
10	BENGAL	23	GOA
11	SIKKIM	24	KARNATAKA
12	ASSAM	25	KERALA
13	ARUNACHAL PRADESH	26	TAMIL-NADU

FIG. 24.27. *The states of present-day India*

Afghanistan to Bengal and from Nepal to the south of Deccan [see L. Frédéric (1987)]. These inscriptions are mainly edicts carved on rocks or columns for which diverse styles of writing were used: Greek and Aramaean in Kandahar and Jalalabad in Afghanistan; the *Kharoshthî* system in Manshera and Shahbasgarhi to the north of the Indus; and *Brâhmî* writing in all the other regions of the Empire.

Kharoshthî comes directly from the old Aramaean alphabet and is similarly written from right to left. This is why it is also labelled "Aramaeo-Indian" writing. Probably introduced in the fourth century BCE, it remained in use in the northwest of India until the end of the fourth century CE.

As for the written form of *Brâhmî*, it was written from left to right and was used to note the sounds of Sanskrit.

The origin of this writing is still not known. Attempts have been made to prove that it comes from *Kharoshthî* writing, but the explanation for this is far from convincing. *Brâhmî* certainly derives from the Western Semitic world, doubtless via some other variety of Aramaean, of which specimens have not yet been found [see M. Cohen (1958); J. G. Février (1959)].

Since the first millennium BCE, India was already open to outside influences, due to long-established ties with the Persians and Aramaean merchants who used the routes which went from Syria and Mesopotamia to the valley of the Indus.

However, the appearance of *Brâhmî* probably pre-dates Emperor Asoka, by whose time it was in widespread use in the different regions of the subcontinent of India.

This language outlived all the others, becoming the unique source of all the forms of writing that later emerged in India and her neighbouring countries. It was given the name *Brâhmî*, in Hindu religion one of the names of the seven *mâtrikâ* or "mothers of the world": one of the feminine energies (*shakti*) supposed to represent the Hindu divinities. Represented as sitting on a goose, her power was equal to that of Brahma, the "Immeasurable", god of the Sky and the horizons, who "endlessly gives birth to the Creation" and who one day invented *Brâhmî* writing for the well-being and diversity of humankind.

According to the edicts of Asoka, *Brâhmî* appeared, in a slightly modified form, in contemporary inscriptions of the Shunga Dynasty (185 – c. 75 BCE on the Magadha, in the present Bihar state, south of the Ganges, then in those of the Kanva Dynasty (who succeeded the former from 73 to c. 30 BCE).

The following is a more developed exploration of *Brâhmî*, first through the inscriptions of the Shaka Dynasty (Scythians, who reigned over Kabul in Afghanistan, Taxilâ in Punjab and Mathura, from the second century BCE to the first century CE) and through the coins embossed with the sovereigns of the Shaka Dynasty who reigned from the second to the fourth century CE in Maharashtra (under the name of *Kshatrapa*, "Satraps").

Brâhmî evolved a little more in the writing of the Andhra and Satavahana Dynasties which reigned during the first two centuries CE in the northwest of Deccan.

Then the system appeared, in an even more developed form, in the inscriptions of the Kushan emperors (who reigned from the first to the third century CE, and who, at first based in Gandhara and Transoxiana attempted to conquer Northwestern India).

Thus through numerous successive and perceptible modifications, *Brâhmî* gave birth to many highly individual styles of writing; styles which constitute the main groups currently in use (Fig. 24.28):

1. the group of types of writing in Northern and Central India and in Central Asia (Tibet and East Turkestan);
2. the writing of Southern India;
3. oriental writing (Southeast Asia).

The apparently considerable differences between the forms of writing of these various groups is ultimately due either to the specific character of the language and traditions to which they have been adapted, or to the techniques of the scribes of each region and the nature of the material they used.

A parallel evolution: Indian figures

In this context, everything becomes clear: in India and the surrounding regions, the notation of the nine units evolved in much the same way as the styles of writing that were born out of *Brâhmî*. In other words, in the same way as the writing they belong to, the various series of 1 to 9 formerly or currently in use in India, Central and Southeast Asia all derive more or less directly from the *Brâhmî* notation for the corresponding numbers.

The numerical symbols of the original Brâhmî notation

This notation appeared for the first time in the middle of the third century BCE in edicts written in both *Ardha-Mâgadhî* and *Brâhmî* which the

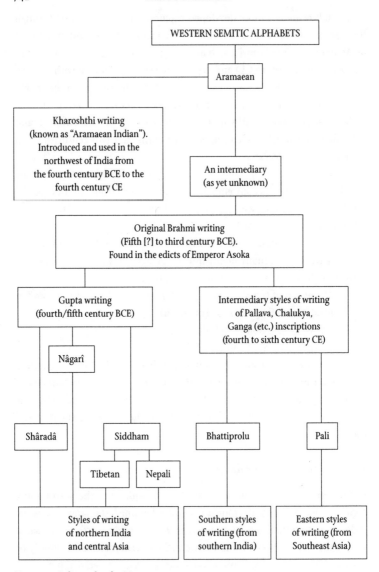

FIG. 24.28. *Indian styles of writing*

emperor Asoka had engraved on rocks, polished sandstone columns and temples hewn out of the rock, in diverse regions of his empire.

But the numerical notation that is found within these edicts is fragmentary, only giving the representations for the numbers 1, 2, 4, and 6:

1	2	3	4	5	6	7	8	9	0	Ref.
ı	**ıı**		**+**		**Ɛ** **ᵠ** **ᵱ**					EI, III p. 134 IA, VI, pp. 155 ff. IA, X, pp. 106 ff. Indrâji, JBRAS XII

Date: third century BCE.
Source: edicts of Asoka written in Brahmi, in various regions of the Empire of the Mauryas, from the regions of Shahbazgarhi, Manshera, Kalsi, Girnar and Sopara (north of Bombay) to Tosali and Jaugada in Kallinga (Orissa), Yerragudi in Kannara, Rampurwa and Lauriya-Araraj in the north of Bihar, Toprah and Mirath north of Delhi, and Rummindei and Nigliva in Nepal (Fig. 24.27).

Fɪɢ. 24.29. *Numerals of the original Brâhmî style of writing: our present-day 6 is already recognisable*

The numerical symbols of intermediate notations

The same system appears in the documents of the eras which followed and this gives a much more precise idea of how *Brâhmî* figures looked.

The following figures appeared at the beginning of the Shunga and Magadha dynasties in the Buddhist inscriptions which adorn the walls of the grottoes of Nana Ghat:

1	2	3	4	5	6	7	8	9	0	Ref.
- **-**	**=** **=**		**Ŧ** **⸸**		**Ƴ** **ᵩ**	**ገ** **ገ**		**ᒣ** **ᒣ**		Datta and Singh Indraji, JBRAS XII Smith and Karpinski

Date: second century BCE.
Source: the caves of Nana Ghat (central India, Maharashtra, c. 150 km from Poona), Buddhist inscriptions written for a sovereign named Vedishri which mainly concern various presents offered during religious ceremonies.

Fɪɢ. 24.30. *Numerals of the intermediary notation of the Shunga: we can already see the prefiguration of our numerals 4, 6, 7 and 9.*

The same series appeared a little later, but in a much more complete form, in the first or second century CE, in the inscriptions of the Buddhist grottoes of Nasik (Fig. 24.31).

Brâhmî figures are also found, in more and more varied forms, in Mathuran inscriptions (Fig. 24.32), Kushana and Andhran inscriptions (Fig. 24.33 and 24.34), western Satrap coins (Fig. 24.35), the inscriptions of Jaggayyapeta (Fig. 24.36), and of the Pallava Dynasty (Fig. 24.37) .

As these numerals derive from *Brâhmî* figures and consequently serve as a go-between with the later forms of the numerals, they shall henceforth be referred to as the *numerical symbols of the intermediate notations*.

1	2	3	4	5	6	7	8	9	0	Ref.

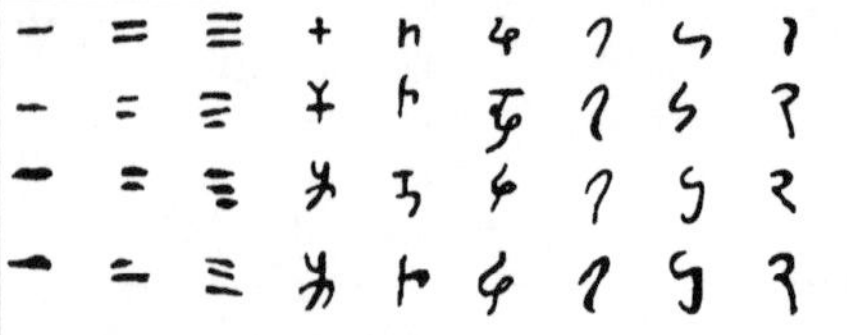

EI, VIII, pp. 59–96
EI, VII, pp. 47–74
Bühler
Datta and Singh
Renou and Filliozat
Smith and Karpinski

Date: first or second century CE.
Source: Buddhist caves of Nasik (in Maharashtra, at least 200 km north of Bombay).

FIG. 24.31. *Numerals of the intermediary system of Nasik: we can see the prefiguration of our numerals 4, 5, 6, 7, 8 and 9.*

1	2	3	4	5	6	7	8	9	0	Ref.

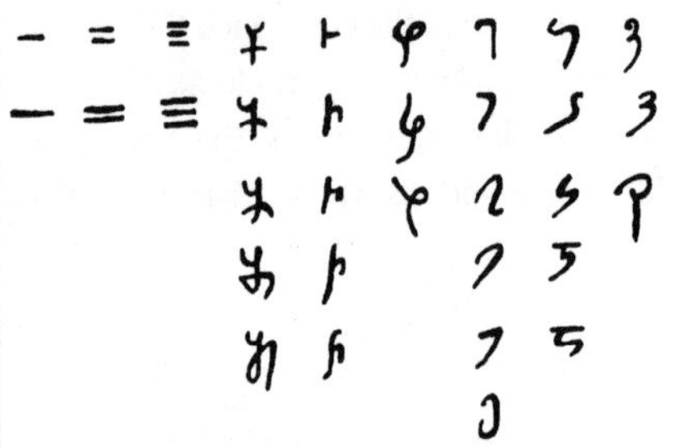

Bühler
Datta and Singh
Ojha

Date: first – third century CE.
Source: inscriptions of Mathura (town of Uttar Pradesh, on the banks of the Yamuna 60 km northwest of Agra), contemporary with a Shaka dynasty.

FIG. 24.32. *Numerals of the intermediary system of Mathura*

1	2	3	4	5	6	7	8	9	0	Ref.

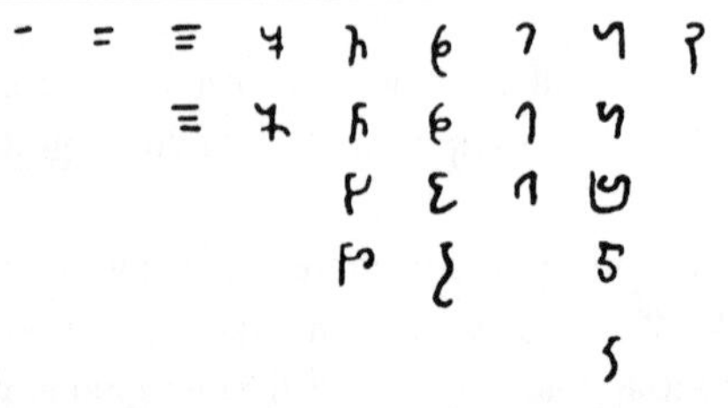

EI, I, p. 381
EI, II, p. 201
Bühler
Datta and Singh
Ojha
Smith and Karpinski

Date: first – second century CE.
Source: contemporary inscriptions of the Kushana dynasty.

FIG. 24.33. *Numerals of the intermediary system of the Kushana*

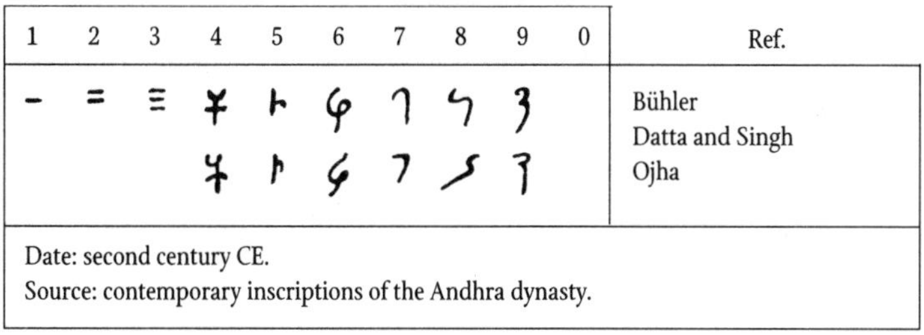
1 2 3 4 5 6 7 8 9 0
Ref.
Bühler
Datta and Singh
Ojha
Date: second century CE.
Source: contemporary inscriptions of the Andhra dynasty.

FIG. 24.34. *Numerals of the intermediary notation of the Andhra*

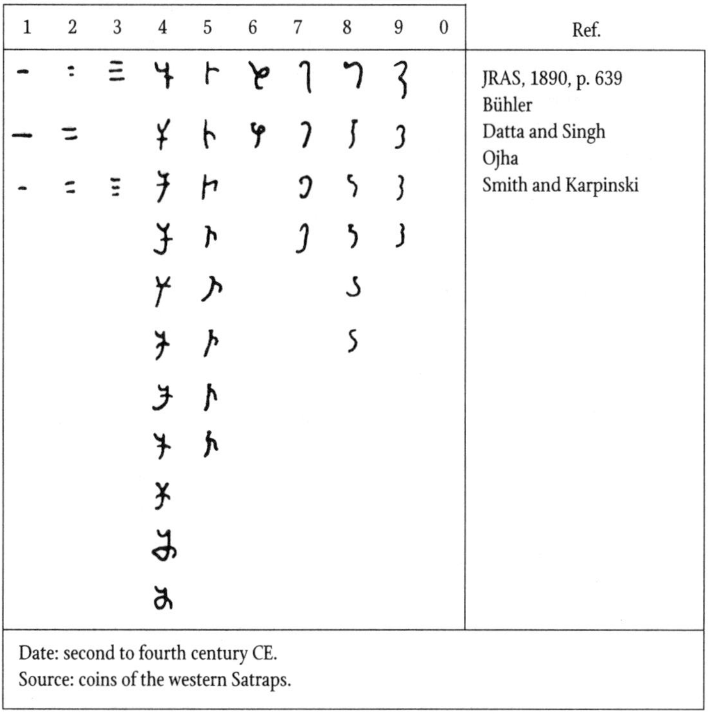
1 2 3 4 5 6 7 8 9 0
Ref.
JRAS, 1890, p. 639
Bühler
Datta and Singh
Ojha
Smith and Karpinski
Date: second to fourth century CE.
Source: coins of the western Satraps.

FIG. 24.35. *Numerals of the intermediary notation of the western Satraps*

These intermediate notations spread over the various regions of India and the neighbouring areas, as did the letters of the corresponding writing, and, over the centuries, they underwent graphical modifications, finally to acquire extremely varied cursive forms, each with a regional style.

The origin of the notations of Northern and Central India

One of the first individual notations to appear was *Gupta* notation, used during the dynasty of the same name (its sovereigns reigned over the Ganges and its tributaries from c. 240 to 535 CE) (Fig. 24.38).

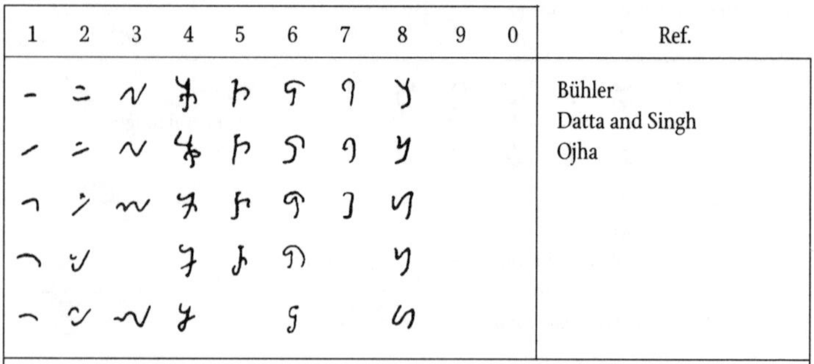
1 2 3 4 5 6 7 8 9 0
Ref.
Bühler
Datta and Singh
Ojha

Date: third century CE.
Source: inscriptions of Jaggayyapeta (site of an ancient Buddhist centre established on the River Krishna, in the present-day state of Andhra Pradesh, in the southeast of the Indian peninsula, opposite Amaravati, capital of the Andhra kingdom during the Shatavahana dynasty).

FIG. 24.36. *Numerals of the intermediary notation of Jaggayyapeta*

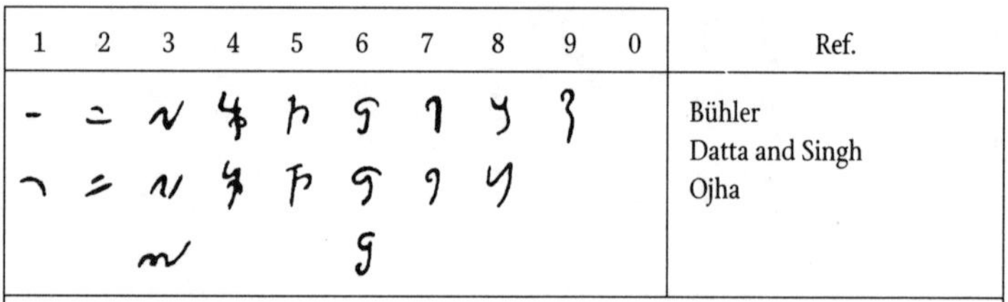
1 2 3 4 5 6 7 8 9 0
Ref.
Bühler
Datta and Singh
Ojha

Date: fourth century CE.
Source: inscriptions of King Skandravarman (c. 75 CE) of the Pallava dynasty, who reigned in the southeast of India at the end of the third century CE, after the fall of the Andhra and Pandya rulers.

FIG. 24.37. *Numerals of the first intermediary notation of the Pallava*

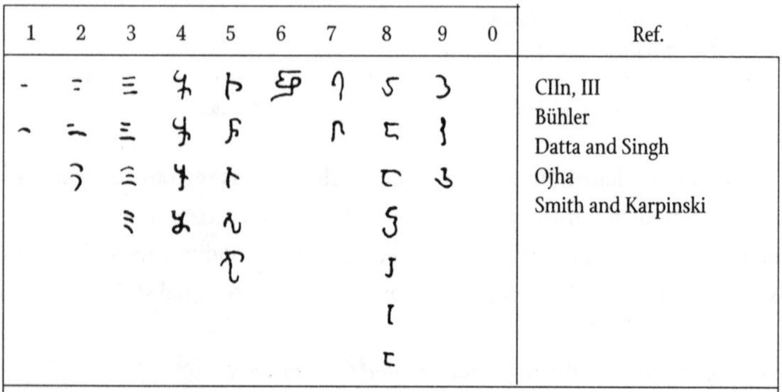
1 2 3 4 5 6 7 8 9 0
Ref.
CIIn, III
Bühler
Datta and Singh
Ojha
Smith and Karpinski

Date: fourth to sixth century CE.
Source: inscriptions of Parivrajaka and Uchchakalpa

FIG. 24.38. *Gupta numerals*

This notation was the origin of all the series of figures in common use in Northern India and Central Asia.

The first developments in Nâgarî notation

As *Gupta* writing became more refined, it gave birth to *Nâgarî* notation (or "urban" writing, the magnificent regularity of which gave it the name of *Devanâgarî*, or "Nâgarî of the gods").

This writing soon acquired great importance, becoming not only the main writing of the Sanskrit language, but also of Hindi, the great language of modern Central India.

As numerical notation experienced a parallel evolution, so *Nâgarî* figures were born out of *Gupta* figures, which later led to the emergence of modern *Nâgarî* figures (see also Fig. 24.3 above):

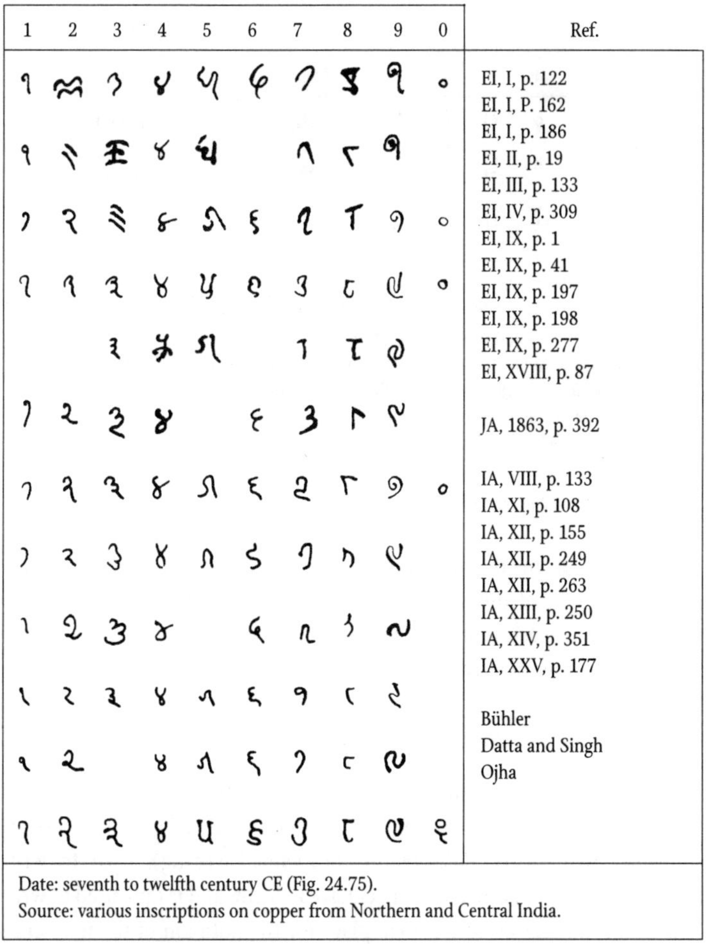

1	2	3	4	5	6	7	8	9	0	Ref.
१	∾	backslash	४	५	६	७	५	९	०	EI, I, p. 122 EI, I, P. 162
१	९	ए	४	५			∧	९		EI, I, p. 186 EI, II, p. 19 EI, III, p. 133
)	२	३	६	५	६	७	⊤	९	०	EI, IV, p. 309 EI, IX, p. 1
१	१	३	४	५	९	३	६	⊍	०	EI, IX, p. 41 EI, IX, p. 197 EI, IX, p. 198
		३	४	५		७	⊤	९		EI, IX, p. 277 EI, XVIII, p. 87
)	२	३	४		६	३	⊦	९		JA, 1863, p. 392
)	२	३	४	५	६	३	⊤	९	०	IA, VIII, p. 133 IA, XI, p. 108 IA, XII, p. 155
)	२	३	४	५	५	७	७	⊍		IA, XII, p. 249 IA, XII, p. 263
१	२	३	४		४	८	३	∿		IA, XIII, p. 250 IA, XIV, p. 351 IA, XXV, p. 177
ι	२	३	४	५	६	९	⊂	३		Bühler
९	२		४	५	६	७	⊏	⊍		Datta and Singh Ojha
१	२	३	४	५	६	७	८	९	०	

Date: seventh to twelfth century CE (Fig. 24.75).
Source: various inscriptions on copper from Northern and Central India.

FIG. 24.39A. *Ancient Nâgarî numerals*

1	2	3	4	5	6	7	8	9	0	Ref.
﹁	﹖	≡	千	﹀	Ɛ	ʔ	ŋ	₹		Datta and Singh
⌐	﹖	≋	↯	⇃		ꞁ	⇂	℉		Ojha
ヽ	﹖	⋑	⅄				⌇	⁇		Smith and Karpinski
ʔ	﹖	⅄	⅄				⊏	℉		
℉			⅄					℃		
								℃		

Date: eighth to twelfth century CE (Fig. 24.3).
Source: various manuscripts from northern and central India (which use neither zero nor the place-value system).

Fɪɢ. 24.39ʙ. *Ancient Nâgarî numerals*

1	2	3	4	5	6	7	8	9	0	Ref.
⇂	₹	₹	𝒴	⅃	₹	ꞁ	⊤	⟩	○	ASI, Rep. 1903–1904, pl. 72
ヽ	⅃	₹	⅄	⅃	⟩	ꞁ	⌐	⟩	○	EI, 1/1892, pp. 155–62
⇂	₹	₹	⅄	⅃	⟩	ꞁ	⌐	⟩	○	Datta and Singh Guitel

Date: 875 to 876 CE (Fig. 24.73).
Source: inscriptions of Gwalior (capital of the ancient princely state of Madhyabharat, situated between the present-day states of Madhya Pradesh and Rajasthan, c.120 km from Agra and over 300 km south of Delhi). The two Sanskrit inscriptions are from the temple of Vaillabhatta-svamim dedicated to Vishnu, and are from the time of the reign of Bhojadeva, dated 932 and 933 of the Vikrama Samvat era, or 875 and 876 CE.

Fɪɢ. 24.39ᴄ. *Ancient Nâgarî numerals*

These are the forms that the Arabs used when they adopted Indian numeration: the proof of this will be seen later on; moreover, in the following tables it can be seen that these figures, if not identical, are very similar to the numerical symbols that we use today.

Notations which are derived from Nâgarî

In Maharashtra, via a southern variant, *Nâgarî* gave birth to *Mahârâshtrî*, which gradually evolved into modern *Marâthî* writing, of which there are currently two forms: *Bâlbodh* (or "academic" writing), used to write Sanskrit, and *Modî*, which is more cursive in form, and is only used to write *Marâthî*. A similar evolution took place for the notation of the nine units (Fig. 24.4 above).

In the state of Rajasthan (bordering Pakistan in the west, Punjab, Haryana and Uttar Pradesh in the north, Madhya Pradesh in the east and Gujarat in the south) *Nâgarî* evolved into *Râjasthanî*. In the northwest of India, however, between the Aravalli Range and the Thar Desert, *Nâgarî* diversified into the cursive forms of *Mârwarî* and *Mahâjanî*, mainly used for commercial purposes.

After the end of the eleventh century, a notation called *Kutilâ* (or "Proto-Bengali") was also born out of *Nâgarî*, from which, in turn, modern *Bengâlî* evolved, sometime after the beginning of the seventeenth century (Fig. 24.10), to which *Oriyâ* (Fig. 24.12), *Gujarâtî* (Fig. 24.8), *Kaîthî* (Fig. 24.9), *Maithilî* (Fig. 24.11) and *Manipurî* can be linked.

The development of Shâradâ notation

After the beginning of the ninth century in Kashmir and Punjab, a northern variant of *Gupta* led to *Shâradâ* notation, which was used in the above parts of India until the fifteenth century at least (Fig. 24.14).

1	2	3	4	5	6	7	8	9	0	Ref.
										IA, XVII, pp. 34–48
										Datta and Singh
										Kaye: *Bakhshâlî manuscript*
										Smith and Karpinski

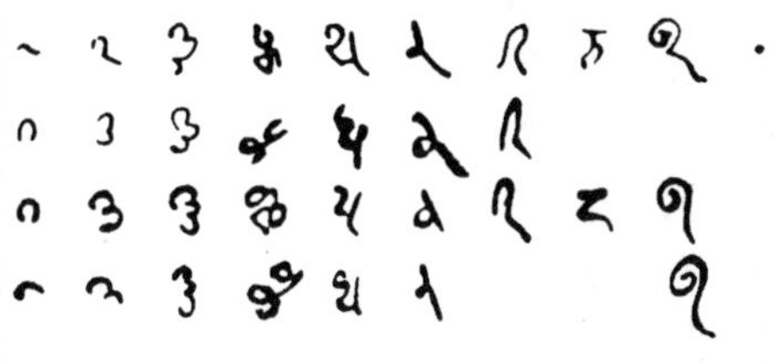

Date: between the ninth and twelfth century CE. (Fig. 24.14).
Source: Manuscript from Bakshali (a village in Gandhara, near Peshawar, in present-day Pakistan, where it was discovered in 1881). The manuscript is written entirely in the *Shâradâ* style, in the Sanskrit language, in both verse and prose, by an anonymous author. It deals with algebraic problems, the numbers being expressed in *Shâradâ* numerals using the place-value system, zero being written as a dot (*bindu*). This manuscript could not have been written earlier than the ninth century CE or later than the twelfth century, but it is possible that it is a copy of – or a commentary on – an earlier document.

FIG. 24.40A. *Ancient Shâradâ numerals*

Notations derived from Shâradâ

It is from this notation that *Tâkarî* (Fig. 24.13), *Dogrî*, *Chamealî*, *Mandealî*, *Kuluî*, *Sirmaurî*, *Jaunsarî*, *Kochî*, *Landa*, *Multânî*, *Sindhî* (Fig. 24.6), *Khudawadî*, *Gurûmukhî* (Fig. 24.7), *Punjâbî* (Fig. 24.5), etc., originated.

1	2	3	4	5	6	7	8	9	0	Ref.
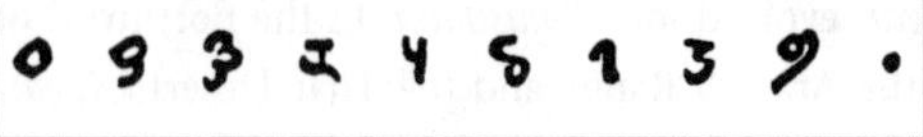 ο	9	3	౩	Y	६	٩	3	9	•	KAV Smith and Karpinski

Date: the fifteenth century CE (approximately).
Source: A Kashmiri document which reproduces the Vedi hymns and texts of the *Atharvaveda* in *Shâradâ* characters (the document is preserved at Tübingen University).

FIG. 24.40B. *Shâradâ numerals (most recent style)*

Nepalese notations

1	2	3	4	5	6	7	8	9	0	Ref.
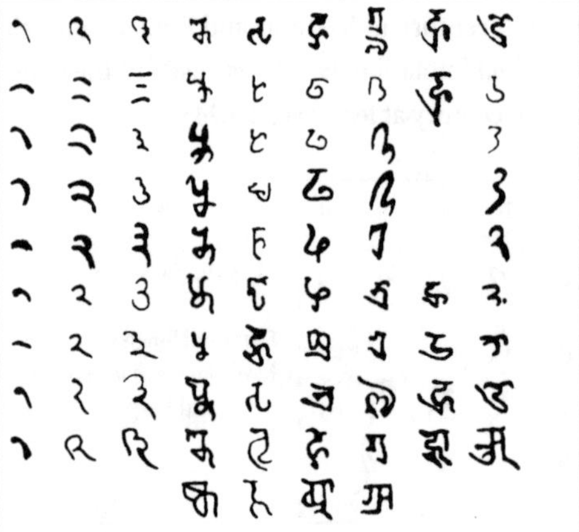 ◌	◌	◌	◌	◌	◌	◌	◌	◌		Bendall Bühler Datta and Singh Ojha Smith and Karpinski

Date: eighth to twelfth century CE (Fig. 24.15).
Source: inscriptions from Nepal and various Buddhist manuscripts from Nepal.

FIG. 24.41. *Ancient Nepali numerals*

Many other systems originated from *Gupta*. After the fifth century CE, one variation evolved into *Siddhamatrikâ* (or Siddam) writing which was used mainly in China and Japan for Sanskrit notation. During its development, some time after the beginning of the ninth century, it gave birth to *Limbu* and modern Nepali (also called *Gurkhali*), specific notations of Nepal whose numerical symbols underwent a parallel evolution (Fig. 24.41).

Notations which originated in India and Central Asia

From the time of the Kushana Empire (first to third century CE) until the Empire of the Guptas, Indian civilisation, along with Buddhism, stretched to Chinese Turkestan, as well as towards northern Afghanistan and Tibet.

Thus one of the notations to be born out of Gupta reached these regions.

Without any radical change, this notation evolved into the writings of Chinese Turkestan, which were used to write Agnean, Kutchean and Khotanese. Each style would have possessed its own figures.

On the other hand, in the various regions of Tibet, the high valleys of the Himalayas and the neighbouring areas of Burma, *Gupta* underwent quite drastic changes to enable spoken languages with very different inflexions to be written down. This is how the Tibetan alphabet came about, the Guptan numerical symbols also being adapted to this graphical style (Fig. 24.16).

Mongolian figures

When the great conqueror Genghis Khan died in 1227, the Mongolian Empire stretched from the Pacific to the Caspian Sea.

J. G. Février (1959) claims that "the Mongolians did not possess any form of writing and that all their conventions were oral; their 'contracts' were alleged to be certain signs carved onto wooden tablets."

But by conquering nearly all of Asia, these half-savage hordes could no longer be contented with such rudimentary methods; so they decided to adopt the writing of the Uighur people of Turfar after they defeated them (the Uighur alphabet constituting a type of Syriac writing, imported by Nestorian monks).

The Mongolians then decided that they wanted an alphabet that was more appropriate for writing their language, mainly because of pressure from the propagators of Buddhism to have their own specific instrument for translating their texts. Their alphabet was created with the collaboration of the Uighurs. They wrote in vertical columns which read from left to right.

However, instead of adopting the non-positional system of the afore-mentioned region, the Mongolians preferred to use Tibetan figures, after the contact that they had had with the latter. Thus "Mongolian" figures were born:

1	2	3	4	5	6	7	8	9	0	Ref.
૧	૨	૩	౮	౧	౬	౨	≮	౯	0	Pihan
Date: thirteenth to fourteenth century CE.										

FIG. 24.42. *Mongolian numerals: the numerals 2, 3, 6 and 0 are recognisable, as well as 9 (or rather its mirror image).*

An evolution from the South to the East

Like *Gupta*, there is another style of writing to come out of *Brâhmî* that is very different from its origins.

1	2	3	4	5	6	7	8	9	0	Ref.
∩	౨	౩	౪	౫	౬	౭	౮			CIIn, III
∩	౩	౩	౪	౫						Bühler Datta and Singh Ojha
௧	౩	౩	౪							
౧	౩		౪							
∩			౫							
			౫							

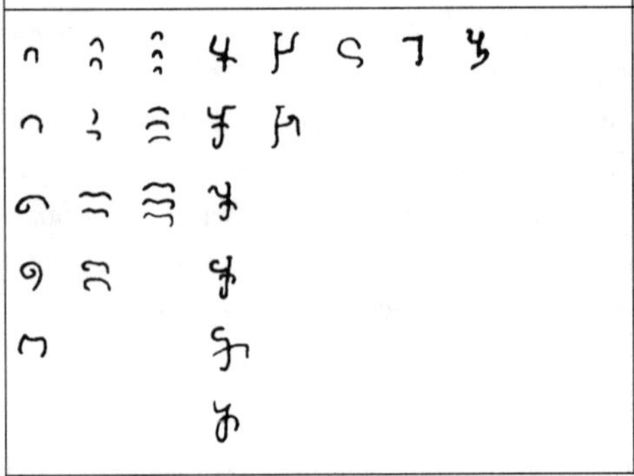

Date: fifth to sixth century CE.
Sources: inscriptions from the Pallava dynasty (who reigned in the southeast of India in the region of the lower Krishna on the Coromandel coast, from the end of the third century CE until the end of the eighth century); Shalankayana inscriptions (a small Hindu dynasty that reigned from 300 to 450 CE, in Vengi and Pedda Vengi, in the region of the Krishna river).

FIG. 24.43. *Numerals of the intermediary counting system of the Pallava*

This is the writing used in inscriptions in Pallava, Shâlankâyana and Valabhî (Fig. 24.43 and 24.44) and the more individualised style of Chalukya and Deccan (Fig. 24.45) and Ganga and Mysore (Fig. 24.46):

1	2	3	4	5	6	7	8	9	0	Ref.
–	꞊	꞉	ꝉ	ꝏ	ꝑ	౨	౫	౯		CIIn, III Bühler
∩	꞉	꞉	౪	ꝏ	ꝑ	౨	౫	౩		Datta and Singh Ojha
⌐	꞉	꞉		౫	ꝑ	౩	౫	౯		
	⌐	꞊		౫	ꝑ	౪	౫			
	꞉			౫		౪				
	꞊			౫						

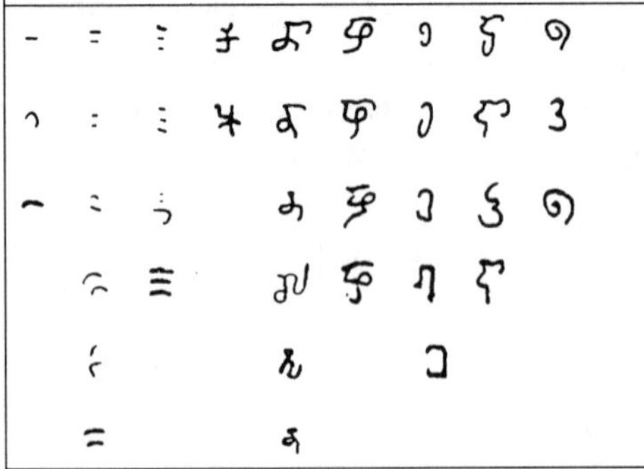

Date: fifth to eighth century CE.
Source: inscriptions from Valabhi (a village in Marathi, capital of the Hindu and Buddhist kingdom which, from 490 to 775, encompassed the present-day States of Gujarat and Maharashtra).

FIG. 24.44. *Numerals of the intermediary counting system of Valabhî*

This is the common basis which would lead progressively on the one hand to the formation of the southern Indian style (attached to Dravidian

1	2	3	4	5	6	7	8	9	0	Ref.
↘	ॐ	ॐ	६	ॐ	ॐ	౨	౪	९		CIIn, III
			౯	౪	ॐ					Bühler
			౯							Datta and Singh
										Ojha

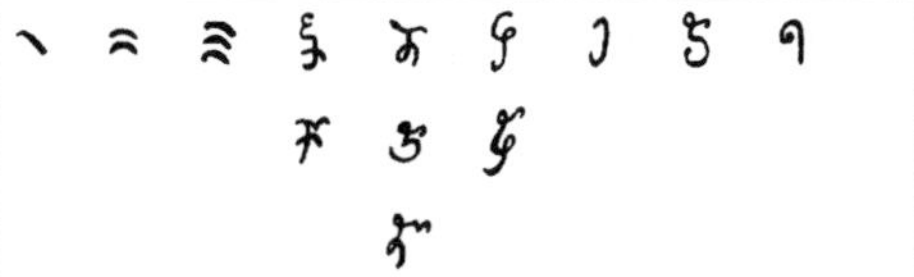

Date: fifth to seventh century CE.
Source: inscriptions of the oldest branch of the Chalukya dynasty of Deccan (known as "de Vatapi", who lived in Badami, in the present-day district of Bijapur, during the sixth century CE).

FIG. 24.45. *Numerals of the intermediary counting system of the Chalukya of Deccan*

1	2	3	4	5	6	7	8	9	0	Ref.
	౩		౨		౮	౭	౧	౫		CIIn, III
					౮			౫		Bühler
								౫		Datta and Singh
										Ojha

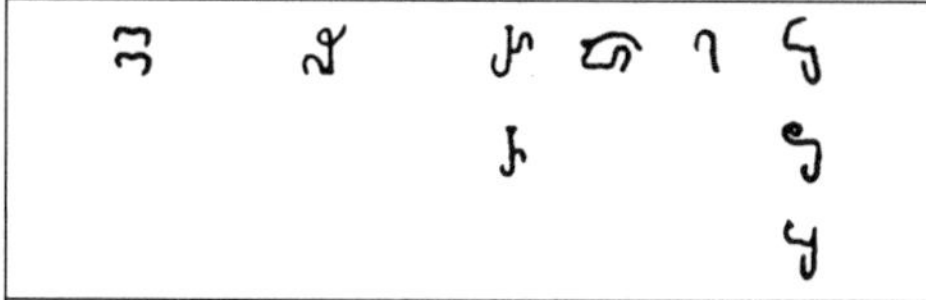

Date: sixth to eighth century CE.
Source: inscriptions of the Ganga dynasty of Mysore (who ruled over a substantial part of the present-day State of Karnataka, from the fifth to the sixteenth century).

FIG. 24.46. *Numerals of the intermediary counting system of the Ganga of Mysore*

styles of writing), and on the other hand to the development of *Pâlî* styles, attached to eastern styles of writing.

Southern (or Dravidian) styles

From one of these systems was derived *Bhattiprolu* writing.

In Telingana, to the east of Andhra Pradesh and the south of Orissa, this gradually became *Telugu* (Fig. 24.20):

In the centre of Deccan, in Karnataka and Andhra Pradesh, it became *Kannara* (Fig. 24.21).

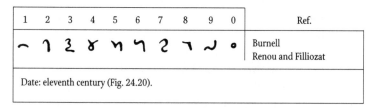

1	2	3	4	5	6	7	8	9	0	Ref.
౧	౧	౩	౪	౫	౧	౨	౧	౨	•	Burnell
										Renou and Filliozat

Date: eleventh century (Fig. 24.20).

FIG. 24.47. *Ancient Telugu numerals*

1	2	3	4	5	6	7	8	9	0	Ref.
∩	⌐	᠀	४	᠈	६	८	᠋	᠌	○	Burnell Renou and Filliozat
Date: sixteenth century CE (Fig. 24.21).										

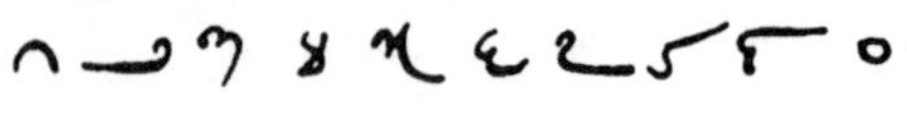

FIG. 24.48. *Ancient Kannara numerals*

To the east of the more southern regions, this became *Grantha* and *Tamil* (Fig. 24.17), as well as *Vatteluttu* (used primarily on the coast of Malabar from the eighth to the sixteenth century), whilst in the west this became the styles known as *Tulu* and *Malayâlam* (Fig. 24.19).

1	2	3	4	5	6	7	8	9	0	Ref.
௮	2	௴	௷	௫	௸	9	ᘉ	ᘃ		Burnell Pihan Renou and Filliozat
	ᘁ	ᘂ	௲	③	௸		ᘄ	ᘆ		
Date: sixteenth century (Fig. 24.19).										

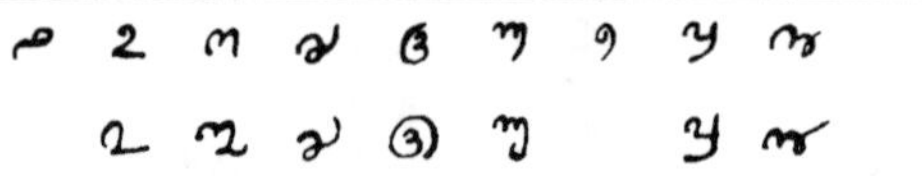

FIG. 24.49. *Ancient Tamil numerals*

Finally, in the extreme south, primarily in Sri Lanka, *Sinhalese* was derived (Fig. 24.22).

The styles of writing of Southeast Asia

At the same time, to the east of India, another variety of intermediate systems developed to lead to the first forms of *Pâli*. Attached to the ancient writing *Ardha-Mâgadhî* (the ancient language spoken in Magadha), these diversified, and led to the characteristic forms of writing used (and still used today) to the east of India and in Southeast Asia.

From this system was derived: *Old Khmer* (developed some time after the beginning of the sixth century CE); *Cham*, used in part of Vietnam, from the seventh century to some time around the thirteenth century; *Kawi* in Java, Bali and Borneo (Fig. 24.50), which dates back to the end of the seventh century, but which has now fallen into disuse; modern *Thai* writing (Shan, Siamese, Laotian, etc.), whose first developments date back to the thirteenth century (Fig. 24.24); *Burmese* (Fig. 24.24 and 24.51), which derived from *Môn* in the eleventh century, used by populations of Pegu before the Burmese invasion; *Old Malay* (Fig. 24.51), from which *Batak* (in the central region of the island of Sumatra), *Redjang* and *Lampong* (in the southeast of the same island), *Tagala* and *Bisaya* in the Philippines, as well as *Macassar* and *Bugi* (from Sulawezi) derived.

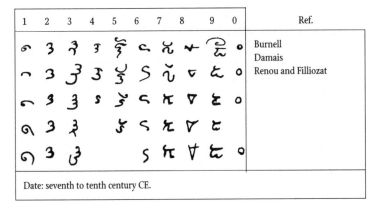

1	2	3	4	5	6	7	8	9	0	Ref.
										Burnell
										Damais
										Renou and Filliozat

Date: seventh to tenth century CE.

FIG. 24.50. *Kawi (ancient Javanese and Balinese) numerals*

1	2	3	4	5	6	7	8	9	0	Ref.
										Latter
										Smith and Karpinski

Date: seventeenth century CE (approx.).

FIG. 24.51. *Ancient Burmese numerals*

Types of numerals that derive from Brâhmî

These fall into three categories, like the types of writing of the same names (Fig. 24.52 and 24.53):

I. The family of writing styles from Northern and Central India and Central Asia, which developed from *Gupta* writing:

    1. Forms of writing derived from *Nâgarî*:

        a. Mahârâshtrî numerals;

        b. Marâthî, Modî, Râjasthânî, Mârwarî and Mahâjani (derived from Mahârâshtrî) numerals;

        c. Kutilâ numerals;

        d. Bengalî, Oriyâ, Gujarâtî, Kaîthî, Maithilî and Manipurî (derived from Kutilâ) numerals.

    2. Forms of writing derived from *Shâradâ*:

        a. Tâkarî and Dogrî numerals;

        b. Chameâlî, Mandealî, Kuluî, Sirmaurî and Jaunsarî numerals;

        c. Sindhî, Khudawadî, Gurumukhî, Punjâbî (etc.) numerals;

        d. Kochî, Landa, Multânî (etc.) numerals.

    3. Types of *Nepalese* writing:

        a. Siddham numerals (influenced by the Nâgarî style);

        b. Modern Nepali numerals (derived from Siddham numerals).

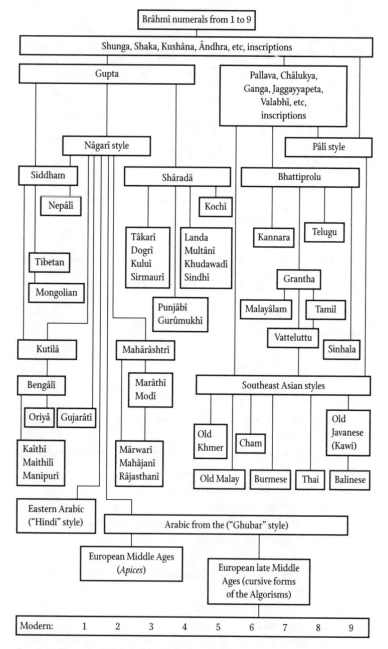

FIG. 24.52. *Numerals which derived from Brâhmî numerals*

4. Types of *Tibetan* writing:

    a. Tibetan numerals (derived from Siddham numerals);

    b. Mongolian numerals (derived from Tibetan numerals).

5. Types of writing from Chinese Turkestan (derived from *Siddham* numerals).

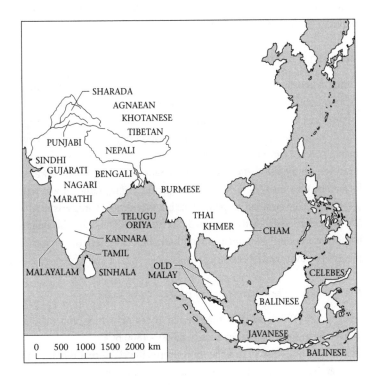

FIG. 24.53. *Geographical areas where writing styles of Indian origin are used*

II. The family of writing styles from southern India, which developed from *Bhattiprolu*, a distant cousin of Gupta:

    1. Telugu and Kannara numerals;

    2. Grantha, Tamil and Vatteluttu numerals;

    3. Tulu and Malayâlam numerals;

    4. Sinhalese (Singhalese) numerals.

III. The family of "oriental" writing styles, which developed from *Pâlî* writing, which itself derives from the same source as Bhattiprolu:

    1. Old Khmer numerals;

    2. Cham numerals;

    3. Old Malay numerals;

    4. Kawi (old Javanese and Balinese) numerals;

    5. Modern Thai-Khmer (Shan, Lao and Siamese) numerals;
    6. Burmese numerals;
    7. Balinese, Buginese, Tagalog, Bisaya and Batak numerals.

As we will see later, Arabic numerals, East and North African alike, derive from the Indian *Nâgarî* numerals, and the numerals that we use today come from the *Ghubar* numerals of the Maghreb. Thus these diverse signs can be placed in the first category of group I.

The mystery of the origin of Brâhmî numerals

Having demonstrated how the above types of numerals all derived from *Brâhmî*, it is now time to explain the origin of the *Brâhmî* numerals themselves.

For some time now, this writing has conserved an ideographical representation for the first three units: the corresponding number of horizontal lines. However, since their emergence, the numerals 4 to 9 have offered no visual key to the numbers that they represent (for example, the 9 was not formed by nine lines, nine dots or nine identical signs; rather, it was represented by a conventional graphic). This is a significant characteristic which has yet to be satisfactorily explained. To try and understand this enigma, let us now examine the principle hypotheses that have been put forward on this subject over the last century.

First hypothesis: The numerals originated in the Indus Valley

S. Langdon (1925) believed that Indian styles of writing and numerals derived from the Indus Valley culture (2500 – 1800 BCE).

The first objection to this theory concerns the claim that there is a link between Indian letters and Proto-Indian pictographic writing.

We have just seen that *Brâhmî* writing actually developed from the ancient alphabets of the western Semitic world through the intermediate of a variety of Aramaean: this link has been satisfactorily established, even if samples of this intermediate writing have not yet been found (Fig. 24.28). Documents from this civilisation are separated from the first texts written in *Brâhmî* and in a purely Indian language by the space of two thousand years. However, the fact that the writing of Indus civilisation has not yet been deciphered does not concern us here.

It is not known whether the ancient civilisations of Mohenjo-daro and Harappâ still existed when India was invaded by the Aryans, or if their writing had developed during this interval.

Moreover, no mention is made of this link in Indian literature, and with good cause: the invaders probably found writing repugnant because, like all Indian European peoples, they attached great importance to oral tradition [see T. V. Gamkrelidze and V. V. Ivanov (1987); A. Martinet (1986)].

It is almost certain that when the Aryans arrived in India they brought no form of writing with them, as happened in Greece and the rest of Europe, whilst various Indo-European peoples came in successive waves to conquer the continent. Their intellectual and spiritual leaders would certainly have had "a knowledge of the great religious poems ; but it seems that their literature was written at a later date, and then the literate men would doubtless have preferred to keep the oral tradition going as long as possible to perpetuate their prestige and their privileges" [M. Cohen (1958)].

Therefore, Langdon's hypothesis has no foundation, because we do not know if any link exists between numerals used by the Indus civilisation and "official" Indian numerals. The theory becomes even more unlikely when one considers that the documentation which survived from the Indus Valley is very fragmentary and does not provide enough information for us to reconstruct the system as a whole.

Second hypothesis: Brâhmî numerals derive from "Aramaean-Indian" numerals

Since Indian letters derive from the Aramaean alphabet, would it not be natural to presume that *Brâhmî* numerals were similarly the offshoot of one of the ancient systems of numerical notation of the western Semitic world? At first glance, this hypothesis seems plausible, since a numerical notation which derives directly from Aramaean, Palmyrenean, Nabataean, etc., can be found in many inscriptions from Punjab and Gandhara. This style is known as "Aramaean-Indian", and is related to *Karoshthî* writing (see Fig. 24.54 and Fig. 18.1 to 12).

However, once we have looked at *Brâhmî* numerals for numbers higher than the first nine units, we will see that this system is too different from Aramaean-Indian for this hypothesis to be taken into consideration.

On the one hand, Aramaean-Indian reads from right to left, whilst *Brâhmî* (and nearly all the styles of writing related to it) reads from left to right.

On the other hand, in the *Karoshthî* system, the numbers 4 to 9 are generally represented by the appropriate number of vertical lines, whilst the *Brâhmî* system gives them independent signs which give no direct visual indication as to their meaning.

Moreover, the original *Brâhmî* system possesses specific numerals for 10, 20, 30, 40, 50, 60, 70, 80, 90, 100, 200, etc., whilst the *Karoshthî* system only has specific figures for 1, 10, 20 and 100.

Finally, the initial Indian system is essentially based on the principle of addition, whilst Aramaean-Indian is based on a hybrid principle combining addition and multiplication.

Thus this hypothesis must be discarded.

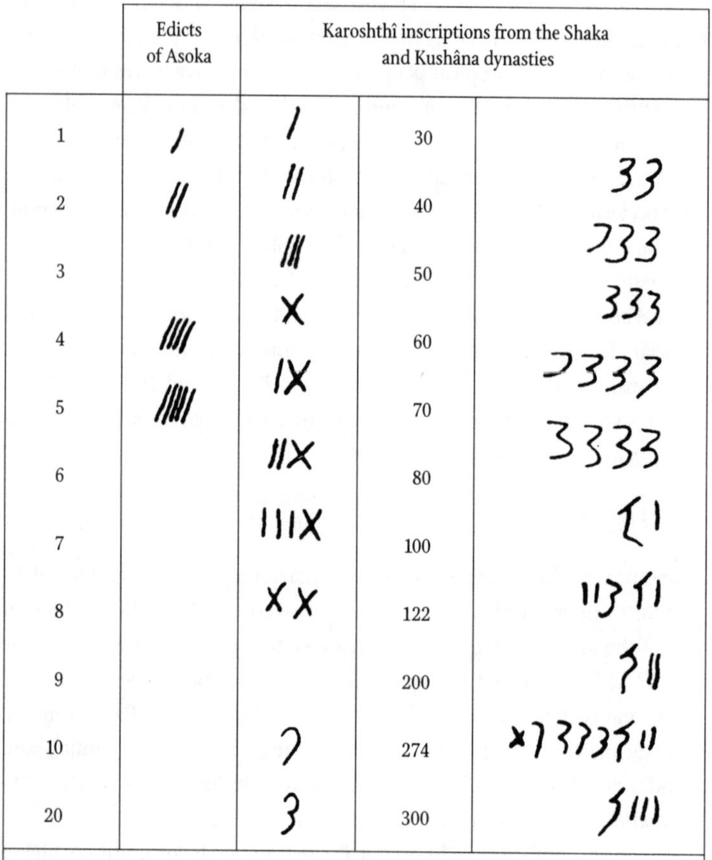

	Edicts of Asoka	Karoshthî inscriptions from the Shaka and Kushâna dynasties		
1	/	/	30	33
2	//	//	40	ㄱ33
3		///	50	333
4	////	X	60	ㄱ333
5	/////	IX	70	3333
6		//X	80	ꝍ
7		///X	100	॥ʒꝍ
8		XX	122	ʒ॥
9			200	Xㄱʒʒʒꝍ॥
10		?	274	
20		3	300	ʒ॥)

Date: third century BCE to fourth century CE.
Sources: inscriptions written in Karoshthî from the edicts of Asoka (3rd c. BCE), where the numerals are partially attested; and Karoshthî inscriptions (2nd c. BCE – 4th c. CE) from the north of Punjab and the former province of Gandhara (region in the north-west of India, the extreme north of Pakistan and the northeast of Afghanistan, which was part of the Persian Empire, before it was conquered in 326 BCE by Alexander the Great), where these numerals are more fully attested.

FIG. 24.54. *"Aramaean-Indian" numerical notations*

Third hypothesis: Brâhmî numerals derive from the Karoshthî alphabet

Another hypothesis (suggested by Cunningham, and later shared by Bayley and Taylor), proposes that *Brâhmî* numerals derived from the letters of the *Brâhmî* alphabet, used as the initials of the Sanskrit names of the corresponding numbers. The following table demonstrates this theory:

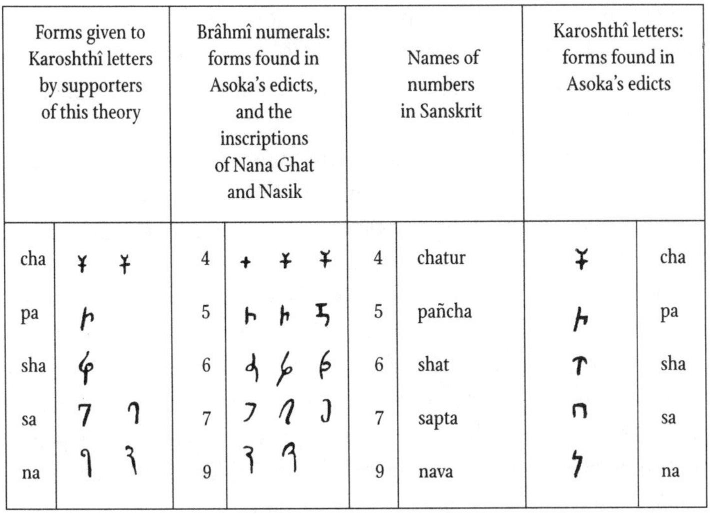

Forms given to Karoshthî letters by supporters of this theory		Brâhmî numerals: forms found in Asoka's edicts, and the inscriptions of Nana Ghat and Nasik		Names of numbers in Sanskrit		Karoshthî letters: forms found in Asoka's edicts	
cha	⅄ ⅄	4	+ ⅄ ⅄	4	chatur	⅄	cha
pa	⍩	5	⍩ ⍩ ⅂	5	pañcha	⍩	pa
sha	φ	6	⏀ φ φ	6	shat	Τ	sha
sa	7 ⏋	7	⏌ ⏋ ⏌	7	sapta	⊓	sa
na	⏋ ⏗	9	⏗ ⏘	9	nava	⅂	na

FIG. 24.55.

The link that has been established here, however, is too tenuous, for at least three reasons.

The first is that the forms given by the supporters of this theory actually come from inscriptions from different eras, most often from later eras, therefore holding little significance for the problem in question, which concerns a graphical system. This is how such evolved forms like those of Kushâna inscriptions in the northwest of Punjab (second to fourth century CE) came to be confused with more ancient styles such as inscriptions from the Shaka era (second century BCE to first century CE) or those from Asoka's time (third century BCE).

The second reason is that the signs which are given for the presumed phonetic values are very similar (if not identical) to letters which are known to represent other numerical values.

Thirdly, the supporters of this hypothesis allowed themselves to get so carried away that they themselves actually added the final touch to the Aramaean-Indian letters which was needed in order to prove their theory.

There is another even more fundamental reason, however, why the above two theories must be rejected: they presume that *Karoshthî* pre-dates *Brâhmî*, whilst today's specialists believe precisely the opposite.

It is certain that *Karoshthî* writing derives from the Aramaean alphabet, because several of its characters are identical (in form and structure) to their Aramaean equivalents; and, like Semitic writing, it reads from right to left. *Karoshthî* remains very different from the latter style of writing, however, because it was adapted to the sounds and inflexions of Indian-European languages. It was probably introduced to the northwest of India in Alexander the Great's time (c. 326 BCE), and was used there until the fourth century CE, and until a slightly later date in Central Asia.

Nevertheless, *Brâhmî* does not derive from this writing. *Brâhmî* stems from another variant of Aramaean, whose characters were adapted to Indian languages whilst the direction of the writing was changed so that it read from left to right.

It is highly probable that *Brâhmî* had been around long before Asoka's time, because by then it was not only fully established, but also and above all it was in use in all of the Indian sub-continent. Therefore, it is very likely that *Karoshthî* was not used in other parts of India except for the regions of Gandhara and Punjab because, as J. G. Février (1959) has already pointed out, it emerged when an Indian style of writing already existed, namely *Brâhmî*, which was in use since roughly the fifth century BCE.

Thus it would seem unlikely that *Karoshthî* could have influenced the formation of *Brâhmî* letters and numerals.

Fourth hypothesis: Brâhmî numerals derive from the Brâhmî alphabet

This hypothesis would initially seem quite feasible when one considers that many kinds of numerals have developed in this way.

The Greeks and the southern Arabic people, for example, gave, as a numerical symbol, the initial letter from their respective alphabets of the name of the number.

We also know that the Assyro-Babylonians, who had no numeral for 100 in their Sumerian cuneiform system, decided to use acrophonics; thus, the syllable *me* was used to denote this number, the initial of the word *me'at*, which means "hundred" in Akkadian.

Ethiopian numerals, which now appear to be completely independent from Ethiopian writing, actually derive from the first nineteen letters of the Greek numeral alphabet; this dates back to the fourth century CE, when the town of Aksum (not far from the modern town of Adoua) was the capital of the ancient kingdom of Abyssinia.

Thus the theory that the numerals of a given civilisation derive from its own alphabet is quite feasible.

In other words, the Indians could quite possibly have used a certain number of the letters of the *Brâhmî* alphabet to create a corresponding numbering system. This is the substance of J. Prinsep's hypothesis (1838); he believed that the prototypes of the Indian numerals constituted the initial, in *Brâhmî* characters, of the Sanskrit names for the corresponding numbers.

However, as this hypothesis has never been confirmed, it remains in the realm of conjecture. Moreover, the author also mixed archaic styles with later ones, and "customised" the characters in question to make his theory appear to hold water.

Fifth hypothesis: Brâhmî numerals derive from a previous numeral alphabet

B. Indrâji (1876) put forward the theory that *Brâhmî* numerals derived from an alphabetical numeral system that was in use in India before Asoka's time.*

If we compare the shapes of the numerals with the letters that appear in Asoka's *Brâhmî* inscriptions of Nana Ghat and Nasik, we can see that there are quite obvious similarities. The numeral for the number 4, a kind of "+" in Asoka's edicts, is identical to the sign used to write the syllable *ka*. Likewise, the 6 is very similar to the syllable (Fig. 24.29). The 7 resembles the syllable *kha*, whilst the 5 has the same appearance as the *ia*, *ña*, etc. (Fig. 24.30).

However, this link which has been established between the original Indian numerals and the letters of the *Brâhmî* alphabet is not convincing.

First, the *Brâhmî* numerals for 1, 2 and 3 do not resemble any letter: they are formed respectively by one, two and three horizontal lines (Fig. 24.29 to 35). Moreover, no phonetic value was assigned to the ancient forms of the *Brâhmî* numeral which represented multiples of 10 (see Fig. 24.70). Even where this relationship has been established, there is too much variation in the attribution of phonetic values to the signs in question. Thus, whilst the numeral 4 has been connected to the syllable *ka*, in its diverse forms it can equally be said to resemble the letter *pka* or the syllables *pna*, *lka*, *tka* or *pkr*. Similarly, the shape of the numeral representing the number 5 resembles the syllable *tr* as well as the following: *ta*, *tâ*, *pu*, *hu*, *ru*, *tr*, *trâ*, *nâ*, *hr*, *hra or ha* [see B. Datta and A. N. Singh (1938), p. 34].

In other words, if such a system did exist in Asoka's time, it is impossible to discover the principle by which it might have functioned.

_____

* Along with the various styles of numerals, the Indians have long known and used a system for representing numbers which involves vocalised consonants of the Indian alphabet which, in regular order, each have numerical value. These are known as *Varnasankhyâ* in Sanskrit, the system of "letter-numbers". The system varies according to the era and region but always follows the method of attributing numerical values to Indian letters, sometimes even following the principle used in numerical representations (the place-value system or the principle of addition). Included in the numerous systems of this kind is the one which the famous astronomer Âryabhata (c. 510 CE) used to record his astronomical data; there is also the system called *Katapayâdi* used (amongst others) by the ninth-century astronomers Haridatta and Shankaranârâyana, as well as *Âksharapallî* frequently used in *Jaina manuscripts. These systems are still in common use today in various regions of India, from Maharashtra, Bengal, Nepal and Orissa to Tamil Nadu, Kerala and Karnataka. They are also used by the Sinhalese, the Burmese, the Khmer, the Siamese and the Japanese, as well as the Tibetans, who often use their letters as numerical signs, mainly to number their registers and manuscripts. Details can be found under the entries *Varnasankhyâ, *Âksharapallî, *Numeral Alphabet, *Âryabhata and *Katapayâdi numeration of the Dictionary.

Despite the shakiness of their explanations, the supporters of Indrâji's theory conjectured that the idea of assigning numerical values to letters of the alphabet dated back to the most ancient of times, their reasoning being that "Indian, Hindu, Jaina and Buddhist traditions attribute the invention of *Brâhmî* writing and its corresponding numeral system to Brahma, the god of creation."

(Of course, such an argument cannot be taken seriously, especially in the case of Indian civilisation, where such traditions were actually only developed relatively recently and are due to two basic traits of the Indian mentality. First, there is the desire of some of these theorists to make such concepts hold more water in the eyes of their readers, disciples or speakers, in attributing their invention to Brahma. There are also those who, convinced of the innate character of the Indian letters and numerals, do not even consider it necessary to give a historical explanation for their existence. In the first instance, the motive was to make these concepts sacred, and in the second, to make them timeless. The latter conveys India's fundamental psychological character; an obsession with the past which always involves wiping out historical events and replacing fact with religious history, which takes no account of archaeology, palaeography or, most importantly, history.)

The pioneer of the above theory even went so far as to claim that the first Indian numeral alphabet dates back to the eighth century BCE. According to Indrâji, it was Pânini (c. 700 BCE) who first had the idea of using the consonants and vowels of the Indian alphabet to represent numbers.

क ka	ख kha	ग ga	घ gha	ङ ṅa
च cha	छ chha	ज ja	झ jha	ञ ña
ट ṭa	ठ ṭha	ड ḍa	ढ ḍha	ण ṇa
त ta	थ tha	द da	ध dha	न na
प pa	फ pha	ब ba	भ bha	म ma
य ya	र ra	ल la	व va	
श śha	ष sha	स sa		
ह ha				

Fig. 24.56. *Consonants of the Devanagari (or Nâgari) alphabet*

Pânini is the famous grammarian of the Sanskrit alphabet: born in Shalatula (near to Attock on the Indus, in the present-day Pakistan), he is considered to be the founder of Sanskrit language and literature; his work, the *Ashtâdhyâyî* (also known as *Pâninîyam*), remains the most famous work on Sanskrit grammar [see L. Frédéric (1987)]. We have no exact dates in Pânini's life, and there is much doubt surrounding the work which is attributed to him.

In other words, the date suggested by Indrâji for the invention of the first Indian alphabetical numeral system has no foundation at all, especially when one considers that there is no known written document, nor specimen of true Indian writing, which dates so far back in Indian history. It goes without saying, then, that this hypothesis must also be rejected.

The origin of Indian alphabetical numerals

So where does the idea of writing numbers using the Indian alphabet come from?

It must be made clear straight away that the idea did not come from Aramaean merchants, who brought their own writing system to India (Fig. 24.28). With a few later exceptions, the northwestern Semites never used their letters for counting; their numerals were of the same kind as the *Karoshthî* (or Aramaean-Indian) system (see Fig. 24.54 and Fig. 18. 1 to 12).

One could attribute the idea to a Greek influence, in the light of Alexander the Great's conquest of the Indus in 326 BCE, and moreover because this kind of system was in use in Greece since the fourth century BCE. However, this hypothesis is not plausible, because no Indian inscriptions, during or after Asoka's reign, show any evidence of alphabetical numeration.

In fact, the first numeral system of this kind was invented in the Indian sub-continent by the famous mathematician and astronomer, *Âryabhata. This system was undeniably unique compared to all the other previous and contemporary systems; not only has it been quoted in numerous works and commentaries over the centuries, but it has also inspired a considerable diversity of authors, commentators and transcribers, in various eras, to draw comparisons with it and both similar and very different systems (see *Âryabhata* and *Katapayâdi* in the Dictionary).

Sixth Hypothesis: Brâhmî numerals came from Egypt

Here are some other hypotheses put forward as to the origin of *Brâhmî* numerals.

Bühler (1896) and A. C. Burnell (1878) believed that the Indians owed their *Brâhmî* writing to Pharaonic Egypt. Bühler claimed that it derived from hieratic writing (see Fig. 23.10), and Burnell believed that *Brâhmî* writing derived from demotic writing.

Indian Civilisation

Bühler's theory is not totally unfounded, because there is a much stronger similarity between *Brâhmî* writing and Egyptian hieratic writing than there is between the former and the demotic writing of the same civilisation. However, is this partial resemblance significant enough to suggest that Egypt had such an influence on India's distant past?

Arabia, the legendary land of "Pount", was a staging post for Egyptian trade. Ships sailed to the Red Sea along the eastern Nile delta and along a canal first to the Bitter Lakes, then to the Gulf of Suez. It is possible that these same merchant ships, in their quest for eastern goods, travelled further than Arabia, not only to the areas around the Persian Gulf, but also as far as the mouth of the Indus [see A. Aymard and J. Auboyer (1961)].

Conversely, during Alexander the Great's time, India communicated with the Caspian and the Black Sea by river navigation, notably along the Amu Darya; overland routes also led from Europe to India through Bactria, Gandhara and the Punjab, giving access to ports on the western coast of India. Commercial relations became more and more firmly established between Egypt and India, and ships even sailed as far as the coast of Malabar, in particular to the port of Muziris (now the town of Canganore) [see Aymard and Auboyer (1961)].

These relations, however, occurred at a relatively late time and do not really prove anything in terms of the transmission of Egyptian hieratic numerals: the amount of time which separates these numerals from their *Brâhmî* counterparts is too great to allow this hypothesis to be taken into consideration.

(It must be remembered that hieratic numerals were almost obsolete in Egypt by the eighth century BCE; therefore, if this system was transmitted to India, the transmission cannot have taken place any later than this time. As we possess no information about India at this time, this hypothesis cannot be proved.)

Moreover, the above comparison only concerns units; there is a clear difference between the other symbols (numerals representing 10 or above, which have not yet been discussed: see Fig. 24.70). Therefore the comparison only concerns a small percentage of the numerals.

The origin of the first nine Indian numerals

There is another hypothesis which seems much more plausible, even in the absence of any documentation.

Basically, we have already proved this hypothesis: different civilisations have had the same needs, living under the same social, psychological, intellectual and material conditions. Independently of one another, they have followed identical paths to arrive at similar, if not identical results.

This explains the existence of certain numerals of other civilisations which resemble and often represent the same numbers as *Brâhmî* numerals, and which generally date back several centuries before Asoka's time.

On consulting Figs. 24.57 and 24.29 to 35, we can see signs which are not Indian, yet which are very similar to the various ways of writing the numbers 1, 2 and 3 in Indian civilisations. We can also see the evident similarity between the Nabataean or Palmyrenean "5" and that of ancient India, as well as the physical resemblance between the Egyptian hieratic or demotic "7" and "9" and their Indian counterparts.

What these analogies actually prove is not the unlikely theory that the first nine Indian numerals came from one of the other civilisations, but rather that there are universal constants caused by the fundamental rules of history and palaeography. These similarities occur because other civilisations used similar writing materials to those of ancient India, for example the calamus (a type of reed whose blunted end was dipped in a coloured substance), and which was used by Egyptian and western Semitic scribes (Aramaeans, Nabataeans, Palmyreneans, etc.) to write on papyrus or parchment, which was long used on tree bark or palm leaf in Bengal, Nepal, the Himalayas and in all of the north and northwest of India.

We know to what extent the nature of the instrument influenced, on the one hand, Egyptian manuscript writing, and on the other hand, the writings of the ancient Semitic world.

In the first case, the use of the calamus turned the hieroglyphics of monumental Egyptian writing into cursive hieratic signs, changing the detailed, pictorial symbols into a shortened, more simplified form, perfectly adapted to the needs of manuscript writing and rapid notation.

In the latter case, the same writing apparatus was used to transform the rigid and angular shape of Phoenician writing into the rounder, more cursive and fluid forms, like that of Elephantine Aramaean scribes.

Thus the superposition of two or three horizontal lines, first transformed into one complete sign by a ligature, gave birth to the same forms as the Indian numerals for 2 and 3, whose palaeographical styles vary considerably according to the era, the region and the habits of the scribe (Fig. 24.58).

This explanation relies on the assumption that horizontal lines formed the first three of the ancient ideographical Indian numerals, and this is what *Brâhmî* inscriptions written after the third century BCE (Shunga, Shaka, Kushâna, Mathurâ, Kshatrapa, etc.) would suggest (Fig. 24.30 to 38). This figurative representation was still in use during the time of the Gupta Dynasty (third to sixth century CE), and even persisted in some areas until the eighth century.

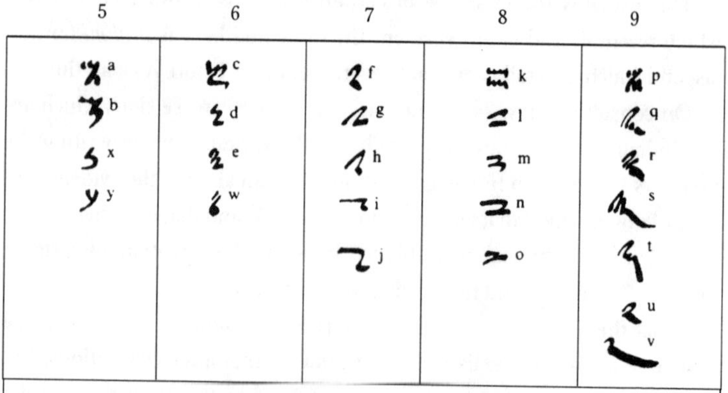

5	6	7	8	9
γ a	**ξ** c	**ζ** f	**≡** k	**χ** p
ξ b	**ξ** d	**ζ** g	**ζ** l	**ζ** q
ζ x	**ξ** e	**ζ** h	**ʒ** m	**ζ** r
y y	**ζ** w	**ʔ** i	**ʒ** n	**ζ** s
		ʔ j	**ʒ** o	**ζ** t
				ζ u
				ζ v

- Egyptian numerals: a (HP I, 618, Abusir); b (HP I, 618, Elephantine); c (HP II, 619, Louvre 3226); d (HP II, Louvre 3226); e (HP II, 619, Gurob); f (HP I, 620, Illahun); g (HP I, 620, Bulaq 18); h (HP II, 620, Louvre 3226); i (HP II, 620, P. Rollin); j (HP III, 620, Takelothis); k (HP I, 621, Elephantine); l (HP I, 621, Illahun); m (HP I, 621, Math); n (HP I, 621, Ebers); o (HP III, 621, Takelothis); p (HP I, 622, Abusir); q (HP I, 622, Illahun); r (HP I, 622, Illahun); s (HP I, 622, Bulaq, Harris); t (HP II, 622, P. Rollin); u (HP II, 622, Gurob); v (HP II, 622, Harris); w (DG, 697, Ptol.); Nabataean numerals: x (CIS, II1, 212); Palmyrenean numerals: y (CIS, II3, 3913).

FIG. 24.57. *Numerals which have the same appearance and numerical value as their Brâhmî equivalents. [Ref. Möller (1911–12); Cantineau; Lidzbarski (1962)]*

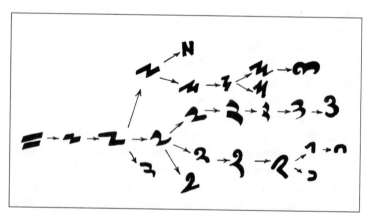

FIG. 24.58. *Evolution of Indian numerals 2 and 3*

However, if we examine Asoka's edicts (c. 260 BCE), we can see that throughout the Mauryan empire, the numbers 1, 2 and 3 were not represented by superposed horizontal lines, but by one, two or three vertical lines (Fig. 24.29).

Why did this change of direction occur? And why did it happen between the third and second century BCE, when the representations had been horizontal since the time of the Buddhist inscriptions of Nana Ghat (Fig. 24.30)?

The second question is difficult to answer, as no documentation has been found from that time on this subject (if indeed anyone took the trouble to write about something which must have seemed so insignificant). However, this is of little importance; we are only interested here in how such a change came about.

Could it have occurred due to aesthetic reasons? This is as unlikely as the possibility that the new notation evolved for practical reasons. To draw a line one, two or three times, whether vertically or horizontally, has no aesthetic value, and involves practically the same amount of exertion, unless it goes against what one is accustomed to writing.

On the other hand, this modification could have been due to the realisation that a vertical representation of the first three units was likely to be confused with the *danda*. This is a punctuation mark in the form of a small vertical line (|), which the Indians have long used in their Sanskrit texts to mark the end of a line or of part of a sentence, which they double (||) to indicate the end of a sentence, couplet or strophe.

The invention of the *danda* in the second century BCE could be responsible for the change in direction of the lines representing the numbers 1 to 3. This is mere conjecture which for the moment remains without proof or confirmation.

Here is another question: why did the Indians conserve these representations of the first three units for so long, when the numerals for 4 to 9 had already graphically evolved into independent numerals, which offered no visual clue as to the numbers they represented (Fig. 24.29 to 38)? This is not only true of the Indians: many other civilisations have offered us similar puzzles over the ages, notably those of China and Egypt.

The explanation lies in a basic human psychological trait, which was discussed in Chapter 1. Whilst it was necessary to have other signs than four or five to nine lines for the numbers 4 or 5 to 9, it was not judged necessary to change the signs for units which were lower or equal to 4; this was not only because these symbols could be drawn quickly and easily, but also and above all because without needing to count, the eye can easily distinguish between a number of lines when they number four or less. Four is the limit, beyond which the human mind has to begin to count in order to determine the exact quantity of a given number of elements.

So what was the reasoning behind the formation of the other six *Brâhmî* numerals? Are they purely conventional signs, created artificially to supply a need? Probably not. Taking the universal laws of palaeography into account, and the evidence surrounding the formation of similar numerals in other cultures, it is more likely that the numerals were born out of prototypes formed by the primitive grouping of a number of lines representing the value of the unit.

In other words, to all appearances, the *Brâhmî* numerals of Asoka's edicts were to their ideographical prototypes as Egyptian hieratic numerals were to their corresponding hieroglyphic numerals.

As the lines representing the numbers 1 to 3 were vertical before they were horizontal, one could reasonably presume that the first nine *Brâhmî* numerals constituted the vestiges of an old indigenous numerical notation, where the nine numerals were represented by the corresponding number of vertical lines; a notation, doubtless older than *Brâhmî* itself*, where, like the Egyptian hieroglyphic system, the Cretan or the Hittite system for example, the vertical lines were set out as in Figure 24.59.

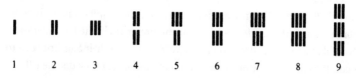

FIG. 24.59. *A plausible reconstruction of the original Indian ideographical notation : the starting point of the evolution which led to the Brâhmî numerals for 4 to 9 (those for 1 to 3 retaining their ideographical form for many centuries, although represented horizontally rather than vertically)*

To enable the numerals to be written rapidly, in order to save time, these groups of lines evolved in much the same manner as those of old Egyptian Pharaonic numerals. Taking into account the kind of material that was written on in India over the centuries (tree bark or palm leaves) and the limitations of the tools used for writing (calamus or brush), the shape of the numerals became more and more complicated with the numerous ligatures (Fig. 24.60), until the numerals no longer bore any resemblance to the original prototypes. Thus a primitive numbering system became one of numerals of distinct forms which gave no visual indication as to their numerical value: the *Brâhmî* numerals of the first three centuries BCE.

Taking into consideration the universal constants of both psychology and palaeography, this is currently the most plausible explanation of the origin of the nine Indian numerals. The summary at the end of this chapter demonstrates the likely stages of the development of *Brâhmî* numerals.

Therefore, it appears that *Brâhmî* numerals were autochthonous, that is to say, their formation was not due to any outside influence. In all probability, they were created in India, and were the product of Indian civilisation alone.

In other words, one could say that the problem of the origin of our present-day numerals has been satisfactorily solved. This is also demonstrated in

---

*This is not at all surprising if one considers that, on the one hand, the ancient civilisation of the Indus which preceded the Aryans used exactly this type of notation (Fig. 1.14), and that on the other hand, Sumerian civilisation developed a numeral system even before creating its own writing system.

the palaeographical tables of Fig. 24.61 to 69, which constitute the complete historical synthesis of the question, and which have been set out taking into account all the details proved both previously and subsequently.

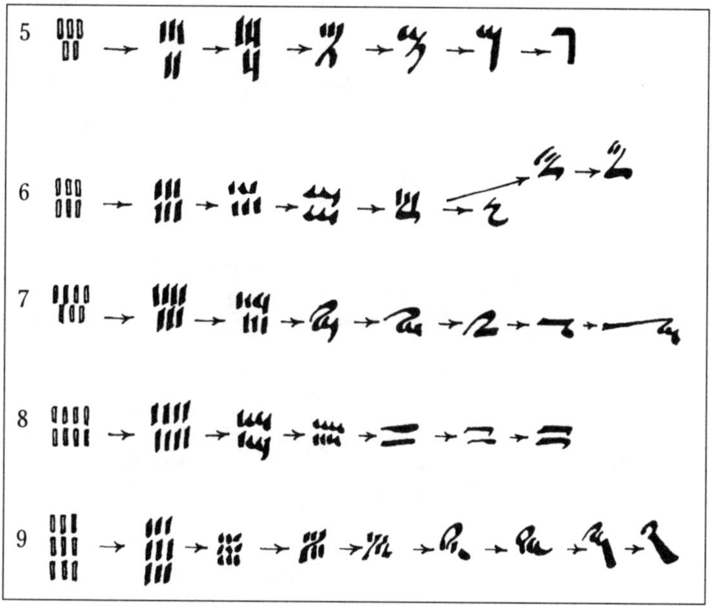

FIG. 24.60. *Results of the graphical evolution of the signs which were originally formed by the juxtaposition or superposition of several vertical lines, these lines being drawn on a smooth surface and written with a calamus with a blunt tip dipped into a coloured substance. This evolution took place among the Egyptians.*

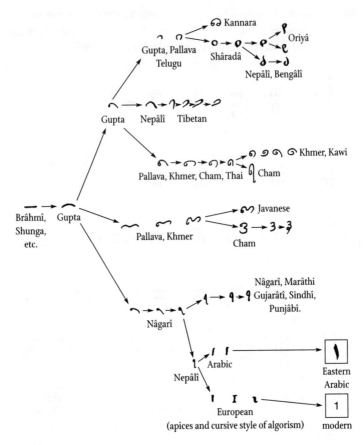

FIG. 24.61. *Origin and evolution of the numeral 1. (For Arabic and European numerals, see Chapters 25 and 26.)*

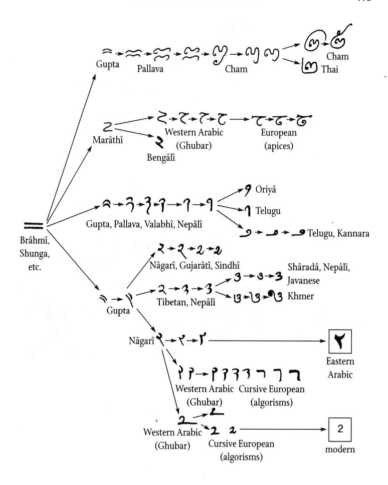

FIG. 24.62. *Origin and evolution of the numeral 2. (For Arabic and European numerals, see Chapters 25 and 26.)*

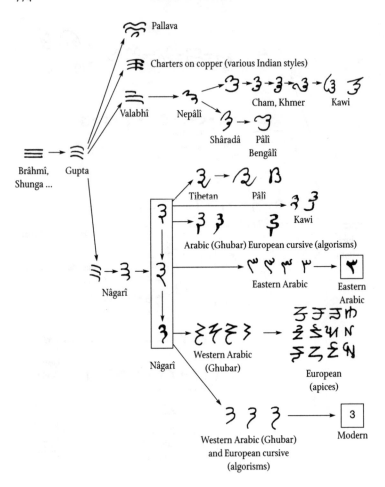

FIG. 24.63. *Origin and evolution of the numeral 3. (For Arabic and European numerals, see Chapters 25 and 26.)*

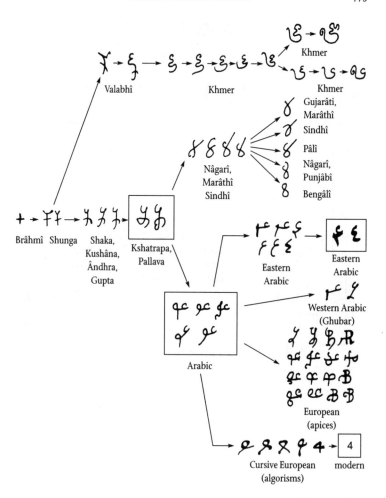

FIG. 24.64. *Origin and evolution of the numeral 4. (For Arabic and European numerals, see Chapters 25 and 26.)*

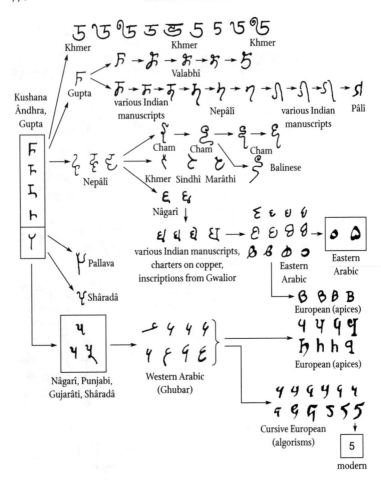

FIG. 24.65. *Origin and evolution of the numeral 5. (For Arabic and European numerals, see Chapters 25 and 26)*

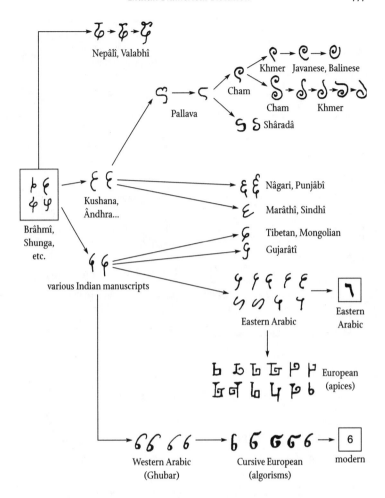

F<small>IG</small>. 24.66. *Origin and evolution of the numeral 6. (For Arabic and European numerals, see Chapters 25 and 26.)*

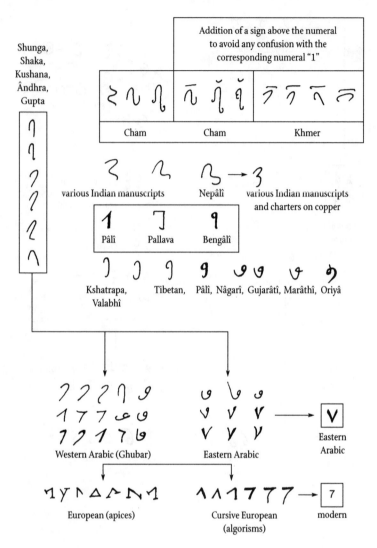

FIG. 24.67. *Origin and evolution of the numeral 7. (For Arabic and European numerals, see Chapters 25 and 26.)*

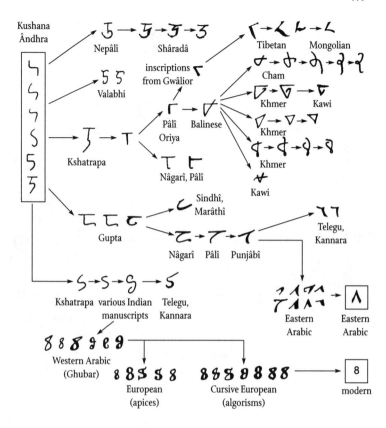

FIG. 24.68. *Origin and evolution of the numeral 8. (For Arabic and European numerals, see Chapters 25 and 26.)*

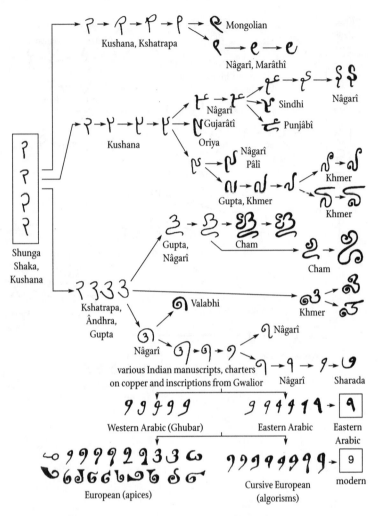

FIG. 24.69. *Origin and evolution of the numeral 9. (For Arabic and European numerals, see Chapters 25 and 26.)*

OLD INDIAN NUMERALS: A VERY BASIC SYSTEM

As the preceding diagrams have shown, Indian numerals, even in their earliest forms, were the forerunners of the nine basic numerals of our present-day system. In other words, it was from these signs that, some centuries later, the numerals that we wrongly call "Arabic" were derived. As we shall see later, modern numerals are the descendants of North African numerals, which themselves are cousins of the eastern Arabic numerals, which in turn are linked to *Nâgarî* numerals, which belong to the family of decimal numeral systems currently in use in India and Southeast and central Asia.

From a graphical point of view, then, the first nine *Brâhmî* numerals shared one of the fundamental characteristics of our present-day numerals. This, however, was the only aspect which they originally had in common.

If we examine *Brâhmî* inscriptions or intermediate Indian inscriptions, from Asoka's edicts to those of the Shungas, Shakas, Kushânas, Ândhras, Kshatrapas, Guptas, Pallavas or even the Châlukyas, that is to say from the third century BCE to the sixth and seventh centuries CE, we can see that the corresponding principle of numerical notation is very rudimentary.

For a decimal base, this system relied largely upon the principle of addition, attributing a specific sign to each of the following numbers (Fig. 24.70):

1	2	3	4	5	6	7	8	9
10	20	30	40	50	60	70	80	90
100	200	300	400	500	600	700	800	900
1,000	2,000	3,000	4,000	5,000	6,000	7,000	8,000	9,000
10,000	20,000	30,000	40,000	50,000	60,000	70,000	80,000	90,000

This written numeral system had special numerals, not only for each basic unit, but also for every ten, hundred, thousand and ten thousand units. To represent a number such as 24,400, one needed only to juxtapose, in this order, the numerals 20,000, 4,000 and 400 (Fig. 24.71):

Ƭᴏ   Tɟ   Hɣ

20,000   4,000   400

Of course, if these numerals had belonged to a place-value system, the number in question would have been written in the following way, using the style of numerals in use at that time, the zero being represented by a little circle, as it appears in later Indian inscriptions:

≈ ɣ ɣ o o

2   4   4   0   0

UNITS

	1	2	3	4	5	6	7	8	9
Third century BCE: Brahmi of Asoka	I	II		+		ϸ ६			
Second century BCE: Inscriptions of Nana Ghat	—	=		¥ ¥		φ	า		₽
First or second century CE: Inscriptions of Nasik	—	≃	≡	¥ ⅄ ⊦ ⊐		𝆙	ๆ	⑁ Ꙅ	₹
First to second century CE: Kushana inscriptions	—	=	≡	⅄ ⅄ ⊦ ⍴	F Y	६ ६ Ɛ	ๆ ๆ ⁊	ๆ ५ ⅁ ७	₹ ₹ ₽
First to third century CE: Andhra, Mathura and Kshatrapa inscriptions	—	=	≊	⅄⅄¥ ⅄⅄⅄	⊦⊦⊦ ⍴⍴	⅄⅄	⅃ ⅂ ⅃⅂	⅃ ⅂ ⅃⅂	⅃₽ ⅌ ⅏⅏⅊
Fourth to sixth century CE: Gupta inscriptions	⌒	⌇ ⅊	⅀ ⅀	⅄ ⅄ ⅄	F ⊦ Ɛ ᴧ	𝆑	⅃ ⊦	ⅉ⅃ ⅀	⅔
Sixth to ninth century CE: Inscriptions of Nepal	⌐	⅊	⅀ ⅊ ₃	⅄ ⅄	𝅗	𝅗𝅥	⅔	⅀⊥	⅔⅄
Fifth to sixth century CE: Pallava inscriptions	⌐ ⌐ ໑ ໑	⅔⅔	⅔ ⅽ	⅊	⅄	ⅽ	⅃	⅙	
Sixth to seventh century CE: Valabhi inscriptions	⌐	⅀	⅀	⅊𝆍	⅔⅊ ⅀⅊	⅁⅁	⅄	⅙	⅗⅄
Various Indian manuscripts	⅃	⅔	⅔	⅄⅔	⅔⅄ ⅊ⅆ ⅀	⅄⅊ ⅀	⅃⅀ ⅔	⅀⅄ ⊤⅊	⅔⅔ ⅊⅊

FIG. 24.70A *Numerical notation linked to Brâhmî writing and its immediate derivatives. There is evidence of the signs formed by straight lines; the others are reconstructions based on a comparative study of forms. (For references, see Fig. 24. 29 to 38 and 24.41 to 46.)*

TENS

	10	20	30	40	50	60	70	80	90
Third century BCE: Brahmi of Asoka					𝟨 𝟛				
Second century BCE: Inscriptions of Nana Ghat	ɑ ɑ̃	O				⊣		⌀	
First or second century CE: Inscriptions of Nasik	ɑ ᴕ	θ		𝟅			ɣ		
First to second century CE: Kushana inscriptions	ᴕᴕɑ ᴕᴕ	θ ◁θ	ν ϣ ᴜ	ϫϫϫ × ×	𝟨 𝟨𝟨	ν ɣ ɣ	× ϫ	ω ⑦ ω	⊕ ⊕
First to third century CE: Andhra, Mathura and Kshatrapa inscriptions	× ɑ ᴕ	𝟠◁ (	ᴎ ᴜ	ᴋ ᴜ ᴋ ᴎ	ᴐ ᴊ	ᴎ ᴎ	⊤ ⊤ ᴎ ⊤ ᴎ	⑦ ⓪ ⑦ ⓪	⊕ 𝟠𝟠
Fourth to sixth century CE: Gupta inscriptions	ᴕ ᴕᴕ ω	O	ᴜ			ᴎ	ᴕᴕ θ	⊕ ⊕ ⑮	
Sixth to ninth century CE: Inscriptions of Nepal	ᴕ 𝟑 ᴎ	𝟪	ᴎᴎ ᴎ	ᴋ ᴋ	𝟨 𝟨		⑪		
Fifth to sixth century CE: Pallava inscriptions									
Sixth to seventh century CE: Valabhi inscriptions	ᴈ ᴈ ᴈ	𝟠	ᴜ	ᴜᴜ ᴜ ᴜ	ᴈ ᴈ	ᴜ	ᴈᴜ ᴈᴜ	ᴃ ᴈ ᴃ	𝟠𝟠
Various Indian manuscripts	ᴈᴈ ᴈᴈ ᴕᴕ	𝟪 ᴈ 𝟪 ᴕᴕ		ᴈ	ᴈ 𝟨	ᴜ	ᴕᴜ ᴕᴜ	ᴎ ᴈ	𝟠𝟠 𝟠𝟠

Fig. 24.70b.

HUNDREDS, THOUSANDS AND TENS OF THOUSANDS

	100	200	300	400	1,000	2,000	3,000	4,000	6,000	8,000	10,000	20,000	70,000
Third century BCE: Brahmi of Asoka		⟨glyph⟩											
Second century BCE: Inscriptions of Nana Ghat	⟨glyph⟩	⟨glyph⟩	⟨glyph⟩	⟨glyph⟩	⟨glyph⟩			⟨glyph⟩	⟨glyph⟩		⟨glyph⟩	⟨glyph⟩	
First or second century CE: Inscriptions of Nasik	⟨glyph⟩	⟨glyph⟩	⟨glyph⟩	⟨glyph⟩	⟨glyph⟩	⟨glyph⟩	⟨glyph⟩	⟨glyph⟩	⟨glyph⟩	⟨glyph⟩	⟨glyph⟩	⟨glyph⟩	⟨glyph⟩
First to second century CE: Kushana inscriptions	⟨glyph⟩	⟨glyph⟩	⟨glyph⟩	⟨glyph⟩									
First to third century CE: Andhra, Mathura and Kshatrapa inscriptions	⟨glyph⟩	⟨glyph⟩	⟨glyph⟩	⟨glyph⟩									
Fourth to sixth century CE: Gupta inscriptions	⟨glyph⟩	⟨glyph⟩	⟨glyph⟩	⟨glyph⟩									
Sixth to ninth century CE: Inscriptions of Nepal	⟨glyph⟩	⟨glyph⟩	⟨glyph⟩	⟨glyph⟩									
Fifth to sixth century CE: Pallava inscriptions													
Sixth to seventh century CE: Valabhi inscriptions	⟨glyph⟩	⟨glyph⟩	⟨glyph⟩	⟨glyph⟩									
Various Indian manuscripts	⟨glyph⟩	⟨glyph⟩	⟨glyph⟩	⟨glyph⟩									
	Special sign	↑ 100+1×100 a vertical line is added to "100"	↑ 100+2×100 two vertical lines are added to "100"	100×4	↑ Special sign	↑ 1,000+1×1,000 a vertical line is added to "1,000"	↑ 1,000+2×1,000 two vertical lines are added to "1,000"	↑ 1,000×4	↑ 1,000×6	↑ 1,000×8	↑ 1,000×10	↑ 1,000×20	↑ 1,000×70

FIG. 24.70C.

Like certain other systems of the ancient world, this numeration was very limited. Arithmetical operations, even simple addition, were virtually impossible. Moreover, the highest numeral represented 90,000: therefore such a system could not be used to record very high numbers.

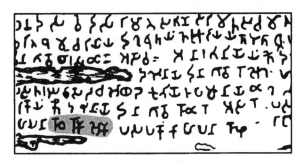

FIG. 24.71. *Detail of a Buddhist inscription in Brâhmî characters adorning one of the walls of the cave at Nana Ghat (second century BCE). The shaded section shows the Brâhmî notation for the number 24,400. [Ref. Smith and Karpinski (1911), p. 24]*

THE PROBLEM OF THE DISCOVERY OF THE INDIAN PLACE-VALUE SYSTEM

Thus the ancestors of our numerals remained static for a long time before acquiring the dynamic and manageable character that they have today thanks to the place-value system.

This leads us to ask two fundamental questions, which we will tackle through an archaeological, epigraphic and philological examination of the mathematical, astrological and astronomical texts of India: *When and how did the first nine numerals of this rudimentary system come to be governed by this essential rule? And when was zero first used?*

The first significant clues

Before we look at archaeology and epigraphy, it is worthwhile investigating whether some clues about zero and the place-value system can be discovered in Indian Sanskrit mathematical literature.

Here, for example, is an extract from the *Ganitasârasamgraha* (Chapter 1, line 27) by the mathematician Mahâvîrâchârya who, giving 12345654321 as the result of a previous calculation, defines this number in the following way [see B. Datta and A. N. Singh (1938)]:

ekâdishadantâni kramena hînâni

which means the quantity "beginning with one [which then grows] until it reaches six, then decreases in reverse order".

This comment is significant because it constitutes a numerical "palindrome": the number reads the same from left to right or from right to left, which is only possible if we are dealing with the place-value system:

$$12345654321$$

$$< \quad \cdot \quad >$$

It should be noted that these types of numbers possess unusual properties; take the following, for example:

$$1^2 = 1$$
$$11^2 = 121$$
$$111^2 = 12321$$
$$1111^2 = 1234321$$
$$11111^2 = 123454321$$
$$111111^2 = 12345654321$$
$$1111111^2 = 1234567654321$$
$$11111111^2 = 123456787654321$$
$$111111111^2 = 12345678987654321$$

These are properties that could not have been worked out using a non-positional system, due to its inconsistencies and the rules that would have governed it.

In other words these types of numbers could only have been discovered after the place-value system was invented. As we know that the *Ganitasârasamgraha* is dated c. 850 CE, we can infer that the place-value system was discovered before the middle of the ninth century.

Here is another piece of evidence which places the discovery of the place-value system at an earlier date: the arithmetician Jinabhadra Gani, who lived at the end of the sixth century, gave to the number 224,400,000,000 the following Sanskrit expression, in his *Brihatkshetrasamâsa* I, 69 (see Datta and Singh, p. 79):

dvi vimshati cha chatur chatvârimshati cha ashta shûnyâni
"twenty-two and forty-four and eight zeros" (=224400000000).

This proves that the Indians knew of zero and the place-value system in the sixth century.

The preceding examples do not constitute "proof" in the strictest sense of the word, but they show that the place-value system must have been in use for some time if its subtleties were understood and appreciated by the contemporary public.

Evidence found in Indian epigraphy

The first known Indian lapidary documents to bear witness to the use of zero and the decimal place-value system actually only date back to the second half of the ninth century CE.

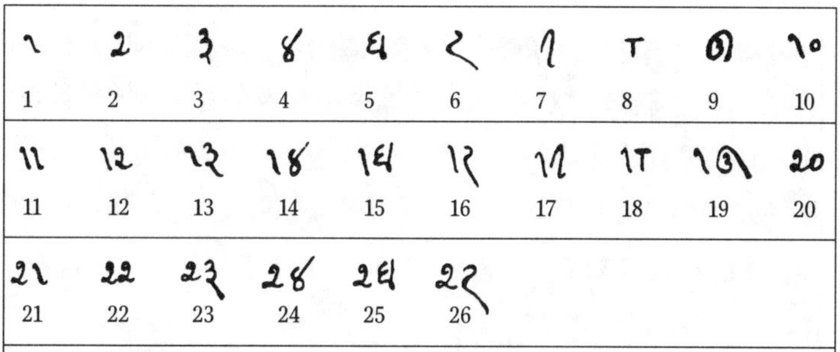

٦	2	३	४	६l	८	٩	T	9	٦o
1	2	3	4	5	6	7	8	9	10
٦٦	٦2	٦३	٦४	٦६l	٦८	٦٩	٦T	٦9	2o
11	12	13	14	15	16	17	18	19	20
2٦	22	2३	2४	2६l	22				
21	22	23	24	25	26				

Ref.: ASI, Rep. 1903-1904, pl. 72; EI, 1/1892, p. 155-162; Datta and Singh (1938); Guitel; Smith and Karpinski (1911).

Fig. 24.72. *Numerals from the first inscription of Gwalior*

These are two stone inscriptions from Bhojadeva's reign, discovered in the nineteenth century in the temple of Vâillabhattasvâmin, dedicated to Vishnu, near the town of Gwalior (capital of the ancient princely state of Madhyabharat, situated approximately 120 kilometres from Agra and a little over 300 kilometres south of Delhi).

The first inscription is quite clearly dated 932 in the *Vikrama* calendar (932 – 57 = 875 CE, see *Vikrama, Dictionary). It is in Sanskrit, and consists of twenty-six stanzas, which are numbered in the following manner using *Nâgarî* numerals (the signs for the numbers 1, 2, 3, 7, 9 and 0 already strongly resemble their modern equivalents) (Fig. 24.72).

The second inscription is dated (in numerals) the year 933 in the *Vikrama* calendar (= 876 CE). Written in Sanskrit prose, it gives an account of the offerings the inhabitants of Gwalior made to Vishnu. It tells mainly of the offering of a piece of land 270 × 187 *hasta*, which was to be turned into a flower garden, and of fifty garlands of seasonal flowers which the gardeners of Gwalior were to bring to the temple as a daily offering. The number denoting the date (933), as well as the three other numbers mentioned, are represented by *Nâgarî* numerals as they appear in Fig. 24.74.

There is no question as to the authenticity of these two inscriptions, and they clearly demonstrate the extent to which the inhabitants of the region were familiar with zero and the place-value system during the second half of the ninth century.

The inscriptions from Gwalior are not the oldest documents to contain evidence of the use of this system. Of the numerous other examples, of which there follows a list in ascending chronological order, there are documents engraved on copper which come from diverse regions of central and western India and date back to the era between the end of the sixth century and the tenth century CE.

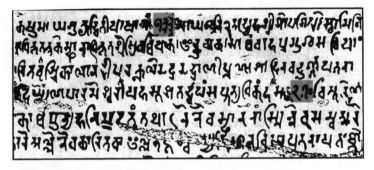

FIG. 24.73. *Detail from the second inscription of Gwalior (876 CE). The shaded section shows the representation of the numbers 933 and 270. [Ref. : EI, I, p. 160]*

୧३३	२१०	१४१	६०
933	270	187	50
Ref.: EI, I, p. 160, lines 1, 4, 5 and 20.			

FIG. 24.74. *Numerals from the second inscription of Gwalior*

These documents are legal charters written in Sanskrit and engraved in ancient Indian characters. They record donations given by kings or wealthy personages to the Brahmans. Each one contains details of the religious occasion when the donation(s) was (or were) offered and gives the name of the donor, the number of gifts plus a description of them, as well as a date which corresponds to one of the Indian calendars (**Chhedi*, **Shaka*, **Vikrama*, etc.; see Dictionary).

These dates are usually expressed in both letters and numerals, with the basic numerals, written in various Indian styles, varying in value according to their decimal position (Fig. 24.75 and 76).

The preceding evidence led historians, in the nineteenth century, to conclude that our present-day numerals were of Indian origin, and that they had been in use at least since the end of the sixth century CE (Fig. 24.75).

The foundations of this evidence seemed to crumble, however, at the beginning of the twentieth century, when three science historians, G. R. Kaye, N. Bubnov and B. Carra de Vaux, who were among those the most opposed to the idea that our numerals originated in India, questioned the authenticity of the copper inscriptions. They claimed that these documents had been re-written, altered or falsified at a much later date than the years given in the lists. It was concluded that, of all the texts which had been thought to be of Indian origin, only the inscriptions carved in stone could be regarded as proof of the existence of the system in question.

	(= 346 + 248 = 594 CE). DOCUMENTS AND SOURCES		
972	Donation charter of Amoghavarsha of the Râshtrakûtas, dated 894 in the *Shaka calendar (= 894 + 78 = 972 CE).	The number 894 is expressed: 	IA, XII, p. 263
933	Donation charter of Govinda IV of the Râshtrakûtas, dated 855 in the *Shaka Samvat calendar (= 855 + 78 = 933 CE).	The number 855 is expressed: 	IA, XII, p. 249
917	Donation charter of Mahîpâla, dated 974 in the *Vikrama Samvat calendar (= 974 – 57 = 917 CE).	The numbers 974 and 500 are expressed: 	IA, XVI, p. 174
837	Bâuka inscription. Dated 894 in the *Vikrama Samvat calendar (= 894 – 57 = 837 CE).	The number 894 is expressed: 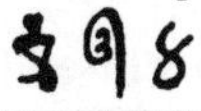	EI, XVIII, p. 87
815	Donation charter of Nâgbhata of Buchkalâ. Dated 872 in the *Vikrama Samvat calendar (= 872 – 57 = 815 CE).	The number 872 is expressed: 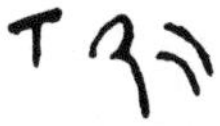	EI, IX, p. 198
793	Donation charter of Shankaragana of Daulatâbâd. Dated 715 in the *Shaka calendar (= 715 + 78 = 793 CE).	The number 715 is expressed: 	EI, IX, p. 197
753	Donation charter of Dantidurga of the Râshtrakûtas. Dated 675 in the *Shaka Samvat calendar (= 675 + 78 = 753 CE).	The number 675 is expressed: 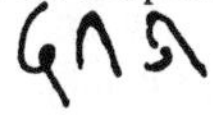	IA, XI, p. 108
753	Inscription of Devendravarman. Dated 675 in the *Shaka calendar (= 675 + 78 = 753 CE).	The number 20 is expressed: 	EI, III, p. 133
737	Donation charter of Dhiniki. Dated 794 in the *Vikrama Samvat calendar (= 794 – 57 = 737 CE).	The number 794 is expressed: 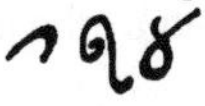	IA, XII, p. 155
594	Donation charter of Dadda III, of Sankheda in Gujarat (Bharukachcha region). Dated 346 in the *Chhedi calendar	The number 346 is expressed: (see Fig. 24.76) 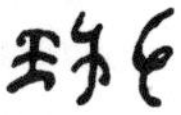	EI, II, p. 19

Fig. 24.75.

Since the inscriptions of Gwalior (875/876 CE) constituted the first evidence of this kind, these authors surmised that in India, zero and the place-value system could not have been used much before the second half of the ninth century CE.

It is true that amongst the charters recorded copper found in India, the authenticity of a certain number of them has been questioned, and quite rightly so, by Indianists (including Torkhede, Kanheri and Belhari, dated respectively 813, 674 and 646 CE [EI, III, p. 53; IA, XXV, p. 345; JA, 1863, p. 392]. Therefore, we have eliminated them from our investigation. As for the other documents of this kind, their authenticity has never been questioned by anyone except for Kaye and others who shared his motives.

The evidence was questioned in the hope of proving that Greek mathematicians were the "real" inventors of our numeral system, and that historians had been mistaken in attributing the invention to the Indians. However, as we have already seen, this hypothesis had no historical foundation, it was simply concocted in order to extend the tradition of the "Greek miracle".

The questioning of the authenticity of the Indian charters has never been satisfactorily justified.

The authors of the controversy would have it that these documents were "fabricated" at a later date, when the opportunity presented itself to a group of dishonest people who wished to take possession of the wealth which had long belonged to religious institutions and which the local authorities had confiscated or requisitioned some time before.

This explanation sounds feasible; however, there is no evidence to prove it, and the event was given a totally arbitrary date (some time during the eleventh century).

It was alleged that on the oldest known dated charter (Fig. 24.76), the numerals 3, 4 and 6, which come at the end of the inscription and which denote the *Chhedi year 346 (594 CE), were added at a later date.

If this were true, then why is the numeral 3 written as three horizontal lines? At the end of the sixth century (which corresponds with the date on the document), this way of writing the number was still used, although it was already becoming obsolete. It had disappeared completely by the next epoch, to be replaced by the non-ideographical sign belonging to the same style as the 4 or 6 which appear in the same document.

Of course, it could be argued that the forger (if there had been a forger) could have studied the palaeography of Indian numerals before imitating the style in question. The date on the legal document which tells of offerings made by someone's ancestor (authentic or not) would have been important to the descendant or person claiming to be so in order for them to prove that they were the rightful owners of the property mentioned on the charter.

But why would someone go to so much trouble, when the date is already given in the text in the form of the names of the numbers in Sanskrit? At that time, this indication was quite acceptable on its own; it was even more reliable than the numerals, whose appearance was susceptible to so many alterations in the hands of scribes and engravers.

What would have been the point of such an addition? And why would the date have been written in keeping with the place-value system when non-positional notation derived from the *Brâhmî* system was still frequently used (at least by the lay person) to write this type of legal document (Fig. 24.70)?

In other words, if the document was forged, why was the place-value system favoured rather than the old non-positional system?

No acceptable answers to these questions have been provided by those who put forward the theory of a forgery. On the other hand, to support his theory, Kaye did not hesitate to cite the charters inscribed on copper containing dates written using the old system and dating back to the era between the end of the sixth century and the ninth century (source: IA, VI, p. 19 ; EI, III, p. 133, etc.).

The most amusing part of this story is that the above dispute only centred on the oldest copper charters containing examples of the use of the place-value system, and not on the numerous other documents of the same nature which were written after or at the same time as the Gwalior inscriptions (876 CE). As for those containing examples of numbers written in the old non-positional Indian numerals whose date oscillates between the sixth and eighth century, their arithmetic was never questioned by Kaye. Thus we can see that these authors had worked out their conclusions far better than their arguments.

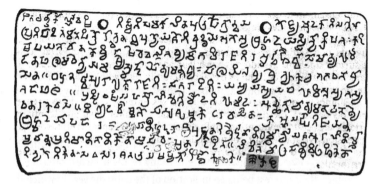

FIG. 24.76. *Donation charter of Dadda III, from Sankheda in Gujarat (region of Bharukachcha). Dated 346 in the *Chhedi calendar (= 346 + 248 = 594 CE), this document is the oldest known formal evidence of the use of the place-value system in India (in the shaded section, the number 346 is expressed according to this system). [Ref : EI, II, p. 19]*

We must be careful, however, because there is no way of ascertaining whether or not any of these copper charters are authentic; it is easy to make forgeries with copper, and we are dealing with a region of the world where counterfeiters, since time immemorial, have been masters at their craft.

The preceding counter-arguments would seem to suggest that these charters could well be authentic.

For the benefit of the doubt, however, we will not use these documents as evidence in our investigation, even though, from a purely graphical point of view, the letters and numerals they contain are indisputably authentic, unless the "forgers" pushed their talents to the limit to make exact copies of the contemporary and regional styles for each of the charters in question.

The fact remains that the history of the Indian decimal place-value system owes much to men such as Kaye. They proved that the subject was a lot more complicated than it seemed at first, and that all the documentation must be scrutinised very closely in establishing the facts. The controversy obliged scholars to go back to square one and apply stricter rules to their analysis of the facts and documents in this very rich and fertile field where they had not always exercised the correct degree of caution. On the other hand, men such as Kaye displayed a certain narrow-mindedness in limiting themselves to the literal frontiers of this civilisation which spread across a geographical area of truly continental dimensions, and which influenced and was witness to the flourishing of many other cultures which were situated beyond the limits of its own territory.

The following demonstrates that there are a great many other (unquestionably authentic) documents, which prove that zero and the place-value system are truly and exclusively Indian inventions, and that their discovery dates back even further than the oldest known inscription on copper.

Proof found in epigraphy from Southeast Asia

The texts that we will consider now are of considerable value to this investigation, for at least two reasons: first, they are all carved in stone, which means that there can be no doubt as to their authenticity; secondly, they are extracts from dated inscriptions, the oldest of which date back into the distant past.

These inscriptions are written either in Sanskrit or in vernacular language, that is to say in the regional language, be it Old Khmer, in Old Malay, in *Cham*, in Old Javanese, etc. Many of them record offerings to temples, their interest being an indication of the date (the year in which the inscription was written) and a detailed description of the offerings.

The way in which the corresponding numbers are expressed gives us the most significant indication of the use of the Indian place-value system.

If we only look at the indigenous inscriptions for the moment (those which are unique to each of the civilisations in question), we can see a very interesting particularity: the commonly-used numbers are not expressed in the same manner as the dates.

For the common numbers (expressing units of length, surface areas or capacities; the number of slaves, objects or animals; the quantity of gifts offered to the divinities and temples, etc.), the engravers usually simply expressed them in the letters of their vernacular language.

However, Cambodia is an exception to this rule; the Khmer engravers often preferred to use their local numeral system, which is immediately identifiable due to its undeniably primitive character (Fig. 24.77). This system uses one, two, three or four vertical lines to represent the first four units, although the fourth is often represented by a sign which gives no ideographical clue to the number it represents. As for the units 5 to 9, these are also represented by independent signs. This system also has a particular sign for 10, 20 and 100. As the system relies on the additive principle to represent numbers below 100, the multiples of 10, from 30 to 90, are expressed by combinations of the numerals for 20 and 10, according to the following rule:

$30 = 20 + 10$	Juxtaposition of the signs 20 and 10
$40 = 20 + 1 \times 20$	A vertical line is added to the sign for 20
$50 = 40 + 10$	Juxtaposition of the signs 40 and 10
$60 = 20 + 2 \times 20$	Two vertical lines are added to 20
$70 = 60 + 10$	Juxtaposition of the signs 60 and 10
$80 = 20 + 3 \times 20$	Three vertical lines are added to 20
$90 = 80 + 10$	Juxtaposition of the signs 80 and 10

The multiples of 100 are expressed in much the same manner, the numeral 100 being accompanied by the corresponding unit:

$200 = 100 + 1 \times 100$	A vertical line is added to the sign 100
$300 = 100 + 2 \times 100$	Two vertical lines are added to 100

The system seems to be limited to numbers below 400: there is no example of a higher number than this; above this quantity the Khmers wrote the names of the numbers in the letters of their language.

Thus, in terms of graphical representation, the ancient Khmer vernacular numeral system derived from the old *Brâhmî* system, as can be seen in the above table.

On the other hand, the structure of the system stems from the counting system of the Old Khmer language, for which we know the base was 20

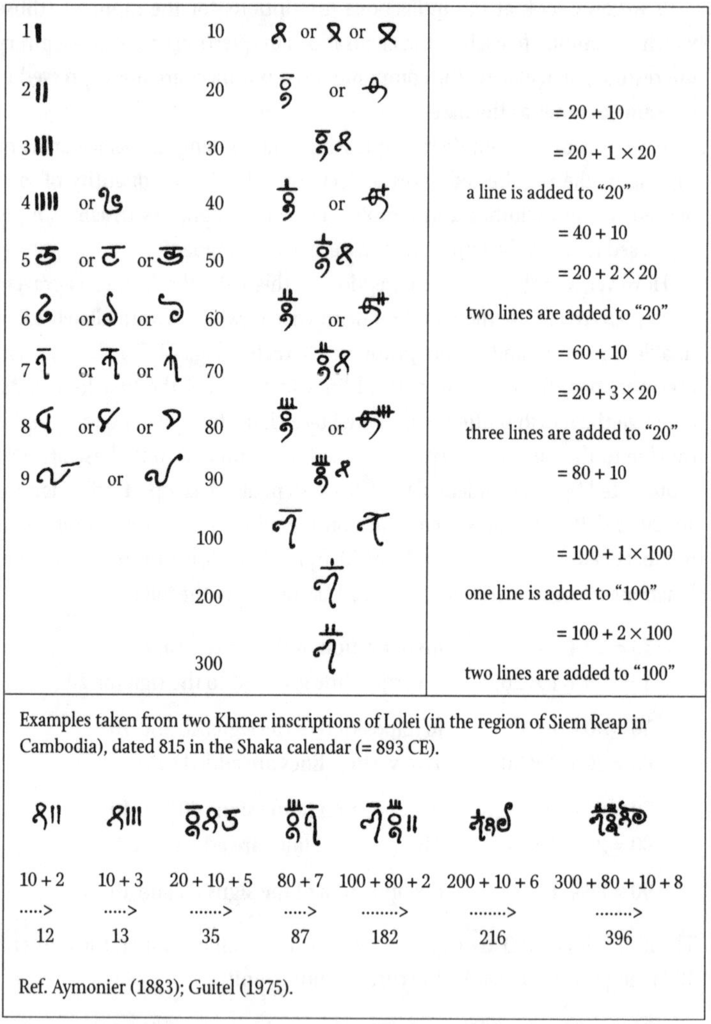

1 **ǀ**		10	𝑔 or 𝑔 or 𝑔	
2 **ǁ**		20		= 20 + 10
3 **ǁǀ**		30		= 20 + 1 × 20
4 **ǁǁ** or		40		a line is added to "20"
				= 40 + 10
5 or or		50		= 20 + 2 × 20
6 or or		60		two lines are added to "20"
				= 60 + 10
7 or or		70		= 20 + 3 × 20
8 or or		80		three lines are added to "20"
9 or		90		= 80 + 10
		100		= 100 + 1 × 100
		200		one line is added to "100"
		300		= 100 + 2 × 100
				two lines are added to "100"

Examples taken from two Khmer inscriptions of Lolei (in the region of Siem Reap in Cambodia), dated 815 in the Shaka calendar (= 893 CE).

10 + 2	10 + 3	20 + 10 + 5	80 + 7	100 + 80 + 2	200 + 10 + 6	300 + 80 + 10 + 8
.....>	.....>	........>	.....>	........>	........>	........>
12	13	35	87	182	216	396

Ref. Aymonier (1883); Guitel (1975).

FIG. 24.77. *The written numeration of the ancient Khmers: a system which uses the additive principle and which contains a curious trace of base 20. Used until the thirteenth century CE in vernacular inscriptions of Cambodia to express everyday numbers.*

(which explains the presence of a special sign for 20 and its multiples). As Coedès observes:

> The numeral system was not decimal, and today, despite the fact that Siamese numerals are used to represent multiples of ten above thirty, and likewise for 100, 1,000, etc., it still is not completely decimal: the names of the numbers from six to nine are expressed as five-one, five-

two, five-three, five-four, and special names for four and many of the multiples of twenty are still in common use. In ancient times, the Khmer people used no more than the names for one, two, three, four, five, ten, twenty and some multiples of twenty to express numbers, no matter how high the numbers were, and they used the Sanskrit word *shata* for "hundred", to which the term *slika* was added, which they also used to express the number 400 (= 20^2).

In other words, the spoken Khmer numeral system constituted a kind of compromise between Indian decimal numeration and a very old and far more primitive indigenous system, based both on 4 and 5 [see BEFEO, XXIV (1924) 3–4, pp. 347–8; JA, CCLXII (1974) 1–2, pp. 176–91].

On the other hand, in order to express dates, the stone-carvers of the diverse civilisations of Southeast Asia never used their vernacular numeral system or wrote the numbers in word form in their own language; as we will see, this fact is of great significance.

They only recorded dates using one of the two following methods: either the names of the numbers in Sanskrit, or, more frequently after a certain date, a decimal numeral system using nine numerals and a zero in the form of a dot or a little circle, strictly adhering to the place-value system (Fig. 24.50 and 24.78 to 80).

There is evidence of this use of the place-value system until the thirteenth century, from the ninth, eighth and even the seventh century CE, depending on the region.

In Champa, it was used consistently, at least since the *Shaka* year 735 (813 CE), which is the date of the oldest known *Cham* inscription of Po Nagar (Fig. 24.80).

In the Indian islands, however, the system appeared much earlier:

• at the end of the eighth century in Java; the oldest vernacular inscription (in *Kawi* writing) to bear witness to the use of the place-value system on this island is from Dinaya, dated the *Shaka* year 682 (760 CE) (Fig. 24.80);

• at the end of the seventh century at Banka; the most ancient vernacular inscription (in Old Malay) which attests its use in this island is that of Kota Kapur, dated 608 Shaka (686 CE) (Fig. 24.80);

• at the end of the seventh century in Sumatra; the oldest vernacular inscriptions (in Old Malay) to bear witness to its use in the region come from Talang Tuwo and Kedukan Bukit in Palembang, dated the respective *Shaka* years 606 and 605 (684 and 683 CE) (Fig. 24.80);

• and also at the end of the seventh century in Cambodia; the oldest vernacular inscription (in Old Khmer) to bear witness to its use in this

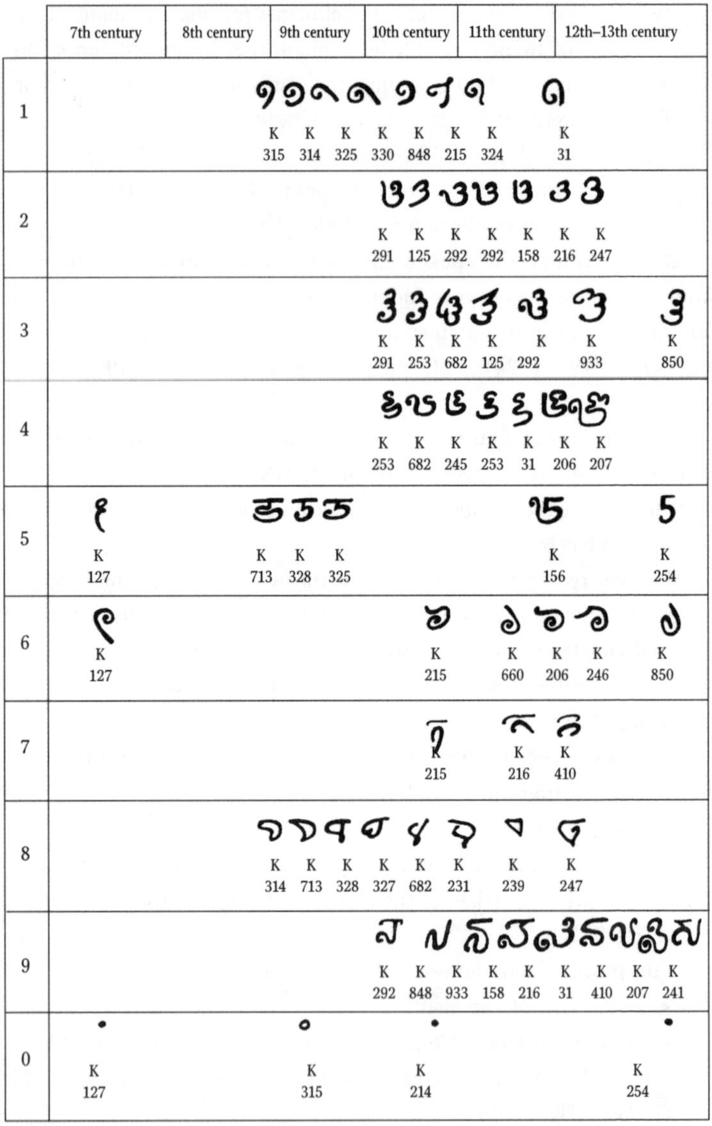

	7th century	8th century	9th century	10th century	11th century	12th–13th century
1		୧ ୨ ଯ ଯ ୨ ୫ ୧ ୧				୧
		K 315 K 314 K 325 K 330 K 848 K 215 K 324				K 31
2				୪୩ ୳୪୩ ୪ ୫ ୫		
				K 291 K 125 K 292 K 292 K 158 K 216 K 247		
3				୫ ୫ ୲୫ ୫ ୴ ୬		୫
				K 291 K 253 K 682 K 125 K 292 K 933		K 850
4				୫୳୪ ୫ ୫ ୲୴		
				K 253 K 682 K 245 K 253 K 31 K 206 K 207		
5	୪	୫ ୪ ୫			୪	5
	K 127	K 713 K 328 K 325			K 156	K 254
6	୧			୬	୬ ୫ ୬	୬
	K 127			K 215	K 660 K 206 K 246	K 850
7				୬	୬ ୬	
				K 215	K 216 K 410	
8			୬ ୬ ୧ ୫ ୪ ୬ ୬ ୬			
			K 314 K 713 K 328 K 327 K 682 K 231 K 239 K 247			
9				୫ ୪ ୫ ୫ ୪ ୫ ୪ ୫ ୫		
				K 292 K 848 K 933 K 158 K 216 K 31 K 410 K 207 K 241		
0	•	○	•			•
	K 127	K 315	K 214			K 254

Fig. 24.78. *A selection of palaeographical variants (dated) of numerals of the place-value system which, in vernacular inscriptions of Cambodia written in Old Khmer, were exclusively used to express dates of the Shaka calendar. (For the K references, see IMCC)*

country is from Trapeang Prei, in the province of Sambor, the respective *Shaka* year 605 (683 CE) (Fig. 24.80).

In Cambodia, however, this is not the oldest existing dated vernacular inscription. There is one which dates back even further, the earliest possible inscription to contain a date; it is from Prah Kuha Luhon, dated the *Shaka* year 596 (674 CE), this date being written in letters using the

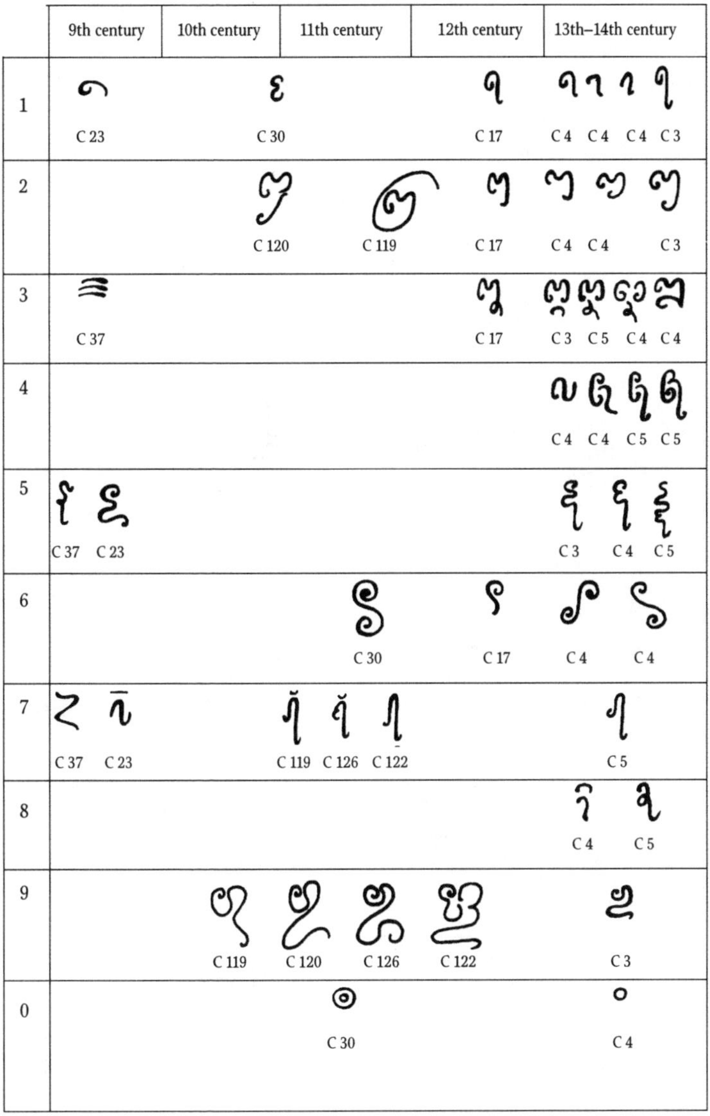

	9th century	10th century	11th century	12th century	13th–14th century
1	C 23	C 30		C 17	C 4 C 4 C 4 C 3
2		C 120	C 119	C 17	C 4 C 4 C 3
3	C 37			C 17	C 3 C 5 C 4 C 4
4					C 4 C 4 C 5 C 5
5	C 37 C 23				C 3 C 4 C 5
6			C 30	C 17	C 4 C 4
7	C 37 C 23		C 119 C 126 C 122		C 5
8					C 4 C 5
9		C 119 C 120 C 126 C 122			C 3
0		C 30			C 4

FIG. 24.79. *A selection of palaeographical variants (dated) of numerals of the place-value system which, in (Cham) vernacular inscriptions of Champa, were exclusively used to express dates of the Shaka calendar. (For the C references, see IMCC)*

Sanskrit names for the numbers (see IMCC, K 44, 1. 6; CIC, IV):

shannavatyuttarapañchashata Shakaparigraha
"the *Shaka* [year] numbering five hundred and ninety-six"

Thus in the vernacular inscriptions of Southeast Asia, the everyday numbers were always expressed through the names of the numbers in the

	DOCUMENTS AND SOURCES	HOW THE DATE IS RECORDED IN THE *SHAKA* CALENDAR	
1084	Cham inscription of Pô Nagar, northern tower. Dated 1006 in the *Shaka* calendar (= 1006 + 78 = 1084 CE).	1　0　0　6	IMCC, C 30 BEFEO, XV, 2, p. 48
1055	Cham inscription of Lai Cham, region of Hanoi. Dated 977 in the *Shaka* calendar (= 977 + 78 = 1055 CE).	9　7　7	IMCC, C 126 BEFEO, XV, 2, pp. 42–3
1055	Cham inscription of Phú-qui, province of Phanrang. Dated 977 in the *Shaka* calendar (= 977 + 78 = 1055 CE).	9　7　7	IMCC, C 122 BEFEO, XV, 2, p. 41 BEFEO, XII, 8, p. 17
1050	Cham inscription of Pô Klaun Garai (first inscription). Dated 972 in the *Shaka* calendar (= 972 + 78 = 1050 CE).	9　7　2	IMCC, C 120 BEFEO, XV, 2, p. 40
1050	Cham inscription of Pô Klaun Garai (second inscription). Dated 972 in the *Shaka* calendar (= 972 + 78 = 1050 CE).	9　7　2	IMCC, C 120 BEFEO, XV, 2, p. 40
1007	Khmer inscription of Phnom Práh Nét Práh (foot of southern tower). Dated 929 in the *Shaka* calendar (= 929 + 78 = 1007 CE).	9　2　9	IMCC, K 216 BEFEO, XXXIV, p. 423
1005	Khmer inscription of Phnom Práh Nét Práh (foot of southern tower). Dated 927 in the *Shaka* calendar (= 927 + 78 = 1005 CE).	9　2　7	IMCC, K 216 BEFEO, XXXV, p. 201
880	Balinese inscription of Taragal. Dated 802 in the *Shaka* calendar (= 802 + 78 = 880 CE).	8　0　2	Damais, p. 148, g.

F‍IG. 24.80A.

	DOCUMENTS AND SOURCES	HOW THE DATE IS RECORDED IN THE *SHAKA* CALENDAR	
878	Balinese inscription of Mamali. Dated 800 in the *Shaka* calendar (= 800 + 78 = 878 CE).	 𝍫 ο ο 8 0 0	Damais, p. 148, f.
877	Balinese inscription of Haliwanghang. Dated 799 in the *Shaka* calendar (= 799 + 78 = 877 CE).	 𐨨 𐨨 𐨨 7 9 9	Damais, p. 148, f.
829	Cham inscription of Bakul. Dated 751 in the *Shaka* calendar (= 751 + 78 = 829).	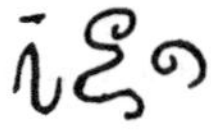 ॶ 𑢻 ௹ 7 5 1	IMCC, C 23 ISCC, p. 238 BEFEO, XV, 2, p. 47
813	Cham inscription of Pô Nagar, northwest tower. Dated 735 in the *Shaka* calendar (= 735 + 78 = 813 CE). *This is the first Cham inscription to contain the date written in numerals.*	 ۲ ≋ ۲ 7 3 5	IMCC, C 37 JA 1891, i, p. 24 BEFEO, XV, 2, p. 47
686	Malaysian inscription of Kota Kapur (isle of Banka). Dated 608 in the *Shaka* calendar (= 608 + 78 = 686 CE).	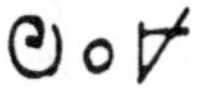 ℮ ο 𝍫 6 0 8	BEFEO, XXX, pp. 29ff. Kern VIII, p. 207
684	Malaysian inscription of Talang Tuwo, Palembang (Sumatra). Dated 606 in the *Shaka* calendar (= 606 + 78 = 684 CE).	 ꠉ ο ꠉ 6 0 6	BEFEO, XXX, pp. 29ff. ACOR, II, p. 19
683	Malaysian inscription of Kedukan Bukit, Palembang (Sumatra). Dated 605 in the *Shaka* calendar (= 605 + 78 = 683). *This is the oldest inscription in Old Malay to be dated in numerals.*	 ℮ ο ꠚ 6 0 5	BEFEO, XXX, pp. 29ff. ACOR, II, p. 13

FIG. 24.80B.

	DOCUMENTS AND SOURCES	HOW THE DATE IS RECORDED IN THE *SAKA* CALENDAR	
683	Khmer inscription of Trapeang Prei, province of Sambór. Dated 605 in the *Shaka* calendar (= 605 + 78 = 683 CE). *This is the first Khmer inscription dated in numerals.*	 ρ • ₹ 6 0 5	IMCC, K 127 CIC, XLVII

Note: In Java, the oldest Kawi inscription (written in Old Javanese) to be dated in numerals is of Dinaya, which bears the date 682 in the *Shaka* calendar = 682 + 78 = 760 CE. [Ref: Tijdschrift, LVII, (1976), p. 411; LXIV, (1924), p. 227].

Fig. 24.80c.

indigenous language or its very rudimentary numerals. For the dates of the *Shaka* calendar, however, either the Sanskrit word, or, more commonly, the decimal place-value system was used, from no later than the end of the seventh century CE.

As we have seen, the different numerals used throughout Southeast Asia were actually nothing more than palaeographical variations of Indian numerals, which themselves derived from the *Brâhmî* form of the first nine units (Fig. 24.52, 24.53, and 24.61 to 69). The only difference between these diverse systems is their complete transformation into local cursive forms (Khmer, Javanese, *Cham*, Malay, Balinese, etc.), according to the habits of the scribes and engravers of the region.*

On the other hand, as well as the use of Sanskrit (the learned language of Indian civilisation) to record the dates, all the vernacular inscriptions reveal that the dates were written exclusively, for many centuries, according to a system whose Indian origin is indisputable: the *Shaka* era of the Indian astronomers [see R. Billard (1971)].

* It should be noted that the interpretation of the *Cham* numerals posed considerable difficulties because of mistakes made as to their values when they were first deciphered. 7 was mistaken for 1, the (more recent form of) 1 was mistaken for the very ancient form of 5, 7 for 9, etc. This caused even graver errors in the nineteenth century when it came to dating the inscriptions, which in turn led to mistakes in the interpretation of historical events and chronology. Thus it appeared that inscriptions of the same person, referring to the same event, were written at completely different points in time. The inscriptions of King Parameshvaravarman, for example, gave the *Shaka* date 972 (1050) in Sanskrit chronograms, whilst other inscriptions, or the same one, gave the date as the *Shaka* year 788 (866). Another example is an inscription of Mî-so'n (BEFEO, IV, 970, 24) which lists a series of religious foundations set up by King Jaya Indravarman of Gramapura; logically, these dates should have been listed in chronological order; however, when the values that were believed to be correct were applied, the dates emerged in the following incoherent manner: 1095, 1096, 1098, 1097, 1070 and 1072. There were many similar enigmas, which seemed to have no solution and were distorting all the acquired data, until L. Finot (BEFEO, XV, 2, 1915, pp. 39–52) discovered the true origin of the *Cham* numerals and with it the solution which had eluded his predecessors. The inaccurate interpretations of the *Cham* numerals were due to the very unusual variations that they had undergone over the centuries because of the whims and aesthetical preoccupations of the corresponding engravers.

These facts are even more significant because they concern the ancient civilisations of Indo-China and Indonesia (Cambodia, Champa, Java, Bali, Malaysia), which were strongly influenced by India in the early centuries CE, partly because of the widespread nature of Shivaism and Buddhism, and also because of the important intermediate role they played in the trading of spices, silk and ivory between India and China [see G. Coedès (1931), (1964)].

Champa is the ancient kingdom that stretched along the coast of what is now Vietnam with the region of Hue at its centre. The native inhabitants of Champa had become Hindu by religion due to their frequent contact with Indian traders. Champa first became a powerful nation at the beginning of the fifth century CE, under the rule of Bhadravarman, who dedicated the shrine of Mî-so'n (which would remain the religious centre of the kingdom) to Shiva, one of the greatest divinities of the Indian Brahmanic pantheon.

Not far from Champa is Cambodia, which previously belonged to the Hindu kingdom of Fu Nan from the first to the sixth century CE, before becoming the centre of Khmer civilisation, which, conserving the foundations of an Indian culture, flourished until the fourteenth century.

In Java, too, which entered into relations with India at the beginning of the second century CE, and all of ancient Indonesia, early developments owe much to Indian civilisation through the influence of Buddhism and Brahmanism.

Thus we can see the great influence that Indian astronomers and mathematicians once exercised over the various cultures of Southeast Asia. All the preceding facts are highly significant for they show how the Khmer, the Cham, the Malaysian, the Javanese, the Balinese, and other races, were profoundly influenced by Indian culture, and borrowed elements of Indian astronomy, in particular the *Shaka* calendar, and conformed to the corresponding arithmetical rules [see F. G. Faraut (1910)].

In these regions, the appearance of zero and the place-value system coincides directly with the appearance of the dates of the *Shaka* calendar:

> The place-value system was used in Indo-China and Indonesia from the seventh century CE, in other words at least two centuries before Kaye claimed to find evidence of its use in India itself. However, unless zero and the Arabic numerals came from the Far East [which is precisely the opposite of what did happen] evidence of their use in the Indian colonies would suggest that they were in use in India at an even earlier date [G. Coedès (1931)].

Thus we have confirmed the words of the Syrian Severus Sebokt, who wrote, in the seventh century, that the Indian place-value system was already known and held in high esteem beyond the borders of India.

THE INDIAN PLACE-VALUE SYSTEM: OUTSIDE INFLUENCE OR INDIAN INVENTION?

The place-value system is unquestionably of Indian origin, and its discovery doubtless dates back much further than the seventh century CE. The question we must now ask is whether this concept was inspired by an outside influence or whether it was a purely Indian discovery.

We know that during the course of history the Babylonians, the Chinese, the Maya and, of course, the Indians, succeeded in inventing a place-value system. If the Indians were influenced by any other civilisation, it would have to have been one of the other three we have just mentioned, either directly or through an intermediary.

Putting to one side the Maya civilisation, which apparently had no contact with the ancient world, this leaves us with the Chinese and Babylonian.

The possibility of Babylonian influence

If the Indian place-value system was derived from that of the Babylonians, it might have been through the intermediary of Greek civilisation.

In 326 BCE, Alexander the Great took possession of the land of the Indus and of the ancient province of Gandhara, from the northeast of Afghanistan and the extreme north of present-day Pakistan to the northwest of India, before these regions were governed by the "Indian-Greek" Satraps c. 30 BCE. We know that many elements and methods of Babylonian astronomy were introduced into India shortly before the beginning of the first millennium CE, probably through the northwest of the Indian sub-continent; no doubt this took place in the eastern part of the present-day state of Gujarat, probably in the region of the port of Bharukachcha, which saw the development of both cultural and maritime activities and trade with the West during the first centuries CE. Thus, as R. Billard (1971) stresses, the period between the third century BCE and the first century CE is characterised by the appearance of the *tithi*, a unit of time used in the Babylonian tablets and corresponding to the thirtieth of a synodic revolution of the Moon, more or less the equivalent of a day or nychthemer: elements which are known to have been transmitted to the Greek astronomers by their Mesopotamian colleagues no later than the Hellenistic era.

We know that Babylonian scholars had invented and used a place-value system with 60 as a base since the nineteenth century BCE, and they had used zero since the fourth or third century BCE. As the sexagesimal system was "one of the elements that the Greeks had acquired from Babylonian astronomy, the mother and wet-nurse of their own astronomy" [F. Thureau-Dangin (1929)], it is possible to suppose that the idea of the place-value system arrived in India at the same time as Babylonian astronomy.

Although this hypothesis cannot be ruled out, we can nevertheless raise one serious objection to it. The Greek astronomers only used the Babylonian sexagesimal system to write the negative values of 60, in other words the sexagesimal fractions of the unit, whilst the system was originally developed in order to express whole numbers as well as fractions. Thus, if a similar influence was exercised over the Indians by the Greeks (for if the Indians were influenced by the Babylonian system it could only have been via the Greeks), how could an incomplete system, only used to record fractions, moreover with a base of 60, have influenced the invention of a decimal place-value system which was originally invented to record whole numbers? This is an obvious flaw, which makes this hypothesis appear rather paradoxical.

The possibility of a Chinese influence

Therefore, at first glance, a Chinese influence would seem more plausible. We know that since the time of the Han (206 BCE to 220 CE), Chinese scholars used a decimal place-value system known as *suan zi* ("calculation using rods"). A regular system which combined horizontal and vertical lines was used to represent the nine basic units, constituting a written transcription of a concrete counting system which used reeds, standing on one end or placed horizontally on a counting board like abacuses in columns.

Thus one could be forgiven for assuming that following the links established between China and India at the beginning of the first millennium CE, Indian scholars were influenced by Chinese mathematicians to create their own system in an imitation of the Chinese counting method.

However, this hypothesis is contradicted by the fact that zero only appeared in the *suan zi* system relatively late. The Chinese scholars overcame the difficulties this caused by expressing a number such as 1,270,000 either in the characters of their ordinary counting system (a non-positional system which did not require the use of zero) or by placing their rod numerals in a series of squares, the missing units being represented by empty squares:

	1	=	π				
	1	2	7	0	0	0	0

It was only after the eighth century CE, and doubtless due to the influence of the Indian Buddhist missionaries, that Chinese mathematicians introduced the use of zero in the form of a little circle or dot (signs that originated in India), thus representing the preceding number in the following manner:

I	**=**	**ㅠ**	**O**	**O**	**O**	**O**
1	2	7	0	0	0	0

A symbol for zero is mentioned in the *Kai yun zhan jing*, a major work on astronomy and astrology published by *Qutan Xida between 718 and 729 CE. The chapter of this work devoted to the *jui zhi* calendar of 718 CE contains a section on Indian calculating techniques [J. Needham (1959)].

After saying that the (Indian) figures are all written in the cursive form in just one stroke, Qutan Xida continues:

> When one or another of the nine numerals has to be used to express the number ten [literally: "when it reaches ten"], it is then written in a preceding column [before the numerals for the units] (*qian wei*). And each time an empty space appears in a column, a dot is always written [to convey the empty space] (*mei gong wei qu heng an yi dian*).

The author of this book on astronomy was not Chinese: *Qutan Xida was actually an adaptation of the Indian name *Gautama Siddhânta, the famous Indian Buddhist mathematician and astronomer living in China and the head of a school of astronomy at Chang'an since approximately 708 CE. According to L. Frédéric (1987), he was the one to introduce the notion of zero in China as well as the division of a circle into sixty sections.

This remarkable account confirms the influence, which has already been proved, of the rapid expansion of the Buddhist movement which accompanied the propagation of Indian science in the Far East. It also adds an important piece of evidence to our investigation of the origin of our modern numeral system:

> Living in China, doubtless knowing all the subtleties of the Chinese language, Qutan Xida insists not only on the fact that Indian numerals were written in a cursive form, but also that each one was written in just one stroke [G. Guitel, (1975)].

In the Chinese place-value system (the "learned" system), the units were written by juxtaposing or superposing one or more vertical or horizontal lines:

I	**II**	**III**	**IIII**	**IIIII**	**T**	**ㅠ**	**ㅠ**	**ㅠㅠ**
1	2	3	4	5	6	7	8	9

The Chinese numerals in common usage are formed by lines, in various positions, and written in a strict order, lifting the writing tool several times, the symbol for 2 being formed by two lines, 4 by six lines, 6 by four lines, and so on (Figure 21. 1 above). This will become clearer later, when we see the succession of lines that forms the Chinese numeral for 100 (see Fig. 21. 8 above):

> This could only have surprised the learned men amongst which Gautama Siddhânta (alias Qutan Xida) lived, because Chinese words are grouped according to the number of lines their drawing requires; for each character, one is taught the order in which the successive lines must be drawn [G. Guitel (1975)].

The nine numerals (of Indian origin) that we use today, on the other hand, are drawn in just one stroke of a pen or pencil. This is one of the characteristics of our numeral system, whose remarkable simplicity we forget because we have been using it all our lives.

This evidence proves that at the beginning of the eighth century, zero and the place-value system had spread as far as China. At the same time, it almost completely rules out any possibility of a Chinese influence over the development of our present-day numerals.

THE AUTONOMY OF THE INDIAN DISCOVERIES

Thus it would seem highly probable under the circumstances that the discovery of zero and the place-value system were inventions unique to Indian civilisation. As the *Brâhmî* notation of the first nine whole numbers (incontestably the graphical origin of our present-day numerals and of all the decimal numeral systems in use in India, Southeast and Central Asia and the Near East) was autochthonous and free of any outside influence, there can be no doubt that our decimal place-value system was born in India and was the product of Indian civilisation alone.

THE NUMERICAL SYMBOLS OF THE INDIAN ASTRONOMERS

We are now going to look at a truly remarkable method of expressing numbers which is frequently found on mathematical and astronomical texts written in Sanskrit; there is no doubt that these texts are of Indian origin.

It is to curious to note that historians of science have not always accorded it the importance it deserves. It constitutes the main piece of evidence of our investigation: added to all the other evidence, it allows us not only to prove beyond doubt that our present-day numeration is of Indian origin, and Indian alone, but also and above all to date the discovery even earlier than the seventh century CE. Moreover, it is even more significant when we consider that the nature of this system is unique in the history of numerals.

By way of introduction, here is a passage from the first modern Indian historian, the Persian astronomer al-Biruni, who wrote the following c. 1010, in his famous work on India [see al-Biruni (1910); F. Woepcke (1863), pp. 283–90]:

> When [the Indian scholars] needed to express a number composed of many orders of units in their astronomical tables, they used certain words for each number composed of one or two orders. For each number, however, they used a certain number of words, so that, if it was difficult to place one word in a certain place, they could choose another from "amongst its sisters" [amongst those which denoted the same number]. Brahmagupta said: If you want to write one, express it through a word which denotes something unique, like the Earth or the Moon; likewise, you can express two with any words which come in pairs, like black and white [this is probably an allusion to the "half black" and "half white" of the month, which corresponds to a division used by the Indians], three by things that come in threes, zero with the names for the sky, and twelve by the names of the sun . . . [Such is the way the system works] as I have understood it. It is an important element in the analysis of their [the Indians'] astronomical tables . . .

Instead of the word *eka, which means "one", the Indian astronomers used names such as *âdi (the "beginning"), *tanu ("the body"), or *pitâmaha ("the Ancestor", which alludes to *Brahma, considered to be the creator of the universe).

Instead of *dvi, which means "two", they used all the words which express ideas, things or people which come in pairs: *Ashvin ("the twin gods"), *Yama ("the Primordial Couple"), *netra ("the eyes"), *bâhu ("the arms"), *paksha ("the wings"), etc.

In other words, rather than using the ordinary Sanskrit names for the numbers 1 to 9 (*eka, *dvi, *tri, *chatur, *pañcha, *shat, *sapta, *ashta, *nava), the Indian scholars expressed them by names which had symbolical value. For each number, there was a wide choice of words, whose literal translation evoked the numerical value they denoted in the reader's mind.

It is difficult to give an exhaustive list of these diverse symbolic words, there being an abundant, if not infinite, number of synonyms. However, the reader will get some idea of the variety of these words from the following examples:

ONE

eka:	Ordinary name for the number 1
pitâmaha:	First father
âdi:	Beginning
tanu:	Body
kshiti, go . . .:	Words meaning "Earth"
abja, indu, soma . . .:	Words meaning "Moon"

TWO

dvi:	Ordinary name for the number 2
Ashvin:	Horsemen
Yama:	Primordial Couple
yamala, yugala . . .:	Words meaning twins or couples
netra:	Eyes
bâhu:	Arms
gulphau:	Ankles
paksha:	Wings

THREE

tri:	Ordinary name for the number 3
guna:	Primordial properties
loka:	[Three] worlds
kâla:	Time
agni, vahni . . .:	Fire
Haranetra:	"Eyes of Hara"

FOUR

chatur:	Ordinary name for the number 4
dish:	The [four] cardinal points
abdhi, sindhu . . .:	The [four] oceans
yuga:	The [four] cosmic cycles
iryâ:	The positions [of the human body]
Haribâhu:	The arms of Vishnu
brahmâsya:	The faces of Brahma

FIVE

pañcha:	Ordinary name for the number 5
bâna, ishu . . .:	Arrows
indriya:	The [five] senses
rudrâsya:	The [five] faces of Rudra
bhûta:	The elements
mahâyajña:	The sacrifices

SIX

shat:	Ordinary name for the number 6
rasa:	The senses
anga:	The [six] limbs [of the human body]
shanmukha:	The [six] faces of Kumara

SEVEN

sapta:	Ordinary name for the number 7
ashva:	Horses
naga:	Mountains
rishi:	The [seven] sages
svara:	The vowels
sâgara:	The [seven] oceans
dvîpa:	The island-continents

EIGHT

ashta:	Ordinary name for the number 8
gaja:	The [eight] elephants
nâga:	Word meaning "serpent"
mûrti:	Forms

NINE

nava:	Ordinary name for the number 9
anka:	Numerals
graha:	Planets
chhidra:	The orifices [of the human body]

ZERO

shûnya:	Ordinary name for 0
bindu:	The point or dot
kha, gagana . . .:	Words meaning "sky"
âkâsha:	Ether
ambara, vyoman . . .:	Atmosphere

The Sanskrit language, which is very learned and rich, lends itself admirably to this system, as it does to poetry and the Indian way of thinking.

These symbols are all taken from nature, human morphology, animal or plant representations, everyday life, legends, traditions, religions, attributes of the divinities of the Vedic, Brahman, Hindu, Jaina or Buddhist pantheons, as well as from the associations of traditional or mythological ideas or from diverse social conventions of Indian civilisation.

With this unique system of numerical notation, we have now entered into the world of symbols of Indian civilisation.

To give the reader a better understanding of the characteristic way of thinking of Indian philosophers, astrologers, cosmographers, astronomers and mathematicians, (the true "inventors" of our present-day counting system), we have included the "Dictionary of Numerical Symbols of Indian Civilisation" at the end of this chapter, the necessity and usefulness of which will become clear in the course of the following pages. *

THE PLACE-VALUE SYSTEM OF THE INDIAN NUMERICAL SYMBOLS

To give us some idea of the principle this system was based on, here is a literal translation of a Sanskrit verse taken from a work on astronomy entitled *Sûrya Siddhânta* (or "Astronomical canon of the Sun"; [see Anon. (1955), I, 33; Burgess and Whitney (1860)]:

> *Chandrochchasyâgnishûnyâshvivasusarpârnavâ yuge*
> *Vâmam pâtasya vasvagniyamâshvishikhidasrakâh*
> "The apsids of the moon in a yuga
> Fire. Vacuum. Horsemen. *Vasu.* Serpent. Ocean,
> and of its waning node
> *Vasu.* Fire. Primordial Couple. Horsemen. Fire. Twins"

This verse is incomprehensible to a reader who does not know that the words "Fire. Void. Horseman. Vasu. Serpent. Ocean" (*âgnishûnyâshviva-susarpârnavâ*) and "Vasu. Fire. Primordial Couple. Horseman. Fire. Horseman" (*vasvagniyamâshvishikhidasra*), in the minds of the Indian astronomers, represented the numbers 488,203 and 232,238 respectively.

Here is a comprehensible translation of the verse:

"[The number of revolutions] of the apsids of the moon in a yuga [is]: 488,203, and [of] its waning node: 232,238."

---

* For each of the word-symbols in question (*Ashvin, *Graha, *Kha, etc.), the reader might find it interesting to consult the corresponding rubric, where the symbol is denoted by [S], then defined in terms of its numerical value and its literal meaning in Sanskrit, before, as far as possible, its implied symbolism is explained. To find the list of Sanskrit word-symbols used (in their abundant synonymy) for a given number, one only need consult the corresponding English word in the Dictionary (*One, *Two, *Zero, etc.).

Thus the author of this text expressed, in his own way, two pieces of astronomical numerical data, concerning a *yuga or "cosmic cycle" (in this case a cosmic cycle named *Mahâyuga and corresponding to a period of time of 4,320,000 years).

The key to the system lies in knowing that, in a number-system which has 10 as its base, the first nine whole numbers, 10 and each multiple of 10 have a specific name; thus one expresses a given number by placing the name for "ten" between that of the units of the first order and that of the units of the second order, then the name for "hundred" between those of the second and third orders, and so on, respecting a previously agreed method of reading.

The number 8,237, for example, might be expressed in the following manner: "eight thousand, two hundred, three times ten and seven", according to this mathematical breakdown of the components:

$$8 \times 10^3 + 2 \times 10^2 + 3 \times 10 + 7 = 8,237.$$

As well as writing the number in terms of decreasing powers of ten, it can also be written in the opposite order, in increasing powers of ten, starting with the smallest unit, for example:

"Seven, three times ten, two hundred, eight thousand".

This is exactly how the Indian astronomers expressed numbers when they used the Sanskrit names of the numbers. Thus the preceding number can be mathematically broken down in the following way:

$$7 + 3 \times 10 + 2 \times 10^2 + 8 \times 10^3 = 8,237.$$

The method of expressing numbers that we are interested in here is the "oral" method, because it uses Sanskrit *words*, the difference being that it simply gives a succession of the corresponding names of the units, in keeping with the method of representation that we have just seen. In other words, there is no mention of the names which indicate the base and its various powers ("ten", "hundred", "thousand", etc.) Thus the preceding number would be expressed in the following manner:

Seven. Three. Two. Eight.

In the same way, two.eight.nine.three corresponds to the value:

$$2 + 8 \times 10 + 9 \times 10^2 + 3 \times 10^3 = 3,982.$$

In other words, the Sanskrit names for the numbers 1 to 10 had a varying value according to their position in the description of numbers of several orders of units. In saying *one, three, nine* for 931 for example, the word one is given the simple value of *one* unit, *three* is given the power of ten and *nine* the value of a multiple of one hundred.

Thus there can be no doubt that we are dealing with a decimal place-value system. This seems even more remarkable when we consider that the Indian scholars were the only ones to invent a system of this kind. The above example, however, poses a fundamental question. We have just seen that in this system, a number such as 931 can be expressed relatively easily, by writing *one, three, nine*. On the other hand, it is difficult to express a number such as 901, where there is an empty space, if you like, in the decimal order (the "ten" column). To write this number, one could obviously not simply write *one, nine*, because this would convey the number 91 (= 1 + 9 × 10), and not 901. How, then, do we communicate that there is nothing in the decimal order?

In other words, when one rigorously applies the place-value system to the nine simple units, the use of a special terminology is indispensable to indicate the absence of units in a certain order.

The Indian astronomers overcame this obstacle by using the Sanskrit word *shûnya* meaning "void" and by extension "zero". Thus they were able to express the number 901 in words which can be translated in the following way:

One. Zero. Nine (= $1 + 0 \times 10 + 9 \times 10^2 = 901$).

The word *shûnya* ("zero") actually became the concept it signified; it played the role of zero in the place-value system, and thus prevented any confusion as to the value of the number expressed.

If we return to the verse quoted above, the Sanskrit numerical expression *agnishûnyâshvivasususarpârnava* (which represents the number 488,203) can be broken down as follows:

agni.shûnya.ashvi.vasu.sarpa.arnava

The words which act as components of this expression, however, are not the ordinary Sanskrit names of numbers. They are word-symbols, the literal translation of which, due to the association of ideas which characterises the Indian way of thinking, evoked a numerical value, rather like the way that the words pair and triad evoke the numbers two and three in our minds, except that the Sanskrit language had a greater choice of synonyms. Indian astronomers nearly always chose to express their numerical data using this almost infinite synonymy.

In order to represent the above number, the word-symbols appeared with the value indicated below:

*agni	= "fire"	= 3
*shûnya	= "void"	= 0
ashvi (= *Ashvin)	= "horsemen"	= 2
*vasu		= 8
*sarpa	= "serpent"	= 8
*arnava	= "ocean"	= 4

Thus one can translate the above expression in the following manner:

Fire.Void.Horsemen.*Vasu*.Serpent.Ocean.

   3   0   2   8   8   4

Remembering the earlier explanation of the system, we can see that the number represented is:

$$3 + 0 \times 10 + 2 \times 10^2 + 8 \times 10^3 + 8 \times 10^4 + 4 \times 10^5 = 488{,}203.$$

The second numerical expression that appears in the verse is *vasvagniyamâshvishikhidasra*, which can also be broken down in the following way:

vasv.agni.yama.ashvi.shikhi.dasra

These are also word-symbols possessing the following numerical values:

*Vasv (= *Vasu)*		= 8
agni	= "fire"	= 3
yama	= "Primordial Couple"	= 2
*ashvi (= *Ashvin)*	= "Horsemen"	= 2
*shiki (= *Shikhin)*	= "fire"	= 3
dasra	= "(one of the) Twins"	= 2

Which is interpreted as:

Vasu.Fire.Primordial Couple.Horsemen.Fire.Twins.

   8   3   2   2   3   2

This corresponds to the number:

$$8 + 3 \times 10 + 2 \times 10^2 + 2 \times 10^3 + 3 \times 10^4 + 2 \times 10^5 = 232{,}238.$$

This method of expressing numbers shows a perfect understanding of zero and the place-value system using 10 as a base.

It is a type of symbolic representation subject to many variations, yet the numerical symbols were always perfectly comprehensible to the Indian astronomers. Even if the value of certain words could vary according to the author, region or the time when they were written, the context always confirmed the intended numerical value.

Dating the Indian numerical word-symbols

When were these word-symbols first used? The answer is highly significant because the concept of zero and the place-value system in India are at least as old as this method of expressing numbers.

Dates on Sanskrit inscriptions from Southeast Asia

In India itself, as well as outside India, many documents exist which prove that this method of counting was, for a great many years, the privileged system of the Indian scholars, from the end of the sixth century at least until a relatively recent date.

The dated Sanskrit inscriptions of Southeast Asia figure very prominently amongst these documents.

It is important to make a clear distinction between the vernacular inscriptions and those written in Sanskrit. Both, however, date back to the *Shaka* era of the Indian astronomers. Primarily, in both types of inscriptions, the dates were recorded in words using the Sanskrit names for the numbers.

In the vernacular inscriptions (according to the region, written in Old Khmer, Old Javanese, Cham, etc.), these dates were then expressed using the nine numerals and zero of the Indian place-value system (Fig. 24.80).

In the Sanskrit inscriptions, however, the dates were recorded exclusively in the Indian word-symbols observing the place-value system and using 10 as the base. Here are some examples, taken from the oldest documents found in each of the regions in question.

The oldest dated Sanskrit inscription from Java is the *Stela* of Changal, the *Shaka* date of which is expressed in the following way [see H. Kern, VII, 118]:

shrutîndriyarasair

This can be broken down into separate words:

shruti.indriya.rasair

On consulting the Dictionary, under the headings **shruti, *indriya* and **rasa* (= *rasair*), the following meanings are obtained:

$$*shruti \ = \text{Veda} \qquad = 4$$
$$*indriya = \text{properties} = 5$$
$$*rasair \ = \text{senses} \quad = 6$$

Bearing in mind that the numbers are always written according to the decimal place-value system, beginning with the smallest unit, in ascending powers of ten (which the Indian astronomers called **ankânâm vâmato gatih*, or the principle of the "movement of the numbers [the numerical symbols] from the right to the left"), we can see that the date in question can be interpreted as:

Veda.Properties.Senses.
4   5   6

This corresponds to the number:

$$4 + 5 \times 10 + 6 \times 10^2 = 654.$$

Thus the inscription in question dates back to the *Shaka* year 654 + 78 (732 CE).

The oldest dated inscription from Champa is the *Stela* of Mî-so'n, the Shaka date of which is written in the following numerical symbols [see G. Coedès and H. Parmentier (1923), C 74 B; BEFEO, XI, p. 266):

ânandâmvarashatshata

which can be translated as follows (bearing in mind that *ânanda* means the "(nine) Nanda"; *amvara* = *ambara* = "space" = 0; and *shatshata* = "six hundreds"):

Nanda.Space.Six.Hundred.

9   0   6 × 100

which corresponds to the date: $9 + 0 \times 10 + 6 \times 100 = 609 + 78$ *Shaka* (687 CE).

The use of the term *shatshata* to denote six hundred shows a certain inexperience in the writing of numerical symbols, because the number 609 can be written *ânandâmvarashat*, which places the symbols for nine (*ânanda*), zero (*amvara*) and six (*shat*) in order.

The oldest dated Sanskrit inscription from Cambodia is that of Prasat Roban Romas, in the province of Kompon Thom. This is also the oldest dated Sanskrit inscription in the whole of Southeast Asia. It contains the following *Shaka* date [see Coedès and Parmentier (1923), K 151; BEFEO, XLIII, 5, p. 6]:

khadvishara

Here is the literal translation (where *kha = "space" = 0; *dvi = "two" = 2; and *shara = arrows = 5):

Space.Two.Arrows.

0   2   5

which corresponds to the date: $0 + 2 \times 10 + 5 \times 10^2 = 520 + 78$ *Shaka* (598 CE).

This proves that the use of Sanskrit word-symbols to express numbers was already widespread in Indo-China and Indonesia at the end of the sixth century CE.

As the civilisations were greatly influenced by Indian astronomers and astrologers, we can quite rightly presume that Indian scholars were using this technique at an even earlier date.

Evidence from the astronomers and mathematicians of India

There is a great deal of evidence pointing to the fact that the system was used by Indian scholars from the sixth century CE until a relatively recent date, as the following (non-exhaustive) list of Indian texts (containing many examples of the word-symbols) shows. The list is written in reverse chronological order (after R. Billard, 1971):

1. *Trishatikâ* by Shrîdharâchârya (date unknown) [B. Datta and A. N. Singh (1938) p. 59]
2. *Karanapaddhati* by Putumanasomayâjin (eighteenth century CE) [K. S. Sastri (1937)]
3. *Siddhântatattvaviveka* by Kamâlakara (seventeenth century CE) [S. Dvivedi (1935)]
4. *Siddhântadarpana* by Nîlakanthaso-mayâjin (1500 CE) [K. V. Sarma (undated)]
5. *Drigganita* by Parameshvara (1431 CE) [Sarma (1963)]
6. *Vâkyapañchâdhyâyi* (Anon., fourteenth century CE) [Sarma and Sastri (1962)]
7. *Siddhântashiromani* by Bhâskarâchârya (1150 CE) [B. D. Sastri (1929)]
8. *Râjamrigânka* by Bhoja (1042 CE) [Billard (1971), p. 10]
9. *Siddhântashekhara* by Shrîpati (1039 CE) [Billard (1971), p. 10]
10. *Shishyadhîvrddhidatantra* by Lalla (tenth century CE) [Billard (1971), p. 10]
11. *Laghubhâskarîyavivarana* by Shankaranârâyana (869 CE) [Billard (1971), p. 8]
12. *Ganitasârasamgraha* by Mahâvîrâchâryâ (850 CE) [M. Rangacarya (1912)]
13. *Grahachâranibandhana* by Haridatta (c. 850 CE) [Sarma (1954)]
14. *Bhâskarîyabhâsya* by Govindasvâmin (c. 830 CE) [Billard (1971), p. 8.]
15. Commentary on the *Âryabhatîya* by Bhâskara (629 CE) [K. S. Shukla and K. V. Sarma (1976)]
16. *Brahmasphutasiddhânta* by Brahmagupta (628 CE) [S. Dvivedi (1902)]
17. *Pañchasiddhântikâ* by Varâhamihîra (575 CE) [O. Neugebauer and D. Pingree (1970)]

Examples taken from the work of Bhâskara I

We will now look at some examples in their original form, taken from some of the oldest texts, which give a clearer indication than the above table of the earliest uses of this system in India.

The first concerns an example of how the number of years (4,320,000) that make up a **chaturyuga* (see also **yuga*) was expressed in word-symbols. It is an extract from the commentary which Bhâskara I wrote in 629 CE on the **Âryabhatîya* [see Shukla and Sarma (1976) p. 197]:

viyadambarâkâshashûnyayamarâmaveda

This can be broken down in the following manner:

viyad. ambara. âkâsha. shûnya. yama. râma. veda

On consulting the Dictionary, the following meanings are obtained:

*viyat (here written *viyad*)	= "sky"	= 0
*ambara	= "atmosphere"	= 0
*âkâsha	= "ether"	= 0
*shûnya	= "void"	= 0
*yama	= "(the) Primordial Couple"	= 2
*râma	= "(the) Râma"	= 3
*veda	= "(the) Veda"	= 4

This gives the following translation, with the corresponding mathematical breakdown:

Sky.Atmosphere.Ether.Void.Primordial Couple.Râma.Veda.

  0   0   0   0   2   3   4

$$= 0 + 0 \times 10 + 0 \times 10^2 + 0 \times 10^3 + 2 \times 10^4 + 3 \times 10^5 + 4 \times 10^6 = 4,320,000.$$

Here are three lines from the same work by Bhâskara (Commentary on the *Âryabhatîya*, manuscript R 14850 of the Government Oriental Manuscript Library, Madras, *Dashagîtika*, [see R. Billard (1971), pp.105–6], in Sanskrit, with the corresponding translation (the numerical word-symbols are underlined to distinguish them from the rest of the text):

tadânayanam idânîm kalpâder adyanirodhâd ayam abdarashir itîritah
khâgnyadrirâmârkarasavasurandhrendavah
te chânkair api 1986123730.

Before we look at the translation, it should be noted that the above word-symbols can be broken down in the following way:

kha.agny.adri.râma.arka.rasa.vasu.randhra.indavah

The Dictionary gives the following meanings for these words:

*kha	= "space"	= 0
agny (= **agni*)	= "fire"	= 3
*adri	= "mountains"	= 7
*râma	= "(the) Râma"	= 3
*arka	= "sun"	= 12
*rasa	= "senses"	= 6
*vasu		= 8
*randhra	= "orifices"	= 9
indavah (= **indu*)	= "moon"	= 1

Thus the following translation is obtained for the preceding extract from the Sanskrit text:

"In order to carry out the translation, here are the number of years which have transpired since the beginning of the [current] **kalpa* until the present day:

Space.Fire.Mountain.*Râma*.Sun.Sense.*Vasu*.Orifice.Moon.

"In figures this reads (*te chânkair api*): 1986123730".

As with the above example, here is the meaning of the word-symbols:

Space	Fire	Mountain	*Râma*	Sun	Sense	*Vasu*	Orifice	Moon
0	3	7	3	12	6	8	9	1

This corresponds to the following number:

$$0 + 3 \times 10 + 7 \times 10^2 + 3 \times 10^3 + 12 \times 10^4$$
$$+ 6 \times 10^6 + 8 \times 10^7 + 9 \times 10^8 + 1 \times 10^9 = 1,986,123,730.$$

One might be surprised to find, in a place-value system, word-symbols denoting values higher than or equal to ten, such as the word **arka* (= "sun" = 12) which is used here to express a number which contains two orders of units. Later, however, we will see why this symbol is used here, which does not constitute an exception to the rule of position where 10 is the base. If in this example, the word *arka*, on its own, expresses the number 12, it only acquires the value of 120,000 (= 12×10^4) because of the place it occupies in the above expression.

Moreover, the value (1,986,123,730) of the preceding word-symbols is clearly indicated "in figures" according to the place-value system, in the third line of the Sanskrit text, accompanied by the words "in figures this reads . . .", evidently in order to prevent any ambiguity as to the intended value. Thus we have a bilingual text of sorts which reinforces the above explanations.

This is not the only instance where Bhâskara felt the need to give the number in its corresponding numerals (using the place-value system of nine units and zero) as well as in astronomical word-symbols. Here is another example, this time involving a much higher number than the previous one [see Shukla and Sarma (1976), pp. 155]:

shûnyâmbarodadhiviyadagniyamâkâshasharasharâdri-
shûnyendurasâmbarângânkâdrishvarendu
ankair api 1779606107550230400.

As in the previous example, this compound word can be literally translated in the following way (given that: **shûnya* = 0, **ambara* = 0, **udadhi* [= *dadhi*] = 4, *viyad* (= **vyant*) = 0, **agni* = 3, **yama* = 2, **âkâsha* = 0, **shara*

= 5, *shara* = 5, *adri* = 7, *shûnya* = 0, *indu* = 1, *rasa* = 6, *ambara* = 0, *anga* = 6, *anka* = 9, *adri* = 7, *Ashva* = 7 and *indu* = 1), where the following two consecutive expressions constitute two ways of writing the same number according to the same principle:

Void.Sky.Ocean.Sky.Fire.Couple.Space.Arrow.Arrow.Mountain.
  0   0   4   0   3   2   0   5   5   7   →

Void.Moon.Sense.Atmosphere.Limb.Numeral.Mount.Horse.Moon
  0   1   6   0   6   9   7   7   1

"In figures this reads: 1,779,606,107,550,230,400."

The number expressed in word symbols is the one expressed "in figures"; according to the text itself:

1,779,606,107,550,230,400.

Here Bhâskara uses the Sanskrit word *anka*, the "numerals", not only to indicate the equivalent of the number concerned in the place-value system using nine numerals (*ankair api*, "in figures this reads . . ."), but also to designate the number 9. This is of great importance, because the basic meaning of *anka* is "a mark" or "a sign", which by extension can mean "numeral", although there is no connection between its other meanings and the number 9. Therefore, Bhâskara's use of *anka* to represent the number 9 proves that nine numerals and the place-value system were already being used to write numbers in India when the commentary was written.

Bhâskara gives the number "in figures" as well as in word-symbols, and this leaves no doubt that he was alluding to the nine basic numerals of the decimal place-value system which was invented in India: which, along with zero, enabled the Indian astronomers not only to represent any number, however high it might have been, but also and above all to carry out any mathematical operation with the minimum of complication.

Thus, in 629, the methods of expressing numbers either in numerals or in word-symbols were widely recognised by the learned men of India.

Examples found in the work of Varâhamihîra

Here are some more examples from the *Pañchasiddhântikâ*, the astronomical work of Varâhamihîra (VIII, lines 2,4 and 5). [See S. Dvivedi and G. Thibaut; O. Neugebauer and D. Pingree] [Personal communication of Billard]:

1) How the number 110 is expressed:
   shûnyaikaika = *shûnya.*eka.eka
   = void. one.one
     0   1   1
   $= 0 + 1 \times 10 + 1 \times 10^2 = 110.$

2) How the number 150 is expressed:

khatithi = **kha.**tithi

= space.day

0   15

= 0 + 15 × 10 = 150.

3) How the number 38,100 is expressed:

khakharûpâshtaguna = **kha.* **kha.***rûpa.***ashta.**guna

= space.space.shape.eight.quality

=   0   0   1   8   3

$= 0 + 0 \times 10 + 1 \times 10^2 + 8 \times 10^3 + 3 \times 10^4 =$
38,100.

This astronomical text was written c. 575 CE. This proves that zero and the place-value system were already in use in India in the second half of the sixth century CE.

THE EARLIEST KNOWN EVIDENCE OF THE INDIAN PLACE-VALUE SYSTEM

We will now look at the most important source of evidence relative to the history of the place-value system: the *Lokavibhâga (or *The Parts of the Universe*), a work on *Jaina cosmology which constitutes the oldest known use of word-symbols.

Besides the fact that the "minus one" is expressed by *rûponaka* (literally: "diminished form", *rûpo* = *rûpa = "shape" or "form" = 1) and that the concept of zero is expressed by *shûnya (void) or by words such as *kha, *gagana or *ambara ("sky", "atmosphere", "space", etc.), we find the following expression used for the number 14,236,713 [source: Anon. (1962), Chapter III, line 69, p. 70] [Personal communication of Billard]:

trîny ekam sapta shat trîni dve chatvâry ekakam

As the words used here are all names of numbers, they can be translated as follows (given that *eka* = 1 [= *ekaka*, the suffix *ka* here being a device used to regulate the metre of the line]; *dve* = 2; *trîni* = 3; *chatvâry* = 4; *shat* = 6; *sapta* = 7):

Three.One.Seven.Six.Three.Two.Four.One

3   1   7   6   3   2   4   1

$(= 3 + 1 \times 10 + 7 \times 10^2 + 6 \times 10^3 + 3 \times 10^4 + 2 \times 10^5 + 4 \times 10^6 + 1 \times 10^7 =$
14,236,713).

The author of this text seems generally to have avoided the abundant synonyms for the numerals and chosen to almost exclusively use the ordinary Sanskrit names of the numbers (*eka, dvi, tri, chatur, pañcha*, etc.).

The reason for this is, perhaps, that the word-symbols were not sufficiently well-known outside "learned" circles. However, there is another probable reason: the author wanted to make his work accessible in order to promote the merits of the philosophy of his religion and the superiority of Jaina science to the public at large, and therefore avoided technical terms.

Nevertheless, at times the author does use certain word-symbols, as in this expression of the number 13,107,200,000 [see Anon. (1962), Chapter IV, line 56, p. 79]:

> *pañchabhyah khalu shûnyebhyah param dve sapta*
> *châmbaram ekam trîni cha rûpam cha . . .*

five voids, then two and seven, the sky, one and three and the form

| 00000 | 2 | 7 | 0 | 1 | 3 | 1 |

$(= 0 + 0 \times 10 + 0 \times 10^2 + 0 \times 10^3 + 0 \times 10^4 + 2 \times 10^5 + 7 \times 10^6 + 0 \times 10^7 + 1 \times 10^8 + 3 \times 10^9 + 1 \times 10^{10} = 13,107,200,000)$.

However, each time the author uses one of these expressions, careful not to confuse his readers, he feels obliged to:

- either be more precise by adding:
  kramât, "in order",
  or *sthânakramâd*, "in positional order (*sthâna*)"
- or, which is even more remarkable, to add the following explanation:
  ankakramena, in the order of the numerals (*anka*)".

In other words, the concept of zero and the place-value system was widespread in India in the fifth century CE and had probably already been known for some time in "learned" circles.

In fact, the *Lokavibhâga* is the oldest known authentic Indian document to contain the use of zero and decimal numeration. As we shall see, it dates back to the middle of the fifth century CE.

We even know the exact year of the document thanks to the following verses [see Anon. (1962), Chapter XI, lines 50–54, pp. 224ff.] [Personal communication of Billard]:

> *vaishve sthite ravisute vrshabhe cha jîve*
> *râjottareshu sitapaksham upetya chandre*
> *grâme cha pâtalikanâmani pânarâshtre*
> *shâstram purâ likhitavân munisarvanandî* (verse 52)
>
> *samvatsare tu dvâvimshe kâñchîshah simhavarmanah*
> *ashîtyagre shakâbdânâm siddham etach chhatatraye* (verse 53).

Here is the translation:

Verse 52: "This work was written long ago by the *Muni* Sarvanandin, in the town called Pâtalika, in the kingdom of Pâna, when Saturn was in

Vaishva, Jupiter in Taurus, the Moon in *Râjottara*, on the first day of the light fortnight."

Verse 53: "Year twenty-two [of the reign] of Simhavarman, king of Kânchî, three hundred and eighty *Shaka* years."

In verse 52, we are told that when the text (or the copy of it) was written, the Moon was in *Râjottara*. This word means the *nakshatra** called *Uttaraphalgunî*: one of the twenty-seven constellations of the sidereal sphere, divided according to the sidereal revolution of the Moon. As it is the tenth constellation which is referred to here, this position corresponds (according to reliable mathematical calculations) to the interval between 146° 40' and 160° of sidereal longitude. We are also told that the Moon was in its phase corresponding to the first day of the "light fortnight": the first half of the month. We can determine that the work was written in the *Shaka* year 380, the corresponding date being written "entirely in letters" using the ordinary Sanskrit names for the numbers.

Looking at the information given in the verses, which has been interpreted according to the elements of Indian history and astronomy, we have:

- the year, namely the *Shaka* year 380;
- the day of the month, in other words, the Moon is in the first day of the first fortnight of the month;
- and the position of the Moon: 146° 40' / 160° of sidereal longitude, which allows us to determine the month.

Without going into too much detail about the methodology used to determine the dates and to study the astronomical data, suffice to say that the information given leaves us in no doubt as to the date expressed here; the date, in the Julian calendar, corresponds exactly to:

Monday, 25 August, 458 CE.

This is the precise date of the Jaina cosmological text, **Lokavibhâga* (or *The Parts of the Universe*).[†]

We can now add the other two pieces of information given in verse 52: the planet Jupiter was in Taurus, in the second sign of the zodiac, thus occupying a position of 30° to 60° of sidereal longitude; at the same time,

---

* Here, this word is used to explain the "lunar mansions" in equal divisions. See **Nakshatra*.

† We also see in verse 53 that this text is dated the 22nd year of the rule of Simhavarman, king of Kânchî (the "Golden Town", sacred place of the Hindus, in Tamil Nadu, approximately 60 km southwest of Madras). According to Frédéric, DCI (1987), pp. 819–20, this king, the son of Skandavarman II, issued from ones of the lines of the Pallava Dynasty, reigned from 436. As this was the 22nd year of his reign, this date corresponds to 436 + 22 = 458 CE. We do not know, however, if the chronology of this sovereign was established by specialists using the text of the **Lokavibhâga*. If this is the case, then this information is of no interest to us. On the other hand, if this is not the case, if the dates of the reign of Simhavarman were established from another inscription, then we have real confirmation of the date we have just determined using the astronomical data in the text in question.

Saturn was in *Vaishva*, the **nakshatra* called *Uttarâshâdha* (the nineteenth constellation of the sidereal revolution), therefore between 266° 40' and 280° of sidereal longitude. As this data agrees with the preceding date, the date is astronomically confirmed.

Whilst this information allows us to date the *Lokavibhâga* with precision, it also irrefutably proves the authenticity of the document, due to the very nature of one of the preceding pieces of astronomical data.

Because Jupiter is situated in the text according to its position in a zodiacal sign, we can also find, for astrological reasons, the position of Saturn in the **nakshatra* system.

This is an irrefutable archaism characterised by the very history of Indian astrology. After this time, there are no more examples where the positions of the planets (with the exception of the Moon) are described in**nakshatra*, they are only expressed in relation to the position of the twelve signs of the zodiac (previously unpublished information given by Billard).

The very existence of this archaism and its almost total disappearance from later Indian texts prove the complete authenticity of its usage, of the document, and all the information it gives us. Moreover, the *Lokavibhâga* as a whole, from an astronomical and cosmological point of view, is undeniably archaic in character in comparison with later texts of the same genre.

Let us now look even more closely at the problem in hand. This text was "written" long before by a *Muni* named Sarvanandin, but the word "written" is ambiguous because in Sanskrit it can mean "copied" as well as "written". The *Lokavibhâga* appears to be the Sanskrit translation of an earlier work written in Prakrit (probably in a Jaina dialect), judging from the translation of verse 51:

> The **Rishi* Simhasûra translated into the Language [= Sanskrit] that which the uninterrupted line of doctors had transmitted [in dialect], which the revered *arhant* Vardhamâna [= the **Jina*] delivered to the saints during the grand assembly of the gods and men, namely all that [the disciples of Jina such as] the Sudharma know about the creation of the universe. Let him be praised by all ascetics.

This could and very probably does mean that the current version of the *Lokavibhâga* is an exact reproduction of an original which was written before 458 CE.

Of course we must be wary of relatively recent Indian texts which are frequently attributed to the **Rishi*, the "Sages" of the Vedic era (twelfth to eighth century BCE) who are said to have received the great "Revelation" from the divinities.

The *Lokavibhâga*, however, is much more modest, as it attributes its writing to a *Muni*. This *Muni* could well have lived one or two generations before the above date.

This seems even more likely when we consider that, on the one hand, the numbers which appear in this text conform totally to the rules of the decimal place-value system, and on the other hand, the care that the author took to popularise the text. As we have already seen, when this text was written, Indian scholars were already familiar with the place-value system.

Who, then, is a *Muni*? The answer to this question is in the text itself, in verse 50:

> *Muni* is he who achieves perfection, and, displaying [the] strength of a lion, escapes the terrible [cycle of renaissance], through obeying the decree of respect to all animal life, the exercises of piety such as the vow of honesty, the holiness which conquers all false doctrine and all futility, dominates the empire of the senses, and even defeats the eternal *Karma* through the fire of fervent austerities.

That, in a nutshell, is the doctrine of Jaina, as well as what became of the *Muni* Sarvanandin to whom the writing of the *Lokavibhâga* is attributed.

When did this *Muni* live? A hundred or two hundred years previously? We will never know. What we do know for certain is that the discovery of our present-day numeral system was made well before that famous Monday 25 August, 458 CE.

HIGHLY CONSISTENT EVIDENCE

Considering the quantity and extreme diversity of the information contained in this chapter, it would seem appropriate to present a summary of all the historical facts which have been established concerning the discovery of zero and the place-value system. The following is a list in reverse chronological order, with references to the Dictionary for those wishing to know more details.

Summary of the historical facts relating to the place-value system

1150 CE. The Indian mathematician *Bhâskarâchârya (known as Bhâskara II) mentions a tradition, according to which zero and the place-value system were invented by the god Brahma. In other words, these notions were so well established in Indian thought and tradition that at this time they were considered to have always been used by humans, and thus to have constituted a "revelation" of the divinities. See *Place-value system.

1010–1030 CE. Date of evidence given by the Muslim scholar of Persian origin, *al-Biruni, about India and in particular her place-value system and methods of calculation; a highly documented piece of evidence to add to the others from the Arabic-Muslim world and the Christian West.

End of the ninth century CE. The philosopher *Shankarâchârya makes a direct reference to the Indian place-value system.

875–876 CE. The dates of the inscriptions of Gwalior: the oldest known "real" Indian inscriptions in stone to use zero (in the form of a little circle) and the nine numerals (in *Nâgarî*) according to the place-value system. See *Nâgarî* numerals, and Figs. 24.72 to 74.

869 CE. The Indian astronomer *Shankaranârâyana frequently uses the place-value system with word-symbols.

c. 850 CE. The Indian astronomer *Haridatta invents a system of numerical notation using letters of the Indian alphabet and based on the place-value system using zero (randomly represented by two different letters): this is the first example of a place-value system which uses letters of the alphabet. See *Katapayâdi numeration.

850 CE. The Indian mathematician *Mahâvîrâchârya frequently uses the place-value system with the nine numerals or with Sanskrit numerical symbols [M. Rangacarya (1912)].

c. 830 CE. The Indian astronomer *Govindasvâmin frequently uses the place-value system [R. Billard (1971), p. 8].

813 CE. This is the date of the oldest known vernacular inscription of Champa (Indianised civilisation of Southeast Asia), the *Shaka* date of which is indicated using the nine Indian numerals and zero. See *Cham numerals, and Fig. 24.80.

760 CE. The date of the oldest known vernacular inscription of Java, the *Shaka* date of which is expressed using the nine numerals and zero from India. See *Kawi numerals, and Fig. 24.80.

732 CE. Date of the oldest known Sanskrit inscription from Java, the *Shaka* date of which is expressed using the place-value system and word-symbols of the Indian astronomers [H. Kern (1913–1929)].

718–729 CE. Date of the *Kai yuan zhan jing*, a work on astronomy and astrology by the Chinese Buddhist *Qutan Xida, who was in fact of Indian origin, real name *Gautama Siddhânta, who lived in China from c. 708 CE, and who, in his work, describes zero, the place-value system of the nine numerals and the Indian methods of calculation.

Seventh century CE. The poet *Subandhu makes direct references to the Indian zero (in the form of a dot) as a mathematical processing device. Thus zero and the place-value system were so well-established in India that the poet could use such subtleties with his metaphors. See *Zero and Sanskrit poetry.

687 CE. Date of the oldest known Sanskrit inscription of Champa, the *Shaka* date of which is expressed using the place-value system and the word-symbols of the Indian astronomers [G. Coedès and H. Parmentier, C 74 B;BEFEO, XI, p. 266].

683 CE. The date of the oldest known vernacular inscription from Malaysia, the *Shaka* date of which is written in the Indian numerals (including zero). See Fig. 24.80.

683 CE. Date of the oldest known vernacular inscription of Cambodia, the *Shaka* date of which is written in Indian numerals (including zero). See *Old Khmer numerals and Fig. 24.80.

662 CE. Syrian bishop Severus Sebokt writes of the nine numerals and Indian methods of calculation.

629 CE. Indian mathematician and astronomer *Bhâskara I frequently uses the place-value system with the word-symbols, often also expressing the number using the nine numerals and zero [K. S. Shukla and K. V. Sarma (1976)].

628 CE. Indian astronomer and mathematician Brahmagupta frequently uses the place-value system with the nine numerals as well as with the word-symbols. He also describes methods of calculation using the nine numerals and zero (very similar to the methods we still use today). He also provides fundamental rules of algebra, where zero is present as a mathematical concept (the number nought), and talks of infinity, defining it as the opposite of zero. See *Zero. *Infinity. *Khachheda.

598 CE. Date of the oldest known Sanskrit inscription of Cambodia, the *Shaka* date of which is written in word-symbols according to the place-value system [Coedès and Parmentier, K 151; BEFEO, XLIII, 5, p. 6].

594 CE. Date of the donation charter engraved on copper of Dadda III of Sankheda, in Gujarat. This is the oldest known Indian text to bear witness to the use of the nine numerals according to the place-value system (see Fig. 24.75). As we saw earlier, there can be no doubt as to the authenticity of this document.

End of the sixth century CE. The arithmetician *Jinabhadra Gani gives several numerical expressions which prove that he was well acquainted with zero and the place-value system [Datta and Singh (1938)].

c. 575 CE. Indian astronomer and astrologer *Varâhamihîra makes frequent use of the place-value system with Sanskrit numerals. See *Indian astrology.

c. 510 CE. *Âryabhata invented a unique method of recording numbers which required perfect understanding of zero and the place-value system. Moreover, he used a remarkable process of calculating square and cube roots, which would have been impossible without the place-value system, using nine different numerals and a tenth sign which performed the functions of zero. See *Âryabhata (Numerical notations of), *Âryabhata's numeration, *Square roots (How Âryabhata calculated his).

(Monday 25 August) 458 CE. The exact date of the *Lokavibhâga, (The Parts of the Universe), the Jaina cosmological text: the oldest known Indian text to use zero and the place-value system with word-symbols.

Thus one can see the impressive amount of evidence proving that our modern number-system is of Indian origin, and that it was invented long before the sixth century CE. All the evidence points to the fact that this invention is entirely Indian, and born out of a very specific context.

Moreover, we are not dealing with one isolated piece of evidence, or even a limited number of documents, but a huge collection of proofs from all the disciplines, dating from the most significant eras, which have been situated through the study of the palaeography, epigraphy and philology of Indian civilisations both within and outside India.

THE MOST LIKELY TIME OF THESE DISCOVERIES

It is most likely that the place-value system and zero were discovered in the middle of the reign of the Gupta Dynasty, whose empire stretched the whole length of the Ganges Valley and its tributaries from 240 to approximately 535, known as the "classic" period.

This period saw the highest forms of Indian art (sculpture, painting, in the caves of Ajanta for example, etc.) reach maturity. It was also a classic period because, as Coomaraswamy said, "almost everything that belongs to the Asian spiritual conscience is of Indian origin and dates back to the Gupta Dynasty."

This era coincides with a kind of rebirth of Brahmanism, before it evolved in the wider sense of Hinduism in the following centuries.

Trade was also flourishing at this time, with the Near East, via Persia, and across the sea with the Roman Empire, particularly through Lâta or the eastern area of the present-day state of Gujarat.

Medicine was also developing at this time, particularly dissection.

In the field of literature, Sanskrit, previously the official language of the court and of Brahmanism, was adopted by the Jainas and Buddhists, who did much for the development of the language. And it was probably in this period too that Sanskrit grew to be a much richer language than it had been in the time of the Vedas. This time also saw the beginnings of the *Mahâbhârata*, one of the greatest Indian epic poems, and of the *Dharmashâstra*, collections of texts, mainly about customs, laws and castes.

It was during this time that the *Darshana* – the six systems of Indian philosophy – were developed.

The stories and fables, such as those of the *Pañchatantra*, (the main source of inspiration for the Persian fable *Kalila wa Dimna*), also appeared for the first time, whilst the theatre knew its first blossoming with the poet Kâlidâsa, considered to be one of the greatest dramatists of Indian history, and the *Navaratna* or "Nine Jewels" of Indian tradition.

As for Indian writing, Gupta constitutes the first notation to be individualised in relation to its *Brâhmî* ancestor. As it became more refined, it gave birth to *Nâgarî* (or *Devanâgarî*), in the seventh century CE, which became the principal style in which Sanskrit and then Hindi were written. From *Nâgarî* came the various styles of northern and central India. Another, more northern variant of *Gupta*, evolved into *Shâradâ* of Kashmir, or its derivatives, and also diversified into *Siddham*, from which the script of Nepal, Chinese Turkestan and Tibet would be derived (Fig. 24.52). (See *Indian numerals).

Thus the Gupta period saw the most spectacular progress in almost all the fields of learning, and was a veritable "explosion" of Indian culture.

This was also the time when *Lalitavistara Sûtra* was written, which tells the legend of Buddha and mentions numbers of the highest orders, following very surprising numerical speculation; speculation which grows rapidly after this period, but for which there is no evidence before this time.

It is doubtless no coincidence that the Gupta era saw the first blossoming of *Indian mathematics.

This was also the time of the first developments of trigonometrical astronomy and "Greek" astrology, which was very different from that which existed previously in India, both in terms of claims and material, and which, being in appearance very systematic, already had the scientific foundations of what would soon become Indian astronomy.

Moreover, this was the time when Âryabhata lived. His work would soon lead to a decisive about-turn in Indian astronomy, breaking once and for all with the old Greek-Babylonian traditions and developing the cosmic cycles called *yuga, devoid of physical value but nonetheless based on a series of unique observations which were more or less precise.

The *Lokavibhâga* is dated 458 CE. This being the oldest known testimony of the use of zero and of the Indian decimal place-value system, the latest possible date of this discovery has to be the middle of the Gupta era. Documents written earlier or at the same time show use of either the ordinary system of the Sanskrit names of the numbers or, as we shall see in the following chronology, that of the old non-positional system derived from the *Brâhmî* system (Fig. 24.70).

Thus the earliest possible date of this discovery is the beginning of the Gupta Dynasty. We must take into consideration, however, the fact that documents bearing witness to the use of word-symbols or the decimal place-value system are only found in abundance after the beginning of the sixth century.

Bearing in mind, on the one hand, the perfect understanding of the place-value system displayed in the *Lokavibhâga* and the clear desire to popularise the text, and on the other hand the fact that the text was more than likely a Sanskrit translation of an earlier document (no doubt written in a Jaina dialect), it would not be unreasonable to suggest *the fourth century CE as the date of the discovery of zero and the place-value system.*

Third to second century BCE. First appearances of *Brâhmî* numerals in the edicts of Emperor Asoka and the inscriptions of Nana Ghat. These are very rudimentary. But the first nine figures already constitute the prefiguration of the nine numerals that we use today (Indian, then Arabic and European). Sanskrit numerals are already worked out and there are particular names for the ascending powers of ten up to 10^8 (= 100,000,000) at least.

First century BCE to third century CE. The numerals found in many inscriptions are derived from *Brâhmî* numerals and constitute a sort of intermediary between *Brâhmî* numerals and later styles, but the place-value system is not yet in use.

The Sanskrit system is extended to include powers of ten up to 10^{12} (= 1,000,000,000,000). [See *Names of numbers]

Fourth to fifth century CE. The numerals derived from *Brâhmî* numerals begin to diversify into specific styles (*Gupta, Pali, Pallava, Châlukya,* etc.)

The Sanskrit system is capable of expressing and using powers of ten up to 10^{421} and above, as we see in the *Lalitavistara* (before 308 CE).

Discovery of zero and the place-value system

458 CE. (*To this day, no document has been found to prove that the nine units were used at this date according to the place-value system.*)

The names of the first nine numbers are used according to the place-value system, as we shall see in the *Lokavibhâga,* dated 458 CE, where the names of the numbers are sometimes replaced by word-symbols and the word *shûnya* ("void") and its synonyms are used as zeros.

From the sixth century onwards. The use of the place-value system and zero begin to appear frequently in documents from India and Southeast Asia (the following list is non-exhaustive):

594 CE. Sankheda charter on copper
628 CE. *Brâhmasputasiddhânta* by Brahmagupta
629 CE. Commentary on the *Âryabhatîya* by Bhâskara
683 CE. Khmer inscription of Trapeang Prei
683 CE. Malaysian inscription of Kedukan Bukit
684 CE. Malaysian inscription of Talang Tuwo
686 CE. Malaysian inscription of Kota Kapur
737 CE. Charter of Dhiniki on copper
753 CE. Inscriptions of Devendravarmana
760 CE. Javanese inscription of Dinaya
793 CE. Charter of Râshtrakûta on copper
813 CE. Cham inscription of Pô Nagar
815 CE. Charter of Buchkalâ on copper

829 CE. Cham inscription of Bakul
837 CE. Inscription of Bauka
850 CE. *Ganitasâramgraha* of Mahâvîrâchârya
862 CE. Inscription of Deogarh
875 CE. Inscriptions of Gwalior
877 CE. Balinese inscription of Haliwanghang
878 CE. Balinese inscription of Mamali
880 CE. Balinese inscription of Taragal
917 CE. Charter on copper of Mahipala, etc.

Seventh century CE. Gupta notation gave birth to *Nâgarî* numerals, which in turn were the forerunners of the numerals of northern and central India (*Bengalî, Gujâratî, Oriyâ, Kaîthî, Maithilî, Manipurî, Marâthî, Mârwarî*, etc.).

Eighth century CE. First appearance of the stylised numerals of Southeast Asia (*Khmer, Cham, Kawi*, etc.).

Ninth century CE. A northern variant of *Gupta* led to the *Shâradâ* numerals of Kashmir, ancestors of the numerals of northwest India (*Dogrî, Tâkârî, Multânî, Sindhî, Punjabî, Gurûmukhî*, etc.).

Eleventh century CE. The first appearances of *Telugu* numerals (southern India).

A CULTURE WITH A PASSION FOR HIGH NUMBERS

The early passion which Indian civilisation had for high numbers was a significant factor contributing to the discovery of the place-value system, and not only offered the Indians the incentive to go beyond the "calculable", physical world, but also led to an understanding (much earlier than in our civilisation) of the notion of mathematical infinity itself.

The Indian love for high numbers can be seen in the *Lalitavistara Sûtra or Development of the Games* [of Buddha] (a Sanskrit text of the Buddhism of Mahâyâna, written in verse and prose, about the life of Buddha, the "Saint of the Shaka family", as he is said to have told his disciples), where high numbers are constantly evoked:

Choosing a few random examples, we find in this text a meeting of ten thousand monks, eighty-four million Apsarâs, thirty-two thousand Bodhisattvas, sixty-eight thousand Brahmas, a million Shakras, a hundred thousand gods, hundreds of millions of divinities, five hundred Pratyeka-Buddhas, eighty-four thousand sons of gods, then thirty-two thousand and thirty-six million other sons of gods, sixty-eight thousand *kotis* [= 680,000,000,000] sons of gods and Bodhisattva, eighty-four hundred thousand *niyuta kotis* [= 8,400,000,000,000,000,000,000] of divinities.

The principal signs of Buddha are given the number thirty-two, secondary signs eighty, signs of his mother thirty-two, those of the dwelling-place and the family where he is said to have been born eight and sixty-four. The queen Mâyâ-Devî is served by ten thousand women; the ornaments of the throne of Buddha are enumerated in hundreds of thousands; the hundreds of thousands of divinities and hundred thousand millions of Bodhisattvas and Buddhas pay homage to this throne which is the result of merits accumulated over one hundred thousand million *kalpas, one kalpa being the equivalent of four billion, three hundred and twenty million years. The lotus flower that blossomed the night that Buddha was conceived has a diameter of sixty-eight million *yojana*. Two hundred thousand treasures appeared when Buddha was born; this filled the three thousand great hosts of worlds, and living creatures came to pay homage to his mother, the queen Mâyâ-Devî, in throngs of eighty-four thousand and sixty thousand [F. Woepcke (1863)].

Likewise, in *The Light of Asia*, Edwin Arnold reproduces this passage from the *Lalitavistara Sûtra*, about the education of Buddha as a child, aged eight, by the Sage Vishvâmitra, who explains, in another passage, that numeration, numbers and arithmetic constitute the most important discipline among the seventy-two arts and sciences that the Bodhisattva must acquire:

And Vishvâmitra said: That's enough [now],
Let us turn to Numbers. Count after me
Until you reach *lakh (= one hundred thousand):
One, two, three, four, up to ten,
Then in tens, up to hundreds and thousands.
After which, the child named the numbers,
[Then] the decades and the centuries, without stopping.
[And once] he reached *lakh*, [which] he whispered in silence,
Then came *koti, *nahut, *ninnahut, *khamba,
*viskhamba, *abab, *attata,
Up to *kumud, *gundhika, and *utpala
[Ending] with *pundarîka [leading]
Towards *paduma, making it possible to count
Up to the last grain of the finest sand
Heaped up in mountainous heights.

Let us interrupt the master for a moment to clarify the numerical values mentioned in the passage:

lakh	is worth	$100,000 = 10^5$
koti	is worth	$10,000,000 = 10^7$
nahut	is worth	$1,000,000,000 = 10^9$
ninnahut	is worth	$100,000,000,000 = 10^{11}$
khamba	is worth	$10,000,000,000,000 = 10^{13}$
viskhamba	is worth	$1,000,000,000,000,000 = 10^{15}$
abab	is worth	$100,000,000,000,000,000 = 10^{17}$
attata	is worth	$10,000,000,000,000,000,000 = 10^{19}$
kumud	is worth	$1,000,000,000,000,000,000,000 = 10^{21}$
gundhika	is worth	$100,000,000,000,000,000,000,000 = 10^{23}$
utpala	is worth	$10,000,000,000,000,000,000,000,000 = 10^{25}$
pundarîka	is worth	$1,000,000,000,000,000,000,000,000,000 = 10^{27}$
paduma	is worth	$100,000,000,000,000,000,000,000,000,000 = 10^{29}$

Thus we are dealing with a centesimal scale, the value of each name being one hundred times bigger than the one preceding it.

> But beyond this counting system,
> There is the *kâtha* which is used to count the stars in the night sky,
> The *kôti-kâtha* for [enumerating] the drops of the ocean,
> *Ingga*, to calculate the circular [movements],
> *Sarvanikchepa*, with which it is possible to calculate
> All the sand of a *Gunga*,
> Until we reach *antahkapa*,
> Which is [made up of ten] *Gungas*.
> [And] if a more intelligible scale is required,
> The mathematical ascensions, through the *asankhya*, which is the sum
> Of all the drops of rain which, in ten thousand years,
> Would fall each day on all the worlds,
> Lead [the arithmetician] to the *mahâkalpa*,
> Which the gods use to calculate their future and their past.

THE LIMITATIONS OF THE (INDIAN) "INCALCULABLE"

The *asankhya* or *asankhyeya*, which was poetically defined as "the sum of all the drops of rain which, in ten thousand years, would fall each day on all the worlds", is actually none other than the Sanskrit term meaning "*incalculable*". It literally means: "number which is impossible to count" (from *sankhya* or *sankhyeya*, "number", accompanied by the privative "a").

This word is used in Brahman cosmogony, where it is sometimes used to denote the length of the "*day of Brahma*", in other words 4,320,000,000 human years.

In *Bhagavad Gîtâ*, however, "incalculable" corresponds to the entire length of Brahma's life, which is 311,040,000,000,000 human years. In one of the commentaries on the work, it is pointed out that "this incredible longevity, for us infinite, represents no more than zero in the stream of eternity."

Naturally, the value given to "the incalculable" varies considerably according to the text, the author, the region and the era. Thus, the *Sankhyâyana Shrauta Sûtra* fixes this limit at 10,000,000,000,000 giving this number the name *ananta*, signifying "infinity" [see Datta and Singh (1938) p. 10)]. The Tibetans and the Sinhalese pushed the limit of *asankhyeya* much further in giving it a value of one followed by ninety-seven zeros. In the *Pâli* Grammar of Kâchchâyana, the same concept is given a value of 10^{140} (ten million to the power of twenty), placing this term at the end of this very impressive nomenclature the scale of which is tens of millions [see JA 6/17 (1871), p. 411, lines 51–2)]:

A hundred times a hundred times a thousand makes a	koti	$= 10^{7}$
A hundred times a hundred times a thousand koti makes a	pakoti	$= 10^{14}$
A hundred times a hundred times a thousand pakoti	kotippakoti	$= 10^{21}$
A hundred times a hundred times a thousand kotippakoti	nahuta	$= 10^{28}$
A hundred times a hundred times a thousand nahuta	ninnahuta	$= 10^{35}$
A hundred times a hundred times a thousand ninnahuta	akkhobhini	$= 10^{42}$
A hundred times a hundred times a thousand akkhobhini	bindu	$= 10^{49}$
A hundred times a hundred times a thousand bindu	abbuda	$= 10^{56}$
A hundred times a hundred times a thousand abbuda	nirabbuda	$= 10^{63}$
A hundred times a hundred times a thousand nirabbuda	ahaha	$= 10^{70}$
A hundred times a hundred times a thousand ahaha	ababa	$= 10^{77}$
A hundred times a hundred times a thousand ababa	atata	$= 10^{84}$
A hundred times a hundred times a thousand atata	sogandhika	$= 10^{91}$
A hundred times a hundred times a thousand sogandhika	uppala	$= 10^{98}$
A hundred times a hundred times a thousand uppala	kumuda	$= 10^{105}$
A hundred times a hundred times a thousand kumuda	pundarîka	$= 10^{112}$
A hundred times a hundred times a thousand pundarîka	paduma	$= 10^{119}$
A hundred times a hundred times a thousand paduma	kathâna	$= 10^{126}$
A hundred times a hundred times a thousand kathâna	mahâkathâna	$= 10^{133}$
A hundred times a hundred times a thousand mahâkathâna	asankhyeya	$= 10^{140}$

The extravagant numbers of the legend of Buddha

Thus we can see the extent to which the Indians took their naming of numbers.

We can get an even clearer idea of this if we return to the *Lalitavistara Sûtra*, where Bodhisattva (Buddha), now an adult, is almost forced to take part in a competition:

When Bodhisattva reached a marriageable age, he was betrothed to Gopâ, the daughter of Shâkya Dandapâni. But Dandapâni refused to let him marry his daughter, unless the son of the king Shuddhodana [Bodhisattva] made a public show of his mastery of the arts. Thus a type of contest, the winner of which would be given Gopâ's hand in marriage, took place between Bodhisattva and five hundred other young Shakyas. This contest included writing, arithmetic, wrestling and archery [F. Woepcke (1863)].

After easily beating all the young Shakyas, Bodhisattva was invited by his father to pit his wits against the great mathematician Arjuna, who had judged the contest:

> "Young man," said Arjuna, "do you know how we express numbers that are higher than a hundred *koti?"
>
> Bodhisattva nodded, but Arjuna impatiently continued:
>
> "So how do we count beyond a hundred *koti in hundreds?"

Here is Bodhisattva's reply, bearing in mind that one *koti is the equivalent of ten million (= 10^7):

> "One hundred *koti* are called an *ayuta, a hundred *ayuta* make a *niyuta, a hundred *niyuta* make a *kankara, a hundred *kankara* make a *vivara, a hundred *vivara* are a *kshobhya, a hundred *kshobhya* make a *vivaha, a hundred *vivaha* make a *utsanga, a hundred *utsanga* make a *bahula, a hundred *bahula* make a *nâgabala, a hundred *nâgabala* make a *titilambha, a hundred *titilambha* make a *vyavasthânaprajñapati, a hundred *vyavasthânaprajñapati* make a *hetuhila, a hundred *hetuhila* make a *karahu, a hundred *karahu* make a *hetvindriya, a hundred *hetvindriya* make a *samâptalambha, a hundred *samâptalambha* make a *gananâgati, a hundred *gananâgati* make a *niravadya, a hundred *niravadya* make a *mudrâbala, a hundred *mudrâbala* make a *sarvabala, a hundred *sarvabala* make a *visamjñagati , a hundred *visamjñagati* make a *sarvajña , a hundred *sarvajña* make a *vibhutangamâ, a hundred *vibhutangamâ* make a *tallakshana."

Thus, in his reply, Bodhisattva had given the following table:

1 *ayuta*	= 100 *koti*	= 10^9
1 *niyuta*	= 100 *ayuta*	= 10^{11}
1 *kankara*	= 100 *niyuta*	= 10^{13}
1 *vivara*	= 100 *kankara*	= 10^{15}
1 *kshobhya*	= 100 *vivara*	= 10^{17}
1 *vivaha*	= 100 *kshobhya*	= 10^{19}
1 *utsanga*	= 100 *vivaha*	= 10^{21}

1 *bahula*	= 100 *utsanga*	= 10^{23}
1 *nâgabala*	= 100 *bahula*	= 10^{25}
1 *titilambha*	= 100 *nâgabala*	= 10^{27}
1 *vyavasthânaprajñapati*	= 100 *titlambha*	= 10^{29}
1 *hetuhila*	= 100 *vyavasthânaprajñapati*	= 10^{31}
1 *karahu*	= 100 *hetuhila*	= 10^{33}
1 *hetvindriya*	= 100 *karahu*	= 10^{35}
1 *samâptalambha*	= 100 *hetvindriya*	= 10^{37}
1 *gananâgati*	= 100 *samâptalambha*	= 10^{39}
1 *niravadya*	= 100 *gananâgati*	= 10^{41}
1 *mudrâbala*	= 100 *niravadya*	= 10^{43}
1 *sarvabala*	= 100 *mudrâbala*	= 10^{45}
1 *visamjñagati*	= 100 *sarvabala*	= 10^{47}
1 *sarvajña*	= 100 *visamjñagati*	= 10^{49}
1 *vibhutangamâ*	= 100 *sarvajña*	= 10^{51}
1 *tallakshana*	= 100 *vibhutangamâ*	= 10^{53}

In modern terms, the value of the *tallakshana* corresponds to the following formula:

$$1\ {}^*tallakshana = (10^7) \times (10^2)^{23} = 10^{7 + 46 \times 1} = 10^{53}.$$

"Having thus reached the **tallakshana*, which we would write today as 1 followed by fifty-three zeros, Bodhisattva added that this whole table forms only one counting system, the **tallakshana* counting system, [from the name of its last term]; but there is, above this system, that of **dhvajâgravati*; beyond that, the counting system **dhvajâgranishâmani*, and beyond that again, six other systems for which he gave the respective names" [Woepcke (1863)].

The **dhvajâgravati* system is also made up of twenty-four terms, and its first term is the **tallakshana* (the largest number in the preceding system, that is 10^{53}). Since its progression increases geometrically by a ratio equivalent to one hundred, its final term therefore has the value:

$$1\ dhvajâgravati = (10^{7 + 46 \times 1}) \times (10^2)^{23} = 10^{7 + 46 \times 2} = 10^{99}.$$

As this is the last term in the preceding system, it becomes the first in the following one, that is to say the third system, the *dhvajâgranishâmani*, the final number of which being equal to:

$$1\ dhvajâgranishâmani = (10^{7 + 46 \times 2}) \times (10^2)^{23} = 10^{7 + 46 \times 3} = 10^{145}.$$

Step by step, we thus arrive at the ninth counting system, of which the name of the last term has the value:

$$(10^{7 + 46 \times 8}) \times (10^2)^{23} = 10^{7 + 46 \times 9} = 10^{421}.$$

(We write this number as 1 followed by 421 zeros).

Arjuna, full of admiration for the superiority of Buddha's knowledge, and wanting nothing more than to learn from him, asked him to explain how one enters into "the counting system which extends to the particles of the first atoms (*Paramânu)"* (literally: "first-atom-particle-penetration-enumeration") and to teach him and the young Shakyas how many first atoms there were in a *yojana* (a unit of length).

Here is Buddha's reply:

> If you want to know this number, use the scale that takes you from the *yojana* to four *krosha* of Mâgadha, from the *krosha* of Mâgadha to a thousand arcs (*dhanu*), from the arc to four cubits (*hasta*), from the cubit to two spans (*vitasti*), from the span to twelve phalanges of fingers (*angulî parva*), from the phalanx of the finger to seven grains of barley (*yava*), from the grain of barley to seven mustard seeds (*sarshapa*), from the mustard seed to seven poppy seeds (*likshâ râja*), from the poppy seed to seven particles of dust stirred up by a cow (*go râja*), from the particle of dust stirred up by a cow to seven specks of dust stirred up by a ram (*edaka râja*), from the speck of dust disturbed by a ram to seven specks of dust stirred up by a hare (*shasha râja*), from the speck of dust stirred up by a hare to seven specks of dust carried off by the wind (*vâyâyana râja*), from the speck of dust carried away by the wind to seven tiny specks of dust (*truti*), from a tiny speck of dust to seven minute specks of dust (*renu*), and from the minute speck of dust to seven particles of the first atoms (*paramânu râja*).

In other words, if we use the modern notation of the exponents and if we use the letter "p" to denote these "first atoms" (*paramânu*), this "scale" can be written in the following manner, starting with the smallest and finishing with the largest quantity:

1 minute speck of dust
= 7 particles of dust of the first atoms . . . . $7\,p$
1 tiny speck of dust
= 7 minute specks of dust . . . . $7^2\,p$
1 speck of dust carried away by the wind
= 7 tiny specks of dust . . . . $7^3\,p$
1 speck of dust stirred up by a hare
= 7 specks of dust carried away by the wind . . . . $7^4\,p$
1 speck of dust stirred up by a ram
= 7 specks of dust stirred up by a hare. . . . . $7^5\,p$
1 speck of dust stirred up by a cow
= 7 specks of dust stirred up by a ram. . . . . $7^6\,p$

1 poppy seed
  = 7 specks of dust stirred up by a cow .... 7^7 p
1 mustard seed
  = 7 poppy seeds .... 7^8 p
1 grain of barley
  = 7 mustard seeds .... 7^9 p
1 phalanx of a finger
  = 7 grains of barley .... 7^{10} p
1 span
  = 12 phalanges of fingers .... 12×7^{10} p
1 cubit
  = 2 spans .... $2 \times 12 \times 7^{10}$ p
1 arc
  = 4 cubits .... $8 \times 12 \times 7^{10}$ p
1 *krosha* from Mâgadha
  = 1,000 arcs .... $1,000 \times 8 \times 12 \times 7^{10}$ p
1 *yojana*
  = 4 *krosha* from Mâgadha ..... $4 \times 1,000 \times 8 \times 12 \times 7^{10}$ p

Carrying out the multiplication $4 \times 1,000 \times 8 \times 12 \times 7^{10}$ which is denoted by the last term in this scale, Buddha gives the sum by expressing in words the number of first atoms contained in the "length" of a *yojana*, namely:

$$108,470,495,616,000.$$

From very high numbers to very small numbers

Using the corresponding Sanskrit terms and taking the scale in reverse order, we have, using the preceding data, the following table which begins with the phalanges of the digits (*angulî parva*) and ends with the atoms (*paramânu râja*):

1 *angulî parva*	= 7 *yava*
1 *yava*	= 7 *sarshapa*
1 *sarshapa*	= 7 *likshâ râja*
1 *likshâ râja*	= 7 *go râja*
1 *go râja*	= 7 *edaka râja*
1 *edaka râja*	= 7 *shasha râja*
1 *shasha râja*	= 7 *vâtayana râja*
1 *vâtayana râja*	= 7 *truti*
1 *truti*	= 7 *renu*
1 *renu*	= 7 *paramânu râja*.

$$1 \text{ } anguli \text{ } parva = 7^{10} \text{ } param\hat{a}nu \text{ } r\hat{a}ja.$$

The *param\hat{a}nu* or "highest atom" constitutes, in Indian thought, the smallest indivisible material particle, which has a taste, a smell and a colour.[*]
In terms of weight, a *param\hat{a}nu* is the equivalent of one seventh of an "atom" (*anu*).

As an *anu* is approximately equal to 1/2,707,200 of a *tola*, which is itself equal to 11.644 grams, the *param\hat{a}nu* weighs the equivalent of 1/18, 950,400 of 11.644 g; thus:

$$1 \text{ } param\hat{a}nu = 0.000000614 \text{ g} = 6.14 \times 10^{-7} \text{ g}.$$

We will now look at the calculation from another angle.

According to the above table, a phalanx of a finger (*anguli parva*) corresponds to 7^{10} "specks of dust of a supreme atom" (*param\hat{a}nu r\hat{a}ja*) ; thus:

$$1 \text{ } *param\hat{a}nu \text{ } r\hat{a}ja = 7^{-10} \text{ } anguli \text{ } parva.$$

Three phalanges of the fingers make an "inch"; therefore a *param\hat{a}nu r\hat{a}ja* is equal to 3.7^{-10} inches. As an inch is equal to 27.06995 mm, we have:

$$1 \text{ } *param\hat{a}nu \text{ } r\hat{a}ja = 0.000000287 \text{ mm} = 2.87 \times 10^{-7} \text{ mm}.$$

These constitute the smallest units of weight and length in India in the early centuries CE.

Thus we have seen how the Indians could easily deal with both "very high" and "very small" numbers.

THE BEGINNINGS OF THESE NUMERICAL SPECULATIONS

The *Shâkyamuni* or "Sage of the Shâkyas", the Indian prince named Gautama Siddhârtha, better known as Buddha, is said to have lived during the fifth century BCE. Does this mean that the Indian passion for high numbers began at this time? We do not know, because no work by Buddha himself has ever been found.

The *Lalitavistara Sûtra* is a collection of stories and ancient legends which was actually only compiled relatively recently.

However, a passage of the *Vâjasaneyî Samhitâ* enumerates the stones needed to construct the sacred altar of fire using the following words [see Weber, in: JSO, XV, pp. 132–40)]:

---

[*]The *param\hat{a}nu* bears no relation to our present-day concept of the "atom", but is more akin to what we would call a molecule: the smallest particle which constitutes a quantity of a compound body.

$$*ayuta = 10,000$$
$$*niyuta = 100,000$$
$$*prayuta = 1,000,000$$
$$*arbuda = 10,000,000$$
$$*nyarbuda = 100,000,000$$
$$*samudra = 1,000,000,000$$
$$*madhya = 10,000,000,000$$
$$*anta = 100,000,000,000$$
$$*parârdha = 1,000,000,000,000.$$

This example, like many others of the same genre, comes from a text belonging to Vedic literature. We know that the texts of the *Vedas* and most of the literature which derives from this civilisation date far back in terms of Indian history, but it is impossible to give an exact date to this era; the texts were first transmitted orally before being transcribed at a later date. As Frédéric explains, "the only chronological order we can give them is a purely internal one. The *Samhitâ* (the three Vedas: *Rigveda*, *Yajurveda*, and *Sâmaveda*) seem to have been composed first; next we have the fourth Veda or *Atharvaveda*, followed by the *Brâhmanas*, the *Kalpasûtras* and finally the *Aranyakas* and the *Upanishads*." What we can say with some certainty is that most of these texts were already in their finished form in the early centuries CE.

The numerical speculation contained in the legend of Buddha cannot have appeared later than the beginning of the fourth century CE, as the *Lalitavistara Sûtra* was translated into Chinese by Dharmarâksha in the year 308 CE.

Thus it would not be unreasonable to place the date of the first developments of these impressive numerical speculations around the third century CE.

The incredible speculations of the Jainas

The members of the Jaina religious movement figure first and foremost amongst the Indian scholars to be well acquainted with such numerical speculations.

There are many examples in the text *Anuyogadvâra Sûtra*, where the sum of the human beings of the creation is given as 2^{96}.

There are other, even older Jaina texts, where numbers containing eighty or even a hundred orders of units are described as "minuscule" in comparison with those under speculation: these numbers are as high as, or greater than *ten to the power of 250*, which we would write today as 1 followed by at least two hundred and fifty zeros.

There is also a period of time called *Shîrshaprahelikâ*, mentioned in several Jaina texts on cosmology, and expressed, according to Hema Chandra

(1089 CE), as "196 positions of numerals of the decimal place-value system", and which corresponds, according to the same source, "to the product of 84,000,000 multiplied by itself twenty-eight times". Thus:

$$\text{the Shîrshaprahelikâ} = (84,000,000)^{28} \approx (8^7)^{28} = 8^{7 \times 28} = 8^{196}.$$

As for the ages of the world, the Jainas used the Brahman classification. Thus the fifth age (which we live in) would have begun in 523 BCE and would be characterised by pain. It would be followed by the sixth and last "age" of 21,000 years, at the end of which the human race would undergo horrific mutations, without the world actually coming to an end. According to Jaina doctrine, the universe is indestructible; this is because it is infinite in terms of both time and space. It was in order to define their vision of this impalpable universe, which is both eternal and limitless, that the Jainas undertook their impressive speculations on gigantic numbers and thus created a "science" which was characteristic of their way of thinking.

Their discovery of *infinity was doubtless due to the fact that they were constantly pushing the limits of the *asamkhyeya* (the "impossible to count", the "innumerable", the "number impossible to conceive") further and further.

THE BIRTH OF MODERN NUMERALS

We can only admire the perfect ease with which the authors and readers of the texts we have just seen were able to write and pronounce these high numbers without ever being struck by a feeling of vertigo at the enormous quantities they were dealing with.

Sanskrit notation had an excellent conceptual quality. It was easy to use and moreover it facilitated the conception of the highest imaginable numbers. This is why it was so well suited to the most exuberant numerical or arithmetical-cosmogonic speculations of Indian culture.

This spoken counting system had a special name for each of the nine simple units:

*eka	*dvi	*tri *	*chatur	*pañcha	*shat	*sapta	*ashta	*nava
1	2	3	4	5	6	7	8	9

There was an independent name for ten, and for each of its multiples, which were used alongside other words in the form of analytical combinations to express intermediate numbers. Like all Indo-European spoken counting systems, the numbers were often expressed – at least in everyday use – in descending order, from the highest to the smallest units.

However, around the dawn of the Common Era (probably from the second century BCE), this order was reversed, particularly in learned and official texts, the numbers being expressed in ascending order, from the

smallest to the highest units. (It has been suggested that this radical transformation was due to the intervention of another civilisation. This idea is
totally without foundation: why and how could this change in direction be
due to an outside influence, bearing in mind that none of the known civilisations, Greek, Babylonian and Chinese included, had reached the same
level as the Indians in terms of numerical concepts and expression? As we
shall see later, the reason for this change has absolutely nothing to do with
any outside influence.)

Where we would say "three thousand seven hundred and fifty-nine",
Indian arithmeticians would have said:

> *nava pañchâshat sapta shata cha trisahasra*
> "nine, fifty, seven hundred and three thousand".

Apart from saying the numbers in the opposite order, the way that numbers
were said in Sanskrit and the way in which we say them are very similar.

However, there is one fundamental difference. When we say the numbers 10,000, 100,000, 10,000,000, 100,000,000, etc., we say *ten thousand, a
hundred thousand, ten million, a hundred million*, etc. In other words, *thousand, million*, etc., play the role of auxiliary bases.

There are no such auxiliary bases in the Sanskrit system, at least none
which were used by learned men; each power of ten had a particular name
which was completely independent of all the others.

These names are discussed in c. 1000 by the Muslim astronomer of
Persian origin, al-Biruni, in his *Kitab fi tahqiq i ma li'l hind* (the book relating to his experiences of Indian civilisation):

> One thing that all nations agree on when it comes to calculations is
> the proportionality of the knots of calculation[*] according to the ratio of
> ten [= the decimal base]. This means that there is no order in which the
> unit is not worth one tenth of the unit which appears in the following
> order and ten times the value of the unit of the preceding order. I care-
> fully researched the names of the different orders of numbers used in
> different languages to the best of my capabilities. I found that the same
> names are repeated once the numbers reach the thousands, as was the
> case with the Arabic system, which is the most appropriate method, and
> the most fitting to the nature of the subject in question. I have also writ-
> ten a whole dissertation on this subject. However, the Indians go beyond
> the thousands in their nomenclature, but not in a uniform manner;
> some use improvised names, others use names which derive from spe-
> cific etymologies; others even mix both these types of names. This

---

[*] According to the contemporary Arabic terminology, the *knot* of calculation is the constituent of a given "order
of units"; thus the knots of units are 1, 2, 3, 4, 5, 6, 7, 8, 9; the knots of tens 10, 20, 30, 40, 50, 60, 70, 80, 90; the
knots of hundreds 100, 200, 300, 400, 500, 600, 700, 800, 900; and so on.

Order of unit	Corresponding name	Numerical value	Power of ten
1	Atmosphere	1	1
2	Ether	10	10
3	Atmosphere	100	10^2
4	Immensity of space	1,000	10^3
5	Atmosphere	10,000	10^4
6	Point (or Dot)	100,000	10^5
7	Canopy of heaven	1,000,000	10^6
8	Voyage on water	10,000,000	10^7
9	Sky, Atmosphere	100,000,000	10^8
10	Sky, Atmosphere	1,000,000,000	10^9
11	Entire, Complete	10,000,000,000	10^{10}
12	Hole	100,000,000,000	10^{11}
13	Void	1,000,000,000,000	10^{12}
14	Point (or Dot)	10,000,000,000,000	10^{13}
15	Foot of Vishnu	100,000,000,000,000	10^{14}
16	Sky	1,000,000,000,000,000	10^{15}
17	Sky, Space	10,000,000,000,000,000	10^{16}
18	Path of the gods	100,000,000,000,000,000	10^{17}

FIG. 24.81. *List of Sanskrit names (translated) for powers of ten according to al-Biruni*

naming reaches as far as the eighteenth order due to certain subtleties which were suggested to the people who use these names, by lexicographers, through the etymologies of these names. I will now describe the differences [which exist in the Indians' usage of these names]. One difference is that some people claim that after the *parârdha* [the name of the eighteenth order of units] there is a nineteenth order, which is called *bhûri*, and that beyond that there is no more need for calculation. But if calculation stops at a certain point, and there is a limit to the order of numbers used, this is only a convention; because this could only occur if one understood nothing besides the names used in the calculations. We also know [according to the same people] that a unit of this order [the nineteenth] is one fifth of the biggest nychthemeron. However, in terms of this method, no mention is made of the influence of any tradition in the work of those who share this opinion. But traditions do exist which shall be explained which mention periods made up of the largest nychthemeron. Adding a nineteenth order is taking the matter to extremes. Another difference lies in the fact that some people claim that the furthest limit of calculation is in the *koti* [10^7] and that beyond this order we return to multiples of tens, hundreds and thousands because the number of the divinities (*Deva*) is included in this order. These people say that the number of divinities is thirty-three *koti* [= 330,000,000], and that on each of the three [gods] Brahma, Nârâyana and Mahâdeva depend eleven *koti* [= 110,000,000] [of these divinities]. As for the names which come after the eighth order, they were created by the grammarians for reasons we shall give below. A further difference is due to the fact that in everyday usage, the Indians use *dasha sahasra* ["ten thousand"] for the fifth order, and *dasha laksha* for the seventh [the tens of millions], because the names of these two orders are hardly ever used. In the work entitled *Arjabhad* [the Arabic name for *Áryabhata*] from the town of Kusumapura, the names of the orders, from the tens of thousands to the tens of *kotis*, are as follows: *ayuta*, *niyuta*, *prayuta*, *koti*, *padma, parapadma*. Yet another difference lies in the fact that some people create names out of pairs. Thus they call the sixth order *niyuta* to follow the name of the fifth [*ayuta*], and they call the eighth *arbuda* so that the ninth order [*vyarbuda*] can follow on, as the twelfth [*nikharva*] follows the eleventh [*kharva*]. They also call the thirteenth order *shankha*, and the fourteenth *mahâshankha* [the "big *shankha*"]; and according to this rule the *mahâpadma* [the thirteenth order] was preceded by the *padma* [twelfth]. These are the differences it is worthwhile knowing. But there are many more which are of no use to us, and only exist because the numbers are taught without the slightest regard for their proper order, or because some people [use them but] claim that [they do not] know [their exact meaning]. This [knowing the precise meanings of all the names] would be difficult for tradesmen. According to the *Pulisha Siddhanta*, after *sahasra*, which is the fourth order, the fifth is *ayuta*, the sixth *niyuta*, the seventh *prayuta*, the eighth *koti*, the ninth *arbuda*, the tenth *kharva*. The names which follow are the same as the ones above [in Fig. 24.81].

These differences apart, the Sanskrit spoken counting system shows the remarkable spirit of organisation of the Indian scholars who, being the good arithmeticians and lexicographers that they were, sought, at an early stage, to give this system an impeccably ordered structure.

This fact is even more remarkable given that the Greeks got no further than ten thousand. As for the Romans, they only had specific names for numbers up to a hundred thousand. In his *Natural History* (XXXIII, 133), Pliny explains that the Romans, scarcely able to name the powers of ten superior to a hundred thousand, contented themselves with expressing "million" as: *decies centena milia*, "ten hundred thousand".

The French had to wait until the thirteenth century for the introduction of the word *million* in their vocabulary which took place c. 1270 [O. Bloch and W. von Wartburg (1968)], and until the end of the fifteenth century for the names of numbers higher than that.

In 1484, Nicolas Chuquet invented the very first set of names for high numbers above a million, using the million 10^6 as the multiplier: "byllion" = 10^{12}, "tryllion" = 10^{18}, "quadrillion" = 10^{24}, "quyllion" = 10^{30}, "sixlion" = 10^{36}, "septyllion" = 10^{42}, "octyllion" = 10^{48}, and "nonyllion" = 10^{54}. Chuquet's work was never published, so that it was not until the middle of the seventeenth century that words like billion, trillion, etc. were found at all commonly. Nowadays, US English has the most regular naming system, using 10^3 as the multiplier, as follows: 10^6 million, 10^9 billion, 10^{12} trillion, 10^{15} quadrillion. 10^{18} quintillion, 10^{21} sextillion, 10^{24} septillion.

In British English, however, the term "billion" is used for 10^{12} (10^9 being just "a thousand million"), and the multiplier used remains 10^6, so that trillion = 10^{18} and quadrillion = 10^{24}. Despite this, the American sense of billion is now used in all financial calculations, and is rapidly displacing the dictionary meaning in British English. French officially uses the same system as the US, except that the older term "milliard" is commonly used for 10^9; "billion", officially given the value of 10^{12} in 1948, is rarely used, and 10^{12} is most often expressed (as in US English) by "trillion".

A comparison between the Arabic, Greek, Chinese and current British systems of expressing high numbers will give a better idea of the impressive conceptual quality of the Sanskrit system.

To make this even clearer, we will use the following number, which will be expressed successively according to the above systems:

$$523\ 622\ 198\ 443\ 682\ 439.$$

As we know, in their nomenclature of the powers of ten, the ancient Arabs always stopped at one thousand, then superposed thousand upon thousand, whilst still using the names of the inferior powers of ten. In other words, in their language, the above number would be expressed rather like this:

Five hundred *thousand thousand thousand thousand thousand* and twenty-three *thousand thousand thousand thousand thousand* and six hundred *thousand thousand thousand thousand* and twenty-two *thousand thousand thousand thousand* and a hundred *thousand thousand thousand* and ninety-eight *thousand thousand thousand* and four hundred *thousand thousand* and forty-three *thousand thousand* and six hundred *thousand* and eighty-two *thousand* and four hundred and thirty-nine.

Equally, in their nomenclature of powers of ten, the ancient Greeks and the Chinese always stopped at the *myriad* (ten thousand); from there, they superposed myriads on top of myriads, mixed with the names of the inferior powers of ten. In other words, in these languages, the above number would have been expressed rather like this [see Daremberg and Saglio (1873); Dedron and Itard (1974); Guitel (1975); Menninger (1957); Ore (1948); Woepcke (1863)]:

Fifty-two *myriads of myriads of myriads of myriads* and three thousand six hundred and twenty-two *myriads of myriads of myriads* and one thousand nine hundred and eighty-four *myriads of myriads* and four thousand three hundred and sixty-eight *myriads* and two thousand four hundred and thirty-nine.

In the United States this would be expressed as:
Five hundred and twenty-three quadrillion, six hundred and twenty-two trillion, one hundred and ninety-eight billion, four hundred and forty-three million, six hundred and eighty-two thousand, four hundred and thirty-nine.

In British English, this number would be expressed as:
Five hundred and twenty-three thousand six hundred and twenty-two billion, one hundred and ninety-eight thousand four hundred and forty-three million, six hundred and eighty-two thousand four hundred and thirty-nine.

All the above methods are rather complicated, and it is difficult to get a clear idea of the positional value of the number.

Since around the time of the *Vedas, the Sanskrit system was much clearer; it possessed names for all the powers of ten up to 10^8 (= 100,000,000). Later, this was extended to 10^{12} (1,000,000,000,000) (probably at the start of the first millennium CE). When the powers of ten were named up to 10^{17} (and sometimes even further, as we saw in the *Jaina texts and in the *legend of Buddha) around 300 CE, it is likely that this was due to the development of the language itself.

Thus the following would have sufficed to express the above number in Sanskrit, using as an example for the base the nomenclature reported by al-Biruni (Fig. 24.81):

nava cha trimshati cha chaturshata cha dvisahasra cha ashtâyuta cha shatlaksha cha triprayuta cha chaturkoti cha chaturvyarbuda cha ashtapadma cha navakharva cha ekanikharva cha dvimahâpadma cha dveshañka cha shatsamudra cha trimadhya cha dvântya cha pañchaparârdha.

In semi-translation, the number reads something like this:

Nine and three *dasha* and four *shata* and two *sahasra* and eight *ayuta* and six *laksha* and three *prayuta* and four *koti* and four *vyarbuda* and eight *padma* and nine *kharva* and one *nikharva* and two *mahâpadma* and two *shankha* and six *samudra* and three *madhya* and two *antya* and five *parârdha*.

In giving each power of ten an individual name, the Sanskrit system gave no special importance to any number.

Thus the Sanskrit system is obviously superior to that of the Arabs (for whom the thousand was the limit), or of the Greeks and the Chinese (whose limit was ten thousand) and even to our own system (where the names thousand, million, etc. continue to act as auxiliary bases).

Instead of naming the numbers in groups of three, four or eight orders of units, the Indians, from a very early date, expressed them taking the powers of ten and the names of the first nine units individually. In other words, to express a given number, one only had to place the name indicating the order of units between the name of the order of units that was immediately below it and the one immediately above it.

That is exactly what is required in order to gain a precise idea of the place-value system, the rule being presented in a natural way and thus appearing self-explanatory. To put it plainly, *the Sanskrit numeral system contained the very key to the discovery of the place-value system.*

In order to grasp this idea, the names of the powers of ten need not always be the same.

In fact, if the mathematical genius of the Indians could embrace variations on the names of the numbers whilst maintaining a clear idea of the series of the ascending powers of ten, this only made it more disposed to understanding the place-value system.

These names need not necessarily have been in everyday use in India. They need only have been familiar to those who were capable of developing the potential ideas behind them, namely to learned men.

We can understand al-Biruni's surprise at seeing grammarians creating these names, and being practically the only ones to use them, for, in the scientific development of Arabic civilization, grammar, lexicography and literature were completely separate movements from the mathematical, medical and philosophical sciences [F. Woepcke (1863)].

However, grammar and interpretation in ancient India were closely linked to the handling of high numbers. Studies relating to poetry and metrics initiated "scientists" to both arithmetic and grammar, and grammarians were just as competent at calculations as the professional mathematicians.

Thus we can see the importance of the role of Indian "scientists", philosophers and cosmographers who, in order to develop their arithmetical-metaphysical and arithmetical-cosmogonical speculations concerning ever higher numbers, became at once arithmeticians, grammarians and poets, and gave their spoken counting system a truly mathematical structure which had the potential to lead them directly to the discovery of the decimal place-value system.

In fact, since a time which was undoubtedly earlier than the middle of the fifth century CE, all mention of the names indicating the base and its diverse powers was suppressed in the body of the numerical expressions expressed by the names of the numbers.

In other words, the Indian scholars quite naturally arrived at the idea of writing numbers without the names *dasha* (= 10), *shata* (= 10^2), *sahasra* (= 10^3), *ayuta* (= 10^4), *laksha* (= 10^5), *prayuta* (= 10^6), *koti* (= 10^7), *vyarbuda* (= 10^8), *padma* (= 10^9), *kharva* (= 10^{10}), *nikharva* (= 10^{11}), *mahâpadma* (= 10^{12}), *shankha* (= 10^{13}), *samudra* (= 10^{14}), *madhya* (= 10^{15}), *antya* (= 10^{16}), *parârdha* (= 10^{17}), etc. From that time on, they simply wrote, in strict order, the names of the units which acted as multiplying coefficients in their numerical expression, according to the order of the ascending powers of ten. Thus they expressed numbers using nothing more than the names of the units.

Instead of writing the number 523 622 198 443 682 439 using the names of the numbers according to the ordinary principle of the Sanskrit language (the complete form of the *Sanskrit numeral system), they only retained the names of the units forming the coefficients of the diverse consecutive powers (abridged form, characteristic of the *simplified Sanskrit numeral system):

Complete form

Nine and three *dasha* and four *shata* and two *sahasra* and eight *ayuta* and six *laksha* and three *prayuta* and four *koti* and four *vyarbuda* and eight *padma* and nine *kharva* and one *nikharva* and two *mahâpadma* and two *shanka* and six *samudra* and three *madhya* and two *antya* and five *parârdha*.

Mathematical breakdown

$$= 9 + 3 \times 10 + 4 \times 10^2 + 2 \times 10^3 + 8 \times 10^4 + 6 \times 10^5$$
$$+ 3 \times 10^6 + 4 \times 10^7 + 4 \times 10^8 + 8 \times 10^9 + 9 \times 10^{10}$$
$$+ 1 \times 10^{11} + 2 \times 10^{12} + 2 \times 10^{13} + 6 \times 10^{14} + 3 \times 10^{15}$$
$$+ 2 \times 10^{16} + 5 \times 10^{17}$$
$$= 523,622,198,443,682,439$$

Abridged form

> Nine.three.four.two.eight.six.three.four.four.
> eight.nine.one.two.two.six.three.two.five

The numbers in the *Jaina* text, the *Lokavibhâga,* (the first document that we know of to make regular use of the place-value system) were expressed in a very similar manner.

In other words, the Indian system of numerical symbols (or at least the ancestor of this unique system) was born out of a simplification of the Sanskrit spoken numeral system.

Such a simplification is not at all surprising when we consider the consistency and potential of the human mind, as well as humankind's intelligence, actions and thoughts upon such matters. When two human beings or two cultures have the same needs and methods due to identical basic (social, psychological, intellectual and material) conditions, they inevitably follow the same paths to arrive at similar, if not identical results.

This is exactly what happened amongst the priest-astronomers of Maya civilisation. Due to a need to abbreviate increasingly high numerical expressions, and also because the units in their system of expressing lengths of time were presented in an impeccable order which was always rigorously followed, they discovered a place-value system, to which they added a sign which performed the function of zero.

As with the Maya, this simplification held no ambiguity for the Indians.

The fact that the successive names of the powers of ten had always followed an invariable order which was firmly established in the mind made the simplification even more comprehensible.

The actual reason for the simplification was doubtless a need for abbreviation. This would have become increasingly necessary as the Indian mathematicians gradually dealt with higher and higher numbers.

To write numbers containing dozens of orders of units according to their names would have taken up whole "pages" of writing. Even expressing one single number could take up several "sheets".

The scholars would have also wanted to be economical with their writing materials. They had to go and pick palm leaves themselves which they used for writing upon. These had to be picked just at the right time, before they opened out entirely, then dried and smoothed out in order to make them fit for the writing of manuscripts. (See *Indian styles of writing). The scholars wanted to give themselves as much time as possible to devote to the more noble task of contemplation, for example studying sacred texts or practising the physical, spiritual and moral exercises of yoga.

The simplification brought about an authentic place-value system which had the Sanskrit names of the nine units as its base symbols. Their value varied according to their relative position in a numerical expression.

Thus *three, two, one* gave the value of simple units to *three*, the value of a multiple of ten to *two* and the value of a multiple of a hundred to *one*.

However, as we can see in the following example, this method could present certain difficulties.

Given that the Sanskrit name for the number three is *tri*, in order to express the number 33333333333, it would be written thus:

tri.tri.tri.tri.tri.tri.tri.tri.tri.tri.tri

Three.Three.Three.Three.Three.Three.Three.Three.Three.Three.Three

3 3 3 3 3 3 3 3 3 3 3

$$(= 3 + 3 \times 10 + 3 \times 10^2 + 3 \times 10^3 + 3 \times 10^4 + 3 \times 10^5 + 3 \times 10^6 +$$
$$3 \times 10^7 + 3 \times 10^8 + 3 \times 10^9 + 3 \times 10^{10}).$$

This expression, which involves the repetition of *tri* eleven times, neither sounds pleasant nor is conducive to the memorisation of the number in question.

Moreover, this number only has eleven orders of units. It would be much worse if it had thirty or a hundred, or even two hundred orders of units.

To avoid this repetition of the same word, the Indian astronomers used synonyms for the Sanskrit names of the numbers. They used all kinds of ideas from traditions, mythology, philosophy, customs and other characteristics of Indian culture in general. This is how they gradually replaced the ordinary Sanskrit names of numbers with an almost infinite synonymy.

Thus the above number would have been expressed by the following kind of symbolic expression:

agni.mûrti.guna.loka.jagat.dahana.kâla.hotri.vâchana.Râma.vahni

Fire.Shape.Quality.World.World.Fire.Time.Fire.Voice.*Râma*.Fire.

3 3 3 3 3 3 3 3 3 3 3

This substitution of the ordinary names of numbers marked the birth of the representation of numbers by *numerical symbols in the place-value system.

Why did Indian astronomers favour this use of numerical symbols over the nine numerals and the sign for zero?

The fact is that they had excellent reasons for this choice.

First and foremost, the concept of zero and the decimal place-value system is totally independent of the chosen style of expressing the numbers (be it conventional graphics, letters of the alphabet or words with or without evocative meaning). All that matters is that there is no ambiguity and that the chosen system of representation contains a perfect concept of zero and the place-value system.

There are other reasons which are specific to the field of Indian astronomy and mathematics, which were generally written in Sanskrit, as were all important erudite Indian texts. The first thing to remember is that in India and Southeast Asia Sanskrit played, and still does play, a role comparable with that of Latin or Greek in Western Europe, with the added virtue of being the only language capable of translating, at the time of the meditations, the mystical transcendental truths said to have been revealed to the Rishi of Vedic times. Bearing in mind the power given to the language (and thus to its written expression), Sanskrit is considered to be the "language of the gods"; whoever masters the language is said to possess divine consciousness and the divine language (see *Mysticism of letters). This explains why the Sanskrit inscriptions of Cambodia, Champa and other indianised civilisations of Southeast Asia do not contain "numerals" for the expression of the *Shaka* dates. These inscriptions were nearly always in verse. As far as the stone-carvers of these regions were concerned, the introduction of numerical signs (considered "vulgar") in Sanskrit texts in verse would have constituted a sort of heresy, not only from an aesthetic point of view, but also and moreover in terms of mysticism and religion. This is why the dates were firstly written in the names of the numbers and then usually in numerical symbols as well. Moreover, the actual name "Sankskrit" is rather significant in this respect, as the word *samskrita (Sanskrit) means "complete", "perfect" and "definitive". In fact, this language is extremely elaborate, almost artificial, and is capable of describing multiple levels of meditation, states of consciousness and psychic, spiritual and even intellectual processes. As for vocabulary, its richness is considerable and highly diversified [see L. Renou (1959)]. Sanskrit has for centuries lent itself admirably to the diverse rules of prosody and versification. Thus we can see why poetry has played such a preponderant role in all of Indian culture and Sanskrit literature.

This explains why the Indian astronomers preferred to use numerical symbols instead of numerals.

Numerical tables, Indian astronomical and mathematical texts, as well as mystical, theological, legendary and cosmological works were nearly always written in verse:

Whilst making love a necklace broke.
A row of pearls mislaid.
One sixth fell to the floor.
One fifth upon the bed.
The young woman saved one third of them.
One tenth were caught by her lover.
If six pearls remained upon the string
How many pearls were there altogether?

This is a mathematical problem posed in the *Lîlâvatî*, a famous mathematical work in the form of poems, written by *Bhâskarâchârya (in 1150), the title of which is the name of the daughter of the mathematician. Here is another example:

> *Of a cluster of lotus flowers,*
> *A third were offered to *Shiva,*
> *One fifth to *Vishnu,*
> *One sixth to *Sûrya.*
> *A quarter were presented to Bhâvanî.*
> *The six remaining flowers*
> *Were given to the venerable tutor.*
> *How many flowers were there altogether?*

From this type of game, the Indian scholars went on to use imagery to express numbers; the choice of synonyms was almost infinite and these were used in keeping with the rules of Sanskrit versification to achieve the required effect. Thus the transcription of a numerical table or of the most arid of mathematical formulae resembled an epic poem.

We need only look at the following lines from a text recording astronomical data to see how poetic and elliptical such documentation could be:

> *The apsids of the moon in a yuga*
> *fire. void. horsemen. Vasu. serpent. ocean*
> *and of its waning node*
> *Vasu. fire. primordial couple. horsemen. fire. horsemen.*

However, aesthetic refinement was not the only motive. This method also offered enormous practical advantages. Billard provides us with the precise fundamental reasons as to why the Indian astronomers chose to use word-symbols to express numbers:

> Indian astronomical texts were always written in Sanskrit. They contained little historical information, were totally devoid of discussions and demonstrations and of the kind of observations which we recognise the value of today – except for the occasional commentary, which was always written in prose – yet they possessed remarkable, even exceptional qualities. *The astronomical data is not only always explicit, but moreover the numerical values are still perfectly conserved after all this time and after so many copies have been made.* Although expressed in a very elliptical manner in the text, where the tradition of versification, used here for mnemonic purposes, led to a synonymy which was often infinite within the technical language – a rather unusual occurrence in the history of astronomy – the astronomical data is very precise and unrivalled in terms of reliability.

The importance of *numerical data* in the Indian astronomical texts is so great because the texts contain so little direct information. All we know of their *astronomical canons, for example, is the terminology by which they were described (average elements, apsids, nodes, eccentricities, exact longitudes, average longitudes, etc.), the terminology being the word-symbols.

It is precisely the study of the numerical data which led to the finding of a given canon in various different texts from very different eras, as well as facilitating the distinction between different canons (see *Indian astronomy).

Thus we can appreciate just how reliable this numerical data had to be in order for it to be transmitted from one generation to the next.

Although initially it might seem puerile, the use of word-symbols was in fact extremely efficient in conserving the exact value of the numbers they expressed, and it was doubtless to this end that the word-symbols were invented. "This conservation of the value of numbers in Sanskrit texts", writes Billard, "is even more impressive when one considers that Indian manuscripts, in material terms, generally do not survive more than two or three centuries [due to climate and above all vermin, which render the conservation of manuscripts extremely difficult], and had the numerical data been recorded in numerals, it would no doubt have reached us in an unusable state."

And Guitel observes that "from a purely mathematical point of view, the use of many different words to express each of the numbers presented no ambiguity; a text written in word-numbers could easily be translated into numerals [and vice versa]. All one would have needed was a glossary of all the words possessing a numerical value, which could be used like a dictionary of rhymes." Whatever the benefits of this system, however, it could not be used to carry out calculations.

The reason for this is obvious: numbers could be expressed using the place-value system with nine numerals and zero, and this system was doubtless invented at the same time as the positional system of the word-symbols.

However, no one would have dreamt of adding *fire, *arrows, *planets and *serpents, or of subtracting *oceans, *orifices or *nâga from *elephants, or multiplying the *faces of Kumâra by the *eyes of Shiva or dividing the *arms of Vishnu by the *great sacrifices!

Since no later than the fifth century, Indian arithmeticians used the place-value system with the nine numerals and zero to carry out complicated mathematical operations. They avoided the use of numerals for recording numerical data, but used them in their rough calculations.

On the other hand, it was very difficult for the Indian astronomers to express their numerical data in numerals, because numerals were far less reliable than the word-symbols. This is because, graphically speaking, the numerals varied according to the style of writing of each region (Fig. 24.3 to 52), and also according to the era and the author or transcriber. A shape which represented the number 2 to one person might well have represented a 3, a 7, or even a 9 to another.

The situation is completely different for us in the twentieth century, because the shapes of the numerals we use and their respective values are the same the world over. For the Indian astronomers, however, the use of numerals could cause confusion. The use of verse and word symbols, on the other hand, was very reliable, because the slightest error could break the rhythm of the verse or verses in question, and therefore would not escape notice. This is why Indian astronomers favoured word-symbols for many centuries.

There is also another, equally fundamental reason. As we have seen, Indian astronomical texts were always in verse: a prosody of long or short syllables was used, as in Graeco-Latin metrics, except that the metre and the number of syllables used in the Indian texts were always perfectly clear and very systematic. Thus the word-symbols not only guaranteed the conservation of the values of the numbers expressed, but also served a mnemonic function. "As well as allowing the writer to find a synonym which gave the required scansion, the word-symbol formed part of the metre, and the number that it expressed was at once firmly established in the text and in the mathematician's memory, who recited the verses as he worked out his calculations" (Billard). The method facilitated and reinforced the Indian scholars' memory: it allowed them to make the best use of their memory through associations of ideas or images contained within rhythms determined by the metre which was dictated by the rules of Sanskrit versification.

When we consider the above conditions, we can understand why the Indian astronomers developed the Sanskrit word-symbols, and continued to use them for such a prolonged length of time.

The same conditions led the astronomer *Âryabhata to develop his famous alphabetical numeral system. He was no doubt familiar with the Sanskrit word-symbols, but needed a system which was more concise whilst meeting the requirements of certain versified Sanskrit compositions. It is likely that he found the word-symbols to be lacking in brevity and perhaps also precision, especially when he wrote his famous sine table.

Similar reasons led astronomers such as *Haridatta or *Shankaranârâyana, at a later date, to use an alphabetical numeral system which was even more efficient then the *katapaya system.

The coexistence of different methods of achieving the same goals is one of the characteristics particular to the highly inventive genius of the Indians, which enjoyed both the finest distinctions and determinations, and the fluctuating wave of an abundant production, and was little inclined towards that precise and rather dry sobriety of the ancient Semites [F. Woepcke (1863)].

The discovery of the place-value system required another, equally basic progression. As soon as the place-value system was rigorously applied to the nine simple units, the use of a special terminology was indispensable to indicate the absence of units in a given order.

The Sanskrit language already possessed the word *shûnya to express "void" or "absence". Synonymous with "vacuity", this word had for several centuries constituted the central element of a mystical and religious philosophy which had become a way of thinking.

Thus there was no need to invent a new terminology to express this new mathematical notion: the term *shûnya ("void") could be used. This is how the word finally came to perform the function of zero as part of this exceptional counting system.

A number such as 301 could now be expressed:

eka.shûnya.tri

one. void. three.

1   0   3

The Sanskrit language, however, being an unrivalled literary instrument in terms of wealth, possessed more than just one word to express a void: there was a whole range of words with more or less the same meaning: words whose literal meaning was connected, directly or indirectly, with the world of symbols of Indian civilisation.

Thus words such as *abhra, *ambara, *antariksha, *gagana, *kha, *nabha, *vyant or *vyoman, which literally meant the sky, space, the atmosphere, the firmament or the canopy of heaven, came to signify not only a void, but also zero. There was also the word *âkâshâ, the principal meaning of which was "ether", the last and the most subtle of the "five elements" of Hindu philosophy, the essence of all that is believed to be uncreated and eternal, the element which penetrates everything, the immensity of space, even space itself.

To the Indian mind, space was the "void" which had no contact with material objects, and was an unchanging and eternal element which defied description; thus the association between the elusive character and very different nature of zero (as regards numerals and ordinary numbers) and the concept of space was immediately obvious to the Indian scholars. The association between ether and "void" is also obvious because *âkâsha* (to the Indian mind) is devoid of all substance, being considered the condition of all corporal extension and the receptacle of all substance formed by one of the other four elements (earth, water, fire and air). In other words, once zero had been invented and put into use, it brought about the realisation that, in terms of existence, *âkâsha* played a role comparable with the one which zero performed in the place-value system, in calculations, in mathematics, and the sciences.

The following are other Indian numerical symbols for zero: *bindu, "point"; *ananta, "infinity"; *jaladharapatha, "voyage on water"; *vishnupada, "foot of Vishnu"; *pûrna, "fullness, wholeness, integrity, completeness"; etc. (See also *Zero.)

The use of one of these words prevented any misunderstanding. Later than the Babylonians, and most likely before the Maya, the Indian scholars invented zero, although for the time being it was little more than a simple word which formed part of everyday vocabulary.

So just how did the place-value system come to be applied to the nine Indian numerals?

Let us now go back to the numeral system of ancient India: the *Brâhmî* system, which, as we have already seen, constituted the prefiguration of the nine basic numerals that we use today (Fig. 24.29 to 52 and 24.61 to 69).

Current documentation suggests that the history of truly Indian numerals began with the *Brâhmî* inscriptions of Emperor Asoka (in the middle of the third century BCE). But the numerals were invented before the Maurya Dynasty, by which time the numerals were highly developed graphically speaking, and widespread throughout the Indian territory.

In fact, as we have already seen, the first nine *Brâhmî* numerals which appear in Asoka's edicts are probably vestiges of an old indigenous system (no doubt older than *Brâhmî* writing itself), where the nine units were represented by the necessary number of vertical lines, similar to the arrangements in Fig. 24.59.

We will now sum up the evidence we have compiled in this chapter on the early stages in the history of these numerals.

Like all the other civilisations of the world, the Indians initially used the required number of vertical lines to write the first nine numbers. However, as a row of vertical lines was not conducive to rapid reading and comprehension, this system was gradually abandoned, at least for the numbers 4 to 9. To overcome the problem, the lines representing the units were split into two groups (two lines on top of two others for 4, three lines on top of two others for 5, etc.; see Fig. 1.26), or a ternary arrangement was used (three lines on top of one line for 4, three above two for 5, etc.; see

Fig. 1. 27). This was how the Sumerians, Cretans and Urarteans proceeded, as well as the Egyptians, Assyrians, Phoenicians, Aramaeans and Lydians. In the long run, however, such groupings of lines did not allow for rapid writing, or time-saving, which was the main preoccupation of the scribes. Thus – due to a combination of circumstances imposed by the very nature of the writing materials used (the scribes wrote upon tree bark or palm leaves with a brush or calamus) depending on the region – a numeral system evolved which was unique to each civilisation and the numerals no longer visually represented their respective values. In each civilisation the change was brought about by both the nature of the writing materials and the desire to save time. Signs were invented which could be written in one single stroke or in short, quick strokes. Ligatures were exploited wherever possible, so that the brush need not be lifted, allowing several lines to be grouped together in one single sign. The initial representations of the numbers were radically modified, as we can see with the *Brâhmî* numerals for the numbers four to nine (Fig. 24.57, 58 and 60).

At the outset, these nine signs were not used in conjunction with the place-value system: the *Brâhmî* system relied on the principle of addition and a specific sign was given to each of the nine units in each decimal order, up to and including tens of thousands (Fig. 24.70).

Mathematical operations, even simple addition, were almost impossible without the invention of a device. The ancestors of our modern numerals remained static for some time before acquiring the dynamic and workable nature of the current numerals. Like certain other systems of the ancient world, this system was also rudimentary whilst it was only used to express numbers.

Mathematicians, philosophers, cosmographers and all others who, for one reason or another, were handling high numbers at that time, resorted to using the Sanskrit names of the numbers.

However, like all the mathematicians of the ancient world, Indian arithmeticians, before discovering the place-value system, used their fingers or, more often, concrete mathematical devices.

It seems that the most common was the abacus: from right to left, the first column represented the units, the next the tens, the third the hundreds, and so on.

Unlike the Greeks, Romans or Chinese, however, who then went on to use pebbles, tokens or reeds, the Indian mathematicians had the idea of using the first nine numerals of their counting system, tracing them in fine sand or dust, inside the column of the corresponding decimal order.[*]

---

[*] This information was obtained from descriptions given by various Indian authors, and later accounts from many Arabic, Persian, European and even Chinese authors [see Allard (1992); Cajori (1980); Datta and Singh (1938); Iyer (1954); Kaye (1908); Levey and Petruck (1965); Waeschke (1878)].

Thus a number such as 7,629 would have been represented in the following manner, with nine in the units column, two in the tens column, six in the hundreds and seven in the thousands:

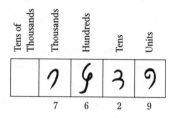

Of course, when a unit within an order of units was missing, one only needed to leave the appropriate column empty; thus the representation of 10,267,000:

The mathematical operation would be carried out by successively erasing the results of the intermediary calculations. (There is a simple example of this in Chapter 25.)

Like us today, the "Pythagorean" tables had to be known by heart, which give the results of the four elementary operations of the nine significant numerals.

Before the beginning of the fifth century BCE, then, all the necessary "ingredients" for the creation of the written place-value system had been amassed by the Indian mathematicians:

- the units one to nine could be expressed by distinct numerals, whose forms were unrelated to the number they represented, namely the first nine *Brâhmî* numerals;
- they had discovered the place-value system;
- they had invented the concept of zero.

A few stages, however, were still lacking before the system could attain perfection:

- the nine numerals were only used in accordance with the additional principle for analytical combinations using numerals higher than or equal to ten, and the notation was very basic and limited to numbers below 100,000;
- the place-value system was only used with Sanskrit names for numbers;
- and zero was only used orally.

In order for the "miracle" to take place, the three above ideas had to be combined.

By using the nine *Brâhmî* numerals in the appropriate columns of the "dust" abacus, the Indian mathematicians had already reached the stage which would inevitably lead them to this major discovery.

This becomes clear when we imagine the Indian mathematicians at work, recording the result of a calculation they had carried out by drawing their abacus in the dust, bearing in mind that they had two methods of expressing numbers: the *Brâhmî* numerals and the Sanskrit names of the numbers.

In the abacus, they would have drawn the numerals in a contemporary style (those from the inscriptions of Nana Ghat, for example, dating from the second century BCE; see Fig. 24.30 and 71), and a calculation might have given the following result:

The figure obtained is 4,769.

As we know, from this time on, the numbers were expressed in their Sanskrit names in the order of ascending powers of ten, from the smallest to the highest.

Therefore this result would have been expressed as follows:

> *nava shashti saptashata cha chatursahasra*
> "nine sixty seven hundred and four thousand".

In numerals, however, the numbers would have been written in the opposite order, reading from left to right:

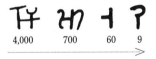

We have evidence of these methods of expressing figures in Indian inscriptions, since the third century BCE, from those of Asoka, Nana Ghat and Nasik to those of the Shunga, Shaka, Kusana, Gupta and Pallava dynasties. The corresponding numerical notations, all issuing from the original *Brâhmî* system, possess a different numeral for each unit of each decimal order (Fig. 24.70 and 71).

When we examine the signs used, we discover that these numerals are not all independent of one another; the only numerals which are really unique are the following:

$$1 \quad 2 \quad 3 \quad 4 \quad 5 \quad 6 \quad 7 \quad 8 \quad 9$$
$$10 \quad 20 \quad 30 \quad 40 \quad 50 \quad 60 \quad 70 \quad 80 \quad 90$$
$$100 \qquad\qquad 1{,}000$$

The numerals for 200 and 300, as well as those for 2,000 and 3,000, derive from the sign for 100 and 1,000 respectively, with the simple addition of one or two horizontal lines (Fig. 24.70 C).

In other words, the four numerals in question conformed to the following mathematical rules:

$$200 = 100 + 1 \times 100 \qquad 2{,}000 = 1{,}000 + 1 \times 1{,}000$$
$$300 = 100 + 2 \times 100 \qquad 3{,}000 = 1{,}000 + 2 \times 1{,}000$$

As for the remaining multiples of one hundred and one thousand, they were represented using the principle of multiplication and placing the numeral for the corresponding unit to the right of the sign for one hundred or one thousand:

$$400 = 100 \times 4 \qquad 4{,}000 = 1{,}000 \times 4$$
$$500 = 100 \times 5 \qquad 5{,}000 = 1{,}000 \times 5$$
$$600 = 100 \times 6 \qquad 6{,}000 = 1{,}000 \times 6$$
$$700 = 100 \times 7 \qquad 7{,}000 = 1{,}000 \times 7$$
$$800 = 100 \times 8 \qquad 8{,}000 = 1{,}000 \times 8$$
$$900 = 100 \times 9 \qquad 9{,}000 = 1{,}000 \times 9$$

It is visibly evident that this rule also applied to the notation of tens of thousands by placing the corresponding number of tens next to the sign for a thousand:

$$10{,}000 = 1{,}000 \times 10$$
$$20{,}000 = 1{,}000 \times 20$$
$$30{,}000 = 1{,}000 \times 30$$
$$40{,}000 = 1{,}000 \times 40$$

Thus, the number 4,769 was written:

$$1{,}000 \ \times \ 4 \ + \ 100 \ \times \ 7 \ + \ 60 \ + \ 9$$

This corresponds exactly, but in the opposite order, to the above Sanskrit expression of the figure:

nava shashti saptashata cha chatur sahasra
$$9 + 60 + 7 \times 100 \qquad + \qquad 4 \times 1{,}000$$

If we look at either way of expressing the sum in the opposite direction from the way it would have been spoken or written, we obtain the arithmetical breakdown of the other. This is what the Indian arithmeticians expressed in the phrase *ankânâm vâmato gatih*, which literally means the "principle of the movement of numerals from the right to the left", which applies to the reading of numbers from the smallest unit to the highest in ascending order of powers of ten.

The inscriptions of Nana Ghat provide the earliest known significant evidence of this principle. Thus we know that from at least as early as the second century BCE, the numbers were expressed in ascending powers of ten in *Brâhmî* numerals; in other words, they were expressed in the opposite order than from left to right. This means that the structure of *Brâhmî* notation had been copied exactly from the Sanskrit system.

Since the highest *Brâhmî* numeral expressed the number 90,000, it was impossible to use this system to express a number that was higher than 99,999.

As the *Brâhmî* numerals constituted an abbreviated written form of the spoken numeration, it had been developed to avoid having to express frequently used numbers through the long-winded Sanskrit names of the numbers.

In other words, the result of a calculation which was equal to or higher than 100,000 could only be written down in the Sanskrit names of the numbers.

The abacus traced in the dust could be used to carry out calculations involving extremely high numbers: each column represented a power of ten, and there was no limit to the number of columns which could be drawn.

Thus there was a very close link between the ability to carry out calculations on this abacus and the level of conception of high numbers and the capacity to express them orally or through writing.

In Indian calculation, the successive columns of the abacus always rigorously corresponded to the consecutive powers of ten. As the Sanskrit counting system possessed the same mathematical structure, these columns corresponded exactly to the impeccable succession of names which the Sanskrit system possessed for the various powers of ten. Thus each system constituted the mirror image of the other.

This is exactly where the potential to discover the place-value system of the nine numerals lay. As with the Sanskrit names of the numbers, the structure of the abacus contained the key to the discovery of the decimal place-value system. This is why the Sanskrit notation was perfect for recording the results of the calculations which were carried out on the abacus.

This becomes even clearer when we take the number 523,622,198,443, 682,439, as it would have appeared when written on the abacus using the nine *Brâhmî* numerals (see adjacent column).

We can see how the close relationship between the representation of the numbers on the abacus and the Sanskrit system led to the change in direc-

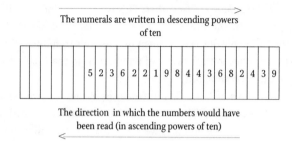

The numerals are written in descending powers
of ten

| | | | | | | | 5 | 2 | 3 | 6 | 2 | 2 | 1 | 9 | 8 | 4 | 4 | 3 | 6 | 8 | 2 | 4 | 3 | 9 | |

The direction in which the numbers would have
been read (in ascending powers of ten)

tion, before the second century BCE, of numerical expressions given using the Sanskrit names of the numbers.

Whilst the numerals read from left to right on the abacus in descending powers of ten, they came to be read from right to left, from the smallest number to the highest.

Bearing in mind the conditions imposed by the very nature of the calculating instrument, the Indian mathematicians had no other choice but to adopt the expression of numbers in the direction described by *ankânâm vâmato gatih*: "the principle of the movement of the numerals from the right to the left". How could they know how to write a high number on the abacus if they began with the highest order? They would have had to work out which column each order had to be placed in by counting each corresponding column beginning with the column for the simple units. This would have wasted time. Thus the best solution was to always start with the column for the simple units.

Thus the old system was abandoned. By beginning with the highest power of ten, the arithmetician immediately knew the size of the number he was dealing with, but this did not facilitate the drawing, on the columns of the abacus, the successive numerals of a number which possessed more than four or five orders of units.

This is why the opposite direction was adopted, the advantage of which being that, no matter how high a number might be, there could be no mistake as to which column each numeral must be written in. It was for the same reason that this direction of expressing the numbers was conserved later on when the positional notation was invented using numerical symbols:

> We must not forget that the numbers which appear in the scientific poems [such as the numerical data given by the Indian astronomers] were destined for mathematical use. Certain lists contain numbers proportional to the differences of sines of angles which ascend in mathematical order; these enabled, with the aid of additions, an almost instant reconstruction of the numbers proportional to the sines of these angles.

The *pandit* dictated the poetic text which the scribe wrote in numerals. How could an addition be carried out if the data consisted of *wing* (= 2) and *fire* (= 3)? If only one number had to be reproduced, even a very high number, it would have been easy to translate it directly onto "paper", but if a series of numbers was involved, how could they be correctly placed on the counting table if they were read out in descending orders of units?

It would have been impossible to transcribe a number in this way unless it was known in advance. The only solution was to read out the numbers in ascending orders of units.

However, when the *pandit* was reading out a high number, the scribe needed to know its highest order; this is why the *pandit* started with the highest powers of the base; this is not possible if one uses the spoken positional numeration, yet this system did enable one to place the number correctly on the counting table, and then one could plainly see the powers of the base which had been recorded [G. Guitel, (1966)].

If we look again at the representation of the number given above, on the "dust" abacus, its mathematical breakdown according to *ankânâm vâmato gatih* was as follows:

$$
\begin{aligned}
&= 9 + 3 \times 10 \quad + 4 \times 10^2 \;\; + 2 \times 10^3 \;\; + 8 \times 10^4 \;\; + 6 \times 10^5 \\
&\quad + 3 \times 10^6 \;\; + 4 \times 10^7 \;\; + 4 \times 10^8 \;\; + 8 \times 10^9 \;\; + 9 \times 10^{10} \\
&\quad + 1 \times 10^{11} \;\; + 2 \times 10^{12} \;\; + 2 \times 10^{13} \;\; + 6 \times 10^{14} \;\; + 3 \times 10^{15} \\
&\quad + 2 \times 10^{16} \;\; + 5 \times 10^{17} \\
&= 523{,}622{,}198{,}443{,}682{,}439
\end{aligned}
$$

This corresponds exactly to the following Sanskrit expression:

"Nine and three *dasha* and four *shata* and two *sahasra* and eight *ayuta* and six *laksha* and three *prayuta* and four *koti* and four *vyarbuda* and eight *padma* and nine *kharva* and one *nikharva* and two *mahâpadma* and two *shankha* and six *samudra* and three *madhya* and two *antya* and five *parârdha*."

Once the Sanskrit numeration was simplified, this number could be expressed in the following manner:

nine.three.four.two.eight.six.three.four.four.
eight.nine.one.two.two.six.three.two.five.

Why was the number written in the names of the numbers instead of in *Brâhmî* numerals, to which the place-value system could have been applied?

This is surely what the Indian mathematicians asked themselves one day, in their continuing desire to economise with time and materials.

They carried out calculations involving high numbers, for which even the intermediate results constituted very high numbers, and which could be difficult to memorise. The results were first recorded in rough. As they needed to be sparing with materials and time, the Indian mathematicians sought a way to write faster and in a more abridged form than the Sanskrit system, even in its simplified form. They realised that the nine *Brâhmî* numerals could provide the "stenography" that they required.

However, bearing in mind the position of the numerals on the abacus, it was necessary to revert to the descending order, thus going from the highest orders of units to the smallest, so as not to cause confusion in the numeral representations; for the results of calculations carried out on the abacus, the numerals had to be placed in the columns in the same positions as they appeared when written in rough.

Thus the number in question acquired the following notation, the numbers reading from left to right in descending order of powers of ten, constituting a faithful reproduction, minus the columns, of its representation on the abacus, as well as the reflection of the abridged form of the corresponding Sanskrit expression:

5 2 3 6 2 2 1 9 8 4 4 3 6 8 2 4 3 9.

Whence came the decimal position values which were given to the first nine numerals of the old notation which originated at the time of the reign of Emperor Asoka.

This was the birth of modern numerals, which signalled the death of the abacus and its columns.

The introduction of a new symbol proved indispensable in order to convey the absence of units in a given decimal order; whilst this sign was not needed when the abacus was used, it was of utmost necessity in the new positional numeral notation.

The Indians, never lacking in resources in these matters, again turned to their unique symbolism.

As we saw earlier, the word-symbols that the Sanskrit language used to express the concept of zero conveyed concepts such as the sky, space, the atmosphere or the firmament.

In drawings and pictograms, the canopy of heaven is universally represented either by a semi-circle or by a circular diagram or by a whole circle. The circle has always been regarded as the representation of the sky and of the Milky Way as it symbolizes both activity and cyclic movements [see J. Chevalier and A. Gheerbrant (1982)].

Thus the little circle, through a simple transposition and association of ideas, came to symbolise the concept of zero for the Indians.

Another Sanskrit term which came to mean zero was the word *bindu, which literally means "point".

The point is the most insignificant geometrical figure, constituting as it does the circle reduced to its simplest expression, its centre.

For the Hindus, however, the *bindu* represents the universe in its non-manifest form, the universe before it was transformed into the world of appearances (*rûpadhâtu*). According to Indian philosophy, this uncreated universe possessed a creative energy, capable of generating everything and anything: it was the causal point.

The most elementary of all geometrical figures, which is capable of creating all possible lines and shapes (*rûpa*) was thus associated with zero, which is not only the most negligible of quantities, but also and above all the most fundamental of all abstract mathematics.

The point was thus used to represent zero, most notably in the *Shârada* system of Kashmir, and in the vernacular notations of Southeast Asia (Fig. 24.82).

From the fifth century CE, the Indian zero, in its various forms, already surpassed the heterogeneous notions of vacuity, nihilism, nullity, insignificance, absence and non-being of Greek-Latin philosophies. **Shûnya* embraced all these concepts, following a perfect homogeneity: it signified not only void, space, atmosphere and "ether", but also the non-created, the non-produced, non-being, non-existence, the unformed, the unthought, the non-present, the absent, nothingness, non-substantiality, nothing much, insignificance, the negligible, the insignificant, nothing, nil, nullity, unproductiveness, of little value and worthlessness (see **Shûnyatâ*, **Zero*, and Fig. 24.D10 and D11 of the latter entry in the Dictionary).

It was also, and above all, an eminently abstract concept: in the simplified Sanskrit system, as well as in the positional system of the numerical symbols, the word **shûnya* and its various synonyms served to mark the absence of units within a given decimal order, in a medial position as well as in an initial or final position; the point or the little circle were used in the same way.

This zero was also a mathematical operator: if it was added to the end of a numerical representation, the value of the representation was multiplied by ten.

By freeing the nine basic numerals from the abacus and inventing a sign for zero, the Indian scholars made significant progress, primarily simplifying quite considerably the rules of a technique which would lead to the birth of our modern written calculation.

The Indian people were the only civilisation to take the decisive step towards the perfection of numerical notation. We owe the discovery of modern numeration and the elaboration of the very foundations of written calculations to India alone.

It is very likely that this important historical event took place around the fourth century CE. It is thanks to the genius of the Indian arithmeticians that three significant ideas were combined:

LIST OF SANSKRIT WORDS FOR ZERO	
SYMBOLS	THEIR MEANINGS
*Abhra	Atmosphere
*Akâsha	Ether
*Ambara	Atmosphere
*Ananta	Immensity of space
*Antariksha	Atmosphere
*Bindu	Point (or Dot)
*Gagana	Canopy of heaven
*Jaladharapatha	Voyage on water
*Kha	Space
*Nabha	Sky, Atmosphere
*Nabhas	Sky, Atmosphere
*Pûrna	Entire, Complete
*Randhra (rare)	Hole
*´Shûnya	Void
*Vindu	Point (or Dot)
*Vishnupada	Foot of Vishnu
*Vyant	Sky
*Vyoman	Sky, Space

Ref.: Al-Biruni; Bühler; Burnell; Datta and Singh; Fleet, in: CIIN, III; Jacquet, in: JA, XVI, 1935; Renou and Filliozat; Sircar; Woepcke (1863).

GRAPHICAL SIGNS FOR ZERO

First sign:
the little circle

Nowadays used in nearly all the notations of India and Southeast Asia (*Nâgarî, Marâthî, Punjabî, Sindhî, Gujarati, Bengâlî, Oriyâ, Nepâlî, Telugu, Kannara, Thai, Burmese, Javanese, etc.*). There is evidence of the use of this sign which dates back many centuries for nearly all these systems.

Second sign:
the point, or dot

Formerly used in the regions of Kashmir (*Shâradâ* numerals). There is also evidence of the use of this sign in the Khmer inscriptions of ancient Cambodia and the vernacular inscriptions of Southeast Asia.

Fig. 24.3 to 51, 24.78 to 80. See also *Numerals "0", *Zero and Fig. 24.D11.

FIG. 24.82. *The various representations of the Indian zero**

- nine numerals which gave no visual clue as to the numbers they represented and which constituted the prefiguration of our modern numerals;
- the discovery of the place-value system, which was applied to these nine numerals, making them dynamic numerical signs;
- the invention of zero and its enormous operational potential.

Thus we can see that the Indian contribution was essential because it united calculation and numerical notation, enabling the democratisation of calculation. For thousands of years this field had only been accessible to the privileged few (professional mathematicians). These discoveries made the domain of arithmetic accessible to anyone.

It still remained for the Indian scholars to perfect the concept of zero and enrich its numerical significance.

Beforehand, the *shûnya* had only served to mark the absence of units in a given order. The Indian scholars, however, soon filled in the gap. Thus, in

---

* The Arabs acquired their signs for zero, as well as the place-value system, from the Indians. This is why we find the point and the little circle used to express zero in Arabic texts. The circle was the sign to prevail in the West, after the Arabs transmitted it to the Europeans some time after the beginning of the twelfth century.

a short space of time, the concept became synonymous with what we now refer to as the "number zero" or the "zero quantity".

The *shûnya* was placed amongst the *Samkhyâ*, which means it was given the status of a "number".

In c. 575, astronomer *Varâhamihira, in *Pañchasiddhântika*, mentioned the use of zero in mathematical operations, as did *Bhâskhara in 629 in his commentary on the *Âryabhatîya*.

In 628, in *Brâhmasphutasiddhânta*, *Brahmagupta defined zero as the result of the subtraction of a number by itself ($a - a = 0$), and described its properties in the following terms:

> When zero (*shûnya*) is added to a number or subtracted from a number, the number remains unchanged; and a number multiplied by zero becomes zero.

Moreover, in the same text, Brahmagupta gives the following rules concerning operations carried out on what he calls "fortunes" (*dhana*), "debts" (*rina*) and "nothing" (*kha*) [see S. Dvivedi (1902), pp. 309–10, rules 31–5)]:

> A debt minus zero is a debt.
> A fortune minus zero is a fortune.
> Zero (*shunya*) minus zero is nothing (*kha*).
> A debt subtracted from zero is a fortune.
> So a fortune subtracted from zero is a debt.
> The product of zero multiplied by a debt or a fortune is zero.
> The product of zero multiplied by itself is nothing.
> The product or the quotient of two fortunes is one fortune.
> The product or the quotient of two debts is one debt.
> The product or the quotient of a debt multiplied by a fortune is a debt.
> The product or the quotient of a fortune multiplied by a debt is a debt.

Modern algebra was born, and the mathematician had thus formulated the basic rules: by replacing "fortune" and "debt" respectively with "positive number" and "negative number", we can see that at that time the Indian mathematicians knew the famous "rule of signs" as well as all the fundamental rules of algebra.

It is clear how much we owe to this brilliant civilisation, and not only in the field of arithmetic; by opening the way to the generalisation of the concept of the number, the Indian scholars enabled the rapid development of algebra, and thus played an essential part in the development of mathematics and exact sciences.

The discoveries of these men doubtless required much time and imagination, and above all a great ability for abstract thinking. The reader will not be surprised to learn that these major discoveries took place within an environment which was at once mystical, philosophical, religious, cosmological, mythological and metaphysical.

Sarasvati, goddess of knowledge and music. From Moor's Hindu Pantheon

CHAPTER 24 PART II

DICTIONARY OF THE NUMERICAL SYMBOLS OF INDIAN CIVILISATION

As we have seen in the course of this chapter, India has always dominated the world in the field of arithmetic, and indeed the art of numbers plays a leading role in Indian culture.

It is precisely this skill in the field of mathematics which led Indian scholars, at a very early stage, to develop their astonishing *arithmetical speculations* which could involve numbers comprising hundreds of decimal places, whilst maintaining a clear idea of the order of the ascending powers of ten within a nomenclature which contained a highly diversified terminology, based as it was upon both specific etymologies and improvised terms born out of a highly creative symbolical imagination.

This same arithmetical genius led to their inevitable discovery of zero and the place-value system and even enabled them to come within touching distance of the concept of mathematical infinity.

Therefore one should not be surprised to learn that, a thousand years earlier than the Europeans, Indian mathematicians already knew that zero and infinity were inverse concepts. They realised that when any number is divided by zero the result is infinity: $a / 0 = \infty$, this "quantity" undergoing no change if it is added to or subtracted from a finite number.[*]

We should not forget that these crucial discoveries were not the fruit of just one genius's individual inspiration nor even that of a group of "mathematicians" as we understand the term today. They were, of course, learned Indians. However, there should be no ambiguity about the meaning of this term: to be termed as learned at that time meant that one was a thinker, a little like the scholarly gentlemen of sixteenth-century Europe, except that an Indian's way of thinking would have been very different from that of the European scholars. The Indian scholar would have been a man of constant reflection whose studies covered the most diverse topics. Moreover, mystical, symbolical, metaphysical and even religious considerations came first and foremost in his learning:

> As India knew nothing of the work of either Aristotle or Descartes, and was ignorant of Jewish or Christian ethics, it would be futile to attempt to draw a parallel between these vastly different civilisations. The foundations are

---

[*] In other words, Indian scholars knew the following properties: $(a/0) \pm k = k \pm (a/0) = (a/0)$, which is to say: $\infty \pm k = k \pm \infty = \infty$.

completely different, as are the customs and ways of thinking. It is impossible to make any comparisons, even if some aspects of Indian culture do seem to coincide with our own. [L. Frédéric, *Dictionnaire de la civilisation indienne* (1987)]

It would also be futile to try and make any comparisons between Indian mathematics and modern mathematics, modern mathematics being the very refined product of contemporary western civilisation: a highly abstract science that has been stripped of any mystical, philosophical or religious influence.

Moreover, the following pages prove that the main preoccupations of Indian scholars had nothing to do with what we in the West today refer to as "hard sciences". In fact, we will see that these major discoveries stem from the incessant study of astronomy, poetry, metric theory, literature, phonetics, grammar, philosophy or mysticism, and even astrology, cosmology and mythology all at once.

In India, an aptitude for the study of numbers and arithmetical research was often combined with a surprising tendency towards metaphysical abstractions: in fact, the latter is so deeply ingrained in Indian thought and tradition that one meets it in all fields of study, from the most advanced mathematical ideas to disciplines completely unrelated to 'exact' sciences.

In short, Indian science was born out of a mystical and religious culture and the etymology of the Sanskrit words used to describe numbers and the science of numbers bears witness to this fact.[†] Together, the discoveries in question represent the culmination of the uniqueness, wealth and incredible diversity of Indian culture.

To give the reader a clearer idea of the circumstances and conditions under which these major discoveries were made, it seems useful now to reiterate the principal notions that have already been explained in this chapter in the form of a Sanskrit and English dictionary and so to define (if need be) these ideas in a more analytical form. This dictionary can serve as a glossary to the numerous ideas which have been covered, each term being marked with an asterisk.

[†] The term for the "Science of numbers" in Sanskrit is *samkhyâna* (also spelt *sankhyâna*) which means "arithmetic": and, by extension, "astronomy" (from the time when the science of the stars was not considered to be a separate discipline from arithmetic). The word is frequently used in this sense in *Jaina* literature, where the science of numbers was considered to be one of the fundamental requirements for the full development of a priest; likewise in later Buddhist literature it was considered to be the most noble of the arts. the word "number" itself is *samkhya* or *samkhyeya*. One should note that this term not only applies to the concept of number but also to the actual numerical symbol. Arithmetician or mathematician is denoted by *sâmkhya*. But *sâmkhya* is also the term used for one of six orthodox (and most ancient) systems of the Hindu philosophy of the six *darshana* ("contemplations"), which teaches "number" as a way of thinking which is connected to the liberation of the soul, and according to which the universe was born out of the union of *prakriti* (nature) and *purusha* (the conscience). It is significant that the word *sâmkhyâ*, which also means a follower of this philosophy, is the term used to refer to the "calculator", but in a mystical sense in this context.

Thus the dictionary can help the reader through the maze of obscure rubric of the Sanskrit language and the complex concepts of Indian science and philosophy.

The dictionary is not only aimed at specialists: the entries, recorded with careful clarity and precision, are concise and easily accessible to the layman. It is not even necessary to have read Chapter 24 (or the preceding chapters) to grasp the concepts it explains.

Its entries are recorded alphabetically, regardless of whether the terminology is in English or Sanskrit.

This dictionary also serves as a thematic index which can clarify the ideas presented in this chapter through an effective reference system of general or specific rubric, giving not only references to Chapter 24 but also to those of the forthcoming Chapters 25 and 26. For example, the reader has only to turn to the entries *Chhedi, *Shaka or *Vikrama to find out about each era.

Under *Asankhyeya, the Sanskrit for "incalculable", we learn that the same term was also sometimes used to express the rather more modest sum of ten to the power 140.

Similarly, the term *Padma, or *Paduma, reveals that the poetic name "pink lotus" was used to denote the number ten to the power of 14 (or 29) as well as ten to the power of 119. The lotus flower was used to represent various numbers in Indian mathematics, the values of which depended on the colour, the number of petals and whether the flower was open, just flowering or still in bud. Thus *white lotus came to mean ten to the power 27 or ten to the power 112, *pink lotus ten to the power 21 or 105, and *blue lotus (half-open) ten to the power 25 or 98. (See *Utpala, *Pundarika, *Kumud and *Kumuda that can all mean "lotus", according to slight characteristic differences of the flower.)

Under entries entitled *High numbers, there are numerous other examples of the unique symbolism which without a doubt was the source of inspiration for the names of these large figures. The same entries also demonstrate how in ancient India, grammar and interpretation were inextricably linked to the handling of high numbers to the extent to which the study of poetry and the Sanskrit metric system inevitably initiated the Indian scholars into the art of arithmetic as well as grammar. Consequently poets, grammarians and astronomers, in fact all learned men, were as skilled at calculation as the arithmeticians and the teachers themselves.

Under the entries *Ananta, we see that the Sanskrit name for infinity was not only used to denote the sum of ten thousand million (ten billion in US English) but also, curiously, it was used as the symbol for zero.

The entries *Infinity and *Serpent will allow the reader to understand the relationship between infinity as we understand the term, and the mythological world of Hinduism, and that *Ananta, often represented as coiled up in a sort of sleeping "8" (like our symbol ∞), is none other than the immense serpent of infinity and eternity, which is linked to the ancient myths of the original serpent.

This dictionary provides an insight into the circumstances under which the Indian scientists invented zero and the place-value system. See *Names of numbers*, *Sanskrit*, *Place-value system*, *Position of numerals*, *Numerical symbols (Principle of the numeration of)*, *Shûnya* and *Zero*.

Under the entries *Shûnya* and *Shûnyatâ*, the Indian philosophical notions of "void" or "emptiness" are very briefly explained, and we can see how these early Indian concepts went far beyond corresponding but very heterogeneous notions of contemporary Graeco-Latin philosophies.

It also shows how, from a very early stage, *shûnya* meant zero as well as emptiness, the central element of the deeply religious and mystical philosophy, *shûnyatâ*. The word came to represent zero when the place-value system was discovered, the two concepts fitting together naturally. The dictionary also explains how, through the use of symbolism, this concept finally came to be graphically represented as the little circle that we all recognise as zero.

The entries *Yuga* and *Kalpa* tell of Indian cosmic cycles, and the speculations developed about them, both in Indian cosmogony and in the learned astronomy introduced by *Âryabhata* at the beginning of the sixth century. See also *Âryabhata*, *Cosmic cycles*, *Day of Brahma* and *Indian astronomy*.

The entry *Âryabhata's number-system* serves as further proof that it was the Indians who discovered zero and who are responsible for our current written number-system. In fact, we will see that this scholar, whilst constructing his own numerical notation (a very clever alphabetical number-system), could not have failed to have known of zero and the place-value system, given the very structure of the system.

If we look at the entries beginning with the expression *numeral alphabet*, we can find information about Indian alphabetical numbering systems. This will also give the reader some idea of the practices which were quite naturally born out of their usage: the composition of chronograms, the invention of secret codes, the preparation of talismans (closely linked to the Indian mysticism of letters and numbers), the development of homiletic or symbolic interpretations, predictive calculations and magical and divinatory practices.

Under the appropriate headings, one can similarly find short biographical notes about the great Indian scientists such as *Âryabhata*, *Bhâskara*, *Bhâskarâchârya*, *Brahmagupta* or *Varâhamihîra*, often accompanied by very precise accounts of the numerical notations they adopted (including bibliographical references).

If we consult the entries *Brâhmî numerals*, *Gupta numerals*, *Nâgarî numerals*, *Shâradâ numerals*, etc., we can also find out all about each style and see the impressive diversity of *Indian written numeral systems*. Alternatively, the reader can consult *Indian numerals*.

Extra details can be found about *Indian arithmetic*, the different *ages* of the sub-continent and *Indian astrology*, *Indian astronomy* and the *Indian mathematics* of this civilisation.

However, it is not necessary to look up all of the references given here: entries are accompanied by references to similar terms.

Entries such as *Algebra*, *Arithmetic*, *High numbers*, *Names of numbers*, *Numerical notation* etc., include either an alphabetical or a numerical list of terms relating to each of the ideas in question.

As for references which seem to have little to do with arithmetic, see: *Astronomical speculations*, *Buddhism*, *Brahmanism*, *Cosmogonic speculations*, *Hinduism*, *Indian cosmogonies and cosmologies*, *Indian divinities*, *Indian mythologies*, *Indian thought*, *Indian religions and philosophies*, *Jaina*, etc.

The main aim behind creating this dictionary has been, however, to give the reader a better idea of the subtle and complicated world of Indian numbers: a world which is largely unknown to Western readers and which is closely linked to Indian legends and cosmogonies. Its symbols, rather than being ordinary graphic signs and names of numbers, derive from a huge wealth of synonyms inspired by nature, human morphology, everyday activities, social conventions and traditions, legends, religion, philosophy, literature, poetry, the attributes of the divinities and by traditional and mythological ideas.

Thus, depending on the context, the idea of *wind* can evoke the number 5, the number 7 or the number 49. This demonstrates a subtlety which Westerners would not grasp if it was not shown from the correct perspective. The reasoning behind these diverse meanings offers a fascinating example of a logic and a way of thinking which is highly characteristic of the Indian mentality, and will help the reader to understand Cartesianism, which can often, due to the very nature of its rationalism, seem in total contradiction. To find out about the logic behind the above values of *wind*, see *Vâyu*, *Pâvana* and *Mount Meru*.

Other examples include: the term *anga*, which literally means "limbs" or "parts" and is often used as the numerical symbol for the number six; the word *rasa*, "sensations", is frequently used to denote the same number; the name of *Rudra*, the ancient Vedic god of the Sky, was used as the numerical symbol for 11, etc. Similarly, the *God of carnal love*, whose name is *Kâma*, was a symbol for the number 13; the *God of water and oceans*, *Varuna*, was the symbol for the 4; *Agni*, the *God of sacrificial fires* meant "three", etc.

These examples (along with many others) show the subtlety of the Indian symbolic system as well as demonstrating one of the most characteristic traits of the Indian cast of mind.

This dictionary contains a considerable amount of symbolism. A term with a symbolic meaning is denoted by an [S], an abbreviation of the actual Indian numerical symbol, and is defined firstly by its numerical value and literal meaning in Sanskrit and then, where possible, its symbolism is explained.

To gain a better understanding of the world in which the Indians created such symbols, the reader might find it useful to read the entries *Symbols* and *Numerical symbols*.

To find the Indian numerical symbol for a given number, one only has to look up the English (or Sanskrit) equivalent: *One, *Two, *Three, etc. (or *Eka, *Dva, *Tri, etc.) Under *Ashta, for example, the normal Sanskrit word for the number 8, there is a list of terms in which the word appears because of a direct link with the idea of the number 8 (for example *ashtadiggaja, the "eight elephants", guardians of the eight horizons of Hindu cosmogony). But if the reader wishes to know about words that have a more symbolical relationship with the same number, he should refer to the entry *Eight, where there is not only a list of numerical symbols which are synonymous with this number, but also a summary explanation of its different symbolical meanings: the serpent (*Ahi, *Naga, *Sarpa), the serpent of the deep (*Ahi), the elephant (*Dantin, *Dvipa, *Gaja), the eight elephants (*Diggaja), a sign that augurs well (*Mangala), etc.

Of course, one could also consult the more detailed rubric either in Sanskrit: *Hastin, *Lokapâla, *Murti, *Tanu, etc., or the English translation of the concepts behind Indian numerical symbols, such as: *Sky, *Space, *Elephant, *Moon, *Earth, *Sun, *Zodiac, *Serpent, etc.

The entry *Numerical symbols (general alphabetic list) contains all the word-symbols of the Sanskrit language that are included in this dictionary, whilst the entry *Symbolism of words with a numerical value gives an alphabetical list of English words which correspond with associated ideas contained within the Sanskrit symbols.

The entry *Symbolism of numbers provides a list of ideas (in arithmetical order) found in the symbolism of ordinary numbers, in high numbers and in the concept of infinity or zero.

This dictionary is the first of its kind to attempt to understand the thought process of the symbolic mind that characterises Indian numerical thinking.

Through a multidisciplinary process, mainly concerned with numbers and the symbols which represent them, the following is a kind of "vertical reading" of literary, philosophical, religious, mystical, mythological, cosmological, astronomical, and of course mathematical elements of this incredibly rich and subtle civilisation: elements which can be found in "horizontal" presentation in a great many wide-ranging publications in the most specialised of libraries.

This dictionary could be said to complement *L'Inde Classique* by L. Renou and J. Filliozat (1953), and also the monumental *Dictionnaire de la civilisation indienne* by L. Frédéric (1987) (the first of its kind to condense, in a remarkably simple yet well-documented manner, the essential facts about the India of both yesterday and today from a historical, geographical, ethnographic, religious, philosophical, literary and linguistic perspective). It also supplements the *Dictionnaire de la sagesse orientale* by K. Friedrichs, I. Fischer-Schreiber, F. K. Erhard and M. S. Diener (1989), which constitutes a vast yet accessible range of references and a very enriching insight to those who are interested in philosophy, mysticism and meditation, or in a general introduction to the doctrines of Hinduism, Buddhism, Taoism and the religion of Zen. It is also the perfect companion to the *Dictionnaire des symboles* by J. Chevalier and A. Gheerbrant (1982), which not only explains the history of symbolic language through the ages and in different civilisations and the indelible yet hidden imprint it has left in our minds, but also opens the doors of the imagination and invites the reader to meditate on the symbols, just as Gaston Bachelard invited us to muse on our dreams in order to discover within them the taste and feel of a living reality. These works have all influenced the writing of this book; without them the following dictionary could not have been compiled because the required research would have taken an inordinate amount of time and would have involved reading analytical works which are inaccessible to the public, and which, moreover, are devoid of any synthesis.

The author warmly thanks Billard, Frédéric, Chevalier and Jacques for their invaluable personal correspondence, especially Billard and Frédéric for reading the rubrics of this dictionary and who offered pertinent and constructive remarks.

The writing of this dictionary had to be handled with utmost caution (especially considering that this field of study was completely new to the author) for several reasons:

- The author was in danger of being carried away by his own enthusiasm; a justified enthusiasm, yet capable of leading to erroneous interpretations.
- The vertiginous world of Indian symbols is highly complex.
- Moreover, the culture in question (whose diverse aspects were studied, notably the countless symbols which are often multivalent) is not only incredibly complex but is also full of pitfalls. See *Indian documentation (Pitfalls of)*.
- Finally, Indian astronomy has played a significant role in this historiography. (The available documentation only offered a relatively simple insight into the literature of Indian astronomy. However, C.

Jacques obligingly recommended Billard's *Astronomie Indienne* (1971)
which is an unprecedented publication on the subject.)

Suffice to say that embarking upon this domain was rather like coming
face to face with one of the many-headed dragons of the legends of
Indian mythology: it was merciless and threatened to devour the author
at any moment, as it had those who had set foot in this territory before
without the necessary amount of precaution and vigilance. Now tamed,
however, the appeased monster offers the reader all the delights of
Eastern subtlety, and the wonder of this ingenious civilisation and its
inestimable contributions.

Shiva's dance of the creation and destruction of the world. After Moor's Hindu Pantheon

A

ABAB. Name given to the number ten to the power seventeen (= a hundred trillion). See *Abhabâgamana*, Names of numbers, High numbers.
Source: *Lalitavistara Sùtra* (before 308 CE).

ABABA. Name given to the number ten to the power seventy-seven. See *Abhabâgamana*, Names of numbers, High numbers.
Source: *Vyâkarana* (Pâlî grammar) by Kâchchâyana (eleventh century CE).

ABBUDA. Name given to the number ten to the power fifty-six. See Names of numbers, High numbers.
Source: *Vyâkarana* (Pâlî grammar) by Kâchchâyana (eleventh century CE).

ABDHI. [S]. Value = 4. "Sea". Four seas were said to surround *Jambudvîpa* (India). See *Sâgara*, Four. See also Ocean.

ABHABÂGAMANA. "Beyond reach". Sanskrit term used to express the uncountable and unlimited. It is possible that the words *abab* (ten to the power seventeen) and *ababa* (ten to the power seventy-seven) were abbreviations of this word. See Names of numbers, High numbers, Infinity. See also *Asamkhyeya*.

ABHRA. [S]. Value = 0. "Atmosphere". The atmosphere represents "emptiness". See *Shûnga*, Zero.

ABJA. Literally: "Moon". Name given to the number ten to the power nine (= a thousand million). See Names of numbers. For an explanation of this symbolism: see High numbers (Symbolic meaning of).
Source: *Lîlâvatî* by Bhâskarâchârya (1150 CE); *Trishatikâ* by Shrîdharâchârya (date uncertain).

ABJA. [S]. Value = 1. "Moon". The moon is unique. Another reason for this symbolism could be that in Indian tradition the moon is considered to be the source and symbol of fertility. It is likened to the primordial waters from whence came the revelation: it is the receptacle of seeds of the cycle of rebirth. It is thus the unity as well as the starting point. See One.

ABLAZE. [S]. Value = 3. See *Shikhin* and Three.

ABSENCE, ABSENT. See *Shûnyatâ*, Zero.

ABSOLUTE. As a symbol representing a large quantity. See High numbers (Symbolic meaning of), Infinity.

ADDITION. [Arithmetic]. *Samkalita* in Sanskrit.

ÂDI. [S]. Value = 1. "Commencement, primordial principle". In Hindu and Brahman philosophy, this principle is said to be found in all things before the creation; thus it is the unity as well as the starting point. See One.

ÂDITYA. [S]. Value = 12. "Children of Âditi". In Brahman and Vedic cosmogony, *Âditi* is the infinite sky, the original space. The *Âditya* are its children. In Vedic times, there were five, then they became seven, and finally twelve and were consequently identified by the twelve months of the year and the course of the sun during this period of time. The same word also signifies Sûrya, the sun god of the *Vedas*. As Sûrya = 12, the children of *Âditi* = 12. See *Sûrya*.

ADRI. [S]. Value = 7. "Mountain". Allusion to *Mount Meru, sacred mountain which, according to ancient Indian cosmological representation, was situated at the centre of the universe and constituted a meeting and resting place for the gods: a representation where we know seven played a significant role. See Seven.

AGA. [S]. Value = 7. "Mountain". See *Adri*. Seven.

AGES (The four). See *Chaturyuga*.

AGNI. [S]. Value = 3. "Fire". In Brahman mythology, Agni is the god of sacrificial fire (the three Vedic fires), which is represented as a man with three bearded heads who appears in three different forms: in the sky as the sun, in the air as lightning and on the earth as fire. Hence: "fire" = 3. See Fire, Three.

AGNIPURÂNA. See *Purâna* and positional numeration.

AHAHA. Name given to the number ten to the power seventy. See Names of numbers, High numbers.
Source: *Vyâkarana* (Pâlî grammar) by Kâchchâyana (eleventh century CE).

AHAR. [S]. Value = 15. "Day". See Tithi, Fifteen.

AHI. [S]. Value = 8. Probably an allusion to Ahirbudhnya (or Ahi Budhnya) who, in Vedic mythology, designates the serpent of the depths of the ocean, born of dark waters. Thus: *Ahi* = 8, because the serpent corresponds symbolically to the number eight. See *Nâga*, Eight. See also Serpent (Symbolism of).

AHIRBUDHNYA. See *Ahi*.

ÂKÂSHA. [S]. Value = 0. "Ether", the "element which penetrates everything", "space". It was considered as emptiness which could not mix with material things, immobile and eternal, beyond description. The association of ideas with the "void" or "emptiness" (*shûnya*) was established well before *shûnya* was identified with the concept of zero. In Indian thought ether was not only the void; it was also and above all the most subtle of the five elements of the revelation. It is certainly devoid of substance, but *âkâsha* is regarded as the condition of all corporeal extension and the receptacle of all matter formed by one of the other four elements (earth, water, fire or air). The association of ideas with zero became even more evident when this fundamental discovery was made: zero not only signified a void and "that which has no meaning", but also played an important role in the place-value system, and in terms of an abstract number, an equally essential role in mathematics and all the other sciences. Hence the symbolism: *âkâsha* = "space" = "void" = "ether" = "element which penetrates everywhere" = 0. See *Shûnya*, *Shûnyatâ*, Zero.

AKKHOBHINI. Name given to ten to the power forty-two. See **Names of numbers. High numbers.**

Source: *Vyâkarana* (Pâli grammar) by Kâchchâyana (eleventh century CE).

AKRITI. [S]. Value = 22. In terms of the poetry of Sanskrit expression, Akriti means the metre of four times twenty-two syllables per verse. See **Indian metric.**

ÂKSHARA. [S]. Value = 1. "Indestructible". A Sanskrit word which, in Hindu philosophy, denotes the "undying" part of the vocal sound corresponding to the revelation of the Brahman. This is a direct reference to the word *ekâkshara*, the "Unique and undying" which is often expressed by the Sacred Syllable *AUM*. See *Trivarna*, Mysticism of letters, One.

AKSHARAPALLÎ. Prakrit word meaning "letter-phoneme, syllable". It denotes a numerical notation of the alphabetical type frequently used in *Jaina manuscripts. See **Numeral alphabet.**

AKSHITI. Name given to the number ten to the power fifteen (= trillion). See **Names of numbers, High numbers.**

Source: *Pañchavimsha Brâhmana* (date uncertain).

AL-BIRUNI (Muhammad ibn Ahmad Abu'l Rayhan) (973–1048). Muslim astronomer and mathematician of Persian origin. After having lived in India for nearly thirty years, and having been initiated into the Indian sciences, he wrote many works, including *Kitab al arqam* ("Book of numerals"), *Tazkira fi'l hisab wa'l mad bi'l arqam al Sind wa'l hind* ("Arithmetic and counting systems using numerals in Sind and the Indias"), and above all *Kitab fi tahqiq i ma li'l hind* which constitutes one of the most important accounts of India in mediaeval times. Al-Biruni described the system of Sanskrit numerical symbols in minute detail, and stressed the importance of the place-value system and zero. He also went into much detail about the Sanskrit counting system, attaching particular importance to the Indian nomenclature of high numbers (see Fig. 24. 81). Here is a list of the principal names of numbers mentioned in *Kitab fi tahqiq i ma li'l hind* [see Woepcke (1863) p. 279]:*Eka* (= 1). *Dasha* (= 10). *Shata* (= 10^2). *Sahasra* (= 10^3). *Ayuta* (= 10^4). *Laksha* (= 10^5). *Prayuta* (= 10^6). *Koti* (= 10^7). *Vyarbuda* (= 10^8). *Padma* (= 10^9). *Kharva* (= 10^{10}). *Nikharva* (= 10^{11}). *Mahâpâdma* (= 10^{12}). *Shankha* (= 10^{13}). *Samudra* (= 10^{14}). *Madhya* (= 10^{15}). *Antya* (= 10^{16}). *Parârdha* (= 10^{17}). See **Indian numerals, Nâgarî numerals, Names of numbers, High numbers, Sanskrit, Numerical symbols.**

ALGEBRA. Alphabetical list of the words relating to this discipline, to which a rubric is dedicated in this dictionary: *Avyaktaganita. *Bîja. *Bîjaganita. *Ghana. *Indian mathematics (History of). *Samîkarana. *Varga. *Varga-Varga. *Varna. *Vyavahâra. *Yâvattâvat.

ALPHABETICAL NUMERATION. See *Aksharapallî*, *Âryabhata's numeration*, Katapayâdi numeration, Numeral alphabet, *Varnasamjña* and *Varnasankhya*.

AMARA. [S]. Value = 33. "Immortal". Allusion to the thirty-three gods. See *Deva*, **Thirty-three.**

AMBARA. [S]. Value = 0. "Atmosphere". See *Abhra*, Zero.

AMBHODHA (AMBHODHI). [S]. Value = 4. "Sea". It was said that four seas surrounded *Jambudvîpa* (India). See *Sâgara*, Four. See also **Ocean.**

AMBHONIDHI. [S]. Value = 4. "Sea". See *Sâgara*, Four. See also *Jala*.

AMBODHA (AMBODHI, AMBUDHI). [S]. Value = 4. "Sea". See *Sâgara*. Four. See also Ocean.

AMBURÂSHI. [S]. Value = 4. "Sea". See *Sâgara*. Four. See also Ocean.

AMRITA. Nectar of "Immortality". See *Soma*, **Serpent (Symbolism of the).**

ANALA. [S]. Value = 3. "Worlds". See **Loka. Three.**

ANANTA. Literally "Infinity". Name given to the number ten to the power thirteen (= ten billion). See *Asamkhyeya*, **Names of numbers.** For an explanation of this symbolism see **High numbers (symbolic meaning of).**

Source: *Sankhyâyana Shrauta Sûtra* (date uncertain), which defines this number as the "limit of the calculable".

ANANTA. Word which literally means "infinity". In Hindu mythology, the *ananta* denotes a huge serpent representing eternity and the immensity of space. It is shown resting on the primordial waters of original chaos (Fig. D. 1). Vishnu is lying on the serpent, between two creations of the world, floating on the "ocean of unconsciousness". The serpent is always represented as coiled up, in a sort of figure eight on its side (like the symbol ∞), and theoretically has a thousand heads. It is considered to be the great king of the **nâgas* and lord of hell (**pâtâla*). Each time the serpent opens its mouth it produces an earthquake because there is a belief that the serpent also supported the world on its back. It is the serpent that at the end of each **kalpa*, spits the destructive fire over the whole of creation. See **Infinity**. See also **Serpent**.

ANANTA. [S]. Value = 0. "Infinity". It seems paradoxical, yet this symbolism comes from the association of *Ananta*, the serpent of infinity, with the immensity of space. As "space" = 0, the name of the serpent became a synonym of zero. See *Ananta* (the second of the above entries). **Zero**.

ANCESTOR. [S]. Value = 1. See **Pitâmaha**. **One**.

ÂNDHRA NUMERALS. Signs derived from **Brâhmî* numerals, through the intermediary of Shunga, Shaka and Kushâna numerals, used in the contemporary inscriptions of the Ândhra dynasty (second – third century CE). These signs are notably found in the inscriptions of Jaggayapeta. The corresponding system did not function according to the place-value system and moreover did not possess zero. See **Indian written numeral systems (Classification of)**. See also Fig. 24.34, 36, 24.52, and 24.61 to 70.

ANGA. [S]. Value = 6. "Limb". The human body consists of six "limbs", or members: the head, the trunk, two arms and two legs. This is not the only reason, however, for this symbolism: there are six appended texts of the *Veda* (a group of Vedic texts called *Vedânga* which deal mainly with Vedic rituals, of their conservation and their perfect transmission). As *Vedânga* means the "members of Veda", we can see how the idea of "member" or "limb" came to signify the number six. See **Veda**, **Vedanga**, **Six**.

ANGULI. [S]. Value = 10. "Digit", because we have ten fingers. See **Ten**.

ANGULI. [S]. Value = 20. "Digit", because we have ten fingers and toes. See **Twenty**.

ANKA. Literally "mark, sign". The term means "numeral", "sign of numeration". See *Anka* [S]. See also all entries beginning with **Numeral**.

ANKA. [S]. Value = 9. "Numerals". Allusion to the nine significant numerals of the Indian place-value system. This symbol was in use no later than the time of **Bhâskara* I (629 CE). See *Anka*. *Ankasthâna*. **Nine**.

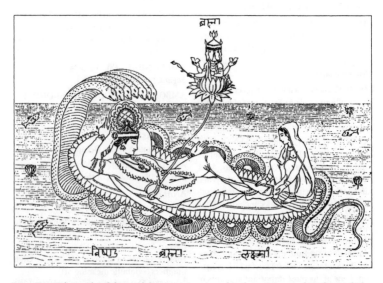

FIG.24.D.1. *Vishnu with Lakshmi and the serpent Ananta and Brahma sitting on a lotus flower which grows out of Vishnu's navel. From Dubois de Jancigny,* L'Univers pittoresque, *Hachette, Paris, 1846*

ANKAKRAMENA. Expression which literally means "in the order of the numerals", and alludes to the principle which the numerals are subjected to in the place-value system. See *Anka. Sthâna.*

Source: *Lokavibhâga* (458 CE).

ANKÂNÂM VÂMATO GATIH. Expression which means "principle of the movement of the figures from the right to the left". The numbers were read out in ascending order, from the smallest units to the highest multiple of ten. This was the reverse of how the numbers were presented in Indian numerical notations (from left to right).

ANKAPALLÎ. Prakrit term which literally means "numerals, representation". It is applied to any system of representing numbers using actual numerals. Thus it denotes "numerical notation".

ANKASTHÂNA. Literally "Numerals in position". The Sanskrit name for positional numeration.

Source: *Lokavibhâga* (458 CE).

ANKLE. [S]. Value = 2. See *Gulpha* and **Two.**

ANTA. Name given to the the number ten to the power eleven (= a hundred thousand million). See **Names of numbers, High numbers.**

Sources: *Vâjasaneyî Samhitâ*, *Taittirîya Samhitâ* and *Kâthaka Samhitâ* (from the start of the first millennium CE); *Pañchavimsha Brâhmana* (date uncertain).

ANTARIKSHA. [S]. Value = 0. "Atmosphere". See *Abhra.* **Zero.**

ANTYA. Literally "(the) last". Name given to the number ten to the power twelve (= a billion). See **Names of numbers, High numbers.**

Source: *Sânkhyâyana Shrauta Sûtra* (date uncertain). An allusion is made to the highest order of units of the ancient Sanskrit numeration at the time of the *Vâjasaneyî Samhitâ*, *Taittirîya Samhitâ* and *Kâthaka Samhitâ* (from the start of the first millennium CE), where the nomenclature stopped at ten to the power twelve.

ANTYA. Literally "(the) last". Name given to the number ten to the power fifteen (= a trillion). See **Names of numbers, High numbers.**

Sources: *Lîlâvatî* by Bhâskarâchârya (1150 CE); *Ganitakaumudî* by Nârâyana (1350 CE), *Trishatikâ* by Shrîdharâchârya (date uncertain). At this later date, when the Sanskrit names for numbers by far surpassed the simple power of fifteen, the name of this number still retained a vestige of the limitation of the spoken numeration of ancient times. See the first entry under *Antya.*

ANTYA. Literally "(the) last". Name given to the number ten to the power sixteen (ten trillion). See *Antya* (the above entries, Sources), **Names of numbers, High numbers.**

Source: *Kitab fî tahqiq i ma li'l hind* by al-Biruni (c. 1030).

ANU. Sanskrit name for "atom". See *Paramânu.*

ANUSHTUBH. [S]. Value = 8. Name given to certain groups of verses of Vedic poetry. This is an allusion to the eight syllables which made up each one of the four elements which constituted the stanzas which were called *anushtubh.* See **Eight, Indian metric.**

ANUYOGADVÂRA SÛTRA. Title of a Jaina cosmological text giving countless examples of extremely high numbers, the corresponding speculations reaching (and even surpassing) easily as high as ten to the power two hundred (one followed by two hundred zeros). Thus, the figure said to express the total number of human beings of the creation is described as the "quantity obtained by multiplying the sixth power of the square of two [= $(2^2)^6 = 2^{12}$] by the third power of two [= $2^3 = 8$], which is equal to the number which can be divided by 2 ninety-six times [= $(2^{12})^8 = 2^{12 \times 8} = 2^{96}$]" [see Datta and Singh (1938), p.12]. There is also the period of time called *Shîrshaprahelikâ*, expressed, according to Hema Chandra (1089 CE), by "196 places of the place-value system" and which corresponds approximately "to the product of 84,000,000 multiplied by itself twenty-eight times" (see Datta and Singh, *op.cit*). This text, amongst many others, shows that the Jainas were amongst the Indian scholars who were most familiar with such arithmetical-cosmogonical speculations. See **Names of numbers, High numbers, Infinity.**

ÂPA. Sanskrit term meaning "water". See *Jala.*

APHORISM. [S]. Value = 3. See *Vâchana.* **Three.**

ÂPTYA. [S]. Value = 3. "Spirit of the Waters". Allusion to the Vedic divinity named *Trita Âptya*, the "Third Spirit of the Waters", who killed Vishvarûpa, the three-headed demon. See **Three.**

ARABIC NUMERATION (Positional systems of Indian origin). See **"Hindi" numerals** and **Ghubar numerals.**

ARAMAEAN-INDIAN NUMERALS. See **Kharoshthî numerals.**

ARAMAEAN-INDIAN NUMERATION. See **Kharoshthî numerals.**

ARBUDA. Name given to the number ten to the power seven (= ten million). See **Names of numbers. High numbers.**

Sources: *Vâjasaneyî Samhitâ* (beginning of Common Era); *Taittirîya Samhitâ* (beginning of Common Era); *Kâthaka Samhitâ* (beginning of Common Era); *Pañchavimsha Brâhmana* (date unknown); *Sankhyâyana Shrauta Sûtra* (date unknown); *Âryabhatîya* (510 CE).

ARBUDA. Name given to the number ten to the power eight (= one hundred million). See **Names of numbers. High numbers.**

Source: *Lîlâvatî* by Bhâskarâchârya (1150 CE); *Ganitakaumudî* by Nârâyana (1350 CE); *Trishatikâ* by Shrîdharâchârya (date unknown).

ARBUDA. Name given to the number ten to the power ten (= ten billion). See **Names of numbers, High numbers.**

Source: *Ganitasârasamgraha* by Mahâvîrâchârya (850 CE).

ARITHMETIC. Here is an alphabetical list of terms relating to this discipline which appear as headings in this dictionary: *Abhabâgamana*, *Addition, *Algebra, *Anka, *Ankakramena, *Ankasthâna, *Arithmetical operations, *Âryabhata, *Âryabhata (Numerical notations of), *Âryabhata's numeration, *Asamkhyeya, *Base 10, *Base of one hundred, *Bhâskara, *Bhâskarâchârya, *Bîja, *Bîjaganita, *Brahmagupta, *Buddha (Legend of), *Calculation (The science of), *Calculation (Methods of), *Calculation on the abacus, *Calculator, *Cube, *Cube root, *Dashaguna, *Dashagunâsamjñâ, *Day of Brahma, *Dhûlikarma, *Digital calculation, *Divi-dend, *Division, *Divisor, *Equation, *Fractions, *High numbers, *Indeterminate equation, *Indian mathematics (The history of), *Infinity, *Kaliyuga, *Kalpa, *Katapayâdi numeration, *Khachheda, *Khahâra, *Mahâvîrâchârya, *Mathematician, *Mathematics, *Mental arithmetic, *Multiplication, *Names of numbers, *Nârâyana, *Numeral alphabet, *Numerals, *Numerical symbols, *Numerical symbols (Principle of the numeration of), *Pâtî, *Quotient, *Remainder, *Rule of five, *Rule of eleven, *Rule of nine, *Rule of seven, *Rule of three, *Sanskrit, *Shatottaragananâ, *Shatottaraguna, *Shatottarasamjñâ, *Shrîdharâchârya, *Square root, *Sthâna, *Subtraction, *Total, *Yuga, *Zero.

ARITHMETICAL OPERATIONS. See **Calculation**, *Dhûlikarma*, **Indian methods of calculation**, *Parikarma*, *Pâtî*, *Pâtîganita*, **Square roots (how Âryabhata calculated his).**

ARITHMETICAL SPECULATIONS. See *Anuyogadvâra Sûtra*, **Asamkhyeya**, **Calculation, Day of Brahma**, *Yuga*, *Kalpa*, **Jaina, Names of numbers, High numbers, and Infinity.**

ARITHMETICAL-COSMOGONICAL SPECULATIONS. See *Anuyogadvâra Sûtra*, *Asamkhyeya*, **Calculation, Cosmic cycles, Day of Brahma,** *Yuga* **(Definition),** *Yuga* **(Systems of calculation of),** *Yuga* **(Cosmogonic speculations on),** **Kalpa, Jaina, Names of numbers, High numbers, Infinity.**

ARJUNÂKARA. [S]. Value = 1,000. "Hands of Arjuna". Allusion to the mythical sovereign Arjunakârtavîrya, leader of the Haihayas and king of the "seven isles", who, according to one of the legends of the *Mahâbhârata*, had a thousand arms. See **Thousand.**

ARKA. [S]. Value = 12. "Bright". An epithet given to Sûrya, the sun god, who, symbolically, represents the number twelve. See **Sûrya. Twelve.**

ARMS. [S]. Value = Two. See *Bâhu* and **Two.**

ARMS OF ARJUNA. [S]. Value = 1,000. See *Arjunâkara* and **Thousand.**

ARMS OF KÂRTTIKEYA. [S]. Value = Twelve. See *Shanmukhabâhu* and **Twelve.**

ARMS OF RÂVANA. [S]. Value = Twenty. See *Râvanabhuja* and **Twenty.**

ARMS OF VISHNU. [S]. Value = Four. See *Haribâhu* and **Four.**

ARNAVA. [S]. Value = 4. "Sea". Four seas were said to surround *Jambudvîpa* (India). See *Sâgara.* **Four.** See also **Ocean.**

ARROW. [S]. Value = 5. See *Bâna, Ishu, Kalamba, Mârgana, Sâyaka, Shara, Vishikha* and **Five.**

ÂRYABHATA. A veritable pioneer of Indian astronomy, Âryabhata is without doubt one of the most original, significant and prolific scholars in the history of Indian science. He was long known by Arabic Muslim scholars as *Arjabhad*, and later in Europe in the Middle Ages by the Latinised name of *Ardubarius*. He lived at the end of the fifth century and the beginning of the sixth century CE, in the town of Kusumapura, near to Pataliputra (now Patna, in Bihar). His work, known as *Âryabhatîya* was written c. 510 CE. It is the first Indian text to record the most advanced astronomy in the history of ancient Indian astronomy. The work also involves trigonometry and gives a summary of the main mathematical knowledge in India at the beginning of the sixth century, bearing witness to the high level of understanding that had been reached in this field at this time. The following rapturous declamation by *Bhâskara (one of Âryabhata's disciples and most fervent of admirers), taken from the Commentary which he wrote on the *Âryabhatîya* in 629 CE gives some indication of the high level of abstract thought achieved by the scholar way ahead of his time [see Billard, IJHS. XII,2, p. 111]: "Âryabhata is the master who, after reaching the furthest shores and plumbing the inmost depths of the sea of ultimate knowledge of mathematics, kinematics and spherics, handed over the three [sciences] to the learned world."

See **Indian astronomy (History of)**. **Indian mathematics (History of)**.

ÂRYABHATA. (Numerical notations of). When referring to numerical data, Âryabhata often used the Sanskrit names of the numbers: at least this is the impression we get if we look at the sections of his work respectively entitled *Ganitapâda*, (which deals with "mathematics"), *Kâlakriyâ* (which talks of "movements", in particular his system of exact longitudes in his **Astronomical canon*) and *Golapâda* (which relates to "spherics" and other three-dimensional problems). Here is a list of the principal names of numbers mentioned in the *Âryabhatîya* [see *Ârya*, II, 2]:

*Eka (= 1). **Dasha* (= 10). **Shata* (= 10^2). **Sahastra* (= 10^3). **Ayuta* (= 10^4). **Niyuta* (= 10^5). **Prayuta* (= 10^6). **Koti* (= 10^7). **Arbuda* (= 10^8). **Vrindâ* (= 10^9).

See **Names of numbers** and **High numbers**.

Âryabhata also used a method of recording numbers which he invented himself: it was a clever (if not terribly practical) alphabetical system. However, he certainly knew the system of *numerical symbols, as we can see if we look at the *Ganitapâda*, which contains two examples of numbers expressed in this way [see *Ârya*, II, line 20; Billard, p. 88]:

sarûpa, "added to the form", and: *râshiguna*, "multiplied by the zodiac".

*Samkalita means addition (literally: "put together") and **gunana* means multiplication. These words can be abbreviated to *sa* ("plus") and *guna* ("multiplied by"). **Rûpa* and **râshi* are the respective numerical symbols for "shape" (or "form") and "zodiac", the numerical values for which are one and twelve. Thus the two above expressions can be translated as follows: *sarûpa*, "added to one", and: *râshiguna*, "multiplied by twelve". This is concrete proof that Âryabhata was familiar with the method of recording numbers using the numerical symbols. These are the only two examples that have been found in his work; however, Billard (pp. 88-89) shows that the *Âryabhatîya*, in its present state, is in fact two different works put together, or rather the result of reorganisation carried out on the original version. Some parts were left unaltered, some were slightly modified, and others still were radically changed in terms of numerical data, basic constants and metre. The text we have today consists of nothing more than the reworked parts, as the original has not been found. It is probable that

Âryabhata used the numerical symbols in the first version of his work and later changed his method of representing numbers as he reorganised his work. Finally, it is extremely likely that Âryabhata knew the sign for zero and the numerals of the place-value system. This supposition is based on the following two facts: first, the invention of his alphabetical counting system would have been impossible without zero or the place-value system; secondly, he carries out calculations on square and cubic roots which are impossible if the numbers in question are not written according to the place-value system and zero. See **Indian written numeration (Classification of)**, **Sanskrit numeration**, **Numerical symbols (Principle of the numeration of)**, **Âryabhata's numeration**. See also **Square roots (How Âryabhata calculated his)**.

ÂRYABHATA'S NUMBER-SYSTEM. This is an alphabetical numerical notation invented by the astronomer Âryabhata c. 510 CE. It is a system which uses thirty-three letters of the Indian alphabet and is capable of representing all the numbers from 1 to 10^{18}.

Âryabhata, it appears, was the first man in India to invent a numerical alphabet. He developed the system in order to express the constants of his **astronomical canons*, and his surprising astronomical speculations about **yugas*, and the system is more elegant and also shorter than that which uses numerical symbols. See **Âryabhata (Numerical notations of)**, **Numeral alphabet**, **Numeration of numerical symbols**, **Yuga (Systems of calculating)** and **Yuga (Astronomical speculations about)**.

The use of this notation is found throughout his work entitled *Dashagîtikapâda*, where he describes it in the following way: *Vargâksharâni varge 'varge 'vargâsha râni kât nmau yah Khadvinavake svarâ nava varge "varge navântya varge vâ*.

Translation: "The letters which are [said to be] classed (*varga*) [are], from [the letter] *ka*, [those which are placed] in odd rows (*varga*); the letters which are [called] unclassed (*avarga*) [are those which are placed] in even rows (*avarga*); [thus, one] *ya* is equal to *nmau* (= *na* + *ma*]; the nine vowels [are used to record] the nine pairs of places (*kha*) in even or odd [rows]. The same [procedure] can be repeated after the last of the nine even rows".

Ref.: JA, 1880, II, p. 440; JRAS, 1863, p. 380; TLSM, I, 1827, p. 54; ZKM, IV, p. 81; Datta and Singh, (1938), p. 65; Shukla and Sarma, *Ganita Section*, (1976), pp. 3ff.

To put it plainly, Âryabhata's method consists of assigning a numerical value to the consonants of the Indian alphabet in the following manner:

1. For the first twenty-five consonants, the order of normal succession is followed for whole numbers starting with 1.

2. The twenty-sixth represents five units more than the twenty-fifth.

3. For the remaining seven, the progression grows by tens.

4. The last consonant of the alphabet receives the value of one hundred.

Thus this notation contains a number of peculiarities which are unique to Indian syllables (which are transcribed below in modern *Nâgarî* characters). For a better understanding of this principle, it should be remembered that this writing uses thirty-three different characters, which represent the consonants, and many other signs representing the vowels in an isolated position (*a, â, i, î, u, û, ri, rî, l, e, o, ai, au*).

An isolated consonant is always pronounced with a short *a*, but when it is combined with another vowel, a special sign is added which graphically has nothing in

common with the sign representing this vowel in an isolated position (a vertical line to the right of the consonant, a line above, a loop below the letter, a horizontal line above the letter, with a loop, and so on).

As for the writing of the word, it is done with a continuous horizontal line called *mâtrâ* (see Fig. D. 2).

It must not be forgotten that the essential phonetic elements of this syllable system are constituted by the association of a consonant with a following vowel (which is either short like *a, i, u, ri, la,* or long like *â, î, û, rî*), or a diphthong (*e, o, ai, au*), which is always long by definition. Thus to a given consonant, a short vowel *a,* or a long vowel *â,* or even any one of remaining vowel or diphthongs can be joined (*i, u, ri, la, î, û, rî, lâ, e, o, ai, au*). Therefore, the following syllables correspond to the consonant *ma* (m) [for example]: *ma, mâ, mi, mî, mu, mû, mri, mrî, mla, mlâ, me, mo, mai.*

Conversely, the thirty-three consonants can be joined to any vowel. Take, for example, *a*:

5 gutturals *ka kha ga gha na*
5 palatals *cha chha ja jha ña*

gutturals	क	ख	ग	घ	ङ
	ka = 1	*kha* = 2	*ga* = 3	*gha* = 4	*na* = 5
palatals	च	छ	ज	झ	ञ
	cha = 6	*chha* = 7	*ja* = 8	*jha* = 9	*ña* = 10
cerebrals	ट	ठ	ड	ढ	ण
	ta = 11	*tha* = 12	*da* = 13	*dha* = 14	*na* = 15
dentals	त	थ	द	ध	न
	ta = 16	*tha* = 17	*da* = 18	*dha* = 19	*na* = 20
labials	प	फ	ब	भ	म
	pa = 21	*pha* = 22	*ba* = 23	*bha* = 24	*ma* = 25
semivowels	य	र	ल	व	
	ya = 30	*ra* = 40	*la* = 50	*va* = 60	
sibilants	श	ष	स		
	śha = 70	*sha* = 80	*sa* = 90		
aspirates	ह				
	ha = 100				

FIG.24D.2. *Alphabetical numeration of Âryabhata: numerical value of consonants in isolated position (vocalisation using a short "a"). Ref. : NCEAM, p. 257. Datta and Singh (1937); Guitel (1966); Jacquet (1835); Pihan (1860); Rodet*

5 cerebrals *ṭa ṭha ḍa ḍha ṇa*
5 dentals *ta tha da dha na*
5 labials *pa pha ba bha ma*
4 semivowels *ya ra la va*
3 sibilants *sha sha sa*
1 aspirated *ha*

The alphabetical notation of numbers invented by this astronomer/phonetician/mathematician is based upon precisely this structure. Starting from the first vowel (*a*):

• the five guttural consonants (*ka, kha, ga, gha, na*) receive the values 1 to 5;

• the five palatals (*cha chha ja jha h*) those of 6–10;

• the 5 cerebrals (*ta tha da dha na*) those of 11 to 15;

• the 5 dentals (*ta ths da dha na*) those of 16 to 20; the five labials (*pa pha ba bha ma*) those of 21 to 25

• the 4 semivowels (*ya ra la va*) receive the values 30, 40, 50 and 60;

• the 3 sibilants (*sha sha sa*) those of 70, 80 and 90;

• and the last letter of the alphabet (the aspirated *ha*) that of 100.

However, if a vertical line is added to the right of a *devanâgarî* consonant (thus vocalising the consonant with a long *a*), the value remains the same: *kâ* = ka; *khâ* = kha; *tâ* = ta, etc.

In other words, this numeration does not distinguish between long and short vowels when attributing numerical values to the letters (*kâ* = ka, *mî* = mi, *ñû* = ñu, *prî* = pri, etc.).

Thus, from this point on, to avoid confusion, only the consonants accompanied by a short vowel will be referred to.

To record numbers above one hundred, Âryabhata came up with the idea of using the rules of the vocalisation of consonants. In keeping with the order of the letters of the alphabet, the first thirty-three consonants with the vowel *a* represent the numbers from 1 to 100 according to the above rule. But if these consonants are vocalised using an *i* or an *î*, (which follow *a* in the Indian syllable system), the value of each is multiplied by a hundred (Fig. D. 3). If they are then accompanied by a *u* or a *û*, their initial values are multiplied by 10,000 (= 10^4). Thus, when either *ri* or *rî* are attached to the successive consonants, they represent the initial

क	खि	गि	धि	डि
ki = 100	*khi* = 200	*gi* = 300	*ghi* = 400	*ṅi* = 500
चि	छि	जि	कि	ञि
chi = 600	*chhi* = 700	*ji* = 800	*jhi* = 900	*ñi* = 1,000
टि	ठि	डि	ढि	णि
ti = 1,100	*ṭhi* = 1,200	*ḍi* = 1,300	*ḍhi* = 1,400	*ṇi* = 1,500
ति	थि	दि	धि	नि
ti = 1,600	*thi* = 1,700	*di* = 1,800	*dhi* = 1,900	*ni* = 2,000
पि	फि	बि	भि	मि
pi = 2,100	*phi* = 2,200	*bi* = 2,300	*bhi* = 2,400	*mi* = 2,500
यि	रि	लि	वि	
yi = 3,000	*ri* = 4,000	*li* = 5,000	*vi* = 6,000	
शि	षि	सि		
shi = 7,000	*shi* = 8,000	*si* = 9,000		
हि				
hi = 10,000				

Fig.24D.3. *Alphabetical numeration of Âryabhata: numerical value of consonants vocalised by a short "i")*

values multiplied by 1,000,000 (= 10^6). And so on with each of the consecutive vowels of the alphabet, multiplying by successive powers of 100 (10^2, 10^4, 10^6, etc.). Using all the possible phonemes, this rule enables numbers to be expressed up to the value of the number that we recognise today in the form of a 1 followed by eighteen zeros (10^{18}).

If we look at the question from another angle, Âryabhata's alphabetical notation follows the successive powers of a hundred: it is thus an additional numeration with a base of 100, where the units and tens (units of the first centesimal order) are expressed by the first thirty-three successive consonants in an isolated position, according to the vocalisation with an *a* (long or short). The units of the second centesimal order (units and tens, multiplied by a hundred = 10^2) are expressed by the same consonants, this time vocalised by an *i* (short or long). Those of the third centesimal order (units and tens multiplied by ten thousand = 10^4) are expressed by the consonants accompanied by the vowel *u* (or *û*). And so on until the units of the ninth centesimal order (units and tens multiplied by 10^{16}) using the thirty-three consonants with *au* (which corresponds to the last vowel. This is how the values for the consonant *ka* are obtained, using the successive vocalisations (Fig. D. 4). The first four orders of Âryabhata's numeration are presented in Fig. D. 5.

As the numbers were set out according to the ascending powers of a hundred, starting with the smallest units, the representation was carried out – at least theoretically – within a rectangle subdivided into several successive rectangles, where the syllables were written from left to right according to the centesimal order (Fig. D. 6). The number 57,753,336 corresponds to the number of synodic revolutions of the moon during a *chaturyuga.

In Âryabhata's language (Sanskrit), this number is expressed as follows (see *Ârya*, II, 2): *Shat thimshati trishata trisahasra pañchâyuta saptaniyuta saptaprayuta pañchakoti.*

This can be translated as follows, where the numbers are expressed in ascending order, starting with the smallest unit:

"Six [= *shat*],
three tens [= *trimshati*],
three hundreds [= *trishata*],
three thousands [= *trisahasra*],
five myriads [= *pañchâyuta*],
seven hundred thousand [= *saptaniyuta*],
seven millions [= *saptaprayuta*],
five tens of millions [= *pañchakoti*]".

See **Âryabhata (Numerical notations of)**, *Ankânâm vâmato gatih*, **Names of numbers**, **Sanskrit**.

Âryabhata's notation conformed rigorously to this order, the only difference being that it functioned according to a base of 100 and not

Centesimal order of units	1st	2nd	3rd	4th	5th	6th	7th	8th	9th
Syllable	क ka	कि ki	कु ku	कृ kri	कॢ kli	के ke	कै kai	कॊ ko	कौ kau
Value	1	10^2	10^4	10^6	10^8	10^{10}	10^{12}	10^{14}	10^{16}

Fig.24D.4. *Consecutive orders of units in Âryabhata's alphabetical numeration (successive values of syllables formed beginning with "ka").*

Vocalisation	with an *a*		with an *i*		with a *u*		with an *r*	
Associated centesimal order	1st		2nd		3rd		4th	
Power of ten	1		10^2		10^4		10^6	
Row of syllable	odd	even	odd	even	odd	even	odd	even

Fig.24D.5.

Centesimal order	←1st →	← 2nd→	← 3rd →	← 4th →
Syllable				
Row	odd even	odd even	odd even	odd even

Fig. 24D.6.

of 10. Thus it was necessary to break down the expression of the number in question (at least mentally) into sections of two decimal orders, as follows:

1st centesimal order: *six, three* tens

2nd centesimal order: *three* hundreds, *three* thousand,

3rd centesimal order: *five* myriads, *seven* hundred thousand,

4th centesimal order: *seven* million, *five* tens of millions.

For the first centesimal order, it was necessary to take the consonants, vocalised by *a*, which correspond respectively to the values 6 and 30 (six and three tens), which gives the syllables

cha and *ya* (see Fig. D. 6A).

For the second centesimal order, the consonants were vocalised by *i*, corresponding respectively to the values 300 and 3,000, the syllables being *gi* and *yi* (see Fig. D. 6B).

For the third centesimal order, the consonants were vocalised by *u*, and corresponded respectively to the values 50,000 and 700,000, the syllables being *nu* and *shu* (see Fig. D.6C).

Finally, for the fourth centesimal order, the consonants were vocalised with *ri*, corresponding respectively to the values 7,000,000 and 50,000,000, the syllables being *chhri* and *lri* (see Fig. D.6D).

	6	30					
36 →	cha	ya					
	odd	even					

Fig. 24D.6A.

			300	3,000			
3,336 →	cha	ya	gi	yi			
			odd	even			

Fig.24D.6B.

					50,000	700,000	
753,336 →	cha	ya	gi	yi	nu	shu	
					odd	even	

Fig.24D.6C.

						7,000,000	50,000,000
57,753,336 →	cha	ya	gi	yi	nu	shu	chhri	lri
						odd	even

Fig.24D.6D.

Thus, the notation for the number in question would be: *chayagiyinushuchhrilri*.

Fig. D. 6E shows the main breakdowns for this value (where the value of a syllable is called *absolute* when the vocalisation accompanying the consonant is ignored, and it is called *relative* where the opposite is the case).

The number 4,320,000 is expressed in the same manner, this figure corresponding to the total number of years in a **chaturyuga* (Fig. D. 6F): *khuyughri*. This notation is proof that inventive genius does not always go hand in hand with simplicity.

Thus, contrary to the opinion of several authors, this notation was not based on the place-value system, and certainly did not use zero. It is in fact an additional numeration of the third category of the classification given in Chapter 23.

However, it is very likely that Âryabhata knew about zero and the place-value system.It is precisely because he already knew about these concepts that he was able to achieve the degree of abstraction that was needed to develop a numerical notation such as this, which is unique

in the whole history of written numerations. Whilst this system is additional in principle, its mathematical structure is full of the purest concepts of zero and the place-value system. This is made clear in Fig. D. 4. The consonant *ka* is the chosen graphical sign upon which everything else is based. Going from left to right, the rule of numerical vocalisation invented by Âryabhata can be resumed as follows: by adding *i* to this sign, in reality two zeros are being added to the decimal representation of the value of the letter *ka* (in other words, the unit); but by adding a *u*, a *ri*, a *li*, an *e*, an *ai*, an *o*, or an *au*, four, six, eight, ten, twelve, fourteen or sixteen zeros would be added.

The Jews, the Syrians and the Greeks certainly used similar conventions, but only for highly specialised usage and without perceiving them from the same angle as Âryabhata. By adding an accent, a dot or even one or two suffixes to a letter, they multiplied its value by 100 or 1,000, but they never managed to generalise their convention from such an abstract angle as Âryabhata.

Syllables	cha	ya	gi	yi	nu	shu	chhri	lri
Absolute values	6	30	3	30	5	70	7	50
Relative total values for each column	36		33×10^2		75×10^4		57×10^6	
Total value	36		$+ \quad 33 \times 10^2 +$		$75 \times 10^4 +$		57×10^6	
57,753,336 →	cha	ya	gi	yi	nu	shu	chhri	lri

Fig. 24D.6E.

Syllables				khu	yu	ghri	
Absolute values				2	30	4	
Relative total values for each column				32×10^4		4×10^6	
Total value				$32 \times 10^4 +$		4×10^6	
57,753,336 →				khu	yu	ghri	

Fig. 24D.6F.

Âryabhata had an advantage, because the phonetic structure which characterises the Indian syllable system is almost mathematical itself. This is confirmed by *Bhâskara I, a faithful disciple separated by a century from Âryabhata. In his Commentary on the *Âryabhatîya* (629 CE), he gives this brief explanation of the rule in question: *nyâsashcha sthânânâm ooooooooooo*. Translation: "By writing in the places (*sthâna*), we have: ooooooooooo [= ten zeros]".

[See: commentary on the *Ganitapâda*, line 2; Shukla and Sarma, (1976), pp. 32–4; Datta and Singh, (1938), pp. 64–7].

In his text, the commentator uses not only the word *sthâna* which means "place" (which the Indian scholars often used in the sense of "positional principle"), but also and above all the little circle, which is the numeral "zero" of the Indian place-value system. See **Sthâna, Numeral 0** and **Zero**.

Later on in his commentary, Bhâskara I writes the following: *khadvinavake svarâ nava varge 'varge khâni shûnyâni, khânâm dvinavakam khadvinavakam tasmin khadvinavake ashtâdashasu shûnyopalakshiteshu*... Translation: "The nine vowels (*nava varge*) [are used to note] the nine pairs of zeros (*khadvinavake*); [because] *kha* means zero (*shûnya*). In the nine pairs of places, that is, in the eighteen (*ashtâdashasu*) [places] marked by zeros (*shûnyopalakshiteshu*)..."

The use of the term *kha* as one of the designations of zero is explained as one of the synonyms of *shûnya*, a word meaning "void" which Indian mathematicians and astronomers have always used, since at least the fifth century CE, in the sense of zero.

This leaves no doubt: even if the master was not very loquacious on the subject, his disciple and commentator explains Âryabhata's system and uses the Indian symbol for zero (the little circle), and also the three fundamental Sanskrit terms (*sthâna*, *kha*, *shûnya*). See **Zero**.

The Sanskrit term *kha*, literally "space", signifies "sky" and "void", and thus by extension "zero" in its mathematical sense. As for "place" (*sthâna*), Âryabhata gave it the meaning of the place occupied by a given syllable; thus, by extension, it meant "order of unit" in his alphabetical numeration. This is due to the "row" occupied by the syllable in a square which is formed by the structure of his notation system (Fig. D. 5). To his mind, it was a completely separate "place", one for the even row, and one for the odd row, within a unit of the centesimal

order (the odd row having a value of a simple unit and the even row a value of a multiple of ten). It is due to the fact that such a "place" can be "emptied" if the units or tens of the corresponding "order" are absent that "place" came to mean both "position" and void. As for the expression *khadvinavake*, for Âryabhata this meant the "nine pairs of zeros", the eighteen zeros added to the decimal positional representation of the initial value of a given consonant, at the end of the successive vocalisation operations.

Moreover, in *Golapâda*, Âryabhata alludes to the essential component of our place-value system when he states that "from place to place (*sthâna*), each [of the numerals] is ten times [greater] than the preceding one" [see Clark (1930), p. 28]. What is more, in the chapter of the *Ganitapâda* on arithmetic and methods of calculation, he gives rules for operations in decimal base for the extraction of square roots and cube roots. Neither of these two operations can be carried out if the numbers are not expressed *in writing*, using the place-value system with nine distinct numerals and a tenth sign which performs the function of zero. See **Pâtîganita, Indian methods of calculation** and **Square roots (How Âryabhata calculated his)**.

This is mathematical proof that, at the beginning of the sixth century CE, Âryabhata had perfect knowledge of zero and the place-value system, which he used to carry out calculations. The question remains: why did he invent such a complicated system when he could have used a much simpler one? It seems that the alphabet offered an almost inexhaustible supply when it came to creating mnemonic words, especially for complicated numbers, and this facilitated the readers' memorising of them. He always wrote in Sanskrit verse, and thus he had found a very convenient way not only to write numbers in a condensed form, but also to meet the demands of the metre and versification of the Sanskrit language.

Luckily, Âryabhata was the only one to make use of the system that he had invented. His successors, including those that referred to his work, generally adopted the method of *numerical symbols*. Even those that opted for an alphabetical numerical notation did not choose to use his system: they used a radically transformed form which was much simpler. See **Katapayâdi numeration**.

When Âryabhata's alphabetical numeration became widespread in the field of Indian astronomy, it no doubt was disastrous for the preservation of mathematical data. Worse yet, it caused the Indian discoveries of the place-value system and zero, which took place before Âryabhata's time, to be irretrievably lost to history.

ÂRYABHATÎYA. Title given to Âryabhata's work by his successors.

ASAMKHYEYA (or ASANKHYEYA). Literally: "number impossible to count" (from *samkhyeya or *sankhyeya, "number", the "highest number imaginable". See **High numbers** and **Infinity.**

ASANKHYEYA. Literally: "non-number". Term designating the "incalculable". See **Asamkhyeya.**

ASANKHYEYA. Literally: "impossible to count". Name given to the number ten to the power 140. See **Names of numbers.** For an explanation of this symbolism, see **High numbers (Symbolic meaning of).**

Source: *Vyâkarana* (Pâlî grammar) by Kâchchâyana (eleventh century CE).

ÂSHÂ. [S]. Value = 10. "Horizons". See *Dish,* **Ten.**

ASHÎTI. Ordinary Sanskrit name for the number eighty.

ASHTA (or ASHTAN). Ordinary Sanskrit name for the number eight. It is used in the composition of several words which have a direct relationship with the idea of this number. Examples: *Ashtadanda, *Ashtadiggaja, *Ashtamangala, *Ashtamûrti, *Ashtânga* and *Ashtavimoksha.*

For words which have a more symbolic relationship with this number, see **Eight** and **Symbolism of numbers.**

ASHTACHATVÂRIMSHATI. Ordinary Sanskrit name for the number forty-eight. For words which have a more symbolic relationship with this number, see **Forty-eight** and **Symbolism of numbers.**

ASHTADANDA. "Eight parts". These are the eight parts of the body that we use to conduct a profound veneration by stretching out face down on the ground. See *Ashtânga.*

ASHTADASHA. Ordinary Sanskrit name for the number eighteen. For words which have a more symbolical relationship with this number, see **Eighteen** and **Symbolism of numbers.**

ASHTADIGGAJA. "Eight elephants". Collective name for the guardians of the eight "horizons" of Hindu cosmogony (these being: *Airâvata, *Pundarika, Vâmana, *Kumuda, Anjana, Pushpadanta, Sarvabhauma* and *Supratîka*). See *Diggaja.*

ASHTAMANGALA. "Eight things that augur well". This concerns the "eight jewels" that Buddhism considers as the witnesses of the veneration of Buddha. See *Mangala.*

ASHTAMÛRTI. "Eight shapes (or forms)". The name of the most important forms of Shiva. See *Ashta* and *Mûrti.*

ASHTAN. A synonym of *Ashta.*

ASHTÂNGA. "Eight limbs (or members)". This term denotes the eight limbs of the human body which are used in prostration (the head, the chest, the two hands, the two feet and the two knees).

ASHTAVIMOKSIIA. "Eight liberations". This refers to a Buddhist meditation exercise, which has eight successive stages of concentration, the aim of which being to liberate the individual from all corporeal and incorporeal attachments.

ASHTI. [S]. Value = 16. In terms of Sanskrit poetry, this refers to the metre of four times sixteen syllables per line. See **Sixteen** and **Indian metric.**

ASHVA. [S]. Value = 7. "Horse". Allusion to the seven horses (or horse with seven heads) of the chariot upon which *Sûrya,* the Brahmanic god of the sun, raced across the sky. See **Seven.**

ASHVIN. [S]. Value = 2. "Horsemen". Name of the twin gods Saranyû and Vivashvant (also called *Dasra and *Nâsatya) of the Hindu pantheon. They symbolise the nervous and vital forces, and are supposed to respectively represent the morning star and the evening star. They are the offspring of horses, hence their name (from *Ashva). These divinities are considered as the "Primordial couple" who appeared in the sky before dawn in a horse-drawn golden chariot. See **Two.**

ASHVINA. [S]. Value = 2. "Horsemen". See **Ashvin** and **Two.**

ASHVINAU. [S]. Value = 2."Horsemen". See **Ashvin** and **Two.**

ASTRONOMICAL CANON. A group of elements conceived as a whole by the author of an astronomical text. These elements are always presented together in a text, commentary or quotation, being effectively (astronomically), or supposedly, interdependent. Often, however, except for historical information, and even though complete, the canons are only in the form of the game of *bîja. Thus, mathematically, a given canon can be placed in any era or represent any unit of time. See **Indian astronomy (The history of).**

ASTRONOMICAL SPECULATIONS. See *Yuga* (Astronomical speculation on).

ASURA. "Anti-god". Name given to the Titans of Indian mythology.

ATATA. Name given to the number ten to the power eighty-four. See **Names of numbers and High numbers**.

Source: *Vyâkarana* (Pâli grammar) by Kâchchâyana (eleventh century CE).

ATIDHRITI. [S]. Value = 19. The metre of four times nineteen syllables per verse. See **Indian metric**.

ÂTMAN. [S]. Value = 1. In Hindu philosophy, this term describes the "Self", the "Individual soul", the "Ultimate reality", even the *Brahman himself, who is said to possess all the corresponding characteristics. The uniqueness of the "Self" and above all the first character of the Brahman as the "great ancestor" explain the symbolism. See *Pitâmaha* and **One**.

ATMOSPHERE. [S]. Value = 0. See **Infinity**, *Shûnya* and **Zero**.

ATRI. [S]. Value = 7. Proper noun designating the seventh of the *Saptarishi* (the "Seven Great Sages" of Vedic India), considered to be the founder of Indian medecine. See *Rishi* and **Seven**.

ATRINAYANAJA. [S]. Value = 1. "Moon". See *Abja* and **One**.

ATTATA. Name given to the number ten to the power nineteen (= ten British trillions). See **Names of numbers and High numbers**.

Source: *Lalitavistara Sûtra* (before 308 CE).

ATYASHTI. [S]. Value = 17. The metre of four times seventeen syllables per line in Sanskrit poetry. See **Indian metric**.

AUM. Sacred symbol of the Hindus. See **Mysticism of letters** and *Ekâkshara*.

AVANI. [S]. Value = 1. "Earth". See *Prithivî* and **One**.

AVARAHAKHA. Generic name of the five elements of the revelation. See *Bhûta* and *Mahâbhûta*.

AVATÂRA. [S]. Value = 10. "Descent". The incarnation of a Brahmanic divinity, birth through transformation, the aim being to carry out a terrestrial task in order to save humanity from grave danger. The allusion here is made to the *Dashâvatâra*, the "ten *Avatâra*", or major incarnations of *Vishnu, attributed to the four "ages" of the world (*yugas*), according to Hindu cosmogony. See *Dashâvatâra* and **Ten**.

AVYAKTAGANITA. Name given to algebra (literally: "science of calculating the unknown"), as opposed to arithmetic, called *vyaktaganita*. See **Vyaktaganita**, **Algebra**, **Arithmetic**.

AYUTA. Name for the number ten to the power four (= ten thousand). See **Names of numbers and High numbers**.

Source: *Vâjasaneyî Samhitâ* (beginning of Common era); *Taittirîya Samhitâ* (beginning of Common era); *Kâthaka Samhitâ* (beginning of Common era); *Pañchavimsha Brâhmana* (date uncertain); *Sankhyâyana Shrauta Sûtra* (date uncertain); *Âryabhatîya* (510 CE); *Kitab fî tahqîq i ma li'l hind* by al-Biruni (1030 CE); *Lîlâvatî* by Bhâskarâchârya (1150 CE); *Ganitakaumudî* by Nârâyana (1350 CE); *Trishatikâ* by Shrîdharâchârya (date uncertain).

AYUTA. Name given to the number ten to the power nine (= a thousand million). See **Names of numbers and High numbers**.

Source: *Lalitavistara Sûtra* (before 308 CE).

B

BÂHU. [S]. Value = 2. "Arms", due to the symmetry of the two arms. See **Two**.

BAHULA. Name given to the number ten to the power twenty-three (= a hundred thousand [British] trillion). See **Names of numbers and High numbers**.

Source: *Lalitavistara Sûtra* (before 308 CE).

BAKSHÂLÎ'S MANUSCRIPT. See **Indian documentation (Pitfalls of)**.

BALINESE NUMERALS. Signs derived from *Brâhmî numerals, through the intermediary of Shunga, Shaka, Kushâna, Ândhra, Pallava, Châlukya, Ganga, Valabhî, "Pâli", Vatteluttu and Kawi numerals. Currently in use in Bali, Borneo and the Celebes islands. The corresponding system functions according to the place-value system and possesses zero (in the form of a little circle). For ancient numerals, see Fig. 24.50 and 80. For modern numerals, see Fig. 24.25. See **Indian written numeral systems (Classification of)**. See also Fig. 24.52 and 24.61 to 69.

BÂNA. [S]. Value = 5. "Arrow". See *Shara* and **Five**. See also *Pañchabâna*.

BASE OF ONE HUNDRED. See *Shatottaragananâ*, *Shatottaraguna* and *Shatottarasamjñâ*.

BASE TEN. See *Dashaguna* and *Dashagunâsamjñâ*.

BEARER. [S]. Value = 1. See *Dharani* and **One.**

BEGINNING. [S]. Value = 1. See *Âdi* and **One.**

BENGALI NUMERALS. Signs derived from *Brâhmî numerals, through the intermediary of Shunga, Shaka, Kushâna, Andhra, Gupta, Nâgarî and Kutila numerals. Currently used in the northeast of India, in Bangladesh, Bengal and in much of the centre of Assam (along the Brahmaputra river). The corresponding system functions according to the place-value system and possesses zero (in the form of a little circle). See **Indian written numeral systems (Classification of).** See also Fig. 24.10, 52 and 24.61 to 69.

BENGALÎ SÂL (Calendar). See *Bengali San.*

BENGALÎ SAN (Calendar). The solar era beginning in the year 593 CE. It is still used today in Bengal. To obtain the corresponding date in Common years, add 593 to a date expressed in this calendar. It is also called *Bengali Sâl.* See **Indian calendars.**

BHA. [S]. Value = 27. "Star". Allusion to the twenty-seven *nakshatra.* See **Twenty-seven.**

BHÂGAHÂRA. Term used in arithmetic to denote division, although the word is most often used to denote the divisor (which is also called *bhâjaka*). See *Chhedana.*

BHAGAVAD GÎTÂ. "Song of the Lord". A long philosophical Sanskrit poem containing the essence of *Vedânta philosophy, explained by Krishna and Arjuna in a dialogue about action, discrimination and knowledge. It is a relatively recent text (c. fourth century CE) and is found in the sixth book of the *Mahâbhârata [see Frédéric; *Dictionnaire* (1987)].

BHÂJAKA. Term used in arithmetic to denote the divisor. See *Bhâgahâra.*

BHÂJYA. Term used in arithmetic to denote the dividend. Synonym: *hârya.* See *Bhâgahâra* and *Chhedana.*

BHÂNU. [S]. Value = 12. An epithet of *Sûrya*, the Sun-god. *Bhânu = Sûrya* = twelve. See **Twelve.**

BHARGA. [S]. Value = 11. One of the names of *Rudra. See **Rudra-Shiva** and **Eleven.**

BHÂSKHARA. Indian mathematician, disciple of *Âryabhata (a century after his death). He was born in the first half of the seventh century. He is known mainly for his Commentary on the *Âryabhatîya, in which examples of the use of the place-value system expressed by

means of the Sanskrit numerical symbols are found in abundance. The translation of the numbers expressed in this manner is often given using the nine numerals and zero (also according to the rules of the place-value system) [see Shukla and Sarma (1976)]. He is usually called "Bhâskara I" so that he is not confused with another mathematician of the same name (*Bhâskarâchârya). See **Âryabhata (Numerical notations of), Âryabhata's numeration, Numerical symbols, Numerical symbols (Principle of the numeration of),** and **Indian mathematics (The history of).**

BHÂSKARA I. See **Bhâskara.**

BHÂSKARA II. See **Bhâskarâchârya.**

BHÂSKARÂCHÂRYA. Indian mathematician, astronomer and mechanic, usually referred to as Bhâskara II. He lived in the second half of the twelfth century CE. He is famous for his work, the *Siddhântashiromani*, an astronomical text accompanied by appendices relating to mathematics, amongst which we find the *Lîlâvatî (the "Player"), which contains a whole collection of problems written in verse. He frequently uses zero and the place-value system, which are expressed in the form of *numerical symbols. He also describes methods of calculation which are very similar to our own and are carried out using the nine numerals and zero. Moreover, he explains the fundamental rules of algebra where the zero is presented as a mathematical concept, and defines *Infinity* as the inverse of zero [see Sastri (1929)].

Here is a list of the main names of numbers given in the *Lîlâvatî* (Lîl, p. 2) [see Datta and Singh (1938), p.13]:

Eka (= 1). *Dasha* (= 10). *Shata* (= 10^2).
Sahasra (= 10^3). *Ayuta* (= 10^4). *Laksha* (= 10^5). *Prayuta* (= 10^6). *Koti* (= 10^7). *Arbuda* (= 10^8). *Abja* (= 10^9). *Kharva* (= 10^{10}). *Nikharva* (= 10^{11}). *Mahâpadma* (= 10^{12}). *Shanku* (= 10^{13}). *Jaladhi* (= 10^{14}). *Antya* (= 10^{15}). *Madhya* (= 10^{16}). *Parârdha* (= 10^{17}).

See **Names of numbers, High numbers, Positional numeration, Numerical symbols (Principle of the numeration of), Zero, Infinity, Arithmetic, Algebra,** and **Indian mathematics (The history of).**

BHÂSKARÎYABHÂSYA. See *Govindasvâmin.*

BHATTIPROLU NUMERALS. Signs derived from *Brâhmî numerals, through the intermediary of Shunga, Shaka, Kushâna, Ândhra, Pallava, Châlukya, Ganga and Valabhî numerals. Used since the eighth century CE by the

Dravidians in southern India. Kannara, Telugu, Grantha, Malayalam Tamil, Sinhalese, etc. numerals derived from these numerals. The corresponding system does not use the place-value system or zero. See **Indian written numeral systems (Classification of)**. See also Fig. 24.52 and 24.61 to 69.

BHAVA. [S]. Value = 11. "Water". One of the names of *Rudra, the etymological meaning of which is related to the tears. See **Rudra, Rudra-Shiva** and **Eleven**.

BHAVISHYAPURÂNA. See *Purâna*.

BHINNA. [Arithmetic]. Sanskrit term used to denote "fractions" in general (literally "broken up"). It is synonymous with *bhâga, amsha*, etc. (literally "portion", "part", etc.).

BHOJA. Indian astronomer who lived in the eleventh century CE. He is known as the author of a text entitled *Râjamrigânka*, in which there are many examples of the place-value system expressed through Sanskrit numerical symbols [see Billard (1971), p. 10]. See **Numerical symbols, Numerical symbols (Principle of the numeration of)**, and **Indian mathematics (The history of)**.

BHÛ. [S]. Value = 1. "Earth". See *Prithivî* and **One**.

BHÛBHRIT. [S]. Value = 7. "Mountain". Allusion to *Mount Meru. See *Adri* and **Seven**.

BHÛDHARA. [S]. Value = 7. "Mountain". Allusion to *Mount Meru. See *Adri* and **Seven**.

BHÛMI. [S]. Value = 1. "Earth". See *Prithivî* and **One**.

BHÛPA. [S]. Value = 16. "King". See *Nripa* and **Sixteen**.

BHÛTA. [S]. Value = 5. "Element". In Brahman and Hindu philosophy, there are five elements (or states) in the manifestation: air (*vâyu), fire (*agni), earth (*prithivî), water (*âpa) and ether (*âkâsha). See *Panchabhûta* and **Five**. See also *Jala*.

BHUVANA. [S]. Value = 3. "World". The "three worlds" (*triloka). See *Loka, Triloka*, and **Three**.

BHUVANA. [S]. Value = 14. "World". According to Mahâyâna Buddhism, the thirteen "countries of election" or "heavens" of Jina and Bodhisattva, to which was added *Vaikuntha. See **Fourteen**.

BÎJA. Word meaning "letters" in terms of *mathematical symbols* (letters used to express unknown values). In algebra, the word is also used in the sense of "element" or even "analy-

sis". See **Algebra** and *Bîjaganita*.

BÎJA. Word meaning "letters" in terms of *religious symbols* (which generally represent the divinities of the Brahman pantheon or the Buddhist tantric) and esoteric symbols (according to a power which is believed to be creative or evocative). See **Mysticism of letters**.

BÎJA. Word used in astronomy to denote corrective terms expressed numerically and applying to the elements of a given text, modifying those of the corresponding *astronomical canon. See **Indian astronomy (The history of)**.

BÎJAGANITA. Word denoting algebraic science or science of analytical arithmetic and the calculation of elements (from *bîja: "letter-symbol", "element", "analysis" and from *ganita: "science of calculation"). The word was used in this sense since Brahmagupta's time (628 CE). However, Indian mathematicians only ever used the first syllable of the word denoting a given operation as their algebraic symbols. See **Indian mathematics (The history of)**.

BILLION. (= ten to the power twelve. US, ten to power of nine). See *Antya, Kharva, Mahâbja, Mahâpadma, Mahâsaroja, Parârdha*, and *Shankha*.

BINDU. Literally "point". This is the name given to the number ten to the power forty-nine. See **Names of numbers**. For an explanation of this symbolism, see **High numbers (The symbolic meaning of)**.

Source: *Vyâkarana (Pâlî grammar) by Kâchchâyana (eleventh century CE).

BINDU. [S]. Value = 0. This word literally means "point". This is the symbol of the universe in its non-manifest form, before its transformation into the world of appearances (*rûpadhâtu*). The comparison between the uncreated universe and the point is due to the fact that this is the most elementary mathematical symbol of all, yet it is capable of generating all possible lines and shapes (*rûpa). Thus the association of ideas with "zero", which is not only considered to be the most negligible quantity, but also and above all it is the most fundamental of mathematical concepts and the basis for all abstract sciences. See **Zero**.

BIRTH. [S]. Value = 4. See *Gati, Yoni* and **Four**.

BLIND KING. [S]. Value = 100. See *Dhârtarâshtra* and **Hundred**.

BLUE LOTUS (half-open). This has represented the number ten to the power twenty-five. See *Utpala* and **High numbers** (The symbolic meaning of).

BLUE LOTUS (half-open). This has represented the number ten to the power ninety-eight. See *Uppala* and **High numbers** (The symbolic meaning of).

BODY. [S]. Value = 1. See *Tanu* and **One**.

BODY. [S]. Value = 6. See *Kâya* and **Six**.

BODY. [S]. Value = 8. See *Tanu* and **Eight**.

BORN TWICE. [S]. Value = 2. See *Dvija* and **Two**.

BOW WITH FIVE FLOWERS. See *Pañchabâna*.

BRAHMA. Name of the "Universal creator", the first of the three major divinities of the Brahman pantheon (Brahma, *Vishnu, *Shiva). See *Pitâmaha, Âtman* and *Parabrahman*.

BRAHMA. [S]. Value = 1. See *Âtman, Pitâmaha,* Parabrahman and **One**.

BRAHMAGUPTA. Indian astronomer who lived in the first half of the seventh century CE. His best-known works are *Brahmasphutasiddhânta* and *Khandakhâdyaka*, where there are many examples of the place-value system using the nine numerals and zero, as well as the *Sanskrit numerical symbols. He also describes methods of calculation which are very similar to our own using the nine numerals and zero. Moreover, he gives basic rules of algebra where zero is presented as a mathematical concept, and he defines *Infinity* as the number whose denominator is zero [see Dvivedi (1902)]. See **Numerical symbols (Principle of the numeration of), Zero, Infinity, Arithmetic, Algebra,** and **Indian mathematics (The history of)**.

BRAHMAN. See *Âtman, Pitâmaha, Parabrahman, Paramâtman,* **Day of Brahma** and *High numbers (The symbolic meaning of)*.

BRAHMANICAL RELIGION. See **Indian philosophies and religions**.

BRAHMANISM. See **Indian religions and philosophies**.

BRAHMASPHUTASIDDHÂNTA. See Brahmagupta and **Indian mathematics (The history of)**.

BRAHMÂSYA. [S]. Value = 4. "Faces of *Brahma". In representations, this god generally has four faces. He also has four arms and he is often depicted holding one of the four *Vedas* in each hand. See **Four**.

BRÂHMÎ ALPHABET. See Fig. 24. 28.

BRÂHMÎ NUMERALS. The numerals from which all the other series of 1 to 9 in India Central and Southeast Asia are derived. These are found notably in Asoka's edicts and in the Buddhist inscriptions of Nana Ghat and Nasik. The corresponding system does not function according to the place-value system, nor does it possess zero. See Fig. 24.29 to 31 and 70. For notations derived from Brâhmî, see Fig. 24.52. For their graphic evolution, see Fig. 24.61 to 69. See **Indian written numeral systems (Classification of)**.

BREATH. [S]. Value = 5. See *Prâna, Pâvana* and **Five**.

BRILLIANT. [S]. Value = 12. See *Arka* and **Twelve**.

BUDDHA (The legend of). Legend recounted in the *Lalitavistara Sûtra*, which is full of examples of immense numbers. See **Indian mathematics (The history of)**.

BUDDHASHAKARÂJA (Calendar). Buddhist calendar which is hardly used outside of Sri Lanka and the Buddhist countries of Southeast Asia. It generally begins in 543 BCE, thus by adding 543 to a date in this calendar we obtain the corresponding date in our own calendar. See **Indian calendars**.

BUDDHISM. Here is an alphabetical list of all the terms related to Buddhism which can be found as headings in this dictionary: *Ashtamangala, *Ashtavimoksha, *Bhuvana, *Buddha (Legend of), *Chaturmukha, *Chaturyoni, *Dashabala, *Dashabûmi, *Dharma, *Dvâdashadvârashâstra, *Dvâtrimshadvaralakshana, *Gati, *High numbers (The symbolic meaning of), *Indriya, *Jagat, *Kâya, *Loka, *Mangala, *Pañchâbhijñâ, *Pañcha Indryâni, Pañchachaksus, *Pañchaklesha, *Pañchânantara, *Ratna, *Saptabuddha, *Shûnya, *Shûnyatâ, *Tallakshana, *Trikâya, *Tripitaka, *Vajra and *Zero.

BUDDHIST RELIGION. See **Buddhism** and **Indian religions and philosophies**.

BURMESE NUMERALS. Signs derived from *Brâhmî numerals, through the intermediary of Shunga, Shaka, Kushâna, Ândhra, Pallava, Châlukya, Ganga, Valabhî, "Pâli", Vatteluttu and Môn numerals. Used since the eleventh century CE by the people of Burma. The corresponding system uses the place-value system and zero (in the form of a little circle). For ancient numerals, see Fig. 24.51. For modern numerals, see Fig. 24.23. See **Indian written numeral systems (Classification of)**. See also Fig. 24.52 and 24.61 to 69.

C

CALCULATING BOARD. See *Pâti, Pâtiganita.*

CALCULATING SLATE. See *Pâtiganita.*

CALCULATION (Methods of). See *Dhûlikarma, Pâti, Pâtiganita* and Indian methods of calculation.

CALCULATION (The science of). See *Ganita.*

CALCULATION ON THE ABACUS. See *Dhûlikarma.*

CALCULATOR. [Arithmetic]. See *Samkhya.*

CANOPY OF HEAVEN. [S]. Value = 0. See Zero, Zero (Indian concepts of) and Zero and Infinity.

CARDINAL POINT. [S]. Value = 4. See *Dish* and Four.

CAUSAL POINT. See *Paramabindu* and Indian atomism.

CELESTIAL YEAR. See *Divyavarsha.*

CENTESIMAL NUMERATION. See *Shatottaragananâ, Shatottaraguna* and *Shatottarasamjñâ.*

CHAITRA. Lunar-solar month corresponding to March / April.

CHAITRÂDI. "The beginning of *Chaitra*". This is the name of the year beginning in spring with the month of *Chaitra*, the first lunar-solar month.

CHAKRA. [S]. Value = 12. "Wheel". This refers to the zodiac wheel. See *Râshi* and Twelve.

CHAKSHUS. [S]. Value = 2. "Eye". See *Netra* and Two.

CHÂLUKYA (Calendar). Calendar of the dynasty of the eastern Châlukyas, beginning in the year 1075 CE. This calendar was used until the middle of the twelfth century (until c. 1162). To obtain the corresponding date in our own calendar, add 1075 to a date expressed in this calendar. See Indian calendars.

CHÂLUKYA NUMERALS. Signs derived from *Brâhmî numerals, through the intermediary of Shunga, Shaka, Kushâna, Ândhra and Pallava numerals. Contemporaries of the "Vatapi" dynasty of the Châlukyas of Deccan (fifth to seventh century CE). The corresponding system does not use the place-value system or zero. See Fig. 24.45 and 70. See Indian written numeral systems (Classification of). See also Fig. 24.52 and 24.61 to 69.

CHAM NUMERALS. Signs derived from *Brahmi numerals, through the intermediary of Shunga, Shaka, Kushâna, Ândhra, Pallava, Châlukya, Ganga, Valabhi, "Pâli" and Vatteluttu numerals. Used from the eighth to the thirteenth century CE to express dates of the Shaka calendar in the vernacular inscriptions of Champa (in part of Vietnam). The corresponding system uses the place-value system and zero. See Indian written numeral systems (Classification of). See Fig. 24.79 and 80. See also Fig. 24.52 and 24.61 to 69.

CHANDRA. [S]. Value = 1. "Luminous". An attribute of the *Moon as a (male) divinity of the Brahmanic pantheon. See *Abja, Soma* and One.

CHARACTERISTIC. [S]. Value = 5. See *Purânalakshana* and Five.

CHATUR. Ordinary Sanskrit name for the number four, which forms part of the composition of many words which have a direct relationship with the idea of this number.

Examples: *Chaturânanavadana, *Chaturyuga, *Chaturanga, *Chaturangabalakâya, *Chaturâshrama, *Chaturdvîpa, *Chaturmahârâja,*Chaturmâsya, *Chaturmukha, *Chaturvarga, *Chaturyoni. For words which have a more symbolic relationship with the number four, see Four and Symbolism of numbers.

CHATURÂNANAVÂDANA. [S]. Value = 4. The "four oceans". See *Chatur, Sâgara,* Four. See also Ocean.

CHATURANGA. "Four parts". Name given to an ancient Indian game, the ancestor of chess: there were four players and the board consisted of eight by eight squares and eight counters (the king, the elephant, the horse, the chariot and four soldiers). See *Chatur.*

CHATURANGABALAKÂYA. "Four corps". Name given to the ancient organisation of the Indian army, which consisted of elephants (*hastikâya*), the cavalry (*ashvakâya*), the chariots (*râthakâya*) and the infantry (*pattikâya*). See *Chatur.*

CHATURÂSHRAMA. "Four stages". According to Hindu philosophy, there were four stages to a man's life, in keeping with the Vedic concept: in the first (called *brahmâchârya*), intellectual capacities are developed, profane and religious instruction are received and the virtues of spiritual life are cultivated; in the second (*grihastha*), marriage and home-making take place; in the third (*hânaprastha*), having fulfilled his role as master of the house and having served his community, he goes alone into the forest to devote himself to intensive meditation, philosophical studies and the Scriptures; finally, in the fourth stage (*sannyâsa*), he gives up all his possessions and no longer cares for earthly things. See *Chatur.*

CHATURDASHA. Ordinary Sanskrit name for the number fourteen. For words with a symbolic relationship with this number, see: **Fourteen** and **Symbolism of numbers**.

CHATURDVÎPA. "Four Islands". In Brahmanic mythology and Hindu cosmology this is the name given to the four island-continents said to surround India (*Jambudvîpa*). See *Chatur*. For an explanation of this choice of number, see **Ocean**, which gives the same explanation about the four seas (**chatursâgara*).

CHATURMAHÂRÂJA. "Four great kings". These are the four guardian divinities of the cardinal points, who are said to live on the peaks of **Mount Meru (Vaishravana in the North, Virûpâksha in the West, Virûdhaka in the South and Dhritarâshtra in the East). See *Chatur*.

CHATURMÂSYA. "Four months". Name of an Indian ritual which takes place every four months, once at the start of spring, once at the start of the rain season, and once at the start of autumn. See *Chatur*.

CHATURMUKHA. "Four faces". Name given to all the Brahmanic or Buddhist divinities who are represented as having four faces (**Brahma, **Shiva, etc.). See *Chatur*.

CHATURSÂGARA. "Four oceans". These are the four seas said to surround India (*Jambudvîpa*). See *Sâgara*. For an explanation of this choice of number, see **Ocean**.

CHATURVARGA. "Four aims". These are the **trivarga* of Hindu philosophy (the three objectives of human existence), namely: *artha*, (material wealth), **kâma* (passionate love), and **dharma* (duty), to which sometimes a fourth is added, *moskha*, the liberation of the soul. See *Chatur*.

CHATURVIMSHATI. Ordinary Sanskrit name for the number twenty-four. For words which have a symbolic relationship with this number, see: **Twenty-four** and **Symbolism of numbers**.

CHATURYONI. The "four types of reincarnation". According to Hindus and Buddhists, there are four different ways to enter the cycle of rebirth (**samsâra*): either through a viviparous birth (*jarâyuva*), in the form of a human being or mammal; or an oviparous birth (*andaja*), in the form of a bird, insect or reptile; or by being born in water and humidity (*samsvedaja*), in the form of a fish or a worm; or even through metamorphosis (*aupapâduka*), which means there is no "mother" involved, just the force of *Karma* [see Frédéric (1994)]. See *Chatur*.

CHATURYUGA. The "four periods". Cosmic cycle of 4,320,000 human years, subdivided into four periods. Synonymous with **mahâyuga*. See *Chatur* and *Yuga* (**Definition**).

CHATURYUGA. [Astronomy]. According to speculations about **yugas*, the *chaturyuga*, or cycle of 4,320,000 years, is defined as the period at the beginning and end of which the nine elements (namely the Sun, the Moon, their apsis and node and the planets) are in average perfect conjunction at the starting point of the longitudes. See *Chaturyuga* (previous entry) and *Yuga* (**Astronomical speculation on**).

CHATVÂRIMSHATI. Ordinary Sanskrit name for the number **forty.

CHHEDANA. [Arithmetic]. Term meaning division (literally: "to break into many pieces"). Synonyms: *bhâgahâra*, *bhâjana*, etc.

CHHEDI (Calendar). Calendar beginning 5 September, 248 CE, which was used in the region of Malva and in Madhya Pradesh. To obtain the corresponding date in our own calendar, add 248 to a given *Chhedi* date. Sometimes called *kalachurî*, it was in use until the eighteenth century CE. See **Indian calendars**.

CHHIDRA. [S]. Value = 9. "Orifice". The nine orifices of the human body (the mouth, the two eyes, the two nostrils, the two ears, the anus and the sexual orifice). See **Nine**.

CHRONOGRAM. A short phrase or sentence whose numerical value amounted to the date of a given event. There are many methods of composing chronograms in India.

CHRONOGRAMS (Systems of letter numerals). One of the processes of composing chronograms involves the use of a **numeral alphabet. The hidden date is revealed by evaluating the various letters of each word of the sentence in question, then totalling the value of each word. This requires a mixture of mathematical and poetical skill, using the imagination to create sentences which have both literal and mathematical meaning. These types of chronograms (for which the system of evaluation clearly varies according to the system of attribution of numerical values to the letters of the alphabet) were not only written in Sanskrit, but also in various **Prakrits (local dialects). Many examples have been found throughout India, from Maharashtra, Bengal, Nepal or Orissa to Tamil Nadu, Kerala or Karnataka. They were also used by the Sinhalese, the Burmese, the Khmers, and in

Thailand, Java and Tibet. Many other examples also exist in Muslim India and in Pakistan, but these are many chronograms which employ numeral letters of the Arabic-Persian alphabet using a process called *Abjad*. See **Numeral alphabet and composition of chronograms.**

CHRONOGRAMS (Systems of numerical symbols). Another method of composing chronograms is only used for expressing the dates of the Shaka calendar: the language used is always Sanskrit and the dates are always expressed metaphorically, using Indian *numerical symbols ruled by the place-value system. This process was used for many centuries in India and in all the Indianised civilisations of Southeast Asia (Khmer, Cham, Javanese, etc., kingdoms). See **Numerical symbols, Numerical symbols (principle of the numeration of).**

CIRCLE. As a symbolic representation of the sky. See **Serpent (Symbolism of the).**

CIRCLE. As the graphic representation of zero. See **Shûnya-chakra, The numeral 0, Zero.**

CITY-FORTRESS. [S]. Value = 3. See *Pura*, *Tripura* and **Three.**

COBRA (Cult and symbolism of). See **Serpent (Symbolism of)** and *Nâga*.

CODE (secret writing and numeration). See **Numeral alphabet and secret writing.**

COLOUR. [S]. Value = 6. See *Râga* and **Six.**

COMPLETE. As a synonym of a large quantity. See **High numbers (Symbolic meaning of).**

COMPLETE. As a synonym of zero. See *Pûrna*.

CONSTELLATION. [S]. Value = 27. See *Nakshatra* and **Twenty-seven.**

CONTEMPLATION. [S]. Value = 6. See *Darshana* and **Six.**

COSMIC CYCLES. The division and length of cosmic cycles has always been of great importance in terms of Brahmanism: These periods represented the successive sections of cosmic life, conceived as cyclical and eternally revolving. The divisions of time were naturally the key elements of these cycles. The temporal dimension was meant to correspond to the duration of the creative and animating power of the cosmos, the "Word" (*vâchana*), which was uttered by the supreme progenitor of the world, Brahman-Prajâpati, and that which assimilates "knowledge" *par excellence*, the Veda. Thus the progenitor resembles the year which is taken as a unit of measurement of its cyclical activity, and the *Veda*, a collection of

lines, is divided into as many metric elements as there are moments in the "year" (see HGS, I, pp. 157–8). Of course, the "year" in question here is a "divine" year as opposed to a human year. See **Divine Year,** *Yuga* **(Definition),** *Yuga* **(Systems of calculation of),** *Yuga* **(Cosmogonic speculations on),** *Kalpa*, **Day of Brahma.**

COSMIC ERAS. See **Cosmic cycles** and *Yuga* **(Definition).**

COSMOGONIC SPECULATIONS. See *Yuga* **(Cosmogonic speculations on),** *Kalpa*, **Jaina.**

COURAGE. [S]. Value = 14. See *Indra* and **Fourteen.**

COW. [S]. Value = 1. See *Go* and **One.**

COW. [S]. Value = 9. See *Go* and **Nine.**

CUBE ROOT. [Arithmetic]. See *Ghanamûla*.

CUBE. [Arithmetic]. See *Ghana*.

D

DAHANA. [S]. Value = 3. "Fire". See *Agni* and **Three.**

DANTA. [S]. Value = 32. "Teeth". Humans have thirty-two teeth. See **Thirty-two.**

DANTIN. [S]. Value = 8. "Elephant". See *Diggaja* and **Eight.**

DARSHANA. [S]. Value = 6. "Vision", "contemplation", "system", and by extension "demonstration" and "philosophical point of view". This concerns the six principal systems of Hindu philosophy: mental research (*mîmâmsâ*); method (*nyâya*); the study and description of nature (*vaisheshika*); number as a way of thinking applied to the liberation of the soul (*sâmkhya*); the philosophies and practices of the liberation of the spirit from material ties (*yoga*); and studies based on the *Vedânta Sûtras* which deal with the basic identity of the soul and the *Brahman (vedânta)*. See *Shaddarshana* and **Six.**

DASHA (or DASHAN). Ordinary Sanskrit name for the number ten, which appears in the composition of many words which have a direct relationship with the idea of this number. Examples: *Dashabala, *Dashabhûmi, *Dashaguna, *Dashagunâsamjñâ, *Dashagunottarasamjñâ, *Dashaharâ, *Dashâvatâra.

For words which have a more symbolic relationship with this number, see **Ten** and **Symbolism of numbers.**

DASHABALA. "Ten powers". This refers to the ten faculties possessed by a Buddha, which give him ten powers, namely: the intuitive knowledge of the possible and the impossible, whatever the situation; the development of actions; the superior and inferior faculties of living beings; the diverse elements of the world; the paths which lead to purity or impurity; contemplation, concentration, meditation and the three deliverances; death; and the purification of all imperfections .

DASHABHÛMI. "Ten lands", "ten paradises". This refers to the "ten stages" of the Buddha Shâkyamuni.

DASHAGUNA. "Ten, primordial property". Sanskrit name for the decimal base. This word can be found in such works as the *Trishatikâ* by Shrîdharâchârya [see TsT, R. 2 – 3] and in the *Lîlavâtî* by Bhâskarâchârya [see Lîl, p. 2].

DASHAGUNÂSAMJÑÂ. "Words representing powers of ten". Term which applies to names of numbers of the Sanskrit numeration, distributed according to a decimal scale (base 10). See *Dashaguna*, **Names of numbers** and **High numbers**. This word can be found in such works as the *Trishatikâ* by Shrîdharâchârya [see TsT, R. 2 – 3].

DASHAGUNOTTARASAMJÑÂ. "Words representing powers of ten". Term which applies to names of numbers of the Sanskrit numeration, distributed according to a decimal scale (base 10). The contrast is made here with the word *shatottarasmjñâ* which applies to the centesimal scale (base 100). See **Dashaguna**, **Names of numbers**, **High numbers** and *Shatottarasamjñâ*.

DASHAHARÂ. Name of the Feast of the tenth day. See *Dasha* and *Durgâ*.

DASHAKOTI. Literally "ten *kotis*". Name given to the number ten to the power eight (= a hundred million). See **Names of numbers** and **High numbers**.

Source: *Ganitasârasamgraha* by Mahâvîrâchârya (850 CE).

DASHALAKSHA. Literally "ten *lakshas*". This is the name given to the number ten to the power six (one million). See **Names of numbers** and **High numbers**.

Source: *Ganitasârasamgraha* by Mahâvîrâchârya (850 CE).

DASHAN. Ordinary Sanskrit name for the number ten. See *Dasha*.

DASHASAHASRA. Literally "ten *sahastras*". Name given to the number ten to the power four (ten thousand). See **Names of numbers** and **High numbers**.

Source: *Ganitasârasamgraha* by Mahâvîrâchârya (850 CE).

DASHÂVATÂRA. Name of the "ten major incarnations" of *Vishnu, which are as follows: *Matsya* (incarnation as a fish); *Kûrma* (incarnation as a tortoise); *Varâha* (as a boar); *Narasimha* (as a lion-man); *Vâmana* (as a dwarf); *Parashu-Râma* (as Râma of the axe); *Râma* (the hero of *Râmâyana*); *Krishna* (the god); *Budha* (the god); and *Kalki*. See *Avatâra*.

DASRA. [S]. Value = 2. Name of one of the two twin gods Saranyû and Vivashvant of the Hindu pantheon (also called Dasra and Nâsatya). Symbolism through association of ideas with the "Horsemen". See *Ashvin* and **Two**.

DAY. [S]. Value = 15. See *Tithi*, *Ahar* and **Fifteen**.

DAY OF BRAHMA (Arithmo-cosmogonical speculations about the). According to Brahman cosmogony, the lifespan of the material universe is limited, and it manifests itself by *kalpa* cycles: "All the planets of the universe, from the most evolved to the most base, are places of suffering, where birth and death take place. But for the soul that reaches my Kingdom, O son of Kunti, there is no more reincarnation. One day of *Brahma is worth a thousand of the ages [*yuga] known to humankind; as is each night" (*Bhagavad Gîtâ*, VIII, lines 16 and 17). Thus each *kalpa* is worth one day in the life of Brahma, the god of creation. In other words, the four ages of a *mahâyuga* must be repeated a thousand times to make one "day of Brahma", a unit of time which is the equivalent of 4,320,000,000 human years.

According to this cosmogony, this is the total length of one created universe. The *kalpa* or "day of Brahma" is meant to correspond to the appearance, evolution and disappearance of a world, and this cycle is followed by a period of "cosmic repose" of equal length, which is followed by a new *kalpa*, and so on indefinitely. In other words, each *kalpa* ends with the total destruction (*pralaya*) of the universe which is followed by a period of reabsorption which is equivalent to a "night of Brahma", of equal length to the corresponding "day", before life is breathed into a new universe. It is precisely during this period of non-creation that *Vishnu, lying on *Ananta, the serpent of Infinity and Eternity, rests while he waits for Brahma to accomplish his work of Creation. This philosophy was developed as far as to speculate on the "length of the life of the god Brahma". A Commentary on the *Bhagavad Gîtâ* says: ". . . nothing in the material universe, not even Brahma can escape birth, ageing and death . . . the Causal Ocean contains countless Brahmas, who, being in a constant state of flux, appear and disappear like bubbles of air".

Here are some calculations relating to this this subject. Given that one whole "twenty-four hour day" in this god's life is the sum of one of his "days" and one of his "nights", "twenty-four hours in the life of Brahma" corresponds to: 4,320,000,000 + 4,320,000,000 = 8,640,000,000 (= eight thousand, six hundred and forty million) human years. One "year of Brahma" is made up of 360 of these "days". Thus it corresponds to 8,640,000,000 × 360 = 3,110,400,000,000 (= three billion, one hundred and ten thousand, four hundred million) human years. As this god is said to live for one hundred of his "years", the total length of his existence is equal to: 3,110,400,000,000 × 100 = 311,040,000,000,000 (= three hundred and eleven billion, forty thousand million) human years. According to certain traditions reported notably by al-Biruni, the "day of Brahma" does not correspond to a simple *kalpa*, but to a *parârdha* of kalpas, which is the length of a kalpa multiplied by *ten to the power seventeen*.

Thus: 1 "day of Brahma" = 100,000,000, 000, 000, 000 (= one hundred trillion) *kalpas*. As one *kalpa* is 4,320,000,000 years long, one "day" of this god corresponds to: 432,000,000, 000,000,000,000,000,000,000 (= four hundred and thirty-two sextillions) human years. Thus the complete "day" = 864,000,000,000,000,000, 000,000,000 (= eight hundred and sixty-four sextillions) human years. If we multiply this number by 36,000, the "life of Brahma" lasts thirty-one octillion and one hundred and four septillion human years. Childish at first sight, such speculations are very revealing of the Indian tendency towards metaphysical abstraction and of the high conceptual level achieved at an early stage by this civilisation. See *Ananta*, *Asamkhyeya*, **Calculation**, **High numbers**, **Infinity**, **Speculative arithmetic**, **Sanskrit**, *Sheshashîrsha*, **Indian mathematics (The history of)** and *Yuga* (**Cosmogonical speculations on**).

DAY OF BRAHMA. Cosmic period corresponding to the total length of one creation of the universe. According to Brahman cosmogony, this "day" is equal to 12,000,000 divine years (*divyavarsha*); and as one divine year is equal to 360 human years, the "day of Brahma" is equal to 4,320,000,000 human years. See *Divyavarsha*, *Mahâyuga* and *Yuga*.

DAY OF THE WEEK. [S]. Value = 7. See *Vâra* and **Seven**.

DECIMAL NUMERATION. See *Dashaguna* and *Dashagunâsamjñâ*.

DELECTATION. [S]. Value = 6. See *Rasa* and **Six**.

DEMONSTRATION. [S]. Value = 6. See *Darshana* and **Six**.

DESCENT. [S]. Value = 10. See *Avatâra* and **Ten**.

DEVA. [S]. Value = 33. "Gods". This is the general name given to all the divinities of the Hindu, Brahmanic, Vedic and Buddhist pantheons. These are the inhabitants of Mount Meru (mythical mountain, situated at the axis of the universe), who are ruled by a god. Unlike the great divinities (*Mahâdeva*) such as *Brahma, *Vishnu and *Shiva, these divinities have neither strength nor creative power. Theoretically numbering thirty-three million, they are reduced to thirty-three in Hindu cosmogony, which also gives their group the name *Trâiyastrimsha* ("thirty-three"). See **Thirty-three**. See also **Mount Meru.**

DEVANAGARI NUMERALS. See **Nagari numerals.**

DEVAPÂRVATA. "Mountain of the gods". One of the names of Mount Meru, the home of the gods in Brahmanic mythology and Hindu cosmology. See **Mount Meru**, *Adri, Dvîpa, Pûrna, Pâtâla, Sâgara, Pushkara, Pâvana* and *Vâyu*.

DHARÂ. [S]. Value = 1. "Earth". See *Prithivî* and **One.**

DHARANÎ. [S]. Value = 1. Literally "Bearer". This is synonymous here with the "earth", in the sense of "the bearer". See *Prithivî* and **One.**

DHARMA. In Indian philosophies, the *Dharma* is the general law, the Duty, the thing which is permanently fixed, the ensemble of rules and natural phenomena which rule the order of things and of men. In Buddhist philosophy in particular, the *dharma* is considered to be one of the three Treasures (**Triratna*) and one of the three refuges of the faithful. It is thus the social duty of the disciple. It represents the ultimate Only Reality, Virtue, Natural Order of all that exists, the Doctrine of Buddha as well as all the perceptions (ideas) hidden in the Manas [see Frédéric, *Dictionnaire*]. See *Shûnyatâ*.

DHÂRTARÂSHTRA. [S]. Value = 100. There is a legend related in the *Mahâbhârata* about the blind king Dhritarâshtra, son of Ambîkâ and the king Vichitravîrya, who married Gandhârî, with whom he had a hundred sons, called *Dhârtarâshtra*. During the Great Battle against the sons of Pându, the latter were all killed and became demons [see Frédéric, *Dictionnaire*]. See *Pândava* and **Hundred.**

DHÂTRÎ. [S]. Value = 1. "Earth". See *Prithivî* and **One.**

DHRITI. [S]. Value = 18. This refers to the metre of four times eighteen syllables per verse in Sanskrit poetry. See **Indian metric.**

DHRUVA. [S]. Value = 1. In Hindu mythology, this was the son of a king called Uttânapâda and his queen Sunîtî, who, through the power of his will, became the *Sudrishti*, the "divinity who never moves": the Pole star, whose uniqueness and fixedness are doubtless at the root of this symbolism. See **One.**

DHÛLÎKARMA. Literally "work on dust" (from *Dhûlî*, "sand", "dust", and *karma*, "act"). Term used in ancient Sanskrit literature to denote the "act of carrying out mathematical operations", in allusion to the ancient Indian practice of carrying out calculations on a board covered in sand. Today, the word is only used in the abstract sense of "superior mathematics". See **Calculation (Methods of).**

DHVAJÂGRANISHÂMANI. Name given to the number ten to the power 145. See **Names of numbers** and **High numbers.**

Source: *Lalitavistara Sûtra* (before 308 CE).

DHVAJÂGRAVATI. Name given to the number ten to the power ninety-nine. See **Names of numbers** and **High numbers.**

Source: *Lalitavistara Sûtra* (before 308 CE).

DIAMOND. A representation of the number ten to the power thirteen. See *Shanku.*

DIGGAJA. [S]. Value = 8. In Hindu cosmogony, the collective name given to the *Ashtadiggaja*, the "eight Elephants", who are said to guard the eight horizons of space. See *Dish* and **Eight.**

DIGITAL CALCULATION. See *Mudrâ.*

DIKPÂLA. [S]. Value = 8. "Guardian of the points of the compass". In Hindu cosmogony, this is the collective name given to the eight divinities considered to be the guardians of the horizons and the points of the compass (*Indra in the east, *Agni in the southeast, *Yama in the south, Nirritî in the southwest, *Varuna in the west, Kuvera in the north, *Vâyu in the northwest and Îshâna in the northeast). See *Diggaja, Dish, Lokapâla* and **Eight.**

DISH. [S]. Value = 4. "Horizon". The four cardinal points (north, south, east and west). See **Four.**

DISH. [S]. Value = 8. "Horizon". The horizons corresponding to the eight points of the compass: the north, the northwest, the west, the

southwest, the southeast, the south, the east and the northeast. See **Eight.**

DISH. [S]. Value = 10. "Horizon". The ten horizons of space: the eight normal horizons, plus the *nadir* and the *zenith*. See **Ten.**

DISHA. [S]. Value = 4. "Horizon". See *Dish* and **Four.**

DISHÂ. [S]. Value = 10. "Horizon". See *Dish* and **Ten.**

DIVÂKARA. [S]. Value = 12. "Sun". See *Sûrya* and **Twelve.**

DIVIDEND. [Arithmetic]. See *Bhâjya.*

DIVINATION. See **Numeral alphabet, magic, mysticism and divination, Indian astrology,** and **Indian astronomy (The history of).**

DIVINE MOTHER. [S]. Value = 7. See *Mâtrikâ* and **Seven.**

DIVINE PERFECTION. As a symbol for a large quantity. See **High numbers (Symbolic meaning of).**

DIVINE YEAR. See *Divyavarsha.*

DIVISION. [Arithmetic]. See *Chhedana, Bhâgahâra, Labdha, Shesha* and *Bhâjya.*

DIVISOR. [Arithmetic]. See *Bhâgahâra.*

DIVYAVARSHA. "Celestial or divine year". To convert a number of divine years into human years, it must be multiplied by 360.

DOGRI NUMERALS. Signs derived from *Brâhmî numerals, through the intermediary of Shunga, Shaka, Kushâna, Andhra, Gupta and Sharada numerals, and constituting a variation of Takari numerals. These are currently used in the Indian part of Jammu (in the southwest of Kashmir). The corresponding system uses the place-value system and possesses zero (in the form of a little circle). See **Indian written numeral systems (Classification of).** See Fig. 24.13, 52 and 24.61 to 69.

DOT. A graphical sign representing zero. see **Numeral 0,** *Bindu, Shûnya-bindu,* **Zero.**

DOT. A name for ten to the power forty-nine. See *Bindu,* **High numbers, High numbers (Symbolic meaning of).**

DOT. [S]. Value = 0. See *Bindu,* **Indian atomism** and **Zero.**

DRAVIDIAN NUMERALS. Numerals used in the southern regions of India, namely Tamil Nadu, Karnataka, Andhra Pradesh and Kerala, where the people are referred to as "Dravidian", and who, unlike the people from northern and central India, do not speak Indo-European languages. These signs are derived from *Brâhmî numerals, through the intermediary of Shunga, Shaka, Kushâna, Andhra, Pallava, Chalukya, Ganga, Valabhi and Bhattiprolu numerals. The corresponding system has not always used the place-value system or possessed zero. See **Tamil numerals**, **Malayalam numerals**, **Telugu numerals**, **Kannara numerals** and **Indian written numeral systems (Classification of)**.

DRAVYA [S]. Value = 6. "Substances". The six "bodies", or "substances" which make up existence according to *Jaina philosophy (these are: *dharmâshtikâya*, which constitutes the means of movement; *adharmâshtikâya*, which allows the animate to become inanimate; *akshatikâya*, which creates the space in which the animate and the inanimate live; *pudgalâshtikâya*, which enables the very existance of matter; *jîvâshtikâya*, which allows the mind to exist through inferences; and *kâla*, which is nothing other than time [see Frédéric (1987)]. See **Six**. This symbol is found in *Ganitasârasamgraha* by Mahâvîrâchârya [see Datta and Singh (1938), p. 55].

DRIGGANITA. See *Parameshvara*.

DRISHTI. [S]. Value = 2. This term is generally used in the sense of "vision", "contemplation", "revelation", "conception of the world" and "theory". Its primary sense, however, is "eye"; hence *drishti* = 2. See **Netra** and **Two**.

DROP. [S]. Value = 1. See *Indu* and **One**.

DUALITY. See *Dvaita*.

DURGÂ. [S]. Value = 9. "Inaccessible". This is the name of a Hindu divinity, the "Divine Mother", wife of Shiva, who is worshipped during the "Feast of the nine days" (*navarâtri*), which is celebrated at the end of the rain season in the month of *Ashvina* (September – October). The association of ideas which led to *Durgâ* becoming the numerical symbol equivalent to nine is obvious, but the choice of this value for the number of days of the feast is difficult to explain. This divinity, who is said to possess great powers, is often represented as having ten arms; moreover, she is depicted standing on a lion, which symbolises her power. The "Feast of nine days", which marks the end of the monsoon, ends on the tenth day with the grand feast of the *dashaharâ* (from *dasha*, "ten"), which is dedicated to Durgâ. The Hindus commemorate the victory of their divinities over the forces of Evil.

In this double religious symbolism, it is possible that, in accordance with the characteristic Indian way of thinking, these nine days were associated with the nine numerals of the place-value system, with which it is possible to write all numbers. The tenth day might then be associated with the tenth sign in this system: the zero, which corresponds to the most elusive, "inaccessible" and abstract concept; a concept whose invention is attributed to *Brahma, and which certainly constituted a great victory over the difficulties presented by numerical calculation. As for the tenth whole number, which in this system is written using a 1 and a 0, this would have corresponded, in the Indian symbolic mind, to an achievement, followed by the return to the unit at the end of the development of the cycle of the first nine numbers. However, this is mere conjecture for which there is no proof or foundation, it is simply based on one of the possible attitudes which characterise Indianity so well. See *Shûnya*, *Shûnyatâ*, **Zero** and **Nine**.

DUST BOARD. See *Pâtî*, *Pâtiganita*.

DVA (or **DVE**, **DVI**). Ordinary Sanskrit name for the number two, which forms a component of many words which have a direct relationship with the concept of duality, opposition, complementarity, etc. Examples: *Dvaipâyanayuga*, *Dvaita*, *Dvandva*, *Dvandvamoha*, *Dvâparayuga*, *Dvaya*, *Dvija*, *Dvivâchana*.

For words which have a more symbolic link with this number, see **Two** and **Symbolism of numbers**.

DVÂDASHA. Literally "twelve". This term is used symbolically in the *Rigveda* to mean "year", in allusion to the twelve months of the year.

Ref: *Rigveda*, VII, 103, 1; Datta and Singh (1938), p. 57.

DVÂDASHA. Ordinary Sanskrit name for the number twelve. For words which have a symbolic link with this number, see **Symbolism of numbers**.

DVÂDASHADVÂRASHÂSTRA. "Tract of the twelve doors". Title of a work by Nâgârjuna, one of the principal Buddhist philosophers, founder of the school of *Mâdhyamika*. See *Dvâdasha* and *Shûnyatâ*.

DVAIPÂYANAYUGA. Synonym of *dvâparayuga*.

DVAITA. "Duality". Term applied to a dualist philosophy, according to which a human creature is different from the *Brahman, its creator. This philosophy opposes the pure doctrine of the *Vedântas*, which is monistic (*Advaitavedânta*, "non dualist *Vedânta*").

DVANDVA [S]. Value = 2. "Couple, contrast". The symbolism is self-explanatory. See **Two**.

DVANDAMOHA. From *dvandva*, "couple, contrast", and *moha*, "illusion". This is the name given by the Hindus to what they consider to be the illusory impression that couples composed of opposites exist, such as shadow and light, joy and pain, etc.

DVAPARAYUGA (or **DVAIPÂYANAYUGA**). Name of the third of the four cosmic ages which make up a **mahâyuga*. This cycle, which is meant to be the equivalent of 864,000 human years, is regarded as the age during which humans have only lived for half of their lives, and where the forces of good have balanced out those of evil. See *Yuga* (**Definition**).

DVÂTRIMSHADVARALAKSHANA. "Thirty-two distinctive signs of perfection". According to Buddhism, these are the signs which allow Buddha to differentiate between ordinary humans from a moral, physical or spiritual perspective. See *Dvatrimshati*.

DVATRIMSHATI (or **DVITRIMSHATI**). Ordinary Sanskrit name for the number thirty-two. For words which have a symbolic connection with this number, see **Thirty-two** and **Symbolism of numbers**.

DVAVIMSHATI (or **DVIVIMSHATI**). Ordinary Sanskrit name for the number twenty-two. For words which have a symbolic link with this number, see **Twenty-two** and **Symbolism of numbers**.

DVAYA. [S]. Value = 2. Word meaning "pair". The symbolism is self-explanatory. See **Two**.

DVE. Ordinary Sanskrit name for the number two. See **Dva**.

DVI. Ordinary Sanskrit name for the number two. See **Dva**.

DVIJA. [S]. Value = 2. "Twice born". Epithet given to people belonging to the first three Brahmanic casts having the right to wear the sacred sash and who, during the ceremony of the handing over of the sash, are considered to be beginning a second life, this time of a spiritual nature [see Frédéric (1987)]. See *Dva* and **Two**.

DVÎPA. [S]. Value = 7. "Island-continent". Allusion to the seven island-continents which, in Hindu cosmology, are meant to radiate out from **Mount Meru. See *Adri* and **Seven**. See also *Sapta Dvîpa*.

DVIPA. [S]. Value = 8. "Elephant". See *Diggaja* and **Eight**.

DVITRIMSHATI. Synonym of **dvatrimshati*.

DVIVÂCHANA. Name of the dual of Sanskrit verbs.

DVIVIMSHATI. Synonym of **dvavimshati*.

DYUMANI. [S]. Value = 12. "Sun". See *Sûrya* and **Twelve**.

E

EARTH. As a mystical symbol for the number four. See *Nâga, Jala,* Ocean, Serpent (**Symbolism of the**).

EARTH. As a name for the number ten to the power sixteen, ten to the power seventeen, ten to the power twenty, ten to the power twenty-one. See *Kshiti, Kshoni, Mahâkshiti, Mahâkshoni* and **High numbers**.

EARTH. [S]. Value = 1. See *Avani, Bhû, Bhûmi, Dharâ, Dharanî, Dhâtrî, Go, Jagatî, Kshaunî, Kshemâ, Kshiti, Kshoni, Ku, Mahî, Prithivî, Vasudhâ, Vasundharâ* and **One**.

EARTH. [S]. Value = 9. See *Go* and **Nine**.

EASTERN ARABIC NUMERALS. Signs derived from **Brâhmî numerals, through the intermediary of Shunga, Shaka, Kushâna, Andhra, Gupta and Nâgarî numerals. Currently in use in Near and Middle East and in Muslim India, Malaysia and Indonesia. The corresponding system functions according to the place-value system and possesses zero (formerly either in the form of a little circle or dot but today exclusively represented by a dot). See **Indian written numeral systems** (**Classification of**). See Fig. 24.2, 24.52 and 24.61 to 24.69.

EIGHT. Ordinary Sanskrit names for the number eight: **ashta, *ashtan*. Here is a list of the corresponding numerical symbols: **Ahi, Anîka, *Anushtubh, Bhûti, *Dantin, *Diggaja, Dik, *Dikpâla, *Dish, Durita, *Dvipa, Dvirada, *Gaja, *Hastin, Ibha, Karman, *Kuñjara, *Lokapâla, Mada, *Mangala, *Matanga, *Mûrti, *Nâga, Pushkarin, *Sarpa, *Siddhi, Sindhura, *Takshan, *Tanu, *Vasu* and *Yâma*.

These words can either be translated by the following words or have a symbolic relationship with them: 1. The serpent (*Ahi, Nâga, Sarpa*). 2. The serpent of the deep (*Ahi*). 3. The elephant (*Dantin, Dvipa, Diggaja*). 4. The eight elephants (*Diggaja*). 5. That which augurs well (*Mangala*). 6. The jewel (*Mangala*). 7. The shapes, or forms (*Mûrti*). 8. The horizons (*Dish*). 9. The guardians of the horizons and of the points of the compass (*Lokapâla*). 10. The guardians of time (*Dikpâla*). 11. Supernatural powers (*Siddhi*). 12. Certain groups of lines of Vedic poetry (*Anushtubh*). 13. A group of eight divinities (*Vasu*). 14. The spheres of existence of *Adibhautika* (*Vasu*). 15. The "acts" (*Karman*) (only in *Jaina philosophy). 16. The "body" (*Tanu*). See **Numerical symbols**.

EIGHTEEN. Ordinary Sanskrit name: *ashtadasha*. The corresponding numerical symbol is *Dhriti*.

EIGHTY. See *Ashîti*.

EKA. Ordinary Sanskrit word for the number one, which appears in the composition of many words which have a direct relationship with the concept of unity. Examples: *Ekachakra, *Ekadanta, *Ekâgratâ, Ekâkshara, *Ekântika, *Ekatva.

For words which have a symbolic connection with the concept of this number, see **One** and **Symbolism of numbers**.

EKACHAKRA. "Who has only one wheel". Attribute of *Sûrya (the Sun-god).

EKADANTA. "Who has only one tooth". Attribute of *Ganesha, son of *Shiva and Pârvati, who is represented as having the body of a man and the head of an elephant, endowed with a unique defence. He is the Hindu divinity of wisdom, guaranteeing success in terrestrial existence and spiritual life.

EKADASHA. The ordinary Sanskrit name for the number eleven. For words which have a symbolic link with this number, see **Eleven** and **Symbolism of numbers**.

EKADASHARÂSHIKA. [Arithmetic]. Sanskrit name for the Rule of Eleven.

EKÂDASHÎ. Name of the eleventh day after the new moon, which orthodox Hindus spend fasting and meditating.

EKÂGRATÂ. In Hindu philosophy, a term which denotes a particular type of esoteric yoga, consisting in concentrating all of one's attention on a single point or object, which allows one to achieve *dhyâna* or "active contemplation".

EKÂKSHARA. "Unique and indestructible". Name of the sacred syllable of the Hindus (*AUM).

EKÂNNACHATVÂRIMSHATI. "One away from forty". The ancient form of the Sanskrit name for the number thirty-nine (in Vedic times). See **Names of numbers**.

EKÂNNATRIMSHATI. "One away from thirty". The ancient form of the Sanskrit name for the number twenty-nine (in Vedic times). See **Names of numbers**.

EKÂNNAVIMSHATI. "One away from twenty". The ancient from of the Sanskrit name for the number *nineteen (during Vedic times). See **Names of numbers**.

EKÂNTIKA. Name of the monotheistic doctrine of the Vishnuite tradition.

EKATVA. In Hindu philosophical systems, a term denoting Unity, the contemplation of Everything. This is the ability to see the Self or the Divine in everything, and everything in the Self or the Divine.

EKAVIMSHATI. Ordinary Sanskrit name for the number twenty-one. For words which have a symbolic connection with this number, see **Twenty-one** and **Symbolism of numbers**.

ELEMENT. [S]. Value = 5. See *Bhûta*, Five and *Pañchabhûta*.

ELEMENTS OF THE REVELATION. See *Bhûta, Pañchabhûta, Jala*, Five, Numeral alphabet, magic, mysticism and divination and Ocean.

ELEPHANT. A symbol for ten to the power twenty-one, ten to the power twenty-seven, ten to the power 105 or ten to the power 112. See *Kumud, Kumuda, Pundarîka* and **High numbers**.

ELEPHANT. [S]. Value = 8. See *Dantin, Diggaja, Gaja, Hastin, Kuñjar*, Eight and *Ashtadiggaja*.

ELEVEN. Ordinary Sanskrit name: *ekâdasha*. Here is a list of corresponding numerical symbols: *Akshauhinî, *Bharga, *Bhava, *Hara, *Îsha, *Îshvara, Lâbha, *Mahâdeva, *Rudra, *Shiva, *Shûlin, Trishtubh*.

These words have the following translation or symbolic meaning: 1. A name or attribute of Rudra-Shiva (*Bharga, Bhava, Hara, Îsha, Îshvara, Mahâdeva, Rudra, Shiva, Shûlin*). 2. The "Supreme Divinity" (*Îshvara*). 3. The "Lord of the Universe" (*Îshvara*). 4. The "Great God" (*Mahâdeva*). 5. "Grumbling" (*Rudra*). 6. The "Lord of tears" (*Rudra*). 7. "Violent" (*Rudra*). 8. The "Master of the animals" (*Shûlin*). See **Numerical symbols**.

ENERGY (feminine). [S]. Value = 3. See *Shakti* and **Three**.

EQUATION. [Algebra]. See *Ghana, Varga, Vargavarga, Samîkarana, Vyavahâra, Yâvattâvat* and **Indian mathematics (The history of)**.

ERAS (of Southeast Asia). See Shaka, *Buddhashakarâja* and **Indian calendars**.

ESOTERICISM. See *Âkshara*, **Numeral alphabet and secret writing, Numeral alphabet, magic, mysticism and divination**, *Âtman*,

AUM, Bîja, Ekâgratâ, Ekâkshara, Kavacha, Mantra, *Trivarna, Vâchana* and Serpent.

ETERNITY. See *Ananta* and Infinity.

ETHER. [S]. Value = 0. See *Âkâsha, Shûnya* and Zero.

EUROPEAN NUMERALS (Algorisms). Numerals used after the twelfth century by European mathematicians (written calculation). The corresponding system functioned according to the place-value system and possessed a zero (in the form of a little circle). These signs derived from *Brâhmî numerals, firstly through the intermediary of types of Indian numerals such as Shunga, Shaka, Kushâna, Andhra, Gupta and Nâgarî, and then via the numerals used by the Arabs. The appearance of the numerals varied greatly from one school to another. Some styles derived from "Hindi" numerals, but most came from Arabic numerals. One such style, standardised due to the requirements of typography, became the origin of the numerals we use today: 1 2 3 4 5 6 7 8 9 0. See Indian written numeral systems (Classification of). See also Fig. 24.52 and 24.61 to 69.

EUROPEAN NUMERALS (Apices of the Middle Ages). Numerals used by European mathematicians in the Middle Ages (who carried out their calculations on an abacus). They derive from *Brâhmî numerals, first through the intermediary of types of Indian numerals such as Shunga, Shaka, Kushâna, Andhra, Gupta and Nâgarî, and then via Ghubar numerals of North African Arabs. The appearance of the numerals varied greatly from one school to another. The corresponding system did not possess zero because calculations were carried out on the abacus. See Indian written numeral systems (Classification of). See also Fig. 24.52 and 24.61 to 69.

EYE. [S]. Value = 2. See *Netra, Drishti* and Two.

EYE. [S]. Value 3. See *Netra* and Three.

EYE OF SHUKRA. [S]. Value = 1. See *Shukranetra* and One.

EYES. [S]. Value = 2. See *Lochana* and Two.

EYES OF INDRA. [S]. Value = 1,000. See *Indradrishti* and Thousand.

EYES OF SENÂNÎ. [S]. Value = 12. See *Senâninetra* and Twelve.

EYES OF SHIVA. [S]. Value = 3. See *Haranetra* and Three.

F

FACE. [S]. Value = 4. See *Mukha* and Four.

FACES OF BRAHMA. [S]. Value = 4. See *Brahmâsya* and Four.

FACES OF KÂRTTIKEYA. [S]. Value = 6. See *Kârttikeyâsya* and Six.

FACES OF KUMÂRA. [S]. Value = 6. See *Kumâravadana* and Six.

FACES OF RUDRA. [S]. Value = 5. See *Rudrâsya* and Five.

FACULTY. [S]. Value = 5. See *Indriya* and Five.

FIFTEEN. Ordinary Sanskrit name: *panchadasha. Here is a list of corresponding numerical symbols: *Ahar, Dina, Ghasra, *Paksha, *Tithi. These words have the following translation or symbolic meaning: 1. "Wing", in allusion to the number of days in one of the two "wings" of the month (*Paksha*). 2. "Day", in allusion to the number of days in one of the two "wings" of the month (*Ahar, Tithi*). See Numerical symbols.

FIFTY. See *Panchâshat* and Names of numbers.

FINGER. [S]. Value = 10. See *Anguli* and Ten.

FINGER (or Digit). [S]. Value = 20. See *Anguli* and Twenty.

FINITE (Number). See Infinity and Indian mathematics (The history of).

FIRE. [S]. Value = 3. See *Agni, Anala, Dahana, Hotri, Hutâshana, Jvalana, Krishânu, Pâvaka, Shikhin, Tapana, Udarchis, Vahni, Vaishvânara* and Three.

FIRE [S]. Value = 12. See *Tapana.* Twelve.

FIRMAMENT. [S]. Value = 0. See *Shûnya,* Zero and Infinity.

FIRST FATHER. [S]. Value = 1. See *Pitâmaha* and One.

FIVE. Ordinary Sanskrit name: *pancha. Here is a list of corresponding numerical symbols: *Artha, *Bâna, Bhâva, *Bhûta, *Gavyâ, *Indriya, *Ishu, *Kalamba, *Karanîya, Kshâra, Lavana, *Mahâbhûta, *Mahâpâpa, *Mahâyjña, *Mârgana, Pallava, *Pândava, Parva, Parvan, *Pâtaka, *Pâvana, *Prâna, *Purânalakshana, *Putra, *Ratna, *Rudrâsya, *Sâyaka, *Shara, Shastra, *Suta, Tanmâtra, Tata, *Tattva, *Tryakshamukha, *Vishaya, *Vishikha.

The translation, or symbolic meaning of these words is as follows: 1. Arrows (*Bâna, *Ishu, *Kalamba, *Mârgana, *Sâyaka, *Shara, *Vishikha). 2. Statistics (*Purânalakshana).

3. "That which must be done" (*Karaniya).
4. Purification (*Pâvana). 5. The gifts of the Cow (*Gavya). 6. The elements, in allusion to the five elements of the revelation (*Bhûta). 7. The Great Elements, in allusion to the five elements of the revelation (*Mahâbhûta). 8. The faculties (*Indriya). 9. The worst sins (*Mahâpâpa). 10. The great sacrifices (*Mahâyajñe). 11. The main observances (*Karaniya). 12. The fundamental principles, realities, truths, the "true natures" (Tattva). 13. The Jewels (*Ratna). 14. The breaths (*Prâna). 15. The senses, or the sense organs (*Vishaya). 16. The Sons of Pându (*Pândava). 17. The Sons, in allusion to the sons of Pându (*Putra). 18. The faces of Rudra (*Rudrâsya, *Tryakshamukha). See Numerical symbols.

FIVE ELEMENTS (philosophy of the). See Bhûta, Pañchabhûta, Jala, Five, Numeral alphabet, Magic, Mysticism and Divination.

FIVE SUPERNATURAL POWERS. See Pañchabhijñâ.

FIVE VISIONS OF BUDDHA. See Pañchachakshus.

FORM. [S]. Value = 1. See Rûpa and One.

FORM. [S]. Value = 3. See Mûrti, Trimûrti and Three.

FORM. [S]. Value = 8. See Mûrti and Eight.

FORTY. Ordinary Sanskrit name: *chatvârimshati. Corresponding numerical symbol: Naraka.

FORTY-EIGHT. Ordinary Sanskrit name: *ashtachatvârimshati. Corresponding numerical symbol: *Jagatî.

FORTY-NINE. Ordinary Sanskrit name: *navachatvârimshati. Corresponding numerical symbols: *Tâna and *Vâyu.

FOUR. Ordinary Sanskrit name for this number: *chatur. Here is a list of the corresponding numerical symbols: *Abdhi, *Ambhodha, Ambhodhi, *Ambhonidhi, Ambudhi, *Amburâshi, *Arnava, Âshrama, Aya, Âya, Bandhu, *Brahmâsya, *Chaturânanavadana, Dadhi, *Dish, *Disha, *Gati, Gostana, *Haribâhu, *Îryâ, *Jala, *Jaladhi, *Jalanidhi, Jalâshaya, Kashâya, Kendra, Khatvâpâda, Koshtha, *Krita, *Mukha, Payodhi, Payonidhi, Purushârtha, *Sâgara, Salilâkara, *Samudra, Senânga, *Shruti, *Sindhu, *Turîya, *Udadhi, Vanadhi, *Varidhi, *Vârinidhi, *Veda, Vishanidhi, Vyûha, *Yoni, *Yuga.

These words have the following translation or symbolic meaning: 1. Water (Jala). 2. Sea or ocean (*Abdhi, *Ambhonidhi, *Ambudhi,

*Amburâshi, *Arnava, *Jaladhi, *Jalanidhi, *Jalâshaya, *Sâgara, *Samudra, *Sindhu, *Udadhi, *Vâridhi, *Vârinidhi). 3. The four oceans (Chaturânanavadana). 4. The "horizons", in the sense of the cardinal points (Dish, Disha). 5. The conditions of existence (Gati). 6. The "Fourth" as an epithet of the Brahman (Turîya). 7. The "revelations" (Shruti). 8. The "positions" (Îryâ). 9. The arms of Vishnu (Haribâhu). 10. The births (Gati, Yoni). 11. The vulva (Yoni). 12. The Vedas (Veda). 13. The faces of Brahma (Brahmâsya). 14. The "faces" (Mukha). 15. The four ages of a mahâyuga (Yuga). 16. The last of the four ages of a mahâyuga (Krita). See Numerical symbols. See also Ocean.

FOUR CARDINAL POINTS. [S]. Value = 4. See Dish and Four.

FOUR ISLAND-CONTINENTS. See Chaturdvîpa and Ocean.

FOUR OCEANS (or FOUR SEAS). See Chatursâgara, Sâgara (= 4) and Ocean.

FOUR STAGES. See Chaturâshrama.

FOURTEEN. Ordinary Sanskrit name: *chaturdasha. Here is a list of corresponding numerical symbols: Bhuvana, *Indra, *Jagat, *Loka, *Manu, Pûrva, *Ratna, *Shakra, *Vidyâ.

These words have the following translation or symbolic meaning: 1. The god Indra (Indra). 2. "Courage", "strength", "power" (Indra). 3. Powerful (Shakra). 4. "Human", in the sense of progenitor of the human race (Manu). 5. The worlds (Bhuvana, Jagat, Loka). 6. The Jewels (Ratna). See Numerical symbols.

FOURTH (The). [S]. Value = 4. Word used as an epithet for *Brahma. See Turîya and Four.

FRACTIONS. [Arithmetic]. See Bhinna, Kalâvarna, Pañcha Jâti.

FUNDAMENTAL PRINCIPLE. [S]. Value = 1. See Âdi and One.

FUNDAMENTAL PRINCIPLE. [S]. Value = 5. See Tattva and Five.

FUNDAMENTAL PRINCIPLE. [S]. Value = 7. See Tattva and Seven.

FUNDAMENTAL PRINCIPLE. [S]. Value = 25. See Tattva and Twenty-five.

G

GAGANA. [S]. Value = 0. Word meaning "the canopy of heaven", "firmament". This symbolism is explained by the fact that the sky is nothing but a "void". See Zero and Shûnya.

GAJA. [S]. Value = 8. "Elephant". See *Diggaja* and **Eight**.

GAME OF CHESS. See **Chaturanga**.

GANANÂ. Word meaning "arithmetic" in ancient Buddhist literature. More commonly, however, it has been used in the sense of "mental arithmetic" (which was and still is particularly developed in the art of Indian calculation).

GANANÂGATI. From *gananâ*, "arithmetic", and *gati*, "condition of existence". Name given to the number ten to the power thirty-nine. See **Names of numbers**. For an explanation of this symbolism, see **High numbers (Symbolic meaning of)**.

Source: *Lalitavistara Sûtra* (before 308 CE).

GANESHA. Hindu divinity of wisdom, also called *Ekadanta*. See **Eka**.

GANESHA. Indian mathematician who lived around the middle of the sixteenth century. Notably his works include a work entitled *Ganitamañjarî*.

GANGA NUMERALS. Signs derived from *Brâhmî numerals, through the intermediary of Shunga, Shaka, Kushâna, Ândhra, Pallava and Chalukya numerals. These were contemporaries of the beginnings of the dynasty of the Gangas of Mysore (sixth to eighth century CE) The corresponding system did not use the place-value system or zero. See **Indian written numeral systems (Classification of)**. See Fig. 24.46, 52 and 24.61 to 69.

GANITA. Sanskrit name for mathematics. In Vedic literature, this word is used to mean "the science of calculation", which is no doubt its original meaning. By extension, this word later acquired the meaning "science of measuring". In ancient Buddhist literature, there are three types of *ganita*: *mudrâ* or "manual arithmetic"; *gananâ* or "mental arithmetic"; and *samkhyâna* or "high arithmetic". Note that the word *ganita* was often used in ancient times to mean astronomy and even geometry (*kshetraganita*). See **Arithmetic, Calculation** and **Indian mathematics (The history of)**.

GANITAKAUMUDÎ. See **Nârâyana**.

GANITÂNUYOGA. Word meaning "explanation of mathematical principles". Term used mainly in *Jaina texts.

GANITASÂRASAMGRAHA. See **Mahâvîrâchârya**.

GATI. [S]. Value = 4. Literally "condition of existence". This word denotes the different forms of existence that reincarnation can assume (*samsâra*). The word became the numerical symbol for 4, synonymous with *yoni*, "birth" [see Frédéric (1994)]. See *Chaturyoni* and **Four**.

GAUTAMA SIDDHÂNTA. (Not to be confused with Gautama Siddhârtha, the Buddha). Chinese Buddhist astronomer of Indian origin, author of a work on astronomy and astrology entitled *Kai yuan zhan jing* (718 – 729 CE), where he describes zero, the place-value system and Indian methods of calculation. See **Place-value system**, and **Zero**.

GAVYÂ. [S]. Value = 5. "Gifts of the Cow". These are the *Pañchagavyâ*, the "five gifts of the Cow" (namely: milk, curds, dung, *ghî* and urine), which make up the sacred drink *gavyâ*, used by certain *samnyâsin* ascetics for its supposedly curative and purifying properties [see Frédéric (1994)]. See **Five**.

GÂYATRÎ. [S]. Value = 24. In expressive Sanskrit poetry, this is a stanza composed of three times eight syllables. See **Indian metric**.

GEOMETRY. See *Kshetraganita* and **Indian mathematics (The history of)**.

GHANA. "Cube". Sanskrit term used in arithmetic and algebra to denote the operation of cubing a number.

GHANA. Word used in algebra to denote "cube", in allusion to the third degree of equations of this order. See *Varga, Varga-Varga* and *Yâvattâvat*.

GHANAMÛLA. Sanskrit term used in arithmetic and algebra to denote the operation of the extraction of the cubic root.

GHUBAR NUMERALS. Signs derived from *Brâhmî numerals, through the intermediary of Shunga, Shaka, Kushâna, Ândhra, Gupta and Nâgarî numerals. Formerly used by the Arabic mathematicians of North Africa (for calculations carried out on the "dust" abacus). The corresponding system did not always possess zero. See **Indian written numeral systems (Classification of)**. See Fig. 24.52 and 24.61 to 69.

GIFTS OF THE COW. [S]. Value = 5. See *Gavyâ* and **Five**.

GIRI. [S]. Value = 7. "Mountain, hill". See *Adri* and **Seven**.

GO. [S]. Value = 1. "Cow", "Earth". This is the name of the sacred cow worshipped by the Hindus. This cow is said to have been created by *Brahma on the first day of the month of

Vaishâkha (April–May). The word forms part of the composition of the name *Govinda* ("Cowherd") attributed to *Vishnu as "Saviour of the earth". This is also an allusion to the fact that the earth (*Prithivî*) is often symbolically associated with a cow named *Prishnî*. This relationship (which also explains the veneration of the cow in Hindu religion) stems from the fact that the cow, like the earth, gives life [see Frédéric (1994)]. See **One.**

GO. [S]. Value = 9. "Cow, Earth". Another meaning of this word is "radiance", and by extension "star". This is why the word became synonymous with *graha*, "planets" (in the sense of *navagraha, the "nine planets of the Hindu cosmological system"). Thus *Go* = 9. See **Nine.**

GOAL (The three). See *Trivarga.*

GOAL (The four). See *Chaturvarga.*

GOD OF CARNAL LOVE. [S]. Value = 13. See *Kâma* and **Thirteen.**

GOD OF COSMIC DESIRE. [S]. Value = 13. See *Kâma* and **Thirteen.**

GOD OF SACRIFICIAL FIRES. [S]. Value = 3. See *Agni* and **Three.**

GOD OF WATER AND OCEANS. See *Varuna.*

GODS. [S]. Value = 33. See *Deva* and **Thirty-three.**

GOOGOL. This term is of English origin. It was invented by the American mathematician Edward Kastner in the 1940s. It denotes the number ten to the power 100. This number, which no longer represents anything palpable, surpasses all that is possible to count or measure in the physical world. See **Infinity** and **High numbers.**

GOVINDASVÂMIN. Indian astronomer c. 830 CE. Notably, his works include *Bhâskarîyabhâsya,* in which there are many examples of the use of the place-value system using Sanskrit numerical symbols [see Billard (1971), p. 8]. See **Numerical symbols,** and **Numeration of numerical symbols.**

GRAHA. [S]. Value = 9. "Planet". This alludes to the *navagrahas, the "nine planets" of the Hindu cosmological system (namely: *Sûrya, the Sun; *Chandra, the Moon; *Angâraka, Mars; *Budha, Mercury; *Brihaspati, Jupiter; *Shukra, Venus; *Shani, Saturn; and the two demons of the eclipses *Râhu and *Ketu. See *Paksha* and **Nine.**

GRAHA. "Planet". See previous entry, *Saptagraha* and *Navagraha.*

GRAHACHÂRANIBANDHANA. See *Haridatta.*

GRAHÂDHÂRA. "Axis of the planets". Name given to the Pole star. See **Dhruva** and *Sudrishti.*

GRAHAGANITA. Name given to astronomy by Brahmagupta (628 CE). Literally: "calculation of the planets", and, by extension, "mathematics of the stars". See **Indian astronomy (The history of)** and *Ganita.*

GRAHAPATI. "Master of the planets". Name sometimes given to *Sûrya, the Sun-god. See *Graha.*

GRAHARÂJA. "King of the planets". Name sometimes given to *Sûrya, the Sun-god. See *Graha.*

GRANTHA NUMERALS. Symbols derived from *Brâhmî numerals and influenced by Shunga, Shaka, Kushana, Andhra, Pallava, Chalukya, Ganga, Valabhi and Bhattiprolu numerals. Formerly used by the Dravidian peoples of Kerala and Tamil Nadu. The symbols corresponded to a mathematical system that was not based on place-values and therefore did not possess a zero. See: **Indian written numeral systems (Classification of).** See also Fig. 24.52 and 24.61 to 69.

GREAT ANCESTOR. [S]. Value = 1. See *Pitâmaha* and **One.**

GREAT ELEMENT. See *Mahâbhûta.* Value = 5.

GREAT GOD. [S]. Value = 11. See *Mahâdeva,* **Eleven** and *Rudra-Shiva.*

GREAT KINGS (The four). See *Chaturmahârâja.*

GREAT SACRIFICE. [S]. Value = 5. See *Mahâyajña* and **Five.**

GREAT SIN. [S]. Value = 5. See *Mahâpâpa* and **Five.**

GREAT YEAR OF BEROSSUS. Cosmic period mentioned in the work of the Babylonian astronomer Berossus (fourth – third century BCE), 432,000 years long. There is an "arithmetical" relationship between this "Great year" and the Indian cosmic cycles called *yugas, because it corresponds: to a *kaliyuga, to 1/10 of a *mahâyuga, and to 2/5 of a *yugapâda. However, it is not known if there is a historical link between this "year" and the Indian *yugas. See **Great year of Heraclitus** and *Yuga* **(Astronomical speculations).**

GREAT YEAR OF HERACLITUS. Cosmic period of the ancient Mediterranean world which, according to Censorinus, is 10,800 years long. There is a mathematical relationship between this "Great year" and the Indian cosmic cycles known as *yugas, because it corresponds: to 1/40 of a *kaliyuga, to 1/100 of a *yugapâda and to 1/400 of a *mahâyuga. However, it is not known if there is a historical link between this "year" and the Indian *yugas. See **Great year of Berossus**.

GUARDIAN OF THE HORIZONS. [S]. Value = 8. See *Lokapâla* and **Eight**.

GUARDIAN OF THE POINTS OF THE COMPASS. [S]. Value = 8. See *Lokapâla, Dikpâla* and **Eight**.

GUJARATI NUMERALS. Signs derived from *Brâhmî numerals, through the intermediary of Shunga, Shaka, Kushâna, Ândhra, Gupta, Nâgarî and Kûtilâ numerals. Currently in use in Gujarat State, on the Indian Ocean, between Bombay and the border of Pakistan. The corresponding system functions according to the place-value system and possesses zero (in the form of a little circle). See **Indian written numeral systems (Classification of)**. See also Fig. 24.8, 52 and 24.61 to 69.

GULPHA. [S]. Value = 2. "Ankle". This symbolism is due to the symmetry of this part of the body. See **Two**.

GUNA. [S]. Value = 3. "Merit", "Quality", "primordial property". Philosophically, the *gunas* are the qualities or conditions of existence which make up Nature. They are in a state of rest when the qualities are in perfect equilibrium, and in a state of evolution when one or more of them prevail over the others. According to the philosophy of the *Sâmkhya*, these qualities are composed of three natural "materials": *Sattva* (representing kindness, the pure essence of things), *Rajas* (active energy, passion), and *Tamas* (passivity, apathy). Here the word is synonymous with *Triguna*, "three qualities", "three primordial properties" [see Frédéric, *Dictionnaire* (1987)]. See *Triguna* and **Three**.

GUNA. [S]. Value = 6. "Merit", "quality", "primordial property". The allusion here is to *shadâyatana*, the "six *gunas*" of Buddhist philosophy. This value was only acquired relatively recently. See *Shâdayatana* and **Six**.

GUNANA. Term used in arithmetic to mean multiplication. Other synonyms: *hanana*, *vadha*, *kshaya*, etc. (which literally mean: "destroy", "kill", etc., in allusion to the successive erasing of the results of the partial products whilst carrying out calculations on sand or using chalk on a board). See **Calculation, *Pâtiganita*, and Indian methods of calculation**. See also Chapter 25.

GUNDHIKA. Name given to the number ten to the power twenty-three. See **Names of numbers** and **High numbers**.

Source: *Lalitavistara Sûtra* (before 308 CE).

GUPTA (Calendar). A calendar (with normal years) established by Chandragupta I beginning in 320 CE. To find the date in the universal calendar which corresponds to one expressed in Gupta years, add 320 to the Gupta date. Sometimes the first year of this calendar is given as 318 or 319. It was used during the Gupta dynasty. In Central India and Nepal, it persisted until the thirteenth century. See **Indian calendars**.

GUPTA NUMERALS. Signs derived from *Brâhmî numerals, through the intermediary of Shunga, Shaka, Kushâna and Ândhra numerals. Contemporaries of the Gupta dynasty (inscriptions of Parivrajaka and Uchchakalpa). The corresponding system does not use the place-value system or zero. These numerals were the ancestors of Nâgarî, Shâradâ and Siddham notations. See Fig. 24.38 and 24.70. For notations derived from Gupta numerals, see Fig. 24.52. For their graphical evolution, see Fig. 24.61 to 69. See **Indian written numeral systems (Classification of)**.

GURKHALI NUMERALS. See **Nepali numerals**.

GURUMUKHI NUMERALS. Signs derived from *Brâhmî numerals, through the intermediary of Shunga, Shaka, Kushâna, Ândhra, Gupta and Shâradâ numerals, and constituting a sort of mixture of Sindhi and Punjabi numerals. Once used by the merchants of Shikarpur and Sukkur. (These merchants also used Sindhi or Punjabi numerals, as well as the eastern Arabic "Hindi" numerals.) The corresponding system functions according to the place-value system and possesses zero (in the form of a little circle). See Fig. 24.7. See also **Indian written numeral systems (Classification of)** and Fig. 24.52 and 24.61 to 69.

H

HALF OF THE BEYOND. As a representation of the numbers ten to the power twelve, ten to the power seventeen and ten to the power eighteen. See *Parârdha*.

HALF OF THE MONTH. [S]. Value = 2. See *Paksha*.

HALF OF THE MONTH. [S]. Value = 15. See *Paksha*.

HAND. [S]. Value = 2. See *Kara* and **Two**.

HARA. [S]. Value = 11. One of the names of *Shiva who is an emanation of *Rudra, the symbolic value of which is eleven. See *Rudra-Shiva* and **Eleven**.

HARANAYANA. [S]. Value = 3. The "eyes of *Hara". See *Haranetra*.

HARANETRA. [S]. Value = 3. "Eyes of *Hara". *Shiva, who has a multitude of names and attributes, one of which is *Hara*, often represented with a third eye in his forehead, which is meant to symbolise perfect knowledge. From which: *Haranetra* = 3. See **Three**.

HARIBÂHU. [S]. Value = 4. "Arms of Hari". *Hari* (literally "he who removes sin") is one of the names for *Vishnu, who is always represented as having four arms.

HARIDATTA. Indian astronomer c. 850 CE. Notably, his works include *Grahachâranibandhana*, in which he tells of the fruit of his invention: a system of numerical notation which uses the letters of the Indian alphabet. This is based on the place-value system and a zero (always expressed by one of two letters). This system is called *katapayâdi*: the first ever alphabetical positional number system [see Sarma (1954)]. See **Katapayâdi numeration**, and **Indian Mathematics (The history of)**.

HARSHAKÂLA (Calendar). Calendar beginning in the year 606 CE, created by Harshavardhana, King of Kanauj and Thaneshvar. To find the date in the universal calendar which corresponds to one expressed in Harshakâla years, add 606 to the Harshakâla date. This calendar was only used during the reign of Harshavardhana and for a short time afterwards in Nepal. See **Indian calendars**.

HÂRYA. "Dividend" (in the mathematical sense). See *Bhâjya*.

HASTIN. [S]. Value = 8. "Elephant". See *Diggaja* and **Eight**.

HEADS OF RÂVANA. [S]. Value = 10. See *Râvanashiras* and **Twenty**.

HEADS OF RUDRA. See *Rudrâsya*.

HEGIRA (Calendar of the). See *Hijra*.

HELL. Value = 7. See *Pâtâla*.

HEMADRÎ. One of the names of *Mount Meru.

HETUHILA. Name given to the number ten to the power thirty-one. See **Names of numbers** and **High numbers**.
Source: *Lalitavistara Sûtra* (before 308 CE).

HETVINDRIYA. Name given to the number ten to the power thirty-five. See **Names of numbers** and **High numbers**.
Source: *Lalitavistara Sûtra* (before 308 CE).

HIGH NUMBERS. Early in Indian civilisation, there was a sort of "craze" for high numbers. *Sanskrit numeration lent itself admirably to the expression of high numbers because it possessed a specific name for each power of ten. There are numerous examples to be found, not only in works on mathematics, but also in those concerning astronomy, cosmology, grammar, religion, legends and mythology. This proves that these names were not in everyday use in India, but rather they were familiar in learned circles, at least as early as the beginning of the Common Era. See **Names of numbers**.

In the naming of high numbers, these texts generally reached the highest numbers that were used in calculations. Thus each of the ascending powers of ten up to a *billion (ten to the power twelve), or even up to *quadrillion (ten to the power 18) were named. In cosmological texts, however (especially those developed by members of the religious cult of *Jaina, such as the *Anuyogadvâra Sûtra*), this limit was pushed much further, bearing witness to the extraordinary fertility of Indian imagination. The Jainas attempted to define their vision of an eternal and infinite universe; thus they undertook impressive arithmetical speculations, which always involve extremely high numbers, equal to or higher than numbers such as ten to the power 190 or ten to the power 250.

This obsession with high numbers is also found in *Vyâkarana, a famous Pali grammar of Kâchchâyana, and in the legend of Buddha, related in the *Lalitavistara Sûtra*, which juggles with numbers as high as ten to the power 421. At first glance childish, this passion for high numbers can tell us something about the high conceptual level achieved early on by Indian arithmeticians. It led the Indians not only to expand the limits of the "calculable", physical world, but also and above all to conceive of the notion of infinity, long before the Western world. See **Googol** and all other entries entitled **High numbers** as well as those entitled **Infinity**.

HIGH NUMBERS. Here is a (non-exhaustive) alphabetical list of Sanskrit words which represent high numbers:*Abab (= 10^{17}), *Ababa (= 10^{77}), *Abbuda (= 10^{56}), *Abja (= 10^{9}), *Ahaha (= 10^{70}), *Akkhobhini (= 10^{42}), *Akshiti (= 10^{15}), *Ananta (= 10^{13}), *Anta (= 10^{11}), *Antya (= 10^{12}), *Antya (= 10^{15}), *Antya (= 10^{16}), *Arbuda (= 10^{7}), *Arbuda(= 10^{8}), *Arbuda (= 10^{10}), *Asankhyeya (= 10^{140}), *Atata (= 10^{84}), *Attata (= 10^{19}), *Ayuta (= 10^{4}), *Ayuta (= 10^{9}), *Bahula (= 10^{23}), *Bindu (= 10^{49}), *Dashakoti (= 10^{8}), *Dashalaksha (= 10^{6}), *Dashasahasra (= 10^{4}), *Dhvajâgravati (= 10^{99}), *Dhvajâgranshâmani (= 10^{145}), *Gananâgati (= 10^{39}), *Gundhika (= 10^{23}), *Hetuhila (= 10^{31}), *Hetvindriya (= 10^{35}), *Jaladhi (= 10^{14}), *Kankara (= 10^{13}), *Karahu (= 10^{33}), *Kathâna (= 10^{119}), *Khamba (= 10^{13}), *Kharva (= 10^{10}), *Kharva (=10^{12}), *Kharva (= 10^{39}), *Koti (= 10^{7}), *Kotippakoti (= 10^{21}), *Kshiti (= 10^{20}), *Kshobha (= 10^{22}) *Kshobhya (= 10^{17}), *Kshoni (= 10^{16}), *Kumud (=10^{21}), *Kumuda (= 10^{105}), *Lakh (= 10^{5}), *Lakkha (= 10^{5}), *Laksha (= 10^{5}), *Madhya (= 10^{10}), *Madhya (= 10^{11}), *Madhya (= 10^{15}), *Madhya (= 10^{16}), *Mahâbja (= 10^{12}), *Mahâkathâna (= 10^{126}), *Mahâkharva (= 10^{13}), *Mahâkshiti (= 10^{21}), *Mahâkshobha (= 10^{23}), *Mahâkshoni (= 10^{17}), *Mahâpadma (= 10^{12}), *Mahâpadma (=10^{15}), *Mahâpadma (= 10^{34}), *Mahâsaroja (= 10^{12}), *Mahâshankha (= 10^{19}), *Mahâvrindâ (= 10^{22}), *Mudrâbala (= 10^{43}), *Nâgabala (= 10^{25}), *Nahut (= 10^{9}), *Nahuta (= 10^{28}), *Nikharva (= 10^{9}), *Nikharva (= 10^{11}), *Nikharva (= 10^{13}), *Ninnahut (= 10^{11}), *Ninnahuta (= 10^{35}), *Nirabbuda (= 10^{63}), *Niravadya (= 10^{41}), *Niyuta (= 10^{5}), *Niyuta (= 10^{6}), *Niyuta (= 10^{11}), *Nyarbuda (10^{8}), *Nyarbuda (= 10^{11}), *Padma (= 10^{9}), *Padma (= 10^{14}), *Padma (= 10^{29}), *Paduma (= 10^{119}), *Pakoti (= 10^{14}), *Parârdha (= 10^{12}), *Parârdha (= 10^{17}), *Pârâvâra (= 10^{14}), *Prayuta (= 10^{5}), *Prayuta (= 10^{6}), *Pundarîka (= 10^{27}), *Pundarîka (= 10^{112}), *Salila (= 10^{11}), *Samâptalambha (= 10^{37}), *Samudra (= 10^{9}), *Samudra (= 10^{10}), *Samudra (= 10^{14}), *Saritâpati (= 10^{14}), *Saroja (= 10^{9}), *Sarvabala (= 10^{45}), *Sarvajña (= 10^{49}), *Shankha (= 10^{12}), *Shankha (= 10^{13}), *Shankha (= 10^{18}), *Shanku (= 10^{13}), *Shatakoti (= 10^{9}), *Sogandhika (= 10^{91}), *Tallakshana (= 10^{53}), *Titilambha (= 10^{27}), *Uppala (= 10^{98}), *Utpala (= 10^{25}), *Utsanga (= 10^{21}), *Vâdava (= 10^{9}), *Vâdava (= 10^{14}), *Vibhutangamâ (= 10^{51}), *Visamjñagati (= 10^{47}), *Viskhamba (= 10^{15}), *Vivaha (= 10^{19}), *Vivara (= 10^{15}), *Vrindâ (= 10^{9}), *Vrindâ (= 10^{17}), *Vyarbuda (= 10^{8}), *Vyavasthânaprajñapati (= 10^{29}).

HIGH NUMBERS (SYMBOLIC MEANING OF). The preceding list is enough to give the reader some idea of the arithmetical genius of the Indian scholars. However, it only gives the mathematical value of the words in question, and neither their literal nor their symbolic meaning. The following (summary) explanations should give the reader a precise idea of the associations of ideas and the symbolism which is implied in this unique terminology.

Firstly, the word *padma (which represents the number ten to the power nine, ten to the power fourteen or ten to the power twenty-nine) literally means "*lotus". However, there is another word *paduma (which can represent ten to the power twenty-nine as well as ten to the power 119), as well as the terms *utpala (ten to the power twenty-five), *uppala (ten to the power ninety-eight), *pundarîka (ten to the power twenty-seven or ten to the power 112), *kumud (ten to the power twenty-one) and *kumuda (ten to the power 105), which also mean "lotus". The reasoning behind this metaphor lies in the fact that the lotus flower is the best-known and most widespread symbol in the whole of Asia. "Born of miry waters, this flower maintains a flat and immaculate purity above the water in all its splendour. Thus it became the symbol of a pure spirit leaving the impure matter of the body. Nearly all Indian philosophies and religions adopted the flower as their symbol, and its diffusion throughout Asia took place due to the spread of Buddhism, even though it is almost certain that the lotus flower was already used as a symbol by many peoples before the advent of Buddhism. Indian philosophers saw it as the very image of divinity, which remains intact and is never soiled by the troubled waters of this world. The closed flower of the lotus, in the shape of an egg, represents the seed of creation which rose out of the primordial waters, and as it opens all the latent possibilities contained within the seed develop in the light. This is why, in Hindu iconography, *Brahma is seen to be born from a lotus flower growing out of the navel of *Vishnu who lies upon the serpent *Ananta who is coiled up on the primordial waters which represent infinity [see Fig. D. 1, p. 446, of the entry entitled **Ananta**]. Similarly, this flower is the 'throne' of Buddha and most of his manifestations: here the lotus represents the *bodhi*, the 'nature of Buddha' which remains pure when it leaves the *samsâra, the cycle of rebirth of this world. A whole symbolic system developed around the lotus flower, which takes into account its colour, the number of petals it possesses, and whether it is open, half-closed or in bud. In the *Kundalinî Yoga*, it is the stem of the lotus which forms the spiritual axis of the world and upon which the lotus flowers become steadily more fully open and the number of petals becomes greater and greater up until the highest illumination where the corolla, which has become divine and of unequalled brilliance, possesses a thousand petals," [see Frédéric, *Le Lotus*, (1987)].

Indian art also seized upon this flower and it has been widely represented in painting as well as in sculpture. We can appreciate why Indian mathematicians, with their perfect command of symbolism, also adopted the lotus flower and all its corresponding mysticism in order to express in Sanskrit gigantic quantities. The *padma* (or *paduma*), is the pink lotus. As well as the purity which it represents, this flower, to the Indian mind, symbolises the highest divinity, as well as innate reason. Written *Padmâ* (with a long a), the pink lotus flower figures amongst the names of Lakshmî as feminine energy (*shakti*) of *Vishnu. In the word *sahasrapadma (the "thousand-petalled lotus"), it represents the "third eye", that of perfect Knowledge; it also represents the superior illumination, of the divine corolla, of unequalled brilliance, flowering on the axis of the spirituality of the world as a thousand-petalled pink lotus [see Frédéric, *Le Lotus* (1987).

It is probably the idea of absolute and divine perfection which gave *padma* a value as elevated as ten to the power 119. However, it did not initially represent such a quantity. Initially, as the Indian mathematicians were gradually becoming more accustomed to dealing with large quantities, and with the idea of perfection and absolution, they probably gave the lotus the value of ten to the power nine. Its value gradually increased as it was successively attributed the values of ten to the power fourteen, then ten to the power twenty-nine and finally ten to the power 119. The flower in question here possesses a thousand petals.

Padmâ is also the name of one of the branches of the Ganges at its mouth. It is interesting to note that this swampy delta with branches radiating from it, like the petals of a lotus flower, is often referred to as *jahnavîvaktra*, literally the "mouths of *Jâhnavî* (the Ganges)". The name, as an ordinary numerical symbol, corresponds to the number one *thousand, precisely because of the hundreds of branches which characterise it. Moreover *Vishnu is associated with the thousand-petalled lotus and has a thousand attributes,

amongst which are: *Sahasranâma, "the thousand names (of Vishnu)".What is more, in Hindu mythology, it is from the "feet of *Vishnu" (*Vishnupada) that the celestial *Gangâ* (the Ganges) sprang, considered to be the "divine mother of India". Thus this flower was associated with both the name and the concept of thousand (*sahasra). However, the terminology which was used had recourse to a secondary symbolism: *thousand was no longer really a numerical concept; its figurative sense was the idea of plurality, of "vast number". Like the word *padma* which initially only meant ten to the power nine, the name of this vast number grew to have the value of ten to the power twelve; which then gave ten to the power fifteen the name *mahâpadma, which means "great lotus". Through a similar association of ideas, the word *shankha, which means "sea conch", was assigned to the numbers ten to the power thirteen and ten to the power eighteen. This symbolises certain Buddhist or Hindu divinities (such as *Vishnu or *Varuna for example). In India, the conch represents riches, good fortune and beauty. This can be associated with the image of a diamond which is pure and beautiful in equal measures. As the diamond is everlasting and shines with a thousand fires, the beauty represented by the conch can be compared to this precious stone. Thus, for some Indian arithmeticians, *shanku ("diamond") is equal to ten to the power thirteen.

Returning to the conch, one of the attributes of Vishnu is expressed by the Sanskrit word for conch (*shankha*), which symbolises the conservative principle of the revelation, due to the fact that the sound and the pearl are conserved within the shell. The conch is also the symbol of abundance, fertility and fecundity, which are precisely the characteristics of the sea from which the shell comes. The shell is also related to the water. This explains the connection with *Varuna, the lord of the Waters. Here there is also the connection with the lotus, which also symbolises not only abundance, but also and above all, in the eyes of humans, a limitless expanse. This is why the word *samudra, which means "ocean", came to mean, in this symbolic terminology, the number ten to the power nine, ten to the power ten or ten to the power fourteen. This is the reason why *Bhâskarâchârya used the word *jaladhi, which also means *ocean, to denote the number ten to the power fourteen. The mathematician must have chosen this word because he was writing in verse and in Sanskrit and he chose his words in order to achieve the desired effect, using an almost limitless choice of synonyms, following the exacting rules of Sanskrit versification. See **Poetry and writing of numbers**.

The Indians also see *samudra* as the waves of superior consciousness which bring immortality to mere mortals; eternal existence and infinity. This explains the connection with *soma*, which is the *amrita*, the "drink of immortality". *Soma* can also mean *moon*, which became a metaphor for a goblet full of the heady brew. Thus it was quite natural that this star should also be associated with incalculably vast quantity. So *abja* ("moon") and *mahâbja* ("great moon"), came to represent numbers such as ten to the power nine or a billion (ten to the power twelve). As well as being connected to water, the conch is symbolically related to the moon, as it is white, the colour of the full moon. This gives double justification to this association of ideas. The apparently limitless expanse of the sea is the most immense thing in the "terrestial world". As the *earth is called *kshiti* or *kshoni*, (also referred to as *mahâkshiti* and *mahâkshoni*, meaning "great earth") we can see how these words came to represent such immense values as ten to the power sixteen, ten to the power seventeen, ten to the power twenty or ten to the power twenty-one.

The Sanskrit word *abhabâgamana* means "the unachievable". The term *ababa* is used in the *Lalitavistara Sûtra* (before 308 CE), to express the quantity ten to the power seventy, and it is possible that this is an abbreviation of *abhabâgamana*. The word *ahaha*, used in the same text to express ten to the power seventy-seven, is almost definitely an abbreviation of the word *abaharaka*, which means extravagant and is similar to our word "abracadabra".

Pundarîka means white lotus with eight petals and is the symbol of spiritual and mental perfection. The term is generally reserved for esoteric divinities, and was dedicated to Shikhin, the second Buddha of the past. This lotus has the same number of petals as the eight directions of space, the eight points of the compass and the eight elephants (*diggaja*) of Hindu cosmogony. Amongst these elephants figures *Pundarîka* who guards the "southeastern horizon" of the universe for the god of fire *Agni*. The "southwestern horizon" is guarded by the elephant *Kumuda*, whose real name also means "lotus", but this time refers to the white-pink flower. The sun is not far from this lotus, as it is situated at the axis of the eight horizons. The elephant *Kumuda* also

symbolises the Sun-god *Sûrya*, who is often denoted by names which evoke the idea of a thousand or the lotus flower: *Sahasrâmshu* ("Thousand of the Shining", in allusion to its rays), *Sahasrakirana* ("Thousand rays"), *Sahasrabhûja* ("Thousand arms") and *Padmapâni* ("Lotus carrier"). Thus the Indians expressed the omniscience of this god and his incalculable powers.

If the sun is a source of light, warmth and life, then like the petals of a lotus, its rays must also contain the spiritual influences received by all things on earth. This is why the names of *Pundarîka*, *Kumud* and *Kumuda* came to represent such vast quantities as ten to the power twenty-one, ten to the power 105 and ten to the power 112. Indian mathematicians soon took the step from the Sun to the canopy of heaven.*Parârdha*, one of the names attributed to ten to the power twelve or ten to the power seventeen, comes from *para*, "beyond", and from *ardha*, "half of beyond". Due to a similar association of ideas, *Madhya*, "middle" (representing the "middle of the beyond") was used to express such numbers as ten to the power ten, ten to the power eleven, ten to the power fifteen or ten to the power sixteen.

According to the Indians, *parârdha* is the spiritual half of the path which leads to death. It is the same as *devayâna*, the "path of the gods", which, according to the *Vedas*, was one of the two possibilities offered to human souls after death, *parârdha* leading to deliverance from *samsâra* or cycle of reincarnation. On reaching the sky, one cannot fail to achieve divine transcendence, power, durability and sanctity, thus touching upon the incalculable in terms of mental and physical perfection, represented by the word *pundarîka*. Intelligence, wisdom and the triumph of the mind over the senses is represented by *utpala*, the blue, half open lotus. This is why these words came to be worth such quantities as ten to the power twenty-five, ten to the power twenty-seven, ten to the power ninety-eight or even ten to the power 112.

No living being can attain divine transcendence, which is conveyed by the Inaccessible, the Absolute, and the Ineffable. This is similar to the "incalculable", *Asamkhyeya* (or *Asankhyeya*), "that which cannot be counted". According to the *Lalitavistara Sûtra*, this word means "the sum of all the drops of rain which, in ten thousand years, would fall each day on all the worlds". In other words, this is the "highest number imaginable". *Asankhyeya* is the term used to express the number ten to the power 140. The terminology used here deals symbolically with the notion of eternity. This is explained by al-Biruni in the thirty-third chapter of his work on India, where he gives this word the value of ten to the power seventeen (Fig. 24. 81):

"The name of the eighteenth order is *parârdha*, which means half of the sky, or more precisely, half of what is above. The *reason* for this is that if a period of time is made up of *kalpas* (cycles of 4,320,000,000 years), a unit of this order is a day of the *purusha* (= one day of the Supreme Being, namely *Brahma). As there is nothing beyond the sky, this is the largest body. Half of the biggest nychthemer (= the longest possible day) is similar to the other; in doubling it, we obtain a "night" with a "day", and thus complete the biggest nychtemer. It is certain that *parârdha* is from *para*, which means the whole sky". Ref: *Kitab fi tahqiq i ma li'l hind* (1030 CE).

Al-Biruni also tells that "according to some, the day of *purusha* (the day of Brahma) is made up of a *parârdha* and a *kalpa*". As a *kalpa* is 4,320,000,000,000 years, this "day" corresponds to: 432,000,000,000,000,000,000,000,000 (four hundred and thiry-two sextillion) years. See **Day of Brahma**.

Traditional brahmanic cosmogony more modestly attributes the length of 4,320,000,000 human years to the "*day of Brahma". This is also what it refers to as *asankhyeya*, the "incalculable". The *Bhagavad Gîtâ* assigns 311,040,000,000,000 years to this word. In a commentary on the text is written: "This formidable longevity, to us infinite, only represents a zero in the tide of eternity." The word *Padmabhûta*, "born from the lotus (with a thousand petals)" is an attribute of *Brahma. Brahma is said to have been born from the lotus which grew out of Vishnu's navel as he lay on the serpent *Ananta floating on the primordial waters (see Fig. D. 1 and **Ananta**). Another attribute of Brahma is *Padmanâbha*, which means "the one whose navel is the lotus (Vishnu)". This is why the word *ananta*, which means "infinity" and "eternity", has sometimes been used to express the number ten to the power thirteen, in memory of distant times when the Sanskrit names for numbers went no further than ten to the power twelve. See **Antya** (first entry, note in the reference).

Ananta is another name for *Shesha, the king of the *nâga who resides in the lower regions of the *Pâtâla. It is an immense serpent with a thousand heads, who serves as a seat for *Vishnu as he rests amongst creatures between two periods of creation. At the end of each *kalpa*, he spits the fire which destroys creation. Considered as the "Remainder" (*Shesha), the "Vestige" of destroyed universes and as the seed of all future creations, he represents immensity and space, and infinity and eternity all at once. See **Serpent (Symbolism of the)**.

The words *eternal* and *infinity* mean "that which has no end, that which never ends, that which can never be reached". This leads to ideas of absoluteness and totality, in the strongest sense of the terms. Words such as *Sarvabala* and *Sarvajña*, formed with the Sanskrit adjective *sarva*, meaning "everything" or "totality", have respectively been associated with numbers as high as ten to the power forty-five or ten to the power forty-nine. Moreover, *Sarvajñata* expresses omniscience in Buddhism, the knowledge of Buddha, one of his fundamental attributes. In the Buddhism of Mahâyâna, this word has even acquired the meaning "the knowledge of all the *dharmas* and of their true nature"; a nature which, in essence, is *shûnyatâ*, vacuity. According to the Indians, vacuity materialises in the centre of the *vajra*, the "diamond", symbol of what remains once appearances have disappeared. The *vajra* is also the projectile "of a thousand points", reputed to never miss its mark, and made of bronze by Tvashtri for *Indra; but this is above all a religious instrument, symbol of the *linga* and divine power, indicating the stability and resoluteness of mind as well as its indestructible character. And as vacuity also means the void for Indians (also caused as much by nothingness, absence or insignificance as by the unthought, immateriality, insubstantiality and non-being), this explains why the *bindu*, the "point" (destined to become a numerical symbol and a graphical representation for zero), represented, for Indian arithmetician-grammarians, a number as high as ten to the power forty-nine.

Before the discovery of infinity or zero, the *bindu* (the "point"), was the Indian symbol for the universe in its non-manifest form, thus that of a universe before its transformation into *rûpadhâtu*, the "world of appearances". For Indian scholars, "nothing" could be united with "everything", even before mathematics made these two concepts inverse notions of one another. See **Zero, Infinity** and **Indian Mathematics (The history of)**. See **Names of numbers, Numerical symbols, Arithmetical speculations, Cosmogonical speculations, *Sheshashîrsha*,**

Shûnya, Shûnyatâ, Indian atomism. See also Serpent (Symbolism of the).

HIJRA (Calendar). Arabic name for the Islamic calendar, which, according to tradition, begins on the 16 July 622 CE, day of the "Escape" or "Flight" (*hijra*, "Hegira") of the prophet Mohammed of Mecca to Medina. As the Muslim year has twelve lunar months each twenty-nine or thirty days long, making up a year of 354 days, this calendar must be rectified by the addition of eleven intercalary years of 355 days every thirty years to catch up with normal solar years. To obtain a date in the universal calendar from one in the Hegira, multiply the latter by 0.97 and add 625.5 to the result. For example: the start of the year of the Hegira 1130 corresponds to July 1677:

1130 *Hegira* = (1130 × 0.97) + 625.5 = 1721.6. Inversely, to find a date of the Hegira from a date in the universal calendar, subtract 625.5 from the latter, then multiply the result by 1.0307 and add 0.46. If there are decimals remaining, add a unit.

For example, to convert the year 1982 into the Hegira calendar, proceed as follows: 1st stage: 1982 − 625.5 = 1356.5.

2nd stage: 1356.5 × 1.0307 = 1398.14.
3rd stage: 1398.14 + 0.46 = 1398.6.

In rounding off this result, the year of the Hegira 1399 is obtained [see Frédéric, *Dictionnaire* (1987)]. See **Indian calendars.**

HINDI NUMERALS. See **Eastern Arabic numerals.**

HINDU RELIGION. See **Indian religions and philosophies.**

HINDUISM. See **Indian religions and philosophies.**

HOLE. [S]. Value = 0. See *Randhra*.

HOLE. [S]. Value = 9. See *Chhidra, Randhra* and **Nine.**

HORIZON. [S]. Value = 4. See *Dish* and **Four.**

HORIZON. [S]. Value = 8. See *Dish* and **Eight.**

HORIZON. [S]. Value = 10. See *Dish* and **Ten.**

HORSE. [S]. Value = 7. See *Ashva* and **Seven.**

HORSEMEN. [S]. Value = 2. See *Ashvin* and **Two.**

HOTRI. [S]. Value = 3. "Fire". See *Agni* and **Three.**

HUMAN. [S]. Value = 14. In the sense of progenitor of the human race. See *Manu* and **Fourteen.**

HUNDRED. Ordinary Sanskrit name: **shata.* Corresponding numerical symbols: *Abjadala,*

*Dhârtarâshtra, Purushâyus and *Shakrayajña.* See **Numerical Symbols.**

HUNDRED BILLION (= ten to the power fourteen). See *Jaladhi, Padma, Pakoti, Pârâvâra, Samudra, Saritâpati, Vâdava.* See also **Names of numbers.**

HUNDRED MILLION (= ten to the power eight). See *Arbuda, Dashakoti, Nyarbuda, Vyarbuda.* See also **Names of numbers.**

HUNDRED THOUSAND BILLION (UK) (= ten to the power seventeen). See *Abab, Kshobhya, Mahâkshoni, Parârdha, Vrindâ.* See also **Names of numbers.**

HUNDRED THOUSAND MILLION (UK) (= ten to the power eleven). See *Anta, Madhya, Nikharva, Ninnahut, Niyuta, Nyarbuda, Salila.* See also **Names of numbers.**

HUNDRED THOUSAND TRILLION (UK) (ten to the power twenty-three). See *Bahula, Gundhika, Mahâkshobha.* See also **Names of numbers.**

HUNDRED TRILLION (UK) (= ten to the power twenty). See *Kshiti.* See also **Names of numbers.**

HUTÂSHANA. [S]. Value = 3. "Fire". See *Agni* and **Three.**

I

IMMENSE. [S]. Value = 1. See *Prithivî* and **One.**

IMPURITIES (The five). See *Pañchaklesha.*

IMRAJÎ (Calendar). See *Kristâbda.*

INACCESSIBLE. [S]. Value = 9. See *Durgâ* and **Nine.**

INCARNATION. [S]. Value = 10. See *Avatâra* and **Ten.**

INDESTRUCTIBLE. [S]. Value = 1. See *Âkshara* and **One.**

INDETERMINATE EQUATION (Analysis of an). See *Kuttaganita* and **Indian mathematics (History of).**

INDETERMINATE. See **Infinity.**

INDIA (States of the present-day Indian union). See Fig. 24. 27.

INDIAN ARITHMETIC. Alphabetical list of words related to this discipline which appear as entries in this dictionary:*Arithmetic. *Bhâgahâra, *Bhâjya, *Bhinna, *Chhedana, *Dashaguna, *Dashagunâsamjña, *Dhûlikarma,

*Ekadasharâshika, *Gananâ, *Ganita, *Ghana, *Ghanamûla, *Gunana, *High numbers, *Indian mathematics (history of), *Indian methods of calculation, *Infinity, *Kalâsavarnana, *Labhda, *Mudrâ, *Names of numbers, *Navarâshika, *Pañcharâshika, *Pañcha jâti, *Parikarma, *Pâti, *Pâtiganita, *Râshi, *Râshividyâ, *Samkalita, *Samkhyâna, *Samkhyeya, *Saptarâshika, *Sarvadhana, *Shatottaragananâ, *Shatottaraguna, *Shatottarasamjñâ, *Shesha, *Square roots (How Âryabhata calculated his), *Trairâshika, *Varga, *Vargamûla, *Vyastrairâshika, *Vyavakalita.

INDIAN ASTROLOGY. This discipline used to go by the name of *Jyotisha, which literally means the "science of the stars". But today this term is more commonly used to describe astronomy. This naming dates back to the time when astronomy was not yet considered to be a separate discipline from arithmetic and calculation.

Until early in the Common Era, astrology was often confused with astronomy, the latter at that time having no other objective than to serve the former. Knowledge of the stars and their movements was a method of predicting the future and determining the favourable dates and times of any given human action: consecration of a ritual sacrifice, commercial transactions, setting out on a voyage, etc. [see Frédéric (1994)]. Thus, when an individual was born, the astrologers, having determined the exact time and the position of the planets and the sun, established the horoscope of the newly-born infant, which was considered indispensable in ascertaining the child's birth chart.

*Varâhamihîra stands out as one of the most famous astrologers of Indian history. He lived in the sixth century CE, and is known principally for his work, *Pañchasiddhântikâ* (the *"Five *Astronomical Canons"*), which is dated c.575 CE. But he also wrote many works on astrology, divination and practical knowledge. The most important of these is *Brihatsamhitâ* (the "great compilation") which covers many subjects: descriptions of heavenly bodies, their movements and conjunctions, meteorological phenomena, indications of the omens these movements, conjunctions and phenomena represent, what action to take and operations to accomplish, signs to look for in humans, animals, precious stones, etc. [see Filliozat, in: HGS, I, pp. 167–8]. See **Indian astronomy (The history of), Ganita, Râshi, Tanu and Yuga**.

INDIAN ASTRONOMY (The history of). Here is an alphabetical list of terms related to this discipline which appear in this dictionary: *Âryabhata, *Astronomical canon, *Bhâskara, *Bhâskarâchârya, *Bhoja, *Bîja, *Brahmagupta, *Chaturyuga, *Ganita, *Govindasvâmin, *Great year of Berossus, *Great year of Heraclitus, *Haridatta, *Indian astrology, *Indianity (fundamental mechanisms of), *Indian mathematics (The history of), *Jyotisha, *Kaliyuga, *Kalpa, *Kamâlakara, *Karana, *Lalla, *Mahâyuga, *Nakshatra, *Nakshatravidyâ, *Nîlakanthasomayâjin, *Numerical symbols, *Parameshvara, *Putumanasomayâjin, *Samkhyâna, *Shankaranârâyana, *Shrîpati, *Siddhânta, *Tithi, *Varâmihira, *Yuga (definition), *Yuga (systems of calculating), *Yuga, (astronomical speculation on), *Yugapâda.

INDIAN ASTRONOMY (The history of). If we take the word "astronomy" in its wider sense, we can traditionally distinguish three principal periods. The first corresponds to the astronomy of era and ritual: a lunar-solar era devoid of any time-scale or era. The corresponding "material" is characterised by the *nakshatra, division of the sidereal sphere in twenty-seven or twenty-eight constellations or asterisms according to the twenty-seven or twenty-eight days of the sidereal revolution of the Moon. The planets (it is unlikely that they had all been discovered at this time) only played a very small part in divination. Between the third century BCE and the first century CE, elements and procedures of Babylonian astronomy appeared in Indian astronomy. A unity of time appeared called the *tithi, which is approximately the length of a day or nychthemer, and corresponds to one thirtieth of the synodic revolution of the Moon. It was at this time that the planets came to the fore in divination and were subjected to arithmetical calculations, based on their synodic revolutions. However, it was at the beginning of the sixth century CE that Indian astronomy underwent its most spectacular developments: *scientific astronomy* was inaugurated by the work of *Âryabhata, which dates back to c. 510 CE. From the outset, this astronomy was based on an astonishing speculation about the cosmic cycles called *yugas, of a very different nature from arithmetical cosmogonic speculations. This speculation involves astronomical elements, where the *mahâyuga (or *chaturyuga), a cycle of 4,320,000 years is presented as the period at the beginning and the end of which the nine elements (the sun, the moon, their apsis and node, and the planets) should be found in average perfect conjunction with the starting point of the longitudes. Thus the length of the revolutions, which had hitherto been considered commensurable, were subjected to common multiplication and general conjunctions.

It is precisely this which makes the speculation so surprising and audacious, because this fact is totally devoid of any physical value. These supposed general conjunctions confer absurd values upon average elements even by the approximative standards of ancient astronomy. However, thanks to a veritable paradox, it is this strange coupling of speculation and reality that enabled Billard to develop a powerful method of determining (with precision) hitherto unknown facts and a chronology which had been despaired of due to the unique conditions of the Indian astronomical text. It is interesting to note that the speculative elements of this astronomy have been as useful to contemporary historical science as they were once harmful. For more details, see: *Yuga* (cosmogonical speculation on), Indian astrology, Yuga (astronomical speculation about), **Indian mathematics (The history of)**, **Indianity (Fundamental mechanisms of)**. [See Billard (1971)].

INDIAN ATOMISM. See *Paramânu, Paramânu Raja* and *Paramabindu.*

INDIAN CALENDARS. India (which only adopted the universal calendar in 1947) has known a great many different eras during the course of its history. Certain eras, mythical or local, have existed on a very limited scale. Others, however, have become so widely used that they still exist today. The (non-exhaustive) list in the following entry gives an idea of the vast number of eras which have been used, and also allow for a better understanding of the elements of Chapter 24, where the documentation is often dated in one of these eras.

INDIAN CALENDARS. An alphabetical list of terms relating to these eras which can be found in this dictionary: *Bengalî San,* *Buddhashakarâja,* *Châlukya,* *Chhedi,* *Gupta,* *Harshakâla,* *Hijra,* *Kaliyuga,* *Kollam,* *Kristabda,* *Lakshamana,* *Laukikasamvat,* *Marâtha,* *Nepâlî,* *Parthian,* *Samvat,* *Seleucid,* *Shaka,* *Simhasamvat,* *Thâkurî,* *Vikrama,* *Vilâyatî,* *Vîrasamvat.* [See Cunningham (1913); Frédéric, *Dictionnaire,* (1987); Renou and Filliozat (1953)].

INDIAN COSMOGONIES AND COSMOLOGIES. Here is an alphabetical list of terms relating to these subjects which appear as entries in this dictionary: *Âditya,* *Adri,* *Anuyogadvâra Sûtra,* *Arithmetical-cosmogonic speculations,* *Ashtadiggaja,* *Avatâra,* *Bhuvana,* *Chaturânanavâdana,* *Chaturdvîpa,* *Chaturmahârâja,* *Chaturyuga,* *Cosmognic speculations,* *Dantin,* *Day of Brahma,* *Diggaja,* *Dikpâla,* *Dvaparayuga,* *Dvîpa,* *Gaja,* *Go,* *Graha,* *Hastin,* *High numbers* (the symbolic meaning of), *Indra,* *Jaina,* *Kala,* *Kaliyuga,* *Kalpa,* *Krita,* *Kritayuga,* *Kuñjara,* *Lokapâla,* *Mahâkalpa,* *Mahâyuga,* *Manu,* *Mount Meru,* *Ocean,* *Paksha,* *Râhu,* *Sâgara,* *Serpent* (Symbolism of the), *Shîrshaprahelikâ,* *Takshan,* *Tretâyuga,* *Tribhuvana,* *Triloka,* *Vaikuntha,* *Vasu,* *Vishnupada,* *Vishva,* *Vishvadeva,* *Yuga.*

INDIAN DIVINITIES. Here is an alphabetical list of terms relating to this theme, which can be found as entries in this dictionary:*Âditya,* *Agni,* *Amara,* *Âptya,* *Arka,* *Ashtamangala,* *Ashtamûrti,* *Ashva,* *Ashvin,* *Âtman,* *Avatâra,* *Bhânu,* *Bharga,* *Bhava,* *Bhuvana,* *Bîja,* *Brahma,* *Brahmâsya,* *Buddha* (Legend of), *Chandra,* *Chaturmahârâja,* *Chaturmukha,* *Dashabala,* *Dashabhûmi,* *Deva,* *Dhruva,* *Dikpâla,* *Divâkara,* *Durgâ,* *Dvâtrimshadvaralakshana,* *Dyumani,* *Ekachakra,* *Ekadanta,* *Ganesha,* *Go,* *High numbers* (Symbolic meaning of), *Hara,* *Haranayana,* *Haranetra,* *Haribâhu,* *Indra,* *Infinity* (Indian mythological representation of), *Îsha,* *Îshadrish,* *Îshvara,* *Jaina,* *Kâma,* *Kârttikeya,* *Kârttikeyâsya,* *Kâya,* *Keshava,* *Krishna,* *Kumârâsya,* *Kumâravadana,* *Lokapâla,* *Lotus,* *Mahâdeva,* *Mârtanda,* *Netra,* *Pañchabâna,* *Pañchânana,* *Parabrahman,* *Pâvana,* *Pinâkanayana,* *Pitâmaha,* *Ravi,* *Ravibâna,* *Rudra,* *Rudra-Shiva,* *Rudrâsya,* *Sâgara,* *Sahasrâmshu,* *Sahasrakirana,* *Sahasrâksha,* *Sahasranâma,* *Shakra,* *Shakti,* *Shatarûpâ,* *Shikhin,* *Shiva,* *Shukranetra,* *Shûla,* *Shûlin,* *Sura,* *Sûrya,* *Tapana,* *Triambaka,* *Tribhuvaneshvara,* *Tripurasundarî,* *Trishûlâ,* *Trivarna,* *Tryakshamukha,* *Tryambaka,* *Turîya,* *Vaikuntha,* *Vaishvanara,* *Varuna,* *Vasu,* *Vâyu,* *Vrindâ,* *Yama.*

INDIAN DOCUMENTATION (Pitfalls of). The purpose of this entry is to warn readers about texts which have absolutely no historical worth whatsoever, which contemporary Indianists – doubtless through bias towards material or excessive admiration of Indian culture – have put forward, due to shocking carelessness and lack of objectivity, in order to claim that the invention of zero and Indian positional numeration date back to the most ancient times. These documents are either fakes, or works resulting from recent compilations, or even ancient texts which successive generations reproduced whilst constantly correcting and revising them over the course of time.

Amongst these documents figures the manuscript of Bakshâlî, discovered in 1881 in the village of Bakshâlî in Gandhara, near Peshawar, in present-day Pakistan. The author of this mathematical text is unknown. It is written in Sanskrit (in verse and prose) and is mainly concerned with algebraic problems. This document is interesting from the point of view of the history of Indian numeration because it contains many examples of numbers written using the sign zero and the *place-value system, as well as several numerical entries expressed in *numerical symbols. According to certain historians of mathematics, this manuscript dates back to "the fourth, or possibly even the second century CE". This document undeniably constitutes an invaluable source of information about *Indian mathematics, but the manuscript itself, in its present form, could not possibly be as old as is claimed. The reason for this is that the numerals, like the characters used for writing, are written in the *Shâradâ* style, of which we know both the origin and the date of its first developments. See **Indian styles of writing** and **Shâradâ numerals**. See also Fig. 24. 38 and 40A.

To give the second or fourth century CE as the date of this document would be an evident contradiction; it would mean that a northern derivative of Gupta writing had been developed two or three centuries before Gupta writing itself appeared. Gupta only began to evolve into the Shâradâ style around the ninth century CE. In other words, the Bakshâlî manuscript cannot have been written earlier than the ninth century CE. However, in the light of certain characteristic indications, it could not have been written any later than the twelfth century CE. Nevertheless, when certain details are considered, which probably reveal archaisms of styles, terminologies and mathematical formulations, it seems likely that the manuscript in its present-day form constitutes the commentary or the copy of an anterior mathematical work. See **Shâradâ numerals**, **Indian styles of writing** and **Indian mathematics (The history of)**.

Other so-called "proofs" put forward to demonstrate that zero and the place-value system were discovered well before the Common Era include the texts of the *Purânas (particularly *Agnipurâna*, *Shivapurâna* and *Bhavishyapurâna*).

In the *Purânas*, great importance is placed in decimal numeration. Thus, in *Agnipurâna*, the eighth text, during an explanation of the place-value system, it is written that "after the place of the units, the value of each place (*sthana*) is ten times that of the preceding place". Similarly, in the *Shivapurâna*, it is explained that usually "there are eighteen positions (*sthana*) for calculation", the text also pointing out that "the Sages say that in this way, the number of places can also be equal to hundreds". These cosmological-legendary texts have often been dated from the fourth century BCE, and some have even been dated as far back as 2000 years BCE. These dates, however, are totally unrealistic, because these texts are from diverse sources and they are the fruit of constant reworkings carried out within an interval of time oscillating between the sixth and the twelfth centuries CE.

In fact, the *Purânas* only seem to have become part of traditional Indian writings after the twelfth century CE. This is a characteristic trait of the Indian mentality which, in order to give more weight to explanations of mythology and legends and to support its tendency to sanctify, immortalise and distort certain elements of knowledge, often invokes an authority from scripture which assumes antiquity. See **Indianity**. Of course these texts do stem from a relatively ancient source; but this source, which has accrued diverse rubrics in quite recent times, has been steadily and constantly reworked.

Here is a typical passage: *ravivâre cha sande cha phâlgune chaiva pharvarî shashtish cha siksatî jñeyâ tad udâhâram îdrisham*

Translation: Namely, for example, that *ravivâre* (Sunday) means *sande*, [the month of] *phâlguna* (February) *pharvâri*, and sixty *siksatî*. Ref: *Bhavishyapurâna* (III, *Pratisargaparvan*, I, line 37); ed. Shrivenkateshvar, Bombay, 1959, p. 423) (Personal communication of Billard). The text refers to a "barbaric language" which is none other than Old English! This would suggest that the English race already existed four thousand years ago, and were contemporaries of the Sumerians. This demonstrates just how far biased authors can go with their unreliable dating of documents. The above line (which was doubtless added at the time of the British domination of India) is clearly out of context. Thus we can see the difficulties we are faced with when dealing with Indian documentation.

[See Datta, in BCMS, XXI, pp. 21ff.; IA, XVII, pp. 33–48; Datta and Singh (1938); Kaye, *Bakhshâlî* (1924); Smith and Karpinski (1911)].

INDIAN MANUSCRIPTS (First material of). See *Pâtîganita* and **Indian styles of writing (Material of).**

INDIAN MATHEMATICS (The history of). What we know of this discipline only really dates back to the beginning of the Common Era, as no documents written in Vedic times survive. Of course this does not mean that Indian mathematical activities only commenced at this time. However, vital information about geometry can be found from this time in the *kalpasûtra*, a collection of Brahmanic rites including the *Shulvasûtra* (or "Aphorisms on lines"), dedicated to the description of the rules of construction for altars and the measurements of sacrificial altars.

Three versions of these texts exist: these are called *Baudhâyana*, *Apastamba* and *Kâtyâyana*. The best known is the *Apastamba* version, in which a similar statement to Pythagoras's theorem can be found as well as some problems similar to those in *Elements* by Euclid. Thus, to build altars of a predetermined size, a square equal to the sum or to the difference of two others had to be built. The altars, which were constructed out of bricks, had to be constructed in certain dimensions and with a determined number of bricks, and in certain cases had to undergo transformations to increase their surface area by a quantity specified in advance.

Some historians think that Indian mathematics of this era only constituted a utilitarian science. However, there is no evidence to prove this theory. As documentation currently stands, only the obtained mathematical results are known. The concise and essentially technical style of the texts in question did not leave room for even a summary description of the corresponding reasoning and methodology.

During the "classic" era (third to sixth century CE), there was a veritable renaissance in all fields of learning, especially in arithmetic and calculation, which underwent rapid expansion at this time. Moreover, it is probably at the beginning of this period that the impressive Indian speculations on high numbers were developed.

In Vedic literature, there is already evidence of skilful handling of quantities as large as ten to the power seven or ten to the power ten, the *Vedas* mentioning names of numbers from *eka* (= 1) to *arbuda* (= ten million). In the texts *Vâjasaneyi Samhitâ*, *Taittirîya Samhitâ* and *Kâthaka Samhitâ* (written at the

start of the Common Era), there are numbers as high as *parârdha*, which, according to contemporary values, represents a billion.

Before the third century CE, however, there was no known text as long as the *Lalitavistara Sûtra*. It is a text about the life of the prince Gautama Siddhârtha (founder of Buddhist doctrine and thereafter named Buddha), which tells of how Buddha, whilst still a boy, becomes master of all sciences. It tells of the evaluation of the number of grains of sand in a mountain, which evokes the famous problem of Archimedes's *Sand-Reckoner*. What is significant is that the speculation goes way beyond the limits of numbers considered by Greek mathematicians: in one passage, when Buddha arrives at the number which today we write as "1" followed by fifty-three zeros, he adds that the scale in question is only one counting system, and that beyond it there are many other counting systems, and cites all their names without exception. See **Buddha (Legend of), Names of numbers** and **High numbers.**

When the *Lalitavistara Sûtra* was written (before 308 CE), Indian arithmetical speculation had reached and surpassed the number ten to the power 421! After this time, equally vast quantities are found in *Jaina cosmological texts, which, speculating on the dimensions of a universe believed to be infinite in terms of both time and space, easily reach and surpass numbers equivalent to ten to the power 200. See **Jaina**, *Shîrshaprahelikâ*, *Anuyogadvâra Sûtra*. This means that, since the third century CE, the Indian mind had an extraordinary penchant for calculation and handling numbers which no other civilisation possessed to the same degree. See **Arithmetic, Calculation, Arithmetical speculations** and **Arithmetical-cosmogonical speculations.**

In fact, long before the *Lalitavistara Sûtra* and Jaina speculations, astrological and astronomical considerations had led the Indians to be deeply interested in mathematics. This took place between the third century BCE and the first century CE, under the influence of Greek astronomers and after the deployment of India's cultural, maritime and commercial activities with the West during this period. This was the time when astronomical procedures of Babylonian origin were introduced to Indian astronomy: a period characterised notably by the appearance of the unit of time called *tithi*, of similar length to the nychthemer (the "day") and consequently corresponding to 1/30 of the synodic revolution of the moon. It was also at this time that planets came into divination and became subjects for arithmetical calculation, based on their apparent synodic revolutions. This is how mathematics, which was essentially applied to religion, came to be used in astronomy at the time of the Gupta dynasty.

The beginning of the third period of this history roughly corresponds to the end of the "classic" era, around the end of the fifth century CE and the beginning of the sixth century, thus coinciding with the epoch of the *Āryabhatīya*. The *Āryabhatīya* is the name of the work by the mathematician and astronomer *Āryabhata, one of the most original, productive and significant scholars in the history of Indian science. This work (written c. 510 CE) is the first Indian text to display deep knowledge of astronomy, and is arguably the most advanced in the history of ancient Indian astronomy, which at this time developed amazing speculations about cosmic cycles called *yugas. See *Yuga* (Astronomical speculations about) and Indian astronomy (The history of).

This work also deals with trigonometry and gives a summary of the principal Indian mathematical knowledge at the beginning of the sixth century: rules for working out square and cubic roots; rules of measurement; elements and formulas of geometry (triangle, circle, etc.); rules of arithmetical progressions; etc. Here is another important particularity of the *Āryabhatīya*: whilst Ptolemy's trigonometry was based principally on a relationship between the chords of a circle and the angle at the centre which subtends each one of them, Āryabhata's trigonometry established a relationship of a different nature between the chord and the arc of the centre, which is the *sine function* as a trigonomic ratio. That is one of the fundamental contributions of Āryabhata. The work also gives a sine table with the "approximate value" (*āsanna*) of the number π (*pi*):$\pi \approx 62,832/20,000 \approx 3.1416$ [see Shukla and Sarma, (1976) and Sarma (1976)]. See Āryabhata.

Āryabhata invented a unique numerical notation, the conception of which required a perfect knowledge of zero and the place-value system. He also used a remarkable procedure for calculating square and cube roots, which would be impossible to carry out if the envisaged numbers were not expressed in written form, according to the place-value system, using nine distinct numerals and a unique tenth sign performing the function of zero. See Āryabhata (Numerical notations of), Āryabhata's numeration, *Pâtiganita*, Indian methods of calculation and Square roots (How Āryabhata calculated).

Of course, Āryabhata was not the first to use zero and the place-value system: the *Lokavibhâga*, or "Parts of the universe", contains numerous examples of them more than fifty years before the *Āryabhatīya* was written: it is a *Jaina cosmological work, which is very precisely dated Monday 25 August 458 CE in the Julian calendar. Moreover, this is the oldest known Indian text to use zero and the place-value system, except that the text only uses the system of *Sanskrit numerical symbols. See Numerical symbols (Principle of the numeration of). However, bearing in mind the perfect conception of the examples taken from the *Lokavibhâga* and the concern about vulgarisation which is clearly expressed, and moreover taking into consideration the fact that this text was probably the Sanskrit translation of an earlier document (most likely written in a Jaina dialect), it seems very likely that these major discoveries were made in the fourth century CE. This system had become widespread amongst the learned in India by the end of the sixth century. After the beginning of the seventh century, it had gone beyond the frontiers of India into the Indian civilisations of Asia.

As a consequence, calculation and the science of mathematics made substantial progress, as did astronomy, the most spectacular developments of which took place after the start of the sixth century CE. See Indian astronomy (The history of). Amongst the many successors of Āryabhata was Bhâskara I, his faithful disciple and fervent admirer, who wrote a Commentary on the *Āryabhatīya* in 629 CE. This gives invaluable information about the events which took place during the century which separated him from his preceptor. Moreover, this work reveals that Bhâskara had fully mastered mathematical operations which employed the nine numerals and zero using, for example, the Rule of Three. He dealt with arithmetical fractions with ease, expressing them in a very similar manner to our own, although he did not use the horizontal line (which was introduced by the Arabic-Muslim mathematicians). Bhâskhara's work also gives many clues about the development of algebra during that time.

*Brahmagupta was a contemporary of Bhâskhara. In 628 CE he wrote an astronomical text called *Brahmasphutasiddhânta* ("Revised system of Brahma"). In 664 CE he wrote a text on astronomical calculation called *Khandakhâdyaka*. He contradicted some of

Âryabhata's accurate ideas about the rotation of the earth, thus delaying the progress of certain aspects of astronomy. However, his work marked progress in fields such as algebra. He is without a doubt the greatest astronomer and mathematician of the seventh century CE. He made frequent use of the place-value system, and described methods of calculation which are very similar to our own using nine numerals and zero. Amongst his most important contributions are his systematisation of the science of negative numbers, his generalisation of Hero of Alexandria's formula for calculating the area of a quadrilateral, as well as his explanations of general solutions of quadratic equations. The progress of Indian mathematics was stimulated by the development of astronomy initiated by Âryabhata. Indian astronomers used all sorts of mathematical techniques and theories in this discipline. See **Indian astronomy (The history of)** and *Yuga* **(Astronomical speculations on).**

Through resolving indeterminate equations, which depend entirely upon knowing the properties of whole numbers, the Indians arrived at discoveries which went far beyond those of other races of Antiquity or the Middle Ages, and which modern science only arrived at through the efforts of Euler [Woepcke (1863)].

Indian algebra never took the decisive step which would have elevated it to the same level as modern algebra. Instead of using symbols such as *a, b, x, y*, etc., which are independent of the real quantities that they represent, it never occurred to the Indian mathematicians to use symbols other than the first syllables of words which denoted the operations concerned. Moreover, the presentation of and solutions to their various mathematical problems were usually written in verse and consequently subjected to the rules of Sanskrit metric. This explains why their algebraic symbols remained for so long wrapped up in verbiage which was subject to diverse interpretations. See **Sanskrit** and **Poetry and the writing of numbers.**

Brahmagupta's successors included the *Jaina mathematician *Mahâvîrâchârya c. 850 CE, in Kannara in southern India. He wrote a work entitled *Ganitasârasamgraha*: "This work deals with the teachings of Brahmagupta but contains both simplifications and additional information. First he explains the mathematical terminology that he uses, then he deals with arithmetical operations, fractions, the Rule of Three, areas, volumes and in particular calculations with practical applications. He gives examples of solutions to problems. Although like all Indian versified texts it is extremely condensed, this work, from a pedagogical point of view, has a significant advantage over earlier texts" [Filliozat (1957–64)].

Other astronomers or mathematicians who corrected or significantly improved the work of their predecessors include *Govindasvâmin (c. 830), *Shankaranârâyana (c. 869), *Lalla (ninth century CE), *Shrîpati (c. 1039), *Bhoja (c. 1042), *Nârâyana (c. 1356), *Parameshvara (c. 1431), *Nîlakanthasomayâjin (c. 1500) and *Shrîdharâchârya (date uncertain). After Âryabhata and Brahmagupta, however, one of the greatest Indian mathematicians of the Middle Ages was without a doubt *Bhâskarâchârya, who is usually referred to as Bhâskara II, to avoid confusion with Bhâskara I. Born in 1114, the son of Chûdâmani Maheshvar, the astronomer in charge of the observatory of *Ujjain, he finished writing his *Siddhântashiromani* in 1150. This work is divided into four parts, the first two being devoted to mathematics and the second two to astronomy. These are respectively: the *Lîlâvatî (named after his daughter), in which he explains the principle rules of arithmetic; the *Bîjaganita (*bîja means "letter" or "symbol", and *ganita means "calculation"), which is about algebra; *Grahganita*, or "Calculation of the Planets"; and finally the *Golâdhyâya*, or "Book of the Spheres". In the field of astronomy, Bhâskarâchârya "repeats his predecessors but criticises them, even Brahmagupta, who he agrees with most often . . . he compares the gravitational forces of the stars to the winds, in distinguishing these winds from the atmosphere and its deplacements. Mathematically, he accounts for the movements by a developed theory of epicycles and eccentrics. One of the most interesting aspects of his work is that he analyses the movement of the sun, not only in considering the difference of the longitudes from one day to another, but even dividing the day into several intervals, and considers the movement in each one of them to be uniform" [Filliozat in: HGS, I]. The mathematical sections are mainly the study of linear or quadratic equations, indeterminate or otherwise; measurements, arithmetical and geometrical progressions, irrational numbers, and many other arithmetical questions of an algebraic, trigonomic or geometric nature.

Thus we have the names and the principle contributions of some of the most renowned figures in the history of Indian science. See **Infinity, Infinity (Indian concepts of)** and **Infinity (Indian mythological representation of).**

INDIAN METHODS OF CALCULATION.

When they first started out, Indian mathematicians carried out their arithmetical operations by drawing the nine numerals of their old *Brahmi numeration on the soft soil inside a series of columns of an abacus drawn in advance with a pointed stick. If a certain order lacked units the corresponding column was simply left empty. See *Dhûlîkarma*. This archaic method was used later by the Arabic arithmeticians, particularly those of the Maghreb and Andalusia, who had adopted the nine Indian numerals but who did not tend to use zero or carry out their calculations without the aid of columns. However, these mathematicians did not only write out their calculations on the ground: they normally used a wooden board covered in dust, fine sand, flour or any kind of powder, and wrote with the point of a stylet, the flat end of which was used to erase mistakes. This board might be placed on the ground, on a stool or a table, or sometimes the board was equipped with its own legs, like the "counting tables" which were later used in Arabic, Turkish and Persian administrations. Sometimes this board constituted part of a type of kit, and was thus smaller and could be carried in a case. See *Pâtî* and *Pâtîganita*.

In the fifth century CE, the first nine Indian numerals taken from the *Brâhmî* notation began to be used with the place-value system and were completed by a sign in the form of a little circle or dot which constituted zero: this system was to be the ancestor of our modern written numeration. See **The place-value system, Position, Zero (Indian concepts of).**

Thus the Indian mathematicians radically transformed their traditional methods of calculation by suppressing the columns of their old abacus and using the place-value system with their first nine numerals whilst continuing to use the board covered in dust. This step thus marked the birth of our modern written calculation.

To start with, although the corresponding techniques had been liberated from the abacus, they were still a faithful reproduction of the old methods of calculation: they were still carried out, as always, by successive corrections, continually erasing the results at each stage of the calculation, and this limited human memory whilst also preventing them from finding out the errors they had committed on the way to arriving at the final result. This method was used with various variations by *Mahâvîrâchârya (850 CE), *Shrîpati (1039 CE), *Bhâskarâchârya (1150 CE) and even by *Nârâyana (1356 CE). Alongside this technique, the Indian mathematicians (and the Arabic arithmeticians after them) developed a way of carrying out operations without any erasing, by writing the intermediary results above the final result. This, of course, was a great advantage because they could see if they had made a mistake in their calculations if the final result turned out to be wrong, yet brought with it the inconvenience of a lot more writing and more difficulty in deciphering the results, and this is why this method of calculation using nine numerals and zero remained beyond the understanding of the layman for so long.

Moreover, it was impossible for these methods to progress further without a radical transformation of the writing materials which were being used. The use of chalk and blackboard, long before the use of pen and paper became widespread, made the task much less onerous because the intermediary results could either be preserved or rubbed out with a cloth.*Bhâskarâchârya (1150 CE) used his *pâtî to work out highly advanced methods of calculation (notably multiplication which he referred to as *sthânakkhanda*, which literally means: the procedure of "separating the positions"). Even before him, the mathematician *Brahmagupta, in his *Brahmasphutasiddhânta* (628 CE), had described four methods of multiplication which were even more advanced and are more or less identical to those we use today. See also **Square roots (How Âryabhata calculated his).**

INDIAN METRIC.

Here is an alphabetical list of terms related to this discipline, which appear as entries in this dictionary:*Akriti, *Anushtubh, *Ashti, *Atidhriti, *Atyashti, *Dhriti, *Gâyatrî, *Jagatî, *Kriti, *Poetry and writing of numbers, *Prakriti, *Sanskrit, *Serpent (Symbolism of the), *Numerical symbols, *Vikriti.

INDIAN MYTHOLOGIES.

Here is an alphabetical list of words relating to this theme, which appear as article headings in this dictionary: *Agni, *Ahi, *Ananta, *Âptya, *Arjunâkara, *Ashva, *Ashtadiggaja, *Ashvin, *Atri, *Avatâra, *Buddha (Legend of), *Brahmâsya, *Chaturâvîpa, *Chaturmukha, *Dantin, *Dasra, *Dhârtarâshtra, *Dhruva, *Diggaja, *Dvîpa, *Gaja, *High numbers (The symbolic meaning of), *Haribâhu, *Hastin, *Indrarishti, *Indu, *Infinity (Indian mythological representation of), *Jagat, *Jaladharapatha, *Kâla, *Kârttikeya, *Kârttikeyâsya, *Kumârâsya, *Kumâravadana, *Kumud, *Kumuda, *Kuñjara, *Lokapâla, *Manu, *Mount Meru, *Mukha, *Muni, *Mûrti, *Naga, *Nâga, *Nâsatya, *Nripa, *Paksha, *Pândava, *Pâtâla, *Pâvana, *Pundarîka, *Pûrna, *Pushkara, *Putra, *Râhu, *Râvana, *Râvanabhuja, *Râvanashiras, *Rishi, *Sahasrârjuna, *Saptarishi, *Senâninetra, *Shanmukhabâhu, *Shatarûpâ, *Sheshashîrsha, *Shukranetra, *Soma, *Trimûrti, *Tripura, *Trishiras, *Uchchaishravas, *Vaikuntha, *Vasu, *Vâyu, *Vishnupada.

INDIAN NUMERALS (or numerals of Indian origin). List of the principal series of numerals that originated in India (graphical signs which derived from Brahmi numerals): *Agni, *Andhra, Balbodh, *Balinese, Batak, *Bengali, *Bhattiprolu, Bisaya, *Brahmi, Bugi, *Burmese, *Chalukya, *Cham, Chameali, *Dogri, *Dravidian, *Eastern Arabic, *European (*apices*), *European (*algorisms*), *Ganga, *Ghubar, *Grantha, *Gujarati, *Gupta, *Gurumukhi, Jaunsari, *(Ancient) Javanese, *Kaithi, *Kannara, *Kawi, *Khudawadi, Khutanese, Kochi, *Kshatrapa, Kului, *Kushâna, Kutchean, *Kutila, Landa, (Ancient) Laotian, Mahajani, *Maharashtri, *Maharashtri-Jaina, *Maithili, *Malayalam, Mandeali, *Manipuri, *Marathi, *Marwari, *Mathura, Modi, *Mon, *Mongol, Multani, *Nagari, *Nepali, *Old Khmer, Old Malay, *Oriya, *"Pali", *Pallava, *Punjabi, *Rajasthani, *Shaka, Shan, *Sharada, *Shunga, Siamese, *Siddham, *Sindhi, *Sinhala, Sirmauri, Tagala, *Takari, *Tamil, *Telugu, *Thai-Khmer, *Tibetan, Tulu, *Valabhi, and *Vatteluttu numerals.

For origins and graphical evolution, see Fig. 24.61 to 69. For genealogy, classification and geographical distribution, see Fig. 24.52 and 53. For all numerical notations of both Ancient and Modern India, see **Numerical notation**. See also **Indian styles of writing** and **Indian written numeral systems (Classification of)**.

INDIAN RELIGIONS AND PHILOSOPHIES. Here is an alphabetical list of terms related to this theme which appear as entries in this dictionary: *Abhra, *Abja, *Âdi, *Âditya, *Agni, *Âkâsha, *Amara, *Anala, *Âptya, *Arka, *Ashtadiggaja, *Ashtamangala, *Ashtamûrti, *Ashtavimoksha, *Ashva, *Ashvin, *Âtman, *Avatâra, *Bhânu, *Bharga, *Bhava, *Bhûta, *Bhuvana, *Bîja, *Bindu, *Brahma, *Brahmâsya, *Chandra, *Chaturâshrama, *Chaturmahârâja, *Chaturmukha,

*Chaturvarga, *Chaturyoni, *Dahana, *Dantin, *Darshana, *Dashabala, *Dashabhûmi, *Dashaharâ, *Dashâhavatâra, *Dasra, *Deva, *Dharma, *Dhruva, *Diggaja, *Dikpâla, *Dish, *Divâkara, *Divyavarsha, *Dravya, *Drishti, *Durgâ, *Dvaita, *Dvandvamoha, *Dvâtrimshadvaralakshana, *Dvija, *Dvîpa, *Dyumani, *Ekachakra, *Ekadanta, *Ekâdashî, *Ekâgratâ, *Ekâkshara, *Ekântika, *Ekatattvâbhyâsa, *Ekatva, *Eleven, *Gagana, *Gaja, *Ganesha, *Gati, *Go, *Guna, *Hara, *Haranayana, *Haranetra, *Haribâhu, *Hastin, *High numbers (The symbolic meaning of), *Hotri, *Hutâshana, *Indian atomism, *Indra, *Indradrishti, *Indriya, *Infinity, *Infinity (Indian concepts of), *Infinity (Indian mythological representations of), *Îsha, *Îshadrish, *Ishvara, *Jagat, *Jaina, *Jala, *Jvalana, *Kâla, *Kâma, *Karanîya, Karttikeya, Karttikeyâsya, *Kârttikeya, *Kârttikeyâsya, *Kâya, *Keshava, *Kha, *Krishânu, *Kumârâsya, *Kumâravadana, *Kumud, *Kumuda, *Kuñjara, *Loka, *Lokapâla, *Lotus, *Mahâbhûta, *Mahâdeva, *Mahâpâpa, *Mahâyajña, *Mantra, *Manu, *Mârtanda, *Mâtrikâ, *Nâsatya, *Netra, *Pañchabâna, *Pañchâbhijña, *Pañchabhûta, *Pañchachakshus, *Pañcha Indriyâni, *Pañchaklesha, Pañchalakshana, *Pañchânana, *Parabrahman, *Paramânu, *Pâtâla, *Pâvaka, *Pâvana, *Pinâkanayana, *Pitâmaha, *Prâna, *Prithivî, *Pundarîka, *Purâ, *Pûrna, *Purânalakshana, *Pushkara, *Râga, *Râma, *Rasa, *Ratna, *Ravi, *Ravibâna, *Ravichandra, *Rudra, *Rudra-Shiva, *Rudrâsya, *Sâgara, *Sahasrakirana, *Sahasrâksha, *Sahasrâmshu, *Sahasranâma, *Samkhya, *Samkh yâ, *Sâmkhya, *Sâmkhyâ, *Samkhyeya, *Samsâra, *Saptamâtrika, *Shâdanga, *Shadâyatana, *Shaddarshana, *Shakra, *Shakti, *Shatarûpâ, *Shatapathabrâhmana, *Shatkasampatti, *Shikhin, *Shiva, *Shruti, *Shukranetra, *Shûla, *Shûlin, *Shûnya, *Shûnyatâ, *Siddha, *Siddhi, *Soma, *Sura, *Sûrya, *Tallakshana, *Tapana, *Tattva, *Triambaka, *Tribhûvaneshvara, *Trichîvara, *Triguna, *Tripurâ, *Tripurasundarî, *Triratna, *Trishûla, *Trivarga, *Trivarna, *Trividyâ, *Tryakshamukha, *Turîya, *Udarchis, *Vahni, *Vaikuntha, *Vaishvanara, *Vajra, *Varuna, *Vasu, *Vishaya, *Vrindâ, *Yama, *Yoni.

INDIAN STYLES OF WRITING (The materials of). The Indians have used various materials in the history of their writing, starting with stone, which, like nearly all other civilisations, has served for the writing of official inscriptions and important commemo-

Indian Styles of Writing

921

rative texts. Stone has often been replaced, at later times, with copper and other metals. Parchment has also been used, but only really in central Asia, religious reasons probably preventing its use in India. Tree-bark was used, mainly in Assam and southern regions, upon which scribes wrote in ink.

In Kashmir and the whole northwest of India, as well as in the regions of the Himalalyas, ink and brush were (and still are) used on birch-bark. This manuscript writing was called *bhûrjapattra*, and its use was mentioned by Quintus Curtius in this region at the time of Alexander the Great: *Libri arborum teneri, haud secus quam chartae, litterarum notas capiunt* ("The tender part of the bark of trees can be written on, like papyrus") [quoted in Février (1959)]. Wooden boards were also used, upon which characters were not carved, but written in ink. Cotton was another writing support, as reported in the same region by Nearchus, Alexander's admiral.

As for manuscripts (the oldest known examples dating back to the first century CE), palm-leaves were the most popular supports in India and Southeast Asia. These were used since ancient times, in regions of Nepal, Burma, Bengal, as well as in southern India, Ceylon, Siam, Cambodia and Java. Its popularity was due to its availability and the ease with which it could be used: "The leaves chosen for writing were picked young, when they had not yet unfurled. The middle vein was removed and they were left to dry out under pressure. To join them together, they were placed between two boards or between two big dried nervures. Then a thread passed through all the leaves to join them together. Only one instrument was needed to pierce, slice, prepare, join and write a book. The extraordinary simplicity of such material certainly played an important role in the diffusion of Indian culture" [Février, (1959)]. One of the ways of writing on a palm leaf was to engrave the characters using a pointed instrument: "It is undeniable that the characters traced with a point appear pale and unclear, but when sprinkled with dust, they become black, and the dust does not stick to the rest of the leaf because its surface is naturally smooth." Another writing instrument is the calamus, a type of reed whose blunted end is dipped in a type of dye; this has been used since time immemorial in Bengal, Nepal and all southern regions of India. Thus the writing materials used in India and Southeast and Central Asia are as varied as the styles of writing themselves. These conditions account for the great diversity of India writing styles: this diversity has not come about by chance, as the nature of the writing materials has had a profound influence over the appearance of the corresponding styles of writing. See **Indian styles of writing.**

INDIAN STYLES OF WRITING. The various styles of writing which are currently in use in India, Central and Southeast Asia all derive more or less directly from the ancient *Brâhmî* writing, as it is found in the edicts of Emperor Asoka and in a whole series of inscriptions which are contemporaries of the Shunga, Kushâna, Ândhra, Kshatrapa, Gupta, Pallava, etc., dynasties. This writing underwent many successive and relatively subtle modifications over the course of the centuries, which led to the development of various completely individual styles of writing. The apparently considerable differences are due to either the specific character of the languages and traditions to which they have been adapted, or the regional customs and the writing materials used. See **Indian styles of writing (The materials of the).**

These styles of writing can be put into three groups (Fig. 24. 28): the group of styles of writing of northern and central India and of Central Asia (Tibet and Chinese Turkestan): the group of styles of writing of southern India; and finally the group of styles of writing known as "oriental". Naturally, the writing of the first nine numbers has undergone a similar evolution over the centuries: all the series of numerals from 1 to 9 currently in use in India and Central and Southeast Asia derive from ancient *Brâhmî* notation for the corresponding numbers and can be placed in the same groups as those for the styles of writing (Fig. 24. 52). For all the corresponding varieties, see **Indian numerals.**

INDIAN THOUGHT. Here is an alphabetical list of words related to this theme, which appear as entries in this dictionary: *Abhra, *Abja, *Âdi, *Âditya, *Adri, *Agni, *Ahi, *Âkshara, *Amara, *Anala, *Ananta, *Anga, *Anuyogadvâra Sûtra, *Âptya, *Arithmetical-cosmogonical speculations, *Arjunâkara, *Arka, *Asamkhyeya, *Âshâ, *Ashtadanda, *Ashtadiggaja, *Ashtamangala, *Ashtamûrti, *Ashtânga, *Ashtavimoksha, *Ashva, *Ashvin, *Âtman, *Atrinayanaja, *AUM, *Avatâra, *Bâna, *Bhânu, *Bharga, *Bhava, *Bhûta, *Bhuvana, *Bîja, *Bindu, *Brahma, *Brahmâsya, *Calculation, *Chakskhus, *Chandra, *Chaturânanavâdana, *Chaturdvîpa, *Chaturmahârâja, *Chaturmukha, *Chaturyoni,*

*Chaturyuga, *Cosmogonic speculations, *Dahana, *Dantin, *Darshana, *Dashabala, *Dashabhûmi, *Day of Brahma, *Deva, *Dharanî, *Dharma, *Dhruva, *Diggaja, *Dikpâla, *Dish, *Divâkara, *Dravya, *Drishti, *Durgâ, *Dvâdashadvârashâstra, *Dvaparayuga, *Dvâtrimshadvaralakshana, *Dvija, *Dvipa, *Dvîpa, *Dyumani, *Eight, *Eka, *Ekachakra, *Ekadanta, *Ekâdashî, *Ekâgratâ, *Ekâkshara, *Ekatva, Ekavâchana, *Eleven, *Fifteen, *Five, *Four, *Fourteen, *Gagana, *Gaja, *Ganesha, *Gati, *Gavyâ, *Go, *Graha, *Hara, *Haranayana, *Haranetra, *Haribâhu, *Hastin, *High numbers (The symbolic meaning of), *Hotri, *Hutâshana, *Indian astrology, *Indian atomism, *Indian documentation (Pitfalls of), *Indianity (Fundamental mechanisms of), *Indra, *Indradrishti, *Indriya, *Indu, *Infinity, *Infinity (Indian concepts of), *Infinity (Indian mythological representation of), *Îsha, *Îshadrish, *Ishu, *Îshvara, *Jagat, *Jaladharapatha, *Jvalana, *Jyotisha, Kabubh, *Kâla, *Kalamba, *Kaliyuga, *Kalpa, *Kâma, *Karaniya, *Kârttikeya, *Kârttikeyâsya, *Kavacha, *Kâya, *Keshava, *Kha, *Krishânu, *Krita, *Kritayuga, *Kshapeshvara, *Kumârâsya, *Kumâravadana, *Kumud, *Kumuda, *Kuñujara, *Loka, *Lokapâla, *Lotus, *Mahâdeva, *Mahâkalpa, *Mahâyuga, *Mangala, *Mantra, *Manu, *Mârgana, *Mârtanda, *Mrigânka, *Mukha, *Mûrti, *Mysticism of infinity, *Mysticism of zero, *Nâga, *Netra, *Nine, *Numeral alphabet, magic, mysticism and divination, *Numerical symbols, *Ocean, *One, *Paksha, *Pañchabâna, *Pañchâbhijña, *Pañchachakshus, *Pañcha Indriyâni, *Pañchaklesha, *Pañchalakshana, *Pañchânana, *Parabrahman, *Pâtâla, *Pâvaka, *Pâvana, *Pinâkanayana, *Pitâmaha, *Pundarîka, *Purâ, *Pûrna, *Putra, *Râga, *Râhu, *Rasa, *Ratna, *Râvanabhuja, *Ravi, *Ravibâna, *Rudra, *Rudra-Shiva, *Rudrâsya, *Sâgara, *Sahasrakirana, *Sahasrâksha, *Sahasrâmshu, *Sahasranâma, *Samkhya, *Samkhyâ, *Sâmkhya, *Sâmkhyâ, *Samkhyâna, *Samkhyeya, *Sanctification of a concept, *Sanskrit, *Saptabuddha, *Sarpa, *Sâyaka, *Seven, *Shakra, *Shakti, *Shankha, *Shanku, *Shanmukhabâhu, *Shara, *Shashadhara, *Shashanka, *Shashin, *Shatarûpâ, *Shikhin, *Shîrshaprahelikâ, *Shîtâmshu, *Shîtarashmi, *Shiva, *Shukranetra, *Shûla, *Shûlin, *Shûnya, *Shûnyatâ, *Six, *Sixteen, *Soma, *Sud-hâmshu, *Sura, *Sûrya, *Symbolism of numbers, *Symbols, *Takshan, *Tallakshana, *Tapana, *Ten, *Thirteen, *Thirty-three, *Thousand, *Three, *Tretâyuga, *Triambaka, *Tribhuvana, *Tribhûvaneshvara, *Trikâya, *Triloka, *Trimûrti, *Tripitaka, *Tripurâ, *Tripurasundarî, *Trishûla, *Trivarna, *Tryakshamukha, *Tryambaka, *Turîya, *Twelve, *Twenty, *Twenty-five, *Two, *Udarchis, *Uppala, *Utpala, *Vâchana, *Vahni, *Vaikuntha, *Vaishvanara, *Vajra, *Varuna, *Vasu, *Vidhu, *Vishika, *Vishnupada, *Vishva, *Vishvadeva, *Vrindâ, *Yama, *Yuga, *Yuga (Astronomical speculation on), *Yuga (Cosmogonical speculations on), *Zero.

INDIAN WRITTEN NUMERAL SYSTEMS (Graphical classification of). The aim of this article is to give a quick recapitulation of the various numerical notations formerly or currently used in the Indian sub-continent, in order to identify the palaeographic type of each one. The following references to figures are mainly the ones which can be found in Chapter 24. More or less all of the numerical notations which are currently in use in India, Central Asia and Southeast Asia (see Fig. 24. 61 to 69) derive from the ancient *Brâhmî* notation (Fig. 24. 29 to 31 and 70), which is found in the edicts of Emperor Asoka and a whole series of contemporary inscriptions of the Shunga, Kushâna, Ândhra, Kshatrapa, Gupta, Pallava, etc. dynasties (Fig. 24. 29 t 38 and 70). This original notation (which surely derives from an earlier ideographical notation) has undergone several subtle graphical modifications over the course of the centuries (Fig. 24. 70), which led to the development of various types of notations which are all highly individual like *Gupta* (Fig. 24. 38), Bhattiprolu and "*Pali*". See also **Ândhra numerals, Bhattiprolu numerals, Brâhmî numerals, Châlukya numerals, Ganga numerals, Gupta numerals, Kshatrapa numerals, Kushâna numerals, Mathurâ numerals, Pâlî numerals, Pallava numerals, Shaka numerals, Shunga numerals and Valabhî numerals.**

For the graphical origin of *Brâhmî* numerals, see Fig. 24. 57 to 24. 59. For notations derived from *Brâhmî*, see Fig. 24. 52. For their graphical evolution, see Fig. 24. 61 to 24. 69.

The apparently considerable differences between these notations are due to either the specific character of the languages and traditions to which they belong to which the corresponding writing would have been adapted, or to the regional habits of the scribes and the nature of the writing material used. See **Indian styles of writing**.

The notations can be divided into three broad groups (see Fig. 24. 52):

1. – The group of notations from Central India, from Northern India, from Tibet and Chinese Turkestan. These notations are the ones which come from Gupta writing. These can be divided in turn into five sub-groups:

1. 1. – *The sub-group of notations derived from Nâgarî*. This group is made up of notations issued from Nâgarî numerals (Fig. 24.3, 39 and 72 to 74), amongst which are: Mahârâshtrâ; Marâthî (Fig. 24.4); Balbodh; Modi; Rajasthani; Mârwarî; Mahajani; Kutilâ; Bengali (Fig. 24.10); Oriya (Fig. 24.12); Gujarati (Fig. 24.8); Maithili (Fig. 24.11); Manipuri; Kaithi (Fig. 24. 9); etc.

The Arabic notations *"Hindi"* and *Ghubar* also belong to this sub-group, as well as the European *Apices* and *Algorisms*: Arabic numerals both from the East and the Maghreb (Fig. 25.3 and 25.5), derive more or less directly from Nâgarî numerals. The numerals that we use today, and the European numerals of the Middle Ages (Fig. 26.3 and 10), derive from the *Ghubar* numerals of the Maghreb (Fig. 25.5). See **Eastern Arabic numerals, Bengali numerals, European numerals (Apices), European numerals (Algorisms), Ghubar numerals, Gujaratî numerals, Kâithî numerals, Kutila numerals, Mahârâshtrî numerals, Mahârâshtrîjaina numerals, Maithilî numerals, Manipuri numerals, Marâthî numerals, Mârwarî numerals, Nâgarî numerals, Oriya numerals and Rajasthani numerals**.

1.2. – *The sub-group of notations derived from Sharada writing*. This is composed of notations derived from the numerals of the same name (Fig. 24. 14 and 40), including: Tâkarî (Fig. 24. 13); Dogri (Fig. 24. 13); Chameali ; Mandeali; Kului; Sirmauri; Jaunsari; Sindhi (Fig. 24. 6); Khudawadi (Fig. 24. 6); Gurumukhi (Fig. 24. 7); Punjabi (Fig. 24. 5); Kochi; Landa; Multani; etc. See **Dogrî numerals, Gurumukhi numerals, Khudawadî numerals, Punjabi numerals, Shârada numerals, Sindhi numerals, Sirmauri numerals and Tâkarî numerals**.

1.3. – *The sub-group of notations from Nepal*. This includes modern Nepali (Fig. 24. 15), which derives from the ancient Siddham notation (Fig. 24. 42) which itself comes from Gupta but under the influnce of Nagari. See Nepali numerals and Siddham numerals.

1.4. – *The sub-group of Tibetan notations*. This contains Tibetan notations (Fig. 24. 16),

which all derive from *Siddham*, and which are notably related to ancient Mongol writing (Fig. 24.42). See **Tibetan numerals and Mongol numerals**.

1.5. – *The sub-group of notations from Central Asia*. This contains notations of Chinese Turkestan, which also all derive from *Siddham*.

2. – The group of notations from Southern India. These are notations which come from *Bhattiprolu* (Fig. 24. 43 to 24. 46), distant cousin of Gupta. They can be subdivided into four groups:

2. 1. – *The sub-group of Telugu notations*. This is made up of Telugu and Kannara notations (Fig. 24. 20, 21, 47 and 48).

2.2. – *The sub-group of Grantha notations*. This contains Grantha, Tamil and Vatteluttu notations (Fig. 24. 17 and 24. 49).

2.3. – *The sub-group of Tulu notations*. This contains Tulu and Malayalam notations (Fig. 224. 19).

2.4. – *The sub-group of Sinhalese notations*. In this group Sinhala notation can be found (Fig. 24. 22).

See **Dravidian numerals, Grantha numerals, Kannara numerals, Malayalam numerals, Sinhala numerals, Tamil numerals, Telugu numerals and Vatteluttu numerals**.

3. – The group of eastern notations. These are the notations of Southeast Asia, which are all derived from "Pali" writing, which itself comes from the same source as Gupta and Bhattiprolu (Fig. 24. 43 to 46). These in turn can be subdivided into seven groups:

3.1. – *The sub-group of Burmese notations*. This contains ancient and modern Burmese notations (Fig. 24. 23).

3.2. – *The sub-group of Old Khmer notations*. In this group there is the ancient notation of Cambodia (Fig. 24. 77, 78 and 80).

3.3. – *The sub-group of Cham notations*. This contains the notation of Champa (Fig. 24. 79 and 80).

3.4. – *The sub-group of Old Malay notations*. This group contains the writing style once used in Malaysia (Fig. 24. 80).

3.5. – *The sub-group of Old Javanese notations*. This group contains "Kawi" writing which was once used in Java and Bali (Fig. 24. 50 and 24. 80).

3.6. – *The sub-group of present-day Thai-Khmer writing*. This includes Shan, Laotian and Siamese, as well as the notation which is currently used in Cambodia, Laos and Thailand (Fig. 24. 24).

3. 7. – *The sub-group of current Balinese notations*. This sub-group is made up of the current

Balinese, Buginese, Tagala, Bisaya and Batak notations (Fig. 24.25). See **Balinese numerals, Burmese numerals, Cham numerals, Ancient Javanese numerals, Kawi numerals, Thai-Khmer numerals and Old Khmer numerals.** For an overview of all these notations, see Fig. 24.52. For their geography, see Fig. 24.27 and 53. For their mathematical classification, see **Indian written numeral systems (The mathematical classification of).**

INDIAN WRITTEN NUMERAL SYSTEMS (The mathematical classification of). Here is a quick summary of the mathematical structure of the various notations which were once used, or are still in use, in the Indian sub-continent. The numerical notations which derive from *Brâhmî* (see **Indian written numeral systems (Graphical classification of)**) are not the only ones to be used in the Indian sub-continent. In northwest India, after Asoka's time until the sixth or seventh century CE, a style of writing was used which was imported by Aramaean traders. This was known as *Karoshthî* (Fig. 24. 54). See **Karoshthî numerals.** There is also the system which was found in Mohenjo-daro and Harappa (in present-day Pakistan), which was used from c. 2500 to 1500 BCE by the ancient Indus civilisation, long before the Aryans arrived on Indian soil. See **Proto-Indian numerals.**

From a mathematical point of view, according to the classification of numerations in Chapter 23, these different systems (which generally have a decimal base) can be divided into three broad categories:

A. – The category of additional numerations. These are systems which are based upon the additional principle, each numeral possessing its own value, independent of its position in numerical representations. This category can be subdivided into three types:

A.1. – *The first type of additional numerations.* These are numerations which (like the Egyptian hieroglyphic system for example) attribute a particular numeral to each of the numbers 1, 10, 100, 1,000, 10,000, etc., and which repeat these signs as many times as necessary to record other numbers (Fig. 23.30). The ancient *Indusian numeration no doubt belonged to this type.

A.2. – *The second type of additional numeration.* These are numerations which (like the Roman system for example) attribute a specific numeral to each of the numbers 1, 10, 100, 1,000, etc., as well as to 5, 50, 500, etc., and which repeat these signs as many times as necessary to record other numbers (Fig.

23.31). There is no known example of this type in India.

A. 3. – *The third type of additional numeration.* These are numerations which (like the Egyptian hieratic system for example) attribute a particular sign to each unit of each decimal order (1, 2, 3, . . . 10, 20, 30, . . . 100, 200, 300, . . ., etc.) and which use combinations of these different signs to write other numbers (Fig. 23.32). This is the type that all notations derived from Brâhmî belong to, at least initially (Fig. 24.70). Thus the following notations belong to this sub-category: Ândhra notation (Fig. 24.34 and 36); Bhattiprolu notation; Chalukya notation (Fig. 24.45); Ganga notation (Fig. 24.46); Gupta notation (Fig. 24.38); Kshatrapa notation (Fig. 24.35); Kushâna notation (Fig. 24.33); Mathura notation (Fig. 24.32); Ancient Nâgarî notation (Fig. 24.39B); Ancient Nepali notation (Fig. 24.41); Pallava notation (Fig. 24.34 and 24.36); Valabhî notation (Fig. 24.44); etc. Alphabetical notations also fall into this category (which use vocalised consonants of the Indian alphabet, to which a numerical value is assigned in a regular order, and which are still used today in various regions of India, from Tibet, Nepal, Bengal or Orissa to Maharashtra, Tamil Nadu, Kerala, Karnataka and Sri Lanka, and from Burma to Cambodia, in Thailand and in Java); notably that of Âryabhata (the difference being that the latter had a centesimal base, not a decimal one). One exception is *Katapayâdi* numeration (which seems to have been invented by Haridatta), which was alphabetical but based on the place-value system. See **Numeral alphabet, Âryabhata's numeration** and **Katapayâdi numeration.**

B. – The category of hybrid numerations. These are numerations which use both multiplication and addition in their representations of numbers. This category can be divided into five types:

B.1. – *The first type of hybrid numeration.* These are numerations which (like the Babylonian system) attribute a particular numeral to each of the numbers 1, 10, 100, 1,000, etc., using an additive notation for numbers inferior to one hundred and a multiplicative notation for the hundreds, the thousands, etc., and which represents other numbers through combinations which use both the additive and multiplicative principles (Fig. 23.33). Aramaean numeration belongs to this group (Fig. 23.17) as well as *Kharoshthî numeration which is derived from the former (Fig. 24.54).

B.2. – *The second type of hybrid numeration.*
These are numerations which function exactly
like the Sinhalese system (Fig. 23.18 and
24.22): a particular numeral is given to each
simple unit, as well as to each power of ten (10,
100, 1,000, etc.), and the notation of hundreds,
thousands, etc., follows the multiplicative rule
(Fig.23.34).

B.3. – *The third type of hybrid numeration.*
These are Mari numerations (Fig. 23.22),
which do not seem to exist in India.

B.4. – *The fourth type of hybrid numeration.*
These are Ethiopian numerations (Fig. 23.36),
which do not seem to exist in India.

B.5. – *The fifth type of hybrid numeration.*
These are the numerations for which Tamil
and Malayâlam numerations provide the models (Fig. 23.20 and 21); these give a specific
numeral to each simple unit (1, 2, 3 . . .), as
well as to diverse multiples of each power of
ten (10, 20, 30, . . . 100, 200, 300, . . ., etc.), and
where the notation of tens, hundreds, thousands, etc., is carried out using the
multiplicative principle (Fig. 23.37).

C. – The category of positional numerations. These are numerations founded on the
principle according to which the basic value of
numerals is determined by their position in
the writing of the numbers, and which thus
requires the use of a zero (Fig. 23.27). This category can be subdivided into two types:

C.1. – *The first type of positional numerations.* These are Babylonian, Chinese or Maya
(Fig. 23. 23, 24, 25 and 38), which are not
found in India.

C.2. – *The second type of positional numerations.* These are the numerations (Fig. 23.28),
which belong to the one which was invented in
India over fifteen centuries ago and which is
the origin of all decimal positional notations
which are currently in use (Fig. 24.3 to 16 and
20 to 26), including our own (Fig. 23.26) and
the one which is still used in Arabic countries
(Fig. 24.3). This system has a decimal base,
and nine distinct numerals which give no visual indication as to their value, which represent
the nine significant units (from which our
signs 1, 2, 3, 4, 5, 6, 7, 8, 9 are) derived; it also
possesses a tenth sign, called *shûnya* (zero),
which is written as a little dot or circle (Fig.
24.82 and *Zero*, Fig. D. 11), and is the ancestor of our zero, whose function is to mark the
absence of units in any given order, and which
possesses a veritable numerical value: that of
"nil" (Fig. 23.27). The fundamental characteristic of this system is that it can express all

numbers in a simple and coherent way,
whether they are whole, fractional, irrational
or transcendental (Fig. 23.28). Thus the Indian
place-value system (for that is what it is) is the
first of the category of the most evolved written numerations in history (Fig. 28.29). The
following are the notations which belong to
this category:

Modern Nâgarî (Fig. 24.3, 39 A and 39 C);
Marathî (Fig. 24.4); Punjabi (Fig. 24.5); Sindhi
(Fig. 24.6); Gurumukhi (Fig. 24.7); Gujarati
(Fig. 24.8); Kâithî (Fig. 24.9); Bengali (Fig.
24.10); Maithilî (Fig. 24.11); Oriya (Fig.
24.12); Tâkarî (Fig. 24.13); Shâradâ (Fig. 24.14
and 40); modern Nepali (Fig. 24.15); Tibetan
(Fig. 24.16); Telugu (Fig. 24.20 and 47);
Kannara (Fig. 24.21 and 48); Burmese (Fig.
24.23 and 51); Thai-Khmer (Fig. 24.24);
Balinese (Fig. 24.25); modern Javanese (Fig.
24.26); ancient Javanese (Fig. 24. 50); Mongol
(Fig. 24.42); the "Hindi" form of Arabic writing (Fig. 24.3 and 25.5); the "Ghubar" form of
Arabic writing, whilst it was used to represent
numbers with zero, without the columns of
the abacus drawn in the dust (Fig. 25.3); the
"Algorism" form of European writing (Fig.
26.10); etc.

Thus the discovery of Indian positional
numeration not only allowed the simple
and perfectly rational representation of
absolutely any number (however large or
complex), but also and above all a very easy
way of carrying out mathematical operation;
this discovery made it possible for anyone to
do sums. The Indian contribution to the
history of mathematics was essential,
because it united calculation with numerical
notation, thus enabling the democratisation
of the art of calculation.

For the graphical classification of the
various numerical notations, see *Indian
written numerations (The graphical classification of).* For the Sanskrit names, usage,
conditions and discovery of positional
numeration, see: **Anka, Sthâna, Ankakramena, Ankasthâna, Sthânakramâd,
Names of numbers, High numbers,
Sanskrit, Numerical symbols, Numeration
of numerical symbols, Katapayâdi numeration, Âryabhata's numeration.**

For zero and its graphic or symbolic representations, see **Zero, Shûnya, Numeral 0.**

For corresponding methods of calculation, see *Patîganita,* **Indian methods of
calculation.**

For the subtleties relating to zero and the place-value system in Sanskrit poetry, see **Poetry, zero and positional numeration.**

INDIVIDUAL SOUL. [S]. Value = 1. See *Âtman.* **One.**

INDO-EUROPEAN NAMES OF NUMBERS. See Chapter 2, especially Fig. 2. 4 A to 4 J and 2. 5, where the Sanskrit names of numbers are compared to those of other languages of Indo-European origin. See Fig. D. 2 of the entry entitled Âryabhata's numeration.

INDRA. [S]. Value = 14. "Strength", "Courage", "Power". The name of one of the principal gods of Vedic times and of the Brahm anic pantheon. He represents the source of cosmic life that he transmits to the earth through the intermediary of rain. His strength lies in the seminal fluid of all beings, this god being said to be "made of all the gods put together". He is eternally young, because he rejuvenates himself at the start of each *manvantara*, which means each of the fourteen "ages" of our world which make up a *kalpa*. Thus *Indra* = 14. See *Yuga,* **Manu** and **Fourteen.**

INDRADRISHTI. [S]. Value = 1,000. "Eyes of Indra". One of this god's attributes is *Sahastâksha*, "of the thousand eyes". See **Indra, Thousand.**

INDRIYA [S]. Value = 5. "Power". This is due to the Buddhist physical and mental powers, which are divided into five groups: the foundations (*âyatana*); the natures (*bhâva*); the senses (*vedâna*); the spiritual powers (*balâ*); and the supramundane powers. The same word also means the "five roots" (*pañcha indriya*), which, as positive agents, enable a person to lead a moral life: faith (*shraddendriya*), energy (*vîyendriya*), memory (*smritîndriya*), meditation (*samâdhîndriya*), and wisdom (*prâjnendriya*) [see Frédéric, *Dictionnaire* (1987)]. See **Five.**

INDU. [S]. Value = 1. "Drop". This represents the moon, and alludes to the "dew" (*chandrakanta*), the mythical pearl said to have been made from concentrated moonbeams. The moon being worth 1, this symbolism is self-explanatory. This word should not be confused with *bindu* ("point") which is a synonym for zero. See **One.**

INDUSIAN NUMERALS. See **Proto-Indian numerals.**

INDUSIAN NUMERATION. See **Proto-Indian numerals.**

INFERIOR WORLD. [S]. Value = 7. See *Pâtâla* and **Seven.**

INFINITELY BIG. See **High numbers,** *Asamkhyeya,* **High numbers (Symbolic meaning of).**

INFINITELY SMALL. See **Low numbers,** *Paramânu, Shûnya, Shûnyatâ,* **Zero** and **Infinity.**

INFINITY (Indian concepts of). Amongst the Sanskrit words for zero is *ananta*, which literally means "infinity": Ananta is an immense serpent, who, in Indian cosmology and mythology, represents the serpent of infinity, eternity and the immensity of space. *Vishnu is said to rest on the serpent in between creations. See **Serpent (Symbolism of the), High numbers** and **Infinity (Indian mythological representation of).**

In Indian mysticism, the concept of zero and that of infinity are very closely linked. Thus words such as *ambara, *kha, *gagana, etc., meaning "space", "sky" or the "canopy of heaven" came to represent zero. See **Zero,** *Shûnya, Âkâsha, Vishnupada* and *Pûrna.*

Of course, Indian mathematicians knew perfectly well how to distinguish between these two notions, which are the inverse of one another, for to their mind, division by zero was equal to infinity. This was the case at least since the time of *Brahmagupta (628 CE), who defined infinity with the term *khachheda, literally "the quantity whose denominator is zero" [see Datta and Singh (1938), pp. 238–44]. In *Lîlâvatî, *Bhâskarâchârya wrote the following about the same concept, which he refers to as *khahara, which literally means "division by zero" [see Datta and Singh (1938), pp. 238–44]: "In this quantity which has zero as divisor, there is no [possible] modification, even though several [quantities] can be extracted or introduced; in the same way, no changes can be carried out on the constant and infinite God [*Brahma] during the period of the destruction or creation of worlds, however many living species are projected forward or are absorbed." This is what Ganesha wrote on the subject in *Grahalâghava* (c. 1558 CE): "The *Khachheda* is an indefinite quantity, unlimited and infinite; it is impossible to know how high this quantity is. It can be modified by neither the addition nor the subtraction of limited [= finite] quantities, for in the preliminary operation of reducing all the fractional expressions to the same denominator, which it is necessary to do beforehand in order to be able to calculate their sum or their difference, the numerator and the denominator of the finite quantity both disappear." So Indian scholars, at least since Brahmagupta's time, knew that division by zero equalled infinity:

$$a/0 = \infty.$$

To their mind, this "quantity" remained unchanged if a finite number was added to it or subtracted from it; thus:

$$a/0 + k = k + a/0 = a/0$$

and

$$a/0 - k = k - a/0 = a/0.$$

This means that the Indians, at least as early as the beginning of the seventh century CE, knew these mathematical formulas that we use today: ˙

$$\infty + k = k + \infty = \infty$$

and

$$\infty - k = k - \infty = \infty.$$

Brahmagupta, however, (and several of his successors) committed the error of thinking that when zero was divided by itself the result was zero, when in reality the result is an "indefinite quantity". Bhâskarâchârya, who made the necessary corrections to the erroneous assertions of his predecessors, and who quite rightly affirmed that a number other than zero divided by zero gives an infinite quotient, himself committed an error by saying that the product of infinity multiplied by zero is a finite number. However, this in no way detracts from the merits of Indian civilisation which was so advanced in comparison with all the other civilisations of Antiquity and the Middle Ages. See **Infinity**, **Infinity (Indian mythological representation of)** and **Indian mathematics (The history of)**.

INFINITY (Indian mythological representation of). It seems that the lemniscate which today represents the concept of infinity, was introduced for the first time in 1655 by the English mathematician John Wallis. Hindu mythological iconography contains a very similar symbol representing the same idea, although it seems that it was never used in the domain of mathematics. This symbol is that of Ananta, the famous serpent of infinity and eternity, which is always represented coiled up in a sort of figure of eight on its side like the symbol ∞. See *Ananta* (in particular Fig. D. 1), *Pûrna* and *Vishnupada*.

This begs the following question: Did Wallis know of the Indian mythological symbol when he introduced this sign into the list of mathematical conventions? The answer is no; this graphism and its many variants (∞, 8, S, etc.) can be found in diverse civilisations and many different epochs and parts of the world, and the symbolism is similar, if not identical, to that of the Indian mythological representations. This symbolism can be found in many ancient astrological, magical, mystical and divinatory representations, for example in ancient and mediaeval talismans, both Eastern and Western, where the S is very common and is meant to express, for the wearer of the amulet, a sign favourable to *eternal* union and *infinite* happiness. The sign which looks like a figure of eight lying on its side can be found painted on the walls of masonic lodges or embroidered upon clothing. It is not there for merely decorative purposes; it symbolises the bonds which unite the members of a social body: the interlacing expresses the sentiment *united until death* [see Chevalier and Gheerbrant (1982)]. This symbol can also be found in the manuscripts of mediaeval alchemists, where three Ss signify the *abundance* of rain water, as well as its *constance*. The S can also be connected to the *celestial wheel* of the Romanised god of Ancient Gaul, and to talismans which have *celestial* meaning in Greek-Roman magical traditions [see Marquès-Rivière (1972)]. For the Assyrians, *hawu* was also in the form of an S, like the serpent of *eternal life*. This symbol was later used by the Hebrews to represent the "bronze serpent" before it was destroyed by Hezekiah (2 Kings XVIII, 4). This is the serpent made by Moses to save the Israelites who had spoken against the Lord, and who had been bitten by the fiery serpents sent by Yahweh (Numbers, XXI, 6)[see GLE, IX, p. 770].

The interlace is often a symbol for water or for the vibration of the air. In many cosmogonies, the interlace symbolises the very nature of creation, energy and all existence. In Celtic art, it symbolises the notion of *ourobouros*: the endless movement of evolution and involution through the muddle of cosmic and human facts. The *ourobouros* is the serpent which bites its own tail ; this symbolises self-evolution, or self-fertilisation and, consequentially, eternal return. This evokes the *samsâra* (or the Indian cycle of "rebirth"), which is an *indefinite* circle of rebirths, of *continual* repetition. Thus the serpent gradually came to be represented by a circular graphism. Sometimes this circle has been dissected by two perpendicular diameters in order to show the inter-relationship between the sky and the earth. The sign which looks like an X or a cross symbolises the earth with its four horizons. Thus the circle dissected by the cross is none other than the celestial-terrestrial opposition of the mysticism of the serpent. See **Serpent (Symbolism of the).**

Palaeography proves that this dissected circle is, cursively speaking, the S or 8 sign denoting a vast quantity or eternity. This is very significant when we look at the shapes of Roman numerals. The Roman numerals that we know today look like Roman letters: I (1), V (5), X (10), L (50), C (100), D (500), M (1,000).

In reality, however, these symbols are not the original ones used to write the numbers. In fact, Roman numerals derive from the ancient practice of counting using a "tally" system which led to numbers being represented by the following symbols:

I	V	X	Ѱ	Ӿ
1	5	10	50	100

Originally, the unit was represented by a vertical line, the number 5 by an acute angle, the number 10 by a cross, 50 by an acute angle dissected by a vertical line and 100 was a cross dissected by a vertical line. We can easily see how the primitive signs for 1, 5 and 10 became the letters I, V and X. The sign for 50 originally looked like an arrow pointing downwards. This evolved into what looked like a T on its head before finally being mistaken for the letter L. As for the representation of 100, this initially evolved into a sign which looked like this: Ӿ Then, in order to save time, this symbol was cut in half so that it looked like the letter C, or its mirror image. This is also the initial of the Latin word for hundred, *Centum*. To create a sign for 1,000, the Romans decided to use the symbol for 10 (the cross) and draw a circle around it. Then, for 500, they cut the sign for a thousand in half: Ð. This sign would later be mistaken for the letter D. The circle dissected by a cross (1,000) evolved into various shapes, which were replaced by the M due to the Latin word *Mile* (see Fig. 16. 26 to 34): Thus we can see how, graphically, the circle dissected by a cross became a sign which was shaped like an S or an 8 on its side. In Latin, the term *Mile*

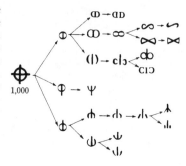

corresponded to the highest number in spoken numeration and, by extension, in everyday language, it meant "vast number" and "the incalculable". In his *Natural History* (XXXIII, 133), Pliny wrote that the Romans had no names for powers of ten superior to a hundred thousand, and so referred to a million as *decies centena milia* ("ten hundred thousand"). The snake as the sign for infinity, in its various forms, has been connected to ideas such as the sky, the universe, the axis of the world, the night of beginnings, the primordial substance, the vital principle, life, eternal life, sexual energy, spiritual energy, vestiges of the past, the seed of times to come, cyclical development and resorption, longevity, extreme fertility, the incalculable quantity, abundance, immensity, totality, absolute stability, endless movement, etc. See **High numbers (The symbolic meaning of).**

INFINITY AND MYSTICISM. See **Infinity (Indian mythological representation of)** and **Serpent (Symbolism of the).**

INFINITY. All confusion must be avoided between *infinity* and *indefinite*. *Indefinite* comes from the Latin *indefinitus*, signifying "vague". This word also has other possible meanings. The first is the opposite of "defined", that which is "unspecified", which remains "undetermined", like death for example: "That which is certain in death is softened a little by that which is uncertain; it is an indeterminate length of time which has something of infinity and of what we call eternity." The second meaning expresses the opposite of that which is "finite"; it is a quantity which, whilst remaining finite, is susceptible to unlimited expansion or growth. This is the meaning of indefinite progress. The third sense can be found in this extract from Descartes: "Each body can be divided into infinitely small parts. I do not know if their number is infinite or not, but certainly, to the best of our knowledge, it is indefinite" [*Oeuvres*, XI, 12]. This is the opposite of that which is infinite: here the indefinite is "that which is only infinite from a certain point of view, because we cannot see its end" [Foulquié (1982)].

On the other hand, a fourth meaning makes this word a synonym of infinity. This is *potential infinity*, illustrated by this quote from Pascal: "The eternal silence of these infinite spaces fill me with fear" [*Pensées*, 428]. This extract inspired the following commentary from Paul Valéry: "This phrase, which is so powerful and magnificent that it is one of the most famous ever to have been uttered, is a *Poem*, and by no means a thought. *Eternal* and *Infinite* are symbols of non-thought. Their value is entirely emotional. They act on our sensitivity. They provoke the peculiar sensation of the inability to imagine" [*Variété, La Pléiade*, I, 458]. Thus potential infinity is "that which, being effectively finite, has the potential for limitless growth" [Foulquié]. In terms of either potential or reality, infinity has posed one of the most serious problems to the human mind in all the history of civilisation. Confronting infinity has been a little like meeting Cerberus at the entrance to the Underworld. *There is one, final number, but it is beyond the power of mortals to reach it; this power belongs only to the gods who are the only ones who may count the stars and the firmament.* Such is the leitmotiv of both ancient and modern religions. It bears witness to humanity's constant obsession with this concept. It demonstrates not only our ability to count numbers "to the end", but also to learn the true meaning of that which conceals the rather vague notion of the unlimited: "We imagine some kind of finite range, then we disregard the limits of the range, and we have the idea of an infinite range. In this way, and perhaps in this way alone, we are able to conceive of an infinite number, an infinite duration, etc. Through this definition, or rather this analysis, we can see to what extent our notion of *infinity* is vague and imperfect; it is only really the notion of the *indefinite*, if we understand by this word a vague quantity to which no limits have been assigned, and not, as one could understand by another meaning of this word, a quantity for which there are limits, yet these limits have not been specified" [D'Alembert, *Essai sur les éléments de philosophie*, Éclairc., XIV].

This explains why comparisons are sometimes made which are reminiscent of religious metaphors and parables. The grains of sand of the desert, the drops of water of the ocean or the stars in the sky are evoked, without the realisation that such comparisons are puerile, as they only involve the domain of the finite. In everyday use, infinity is only understood by its negation. In fact, the word "infinity" derives from the Latin *infinitus*, "that which has no end", "that which never ends". It is the negation of the finite, in the sense that infinity is "that which can never be reached".

[See Blaise (1954); Bloch and von Wartburg (1968); Chantraine (1970); Du Cange (1678); Ernout and Meillet (1959); Estienne (1573); Gaffiot; GLF (1971); Littré (1971); Robert (1985)].

It is precisely this limited conception which prevented the Greek mathematicians from making progress in this domain. Historically, it was in Greece, after Pythagoras's discoveries, that the evolution of this concept began with the undisputed statement that "infinity is something which cannot be measured". The problem, according to Bertrand Russell, represented "in one form or another, the basis of nearly all the theories of space, time and of infinity which persisted from that time up until the present day".

Descartes was one of the first European scholars to establish infinity as a fundamental reality. This notion later became a perfectly precise, objective concept, presenting no basic problems such as those often conferred upon it by the profane. The symbol for infinity (∞) seems to have appeared for the first time in 1655 in a list of mathematical signs compiled by the English mathematician John Wallis.

Mathematically, infinity is that which is bigger than any other quantity, and no finite number can be added to it. Fléchier compared infinity to God, as God is "infinitely powerful and thus infinitely free". Zero is the opposite of infinity: it is infinitely small, the variable quantity which is inferior to all positive numbers, however small they might be. Infinity, or the impossibility of counting all the numbers, remains a *mathematical hypothesis*; it is one of the fundamental axioms upon which contemporary mathematics is based. See **Infinity (Indian concepts of)** and **Infinity (Indian mythological representations of)**.

INFINITY. Term used as a synonym for "potential infinity". See **Infinity**. See also **Indian mathematics (The history of)**.

INFINITY. Term used as a synonym for the "indeterminable". See **Infinity** and **High numbers**. See also **Serpent (Symbolism of the)**.

INFINITY. Term used as a synonym of the "unlimited". See **Infinity, High numbers** and **High numbers (The symbolic meaning of)**. See also **Serpent (Symbolism of the)**.

INFINITY. Term used as a synonym of the eternity and immensity of space. See *Ananta*, **Infinity (The Indian mythological representation of), High numbers (The symbolic meaning of)** and **Infinity**.

INFINITY. Term used to represent the number ten to the power fourteen. See *Ananta* and **High numbers (The symbolic meaning of)**.

INFINITY. [S]. Value = 0. See **Infinity**, *Âkâsha, Ananta, Vishnupada, Shûnya* and **Zero**.

INNATE REASON. As a symbol for a large quantity. See **High numbers (Symbolic meaning of)**.

INNUMERABLE. See *Abhabâgamana, Asamkhyeya*, **High numbers** and **Infinity**.

INNUMERABLE. Term used as the name for the number ten to the power 140. See *Asankhyeya*.

INSIGNIFICANCE. See **Low numbers**, *Shûnyatâ* and **Zero**.

INTERLACING. See **Infinity (Indian mythological representation of)** and **Serpent (Symbolism of the)**.

IRYÂ. [S]. Value = 4. "Position". Allusion to the four principal positions of the human body (positions: lying flat on one's stomach, lying flat on one's back, standing up or sitting down). See **Four**.

ÎSHA. [S]. Value = 11. This is the shortened form of *Îshâna*, one of the names of *Rudra, the symbolic value of which is eleven. See **Rudra-Shiva** and **Eleven**.

ÎSHADRISH. [S]. Value = 3. The "eyes of Hara". See *Îsha, Haranetra* and **Three**.

ISHU. [S]. Value = 5. "Arrow". See *Shara* and **Five**.

ÎSHVARA. [S]. Value = 11. "Lord of the universe", "Supreme divinity". One of the attributes of *Shiva, who is an emanation of *Rudra, whose name has the symbolic value of eleven. See **Rudra-Shiva** and **Eleven**.

ISLAND-CONTINENT. [S]. Value = 7. See *Dvîpa* and **Seven**.

ISLAND-CONTINENTS (The four). See *Chaturdvîpa*.

ISLAND-CONTINENTS (The seven). See *Sapta* and *Dvîpa*.

ÎSVÎ (Calendar). See *Kristâbda*.

J

JAGAT. [S]. Value = 3. "Universe", "Phenomenal world". Here this word is taken in the sense of three "worlds". See *Loka, Triloka* and **Three**.

JAGAT. [S]. Value = 14. "World". Here the word is taken in the sense of the fourteen chosen lands of the Buddhism of the Mahâyana (including *Vaikuntha*). See *Bhuvana* and **Fourteen**.

JAGATÎ. [S]. Value = 1. "Earth". See *Prithivî* and **One**.

JAGATÎ. [S]. Value = 12. In Sanskrit poetry, this is the metre which is made up of a verse of four times twelve syllables. See **Indian metric** and **Twelve**.

JAGATÎ. [S]. Value = 48. In Sanskrit poetry, this is the metre which is made up of a verse of four times twelve syllables. See **Indian metric**.

JÂHNAVÎVAKTRA. [S]. Value 1,000. The "mouths of Jâhnavî". The name Jâhnavî denotes the river Ganges (*Gangâ*), considered to be the daughter of Jâhnu. According to legend, Jâhnu drank the river because it disturbed his prayer, but the water came out of his ears. The Sanskrit name for "thousand" (*sahasra*) often means "multiplicity" and "multitude". The swampy delta of this river is divided into many hundreds of branches, and so these "mouths" came to represent the quantity *thousand* because they are so numerous. See **Thousand** and **High numbers (Symbolic meaning of)**.

JAINA RELIGION. See **Indian religions and philosophies**.

JAINA. This is the name of an Indian religious sect. This religion seems to have been founded around the sixth century BCE by a "sage" (*muni*) named Vardhamâna, better known as *Jina*. Jaina philosophy and logic is accompanied by a very strict moral doctrine, born out of several concepts including *nayavâda* (a highly developed science of the knowledge of the real from its most diverse aspects) and *syâdvâda* (which consists of a relativist vision which is meant to adjust the affirmation and negation of things to their moving reality). Nature is divided into "categories", which are classed in different orders depending on the point of view from which they are considered.

In one of these "categories", there are "principles" and "masses of beings", the most important of which are the soul, matter, the cause of movement, the cause of the halting of movement and space (*âkâsha*). Matter is of atomic structure. Each "atom" of corpororal nature is uncreated, indivisible and indestructible, whilst possessing particular tastes, smells and colours. As for time, it is considered a substance without space, yet according to Jaina philosophy, it is made up of an infinite number of "temporal atoms" (*kâlânu*). These diverse theories are accompanied by a highly developed cosmological vision of the universe (*loka*) in which the universe is represented as a man made up of three worlds, his head forming the superior world, his body the middle world and his legs the inferior world. These three worlds are surrounded by a triple atmospheric cover, made up of air, vapour and ether (*âkâsha*), beyond which is nothing but empty space (*shûnyatâ*). This universe is organised around a hollow vertical axis, inside which live all "mobile" living beings.

Each world is divided into several stages: the inferior world; the middle world, which includes our world and the island-continents (*dvîpa*, *chaturdvîpa*); and finally the superior world, situated above *Mount Meru, the mythical mountain of Hindu and Brahmanic cosmology, which is said to be the centre of the universe where the gods live. The summit, which constitutes the "chignon" of the cosmic man, is said to be occupied by liberated souls. As for the ages of the world, the Jainas accept Brahmanic classification. Thus the fifth age (the age which we are living in) would have begun in 523 BCE and be characterised by pain. This would be followed by a sixth and last "age", 21,000 years long, at the end of which the human race would undergo terrible mutations.

However, the world would not disappear, for, according to Jaina doctrine, the universe is indestructible. This is because it is infinite, in terms of both time and space. It was in order to define their vision of this impalpable universe, situated in the unlimited and the eternal, that the Jainas began their impressive numerical speculations and thus created a science which was characteristic of their way of thinking: a "science" which, by using incredibly high numbers and constantly expanding the limits of *asamkhyeya* (the "incalculable", the "impossible to count") finally allowed them to get within reach of the world of infinite numbers. [See Frédéric, *Dictionnaire* (1987)]. See

Anuyogadvâra Sûtra, Names of numbers, High numbers and Infinity.

JALA. [S]. Value = 4. Synonymous with *âpa*, "water". This symbolism is explained by the Brahmanic doctrine of the "elements of the revelation" (*bhûta*). According to this philosophy, the universe is the result of the interaction of five "powers" (*nritya*) personified respectively by *Brahma, *Vishnu, *Rudra, Maheshvara and *Shiva. These powers are: creation (*shrishti*), conservation (*stithi*), creative emotion (*tirobhava*), destruction (*shangara*) and rest (*anugraha*). On account of these five "powers", the universe is thus the result of transformation and interaction of the "five elements" (*pañchabhûta*). These elements are respectively: ether (*âkâsha*), water (*âpa*), air (*vâyu*), fire (*agni*) and water (*prithivî*). Ether, the most subtle of the five elements, is considered to be the condition of all corporal extension and the receptacle of all matter which manifests itself in the form of any one of the other four elements. Ether is thus space, the "element which penetrates everything", the *shûnya, the "void". In other words, according to this philosophy, *âkâsha (ether) is the immobile and eternal element which is the essential condition for all manifestation, but which, by its very essence, is indescribable, and cannot be mixed with any material thing. Thus this element is not meant to participate directly with the "material order of nature", which comes from *prakriti (the supposed original material substance of the universe).

Hence we are dealing with "natural order" which is very similar to the doctrine of the great philosophers of Ancient Greece (Pythagoras, Plato, Aristotle, etc.). This doctrine states that: the various phenomena of life can be attributed to the manifestations of the elements which determine the essence of the forces of Nature, who carries out her work of generation and destruction using these elements: *water, air, fire*, and *earth*. Each one of these elements is created by the combination of two primordial constituents: water comes from coldness and humidity, air comes from humidity and heat, fire is made by heat and dryness, and earth comes from dryness and cold. Each one of these is representative of a state, liquid, gas, igneous and solid. In each of these groups is a collection of fixed conditions of life, and the groups together form a cycle, which begins with the first element (water) and ends with the last (earth), after passing through the intermediary stages (air and fire). This gives a *quaternary order of nature*, which corresponds to both the human temperaments and to the stages of human life: winter, spring, summer, autumn; midnight to dawn, dawn to midday, midday to dusk, dusk to midnight; phlegmatic, sanguine, bilious and choleric; childhood, youth, maturity and old age; learning, blossoming, culminating, declining; etc. [Chevalier and Gheerbrant (1982)]. It is thus on this basis that water (*jala*) came to symbolise the number four in Sanskrit. This quaternary symbolism is also responsible for the fact that the value of four has often been attributed to the word for "ocean" (*sâgara*). See **Four** and *Sâgara*.

JALADHARAPATHA. [S]. Value = 0. "Voyage on the water". Allusion to *Ananta, the serpent with a thousand heads, who floats on the primordial waters, or the "ocean of unconsciousness", during the space of time which separates two succesive creations of the world. This symbolism thus corresponds to the identification of infinity with zero, because Ananta is none other than the serpent of infinity and eternity. See **Zero**.

JALADHI. [S]. Value = 4. "Sea". See **Sâgara, Four, Ocean.**

JALADHI Name given to the number ten to the power fourteen (= a hundred billion). See **Names of numbers.** For an explanation of this symbolism, see **High numbers (Symbolic meaning of).**

Source: *Lîlâvatî* by Bhâskarâchârya (1150 CE).

JALANIDHI. [S]. Value = 4. "Sea". See **Sâgara, Four.** See also **Ocean.**

JAMBUDVÎPA. "Isle of the Jambu tree". Name in Hindu cosmology for the whole of the Indian subcontinent, which is situated (according to a characterised representation of the structure of the universe) to the south of *Mount Meru.

JAVANESE NUMERALS (Ancient). See **Kawi numerals.**

JAVANESE NUMERALS (Modern). Currently in use in the island of Java, in Bali, Madura and Lombok, as well as in the Sounda islands. The corresponding system functions according to the place-value system and possesses zero (in the form of a little circle). Apart from the numerals 0 and 5 (whose graphical origin is evident), this notation actually corresponds to a relatively recent graphical creation, the shape of the numerals having (curiously) become similar in appearance to some of the letters of the contemporary Javanese alphabet. The

Javanese people formerly used a notation which derived from Brâhmî numerals, which belongs to the group of numerals known as "Pali". See Fig. 24.26 and 52. See also **Indian written numeral systems (Classification of)** and **Kawi numerals.**

JEWEL. [S]. Value = 3. See *Ratna* and **Three.**

JEWEL. [S]. Value = 5. See *Ratna* and **Five.**

JEWEL. [S]. Value = 8. See *Mangala* and **Eight.**

JEWEL. [S]. Value = 9. See *Ratna* and **Nine.**

JEWEL. [S]. Value = 14. See *Ratna* and **Fourteen.**

JINA. Name of the founder of the religious sect of the *Jainas.

JINABHADRA GANI. Indian arithmetician who lived at the end of the sixth century. His works notably include *Brihatkshetrasâmâsa*, where he gives an expression for the number 224,400,000,000 in the simplified Sanskrit system using the place-value system [see Datta and Singh (1938) p. 79]. See **Indian mathematics (The history of).**

JVALANA. [S]. Value = 3. "Fire". See *Agni* and **Three.**

JYOTISHA. Sanskrit name attributed to astronomy, once it was considered to be a separate discipline from arithmetic and calculation. This name, however, (which literally means "science of the stars") was long attributed to astrology. See **Indian astrology, Ganita** and **Indian astronomy (The history of).**

JYOTISHAVEDÂNGA. "Astronomical Element of Knowledge". Name of an ancient text on astrology, notably concerning the determination of the exact dates of the sacrifices of the Brahman cult [see Billard (1971)]. See **Jyotisha, Indian astrology** and **Indian astronomy (The history of).**

K

KÂCHCHÂYANA. Grammarian from Sri Lanka who is believed to have written the *Vyâkarana*, a Pali grammar divided into eight parts. He probably lived during the eleventh century CE. Here is a list of the principal names of numbers mentioned in this work:

*Koti (= 10^7), *Pakoti (= 10^{14}), *Kotippakoti (= 10^{21}), *Nahuta (= 10^{28}), *Ninnahuta (= 10^{35}), *Akkhobhini (= 10^{42}), *Bindu (= 10^{49}), *Abbuda (= 10^{56}), *Nirabbuda (= 10^{63}), *Ahaha (= 10^{70}), *Ababa (=10^{77}), *Atata (= 10^{84}), *Sogandhika

(= 10^{91}), *Uppala (= 10^{98}), *Kumuda (= 10^{105}), *Pundarîka (= 10^{112}), *Paduma (= 10^{119}), *Kathâna (= 10^{126}), *Mahâkathâna (= 10^{133}), *Asankhyeya (= 10^{140}).

See **Names of numbers** and **High numbers**

Source: Vyâkarana by Kâchchâyana [see JA, 6th Series, XVII, 1871, p. 411, line 51–52].

KAÎTHÎ NUMERALS. Signs derived from *Brâhmî numerals, through the intermediary of Shunga, Shaka, Kushâna, Ândhra, Gupta, Nâgarî, Kutila and Bengali numerals. Currently in use in Bihar state, in the east of India, and sometimes used in Gujarat state. The corresponding system functions according to the place-value system and possesses zero (in the form of a little circle). See also Fig. 24.9, 52 and 61 to 69. See **Indian written numeral systems (Classification of)**.

KAKUBH. [S]. Value = 10. "Horizon". See *Dish* and **Ten**.

KÂLA. [S]. Value = 3. "Time". In Brahman mythology, time is personified by the terrible Kâla, Lord of Creation and Destruction. He is often identified as Shiva holding his Trident (*trishûlâ*), which symbolises the three aspects of the revelation (creation, preservation, destruction), as well as the three primordial properties (*guna*) and the three states of consciousness. Here, the word is synonymous with *trikâla*, "three times". See **Guna, Shûla, Triguna** and **Three**.

KALACHURI (Calendar). See *Chhedi*.

KALAMBA. [S]. Value = 5. "Arrow". See *Shara* and **Five**.

KÂLÂNU. "Temporal atom". In *Jaina philosophy, time (*kâla*) is made up of an infinite number of temporal atoms (atom = *anu*).

KALÂSAVARNA. Word used in arithmetic to denote "fundamental operations" carried out on fractions. See *Parikarma*.

KALIYUGA (Calendar). Calendar of fictitious times, which is sometimes referred to in Hindu religious texts and Indian astronomical texts. It begins on the 18 February of the year 3101 BCE. Characteristically, the beginning of this calendar is traditionally related to a theoretical starting point of celestial revolutions corresponding to a supposed general conjunction in average longitude with the starting point of the sidereal longitudes of the sun, the moon and the planets (the ascending apogees and node of the moon being respectively at 90° and 180° of these longitudes). To find the corresponding date in our calendar, simply subtract 3,101

from a date in the *Kaliyuga* calendar. See *Kaliyuga*, **Indian calendars** and *Yuga* **(Astronomical speculation)**.

KALIYUGA. Name of the last of the four cosmic calendars which make up a *mahâyuga*. This cycle, said to be 432,000 human years long, is the "iron age", during which living things only live for a quarter of their existence and the forces of evil triumph over good: we are living in this age now, and it is meant to end with a *pralaya* (destruction by fire and water). See *Yuga* **(Definition)**, *Yuga* **(Systems of calculating)** and *Yuga* **(Cosmogonical speculations about)**.

KALPA. Unit of cosmic time which, according to Indian speculations, corresponds to the length of 1,000 *mahâyugas*. Thus one Kalpa corresponds to 4,320,000,000 human years. See *Yuga* **(Definition)**.

KALPA. [Astronomy]. According to Brahmagupta (628 CE), the *kalpa* cycle, or period of 4,320,000,000 years, is delimited by two perfect conjunctions in real longitude of the totality of elements, each one accompanied by a total eclipse of the sun at exactly six o'clock in *Ujjayinî. See *Kalpa* (first entry above) and *Yuga* **(Astronomical speculations about)**.

KALPAS (Cosmogonical speculations about). In Buddhist cosmogony, the term *kalpa* denotes an infinite length of time. The *kalpa* is made up of four periods: the creation of worlds, the lifespan of existing worlds, the destruction of worlds and the duration of existence of chaos. During the period of creation the different universes are formed with their living beings. The second period sees the appearance of the sun and the moon, the differentiation between the sexes and the development of social life. During the phase of destruction, fire, water and wind destroy everything apart from the fourth *dhyâna*. Chaos represents total annihilation. These four phases make up one "big" *kalpa* (*mahâkalpa*); Each one of them is made up of twenty "little" *kalpas*, which themselves are broken down into fire, bronze, silver and golden ages. During the entire creation phase of a "little" *kalpa*, the life expectancy of humans increases by one year per century until it reaches 84,000 years. In a parallel fashion, the human body grows to a height of 84,000 feet. During the "little" *kalpa's* period of disappearance, which is made up of successive phases of plague, war and famine, human life is shortened to ten years and the human body returns to the height of one foot.

[This article is taken from the *Dictionnaire de la sagesse orientale*, Friedrichs, Fischer-Schreiber, Erhard and Diener (1989)]. See *Kalpa* (first entry) and **Day of Brahma**.

KÂMA. [S]. Value = 13. Name of the Hindu divinity of Cosmic Desire and Carnal Love whose action decides the laws of the reincarnation of living beings (**samsâra*). Kâma presides over the thirteenth lunar day. See **Thirteen** and *Pañchabâna*.

KAMÂLAKARA. Indian astronomer of the seventeenth century CE. His works notably include *Siddhântatattvaviveka*, in which the place-value system with Sanskrit numerical symbols is frequently used [see Dvivedi (1935)]. See **Numerical symbols, Numerical symbols (Principle of the numeration of)**, and **Indian mathematics (The history of)**.

KANKARA. Name given to the number ten to the power thirteen (= ten billion). See **Names of numbers** and **High numbers**.

KANNARA NUMERALS. Signs derived from **Brâhmî* numerals, through the intermediary of Shunga, Shaka, Kushâna, Ândhra, Pallava, Chalukya, Ganga, Valabhî and Bhattiprolu numerals. Currently used by the Dravidians of Karnataka state and part of Andhra Pradesh. They are also called Kannada (or even Karnata) numerals. The corresponding system today uses the place-value system and zero (in the from of a little circle). For ancient numerals, see Fig. 24.48. For modern numerals, see Fig. 24.21. See also Fig. 24.52 and 24.61 to 69. See **Indian written numeral systems (Classification of)**.

KARA. [S]. Value = 2. "Hand". This is because of the symmetry of the two hands. See **Two**.

KARAHU. Name given to the number ten to the power thirty-three. See **Names of numbers** and **High numbers**.

Source: **Lalitavistara Sûtra* (before 308 CE).

KARANA. Name of the astronomical formula employing, for example, in the workings of real longitudes, the interpolation – generally linear – of tabulated values. See **Indian astronomy (The history of)** and *Yuga* (**Astronomical speculation on**).

KARANAPADDHATI. See *Putumanasomayâjin*.

KARANÎYA. [S]. Value = 5. "That which must be done". This refers to the five major observances of **Jaina* religion, which constitute the basic rules of their philosophy: not to harm living beings (*ahimsâ*); not to be false (*sunrita*); not to steal (*asteya*); carnal discipline (*brahmâchârya*); and detachment from earthly possessions (*aparigraha*). See **Five**.

KARNATA NUMERALS. See **Kannara numerals**.

KARNIKÂCHALA. One of the names of **Mount Meru.* See *Adri, Dvîpa, Pûrna, Pâtâla, Sâgara, Pushkara, Pâvana* and *Vâyu*.

KÂRTTIKEYA. Hindu divinity of war and the planet Mars, son of Shiva, often likened to **Kumâra*.

KÂRTTIKEYÂSYA. [S]. Value = 6. "Faces of **Kârttikeya*". Allusion to the six heads of this divinity. See **Six**.

KATAPAYA (Spoken numeration). See **Katapayâdi numeration**.

KATAPAYÂDI NUMERATION. Method of writing numbers using the letters of the Indian alphabet. In this system, the numerical attribution of of syllables corresponds to the following rule, according to the regular order of succession of the letters of the Indian alphabet (see Fig. 24. 56): the first nine letters (*ka, kha, ga, gha, na, cha, chha, ja* and *jha*) represent the numbers 1 to 9, whilst the tenth (*ña*) corresponds to zero; the following nine letters (*ta, tha, da, dha, na, ta, tha, da, dha*) also receive the values 1 to 9, whilst the following letter (*na*) has the value of 0; the next five (*pa, pha, ba, bha, ma*) represent the first five units; and the last eight (*ya, ra, la, va, sha, sha, sa* and *ha*) represent the numbers 1 to 8.

Thus each simple unit is represented by two, three or four different letters: numeral 1 by one of the letters *ka, ta, pa* or *ya* (hence *katapaya* is the name of the system); 2 by *kha, tha, pha* or *ra*; 3 by *ga, da, ba, la*; 4 by *gha, dha, bha* or *va*; 5 by *na, na, ma* or *sha*; 6 by *cha, ta* or *sha*; 7 by *chha, tha* or *sa*; 8 by *ja, da* or *ha*; 9 by *jha* or *dha*; and 0 by *ña* or *na*. This system is infinitely simpler than Âryabhata's.

The complete key is given in the following lines, which are an extract from *Sadratnamâlâ* by Shankaravarman:

Nañâvachashacha shûyâni Samkhyâ katapayâdayah

Mishre tûpânta hal samkhya Na cha chinty-ohalasvarah

Translation: "[The letter] *na* and *a*, as well as the vowels, are zero. [The letters] starting with *ka, ta, pa, ya*, represent the numbers [from 1 to 9]. When two consonants are joined, only the last one corresponds to a number. And a consonant which is not joined to a vowel is insignificant." [See EI, VI, p. 121; Datta and Singh, p. 70]. In other words, in this system, the vowels and the consonants which are not vocalised have no numerical value; and groups such as *ksha*, *tva*, *ktya*, etc., often considered as unitary in Indian alphabets, receive respectively the same value as the letters *sha*, *va*, *ya*, etc. The letters *ña* and *na*, represent zero. Thus the vocalised consonants are the only "numerals" in the system, their numerical value being entirely independent of the vocalisations in question. This means that, unlike Âryabhata's system, there is no difference between syllables such as *sa*, *si*, *su*, *se*, *so*, *sai*, etc. In fact, this system constitutes a simplification of Âryabhata's alphabetical numeration. See **Âryabhata** and **Âryabhata's numeration**.

Historically, the first author who is known to have used this system employing the name of *katapayâdi* is the astronomer Shankaranârâyana, author of a work entitled *Laghubhahâskarîyavivarana* written in 869 CE. This date is given by the author himself, and is expressed as the *Shaka* year 791, which is 791 + 78 = 869 CE.

However, the latter did not invent *katapayâdi*, because the system had already appeared, under the name of *varnasamjñâ*, "from syllables", in *Grahachâranibandhana* by the astronomer Haridatta, for which there is overwhelming evidence to suggest that he was the inventor of this system. First, there is no mention is made of the system by his predecessors; secondly, in his work (where he makes frequent reference to Âryabhata), he takes the trouble to give all the details (like the inventor of a new system who feels obliged to explain it to readers who are used to using a different method); finally, it is Haridatta who is the first and last person to explain the system, which suggests that afterwards it became common knowledge. [Personal communication of Billard]. According to a tradition in Kerala, Haridatta wrote his text in 684 CE [see Sarma (1954), p. v]. However, this date does not seem to correspond to a significant piece of evidence found in the work of astronomer Shankaranârâyana, where he is paying homage

	LETTERS USED			
for the numeral 1	क *ka*	ट *ta*	प *pa*	य *ya*
for the numeral 2	ख *kha*	ठ *tha*	फ *pha*	र *ra*
for the numeral 3	ग *ga*	ड *da*	ब *ba*	ल *la*
for the numeral 4	घ *gha*	ढ *dha*	भ *bha*	व *va*
for the numeral 5	ङ *ṅa*	ण *ṇa*	म *ma*	श *sha*
for the numeral 6	च *cha*	त *ta*		ष *sha*
for the numeral 7	छ *chha*	थ *tha*		स *sa*
for the numeral 8	ज *ja*	द *da*		ह *ha*
for the numeral 9	झ *jha*	ध *dha*		
for the numeral 0	ञ *ña*	न *na*		

FIG. 24D.7 *Letter-numerals of the "Katapayâdi system". Ref. : Datta and Singh (1938); Jacquet (1835); Pihan (1860); Renou and Filliozat (1953); Sarma (1954)*

to his illustrious predecessors, and quotes their respective names, using the word *kramâd*, which means "in the order":

1. *Âryabhata* [c. 510 CE]
2. *Varâhamihîra* [c. 575 CE]
3. *Bhâskara* [c. 629 CE]
4. *Govindasvâmin* [c. 830 CE]
5. *Haridatta.*

Thus Haridatta is placed after Govindasvâmin, of whom Shankaranârâyana was a disciple. Such a list is very rare for an Indian scholar; chronology was not generally of much interest to them. It seems even more remarkable in light of the fact that Indian astronomical texts are usually very sparing with historical facts, and it is very rare to find a reference to another text. If mention is made of earlier authors, the list is usually in order, to aid the rhythm of the versification. This is the only known example of such a list accompanied by a chronological indication. In short, if Haridatta's work was written before 869 (the date of Shankaranârâyana's text), it probably dates back to c. 850 CE. (Personal communication of Billard.) This means that *katapayâdi* numeration was not created until the middle of the ninth century, three centuries after Âryabhata. Through abandoning the method of successively vocalising the consonants of the Indian alphabet, and replacing each value which was equal to or higher than ten with a zero or one of the nine numerals, the inventor of the *katapayâdi* system transformed Âryabhata's system into a place-value system equipped with a zero.

The proof of this is in the following mention in an anonymous text from the tenth century, where there is frequent use of the *katapayâdi* notation: *vibhâvonashakâbdam* . . .

"The *Shaka* date decreased by 444 . . ." Ref.: *Grahachâranibandhanasamgraha*, line 17; Billard (1971), p. 142.

This mention contains the expression *vibhâvona* (= *vibhâva* + *ûna*) which means "444 (= *vibhâva*) decreased by". Bearing in mind the principle of this notation, where the value of a consonant is independent of its vocalisation, and where the numbers are expressed in ascending order starting with the smallest units, the number 444 is written as follows:

vibhâva (= *va.bha.va*).

According to the values of the numeral letters in the *katapayâdi* system, this gives the following (Fig. D. 7):

$$(va) (bhâ) (va) = 4 + 4 \times 10 + 4 \times 10^2$$
$$4 \quad 4 \quad 4 \quad = 444.$$

Thus the numeral letters are combined and are never modified by vowels; these can be inserted wherever necessary, as they have no numerical value. As for the principle of the notation, which is the rule of position applied to any of the nine letter-units and the two letter-zeros, it follows the ascending powers of ten starting with the smallest unit, as it does with the numerical symbols. Here is another example found in an astronomical table of Haridatta's *Grahachâranibandhana* (II, 14), giving a trigonometric function for Saturn [see Sarma (1954), p. 12]:

dhanâdhya	*dha.nâ.dhya*
= *dhîrâdhya*	= *dhî.râ.dhya*
= *dha.nâ.ya*	= *dha.ra.ya*
9 0 1	9 2 1
= 109	= 129

This is more proof that Âryabhata was fully aware of zero and the place-value system, but by confining himself as he did to his system of vocalisation, he made it impossible for his numeration to be positional. (See **Ankânâm vâmato gatih**.)

It is surprising to note the numbers of letters that could be used to record the same numeral. In fact, this system, like the notation which inspired it, offered many possibilities to the mnemonics of numbers. Moreover, it was perfectly capable of meeting the needs of the rules of versification or prosody, with the advantage of being especially useful when reproducing abundant tables of trigonometric functions in a much shorter form than the system of *numerical symbols. Added to the possibility of expressing a given numeral with many different letters was the ability to vocalise these letters without changing the values they expressed. Thus it was always possible to find several intelligible words to express a number. This is doubtless the reason why this system came to be used, in various forms, in southern India (the notation in this case being applied to letters of the Grantha, Tulu, Telugu (etc.) alphabets).

KÂTHAKA SAMHITÂ. Text derived from the *Yajurveda* "black". It figures amongst the texts of Vedic literature. Passed down through oral transmission since ancient times, it only found its definitive form at the beginning of Christianity. See **Veda**. Here is a list of the main names of numbers mentioned in this text:

Eka (= 1), *Dasha* (= 10), *Sata* (= 10^2), *Sahasra* (= 10^3), *Ayuta* (= 10^4), *Prayuta* (= 10^5), *Niyuta* (= 10^6), *Arbuda* (= 10^7),

*Nyarbuda (= 10^8), *Samudra (= 10^9), *Vâdava (= 10^9), *Madhya (= 10^{10}), *Anta (= 10^{11}), *Parârdha (=10^{12}).

See **Names of numbers** and **High numbers**.

Ref.: *Kâthaka Samhitâ*, XVII, 10 [see Datta and Singh (1938), p. 10].

KATHÂNA. Name given to the number ten to the power 119 See **Names of numbers** and **High numbers**.

Ref.: *Vyâkarana* (Pali grammar) by Kâchchâyana (c. eleventh century).

Gutturals	क	ख	ग	घ	ङ
	ka	kha	ga	gha	ṅa
Âryabhata's system[1]	1	2	3	4	5
Katapayâdi system[2]	1	2	3	4	5
Palatals	च	छ	ज	झ	ञ
	cha	chha	ja	jha	ña
Âryabhata's system	6	7	8	9	10
Katapayâdi system	6	7	8	9	0
Cerebrals	ट	ठ	ड	ढ	ण
	ta	tha	da	dha	ṇa
Âryabhata's system	11	12	13	14	15
Katapayâdi system	1	2	3	4	5
Dentals	त	थ	द	ध	न
	ta	tha	da	dha	na
Âryabhata's system	16	17	18	19	20
Katapayâdi system	6	7	8	9	0
Labials	प	फ	ब	भ	म
	pa	pha	ba	bha	ma
Âryabhata's system	21	22	23	24	25
Katapayâdi system	1	2	3	4	5
Semivowels	य	र	ल	व	
	ya	ra	la	va	
Âryabhata's system	30	40	50	60	
Katapayâdi system	1	2	3	4	
Sibilants	श	ष	स		
	sha	sha	sa		
Âryabhata's system	70	80	90		
Katapayâdi system	5	6	7		
Aspirates	ह				
	ha				
Âryabhata's system	100				
Katapayâdi system	8				

1. See Fig. D.2, p. 448
2. See Fig. D.7, p. 474

FIG. 24D.8

KAVACHA. Literally "Charm, armour". This is the name for Tantric talismans and amulets. See **Numeral alphabet, magic, mysticism and divination.**

KAWI NUMERALS. Signs derived from *Brâhmî numerals, through the intermediary of Shunga, Shaka, Kushâna, Ândhra, Pallava, Chalukya, Ganga, Valabhî, "Pali" and Vatteluttu numerals. Formerly used (since the seventh century CE) in Java and Borneo. These are the numerals of Old Javanese writing. The corresponding system uses the place-value system and zero (in the form of a little circle). See Fig. 24.50, 52, 61 to 69 and 80. See also **Indian written numeral systems (Classification of).**

KÂYA. [S]. Value = 6. "Body". Allusion to the *trikâya*, the "three bodies" that a Buddha can assume simultaneously, and which are often associated with the "three spheres" of Buddhas' existence. Thus the symbolic sum: 3 + 3 = 6. See **Six.**

KESHAVA. [S]. Value = 9. This concerns one of the epithets of *Vishnu (and *Krishna). The symbolism is due to the fact that *keshava* is another name for the month of *mârgâshirsha*, the ninth month of the *chaitrâdi* year. See **Nine.**

KHA. [S]. Value = 0. Word meaning "space". This symbolism is explained by the fact that space is nothing but a "void". See *Shûnya* and **Zero.**

KHACHHEDA. Sanskrit term used to denote infinity. Literally "divided by zero" (from *kha*, "space" as a symbol for zero, and *chheda*, "the act of breaking into many parts", "division"). Thus it is the "quantity whose denominator is zero". The term is used notably in this sense by *Brahmagupta in his *Brahmasphutasiddânta* (628 CE). See *Chhedana*, **Infinity** (entries beginning with), **Zero** and **Indian mathematics (The history of).**

KHAHÂRA. Sanskrit word for infinity. Literally "division by zero". Notably used by *Bhâskarâchârya. See *Khachheda*.

KHAMBA. Name given to the number ten to the power thirteen (= ten billion). See **Names of numbers** and **High numbers.**
Source: *Lalitavistara Sûtra* (before 308 CE).

KHAROSHTHÎ ALPHABET. See Fig. 24. 28.

KHAROSHTHÎ NUMERALS. Numerals derived from the numerical notations of western Semitic civilisations. This is attested notably in the edicts of Asoka written in Aramaean Indian. The corresponding system does not use the place-value system, nor does it possess zero. See **Indian written numeral systems (Classification of).** See Fig. 24.54.

KHARVA. Name given to the number ten to the power ten (ten thousand million). See **Names of numbers** and **High numbers.**
Sources: *Kitab fi tahqiq i ma li'l hind by al-Biruni (c. 1030 CE); *Lîlâvatî by Bhâskarâchârya (1150 CE); *Ganitakaumudî by Nârâyana (1350 CE); *Trishatikâ by Shrîdharâchârya (date unknown).

KHARVA. Name given to the number ten to the power twelve (= one billion). See **Names of numbers** and **High numbers.**
Source: *Ganitasârasamgraha by Mahâvîrâchârya (850 CE).

KHARVA. Name given to the number ten to the power thirty-nine. See **Names of numbers** and **High numbers.**
Source: *Râmâyana by Vâlmîki (in the first centuries of the Common Era).

KHMER NUMERALS. For modern numerals, see **Thai-Khmer numerals.** For ancient numerals, see **Old Khmer numerals.**

KHUDAWADI NUMERALS. Signs derived from *Brâhmî numerals, through the intermediary of Shunga, Shaka, Kushâna, Ândhra, Gupta and Shâradâ numerals, and constituting a slight variation of Sindhi numerals. Once used by the merchants of Hyderabad (a town of Sind, built on the delta of the Indus, to the east of Karachi, not to be confused with the other Hyderabad, capital of Andhra Pradesh). The corresponding system functions according to the place-value system and possesses zero (in the form of a little circle). See **Indian written numeral systems (Classification of)** and Fig. 24.6, 52 and 61 to 69.

KING. [S]. Value = 16. See *Bhûpa*, *Nripa* and **Sixteen.**

KITAB FI TAHQIQ I MA LI'L HIND. Arabic work by al-Biruni, which constitutes one of the most important pieces of evidence about Indian civilisation at the beginning of the eleventh century. See **al-Biruni.**

KOLLAM (Calendar). Beginning in 825 CE, created by the sovereign of the town of the same name situated in Kerala near to Travancore, on the Malabar coast. To find the corresponding date in the Common Era, add 825 to a date expressed in Kollam years. This calendar is also called *Parashurâma*. It is rarely used. See **Indian calendars.**

KOTI. Name given to the number ten to the power seven (= ten million). See **Names of numbers** and **High numbers**.

Sources: **Râmâyana* by Vâlmîki (in the first centuries CE); **Lalitavistara Sûtra* (before 308 CE); **Âryabhatîya* (510 CE); **Ganitasârasamgraha* by Mahâvîrâchârya (850 CE); **Kitab fi tahqiq i ma li'l hind* by al-Biruni (c. 1030 CE); **Vyâkarana* (Pali grammar) by Kâchchâyana (eleventh century CE); **Lilâvati* by Bhâskarâchârya (1150 CE); **Ganitakaumudi* by Nârâyana (1350 CE); **Trishtikâ* by Shrîdharâchârya (date unknown).

KOTIPPAKOTI. Name given to the number ten to the power twenty-one (= quintillion). See **Names of numbers** and **High numbers**.

Source: **Vyâkarana* (Pali grammar) by Kâchchâyana (eleventh century CE).

KRAMÂD (KRAMÂT). Word meaning "in the order". See *Sthâna, Sthânakramâd* and *Ankakramena.*

KRISHÂNU. [S]. Value = 3. "Fire". See *Agni* and **Three.**

KRISHNA. See *Avatâra.*

KRISTÂBDA (Calendar). Name given to the Christian calendar. It is also referred to as *îsvî* or *imrâjî.* See **Indian calendars.**

KRITA. [S]. Value = 4. The name of the first of four cosmic cycles (**krityuga*) which make up a **chaturyuga* (or **mahâyuga*) in Brahman cosmogony. The symbolism is not due to the fact that the *krityuga* was the "first age" of the world, but because it inaugurated a new *chaturyuga.* Thus it began a new cosmic cycle composed of four periods corresponding to the life of a universe. See *Yuga* and **Four.**

KRITAYUGA. Name of the first of the four cosmic eras which make up a **mahâyuga* (or **chaturyuga*). This cycle, said to last 1,728,000 human years, is regarded as the "golden age" during which humans have an extremely long life and everything is perfect. See *Yuga.*

KRITI. [S]. Value = 20. In Sanskrit poetry, this is a metre of four times twenty syllables. See **Indian metric** and **Twenty.**

KSHAPESHVARA. [S]. Value = 1. "Moon". See *Abja* and **One.**

KSHATRAPA NUMERALS. Signs derived from **Brâhmî numerals, through the intermediary of Shunga, Shaka and Kushâna numerals. Contemporaries of the western Satraps (second to fourth century CE). The corresponding system did not function according to the place-value system and moreover did not possess zero. See **Indian written numeral systems**

(Classification of). See also Fig. 24.35, 52, 24.61 to 69 and 70.

KSHAUNÎ. [S]. Value = 1. "Earth". See *Prithivî* and **One.**

KSHEMÂ. [S]. Value = 1. "Earth". See *Prithivî* and **One.**

KSHETRAGANITA. Term used in early times meaning geometry. See *Ganita.*

KSHITI. Literally "earth". Name given to the number ten to the power twenty (= a hundred quadrillions). See **Names of numbers.** For an explanation of this symbolism, see **High numbers (Symbolic meaning of).**

Source: **Ganitasârasamgraha* by Mahâvîrâchârya (850 CE).

KSHITI. [S]. Value = 1. "Earth". See *Prithivî* and **One.**

KSHOBHA. Name given to the number ten to the power twenty-two (= ten quintillions). See **Names of numbers** and **High numbers.**

Source: **Ganitasârasamgraha* by Mahâvîrâchârya (850 CE).

KSHOBHYA. Literally "Movement". Name given to the number ten to the power seventeen. This name might have been attributed to such a high number because of the "endless movement" of the waves, since "ocean" (**samudra, *jaladhi*) was also sometimes used to represent large quantities. See **Names of numbers** and **High numbers.**

Source: **Lalitavistara Sûtra* (before 308 CE).

KSHONI. Literally "earth". Name given to the number ten to the power sixteen. See **Names of numbers.** For an explanation of this symbolism, see **High numbers (Symbolic meaning of).**

Source: **Ganitasârasamgraha* by Mahâvîrâchârya (850 CE).

KSHONI. [S]. Value = 1. "Earth". See *Prithivî* and **One.**

KU. [S]. Value = 1. "Earth". See *Prithivî* and **One.**

KUMÂRA. See *Kumârâsya, Kumâravadana* and *Kârttikeya.*

KUMÂRÂSYA. [S]. Value = 6. "Faces of **Kumâra". Allusion to the six heads of **Kârttikeya. See *Kumâra* and **Six.**

KUMÂRAVADANA. [S]. Value = 6. "Faces of **Kumâra". Allusion to the six heads of **Kârttikeya. See *Kumâra* and **Six.**

KUMUD. Literally "(pink-white) lotus". Name given to the number ten to the power twenty-one (= quintillion). See **Names of numbers.** For an explanation of this symbolism, see **High numbers (Symbolic meaning of).**

Source: **Lalitavistara Sûtra* (before 308 CE).

KUMUDA. Literally "pink-white lotus". Name given to the number ten to the power 105. See **Names of numbers.** For an explanation of this symbolism, see **High numbers (Symbolic meaning of).**

Source: *Vyâkarana* (Pali grammar) by Kâchchâyana (eleventh century CE).

KUÑJARA. [S]. Value = 8. "Elephant". See *Diggaja* and *Eight.*

KUSHÂNA NUMERALS. Signs derived from *Brâhmî numerals, through the intermediary of Shunga and Shaka numerals. Contemporaries of the Kushâna dynasty (first to second century CE). The corresponding system did not function according to the place-value system and moreover did not possess zero. See **Indian written numeral systems (Classification of).** See also Fig. 24.33, 52, 24.61 to 69 and 70.

KUTILA NUMERALS. Signs derived from *Brâhmî numerals, through the intermediary of Shunga, Shaka, Kushâna, Ândhra, Gupta and Nâgarî numerals. Formerly used in Bengal and Assam. The corresponding system was based on the place-value system and possessed zero (in the form of a little circle). These numerals were the ancestors of Bengali, Oriyâ, Gujarâtî, Kaîthî, Maithilî and Manipuri numerals. See **Indian written numeral systems (Classification of).** See Fig. 24.52 and 24.61 to 69.

KUTTAKAGANITA. In algebra, the name given to the part related to the analysis of indeterminate equations of the first degree. See **Indian mathematics (The history of).**

L

LABDHA. Term used in arithmetic to denote the quotient of a division. Synonym: *labdhi*. See *Bhâgahâra, Chhedana* and *Shesha.*

LAGHUBHÂSKARÎYAVIVARANA. See Shankaranârâyana.

LAKH. Name given to the number ten to the power five (= a hundred thousand). See **Names of numbers** and **High numbers.**

Source: *Lalitavistara Sûtra* (before 308 CE).

LAKKHA. Name given to the number ten to the power five (= a hundred thousand). See **Names of numbers** and **High numbers.**

Source: *Vyâkarana* (Pali grammar) by Kâchchâyana (eleventh century CE).

LAKSHA. Name given to the number ten to the power five (= a hundred thousand). See **Names of numbers** and **High numbers.**

Source: *Ganitasârasamgraha* by Mahâvîrâchârya (850 CE); *Kitab fi tahqiq i ma li'l hind* by al-Biruni

(c. 1030 CE); *Lîlâvatî* by Bhâskarâchârya (1150 CE); *Ganitakaumudî* by Nârâyana (1350 CE); *Trishtikâ* by Shrîdharâchârya (date unknown).

LAKSHAMANA (Calendar). This calendar begins in the year 1118 CE. To find the corresponding date in the Common Era, add 1118 to the date expressed in the *Lakshamana* calendar. Formerly used in the region of Mithila (north of Bihar). See **Indian Calendars.**

LALITAVISTARA SÛTRA. "Development of games'. Sanskrit text on the Buddhism of the Mahâyâna, written in verse and prose, about the life of Buddha, as he is said to have recounted it to his own disciples, where there is constant reference to numbers of gigantic proportions. This text is in fact a relatively recent compilation of ancient stories and legends. It is clearly later than the *Vâjasaneyî Samhitâ* (written at the start of the Common Era) but not later than the beginning of the fourth century, because the *Lalitavistara Sûtra* was translated into Chinese by Dharmarâksha in the year 308 CE. Here is a list of some of the names of high numbers mentioned in the text:

Lakh (= 10^5), *Koti* (= 10^7), *Nahut* (= 10^9), *Ninnahut* (= 10^{11}), *Khamba* (= 10^{13}), *Viskhamba* (= 10^{15}), *Abab* (= 10^{17}), *Attata* (= 10^{19}), *Kumud* (= 10^{21}), *Gundhika* (= 10^{23}), *Utpala* (= 10^{25}), *Pundarîka* (= 10^{27}), *Paduma* (= 10^{29}).

Here is another list of high numbers mentioned in the text (legend of Buddha):

Ayuta (= 10^9), *Niyuta* (= 10^{11}), *Kankara* (= 10^{13}), *Vivara* (= 10^{15}), *Kshobhya* (= 10^{17}), *Vivaha* (= 10^{19}), *Utsanga* (= 10^{21}), *Bahula* (= 10^{23}), *Nâgabala* (= 10^{25}), *Titilambha* (= 10^{27}), *Vyavasthânaprajñapati* (= 10^{29}), *Hetuhila* (= 10^{31}), *Karahu* (= 10^{33}), *Hetvindriya* (= 10^{35}), *Samâptalambha* (= 10^{37}), *Gananâgati* (= 10^{39}), *Niravadya* (= 10^{41}), *Mudrâbala* (= 10^{43}), *Sarvabala* (= 10^{45}), *Visamjñagati* (= 10^{47}), *Sarvajña* (= 10^{49}), *Vibhutangamâ* (= 10^{51}), *Tallakshana* (= 10^{53}), *Dhvajâgravati* (= 10^{99}), *Dhvajâgranishâmani* (= 10^{145}), etc.

See **Names of numbers** and **High numbers.** [See Lal Mitra (1877); Datta and Singh (1938), pp. 10–11; Woepcke (1863)].

LALLA. Indian astronomer who lived in the ninth century CE. His works notably include an interesting astronomical text entitled *Shishyadhîvriddhidatantra*, in which there is abundant usage of the place-value system recorded by means of *Sanskrit numerical symbols [see Billard (1971), p. 10]. See **Numerical symbols (Principle of the numeration of)**, and **Indian mathematics (The history of).**

LAUKIKAKÂLA (Calendar). See **Laukikasamvat.**

LAUKIKASAMVAT (Calendar). Beginning in 3076 BCE, this calendar was formerly used in Punjab and Kashmir. To find the corresponding date in the Common Era, take away 3076 from a date expressed in the *Laukikasamvat* calendar. This calendar also goes by the names of *Laukikakâla, Lokakâla, Saptarishikâl*a, etc. See **Indian calendars.**

LEGEND OF BUDDHA. See **Buddha (Legend of)** and *Lalitavistara Sûtra.*

LÎLÂVATÎ. "The (female) player". Name of a mathematical work from the twelfth century CE written in a highly poetic style. See **Bhâskarâchârya** and **Indian mathematics (The history of).**

LINGA (LINGAM). Literally "sign". Erected stone, in the shape of a prism or cylinder, phallic in appearance, which, in Hinduism, represents the universe and fundamental nature, complement of the **yoni,* the "feminine vulva", which is symbolised by a stone lying on its side and represents manifest energy [see Frédéric, *Dictionnaire* (1987)].

LOCHANA. [S]. Value = 2. The (two) "eyes". See **Two.**

LOKA. [S]. Value = 3. "World". Division of the Hindu universe. There are three *loka*: the earth (*bhurloka*), the space between the earth and the sun (*bhuvarloka*), and the space between the sun and the pole star (*svarloka*). In Buddhism, there are also three *lokas* and these represent the "spheres" of existence which make up the universe: *kâmaloka* (the world of sensations), *rûpaloka* (world of shapes or forms), and *arûpaloka* (the formless, immaterial world) [see Frédéric *Dictionnaire* (1987)]. See *Triloka* and **Three.**

LOKA. [S]. Value = 7. "World". Here the allusion is to another classification which tells of the existence of seven superior worlds: *bhurloka* (the earth); *bhuvarloka* (the space between the earth and the sun, supposedly the home of the **muni*, the **siddha*, etc.); the *svarloka* (the sky of **Indra); maharloka* (where Bhrigu and many other "saints" are said to reside); *janaloka* (the land of the three sons of **Brahma); taparloka* (home of the *vairâja*); and *satyaloka* or *brahmaloka* (the domain of Brahma). These seven superior worlds defend themselves against seven **pâtâla* ("inferior worlds") [see Frédéric *Dictionnaire*, (1987)]. See **Seven.**

LOKA. [S]. Value = 14. "world". See *Bhuvana* and **Fourteen.**

LOKAKÂLA (Calendar). See *Laukikasamvat.*

LOKAPÂLA. [S]. Value = 8. "Guardian of the horizons". In Hindu mythology, this is the name of the eight divinities who are guardians of the eight "horizons" and the eight points of the compass, who are represented as warriors in armour riding elephants. See *Diggaja, Dikpâla, Dish* and **Eight.**

LOKAVIBHÂGA. "Parts of the Universe". A **Jaina cosmological text which possesses the very exact date of Monday, August 25th of the year 458 CE in the Julian calendar. It is the oldest Indian text known to be in existence which contains zero and the place-value system expressed in numerical symbols [see Anon. (1962), chapter IV, line 56, p. 79]. See *Anka, Ankakramena, Sthâna, Sthânakramâd* and *Ankasthâna.*

LORD OF THE UNIVERSE. [S]. Value = 11. See **Îshvara, Rudra-Shiva** and **Eleven.**

LOTUS. This flower is the most famous symbol in all of Asia. It symbolises the pure spirit leaving the impure vessel of the body. It is the very image of divinity, which remains intact and is never soiled by the troubled waters of this world. A whole symbolism has developed around the lotus, according to its colour, the number of petals, and whether it is open, fresh-blown or in bud [see Frédéric, *Le Lotus* (1987)]. Thus it is not surprising that Indian arithmetic is full of related vocabulary and that such symbolism was often used to express very high numbers. In many texts, the words **padma, *paduma, *utpala, *pundarika* (also spelt **pundarika), *kumud* and **kumuda* (which all literally mean "lotus") express numbers such as: ten to the power four, ten to the power nine, ten to the power fourteen, ten to the power twenty-one, ten to the power twenty-five, ten to the power twenty-seven, ten to the power twenty-nine, ten to the power ninety-eight, ten to the power 105, ten to the power 112 or ten to the power 119. See **High numbers (The symbolic meanings of).**

LOW NUMBERS. See *Paramânu.*

LUMINOUS. [S]. Value = 1. See *Chandra* and **One.**

LUNAR MANSION. [S]. Value = 27. See *Nakshatra* and **Twenty-seven.**

M

MADHYA. Literally "Milieu". Name given to the number ten to the power ten (= ten thousand million). See **Names of numbers**. For an explanation of this symbolism, see **High numbers (The symbolic meaning of)**.

Sources: *Vâjasaneyî Samhitâ* (beginning of the Common Era); *Taitirîya Samhitâ* (beginning of the Common Era); *Pañchavimsha Brâhmana* (date unknown).

MADHYA. Literally "Milieu". Name given to the number ten to the power eleven (= thousand million). See *Names of numbers*. For an explanation of this symbolism, see **High numbers (The symbolic meaning of)**.

Sources: *Kâthaka Samhitâ* (beginning of the Common Era).

MADHYA. Literally "Milieu". Name given to the number ten to the power fifteen (= trillion). See **Names of numbers**. For an explanation of this symbolism, see **High numbers (The symbolic meaning of)**.

Source: *Kitab fi tahqiq i ma li'l hind* by al-Biruni (c. 1030 CE).

MADHYA. Literally "Milieu". Name given to the number ten to the power sixteen (= ten trillion). See **Names of numbers**. For an explanation of this symbolism, see **High numbers (The symbolic meaning of)**.

Source: *Lîlâvatî* by Bhâskarâchârya (1150 CE); *Ganitakaumudî* by Narâyana (1350 CE); *Trishatikâ* by Shrîdharâchârya (date uncertain).

MÂDHYAMIKA. Name given to the adepts of the Buddhist doctrine called the "Middle Path". This doctrine does not separate the reality and non-reality of things, and even considers the latter as a type of "vacuity" (*shûnyatâ*). This is why its adepts are sometimes called the *shûnyavâdin*, the "vacuists". See *Shûnya* and **Zero**.

MAGIC. See **Numeral alphabet, magic, mysticism and divination**.

MAHÂBHÂRATA. Name of a great Indian epic. See *Arjunakara*, *Dhârtarâshtra*, *Nripa* and *Vasu*.

MAHÂBHÛTA. [S]. Value = 5. "Great element". This term is used by Hindus to denote collectively the five elements of the revelation. It is thus synonymous with the word *bhûta*, which can denote any one of the elements. Another generic term for the five elements is *Avarahakha*, made up of the five letters which symbolise each one of them: *A* (earth); *Va* (water); *Ra* (fire); *Ha* (wind); and *Kha* (ether). See *Pañchabhûta* and **Five**.

MAHÂBJA. Literally "great moon". Name given to the number ten to the power twelve (= billion). See *Abja* and **Names of numbers**. For an explanation of this symbolism, see **High numbers (The symbolic meaning of)**.

Source: *Ganitakaumudî* by Narâyana (1350 CE).

MAHÂDEVA. [S]. Value = 11. "Great god". One of the names for *Rudra*, whose symbolic value is eleven. See *Rudra*, *Shiva* and **Eleven**.

MAHÂKALPA. "Great *kalpa*". According to arithmetical-cosmogonical speculations, this term denotes a unit of cosmic time which is even bigger than the *kalpa* (= 4,320,000,000 human years). It is the equivalent of twenty "little" *kalpa* or ordinary *kalpa*. Thus one *mahâkalpa* = 86,400,000,000 human years. See *Yuga*.

MAHÂKATHÂNA. Literally "great *kathâna*". Name given to the number ten to the power 126. See **Names of numbers** and **High numbers**.

Source: *Vyâkarana* (Pali grammar) by Kâchchâyana (eleventh century CE).

MAHÂKHARVA. Literally "great *kharva*". Name given to the number ten to the power thirteen (= ten billion). See **Names of numbers** and **High numbers**.

Source: *Ganitasârasamgraha* by Mahâvîrâchârya (850 CE).

MAHÂKSHITI. Literally: "great earth". Name given to the number ten to the power twenty-one (= quintillion). See **Names of numbers**. For an explanation of the symbolism, see **High numbers (Symbolic meaning of)**.

Source: *Ganitasârasamgraha* by Mahâvîrâchârya (850 CE).

MAHÂKSHOBHA. Literally "great earth". Name given to the number ten to the power twenty-three (= a hundred quintillions). See **Names of numbers** and **High numbers**.

Source: *Ganitasârasamgraha* by Mahâvîrâchârya (850 CE).

MAHÂKSHONI. Literally "great earth." Name given to the number ten to the power seventeen (= a hundred trillion). See **Names of numbers** and **High numbers**.

Source: *Ganitasârasamgraha* by Mahâvîrâchârya (850 CE).

MAHÂPADMA. Literally "great (pink) *lotus*". Name given to the number ten to the power twelve (= billion). See *Padma*, **High numbers** and **Names of numbers**.

Sources: *Kitab fi tahqiq i ma li'l hind* by al-Biruni (c. 1030); *Lîlâvatî* by Bhâskarâchârya (1150 CE).

MAHÂPADMA. Literally "great (pink) *lotus*". Name given to the number ten to the power fifteen (= trillion). See *Padma*, **High numbers** and **Names of numbers**.

Source: *Ganitasârasamgraha* by Mahâvîrâchârya (850 CE).

MAHÂPADMA. Literally "great (pink) *lotus*". Name given to the number ten to the power thirty-four. See *Padma*, **High numbers** and **Names of numbers**.

Source : *Râmayana* by Vâlmîki (in the early centuries CE).

MAHÂPÂPA. [S]. Value = 5. "Great sin". Allusion to the *Pañchaklesha*, the "five impurities", which, in Hindu and Buddhist philosophies, constitute the five main obstacles denying the faithful the Way of Realisation (*bodhi*) : greed, anger, thoughtlessness, insolence and doubt. See **Five**.

MAHARASHTRI NUMERALS. Signs derived from *Brâhmî numerals, through the intermediary of Shunga, Shaka, Kushâna, Ândhra, Gupta and Nâgarî numerals. Formerly used in Maharashtra State. The corresponding system was based on the place-value system and possessed zero (in the form of a little circle). These numerals were the ancestors of Marathi, Modi, Marwari, Mahajani and Rajasthani numerals. See **Indian written numeral systems (Classification of)**. See Fig. 24.52 and 24.61 to 69.

MAHARASHTRI-JAINA NUMERALS. Signs derived from *Brahmi numerals, through the intermediary of Shunga, Shaka, Kushâna, Ândhra, Gupta and Nâgarî numerals. Formerly used by the *Jainas (*Shvetâmbara*). The corresponding system was based on the place-value system and possessed zero (in the form of a little circle). See **Indian written numeral systems (Classification of)**. See Fig. 24.52 and 24.61 to 69.

MAHÂSAROJA. Literally "great *saroja*". Name given to the number ten to the power twelve (= billion). See **Names of numbers** and **High numbers**.
Source : *Trishatikâ by Shrîdharâchârya (date uncertain).

MAHÂSHANKHA. Literally "great conch". Name given to the number ten to the power nineteen (= ten quadrillions). See ***Shankha***, **Names of numbers** and **High numbers**.
Source : *Ganitasârasamgraha by Mâhavîrâchârya (850 CE).

MAHÂVÎRÂCHÂRYA. *Jaina mathematician who lived in the ninth century. His works notably include *Ganitasârasamgraha*, where there is frequent use of the place-value system, written not only in numerical symbols, but also with nine numerals and the sign for zero. Here is a list of the principle names for numbers mentioned in *Ganitasârasamgraha* :
*Eka (= 1), *Dasha (= 10), *Shata (= 10^2), *Sahasra (= 10^3), *Dashasahasra (= 10^4), *Laksha (= 10^5), *Dashalaksha (= 10^6), *Koti (= 10^7), *Dashakoti (= 10^8), *Shatakoti (= 10^9), *Arbuda (= 10^{10}), *Nyarbuda (= 10^{11}), *Kharva (= 10^{12}), *Mahâkharva (= 10^{13}), *Padma (= 10^{14}), *Mahâpadma (= 10^{15}), *Kshoni (= 10^{16}), *Mahâkshoni (= 10^{17}), *Shankha (= 10^{18}), *Mahâshankha (= 10^{19}), *Kshiti (= 10^{20}), *Mahâkshiti (= 10^{21}), *Kshobha (= 10^{22}), *Mahâkshobha (= 10^{23}).

See **Names of numbers, High numbers, Numeration of numerical symbols, Zero** and **Indian mathematics (The history of)**.
Source : GtsS, I, p. 63–68 [See Datta and Singh (1938), p. 13; Rangacarya (1912)].

MAHÂVRINDÂ. Literally "great *vrindâ*". Name given to the number ten to the power twenty-two (= ten quintillions). See **Names of numbers, *Vrindâ*** and **High numbers (The symbolic meaning of)**.
Source : *Râmâyana by Vâlmîki (first centuries CE).

MAHÂYAJÑA. [S]. Value = 5. "Great sacrifice". This is the name for the five daily sacrifices that all orthodox Hindus must make : prayer and devotion (*hapûjâ*), the placing of offerings in various places (*baliharana*), offerings to the shades of ancestors (*pitriyajña*), the offering of a ritual meal (*manushyayajña*) and a sacrifice in honour of the fire which cooks the food [see Frédéric *Dictionnaire* (1987)]. See **Five**.

MAHÂYUGA. "Great period". This is the largest cosmic cycle of Indian speculations. Considered as the "Great age", this cycle is made up of four successive periods (*kritayuga*, *tretâyuga*, *dvâparayuga*, *kaliyuga*); this is why it is also called the *chaturyuga* ("four ages"). It is said to be made up of 4,320,000 human years. See **Yuga**.

MAHÎ. [S]. Value = 1. This term (which is also the name of a river in Rajasthan) means "curds", the first product derived from milk. Milk itself is the first and most important of the "gifts of the Cow" (*gavyâ*), which is the first nourishment by which all others potentially exist. Thus this symbolism embraces the idea of the cow as a whole and even, in an esoterical sense, the sacred Cow of the Hindus, which is identified with the whole world, because the Cow dispenses life. As "Earth" is a numerical symbol for the value 1, *Mahi* = "curds" = 1. See **Prithivî, Go** and **One**.

MAHÎDHARA. [S]. Value = 7. "Mountain". See **Mount Meru, Adri** and **Seven**.

MAHORÂGA. "Great serpents". Category of demons in the form of cobras. See **Serpent (Symbolism of the)**.

MAIN OBSERVANCE. [S]. Value = 5. See ***Karanîya*** and **Five**.

MAITHILI NUMERALS. Signs derived from *Brâhmî numerals, through the intermediary of Shunga, Shaka, Kushâna, Ândhra, Gupta, Nâgarî and Bengali numerals. Currently found mainly in the north of Bihar State. The corresponding system is based on the place-value system and possesses zero (in the form of a little circle). See **Indian written numeral systems (Classification of)**. See Fig. 24.11, 52 and 24.61 to 69.

MALAYALAM NUMERALS. Signs derived from *Brâhmî numerals, through the intermediary of Shunga, Shaka, Kushâna, Ândhra, Pallava, Chalukya, Ganga, Valabhî, Bhattiprolu and Grantha numerals. Currently used by the Dravidians of Kerala State on the ancient coast of Malabar, to the southwest of India. The corresponding system is not based on the place-value system and has only possessed zero since a relatively recent date. See **Indian written numeral systems (Classification of)**. See also Fig. 24.19, 52 and 24.61 to 69.

MANDARA. One of the names for Mount Meru. See **Mount Meru**, *Adri, Dvîpa, Pûrna, Pâtâla, Sâgara, Pushkara, Pâvana* and *Vâyu*.

MANGALA. [S]. Value = 8. "Jewel", "thing which augurs well". Here the allusion is to *ashtamangala*, the eight "things which augur well", Buddhist symbols which represent the veneration of the "Master of the world" (and, by extension, Buddha). These are: the parasol (symbol of royal dignity meant to protect against misfortune); the two fish (signs of the Indian master of the universe); the conch (symbol of victory in combat); the lotus flower (symbol of purity); the container of lustral water (filled with Amrita, the nectar of immortality); the rolled flag (sign of victorious faith); the knots of eternal life; and the wheel of the Doctrine (*Dharmachakra*). See **Eight**.

MANIPURI NUMERALS. Signs derived from *Brâhmî numerals, through the intermediary of Shunga, Shaka, Kushâna, Ândhra, Gupta, Nâgarî, Kutila and Bengali numerals. Currently in use in Manipur State, to the east of Assam and next to the border of Burma. The corresponding system functions according to the place-value system and possesses zero (in the form of a little circle). See **Indian written numeral systems (Classification of)**. See also Fig. 24.52 and 24.61 to 69.

MANTRA. Sacred formula which constitutes the digest, in a material from, of the divinity which it is meant to invoke. See **Numeral alphabet, magic, mysticism and divination** and **Mysticism of letters**.

MANU. [S]. Value = 14. Literally "human". This is the name given in traditional legends to the Progenitor of the human race as a symbol of the thinking being and considered as the intermediary between the Creator and the human race. According to the *Vedas, the manus* constituted the first divine legislators who fixed the rules of religious ceremonies and ritual sacrifices. According to the *purânas, there were fourteen successive *manus*, sovereigns living in ethereal worlds where they are meant to direct the conscious life of humankind and its ability to think. Thus: *manu* = 14. (The *manu* of the present era is the seventh: named *Vaivashvata*, "Born of the Sun") See **Fourteen**.

MANUAL ARITHMETIC. See *Mudrâ*.

MANUSMRITI. Important religious work considered to be the foundation of Hindu society.

MARÂTHA (Calendar). This calendar begins in the year 1673 CE, and was founded by Shivâjî. To find the corresponding date in the Common Era, add 1673 to a date expressed in the *Marâtha* calendar. Formerly used in Maharashtra. See **Indian calendars**.

MARATHI NUMERALS. Signs derived from *Brâhmî numerals, through the intermediary of Shunga, Shaka, Kushâna, Ândhra, Gupta, Nâgarî and Mahârâshtrî numerals. Currently used in the west of India, in the state-province of Maharashtra. The corresponding system is based on the place-value system and possesses zero (in the form of a little circle). See **Indian written numeral systems (Classification of)**. See Fig. 24.4, 52 and 24.61 to 69.

MÂRGANA [S]. Value = 5. "Arrow". See *Shara* and **Five**.

MÂRTANDA. [S]. Value = 12. One of the names of *Sûrya. See **Twelve**.

MARWARI NUMERALS. Signs derived from *Brâhmî numerals, through the intermediary of Shunga, Shaka, Kushâna, Ândhra, Gupta, Nâgarî and Mahârâshtrî numerals. Currently used in the northwest of India (Rajasthan) and in the Aravalli mountains, and between the afore-mentioned mountains and the Thar desert (Mârusthali). The corresponding system is based on the place-value system and possesses zero (in the form of a little circle). See **Indian written numeral systems (Classification of)**. See Fig. 24.52 and 24.61 to 69.

MÂSA. [S]. Value = 12. "Month". Allusion to the twelve months of the year. See **Twelve**.

MÂSÂRDHA. [S]. Value = 5. "Season". See *Ritu* and **Five**.

MATANGA. [S]. Value = 8. "Elephant". See *Diggaja* and **Eight**.

MATHEMATICIAN. See *Samkhyâ*.

MATHEMATICS. See *Ganita, Ganitânuyoga*, **Arithmetic, Calculation** and **Indian mathematics (The history of)**.

MATHURA NUMERALS. Signs derived from *Brâhmî numerals, through the intermediary of Shunga, Shaka, Kushâna and Ândhra numerals. Contemporaries of a Shaka dynasty (first to third century CE). These are attested mainly in the inscriptions of Mathura (in Uttar Pradesh). The corresponding system did not use the place-value system or zero. See Fig. 24.32, 52 and 24.61 to 69, and 70. See also **Indian written numeral systems (Classification of)**.

MÂTRIKÂ. [S]. Value = 7. "Divine Mother". Name given in Hinduism to the *saptamâtrikâ*, the seven aspects of *shakti*, "feminine energy" of the divinities : aspects which are considered to be the "mothers of the world". Thus: *mâtrikâ* = 7. See **Seven**.

MATTER (Indian concept of). See **Indian atomism, Jaina** and *Jala*.

MEMBER. [S]. Value = 6. See *Anga* and **Six**.

MENTAL ARITHMETIC. See *Gananâ*.

MERIT. [S]. Value = 3. See *Guna, Triguna* and **Three**.

MERIT. [S]. Value = 6. See *Guna, Shadâyatana* and **Six**.

MILIEU. As a name of a high number. See *Madhya*.

MILLION (= ten to the power six). See *Dashalaksha, Niyuta, Prayuta* and **Names of numbers**.

MON NUMERALS. Signs derived from *Brâhmî numerals, through the intermediary of Shunga, Shaka, Kushâna, Ândhra, Pallava, Châlukya, Ganga, Valabhî, "Pâlî" and Vateluttu numerals. Formerly used by the people of Pegu before the Burmese invasion. The corresponding system was not based upon the place-value system and did not possess zero. See Fig. 24.52 and 24.61 to 69. See also **Indian written numeral systems (Classification of)**.

MONGOL NUMERALS. Signs derived from *Brâhmî numerals through the intermediary notations of Shunga, Shaka, Kushâna, Ândhra, Gupta, Siddham and Tibetan numerals. Used by the Mongols during the thirteenth and fourteenth centuries. The corresponding system functioned according to the place-value system and possessed zero (in the form of a little circle). See Fig. 24. 42, 52 and 24.61 to 69. See **Indian written numeral systems (Classification of)**.

MONTH (Rite of the four). See *Chaturmâsya*.

MONTH. [S]. Value = 12. See *Mâsa* and **Twelve**.

MOON. Used as a name for ten to the power nine or ten to the power twelve. See *Abja* and *Mahâbja*. See also **High numbers (The symbolic meaning of)**.

MOON. [S]. Value = 1. See *Abja, Atrinayanaja, Chandra, Indu, Kshapeshvara, Mriganka, Shashadhara, Shashanka, Shashin, Shîtâmshu, Shîtarashmi, Soma, Sudhâmshu, Vidhu* and **One**.

MORTAL SINS (The Five). See *Pañchânantarya*.

MOUNT MERU. Mythical mountain in Hindu cosmology and Brahman mythology. It has many Sanskrit names : *Ratnasanu, Sumeru, Hemadrî, Mandara, Karnikâchala, Devapârvata*, etc. Mount Meru was meant to be the place where the gods lived and met. It was said to be situated at the centre of the universe, under the Pole star, and also constituted the "axis of the world". *Indra lived at the summit, the head of the *deva*, whilst the slopes were peopled with the *Trâyastrimsha*, the thirty-three *deva* (gods).

Mount Meru plays an important role in mythology and Brahman and Hindu cosmological texts. Thus this mountain was said to act as a pivot between the *deva* and the *asura* ("anti-gods") during the churning of the sea of milk.

In corresponding representations, Mount Meru, and all that is connected to it, is always associated with the number seven. First there is the concept of the "mountain" and, by extension, that of "hill", which is generally symbolically connected with this number. There are also the "seven oceans" (*sapta sâgara*). Then there are the "island-continents" (*dvîpa*), each one flooded by one of the seven oceans, which surround Mount Meru. As for Mount Meru itself, it has seven faces, each facing one of the seven seas and one of the seven continents. It is above the *pâtâla*, the seven underworlds or "inferior worlds", where the *nâga* live, the master of whom is the king Muchalinda, the chthonian genie in the form of a cobra, depicted with seven heads.

Fig. 24D.9. *Mount Meru, centre of the universe in Hindu and Brahmanic cosmology. Ref.: Dubois de Jancigny,* L'Univers pittoresque, *Hachette, Paris, 1846*

Thus Mount Meru represents total stability and the absolute centre of the universe, around which the universe and the firmament revolve. This image of Mount Meru connects it to one of the universal images of the Pole star. According to the legend, Mount Meru is directly underneath the Pole star and is "on the same axis". The symbolic correspondence between Mount Meru and the number seven comes from the fact that the Pole star, the *Sudrishti*, "That which never moves", is the last of the seven stars of the constellation: the Bear. According to Indian tradition, this constellation is the personification of the seven "great Sages" of Vedic times, the *Saptarishi*, who are thought to be the authors of both the hymns and invocations of the *Rigveda*, and of the most important Vedic texts.

In Sanskrit, the word for "mountain" is *pârvata*, which appears in one of the names of Mount Meru : *Devapârvata*, "mountain of the gods". Because this sacred mountain was associated with the number seven, the mountain, daughter of Himâlaya, sister of Vishnu and wife of *Shiva also came to be synonymous with this number: she was *Kâlî*,

the "Black", who represented the destructive power of time (*Kâla*) and was considered in the *Veda* to be the seventh tongue of Agni, "Fire". It is perhaps not by chance that Manasâ, the Hindu tantric divinity, who symbolises the destructive and regenerative aspects of Pârvati, has been considered as one of the sisters of Muchalinda (the king of the *nâga* with seven cobra heads) and as the *Pâtâla Kumârâ*, divinity of the serpents and "princess of the (seven) Underworlds".

Even *Sûrya, the Sun-god, traditionally associated with the number twelve in the system of *numerical symbols, has represented the number seven : he is often represented as a warrior flying through the sky on a chariot pulled either by seven horses (*ashva), or by *Aruna*, the horse with seven heads. See *Adri*, *Dvîpa*, *Sâgara*, *Pâtâla*, *Pushkara*, *Pâvana*, *Vâyu*, *Loka* (= 7) and **Seven**.

MOUNTAIN. [S]. Value = 7. See *Adri*, *Pârvata*, **Seven** and **Mount Meru**.

MOUTHS OF JÂHNAVÎ. [S]. Value = 1,000. "Mouths of the Ganges". See *Jâhnavîvaktra* and **Thousand**.

MRIGÂNKA. [S]. Value = 1. "Moon". See **Abja** and **One**.

MUCHALINDA (MUCHILINDA). Name of the king of the *nâga*. See **Serpent (Symbolism of the)**.

MUDRÂ. "Mark, sign". In Indian mysticism, mainly in esoteric Buddhism, this word denotes the gestures made by the hands and is meant to symbolise a mental attitude of the divinities. They are mainly used during ceremonies and prayers to invoke Buddha and the power of his divinites.

MÛDRÂ. Term denoting manual arithmetic and digital calcultaion in Ancient Sanskrit literature. See Chapter 3.

MUDRÂBALA. Literally : "Power of the *mudrâ*". Name given to the number ten to the power forty-three. To those using it, such a high number must have symbolically represented a quantity which was as incalculable as the powers concealed within the mystical gestures called *mudrâ*. See **Mudrâ** (first article), **Names of numbers** and **High numbers**.

Source : *Lalitavistara Sûtra* (before 308 CE).

MUKHA. [S]. Value = 4. "Face". Allusion to the *chaturmukha* ("Four Faces"), which refers to all the of the Brahmanic (or Buddhist) divinities who are represented as having four faces (*Brahma, *Shiva, etc.). See **Four**.

MULTIPLICATION. [Arithmetic]. See *Gunana, Pâtiganita* and **Indian methods of calculation**.

MUNI. [S]. Value = 7. "Sage". This is an allusion to the seven mythical sages of Vedic times. Strictly speaking, the word *muni*, "sage", is much less strong than *Rishi*, which denotes the seven "Sages". But the name began to be used as a symbol for the number seven because of the desired effect in the versification of expressions using numerical symbols. See **Seven, Sanskrit** and **Poetry and the writing of numbers**.

MÛRTI. [S]. Value = 3. "Form". Allusion to the "three forms" of Hindu triads (*trimûrti*), constituted by either three different divinities (usually *Brahma, *Shiva and *Vishnu), or three aspects of one single divinity. See **Three**.

MÛRTI. [S]. Value = 8. "Form". Allusion to the *ashtamûrti*, the "eight" most important "forms" of *Shiva : *Rudra, who represents the power of fire; *Bhava, water; *Sharva*, the earth; *Îshâna*, the sun; *Pashupati*, sacrifice; *Bhîma*, the terrible; and *Ugra* and *Mahâdeva*. See **Rudra-Shiva** and **Eight**.

MUSICAL MODE. [S]. Value = 6. See *Râga* and **Six**.

MUSICAL NOTE. [S]. Value = 7. See *Svara* and **Seven**.

MUSLIM INDIA. See **Numeral alphabet and composition of chronograms** and **Eastern Arabic numerals**.

MYSTICISM AND POSITIONAL NUMERATION. See *Durgâ*.

MYSTICISM OF HIGH NUMBERS. See **High numbers (The symbolic meaning of)**.

MYSTICISM OF INFINITY. See **Infinity (Mythological representation of)** and **Serpent (Symbolism of the)**.

MYSTICISM OF LETTERS. See *Akshara*, **Numeral alphabet, magic, mysticism and divination**, *Bîja*, **Mantra**, *Trivarna, Vâchana*.

MYSTICISM OF NUMBERS. See **Numerical symbols, Symbolism of words with a numerical value, Symbolism of numbers (Concept of large quantity), Symbolism of zero and High numbers (The symbolic meaning of)**.

MYSTICISM OF THE NUMBER FOUR. See *Nâga, Jala*, **Ocean** and **Serpent (Symbolism of the)**.

MYSTICISM OF THE NUMBER SEVEN. See **Mount Meru** and **Ocean**.

MYSTICISM OF ZERO. See *Shûnya, Shûnyatâ*, **Zero, Zero (Indian concepts of), Zero and Sanskrit poetry** and **Symbolism of Zero**.

MYTHICAL PEARL. [S]. Value = 1. See **Indu**.

N

NABHA. [S]. Value = 0. "Sky, atmosphere". This symbolism is due to the fact that the sky is considered to be the "void". See **Zero** and *Shûnya*.

NABHAS. [S]. Value = 0. "Sky, atmosphere". See *Nabha*, **Zero** and *Shûnya*.

NÂDÎ. Hindu word denoting the arteries of the human body. See **Numeral alphabet, magic, mysticism and divination**.

NAGA. [S]. Value = 7. "Mountain". Literally, "That which does not move". This is an allusion to *Mount Meru, the mythical mountain of Hindu cosmology and Brahman mythology, the dwelling and meeting place of the gods, which is said to be situated at the centre of the universe and thus constitute the axis of the world. This symbolism comes from the fact that the number seven plays an important role in mythological representations related to

Mount Meru, and because the Pole star, situated directly above this mountain, is the *Sudrishti*, the divinity "who never moves". See **Adri**, **Mount Meru**, **Seven** and *Dhruva*.

NÂGA. [S]. Value = 8. "Serpent". This symbolism is due to the fact that the serpent (especially the *nâga*) is considered to be not only a chthonic genie who owns the earth and its treasures, but also an aquatic symbol. It is a "spirit of the waters" that lives in the *pâtâla* or "underworlds". In Sanskrit, water is *jala*, and this word is used as a numerical symbol for the number four. In their subterranean kingdom, the *nâga* reproduce in couples and evolve in the company of the *nâginî* (the females), so "water", in this case, has been symbolically multiplied by two, to give their generic name the symbolic value of : $4 \times 2 = 8$. In traditional Indian thought, the earth (to which the serpent is also associated) corresponds symbolically to the number four, being associated with the square and its four horizons (or cardinal points). As the *nâga* is also aquatic, water (= 4) has been symbolically added to give the serpent its generic designation as a numerical symbol with a value equal to eight (*nâga* = earth + water = $4 + 4 = 8$). See **Eight**, **Serpent** (Symbolism of the) and **Infinity**.

For a documented example of this : see EI, XXXV, p. 140.

NÂGABALA. Literally, "Power of the *nâga*". Name given to the number ten to the power twenty-five. See **Names of numbers**, **High numbers** and **Serpent (Symbolism of the)**.

Source : *Lalitavistara Sûtra* (before 308 CE).

NÂGARÎ ALPHABET. See Fig. 24.56. See also **Âryabhata's numeration**.

NÂGARÎ NUMERALS. Signs derived from *Brâhmî* numerals through the intermediary notations of Shunga, Shaka, Kushâna, Âdhra and Gupta. Today these are the most widely used numerals in India, from Madhya Pradesh (central province) to Uttar Pradesh (northern province), Rajasthan, Haryana, Himachal Pradesh (the Himalayas) and Delhi. These numerals are also called *Devanagari* because they are the most regular numerals of India. The corresponding system is based on the place-value system and possesses zero (in the form of a little circle). However, this was not always the case, as a considerable number of documents written before the eighth century CE prove. These signs were the ancestors of Siddham, Nepali, Tibetan, Mongol, Kutilâ, Bengali, Oriya, Kaîthî, Maithilî, Manipurî, Gujarâtî,

Mahârâshtrî, Marâthî, Modî, Mârwarî, Mahâjani, Râjasthanî, etc. numerals, as well as the "Hindi" numerals of the eastern Arabs, the Ghubar numerals of North Africa, the *apices* and algorisms of mediaeval Europe, not to mention our own modern numerals. For ancient Nâgarî numerals recorded on copper charters, see Fig. 24.39 A and 75; for those recorded on manuscripts, see Fig. 24.39 B; for inscriptions of Gwalior, see Fig. 24.39 C and 24. 72 to 74. For modern Nâgarî numerals, see Fig. 24.3. For notations which derived from Nâgarî, see Fig. 24.52. For the corresponding graphical evolution, see Fig. 24.61 to 69. See also **Indian written numeral systems (Classification of)**.

NÂGINÎ. Female of the *nâga*. See *Nâga* and **Serpent (Symbolism of the)**.

NAHUT. Name given to the number ten to the power nine. See **Names of numbers** and **High numbers**.

Source : *Lalitavistara Sûtra* (before 308 CE).

NAHUTA. Name given to the number ten to the power twenty-eight. See **Names of numbers** and **High numbers**.

Source : *Vyâkarana* (Pâli grammar) by Kâchâyana (eleventh century CE).

NAIL. [S]. Value = 20. See *Nakha* and **Twenty**.

NAKHA. [S]. Value = 20. "Nail". This is because of the nails of the ten fingers and ten toes. See **Twenty**.

NAKSHATRA. [S]. Value = 27. "Lunar Mansion". This refers to the houses occupied successively by the moon in its monthly cycle, which in solar days lasts between twenty-seven and twenty-eight days. For the representation of the sidereal movements of the moon, however, Indian astronomers usually used the system of twenty-seven *nakshatra* marking twenty-seven ideal equal divisions of the ecliptic zone (each one equal to 13° 20'). This is why the word came to symbolically signify the number twenty-seven. See **Twenty-seven**.

NAKSHATRAVIDYÂ. Literally : "Knowledge of the *nakshatra*". Name given to "astronomy" in the *Chândogya Upanishad*.

NAMES OF NUMBERS (up to thousand). Here is a list of ordinary Sanskrit names of numbers :

Eka (= 1); *Dva* (= 2); *Dve* (= 2); *Dvi* (= 2); *Trai* (= 3); *Traya* (= 3); *Tri* (= 3); *Chatur* (= 4); *Pañcha* (= 5); *Shad* (= 6); *Shash* (= 6); *Shat* (= 6); *Sapta* (= 7); *Saptan* (= 7); *Ashta* (= 8); *Ashtan* (= 8); *Nava* (= 9); *Navan* (= 9).

Dasha (= 10); *Dashan* (= 10); *Ekadasha* (= 11); *Dvâdasha* (= 12); *Trayodasha* (= 13);

Chaturdasha (= 14); *Pañchadasha* (= 15); *Shaddasha* (= 16); *Saptadasha* (= 17); *Ashtadasha* (= 18); *Navadasha* (= 19).

Vimshati (= 20); *Ekavimshati* (= 21); *Dvavimshati* (= 22); *Trayavimshati* (= 23); *Chaturvimshati* (= 24); *Pañchavimshati* (= 25); *Shadvimshati* (= 26); *Saptavimshati* (= 27); Ashtavimshati (= 28); Navavimshati (= 29).

Trimshat (= 30); *Chatvarimshat* (= 40); *Pañchashat* (= 50); *Shashti* (= 60); *Saptati* (= 70); *Ashiti* (= 80); *Navati* (= 90).

At the start of the Common Era, the subtractive forms were also used for the numbers 19, 29, 39, 49, etc. : *ekânnavimshati (= 20 − 1 = 19); *ekânnatrimshati (= 30 − 1 = 29); etc.

[See *Taittirîya Samhitâ*, VII, 2, 11; Datta and Singh (1938), pp. 14–15].

Shata (= 100). This is the classical Sanksrit form of this number. However, at the beginning of the Common Era, the Indo-European form *Sata* was still used.

Ref.: There is evidence of the use of this form in *Vâjasaneyî Samhitâ*, *Taittirîya Samhitâ*, *Kâthaka Samhitâ*, *Pañchavimsha Brâhmana* and *Sankhyâyana Shrauta Sûtra*.

Dvashata (= 200); Trishata (= 300); Chatuhshata (= 400); etc. *Sahasra (= 1,000); Dvasahasra (= 2,000); Trisahasra (= 3,000); Chatursahasra (= 4,000); etc.

See **Sanskrit**.

NAMES OF NUMBERS (Powers of ten above thousand). After ten thousand, Sanskrit spoken numeration assigns names to the various powers of ten which differ considerably from one author to another and from one era to another; thus the same word can have several numerical values depending on the source in question. The use of these names was not commonplace in India. However, they were very familiar to scholars, since the following terms are found in astronomical, mathematical, cosmological, grammatical and religious texts, as well as in legend and mythology.

In the following list, the letters in brackets indicate the source of each word in question; here are the letters, the sources they represent, and the era in which they were written :

(a) *Vâjasaneyî Samhitâ*. (b) *Taittirîya Samhitâ*. (c) *Kâthaka Samhitâ*. (d) *Râmâyana* by Vâlmîki. (e) *Lalitavistara Sûtra*. (f) *Pañchavimsha Brâhmana*. (g) *Sankhyâyana Shrauta Sûtra*. (h) *Âryabhatîya* by Âryabhata. (i) *Ganitasârasamgraha* by Mahâvîrâchârya. (j) *Kitab fî tahqiq i ma li'l hind* by al-Biruni. (k) *Vyâkarana*, Pâli grammar, by Kâchchâyana. (l) *Lilâvatî* by Bhâskarâchârya. (m) *Ganitakaumudî*

by Nârâyana. (n) *Trishatikâ* by Shrîdharâchâryâ.

(a, b, c : beginning of the Common Era; d : early centuries of the Common Era; e : before 308 CE; f, g : date uncertain; h : c. 510 CE; i : 850 CE; j : c. 1030 CE; k : eleventh century CE; l : 1150 CE; m : 1356 CE; n : date uncertain).

Here is a (non-exhaustive) arithmetical list of the Sanskrit names of high numbers :

TEN TO THE POWER 4: *Ayuta (a, b, c, f, g, h, j, l, m, n); *Dashasahasra (i).

TEN TO THE POWER 5: *Lakh (e); *Lakkha (k); *Laksha (i, j, l, m, n); *Niyuta (a, b, f, h); *Prayuta (c).

TEN TO THE POWER 6: *Dashalaksha (i); *Niyuta (c); *Prayuta (a, b, f, g, h, j, l, m, n).

TEN TO THE POWER 7: *Arbuda (a, b, c, f, g, h); *Koti (d, e, h, i, j, k, l, m, n).

TEN TO THE POWER 8: *Arbuda (l, m, n); *Dashakoti (i); *Nyarbuda (a, b, c, f, g); *Vyarbuda (j).

TEN TO THE POWER 9: *Abja (l, n); *Ayuta (e); *Nahut (e); *Nikharva (g); *Padma (j); *Samudra (a, b, c, f); *Saroja (m); *Shatakoti (i); *Vâdava (c); *Vrindâ (h).

TEN TO THE POWER 10: *Arbuda (i); *Kharva (j, l, m, n); *Madhya (a, b, f); *Samudra (g).

TEN TO THE POWER 11: *Anta (a, b, c, f); *Madhya (c); *Nikharva (j, l, m, n); *Ninnahut (e); *Niyuta (e); *Nyarbuda (i); *Salila (g).

TEN TO THE POWER 12: *Antya (g); *Kharva (i); *Mahâbja (m); *Mahâpadma (j, l); Mahâsaroja (n); *Parârdha (a, b, c, f); *Shankha (d).

TEN TO THE POWER 13: *Ananta (g); *Kankara (e); *Khamba (e); *Mahâkharva (i); *Nikharva (f); *Shankha (j); *Shanku (l, m, n).

TEN TO THE POWER 14: *Jaladhi (l); *Padma (i); *Pakoti (k); *Pârâvâra (m); *Samudra (j); *Saritâpati (n); *Vâdava (f).

TEN TO THE POWER 15: *Akshiti (f); *Antya (l, m, n); *Madhya (j); *Mahâpadma (i); *Viskhamba (e); *Vivara (e).

TEN TO THE POWER 16: *Antya (j); *Madhya (l, m, n); *Kshoni (i).

TEN TO THE POWER 17: *Abab (e); *Kshobhya (e); *Mahâkshoni (i); *Parârdha (j, l, m, n); *Vrindâ (d).

TEN TO THE POWER 18: *Shankha (i).

TEN TO THE POWER 19: *Attata (e); *Mahâshankha (i); *Vivaha (e).

TEN TO THE POWER 20: *Kshiti (i).

TEN TO THE POWER 21: *Kotippakoti (k); *Kumud (e); *Mahâkshiti (i); *Utsanga (e).

TEN TO THE POWER 22: *Kshobha (i); *Mahâvrindâ (d).

TEN TO THE POWER 23: *Bahula* (e); *Gundhika* (e); *Mahâkshobha* (i).

TEN TO THE POWER 25: *Nâgabala* (e); *Utpala* (e).

TEN TO THE POWER 27: *Pundarîka* (e); *Titilambha* (e).

TEN TO THE POWER 28: *Nahuta* (k).

TEN TO THE POWER 29 : *Padma* (d); *Paduma* (e); *Vyavasthânaprajñapati* (e).

TEN TO THE POWER 31: *Hetuhila* (e).

TEN TO THE POWER 33: *Karahu* (e).

TEN TO THE POWER 34: *Mahâpadma* (d).

TEN TO THE POWER 35: *Hetvindriya* (e); *Ninnahuta* (k).

TEN TO THE POWER 37: *Samâptalambha* (e).

TEN TO THE POWER 39: *Gananâgati* (e); *Kharva* (d).

TEN TO THE POWER 41: *Niravadya* (e).

TEN TO THE POWER 42: *Akkhobhini* (k).

TEN TO THE POWER 43: *Mudrâbala* (e).

TEN TO THE POWER 45: *Sarvabala* (e).

TEN TO THE POWER 47: *Visamjñagati* (e).

TEN TO THE POWER 49: *Bindu* (k); *Sarvajña* (e).

TEN TO THE POWER 51: *Vibhutangamâ* (e).

TEN TO THE POWER 53: *Tallakshana* (e).

TEN TO THE POWER 56: *Abbuda* (k).

TEN TO THE POWER 63: *Nirabbuda* (k).

TEN TO THE POWER 70: *Ahaha* (k).

TEN TO THE POWER 77: *Ababa* (k).

TEN TO THE POWER 84: *Atata* (k).

TEN TO THE POWER 91: *Sogandhika* (k).

TEN TO THE POWER 98: *Uppala* (k).

TEN TO THE POWER 99: *Dhvajâgravati* (e).

TEN TO THE POWER 105: *Kumuda* (k).

TEN TO THE POWER 112: *Pundarîka* (k).

TEN TO THE POWER 119: *Kathâna* (k); *Paduma* (k).

TEN TO THE POWER 126: *Mahâkathâna* (k).

TEN TO THE POWER 140: *Asankhyeya* (k).

TEN TO THE POWER 145: *Dhvajâgranishâmani* (e). And so on until ten to the power 421 (e).

See **Sanskrit** and **Poetry and writing of numbers**.

Indian scholars did not specialise in just one field of study; they embraced diverse disciplines all at once, such as mathematics, astronomy, literature, poetry, phonetics or philosophy, and even mysticism, divination and astrology. Thus it is not surprising that in arithmetic, their fertile imaginations led them to use subtle symbolism to name high numbers. They gave a unique name to each power of ten up to at least as high as ten to the

power 421. This is why their spoken numeration had a mathematical structure with the potential to lead them to the discovery of the place-value system and consequently the "invention" of zero. See **High numbers**. For an explanation of the symbolism of these diverse words, see: **High numbers (The symbolic meaning of)**, **Zero**, **Numeration of numerical symbols** and **Numerical symbols (Principle of the numeration of)**.

NÂRÂYANA. Indian mathematician c. 1356. His works notably include *Ganitakaumudî*.

Here is a list of the principal names of numbers mentioned in that work: *Eka* (= 1), *Dasha* (= 10), *Shata* (= 10^2), *Sahasra* (= 10^3), *Ayuta* (= 10^4), *Laksha* (= 10^5), *Prayuta* (= 10^6), *Koti* (= 10^7), *Arbuda* (= 10^8), *Saroja* (= 10^9), *Kharva* (= 10^{10}), *Nikharva* (= 10^{11}). *Mahâpadma* (= 10^{12}), *Shanku* (= 10^{13}). *Pârâvâra* (= 10^{14}), *Madhya* (= 10^{15}), *Antya* (= 10^{16}), *Parârdha* (= 10^{17}).

See **Names of numbers** and **High numbers**. [See Datta and Singh (1938), p. 13]

NÂSATYA. [S]. Value = 2. Name of one of the two twin gods Saranyû and Vivashvant of the Hindu pantheon (also called *Dasra* and Nâsatya). The symbolism is through an association of ideas with the "Horsemen". See *Ashvin* and **Two**.

NAVA (NAVAN). Ordinary Sanskrit names for the number nine, which appear in the composition of many words which have a direct relationship with the concept of this number. Examples: *Navagraha*, *Navaratna*, *Navarâshika* and *Navarâtrî*. For words which have a more symbolic relationship with this number, see: **Nine** and **Symbolism of numbers**.

NAVACHATVÂRIMSHATI. Ordinary Sanskrit name for the number forty-nine. For words having a symbolic link to this number, see **Forty-nine** and **Symbolism of numbers**.

NAVADASHA. Ordinary Sanskrit name for the number nineteen. For words which have a symbolic link to this number, see : **Nineteen** and **Symbolism of numbers**.

NAVAGRAHA. Literally : "nine planets". This relates to the nine planets of the Hindu cosmological system: the seven planets (*saptagraha*) plus the demons of the eclipses *Râhu* and *Ketu*. See *Graha* and *Paksha*.

NAVAN. Ordinary Sanskrit name for the number nine. See *Nava*.

NAVARÂSHIKA. [Arithmetic]. Sanskrit name for the Rule of Nine. See *Nava*.

NAVARATNA. "Nine jewels", "Nine precious stones". Collective name given to the nine famous poets of Sanskrit expression who are said to have lived in the court of King Vikramâditya (namely : *Dhavantari*, the pearl; *Kshapanaka*, the ruby; *Amarasimah*, the topaz; *Shanku*, the diamond; *Vetâlabhatta*, the emerald; *Ghatakarpara*, the lapis-lazuli; *Kâlidâsa*, the coral; *Varâhamîhira*, the sapphire; and *Vararuchi*, not identified to any specific stone). See *Nava* and *Ratna* (= 9).

NAVARÂTRÎ. Name of the nine-day Feast. See *Durgâ*.

NAVATI. Ordinary Sanskrit name for the number ninety.

NAYANA. [S]. Value = 2. "Eye". See *Netra* (= 2) and Two.

NEPÂLÎ (Calendar). Beginning in 879. To find the corresponding date in the Common Era, simply add 879 to a date expressed in this calendar. Still used occassionally in Nepal. Also called *Newârî*. See Indian calendars.

NEPÂLÎ NUMERALS. Signs derived from *Brahmi numerals through the intermediary notations of Shunga, Shaka, Kushâna, Adhra, Gupta, Nâgari and Siddham numerals. Currently used mainly in the independent state of Nepal. They are also called *Gurkhali* numerals. The corresponding system is based on the place-value system and has a zero (in the form of a little circle). For ancient numerals, see Fig. 24.41. For modern numerals, see Fig. 24.15. See Fig. 24.52 and 24.61 to 69. See also Indian written numeral systems (Classification of).

NETHER WORLD. [S]. Value = 7. See *Pâtâla*. Seven.

NETRA. [S]. Value = 2. "Eye". See Two.

NETRA. [S]. Value 3. "Eye". Symbol used only in regions of Bengal, where this word is generally used to denote the three eyes of *Shiva. See Three.

NEWARÎ (Calendar). See *Nepâlî*.

NIHILISM. See *Shûnyatâ* and Zero.

NIKHARVA. Name given to the number ten to the power nine. See Names of numbers and High numbers.
Source: *Sankhyâyana Shrauta Sûtra (date uncertain).

NIKHARVA. Name given to the number ten to the power eleven. See Names of numbers and High numbers.
Sources : *Kitab fi tahqiq i ma li'l hind by al-Biruni (c. 1030 CE); *Lîlâvatî by Bhâskarâchârya (1150 CE); *Ganitakaumudi by Nârâyana (1350 CE); *Trishatikâ by Shrîdharâchârya (date unknown).

NIKHARVA. Name given to the number ten to the power thirteen. See Names of numbers and High numbers.
Source: *Pañchavimsha Brâhmana (date uncertain).

NIL, NULLITY. See *Shûnyatâ* and Zero.

NÎLAKANTHASOMAYÂJIN. Indian astronomer c. 1500 CE. His works notably include *Siddhântadarpana*, in which the place-value system with Sanskrit numerical symbols is used frequently [see Sarma, *Siddhântadarpana*]. See Numerical symbols, Numeration of numerical symbols and Indian mathematics (The history of).

NINE. Ordinary Sanskrit names: *nava, *navan. Here is a list of the corresponding numerical symbols: *Abjagarbha, Aja, *Anka, Brihatî, *Chhidra, *Durgâ, Dvâra, *Go, *Graha, *Keshava, Khanda, Laddha, Labdhi, Nanda, Nidhâna, Nidhi, Padârtha, *Randhra, *Ratna, Târkshyadhvaja, Upendra, Varsha. These words have the following literal or symbolic meaning: 1. The Brahman (*Abjagarbha, Aja*). 2. The name of the ninth month of the *chaitradi* year (*Keshava*). 3. The numerals of the place-value system (*Anka*). 4. The "Inaccessible", the "Divine Mother", in allusion to a divinity of the same name (*Durgâ*). 5. The Jewels (*Ratna*). 6. The holes, the orifices (*Chhidra, Randhra*). 7. The planets (*Graha*). 8. The radiance (*Go*). 9. The "Cow" to denote the earth (*Go*). See Numerical symbols.

NINETEEN. Ordinary Sanskrit name : *navadasha. The corresponding numerical symbol is *Atidhriti. Note that at the beginning of the Common Era, and probably since Vedic times, this number was also called *ekânnavimshati*, which literally means "one away from twenty" [see *Taittirîya Samhitâ*, VII, 2. 11]; but it is also used in its normal form from this time [See *Taittirîya Samhitâ*, XIV, 23; Datta and Singh (1938), pp. 14–15].

NINETY. See *Navati*.

NINNAHUT. Name given to the number ten to the power eleven. See Names of numbers and High numbers.
Source : *Lalitavistara Sûtra (before 308 CE).

NINNAHUTA. Name given to the number ten to the power thirty-five. See Names of numbers and High numbers.
Source: *Vyâkarana (Pâlî grammar) by Kâchchâyana (eleventh century CE).

NIRABBUDA. Name given to the number ten to the power sixty-three. See Names of numbers and High numbers.
Source: *Vyâkarana (Pâlî grammar) by Kâchchâyana (eleventh century CE).

NIRAVADYA. Name given to the number ten to the power forty-one. See **Names of numbers** and **High numbers.**

Source : **Lalitavistara Sûtra* (before 308 CE).

NÎRVANA. According to Indian philosophers, this is the supreme state of non-existence, reincarnation and absorption of the being in the Brahman. See *Shûnyatâ* and **Zero.**

NIYUTA. Name given to the number ten to the power five. See **Names of numbers** and **High numbers.**

Sources : **Vâjasaneyî Samhitâ*, **Taittirîya Samhitâ* and **Kâthaka Samhitâ* (from the start of the first millennium CE); **Pañchavimsha Brâhmana* (date uncertain); **Âryabhatîya* (510 CE).

NIYUTA. Name given to the number ten to the power six. See **Names of numbers** and **High numbers.**

Source : **Kâthaka Samhitâ* (start of the Common Era).

NIYUTA. Name given to the number ten to the power eleven. See **Names of numbers** and **High numbers.**

Source : **Lalitavistara Sûtra* (before 308 CE).

NON-BEING. See *Shûnyatâ* and **Zero.**

NON-EXISTENCE. See *Shûnyatâ* and **Zero.**

NON-PRESENT. See *Shûnyatâ* and **Zero.**

NON-PRODUCT. See *Shûnyatâ* and **Zero.**

NON-SUBSTANTIALITY. See *Shûnyatâ* and **Zero.**

NON-VALUE. See *Shûnyatâ* and **Zero.**

NOTHING. See *Shûnya* and **Zero.**

NOTHINGNESS. See *Shûnyatâ* and **Zero.**

NRIPA. [S]. Value = 16. "King". This is an allusion to the sixteen kings of the epic poems of the **Mahâbhârata* (Brihadbala, king of Koshala; Chitrasena, king of the Gandharva; Dhritarâshtra, the blind king of Indraprastha; Drupada, king of the Panchala; Jayadrâtha, king of the Sindhu; Kartavîrya, king of the Haihaya; Kâshîpati, king of the Kâshî; Madreshvara, king of the Madra; king Pradîpa; Shatayûpa, ascetic king; Shishupâla, king of the Chedi; Subala, king of Gandhâra; Vajra, king of Indraprastha; Virâta, king of the Matsya; Yavanâdhipa, king of the Yavana; and Yudhisthira, king of Indraprastha). See **Sixteen.**

NUMBERS (Philosophy and science of). See *Samkhya, Samkhyâ, Sâmkhya, Sâmkhyâ*, Numerical symbols, Symbolism of words with a numerical value, Symbolism of numbers, *Shûnya, Shûnyatâ*, Zero, Infinity and Mysticism of infinity.

NUMBERS (The science of). See *Samkhyâna*. See also **Numbers (The philosophy and science of).**

NUMERAL "0" (in the form of a little circle). Currently the symbol used in nearly all the numerical notations of India (the following types of modern numerals : Nâgarî, Gujarâtî, Marâthî, Bengali, Oriyâ, Punjabi, Sindhi, Gurûmukhî, Kâithî, Maithilî, Tâkarî, Telugu, Kannara, etc.), of central Asia (Nepali and Tibetan numerals) and of Southeast Asia (Thai-Khmer, Balinese, Burmese, Javanese, etc. numerals). There is evidence of the use of this sign since the seventh century CE in the Indianised civilisations of Southeast Asia (Champa, Cambodia, Sumatra, Bali, etc.). See Fig. 24.3 to 13, 24.15, 16, 21, 24, 25, 26, 28, 39, 41, 42, 50, 51, 52, 78, 79 and 24.80. See **Indian written numeral systems (Classification of).** See also **Circle** and **Zero.**

NUMERAL "0" (in the shape of a point or dot). This was formerly in use in the regions of Kashmir and Punjab (Sharada numerals). There is evidence of the use of this sign since the seventh century CE in the Khmer inscriptions of ancient Cambodia. Today, this sign is still used in Muslim India in eastern Arabic numeration ("Hindi" numerals). See Fig. 24.2, 14, 40, 78 and 80. See **Indian written numeral systems (Classification of), Eastern Arabic numerals, Dot** and **Zero.**

NUMERAL "1". (The origin and evolution of the). See Fig. 24. 61.

NUMERAL "2". (The origin and evolution of the). See Fig. 24. 62.

NUMERAL "3". (The origin and evolution of the). See Fig. 24. 63.

NUMERAL "4". (The origin and evolution of the). See Fig. 24. 64.

NUMERAL "5". (The origin and evolution of the). See Fig. 24. 65.

NUMERAL "6". (The origin and evolution of the). See Fig. 24. 66.

NUMERAL "7". (The origin and evolution of the). See Fig. 24. 67.

NUMERAL "8". (The origin and evolution of the). See Fig. 24. 68.

NUMERAL "9". (The origin and evolution of the). See Fig. 24. 69.

NUMERAL (as a sign of written numeration). See *Anka* and **Signs of numeration.**

NUMERAL ALPHABET AND COMPOSITION OF CHRONOGRAMS. Chronograms can be found on certain monuments. These are short phrases written in Sanskrit (or Prakrit), the words of which, when evaluated then totalled according to the numerical value of their letters, give the date of an event which has already taken place or will take place in the future. In Muslim India, the same procedure was used frequently, this time using the numeral letters of the Arabic-Persian alphabet. They are commonly found on epitaphs to express the date of death of the person buried in the tomb. See **Numeral alphabet, Chronogram. Chronograms (System of letter numerals).**

NUMERAL ALPHABET AND SECRET WRITING. Like all those who have used an alphabetical numeration, the Indians, Sinhalese, Burmese, Khmers, Thais, Javanese and Tibetans alike have used it to write in a secret code. We still use such systems today to write information or incantatory or magic formulas. In this way numerical series are written to hide their meanings should they fall into the hands of the profane or uninitiated. Likewise, if the order of pages are numbered in this way, it prohibits the profane from reading the texts, thus keeping them secret in a coherent manner; the initiated only has to put the pages in the correct order before he reads the text. See **Numeral alphabet.**

NUMERAL ALPHABET, MAGIC, MYSTICISM AND DIVINATION. As with the Greeks, the Jews, the Syrians, the Arabs and the Persians, the Indian mystics, Magi and soothsayers used their numeral alphabets as the basic instruments of their magical, divinatory or numerological interpretations or practices. A whole mystical-religious practice, just like gnosis, Judaeo-Christian Cabbala or Muslim Sufism was created in this manner. This led to all kinds of homilectic and symbolic interpretations, to various predictive calculations and to the creation of certain *kavachas*, talismans curiously resembling Hebrew Cabbalistic pentacles and Muslim *herz* from North Africa. The practice was based on a doctrine of sound and the Sanskrit alphabet: *bîjas* or "letter-seeds", where each syllable of the alphabet characterised a divinity of the Brahmanic pantheon (or of the pantheon of tantric Buddhism in the schools in the North), whom it was believed that one could evoke just by pronouncing the letter. The sound, by definition, was considered to be the creative and evocative element *par excellence*. Hence the mystical value attached to each letter in association with the esoterical meaning of its numerical equivalent. The external sound of the voice is born in the secret centre of the person in the form of the essence of the sound, and passes through three vibratory processes

before becoming audible: *parâ, pashyantî* and *madhyamâ*. Beginning subtly, the sound turns into one of the forty-six letters of the Sanskrit alphabet. As the sound is transmitted by the *nâdî*, it becomes one or another of the Sanskrit alphabetical letters. The matter, in Hindu cosmology, is divided into five states of manifestation : air, fire, earth, water, ether. Each state corresponds to a Sanskrit letter as is shown in the following table:

Air	(*Vâyu*): *ka, kha, ga, gha, na, a, â, ri, ha, sha, ya.*
Fire	(*Agnî*): *cha, chha, ja, jha, ña, i, î, rî, ksa, ra.*
Earth	(*Prithivî*): *ta, tha, da, dha, na, u, û, li, sha, va, la.*
Water	(*Âpa*): *ta, tha, da, dha, na, ê, ai, li, sa.*
Ether	(*Âkâsha*): *pa, pha, ba, bha, ma, o, au, am, ah.*

We can now understand the principle of the creation of a *mantra*, which is a combination of sounds which have been carefully studied in terms of their secret values. It is not worth trying to make intelligible sense of a mantra because this is not its aim; just as certain numerical combinations enabled Cabbalists to invent ingenious secret names, names which are impossible to translate (they are artificial creations), the mantra is a precise combination of sounds created with some secret aim in mind [Marquès-Rivière (1972)] See *Âkshara*, **Numeral alphabet, *Bhûta, Mahâbhûta, Trivarna, Vâchana.*** See also Chapter 20, for similar practices in other cultures.

NUMERAL ALPHABET. This denotes any system of representing numbers which uses vocalised consonants of the Indian alphabet, to which a numerical value is assigned, in a predetermined, regular order. In keeping with their diverse systems of recording numbers (in numerals, in symbols or spoken), the Indians knew and used different systems of this kind. This is what is conveyed by the collective name *varnasankhya*, or systems of "letter-numbers".

The inventor of the first numerical alphabet in Indian history was the astronomer *Âryabhata*, who, c. 510 CE, had the idea of using the thirty-three letters of the Indian alphabet to represent all the numbers from 1 to 10^{18}. His aim in creating this system was to express the constants of his *astronomical canon*, as well as the numerical data of his diverse speculations on *yugas*. See **Âryabhata (Numerical notations of),**

Âryabhata's numeration, Yuga (astronomical speculation about).

After Âryabhata, many other numeral alphabets were invented using Indian letters. These vary both according to the numerical value of the letters and the period and region, and sometimes even the principle employed in the numerical representations.

One such system is the *katapayâdi* system, which is still called *varnasamjña* (or "proceeding from syllables"); it was almost certainly created by the astronomer *Haridatta in the ninth century CE. and later adopted by many astronomers, including Shankaranârâyana (c. 869 CE). It is a simplified version of Âryabhata's system; the successive vocalisations of the consonants of the Indian alphabet are suppressed. Each value which is superior or equal to ten is replaced with a zero or one of the first nine units. The author of the system thus transformed the earlier system into an alphabetical numeration which used the place-value system and zero. See **Katapayâdi numeration.**

Amongst the diverse alphabetical notations, it is also worth mentioning the *aksharapallî* system, which is frequently used in Jaina manuscripts. Such systems are still in use today in various regions of India, from Maharashtra, Bengal, Nepal and Orissa to Tamil Nadu, Kerala and Karnataka. They are also found amongst the Sinhalese, the Burmese, the Khmers, the Thais and the Javanese. They can also be found amongst the Tibetans, who have long used their letters as numerical signs, particularly when numbering their registers and the pages of their manuscripts. See Chapters 17 to 20 for similar uses in other cultures.

NUMERAL. [S]. Value = 9. See *Anka* and **Nine.**

NUMERATION OF NUMERICAL SYMBOLS. Name given here to the place-value system written using Sanskrit numerical symbols, used by Indian astronomers and mathematicians since at least the fifth century CE. In Sanskrit, this is often called *samkhya* (or *sankhya*). See **Numerical symbols (Principle of the numeration of).**

NUMERICAL NOTATION. Here is an alphabetical list of terms relating to this notion, which appear as headings in this dictionary: *Aksharapallî, *Ândhra numerals, *Anka, *Ankakramena, *Ankânâm Vamâto Gatih, *Ankapallî, *Ankasthâna, *Arabic numeration (Positional systems of Indian origin), *Âryabhata's numeration, *Brâhmî numerals, *Eastern Arabic numerals, *High numbers, *Indian numerals, *Indian written numeral systems (Classification of), *Indusian numeration, *Katapayâdi numeration, *Kharoshthî numeration, *Numeral alphabet, *Numeral 1, *Numeral 2, etc., *Numerical symbols (Principle of the numeration of), *Sanskrit *Sthâna, *Sthânakramâd, *Varnasamjnâ and *Zero.

NUMERICAL SYMBOLS. These are words which are given a numerical value depending what they represent. They can be taken from nature, the morphology of the human body, representations of animal or plants, acts of daily life, any types of tradition, philosophical, literary or religious elements, attributes and morphologies connected to the divinities of the Hindu, Jaina, Vedic, Brahmanic, Buddhist, etc. pantheons, legends, traditional associations of ideas, mythologies or social conventions of Indian culture. See **Symbols.** See also all entries entitled **Numerical symbols or Symbolism of numbers.**

NUMERICAL SYMBOLS (General alphabetic list). These are Sanskrit numerical symbols which are found in texts on mathematics or astronomy, as well as in various Indian epigraphic inscriptions (this list is not exhaustive):

*Abdhi (= 4), *Abhra (= 0), *Abja (= 1), Abjadala (= 100), Abjagarbha (= 9), Achala (= 7), *Adi (= 1), *Âditya (= 12), *Adri (= 7), *Aga (= 7), Aghosha (= 13), *Agni (= 3), *Ahar (= 15), *Ahi (= 8), Airâvata (= 1), Aja (= 9), *Âkashâ (= 0), *Akriti (= 22), *Âkshara (= 1), Akshauhinî (= 11), Akshi (= 2), *Amara (= 33), Ambaka (= 2), *Ambara (= 0), *Ambhodha (= 4), Ambhodhi (= 4), Ambhonidhi (= 4), *Ambodha (= 4), Ambodhi (= 4), Ambudhi (= 4), *Amburâshi (= 4), *Anala (= 3), *Ananta (= 0), *Anga (= 6), *Anguli (= 10), *Anguli (= 20), Anîka (= 8), *Anka (= 9), *Antariksha (= 0), *Anushtubh (= 8), *Âptya (= 3), Arhat (= 24), Ari (= 6), *Arjunâkara (= 1,000), *Arka (= 12), *Arnava (= 4), Artha (= 5), *Âshâ (= 10), Âshrama (= 4), *Ashti (= 16), *Ashva (= 7), *Ashvin (= 2), *Ashvina (= 2), *Ashvinau (= 2), *Atidhriti (= 19), Atijagatî (= 13), *Âtman (= 1), *Atri (= 7), *Atrinayanaja (= 1), *Atyashti (= 17), *Avani (= 1), *Avatâra (= 10), Aya (= 4), Âya (= 4), Ayana (= 2).

*Bâhu (= 2), *Bâna (= 5), Bandhu (= 4), *Bha (= 27), *Bhânu (= 12), *Bharga (= 11), Bhâva (= 5), *Bhava (= 11), Bhaya (= 7), *Bhû (= 1), *Bhûbrit (= 7), *Bhûdhara (= 7), *Bhûmi (= 1), *Bhûpa (= 16), *Bhûta (= 5), Bhûti (= 8), *Bhuvana (= 3), *Bhuvana (= 14), *Bindu (= 0), *Brahmâsya (= 4), Brihatî (= 9).

*Chakra (= 12), *Chakshus (= 2), Chandah (= 7), Chandas (= 7), *Chandra (= 1), *Chaturânanavâdana (= 4), *Chhidra (= 9).

Dadhi (= 4), *Dahana (= 3), *Danta (= 32), *Dantin (= 8), *Darshana (= 6), *Dasra (= 2), *Deva (= 33), *Dharâ (= 1), *Dharanî (= 1), *Dhârtarâshtra (= 100), *Dhâtrî (= 1), Dhâtu (= 7), Dhî (= 7), *Dhriti (= 18), *Dhruva (= 1), *Diggaja (= 8), Dik (= 8), *Dikpâla (= 8), Dina (= 15), *Dish (= 4), *Dish (= 8), *Disha (= 10), *Dishâ (= 4), * Disha (= 10), *Divâkara (= 12), Dosha (= 3), *Dravya (= 6), *Drishti (= 2), *Durgâ (= 9), Durita (= 8), *Dvandva (= 2), Dvâra (= 9), *Dvaya (= 2), *Dvija (= 2), *Dvipa (= 8), *Dvîpa (= 7), Dvirada (= 8), Dyumani (= 12).

*Gagana (= 0), *Gaja (= 8), Gangâmarga (= 3), *Gati (= 4), *Gavyâ (= 5), *Gâyatrî (= 24), Ghasra (= 15), *Giri (= 7), *Go (= 1), *Go (= 9), Gostana (= 4), *Graha (= 9), Grahana (= 2), *Gulpha (= 2), *Guna (= 3), *Guna (= 6).

*Hara (= 11), *Haranayana (= 3), *Haranetra (= 3), *Haribâhu (= 4), *Hastin (= 8), Haya (= 7), Himagu (= 1), Himakara (= 1), Himâmshu (= 1), *Hotri (= 3), *Hutâshana (= 3).

Ibha (= 8), Îkshana (= 2), Ilâ (= 1), *Indra (= 14), *Indradrishti (= 1,000), *Indriya (= 5), *Indu (= 1), *Iryâ (= 4), *Îsha (= 11), *Îshadrish (= 3), *Ishu (= 5), *Îshvara (= 11).

*Jagat (= 3), *Jagat (= 14), *Jagatî (= 1), *Jagatî (= 12), *Jagatî (= 48), *Jahnavîvaktra (= 1,000), *Jala (= 4), *Jaladharapatha (= 0), *Jaladhi (= 4), *Jalanidhi (= 4), Jalâshaya (= 4), Jana (= 1), Janghâ (= 2), Jânu (= 2), Jâti (= 22), Jina (= 24), *Jvalana (= 3).

*Kakubh (= 10), *Kâla (= 3), Kalâ (= 16), Kalamba (= 5), Kalatra (= 7), *Kâma (= 13), *Kara (= 2), Kâraka (= 6), *Karanîya (= 5), Karman (= 8), Karman (= 10), Karna (= 2), *Kârttikeyâsya (= 6), Kashâya (= 4), *Kâya (= 6), Kendra (= 4), *Keshava (= 9), *Kha (= 0), Khanda (= 9), Khara (= 6), Khatvâpâda (= 4), Koshtha (= 4), *Krishânu (= 3), *Krita (= 4), *Kriti (= 20), Kritin (= 22), Kshapâkara (= 1), *Kshapeshvara (= 1), Kshâra (= 5), *Kshaunî (= 1), *Kshemâ (= 1), *Kshiti (= 1), *Kshoni (= 1), *Ku (= 1), Kucha (= 2), *Kumârâsya (= 6), *Kumâravadana (= 6), *Kuñjara= 6), (= 8), Kutumba (= 2).

Labdha (= 9), Labdhi (= 9), Lâbha (= 11), Lakâra (= 10), Lavana (= 5), Lekhya (= 6), *Loka (= 3), *Loka (= 7), *Loka (= 14), *Lokapâla (= 8), *Lochana (= 2).

Mada (= 8), *Mahâbhûta (= 5), *Mahâdeva (= 11), *Mahâpâpa (= 5), *Mahâyajña (= 5), *Mahî (= 1), *Mahîdhara (= 7), Mala (= 6),

*Mangala (= 8), Manmatha (= 13), *Manu (= 14), *Mârgana (= 5), *Mârtanda (= 12), *Mâsa (= 12), *Mâsârdha (= 6), *Matanga (= 8), *Mâtrika (= 7), *Mrigânka (= 1), *Mukha (= 4), Mûlaprakriti (= 1), *Muni (= 7), *Mûrti (= 3), *Mûrti (= 8).

*Nabha (= 0), *Nabhas (= 0), Nâdî (= 3), Nadîkûla (= 2), *Naga (= 7), *Nâga (= 8), *Nakha (= 20), *Nakshatra (= 27), Nanda (= 9), Naraka (= 40), *Nâsatya (= 2), Naya (= 2), Nâyaka (= 1), *Nayana (= 2), *Netra (= 2), *Netra (= 3), Nidhâna (= 9), Nidhi (= 9), *Nripa (= 16).

Oshtha (= 2).

Padârtha (= 9), *Paksha (= 2), *Paksha (= 15), Pallava (= 5), *Pândava (= 5), Pankti (= 10), *Parabrahman (= 1), Parva (= 5), Parvan (= 5), *Parvata (= 7), *Pâtaka (= 5), *Pâtâla (= 7), *Pâvaka (= 3), *Pâvana (= 5), *Pâvana (= 7), Payodhi (= 4), Payonidhi (= 4), *Pinâkanayana (= 3), *Pitâmaha (= 1), *Prakriti (= 21), Prâleyâmshu (= 1), *Prâna (= 5), *Prithivî (= 1), *Pura (= 3), *Purâ (= 3), *Purânalakshana (= 5), *Pûrna (= 0), Purushârtha (= 4), Purushâyus (= 100), Pûrva (= 14), *Pushkara (= 7), Pushkarin (= 8), *Putra (= 5).

*Rada (= 32), *Râga (= 6), Rajanîkara (= 1), *Râma (= 3), Râmanandana (= 2), *Randhra (= 0), *Randhra (= 9), *Rasa (= 6), *Râshi (= 12), Rashmi (= 1), *Ratna (= 3), *Ratna (= 5), *Ratna (= 9), *Ratna (= 14), *Râvanabhuja (= 20), *Râvanashiras (= 10), *Ravi (= 12), *Ravibâna (= 1,000), *Ravichandra (= 2), Ripu (= 6), *Rishi (= 7), *Ritu (= 6), *Rudra (= 11), *Rudrâsya (= 5), *Rûpa (= 1).

*Sâgara (= 4), *Sâgara (= 7), *Sahasrâmshu (= 12), Sahodarâh (= 3), Salilâkara (= 4), *Samudra (= 4), *Samudra (= 7), Sankrânti (= 12), *Sarpa (= 8), *Sâyaka (= 5), Senânga (= 4), *Senâninetra (= 12), *Shadâyatana (= 6), *Shaddarshana (= 6), *Shâdgunya (= 6), *Shaila (= 7), *Shakra (= 14), Shakrayajña (= 100), *Shakti (= 3), *Shankarâkshi (= 3), *Shanmukha (= 6), *Shanmukhabâhu (= 12), *Shara (= 5), *Shashadhara (= 1), *Shashanka (= 1), *Shashin (= 1), Shâstra (= 5), Shâstra (= 6), *Sheshashîrsha (= 1,000), *Shikhin (= 3), *Shîtâmshu (= 1), *Shîtarashmi (= 1), *Shiva (= 11), *Shruti (= 4), *Shukranetra (= 1), *Shûla (= 3), *Shûlin (= 11), *Shûnya (= 0), Shveta (= 1), Siddha (= 24), *Siddhi (= 8), *Sindhu (= 4), Sindhura (= 8), *Soma (= 1), *Sudhâmshu (= 1), *Sura (= 33), *Sûrya (= 12), *Suta (= 5), Svagara (= 21) *Svara (= 7).

Takshan (= 8), *Tâna* (= 49), *Tanmâtra* (= 5), *Tanu* (= 1), *Tanu* (= 8), *Tapana* (= 3), *Tapana* (= 12), *Tarka* (= 6), *Târkshadhvaj* (= 9), *Tata* (= 5), *Tattva* (= 5), *Tattva* (= 7), *Tattva* (= 25), *Tithi* (= 15), *Trailokya* (= 3), *Trayî* (= 3), *Tridasha* (= 33), *Trigata* (= 3), *Triguna* (= 3), *Trijagat* (= 3), *Trikâla* (= 3), *Trikâya* (= 3), *Triloka* (= 3), *Trimûrti* (= 3), *Trinetra* (= 3), *Tripurâ* (= 3), *Triratna* (= 3), *Trishiras* (= 3), *Trishtubh* (= 11), *Trivarga* (= 3), *Trivarna* (= 3), *Tryakshamukha* (= 5), *Tryambaka* (= 3), *Turaga* (= 7), *Turangama* (= 7), *Turîya* (= 4).

Uchchaishravas (= 1), *Uda* (= 27), *Udadhi* (= 4), *Udarchis* (= 3), *Upendra* (= 9), *Utkriti* (= 26), * *Urvarâ* (= 1).

Vâchana (= 3), *Vahni* (= 3), *Vaishvânara* (= 3), *Vâjin* (= 7), *Vanadhi* (= 4), *Vâra* (= 7), *Vâridhi* (= 4), *Vârinidhi* (= 4), *Varsha* (= 9), *Vasu* (= 8), *Vasudhâ* (= 1), *Vasundharâ* (= 1), *Vâyu* (= 49), *Veda* (= 3), *Veda* (= 4), *Vidhu* (= 1), *Vidyâ* (= 14), *Vikriti* (= 23), *Vindu* (= 0), *Vishanidhi* (= 4), *Vishaya* (= 5), *Vishikha* (= 5), *Vishnupada* (= 0), *Vishtapa* (= 3), *Vishuvat* (= 2), *Vishva* (= 13), *Vishvadeva* (= 13), *Viyata* (= 0), *Vrata* (= 5), *Vyant* (= 0), *Vyasana* (= 7), *Vyaya* (= 12), *Vyoman* (= 0), *Vyûha* (= 4).

Yama (= 2), *Yâma* (= 8), *Yamala* (= 2), *Yamau* (= 2), *Yati* (= 7), *Yoni* (= 4), *Yuga* (= 2), *Yuga* (= 4), *Yugala* (= 2), *Yugma* (= 2).

To gain an idea of the symbolism of these words, see **Symbolism of words with a numerical value** and **Symbolism of numbers**. The first of these two entries gives an alphabetical list of English terms which explain the various corresponding associations of ideas, and the second entry gives a list of the same associations of ideas, set out this time in numerical order (one, two, three, etc.). To understand the principle for using word-symbols to represent numbers, see **Numerical symbols (Principle of the numeration of)**.

Source: Bühler (1896), pp. 84ff; Burnell (1878); Datta and Singh (1938), pp. 54–7; Fleet, in : CIIn, VIII; Jaquet, in : JA, XVI, 1835; Renou and Filliozat (1953), p. 708–9; Sircar (1965), pp. 230–3; Woepcke (1863).

NUMERICAL SYMBOLS (Principle of the numeration of). Procedure used to record numbers by Indian scholars since at least as early as the fifth century CE. This is simply a series of Sanskrit word-symbols (which are used as names of units), which are written in conformity with the "principle of the movement of numerals from the right to the left" (*ankânâm vâmato gatih*). See **Sanskrit** and **Numerical symbols**.

In other words, in this system, numerical symbols have a variable value depending on their position when numbers are written down. The system possesses several different special terms which symbolise zero and which thus serve to mark the absence of units in any given decimal order in this positional notation (*shûnya*, *âkâsha*, *abhra*, *ambara*, *antariksha*, *bindu*, *gagana*, *jaladharapatha*, *kha*, *nabha*, *nabhas*, etc.). An expression such as:

agni. shûnya. ashvi. vasu.

[literally : "fire (= 3). void (= 0). Horsemen (= 2). Vasu (= 8)"] corresponds to the numbers: $3 + 0 \times 10 + 2 \times 10^2 + 8 \times 10^3 = 8{,}203$.

This method of expressing numbers uses the place-value system and zero. What is remarkable about it is that Indian scholars are the only ones to have invented such a system. See **Position of numerals**, and **Zero**.

NUMERICAL SYMBOLS (Sanskrit designation of). The generic term for words used as numerical symbols is *samkhya*, which literally means "number". Also used to refer to the system as a whole, which is the place-value system expressed through numerical symbols.

NUMEROLOGY. See **Numeral alphabet, magic, mysticism and divination.**

NYARBUDA. Name given to the number ten to the power eight (= one hundred million). See **Names of numbers** and **High numbers**.

Sources: *Vâjasaneyî Samhitâ* (beginning of the Common Era); *Taittirîya Samhitâ* (beginning of the Common Era); *Kâthaka Samhitâ* (beginning of the Common Era); *Pañchavimsha Brâhmana* (date uncertain); *Sankhyâyana Shrauta Sûtra* (date uncertain).

NYARBUDA. Name given to the number ten to the power 11. See **Names of numbers** and **High numbers**.

Source: *Ganitasârasamgraha* by Mahâhavîrâchârya (850 CE).

O

OCEAN. Name given to the number ten to the power four, ten to the power nine or ten to the power fourteen. See *Jaladhi*, *Samudra* and **High numbers**.

OCEAN. [S]. The entries entitled *sâgara* or *samudra*, which, as numerical symbols, translate the idea of "sea" or "ocean", can have the value of either 4 or 7. The relation between *sâgara* and 4 can be explained through the allusion to the "four oceans" (*chatursâgara*) which, according to Hindu and Brahmanic mythologies, surround *Jambudvîpa*, (India). However, this explanation does not give the real reason for the choice of the number four for the oceans surrounding India. In reality, it is due to the fact that the mystical symbol for "water" (*jala*) is the number four. According to Brahmanic doctrine of the five elements of the manifestation (*bhûta*), water (which is also called *apa*) forms, along with earth (*prithivî*), air (*vâyu*) and fire (*agnî*), the ensemble of elements which are said to participate directly in the "material order of nature". This order is believed to be quaternary, and the diverse phenomena of life boil down to the manifestations of these four elements in the determination of the essence of the forces of nature as well as in the realisation of the latter in its work of generation and destruction. In traditional Indian thought (and even according to a universal constant), the earth itself corresponds symbolically to the number four, because it is associated with a square due to its four horizons (or cardinal points).

As for the relationship between *sâgara* and the number seven, this can be explained by direct reference to the seven mythical oceans (namely: The ocean of salt water, the ocean of sugar cane juice, the ocean of wine, the ocean of thinned butter, the ocean of whipped cheese, the ocean of milk and the ocean of soft water), which are meant to surround *Mount Meru*. See *Sapta sâgara*.

Mount Meru is the mythical and sacred mountain of Brahman mythology and Hindu cosmology, which constitutes the meeting place and dwelling of the gods. Situated at the centre of the universe, this mountain is placed above seven hells (*pâtâla*), and has seven faces, each one looking at one of the seven "island-continents", themselves each in one of the seven oceans, etc. In this symbolism, Mount Meru represents the total fixedness and the absolute centre around which the firmament and the whole universe pivot in their eternal course. This image is connected to one of the universal symbolic representations of the *Pole star. Mount Meru is said to be situated directly underneath this star, and along exactly the same axis. This symbolic correspondence comes from the fact that the Pole star, the *Sudrishti*, "That which never moves", is the last of the seven stars of the Little Bear, which themselves are considered by Indian tradition to be the personification of the seven "great Sages" (in other words the *saptarishi* of Vedic times, believed to be the authors of hymns and invocations, as well as of the most important texts of the *Veda*). This is why the number seven came to play a preponderant symbolic role in the mythological representations associated with Mount Meru.

It is this symbolism which determined the number of cosmic oceans in the legends about the creation of the universe, and gave words expressing the idea of "ocean" a value of 7. In its representations, India, (*Jambudvîpa*) is considered to be the "centre of the earth", whilst Mount Meru was regarded as the centre of the universe. Ocean has two different numerical values in order to mark the opposition between the human character, essentially terrestrial, of the oceans surrounding India, and the divine character, essentially celestial, of the oceans surrounding Mount Meru. In spite of the apparent paradox, Indian scholars managed to avoid any confusion. The words *samudra and *sâgara*, which both mean "ocean", were both sometimes used as symbols for the number four. But they were usually used (never simultaneously) to express the number seven, words such as *abdhi, *ambhonidhi, ambudhi, *amburâshi, *jaladhi, *jalanidhi, *jalâshaya, *sindhu, *vâridhi or *vârinidhi being reserved for the number four, and which more modestly meant "sea".

OCEAN. [S]. Value = 4. See *Abdhi, Ambhonidhi, Ambudhi, Amburâshi, Arnava, Jaladhi, Jalanidhi, Sâgara, Samudra, Sindhu, Udadhi, Vâridhi* and *Vârinidhi.* See also *Four, Jala.*

OCEAN. [S]. Value = 7. See *Sâgara* and *Samudra.* See also *Seven, Mount Meru.*

OLD KHMER NUMERALS. Symbols derived from *Brâhmî numerals and influenced by Shunga, Shaka, Kushâna, Ândhra, Pallava, Châlukya, Ganga, Valabhî, 'Pâli' and Vatteluttu numerals. Used from the seventh century CE in the ancient kingdom of Cambodia. The notation used for dates in the *Shaka era were based on a place-value system and had a zero (a dot or small circle), whereas vernacular notation was very rudimentary. See: **Indian written numerals systems (Classification of).** See Fig., 24.52, 61 to 69, 77, 78 and 80.

ONE. Ordinary Sanskrit name for this number: *Eka. Here is a list of corresponding numerical symbols: *Abja, *Âdi, Airâvata, *Âkshara, *Âtman, *Atrinayanaja, *Avani, *Bhû, *Bhûmi, *Chandra, *Dharâ, *Dharanî, *Dhâtrî, *Dhruva, *Go, Himagu, Himakara, Himâmshu, Ilâ, *Indu, *Jagatî, Jana, Kshapâkara, *Kshapeshvara, *Kshaunî, *Kshemâ, *Kshiti, *Kshoni,

*Ku, *Mahî, *Mrigânka, Mûlaprakriti, Nâyaka, *Parabrahman, *Pitâmaha, Prâleyâmshu, *Prithivî, Rajanîkara, Rashmi, *Rûpa, *Shashadhara, *Shashanka, *Shashin, Shveta, *Shîtâmshu, *Shîtarashmi, *Shukranetra, *Soma, *Sudhâmshu, *Tanu, *Uch-chaishravas, *Urvarâ, *Vasudha, *Vasundharâ, *Vidhu. These words have the following translation or symbolic meaning: 1. The "Moon". (Abja, Atrinayanaja, Chandra, Indu, Jagatî, Kshapeshvara, Mrigânka, Shashadhara, Shashanka, Shashin, Shîtâmshu, Shîtarashmi, Soma, Sudhâmshu, Vidhu). 2. The drink of immortality (Soma). 3. The "Earth" (Avani, Bhû, Bhûmi, Dharâ, Dharanî, Dhâtrî, Go, Jagatî, Kshaunî, Kshemâ, Kshiti, Kshoni, Ku, Mahî, Prithivî, Urvarâ, Vasudha, Vasundharâ). 4. The "Ancestor", the "First Father", the "Great Ancestor" (Pitâmaha). 5. Individual soul, supreme soul, ultimate Reality, the Self (Âtman). 6. The Brahman (Âtman, Pitâmaha, Parabrahman). 7. The beginning (Âdi). 8. The body (Tanu). 9. The Pole star (Dhruva). 10. The form (Rûpa). 11. The "drop" (Indu). 12. The "immense" (Prithivî). 13. The "Indestructible" (Akshara). 14. The rabbit (Shashin, Shashadhara). 15. The "Luminous", in allusion to the moon as a masculine entity (Chandra). 16. The "cold Rays" of the moon (Shîtamshu, Shîtarashmi). 17. The terrestrial world (Prithivî). 18. The eye of Shukra (Shukranetra). 19. The "Bearer", in allusion to the earth (Dharanî). 20. The primordial principle (Âdi). 21. Rabbit figure (Shashadhara). 22. The Cow (Go, Mahî). 23. Curdled milk (Mahî). See **Numerical symbols.**

OPINION. [S]. Value = 6. See *Darshana* and **Six.**

ORDERS OF BEINGS (The five). See *Panchaparamesthin.*

ORIFICE. [S]. Value = 9. See Chhidra, Randhra and Nine.

ORIGINAL SERPENT (Myth of the). See **Infinity (Indian mythological representation of)** and **Serpent (Symbolism of the).**

ORISSÎ NUMERALS. See Oriyâ Numerals.

ORIYÂ NUMERALS. Symbols derived from *Brâhmî numerals and influenced by Shunga, Shaka, Kushâna, Ândhra, Gupta, Nâgarî, Kutilâ and Bengali. Now used mainly in the state of Orissâ. Also called Orissî numerals. The symbols correspond to a mathematical system that has place values and a zero (shaped like a small circle). See **Indian written numeral systems (Classification of).** See Fig. 24.12, 52 and 24.61 to 69.

OUROBOUROS. See Infinity (Indian mythological representation of) and Serpent (Symbolism of the).

P

PADMA (or PADUMA). This is the name for the pink lotus. As well as the purity it represents, to the Indian mind it symbolises the highest divinity as well as innate reason.

PADMA. Name given to the number ten to the power nine. See **Names of numbers.** See also **High numbers (The symbolic meaning of).**

Source: *Kitab fi tahqiq i ma li'l hind by al-Biruni (c. 1030 CE).

PADMA. Name given to the number ten to the power fourteen. See **Names of numbers.** For an explanantion of this symbolism, see **Padma (or Paduma).** See also **High numbers (The symbolic meaning of).**

Source: *Ganitasârasamgraha by Mahâvîrâchârya (850 CE).

PADMA. Name given to the number ten to the power twenty-nine. See **Names of numbers.** See also **High numbers (The symbolic meaning of).**

Source: *Râmâyana by Vâlmîki (early centuries CE).

PADUMA. Literally, "(pink) lotus". Name given to the number ten to the power twenty-nine. See **Names of numbers.** See also **High numbers (The symbolic meaning of).**

Source: *Lalitavistara Sûtra (before 308 CE).

PADUMA. Name given to the number ten to the power 119. See **Names of numbers.** See also **High numbers (The symbolic meaning of).**

Source: *Vyâkarana (Pali grammar) by Kâchchâyana (eleventh century CE).

PAIR. [S]. Value = 2. See **Dvaya** and **Two.**

PAKOTI. Name given to the number ten to the power fourteen. See **Names of numbers** and **High numbers.**

Source: *Vyâkarana (Pali grammar) by Kâchchâyana (eleventh century CE).

PAKSHA. [S]. Value = 2. "Wing". This is due to the symmetry of this organ. The word can also mean one of the two halves of a month. Thus it is sometimes also used to represent the number fifteen. This double symbolism can be explained by the division of the month (*mâsa) into two periods of fifteen days called paksha, each one corresponding to one phase of the moon. The first, called "shining" (shudi), is progressive, and the second, called "shadow" (*badi*), is degressive. According to Hindu mythology and cosmogony, these two periods formed one whole being (before the churning of the sea of milk); this being was decapitated by Indra when he drank the *amrita (the nectar of eternal life) that he had stolen. This created the "Cut in twos" (*Ashleshâbava*): two beings named *Râhu and *Ketu, who personify the ascending and descending nodes of the moon. See *Mâsa*, *Râhu* and **Two**.

PAKSHA. [S]. Value = 15. See **Fifteen**.

"PÂLÎ" NUMERALS. Symbols derived from *Brâhmî numerals and influenced by Shunga, Shaka, Kushâna, Ândhra, Pallava, Châlukya, Ganga and Valabhî. Formerly used in Magadha (the ancient Hindu kingdon of present-day Bihar, south of the Ganges) from the Mauryan period. All the later numeral symbols of the eastern and southeast Asia (Mon, Burmese, Cham, Ancient Khmer, Thai-Khmer, Balinese, etc.) derive from Pâlî numerals. The symbols corresponded to a mathematical system that was not based on place values and therefore did not possess a zero. See: **Indian written numerals systems (Classification of)**. See Fig. 24.52 and 24.61 to 69.

PALLAVA NUMERALS. Symbols derived from •Brâhmî numerals and influenced by Shunga, Shaka, Kushâna and Ândhra, arising at the time of the Pallava dynasty (fourth to sixth centuries CE). The symbols correspond to a mathematical system that was not based on place values and therefore did not possess a zero. See: **Indian written numeral systems (Classification of)**. See Fig. 24.37, 24.61 to 24.69 and 24.70.

PAÑCHA. Ordinary Sanskrit term for the number five, which appears in many words which have a direct relationship with the idea of this number. Examples:

*Pañchabâna, *Pañchâbhijñâ, *Pañchabhûta, *Pañchachakshus, *Pañchadisha, *Pañchagavyâ, *Pañcha Indriyâni, *Pañcha Jâti, *Pañchaklesha, *Pañchânana, *Pañchanantarya, *Pañchaparameshtin, *Pañcharâshika, *Pañchatantra.

For words which have a more symbolic relationship with this number, see **Five** and **Symbolism of numbers**.

PAÑCHABÂNA. "Bow of five flowers". This is one of the attributes of *Kâma, Hindu divinity of Cosmic Desire and Carnal Love, who is generally invoked in marriage ceremonies. Kâma is often represented as a young man armed with a bow of sugar cane and five arrows covered in or constituted by five flowers.

PAÑCHÂBHIJÑÂ. Name given by the Sinhalese to the "five supernatural powers" of Buddha. The Buddhists of Sri Lanka only recognise five of the six *Abhijña*, or "supernatural powers", which other Buddhist philosophies believe in.

PAÑCHABHÛTA. "Five elements". Collective name for the five elements of the manifestation of Brahman and Hindu philosophies. See *Bhûta* and *Jala*.

PAÑCHACHAKSHUS. "Five visions of Buddha". According to Buddhists, Buddha possesses the five following types of visions: that of the body, of the divine form, wisdom, doctrine and of his eye.

PAÑCHADASHA. Ordinary Sanskrit name for the number fifteen. For words with a symbolic relationship with this number, see **Fifteen** and **Symbolism of numbers**.

PAÑCHADISHA. "Five horizons". These are the four cardinal points plus the zenith. See *Dish*.

PAÑCHAGAVYÂ. "Five gifts of the Cow". See *Gavyâ*.

PAÑCHA INDRIYÂNI. "Five faculties". These are the mental and physical faculties of Buddhist philosophy, which are divided into five groups. See *Indriya*.

PAÑCHA JÂTI. Name of the five fundamental arithmetical rules of the reduction of fractions.

PAÑCHAKLESHA. "Five impurities". According to Hindu and Buddhist philosophies, these are the five major obstacles which keep the faithful off the Way of Realisation. See *Mahâpâpa*.

PAÑCHÂNANA. Name of the five heads of *Rudra. See *Rudrâsya*.

PAÑCHÂNANTARYA. "Five mortal sins" of Buddhism. These are the following sins: parricide; matricide; the killing of an *arhat* (a saint issued from *karma*); causing division in the Buddhist community (*sangham*); and wounding a Buddha.

PAÑCHAPARAMESHTIN. Name of the five orders of beings, considered to be the "five treasures" (*Pañcha Ratna*) of *Jaina religion.

PAÑCHAPARÂSHIKA. [Arithmetic]. Sanskrit name for the Rule of Five.

PAÑCHÂSHAT. Ordinary Sanskrit name for the number fifty.

PAÑCHASIDDHÂNTIKA. "Five astronomical canons". See *Varâhamihira* and **Indian astrology**.

PAÑCHATANTRA. "Five books". Name of the famous collection of moralistic tales and fables, made up of five books. The fables of Aesop and La Fontaine are more or less directly inspired by this collection. See *Pañcha*.

PAÑCHAVIMSHA BRÂHMANA. Text derived from the *Samaveda*, a text of Vedic literature. The contents were transmitted orally since ancient times, but were constantly re-worked and added to, and did not achieve their finished form until relatively recently. Date uncertain. See *Veda*. Here is a list of the main names of numbers mentioned in the text [see Datta and Singh (1938), p. 10]:

*Eka (= 1), *Dasha (= 10), *Sata (= 10^2), *Sahasra (= 10^3), *Ayuta (= 10^4), *Niyuta (= 10^5), *Prayuta (= 10^6), *Arbuda (= 10^7), *Nyarbuda (= 10^8), *Samudra (= 10^9), *Madhya (= 10^{10}), *Anta (= 10^{11}), *Parârdha (=10^{12}), *Nikharva (= 10^{13}), *Vâdava (= 10^{14}), *Akshiti (= 10^{15}). See Names of numbers and High numbers.

PAÑCHAVIMSHATI. Ordinary Sanskrit name for the number twenty-five. For words which are symbolically related to this number, see Twenty-five and Symbolism of numbers.

PÂNDAVA. [S]. Value = 5. "Son of Pându". This refers to one of the five brothers, semi-legendary heroes of the epic *Mahâbhârata* (namely: Yudishtira, Arjuna, Bhîma, Nakula, and Sahadeva), son of the king Pându of Hastinâpura. See Five.

PAPER. See Pâtîganita.

PARÂ. See Numeral alphabet, magic, mysticism and divination.

PARABRAHMAN. [S]. Value = 1. Literally, "Supreme Brahman". Expression synonymous with *Paramâtman*, in terms of "Supreme Soul", and an epithet given to Mahâpurusha (supreme entity of the global spirit of humanity), considered in Hindu philosophy to be the Absolute Lord of the universe and thus identified with the Brahman. See Âtman, Pitâmaha and One.

PARADISE. [S]. Value = 13. See *Vishvadeva* and Thirteen.

PARADISE. [S]. Value = 14. See *Bhuvana* and Fourteen.

PARAMABINDU. "Supreme Point". This is the supreme causal point, which, according to Buddhist philosophy, is both inexistent and identical to all the universe; it is also time considered as a point (*bindu) which lasts no sequential time but gives the impression of having a duration [see Frédéric, *Dictionnaire* (1987)].

PARAMÂNU. "Supreme Atom". This is the smallest indivisible material particle, and has a taste, odour and colour. This is different to our notion of the "atom", and is more like what we call a "molecule", the smallest particle which

constitutes part of a compound body. The *paramânu* and the *paramânu râja* (or "grain of dust of the first atoms") have long been the smallest units of length and weight in India. These are found notably in the Legend of Buddha, told in the *Lalitavistara Sûtra*, where the *paramânu* corresponds to 0.000000287 mm and the *paramânu râja* to 0.000000614 g.

PARAMÂNU RÂJA. "Grain of dust of the first atoms". Name of the smallest Indian unit of weight. At the time of the writing of the *Lalitavistara Sûtra* (before 308 CE), it corresponded to 0.000000614g. See *Paramânu*.

PARAMÂTMAN. "Supreme Soul". Epithet given to the *Brahman. See *Parabrahman*.

PARAMESHVARA. Indian astronomer c. 1431 CE. His works notably include the text entitled *Drigganita*, in which there is abundant use of the place-value system using Sanskrit numerical symbols [see Sarma (1963)]. See Numerical symbols, Numeration of numerical symbols and Indian mathematics (History of).

PARÂRDHA. From *para*, "beyond", and *ardha* "half". This is the spiritual half of the path which leads to death, identical to *devayâna*, the "way of the gods", which, according to the *Vedas*, is one of the two possibilities offered to human souls after death (this path being said to lead to the deliverance from *samsâra* or cycles of rebirth). The symbolism which has led to these words having such high numerical values as ten to the power twelve or ten to the power seventeen comes from an association of ideas, not only with the immeasurable immensity of the sky, but also with the eternity which it represents. For more details, see High numbers (Symbolic meaning of).

PARÂRDHA. Literally "half of the beyond". Name given to the number ten to the power twelve (= billion). See Names of numbers. For an explanation of this symbolism, see *Parârdha* (first entry) and High numbers (Symbolic meaning of).

Sources: *Vâjasaneyî Samhitâ, *Taittirîya Samhitâ and *Kâthaka Samhitâ (from the start of the first millennium CE); *Pañchavimsha Brâhmana (date uncertain).

PARÂRDHA. Literally "half of the beyond". Name given to the number ten to the power seventeen. See Names of numbers. For an explanation of this symbolism, see *Parârdha* (first entry) and High numbers (The symbolic meaning of).

Parârdha

961

Sources: *Kitab fi tahqiq i ma li'l hind by al-Biruni (c. 1030 CE); *Lîlâvatî by Bhâskarâchârya (1150 CE); *Ganitakaumudî by Nârâyana (1350 CE); *Trishatikâ by Shrîdharâchârya (date unknown).

PARASHURÂMA (Calendar). See *Kollam.*

PÂRÂVÂRA. Name given to the number ten to the power fourteen. See **Names of numbers** and **High numbers.**
Source: *Ganitakaumudî by Nârâyana (1350 CE).

PARIKARMA. Word used in arithmetic to denote "fundamental operations" carried out on whole numbers. See *Kalâsavarna.*

PART. [S]. Value = 6. See *Anga* and **Six.**

PARTHIAN (Calendar). Calendar beginning in the year 248 BCE. Formerly used in the northwest of the Indian sub-continent. To find a corresponding date in the Common Era, subtract 248 from a date expressed in Parthian calendar. See **Indian calendars.**

PÂRVATA. [S]. Value = 7. "Mountain". Clearly an allusion to the "Mountain of the gods" (*devapârvata), one of the names for *Mount Meru, which is said to be the home of the gods. This numerical symbolism is due to the preponderance of the number seven in the mythological representations of Mount Meru. See *Adri* and **Seven.**

PÂRVATÎ. See **Mount Meru.**

PASHYANTÎ. See **Numeral alphabet, magic, mysticism and divination.**

PASSION. [S]. Value = 6. See *Râga* and **Six.**

PÂTAKA. [S]. Value = 5. "Great sin". See *Mahâpâpa* and **Five.**

PÂTÂLA. [S]. Value = 7. "Inferior world". This refers to one of the seven "hells" of Hindu and *Jaina mythology (namely: *Atâla, Vitâla, Nitâla, Gabhastimat, Mahâtâla, Sutâla* and *Pâtâla).* These inferior worlds are said to be situated one on top of the other underneath *Mount Meru. They are the dwelling place of the *nâga, who are ruled by *Muchalinda, a chthonian genie in the form of a cobra, depicted as having seven heads. See **Seven.**

PÂTÂLA KUMÂRA. "Princess of the Underworlds". Name given to the daughter of Himalaya, sister of Vishnu and wife of Shiva. See *Pârvati.*

PÂTÎ. Literally "Board", "tablet". Term used for the calculating board or tablet, upon which Indian mathematicians carried out their calculations. See *Pâtiganita* and **Indian methods of calculation.**

PÂTÎGANITA (or GANITAPÂTÎ). In its most general sense, this word is used today to mean "abstract mathematics". In the past, however, it

referred to "arithmetic" and to the "practice of calculation", and appeared in the titles of many works relating to this discipline, for example: *Pâtîsâra* by Munishvara (1658); *Ganitapâtîkaumudî* by *Nârâyana (1356), which deals notably with magic squares; and *Ganitatilaka* by *Shrîpati (1039), the sub-heading of which is *Pâtîganita.* See: Datta and Singh (1938); Kapadia (1935).

Moreover, in his *Brahmasphutasiddhânta* (628), *Brahmagupta describes the ensemble of basic arithmetical operations with the word *pâtîganita.* He writes: "Those that know the twenty logistic operations separately and individually, [these being] addition, multiplication, etc., as well as the eight [methods] of determination, including [in particular measurement by] shadow, is a [true] mathematician." See: BrSpSi; Datta and Singh (1938).

To Brahmagupta's mind, the eight fundamental operations of the Indian mathematicians were the same as the first eight operations of *pâtîganita* (namely: addition, subtraction, multiplication, division, the squaring or cubing of a number, the extraction of the square or cube root), to which the five fundamental rules of the reduction of fractions were added: the *trairâshika* or "Rule of Three", etc. This shows the high level that had been reached by the Indian mathematicians in their calculating techniques at the beginning of the seventh century CE. The methods of calculation which originated in India are known to us today not only because of the information provided by Arabic and European authors, but also by Indian authors themselves. See **Square roots (How Âryabhata calculated his).**

See: Allard (1981); Datta and Singh (1938); Iyer (1954); Kaye (1908); Waeschke (1878).

In some rural regions of India, these processes have been taught through the centuries with hardly any modifications, and calculations are still carried out on the *pâti* (small board) [see Datta and Singh (1938)]. The word *pâtiganita* (or *ganitapâtî)* is composed of *ganita, which means "calculation, arithmetic, science of calculation", and *pâtî, synonymous with *Patta* in the sense of "board" or "tablet". See: AMM, XXXV, p. 526; Datta and Singh, pp. 7–8 and 123. This etymology dates back to the time when Indian mathematicians carried out their calculations on either a board or a tablet. Today, the most natural support for carrying out mathematical operations on is paper. Paper was invented in China, although the circumstances are not fully known. There are texts that attribute the invention of a type of paper made from the pulp created by removing the fibre from rags and fishing nets to Cai-Lun in the year 105 BCE. However, the ideogram used to write the word *paper* in Chinese contains the sign for silk. It seems that paper made from silk preceded paper made from vegetable fibres, the latter quickly replacing the former type because it was cheaper and more resistant. Cai-Lun and other paper makers then went on to use the pulp of vegetable matter, particularly the bark of the mulberry tree. It was probably in the tenth century that they began to use bamboo and, around the fourteenth century, straw. It would be a long time after the Chinese discovery before the West would know about paper. The production of paper began in Samarkand in 751 after the Chinese were taken prisoner by the Arabs at the battle of Talas. Paper was then made by Chinese workers in Baghdad (from 793) and Damascus, which for centuries remained the principal supplier to Europe. From there, methods of fabrication spread to Egypt (c. 900) and Morocco, from where the Arabic invaders introduced it to Spain [see Galiana, (1968)].

Paper was introduced to India by the Persians, who learned the methods of manufacture from the Arabs. It was not until the fourteenth century, however, that the Indians learned the secrets of paper-making. In other words, Indians almost never used paper to carry out their calculations, until very late on in their history. The Arabs and Persians never used paper for this purpose until the twelfth or thirteenth century, because it was such a rare and expensive commodity. The Indians could have used the same material as they used for their manuscripts, carving or writing on palm-leaves or tree-bark (see **Indian styles of writing, The materials of**). However, carrying out calculations was a completely different practice to writing manuscripts: working out sums was "rough work", whilst manuscripts were written last, on durable material and in indelible ink. They used something much more economic than palm-leaves or tree-bark for their calculations: they used chalk and slate, just as most people in the Western world used at school until very recently (or chalk and a blackboard). The mathematician *Bhâskarâchârya (whose favourite instrument was the *pâtî*, or "board", which he wrote upon with a piece of chalk) refers to the use of these materials in his texts, notably in his *Lîlavâti*, where he writes the following:

khatikâyâ rekhâ ucchâdya . . ., "After having drawn the lines [of the numerals for the calculations] on the *pâtî* with chalk . . ."

[See: Datta and Singh (1938), p. 129; Dvivedi (1935), p. 41.]

In other words, the Indian mathematicians began, at some point to use if not slate, then at least a wooden board painted black, and chalk to write their numerals on and cross them out, and a rag to rub them out.

Just as the Arabs and Persians adopted the Indian numerals and methods of calculation, so they began to use the support upon which the Indians carried out their mathematical operations. They gave it the Arabic name *takht* or *luha* (especially in northern Africa) which means "table" or "board" (whether it is made of wood, leather, metal, earth, clay or even slate). As for "arithmetic", this was described by the expression *'ilm al hisab al takht* ("science of calculation on the board"). This support had the advantage of overcoming all the difficulties created when calculations were carried out on boards covered in dust. See **Indian methods of calculation**.

PÂVAKA. [S]. Value = 3. "Fire". See *Agni* and **Three**.

PÂVANA. "Purification", and by extension, "He who purifies". This is another name for *Vâyu, the ancient Brahmanic god of the wind. He is often represented on a mount in the form of an antelope or a deer and holding a fan, an arrow and a standard, respectively symbols of the air (*vâyu*), of speed and of the wind [see Frédéric, *Dictionnaire* (1987)].

PÂVANA. [S]. Value = 5. "Purification". This symbolism is due to this word being associated with one of the attributes of *Vâyu, god of the wind, because the wind itself in Indian cosmologies is considered to be the "cosmic breath". Vâyu is king of the subtle and intermediary domain between the sky and the earth who penetrates, breaks up and purifies. Vâyu is also known by the name *Anila*, which means "breath of life". Thus, according to the Hindus, he is the cosmic energy that penetrates and conserves the body and is manifested most clearly in the form of breath in creatures. Vâyu is also the *prâna*, the "breath" in terms of "vital respiration". Hinduism distinguishes between five types of breath: *prajña*, the very essence of breath, the pure vital force; *vyâna*, the regulator of the circulation of the blood; *samâna*, which regulates the process of absorption and assimilation of food and maintains the balance of the body by looking after the processes of feeding; *apâna*, which looks after secretion; and *udâna*, which acts on the upper part of the organism and facilitates spiritual development by creating a link between the physical part and the spiritual part of the being. Thus *pâvana* = 5. See *Pâvana* (previous entry), *Prâna* and Five.

The use of this numerical symbol can be found in Bhâskarâchârya [see SiShi, I, 27] and in Bhattotpala's Commentary on *Brihatsamhitâ* (chapter II). [See Datta and Singh (1938), p. 55].

PÂVANA. [S]. Value = 7. "Purification", "He who purifies". This is one of the attributes of *Vâyu, god of the wind (see previous article). To understand the reason for this symbolism, it is necessary to be familiar with the relevant episode in Brahmanic mythology. One day Vâyu revolted against the *deva*, or "gods", who live on *Mount Meru. He decided to destroy the mountain, and started a powerful hurricane. But the mountain was protected by the wing of Garuda, the bird-helper of Vishnu, which meant that the assaults of the wind had no effect. One day, however, when Garuda was absent, Vâyu cut off the peak of Mount Meru and threw it into the ocean. This is how Lankâ was born, the island of Sri Lanka. This mythological tale explains how the wind came to have this value. The mythical mountain, *Mount Meru, the living and meeting place of the gods, and centre of the universe, is said to be situated above the seven *pâtâla* (or "inferior worlds"), and has seven faces, each one turned towards one of the seven *dvîpa* (or "island-continents") and one of the seven *sâgara* (or "oceans"); when Vâyu attacked the mountain, he created seven strong winds, one for each face of Mount Meru. Thus: *pâvana* = 7. See Seven.

PERFECT. A synonym for a large quantity. See High numbers (Symbolic meaning of).

PERFECT. [S]. Value = 0. See *Pûrna* and Zero.

PHENOMENAL WORLD. [S]. Value = 3. See Jagat, Loka, Three, Triloka.

PHILOSOPHICAL POINT OF VIEW. [S]. Value = 6. See *Darshana* and Six.

PHILOSOPHY OF VACUITY. See *Shûnya*, *Shûnyatâ*.

PHILOSOPHY OF ZERO. See Symbolism of zero, *Shûnya*, *Shûnyatâ*, and Zero.

PINÂKANAYANA. [S]. Value = 3. This is one of Shiva's names, the third divinity of the Hindu trinity, god of destruction and dissolution. He is often represented with a third eye on his forehead (which symbolises perfect Knowledge). Moreover, his emblem is the *trishûla*, or "trident", symbols of the three aspects of the manifestation (creation, preservation, destruction). See *Haranetra* and Three.

PINK LOTUS. As name of the numbers ten to power nine, ten to power fourteen and ten to power twenty-nine. See: Padma, High Numbers (Symbolic Meaning of).

PINK LOTUS. As name of the numbers ten to power twenty-nine, ten to power 119. See: Padma, High Numbers (Symbolic Meaning of).

PITÂMAHA. [S]. Value = 1. "Great ancestor", "grandfather", "first father". This is an allusion to the god Brahma, first divinity of the trinity of Hinduism; he is the "Director of the sky", the "Master of the horizons", the "One" amongst the diversity. See One.

PLACE-VALUE SYSTEM. The most common Sanskrit term for this is *sthâna*, which literally means "place". See *Sthâna*, *Anka*, *Ankakramena*, *Ankasthâna*, *Sthânakramâd* and Indian written numeral systems (Classification of).

PLANET. [S]. Value = 9. See *Graha* and Nine.

PLANETS. See *Graha*, *Saptagraha* and *Navagraha*.

PLENITUDE. [S]. Value = 0. See *Pûrna* and Zero.

POETRY. See Indian metric, Poetry and writing of numbers, *Nâga*, Serpent (Symbolism of the) and Poetry, zero and positional numeration.

POETRY AND WRITING OF NUMBERS. Like all the scholars of India, astronomers and mathematicians of this civilisation usually wrote in Sanskrit, often writing their numerical tables and texts in verse. These scholars loved to play with and speculate with numbers, and their enjoyment can be seen in the form of their wording which, if not lyrical, is at least in verse. Thus numbers came to be written using words which were connected to them symbolically, and one such word could be chosen from an almost limitless selection of synonyms so that it would fit the rules of Sanskrit versification and give the desired effect. The transcription of a numerical table or of the most arid mathematical formula would often resemble an epic poem. Their language lent itself admirably to the rules of versification, thus giving poetry a significant role in Indian culture and Sanskrit literature. See Sanskrit and Numerical symbols.

POETRY, ZERO AND POSITIONAL NUMERATION. See Zero and Sanskrit poetry.

POINT. [S]. Value = 3. See *Shûla* and Three.

POSITION. [S]. Value = 4. See *Îryâ* and Four.

POSITION OF NUMERALS. See *Sthâna*, *Sthânakramâd*, *Ankasthâna* and *Ankakramena*.

POWER. [S]. Value = 14. See **Indra** and **Fourteen.**

POWERFUL. [S]. Value = 14. See **Shakra** and **Fourteen.**

POWERS OF TEN. See **Ten, Hundred, Thousand, Ten thousand, Million, Ten million, Hundred million, Thousand million, Ten thousand million, Hundred thousand million, Billion, Ten billion, Hundred billion, Trillion, Ten trillion, Hundred trillion, Quadrillion, Quintillion, Names of numbers, High numbers** and **Infinity.**

PRÂKRIT. "Unrefined", "Basic". Generic name commonly used by Indians to refer to the numerous Indo-European dialects of the "Indo-Aryan" category.

PRAKRITI. "Nature, material". According to Indian philosophy, this is the original material that the universe was made from. It is the principal transcendental material, which is associated with terrestrial elements, as opposed to the principal spirit (which is represented by the skies).

PRAKRITI. [S]. Value = 21. In Sanskrit poetry, this is the metre of four times twenty-one syllables per verse. See **Indian metric.**

PRALAYA. Name of the total destruction of the universe in Hindu and Brahman cosmogonies. See **Day of Brahma, Kalpa, Kaliyuga** and **Yuga.**

PRÂNA. [S]. Value = 5. "Breath". In Hindu philosophy, this describes the five breaths which are said to govern the vital functions of the human being (*prajña, apâna, vyâna, udâna* and *samâna*). This term not only applies to respiratory rhythms (like the *prânâyama* physical exercises, which are meant to control breathing and form part of the techniques of *hathayoga*), but also to "subtle breathing" identified with intelligence and wisdom (*prajña*) [see Frédéric, *Dictionnaire* (1987)]. See **Pâvana** and **Five.**

PRAYUTA. Name given to the number ten to the power five (a hundred thousand). See **Names of numbers** and **High numbers.**

Source: *Kâthaka Samhitâ* (beginning of the Common Era).

PRAYUTA. Name given to the number ten to the power six (= million). See **Names of numbers** and **High numbers.**

Sources: *Vâjasaneyî Samhitâ, *Taittirîya Samhitâ and *Kâthaka Samhitâ (from the start of the first millennium CE); *Pañchavimsha Brâhmana (date uncertain); *Âryabhatîya (510 CE). *Lîlâvatî by Bhâskarâchârya (1150 CE); *Ganitakaumudî by Nârâyana (1350 CE); *Trishatikâ by

Shrîdharâchârya (date uncertain); *Kitab fî tahqiq i ma li'l hind by al-Biruni (c. 1030 CE); *Sankhyâyana Shrauta Sûtra (date unknown).

PRECEPT. [S]. Value = 6. See **Six.**

PRIMORDIAL PRINCIPLE. [S]. Value = 1. See **Âdi.** One.

PRIMORDIAL PROPERTY. [S]. Value = 3. See **Guna** and **Triguna.**

PRIMORDIAL PROPERTY. [S]. Value = 6. See **Guna** and **Shadâyatana.**

PRINCIPLE OF THE ENUNCIATION OF NUMBERS. See **Ankânâm vâmato gatih** and **Sanskrit.**

PRINCIPLE OF POSITION. The Sanskrit term usually designating it is *sthâna*, literally: "place". See **Sthâna.**

PRITHIVÎ. [S]. Value = 1. "Immense", "Earth", "terrestrial world". This symbolism primarily refers to the unique nature of the earth, considered to be the spouse of the sky. However, this is also and above all an allusion to the fact that the earth, as principal transcendental material (*prakriti*), as opposed to the principal spirit (represented by the skies), is regarded as the mother of all things. See **One.**

PROGENITOR OF THE HUMAN RACE. [S]. Value = 14. See **Manu** and **Fourteen.**

PROTO-INDIAN NUMERALS. Symbols used from about 2500 to 1500 BCE by people of the Indus civilisation (Mohenjo-daro, Harappa, etc.) who preceded the Aryan settlement of the Indian sub-continent. It is not known how these very different symbols could have evolved into early Brâhmî numerals (nor if indeed there is a connection between them). Only the signs for the nine units have been identified so far; a full understanding of proto-Indian numerals must await further archaeological evidence. See Fig. 1.14.

PUNDARÎKA. Literally "(white) lotus". Name given to the number ten to the power twenty-seven. See **Names of numbers.** For an explanation of this symbolism, see: **High numbers (Symbolic meaning of).** Source: *Lalitavistara Sûtra* (before 308 CE).

PUNDARÎKA. Literally "(white) lotus". Name given to the number ten to the power 112. See **Names of numbers.** For an explanation of this symbolism, see: **High numbers (Symbolic meaning of).** Source: *Vyâkarana* (Pali grammar) by Kâchchâyana (eleventh century CE).

PUNJABI NUMERALS. Symbols derived from *Brâhmî numerals and influenced by Shunga, Shaka, Kushâna, Ândhra, Gupta and Shârada. Currently used in the Punjab (Northwest India). The symbols correspond to a mathematical system that has place values and a zero (shaped like a small circle). See: **Indian written numeral systems (Classification of)**. See Fig. 24.5, 52, and 24.61 to 69.

PURA. [S]. Value = 3. "City". Allusion to the *tripura, the "three cities" of the *Asura (or "anti-gods"), flying iron fortresses from which they directed the war they waged against the *deva. See **Three**.

PURÂ. [S]. Value = 3. "State". Allusion to the three *tripurâ, the "three states of consciousness" according to Hinduism (awake, asleep and dreaming). See **Three**.

PURÂNA. Literally: "Ancients". Traditional Sanskrit texts, dealing with highly diverse subjects, such as the creation of the world, mythology, legends, the genealogy of mythical sovereigns, castes, etc. These texts were written for ordinary people and those of "low caste". Analysis has shown that they are made up of documents written at various times and are from many different sources, and were compiled, revised, added to and corrected in an interval of time oscillating between the sixth and the twelfth century, some even being dated as nineteenth century. Thus the documentation that they contain should be treated with caution, as, from a purely historical point of view at least, they are of no interest. See **Indian documentation (Pitfalls of)**.

PURÂNA AND POSITIONAL NUMERATION. See **Indian documentation (Pitfalls of)**.

PURÂNALAKSHANA. [S]. Value = 5. (Late usage). Allusion to the texts of the *Purâna, which tell of the *Pañchalakshana*, which, in Hindu philosophy, correspond to the "five characteristics" which are said to have defined history: creation (*sarga*), periodical creations (*pratisarga*), divine geneaologies (*vamsha*), the era of a *manu (manvantara) and the genealogies of human sovereigns (*vamshânucharita*) [see Frédéric, *Dictionnaire* (1987)]. See **Five**.

PURIFICATION. [S]. Value = 5. See *Pâvana* and **Five**.

PURIFICATION. [S]. Value = 7. See *Pâvana* and **Seven**.

PÛRNA. [S]. Value = 0. Literally "full, fullness, fulfilled, perfect, finished". To a Western reader, this symbolism might seem paradoxical: how can a word that means "full" represent zero, the void? The allusion is to *Vishnu, the second great divinity of the Hindu and Brahman trinity, whose essential role is to preserve, and cause the evolution of, creation (*Brahma being the creator, *Vishnu the conserver and *Shiva the destroyer). Vishnu is considered to be the internal cause of existence and the guardian of *dharma. Each time the world goes wrong, he hastens (incarnating himself in the form of *avatâra) to show humanity new ways in which to develop. He is often represented as a handsome young man with four arms, holding a conch in the first hand, a bow in the second, a club in the third and a lotus flower in the fourth. The conch represents riches, fortune and beauty, which are the attributes of Vishnu as the principal conserver of the manifestation, because the sound and the pearl are conserved within the shell. As for the *lotus, it symbolises the highest divinity, innate reason and mental and spiritual perfection. It also symbolises the "third eye", that of perfect Knowledge; however, it is also the superior illumination and the divine corolla, the totality of revelation and illumination, as well as intelligence, wisdom and the victory of the mind over the senses. See **High numbers (Symbolic meaning of)**.

Moreover, like the thousand-petalled lotus, Vishnu possesses a thousand attributes and qualities (*sahasranâma*). He represents the innumerable (*thousand* here being treated in its figurative sense). See **Thousand**. Thus Vishnu is associated with the idea of wholeness, integrity, completeness, absoluteness and perfection. The "foot of Vishnu" (*vishnupada*), is the "sky", the "zenith", the "place of the blessed" and the meeting place of the gods; it is, in Hindu cosmology, the summit of *Mount Meru, the mythical mountain situated at the centre of the universe, the source of the celestial *Gangâ* (the sacred Ganges). This makes it easier to understand how "full" came to mean infinity, eternity, and by extension completion and perfection. It is upon *Ananta, the serpent with a thousand heads floating on the primordial waters of the "ocean of unconsciousness", that Vishnu lies to rest during the time separating two creations of the universe. Ananta represents infinity, and has also often represented zero as a numerical symbol. Thus it is clear how *pûrna* came to mean zero. See **Ananta, Jaladharapatha, Shûnya, Zero**. See also **Infinity, Infinity (Indian mythological representation of)** and **Serpent (Symbolism of the)**.

PUSHKARA. [S]. Value = 7. This is a surname attributed to Krishna and Shiva, as well as to

Dyaus (the Sky) considered to be a "reservoir of water". The allusion here is to *Pushkara*, one of the seven mythical continents that surround *Mount Meru. See *Dvîpa, Sapta, Sâgara,* Ocean and Seven.

PUTRA. [S]. Value = 5. "Son". In this symbolism, the word in question is a synonym of *Pândava,* which means the "sons of Pându". See Five.

PUTUMANASOMAYÂJIN. Indian astronomer of the eighteenth century. His works notably include a text entitled *Karanapaddhati,* in which there is frequent use of the place-value system written in the Sanskrit numerical symbols [see Sastri (1929)]. See Numerical symbols, Numeration of numerical symbols, and Indian mathematics (History of).

Q

QUADRILLION (= ten to the power eighteen). See *Shankha* and Names of numbers.

QUALITY. [S]. Value = 3. See *Guna, Triguna* and Three.

QUALITY. [S]. Value = 6. See *Guna, Shadâyatana* and Six.

QUINTILLION (= ten to the power twenty-one). See *Kotippakoti, Kumud, Mahâkshiti* and *Utsanga.* See also Names of numbers.

QUOTIENT. [Arithmetic]. See *Labdha.*

QUTAN XIDA. Chinese astronomer of Indian origin. Qutan Xida is none other than the Chinese rendering of the Indian name *Gautama Siddhânta.

R

RABBIT. [S]. Value = 1. See One, *Shashin, Shashadhara.*

RADA. [S]. Value = 32. "Tooth". See *Danta* and Thirty-two.

RADIANCE. [S]. Value = 9. See *Go* and Nine.

RÂGA. [S]. Value = 6. "Attraction, colour, passion, musical mode". This Sanskrit term describes the moments of emotion provoked by a piece of music (the modes and rhythms causing diverse sensations in the listener) or by a visual work of art. The instants of emotion, which can be provoked by the perception of an exterior agent such as the rain, the wind, a storm, etc., or even by an interior sentiment

such as love, nostalgia, sadness, etc., combine with lines and colours to provoke diverse sensations within the spectator. In the symbolism in question, the allusion is to the *janaka râga,* the six "eastern *râga*", who are male, associated with their six *râgini* (or female *râga*), and with the six "sons" of the latter ones, each of these groups in turn being associated with the *shadâyatana* or "six *guna*" of Buddhist philosophy (in other words the six sense organs: eye, nose, ear, tongue, body and mind) [see Frédéric, *Dictionnaire* (1987)]. See *Rasa,* Six and *Nâga.*

RÂHU. Demon who, according to ancient Indian mythology and cosmology, caused eclipses by "devouring" the sun or the moon, due to a privilege conferred on him by *Brahma. See *Paksha.*

RÂJAMRIGÂNKA. See *Bhoja.*

RÂJASTHANÎ NUMERALS. Symbols derived from *Brâhmî numerals and influenced by Shunga, Shaka, Kushâna, Ândhra, Gupta, Nâgarî and Mahârâshtrî. Currently used in the state of Rajasthan in the west of the sub-continent (bordering on Pakistan, Punjab, Haryana, Uttar Pradesh, Madhya Pradesh and the Gujurat). Râjasthanî numerals are a variant of Mârwarî numerals. The symbols correspond to a mathematical system that has place values and a zero (shaped like a small circle). See: Indian written numeral systems (Classification of). See Fig. 24.52 and 24.61 to 69.

RÂMA. [S]. Value = 3. Allusion to the three Râma of Indian tradition and philosophy: the first, also called Parashu-Râma, or "Râma of the axe", is the sixth incarnation of Vishnu, who came to crush the tyranny of the Kshatriyas, the caste of warriors; the second, called Râma-chandra, seventh incarnation of Vishnu, came to develop *sattva* in humankind, in other words uprightness, equilibrium, serenity and peacefulness; and the third, called simply Râma, was the famous hero of the epic poem *Râmâyana* [see Frédéric, *Dictionnaire* (1987)].

RÂMÂYANA. "The march of Râma". This is an epic Indian poem, written down in Sanskrit by the poet Vâlmîki. It is derived from very ancient legends, but did not find its definitive form until the early centuries of the Common Era. Here is a list of names of the high numbers mentioned in this text (from a passage where, in order to express the number of monkeys that made up Sugriva's army, the author gives the following names successively, which increase each time on a scale of one hundred thousand):

Koti (= 10⁷), *Shanka* (= 10¹²), *Vrindâ* (= 10¹⁷), *Mahâvrinda* (= 10²²), *Padma* (= 10²⁹), *Mahâpadma* (= 10³⁴, *Kharva* (= 10³⁹). See **Names of numbers** and **High numbers**. [See Weber in: JSO, XV, pp. 132–40; Woepcke (1863)].

RANDHRA. [S]. Value = 0. "Hole". Numerical word-symbol used rarely and not until a relatively recent date. The origin of this association of ideas clearly comes from the lack of consideration attached to the anal orifice. See **Zero**.

RANDHRA. [S]. Value = 9. "Hole". This is an allusion to the nine orifices of the human body (the two eye sockets, the two ears, the two nostrils, the mouth, the navel and the anal orifice). See **Chhidra** and **Nine**.

RASA. [S]. Value = 6. "Sensation". In its most general meaning, this word denotes the sensation(s) that a *Shadâyatana* can experience, in other words the "six senses or sense organs" of Indian philosophy (which are: the eye, the nose, the ear, the tongue, the body and the mind). However, the explanation for this symbolism is much more subtle than that. It can only be understood with reference to the Indian aesthetic canons, where this word describes the emotional state of the spectator, listener or reader, in terms of the essence of the evocative power of the musical, pictorial, poetic, theatrical, (etc.) art. This aesthetic distinguishes between nine different types of *rasa*, including the least agreeable, namely: *shringâra* (love or erotic passion); *hâshya* (comedy and humour); *karunâ* (compassion); *vîra* (heroic sentiment); *adbhuta* (amazement); *shânta* (peace and serenity); *raudra* (anger and rage); *bhayânaka* (fear or anguish); and *vîbhatsa* (disgust or repulsion). The first six are the ones which enable enjoyment, and this is what *rasa* refers to in this symbolism: the idea of "savouring". Thus *rasa* = 6. See **Shadâyatana** and **Six**.

RÂSHI. "Rule". Often used in arithmetic to denote the "Rule of Three".

RÂSHI. [S]. Value = 12. "Zodiac". This, of course, refers to the twelve signs of the Indian zodiac: *Mesha* (Aries); *Vrishabha* (Taurus); *Mithûna* (Gemini); *Karka* (Cancer); *Simha* (Leo); *Kanyâ* (Virgo); *Tulâ* (Libra); *Vrishchika* (Scorpio); *Dhanus* (Sagittarius); *Makara* (Capricorn); *Kumbha* (Aquarius); *Mîna* (Pisces). See **Twelve**.

RÂSHIVIDYÂ. Name given to arithmetic in the *Chândogya Upanishad*. Literally: "Knowledge of the rules".

RATNA. [S]. Value = 3. "Jewel". This is probably an allusion to the *triratna*, the "three jewels" of Buddhism, namely: the Community (*sangha*), the Buddhist Law (*dharma*) and Buddha. These "jewels" are represented by a trident. See **Dharma, Shûla** and **Three**.

Note: this symbol is found very rarely representing this value, except for in the *Ganitasârasamgraha* by Mahâvîrâchârya [see Datta and Singh (1938), p. 55].

RATNA. [S]. Value = 5. "Jewel". This is the most frequent value that this word is used for as a numerical symbol. It is probably an allusion to the *panchaparameshtin*, the "five orders of beings" considered to be the "five treasures" of *Jaina religion: the *siddha*, human beings who are omniscient and who became immortal after being freed from the bonds of *karma* and *samsâra*; the *arhat*, sages liberated from the bonds of *karma*, but still subject to the laws of *samsâra*; the *âchârya* or "great masters"; the *upadhya* or "masters"; and the ascetics (*sâdhu*) [see Frédéric, *Dictionnaire* (1987)]. See **Five**.

RATNA. [S]. Value = 9. "Jewel". This allusion could be to the *Navaratna*, "Nine Jewels", the collective name given to the nine famous poets who wrote in Sanskrit who lived in the court of the king Vikramâditya. See **Nine**.

RATNA. [S]. Value = 14. "Jewel". There is no concrete explanation for this symbolism. However, it could have some connection to the *saptaratna* or "seven jewels" of Buddhism, which constitute the seven attributes of the current Buddha ("Golden wheel"; *Chintâmani*, or miraculous pearl said to grant all wishes; "White horse"; "Noble woman"; "Elephant" carrying the sacred Scriptures; "Minister of Finances"; and "Head of war"); these are attributes that would have been associated symbolically with the *saptabuddha*, or seven Buddhas of the past (Vipashyin, Shikhin, Vishvabhû, Krakuchhanda, Kanakamuni and Kâshyapa), including the current Buddha (Shâkyamuni Siddhârtha Gautama); thus, by symbolic addition: *ratna* = 7 + 7 = 14. See **Fourteen**.

RATNASANU. One of the names for *Mount Meru. See **Adri, Dvîpa, Pûrna, Pâtâla, Sâgara, Pushkara, Pâvana** and **Vayu**.

RÂVANA. Name of the king-demon Lankâ who, according to the legends of *Râmâyana*, usurped the throne of his half-brother Kuvera and stole his flying palace (*pushpaka*).

RÂVANABHUJA. [S]. Value = 20. "Arms of *Râvana". Allusion to the twenty arms of this king-demon. See **Twenty**.

RÂVANASHIRAS. [S]. Value = 10. "Heads of *Râvana". This king-demon is said to have had ten heads. See **Ten**.

RAVI. [S]. Value = 12. This is another name for *Sûrya, the divinity of the sun. See **Twelve.**

RAVIBÂNA. [S]. Value = 1,000. "Beams of Ravi". This refers to one of the attributes of *Ravi (= *Sûrya), the divinity of the sun, and expresses the *sahasrakirana or "Thousand Rays" of the sun. See **Thousand.**

RAVICHANDRA. [S]. Value = 2. The couple uniting Ravi and Chandra (named Ravi after *Sûrya, the sun whose other attribute is *Ravi, and *Soma, the moon, the masculine entity also called *Chandra). See **Two.**

REALITY. [S]. Value = 5. See *Tattva* and **Five.**

REALITY. [S]. Value = 7. See *Tattva* and **Seven.**

REALITY. [S]. Value = 25. See *Tattva* and **Twenty-five.**

REMAINDER. [Arithmetic]. See *Shesha.*

RISHI. [S]. Value = 7. "Sage". Allusion to the *Saptarishi, the seven great mythical Sages of Vedic times (Gotama, Bharadvâja, Vishvamitra, Jamadagni, Vasishtha, Kashyapa and *Atri), created by *Brahma and said to be the authors of the hymns and invocations of the *Rigveda* and most of the other *Vedas. They are said to form the seven stars of the Little Bear. See **Seven.**

RITU. [S]. Value = 6. "Season". Allusion to the six seasons, each lasting two "months" in the Hindu calendar: spring (*vasanta*); the hot season (*grishma*); the rain season (*varsha*); autumn (*sharada*); winter (*hemanta*) and the cold season (*shishira*). See **Six.**

RUDRA-SHIVA (Attributes of). [S]. Value = 11. See *Bharga, Bhava, Hara, Îsha, Îshvara, Mahâdeva, Rudra, Shiva, Shûlin* and **Eleven.**

RUDRA. [S]. Value = 11. "Rumbling", "Violent", "Lord of tears". This is the name of the ancient Vedic divinity of the tempest who, according to the *Vedas, was the personification of the vital breaths, which came from *Brahma's forehead, of which there were eleven. Thus: *Rudra* = 11. See **Eleven.**

RUDRÂSYA. [S]. Value = 5. "Faces of *Rudra". This god is said to have had five heads. He is also lord of the "five elements", "the five sense organs", the five "human races" and the five points of the compass (if the zenith is included). See **Five.**

RULE OF THREE. [Arithmetic]. See *Râshi, Trairâshika* and *Vyastatrairâshika.*

RULE OF FIVE. [Arithmetic]. See *Pañchaparâshika.*

RULE OF SEVEN. [Arithmetic]. See *Saptarâshika.*

RULE OF NINE. [Arithmetic]. See *Navarâshika.*

RULE OF ELEVEN. [Arithmetic]. See *Ekâdasharâshika.*

RÛPA. [S]. Value = 1. "Form", "Appearance". This word is synonymous here with "body" as a symbol for the number one. See *Tanu* and **One.**

S

SÂGARA. [S]. Value = 4. "Sea, Ocean". This symbolism can be explained by an allusion to the four "oceans" (*chatursâgara) which surround the four "island-continents" (*chaturdvîpa) which, according to Hindu cosmology, surround *Jambudvîpa* (India). See **Four** and **Ocean.**

SÂGARA. [S]. Value = 7. "Sea, Ocean". This symbolism can be explained by an allusion to the seven "oceans" (*sapta Sâgara) which, according to Hindu cosmology and Brahmanic mythology, surround *Mount Meru. See **Four** and **Ocean.**

SAGE. [S]. Value = 7. See *Atri, Rishi, Saptarishi, Muni* and **Seven.**

SAHASRA. Ordinary Sanskrit name for the number *thousand, the consecutive multiples of which are formed by placing the word *sahasra* to the right of the name of the corresponding unit: *dvasahasra* (two thousand), *trisahasra* (three thousand), chatursahasra (four thousand), *pañchasahasra* (five thousand), etc. This name appears in many words which have a direct relationship with the idea of this number.

Examples: *Sahasrabhûja, *Sahasrakirana, *Sahasrâksha, *Sahasrâmshu, *Sahasranâma, *Sahasrapadma, *Sahasrârjuna.

For words which have a more symbolic relationship with this number, see **Thousand** and **Symbolism of numbers.**

SAHASRABHÛJA. "Thousand arms". This is one of the names of the Sun-god *Sûrya (in allusion to his rays). In the schools of Buddhism of the North, this term refers to an ancient divinity whose thousand arms represented his multiple powers and omniscience.

SAHASRAKIRANA. "Thousand rays". One of the names of the Sun-god *Sûrya.

SAHASRÂKSHA. "Thousand eyes". One of the attributes of *Indra and *Vishnu. See *Indradrishti* and *Sahasra.*

SAHASRÂMSHU. [S]. Value = 12. "Thousand (of the) Shining" (from *sahasra*, "thousand", and *âmshu*, "shining"). This is a metaphorical name for the Sun (the "thousand rays" of its "shining"), and the symbolism has nothing to do with the idea of a thousand, but with the name of the Sun-god as a numerical symbol equal to twelve. See *Sûrya* and **Twelve**.

SAHASRANÂMA. "Thousand names". One of the attributes of *Vishnu and *Shiva.

SAHASRAPADMA. "Lotus of a thousand petals". See **Lotus** and **High numbers (Symbolic meaning of)**.

SAHASRÂRJUNA. "Arjuna's thousand". Name for the thousand arms of Arjunakârtavîrya, mythical sovereign of the *Mahâbhârata. See *Arjunâkara.*

SALILA. Name given to the number ten to the power eleven. See **Names of numbers** and **High numbers**. Source: *Samkhyâyana Shrauta Sûtra* (date uncertain).

SAMÂPTALAMBHA. Name given to the number ten to the power thirty-seven. See **Names of numbers** and **High numbers**.
Source: *Lalitavistara Sûtra* (before 308 CE).

SAMÎKARANA. Term used to denote an "equation". Literally "to make equal" (from *sama* "equal", and *kara* "to make"). Synonyms: *samîkâra, sâmîkriyâ,* etc.

SAMKALITA. Sanskrit term denoting addition. Literally: "put together". Synonyms: *samkalana* (literally: "act of reuniting"); *mishrana* ("act of mixing"); *sammelana; prakshepana; samyojana,* etc.

SAMKHYA (SANKHYA). "Number". Term often used to describe the system of writing numbers using numerical symbols. See **Numerical symbols** and **Numeration of numerical symbols**.

SÂMKHYÂ (SÂNKHYÂ). Literally "calculator". This term describes the adept of the mystical philosophy of *sâmkhya.

SÂMKHYA (SÂNKHYA). Literally "number". This denotes one of the six orthodox systems of Indian philosophies. See *Darshana* and *Tattva.*

SAMKHYÂ (SÂNKHYÂ). Literally "number". Word used to denote "expert-calculator" and, by extension, the arithmetician and mathematician. See *Darshana* and *Samkhyâna.*

SAMKHYÂNA (SANKHYÂNA). "Science of numbers", and by extension "arithmetic" and "astronomy". Word used in this sense in Buddhist and Jaina literature. This science was considered to be one of the fundamental conditions for the development of a Jaina

priest. For the Buddhists, it was also considered (although somewhat later) to be the first and most noble of arts.

SAMKHYEYA (SANKHYEYA). "Number", in the operative and arithmetical sense of the word.

SAMSÂRA. Cycle of rebirth. See *Gati, Kâma* and *Yoni.*

SAMSKRITA. "Complete", "perfect", "definitive". Term used to denote the Sanskrit language. See **Sanskrit**.

SAMUDRA. Literally "ocean". Name given to the number ten to the power nine. See **Names of numbers**. For an explanation of this symbolism, see **High numbers (Symbolic meaning of)**.
Sources: *Vâjasaneyî Samhitâ, *Taittirîya Samhitâ and *Kâthaka Samhitâ (from the start of the first millennium CE); *Pañchavimsha Brâhmana (date uncertain).

SAMUDRA. Literally "ocean". Name given to the number ten to the power ten. See **Names of numbers**. For an explanation of this symbolism, see **High numbers (Symbolic meaning of)**.
Source: *Sankhyâyana Shrauta Sûtra (date uncertain).

SAMUDRA. Literally "ocean". Name given to the number ten to the power fourteen. See **Names of numbers**. For an explanation of this symbolism, see **High numbers (Symbolic meaning of)**.
Source: *Kitab fi tahqiq i ma li'l hind* by al-Biruni (c. 1030 CE).

SAMUDRA. [S]. Value = 4. "Ocean". This is because of the four oceans that are said to surround *Jambudvîpa (India). See *Sâgara,* **Four** and **Ocean**.

SAMUDRA. [S]. Value = 7. "Ocean". This is because of the seven oceans that are said to surround *Mount Meru. See *Sâgara,* **Seven** and **Ocean**.

SAMVAT (Calendar). See *Vikrama.*

SANKHYA, etc. See *Samkhya,* etc.

SANKHYÂNA. See *Samkhyâna.*

SANKHYÂYANA SHRAUTA SÛTRA. Philosophical Sanskrit text (date uncertain). Here is a list of the principal names of numbers mentioned in the text [see Datta and Singh (1938), p. 10]:
*Eka (= 1), *Dasha (= 10), *Sata (= 10^2), *Sahasra (= 10^3), *Ayuta (= 10^4), *Niyuta (= 10^5), *Prayuta (= 10^6), *Arbuda (= 10^7), *Nyarbuda (= 10^8), *Nikharva (= 10^9), *Samudra (= 10^{10}), *Salila (= 10^{11}), *Antya (= 10^{12}), *Ananta (= 10^{13}). See **Names of numbers** and **High numbers**.

SANKHYEYA. See *Samkhyeya.*

SANSKRIT. In India and Southeast Asia, Sanskrit has played, and still does play today, a role comparable with Greek and Latin in Western Europe. This language is capable of translating, through meditation, the mystical transcendental truths said to have been revealed to the **Rishi* in Vedic times. See *Akshara, AUM, Trivarna, Vâchana* and **Mysticism of letters.** Moreover, the name of the Sanskrit language itself is quite significant, because the word **samskrita* ("Sanskrit") means "complete", "perfect" and "definitive". The people who know this Sanskrit are said to speak the divine language and are thus gifted with divine knowledge. Bearing in mind the power accorded to the spoken word (and consequentially its written expression), Sanskrit is considered to be the "language of the gods".

In fact, this language is extremely elaborate, almost artificial. It is capable of describing multiple levels of meditations, states of consciousness and physical, spiritual and even intellectual processes. The inflection of nouns is richly articulated and there are numerous personal forms of the verb, even though the syntax is rudimentary. The vocabulary is very rich and highly diversified according to the means for which it is intended [see Renou (1930); see also Filliozat (1992)].

This shows how, over the centuries, Sanskrit has lent itself admirably to the rules of prosody and versification. This explains why poetry has always played such an important role in Indian culture and Sanskrit literature. It is clear why Indian astronomers favoured the use of Sanskrit numerical symbols, based on a complex symbolism which was extraordinarily fertile and sophisticated, possessing as it did an almost limitless choice of synonyms. See **Poetry and writing of numbers** and **Numerical symbols.**

SAPTA (SAPTAN). Ordinary Sanskrit name for the number seven, which forms part of the composition of many words directly related to the idea of this number. Examples: **Saptabuddha, *Saptagraha, *Saptamâtrikâ, *Saptapadî, *Saptarâshika, *Saptarishi, *Saptarishikâla, *Saptasindhava.* For words which have a more symbolic connection with this number, see **Seven** and **Symbolism of numbers.**

SAPTA. Literally "seven". Term used symbolically in the texts of the *Atharvaveda* as a synonym for each of the following ideas: "sage",

"ocean", "mountain", "island-continent", etc. The allusion here is to the "Seven Sages" of Vedic times (**saptarishi*), to the seven cosmic oceans (**sapta sâgara*), to the seven peaks of Mount Meru, or to the seven "island continents" (*sapta dvîpa*) of Indian mythology and cosmology. See *Saptarishi, Adri, Giri, Sâgara, Dvîpa,* **Mount Meru** and **Ocean.**

For an example, see *Atharvaveda,* I, 1, 1; Datta and Singh (1938), p. 17.

SAPTABUDDHA. Name of the seven Buddhas. See *Sapta* and *Ratna* (= 14).

SAPTADASHA. Ordinary Sanskrit name for the number seventeen. For words which have a symbolic link with this number, see **Seventeen** and **Symbolism of numbers.**

SAPTA DVÎPA. "Seven islands". In Hindu cosmology and Brahmanic mythology, this is the name given to the seven island-continents said to surround **Mount Meru. See *Sapta.* For an explanantion of the symbolism and the choice of this number, see **Ocean.**

SAPTAGRAHA. Literally "seven planets". These are the following: **Sûrya* (the Sun); **Chandra* (the Moon); **Angâraka* (Mars); **Budha* (Mercury); **Brihaspati* (Jupiter); **Shukra* (Venus); and **Shani* (Saturn). See *Graha* and *Paksha.*

SAPTAMÂTRIKÂ. Name for the seven "divine Mothers". See *Mâtrikâ.*

SAPTAN. Ordinary name for the number seven. See *Sapta.*

SAPTAPADÎ. "Seven paces". Name of a Hindu rite which forms part of the nuptial ceremonies, where the bride and groom must take seven paces around the sacred fire in order to consummate the union.

SAPTARÂSHIKA. [Arithmetic]. Sanskrit name for the Rule of Seven.

SAPTARATNA. Name of the "Seven Jewels of Buddhism". See *Ratna* (= 14).

SAPTARISHI. "Seven Sages". These are the seven **Rishi* of Vedic times, who are said to have resided in the seven stars of the Little Bear. See *Atri* and **Mount Meru.**

SAPTARISHIKÂLA. "Time of the seven **Rishi*". Name of an Indian calendar. See *Saptarishi, Kâla* and *Laukikasamvat.*

SAPTA SÂGARA. "Seven oceans". These are the seven oceans which are said to surround **Mount Meru in Hindu cosmology and

Brahmanic mythology: the ocean of salt water, the ocean of sugar cane juice, the ocean of wine, the ocean of thinned butter, the ocean of whipped cheese, the ocean of milk and the ocean of soft water). See *Sâgara*. For an explanation of the choice of this number, see **Ocean**.

SAPTASINDHAVA. "Seven rivers". This is one of the seven sacred rivers of ancient Brahmanism (*Gangâ, Yamunâ, Sarsvatî, Satlej, Parushni, Marurudvridhâ* and *Arjîkîyâ*).

SAPTATI. Ordinary Sanskrit name for the number seventy.

SAPTAVIMSHATI. Ordinary Sanskrit name for the number twenty-seven. For words which have a symbolic relationship with this number, see **Twenty-seven** and **Symbolism of numbers.**

SARITÂPATI. Name given to the number ten to the power fourteen (= hundred billion). See **Names of numbers** and **High numbers.**

Source: *Trishatikâ* by Shrîdharâchârya (date uncertain).

SAROJA. Name given to the number ten to the power nine. See **Names of numbers** and **High numbers.**

Source: *Ganitakaumudî* by Nârâyana (1350 CE).

SARPA. [S]. Value = 8. "Serpent". See **Nâga, Eight** and **Serpent (Symbolism of).**

SARVABALA. Name formed with the Sanskrit adjective *sarva*, which signifies "everything". It is given to the number ten to the power forty-five. See **Names of numbers.** For an explanation of this symbolism, see **High numbers (Symbolic meaning of).**

Source: *Lalitavistara Sûtra* (before 308 CE).

SARVADHANA. [Arithmetic]. Term denoting the "total", or the "whole".

SARVAJÑA. Name formed with the Sanskrit adjective *sarva*, which means "everything". Given to the number ten to the power forty-nine. See **Names of numbers.** For an explanation of this symbolism, see **High numbers (Symbolic meaning of).**

SATA. Ancient Sanskrit form of the name for hundred. See *Shata* and **Names of numbers.** Use of this word is notably found in *Vâjasaneyî Samhitâ, *Taittirîya Samhitâ* and *Kâthaka Samhitâ* (from the start of the first millennium CE); and in *Pañchavimsha Brâhmana* (date uncertain) and *Sankhyâyana Shrauta Sûtra* (date uncertain).

SATYAYUGA. Synonym of *Kritayuga. See **Yuga.**

SÂYAKA. [S]. Value = 5. "Arrow". Allusion to the *Pañchasâyaka*, the "five arrows" of *Kâma. See *Bâna, Pañchabâna, Shara* and **Five.**

SEASON. [S]. Value = 6. See *Ritu* and **Six.**

SELEUCID (Calendar). This calendar began in the year 311 BCE, and was used in the northwest of the Indian subcontinent. To find the corresponding date in the Common Era, subtract 311 from a date expressed in the Seleucid calendar. See **Indian calendars.**

SELF (THE). [S]. Value = 1. See *Abja* and **One.**

SENÂNÎNETRA. [S]. Value = 12. "Eyes of Senânî". This is one of the names of *Kârttikeya, who is often depicted as having six heads. Thus *Senânînetra* = 6 × 2 = 12 eyes. See *Kârttikeyâsya* and **Twelve.**

SENSATION. [S]. Value = 6. See *Rasa* and **Six.**

SENSE. [S]. Value = 5. See *Vishaya* and **Five.**

SERPENT (Cult of the). See **Serpent (Symbolism of the).** See also **Infinity (Indian mythological representation of).**

SERPENT (Symbolism of the). In India and all its neighbouring regions, since the dawn of Indian civilisation, the Serpent has been an object of veneration worshipped by the most diverse of religions. At the beginning of the rain season in Rajasthan, Bengal and Tamil Nadu, the serpent is worshipped through offerings of milk and food. In popular religion, the cobra is very highly considered and these snakes are to be found adorning stones called *Grâmadevata*, or "divinities of the village", which are placed under the banyans. Frédéric (1987) explains that the serpents, in most local religions, are genies of the ground, chthonian spirits who possess the earth and its treasures. The cobras are the most significant type of snake in Indian mythology; they are deified and have their own personality. They are often associated with the cult of *Shiva, and in some pictures of Shiva, he has a cobra wound one of his left arms. In these representations, cobras are actually *nâga*, chthonian divinities with the body of a serpent, considered to be the water spirits in all folklore of Asia, especially in the Far East where they are depicted as dragons.

In fact, in traditional Indian iconography, the *nâga* are usually represented as having the head of a human with a cobra's hood. They live in the *pâtâla*, the underworlds, and guard the treasure which is under the earth. They are said to live with the females, the *nâginî* (renowned for their beauty) and devote themselves to poetry. They are considered to be excellent poets, and are even called the princes of poetry:

first they mastered numbers, which led them naturally to becoming masters of the art of poetic metric. They are also princes of arithmetic because, according to legend, there are a *thousand of them. In other words, due to their considerable fertility, the *nâga* represent the incalculable. Just as metric involves the regulation of rhythm, so they are also sometimes associated with the rhythm of the seasons and the weather cycles.

Coming back to the cobra, this is a long snake which can measure between one metre and one and a half metres. Because of this, the Hindus classified them amongst the demons called *mahorâga* (or "large serpents"). It is the "royal" cobra, however, (which can be up to two metres in length) that was the logical choice of leader of the tribe. This snake, as king of the *nâga*, was given several different names: *Vâsuki, *Muchalinda, Muchilinda, Muchalinga, Takshasa, *Shesha, etc., and there are many corresponding myths. See **Vâsuki**. According to a Buddhist legend, the king Muchilinda protected the Buddha, who was in deep meditation, from the rain and floods, by making his coils into a high seat and sheltering him with the hoods of his seven heads. The name which is used most frequently, however, is *Shesha. He is sometimes depicted as a snake with seven heads, but he is usually represented as a serpent with a thousand heads. This is why the term *Sheshashîrsha (literally "head of Shesha") often means "thousand" when it is used as a numerical word-symbol. Etymologically, the word *shesha* means "vestige", "that which remains". Shesha is also called *Âdi Shesha* (from *Âdi, "beginning"). This is because Shesha is also and most significantly the "original serpent", born out of the union of Kashyapa and Kadru (Immortality). And because he married Anantashîrsha (the "head of *Ananta"), Shesha, according to Indian cosmology and mythology, became the son of immortality, the vestige of destroyed universes and the seed of all future creations all at once.

The king of the *nâga* thus represents primordial nature, the limitless length of eternity and the boundless limits of infinity. Thus Shesha is none other than *Ananta*: that immense serpent that floats on the primordial waters of original chaos and the "ocean of unconsciousness", and *Vishnu lies on his coils when he rests in between two creations of the world, during the birth of *Brahma who is born out of his navel (see Fig. D. 1 in the entry entitled *Ananta). Ananta is also the great prince of darkness. Each time he opens his mouth, he causes an earthquake. At the end of each *kalpa (cosmic cycle of 4,320,000,000 years), Ananta spits and causes the fire which destroys all creation in the universe. He is also *Ahirbudhnya (or *Ahi Budhnya*), the famous serpent of the depths of the ocean who, according to Vedic mythology, is born out of dark waters. Thus, as well as being the genie of the ground and the chthonian spirit who owns the earth and its treasures, the serpent is also a "spirit of the waters" (*âptya), who lives in the "inferior worlds" (*pâtâla).

Some myths clearly indicate this ambivalence surrounding the nature of the reptile, for example the legend which tells the story of Kâliya, the king of the *nâga* of the Yamunâ river; this is a serpent with four heads of monstrous proportions, who defeated by *Krishna, who was then only five years old, went to hide in the depths of the ocean. In this myth, the *four heads* of the king of the *nâga* is significant, because when this serpent goes by the name of Muchalinda, it often has seven heads, or a thousand heads like *Ananta. The choice of these numerical attributions is not simply a question of chance. In fact, in these allegories, the seven heads of Muchalinda represent the subterranean kingdom of the *nâga*, each one being associated with one of the seven hells which constitute the "inferior worlds". These Hells are situated just below *Mount Meru, the centre of the universe, which itself has seven faces, each one facing one of the seven oceans (*sapta sâgara) and one of the "island-continents" (*sapta dvîpa). Muchalinda was the "original serpent" who created primordial nature. *Mount Meru, the sacred and mythical mountain of Indian religions, which is thus associated with the number seven, receives its light from the *Pole star, the last of the seven stars of the Little Bear, situated on exactly the same line as this "axis of the world".

On the other hand, the four heads of Kâliya represent the essentially terrestrial nature of the serpent, which crawls along the ground. In Indian mystical thought, earth corresponds symbolically to the number four, which is linked to the square, which in turn is associated with the four cardinal points. On the other hand, the thousand heads of Shesha-Ananta symbolise both the incalculable multitude and an eternal length of time. As for the battle mentioned above between *Krishna and the king of the *nâga*, this is the mystical expression of the rivalry between man and serpent. This man-snake duality is expressed in a very symbolic manner in Vedic literature (notably in the *Chhândogya*

Upanishad), where Krishna, the "Black", before his deification, is a simple scholar or **asura* (an "anti-god"). After his victory over the snake, he becomes one of the divinities of the Hindu pantheon: he becomes the *eighth* "incarnation" (**avatâra*) of Vishnu, even before becoming the "beneficent protector of humanity".

This duality is also expressed numerically, because Krishna's position as an incarnation of Vishnu is equal to eight, which is exactly the mystical value of the *nâga*. The *nâga* is not only considered to be a genie of the ground, a chthonian spirit who owns the earth and its treasure, but also and above all an aquatic symbol; it is a "spirit of the waters" living in the underworlds. The symbolic value of the earth is 4. In Indian mystical thought, water (see **Jala*) also has the value 4, thus the ambivalence surrounding the serpent is expressed by the relation: *nâga* = earth + water = 4 + 4 = 8. This value is confirmed by the fact that the *nâga* reproduce in couples and always develop in the company of the *nâginî* (their females); this gives the number eight as the result of the symbolic multiplication of two (the *nâga* and his *nâginî*) by 4 (the earth or water). This is why the name of this species became a word-symbol for the numerical value of 8 (see the entry entitled **Nâga*).

As well as its terrestrial character, the serpent symbolises primordial nature: "The underworlds and the oceans, the primordial water and the deep earth form one *materia prima*, a primordial substance, which is that of the serpent. He is spirit of the first water and spirit of all waters, be they below, on the surface of or above the earth. Thus the serpent is associated with the cold, sticky and subterranean night of the origins of life: All the serpents of creation together form a unique primordial mass, an *incalculable* primordial thing, which is constantly in the process of deteriorating, disappearing and being reborn." [Keyserling, quoted in Chevalier and Gheerbrant (1982)]. Thus the serpent symbolises life. The *primordial thing* is life in its latent form. Keyserling says that the Chaldaeans only had one word to express both "life" and "serpent". The symbolism of the serpent is linked to the very idea of life; in Arabic, *serpent* is *hayyah* and *life* is *hayat*. [Guénon, quoted in Chevalier and Gheerbrant (1982)]. The serpent is one of the most important archetypes of the human soul [Bachelard, quoted in Chevalier and Gheerbrant (1982)]. The same images are found in Indian cosmological and mythological representations. Thus in tantric doctrine, the *Kundalinî*, literally the "Serpent" of Shiva, source of all spiritual and sexual energies (energies = **shakti*) is said to be found coiled up at the base of the vertebral column, on the *chakra* of the state of sleep. And when he wakes up, "the serpent hisses and becomes tense, and the successive ascent of the *chakra* begins: this is the arousal of the libido, the renewed manifestation of life" [Frédéric, *Dictionnaire* (1987)]. Moreover, from a macroscopic point of view, the *Kundalinî* is the equivalent of the serpent **Ananta*, who grasps in his coils the very base of the *axis of the universe*. He is associated with Vishnu and Shiva, and symbolises cyclical development and reabsorption, but, as guardian of the nadir, he is the *bearer of the world*, and ensures its continuity and stability. But Ananta is first and foremost the serpent of infinity, immensity and eternity. All these meanings are in fact various applications, depending on the field in question, of the myth of the original Serpent, which represents primordial indifferentiation. The serpent is considered to be both the beginning and the end of all creation. It is not by chance that the Sanskrit language uses the word *Shesha*, the "remainder", to denote the serpent Ananta; to the Indians, the *nâga* with a thousand heads represents the "vestige" of worlds which have disappeared as well as the seed of worlds yet to appear. This explains the importance which so many cosmologies and mythologies place on the eschatological symbolism of the serpent.

In summary, the snake has always been associated with ideas of the sky, celestial bodies, the universe, of the night of origins, *materia prima*, the axis of the world, primordial substance, the vital principle, life, eternal life and sexual or spiritual energy. It is also connected to ideas of the vestige of past creations and the seed of future creations, cyclical development and reabsorption, longevity, an innumerable quantity, abundance, fertility, immensity, wholeness, absolute stability, endless undulating movement, etc.

In other words, since time immemorial, and amongst all the races of the earth, the serpent, as well as being a symbol of the earth and water, personifies the notion of infinity and *eternity*. See **Infinity**, **Infinity (Indian concepts of)** and **Infinity (Indian mythological representation of)**.

SERPENT OF INFINITY AND ETERNITY. See **Ananta**, *Sheshashîrsha*, **Infinity (Indian mythological representation of)** and **Serpent (Symbolism of the)**.

SERPENT OF THE DEEP. [S]. Value = 8. See *Ahi*, **Eight** and **Serpent (Symbolism of the).**

SERPENT WITH A THOUSAND HEADS. [S]. Value = 1,000. See **Sheshashîrsha** and **Thousand**. See also **Infinity (Indian mythological representation of)** and **Serpent (Symbolism of the).**

SERPENT. [S]. Value = 8. See *Nâga, Ahi, Sarpa* and **Eight.**

SEVEN. The ordinary Sanskrit words for the number seven are *sapta* and *saptan*. Here is a list of corresponding numerical symbols: *Abdhi, Achala, *Adri, *Aga, *Ashva, *Atri, Bhaya, *Bhûbhrit, *Bhûdhara, Chandas, Dhâtu, Dhî, *Dvîpa, *Giri, Haya, Kalatra, *Loka, *Mahîdhara, *Mâtrikâ, *Muni, *Naga, *Pârvata, *Pâtâla, *Pâvana, *Pushkara, *Rishi, *Sâgara, *Sâgara, *Samudra, *Shaila, *Svara, *Tattva, *Turaga, *Turangama, *Vâjin, *Vâra, *Vyasana* and *Yati*. These words have the following symbolic meaning or translation:
1. "Purification" and by extension "Purifier" (*Pâvana*). 2. The horses (*Ashva, Turaga, Turangama, Vâjin*). 3. The island-continents (*Dvîpa*). 4. The seas or oceans (*Sâgara, Samudra*). 5. The divine mothers (*Mâtrikâ*). 6. The worlds (*Loka*). 7. The inferior worlds (*Pâtâla*). 8. The mountains or hills (*Adri, Aga, Bhûbhrit, Bhûdhara, Giri, Mahîdhara, Naga, Pârvata, Shaila*). 9. The syllables (*Svara*). 10. The musical notes (*Svara*). 11. The "Sages" of Vedic times (*Muni, Rishi*). 12. The last of the seven Rishi (*Atri*). 13. The days of the week (*Vâra*). 14. "That which does not move" (*Naga*). 15. The seventh "island-continent" (*Pushkara*). 16. The fears (*Bhaya*) (only in *Jaina religion). 17. The winds (*Pâvana*).

See **Numerical symbols.**

SEVENTEEN. Ordinary Sanskrit name: *saptadasha*. Corresponding numerical symbol: *atyashti*.

SEVENTY. See *Saptati*.

SEVERUS SEBOKT. Syrian bishop of the seventh century CE. His works notably include a manuscript dated 662 CE, where he talks of the system of nine numerals and Indian methods of calculation.

SHAD (SHASH, SHAT). Ordinary Sanskrit name for the number six, this word forms part of the composition of many other words which are directly related to the idea of this number.

Examples: *Shâdanga, *Shadâyatana, *Shaddarshana, *Shâdgunya, *Shatkasampatti. For words which are symbolically related to this number, see **Six** and **Symbolism of numbers.**

SHÂDANGA. "Six parts". This is the name for the six aesthetic rules of painting, which are described in a commentary on the *Kâmasutra* by Yashodhara (these six rules being as follows: *rûpabheda*, "shape"; *pramanam*, "size"; *bhava*, "sentiment"; *lavana*, "grace"; *sadrîshyam*, "comparison"; and *varnikabahanga*, "colour").

SHADÂYATANA. [S]. Value = 6. "Six *guna*". These are the "six bases", or "six categories". These are the six senses, objects or sense organs of Buddhist philosophy (namely: the eye, the nose, the ear, the tongue, the body and the mind). See **Six.**

SHADDARSHANA. [S]. Value = 6. "Six visions", "six contemplations", "six philosophical points of view". These are the six principal systems of Hindu philosophy. See *Darshana* and **Six.**

SHADDASHA. Ordinary Sanskrit name for the number sixteen. For words which are symbolically connected to this number see **Sixteen** and **Symbolism of numbers.**

SHÂDGUNYA. [S]. Value = 6. "six *guna*". This is a synonym of *shadâyatana*. See **Six.**

SHADVIMSHATI. Ordinary Sanskrit name for the number twenty-six. For words which are symbolically associated with this number see **Twenty-six** and **Symbolism of numbers.**

SHAILA. [S]. Value = 7. "Mountain". This concept is related to the myth of *Mount Meru, where the numbers seven plays a significant role. See *Adri* and **Seven.**

SHAKA (Calendar). This is the most widely used calendar in Hindu India, as well as in the parts of Southeast Asia influenced by India. It is also known as *Shakakâla, Shakarâja* or *Shakasamvat*. It began in the year 78 of the Common Era. According to certain traditions, this calendar was begun in the first century CE by a Satrap (Kshatrapa) king called Shâlivâhana (or Nahapâna), who then reigned over the city of *Ujjain. To find a corresponding date in the Common Era, add seventy-eight to a date expressed in *Shaka* years. See **Indian calendars.**

SHAKA NUMERALS. Symbols derived from *Brâhmî numerals and influenced by Shunga numerals, arising at the time of the Shunga dynasty (second to first centuries BCE). The symbols corresponded to a mathematical system that was not based on place values and therefore did not possess a zero. See: **Indian written numeral systems (Classification of)**. See Fig. 24.52 and 24.61 to 69.

SHAKASAMVAT (Calendar). See *Shaka*.

SHAKRA. [S]. Value = 14. "Powerful". Allusion to the "strength" of the god *Indra, amongst whose attributes is *Shakradevendra*, "Powerful Indra". This explains the symbolism in question, becuase *Indra* = 14. See **Fourteen**.

SHAKTI. [S]. Value = 3. "Energy". In Brahmanism and Hinduism, this word denotes feminine energy or the active principle of all divinity. The allusion here is to the *shakti* of the most important divinities, namely those of the triad formed by *Brahma, *Vishnu and *Shiva. See **Three**.

For an example of the use of this word-symbol, see: EI, XIX, p. 166.

SHANKARÂCHÂRYA. Hindu philosopher of the late ninth century. His works notably include *Shârîrakamîmâmsâbhâshya* (great commentary on the *Vedântasûtra*), where there is a reference to the place-value system of the Indian numerals.

SHANKARÂKSHI. [S]. Value = 3. Synonym of *Haranetra*, "eyes of *Shiva". See **Three**.

SHANKARANÂRÂYANA. Indian astronomer c. 869 CE. His works notably include a text entitled *Laghubhâskarîyavivarana* in which the place-value system of Sanskrit numerical symbols is used frequently. He also uses the *katapayâdi* method invented by Haridatta [see Billard (1971), p. 8]. See **Numerical symbols, Katapayâdi numeration** and **Indian mathematics (History of)**.

SHANKHA. Word which expresses the sea conch. It is a symbol of riches and of certain Hindu and Buddhist divinities (such as *Vishnu). It is a name given to the number ten to the power twelve. See **Names of numbers.** For an explanation of this symbolism, see **High numbers (Symbolic meaning of)**.

Source: *Râmâyana by Vâlmîki (early centuries CE).

SHANKHA. Word which expresses the sea conch. It is given to the number ten to the power thirteen (ten billion). See **Names of numbers.** For an explanation of this symbolism, see **High numbers (Symbolic meaning of)**.

Source: *Kitab fi tahqiq i ma li'l hind by al-Biruni (c. 1030 CE).

SHANKHA. Word meaning sea conch. It is given to the number ten to the power eighteen. See **Names of numbers.** For an explanation of this symbolism, see **High numbers (Symbolic meaning of)**.

Source: *Ganitasârasamgraha by Mâhavîrâchârya (850 CE).

SHANKU. Literally: "Diamond". Name given to the number ten to the power thirteen (ten billion). See **Names of numbers.** For an explanation of this symbolism, see **High numbers (Symbolic meaning of)**.

Sources: *Lîlâvatî by Bhâskarâchârya (1150 CE); *Ganitakaumudî by Nârâyana (1350 CE), *Trishatikâ by Shrîdharâchârya (date uncertain).

SHANMUKHA. [S]. Value = 6. Synonym of *Kumârâsya, "Faces of *Kumâra (= Shanmukha)". This is an allusion to the six heads of *Kârttikeya. See *Kârttikeyâsya and **Six**.

SHANMUKHABÂHU. [S]. Value = 12. "Arms of *Shanmukha (= *Kumâra = *Kârttikeya)". Kârttikeya is said to have had twelve arms. See *Kârttikeyâsya and **Twelve**.

SHARA. [S]. Value = 5. "Arrow". This is one of the attributes of *Kâma, Hindu divinity of Cosmic Desire and Carnal Love, who is generally invoked during wedding ceremonies, and whose action is said to determine the laws of *samsâra for human beings. The symbolism in question is due to the fact that Kâma is often represented as a young man armed with a bow made of sugar cane shooting five arrows (*pañchabâna) which are either flowers or adorned with flowers. See **Arrow** and **Five**.

SHÂRÂDA NUMERALS. Symbols derived from *Brâhmî numerals and influenced by Shunga, Shaka, Kushâna, Ândhra, and Gupta. Used in Kashmir and the Punjab from the ninth to the fifteenth centuries CE. The symbols correspond to a mathematical system that has place values and a zero (shaped like a dot). The more or less direct ancestor of Tâkarî, Dogrî, Kuluî, Sirmaurî, Kochî, Landa, Maltânî, Khudawadî, Sindhî, Punjabi, Gurûmukhî, etc. numerals. For historic symbols, see Fig. 24.40; for current symbols, see Fig. 24.14; for derived notations, see Fig. 24.52. For the corresponding graphical development, see Fig. 24.61 to 69. See: **Indian written numeral systems (Classification of)**.

SHASH. Ordinary Sanskrit word for the number six. See *Shad*.

SHASHADHARA. [S]. Value = 1. "Which represents a rabbit". This is connected with an attribute of the moon. According to legend, a rabbit, who offered its own flesh to relieve the poor, was rewarded by having its own image impressed on the face of the moon. This explains the symbolism in question, because "Moon" = 1. See *Abja* and **One**.

SHASHANKA. [S]. Value = 1. "Moon". See *Abja* and One.

SHASHIN. [S]. Value = 1. "To the Rabbit". This is the rabbit which, according to legend, was drawn on the face of the moon. See *Shashadhara, Abja* and One.

SHASHTI. Ordinary Sanskrit name for the number sixty.

SHAT. Ordinary Sanskrit name for the number six. See *Shad.*

SHATA. Ordinary Sanskrit name for the number one hundred. Its multiples are formed by placing it to the right of the names of the corresponding units: *dvashtat* (two hundred), *trishata* (three hundred), *chatushata* (four hundred), etc. This name forms part of the composition of several words which are symbolically associated with the idea of this number.

Examples: *Shatapathabrâhmana, *Shatarudriya, *Shatarûpâ, *Shatottaragananâ, *Shatottaraguna, *Shatottarasamjnâ. For words which have a more symbolic link with this number, see Hundred and Symbolism of numbers. See also *Sata* for an ancient form of this number.

SHATAKOTI. Literally: a hundred *koti*. This is the name given to the number ten to the power nine. See Names of numbers and High numbers.

Source: *Ganitasârasamgraha* by Mâhavîrâchârya (850 CE).

SHATAPATHABRÂHMANA. "Brâhmana of the Hundred ways". This is the title of an important work of Vedic literature, divided into a hundred *adhyâya* ("recitations").

SHATARUDRIYA. Name of a Sanskrit hymn which is part of the *Taittirîya Samhitâ* (*Yajurveda*), addressed to *Rudra considered from a hundred different perspectives.

SHATARÛPÂ. "Of a hundred forms". One of the names for the first woman, daughter and wife of *Brahma, who is said to have been gifted with a "hundred bodies". See *Rûpa.*

SHATKASAMPATTI. Literally "six great victories" (from *shatka*, "made up of six", and *sampatti*, "to obtain, achieve, succeed"). In Hinduism, this refers to the "Six Great Victories" of Tattvabodha of Shankara, which constitutes the first of the four conditions that an adept of the philosophy of the *Vedânta* must fulfil.

SHATOTTARAGANANÂ. "Centesimal arithmetic". There is a reference to this in *Lalitavistara Sûtra* [see Datta and Singh (1938), p. 10].

SHATOTTARAGUNA. "Hundred, primordial property". Sanskrit name for the centesimal base. Reference to this is found in the *Lalitavistara Sûtra.*

SHATOTTARASAMJÑÂ. "Names of multiples of a hundred". This term applies to names of numbers in Sanskrit numeration in the centesimal base. There is a reference to this in *Lalitavistara Sûtra* [see Datta and Singh (1938), p. 10]. The equivalent of this word in terms of the decimal base is *Dashagunâsamjnâ*. See *Shatottaragananâ, Shatottaraguna* and *Dashagunâsamjña.*

SHESHA. "Vestige", "that which remains" or "he who remains". In Brahman and Hindu mythologies, this is the name of *Ananta, the king of the *nâga* and serpent of the infinity, eternity and immensity of space. See Serpent (Symbolism of the).

SHESHA. [Arithmetic]. "Vestige". Term describing the remainder in division.

SHESHASHÎRSHA. [S]. Value = 1,000. Literally "heads of *Shesha". Shesha is the king of the *nâga* who lives in the inferior worlds (*pâtâla) and who is considered to be the "Vestige" of destroyed universes as well as the seed of all future creation. This symbolism comes from the fact that Shesha is represented as a serpent with a thousand heads, the number *thousand* here meaning "multitude" and the "incalculable". See Ananta, Thousand, High numbers (Symbolic meaning of) and Serpent (Symbolism of the).

SHIKHIN. [S]. Value = 3. "Ablaze". This is one of the names for *Agni, the god of sacrificial fire, whose name is equal to the number three. See Three.

SHÎRSHAPRAHELIKÂ. From *shîrsha*, "head", and *prahelikâ*, "awkward question, enigma". This term is used in the texts of *Jaina cosmology to denote a period of time which corresponds approximately to ten to the power 196. See *Anuyogadvâra Sûtra*, Names of numbers, High numbers and Infinity.

SHÎTÂMSHU. [S]. Value = 1. "Of the cold rays". This is a synonym of "moon", the opposite of the warm rays of the sun. See *Abja* and One.

SHÎTARASHMI. [S]. Value = 1. "Of the cold rays". This is a synonym of "Moon", the opposite of the warm rays of the sun. See *Abja* and One.

SHIVA. [S]. Value = 11. One of the three main divinities of the Brahmanic pantheon (*Brahma, *Vishnu, *Shiva). There is no mention of Shiva in the *Veda*, and it would seem that Shiva did not become a god until relatively recently. The symbolism in question comes from the fact that Shiva is none other than an incarnation of *Rudra, ancient Vedic divinity of tempests and cosmic anger. As Rudra symbolises the number 11 (because of the eleven vital breaths, born from Brahma's forehead, of which he was the personification), the name of Shiva also came to represent this number. See **Rudra-Shiva** and **Eleven**.

SHRÎDHARÂCHÂRYA. Indian mathematician. The date of his birth is not known. His works notably include *Trishatikâ*. Here is a list of the principal names of numbers mentioned in this work:

Eka (= 1), *Dasha* (= 10), *Shata* (= 10^2), *Sahasra* (= 10^3), *Ayuta* (= 10^4), *Laksha* (= 10^5), *Prayuta* (= 10^6), *Koti* (= 10^7), *Arbuda* (= 10^8), *Abja* (= 10^9), *Kharva* (= 10^{10}), *Nikharva* (= 10^{11}), *Mahâsaroja* (= 10^{12}), *Shanku* (= 10^{13}), *Saritâpati*, (= 10^{14}), *Antya* (= 10^{15}), *Madhya* (= 10^{16}), *Parârdha* (= 10^{17}).

Ref.: TsT, R. 2–3 [see Datta and Singh (1938), p. 13].

See **Names of numbers, High numbers** and **Indian mathematics (History of)**.

SHRÎPATI. Indian astronomer c. 1039 CE. His works notably include a text entitled *Siddhântashekhara*, in which the place-value system of the Sanskrit numerical system is used frequently [see Billard (1987), p. 10]. See **Numerical symbols, Numeration of numerical symbols**, and **Indian mathematics (History of)**.

SHRUTI. [S]. Value = 4. "Recital". Name given to the ancient Brahmanic and Vedic religious texts, which are said to have been revealed by a divinity to one of the seven "Sages" (*rishi*), poets and soothsayers of Vedic times (*Saptarishi*). As this allusion primarily concerns the *Veda*, and there are four of them, *shruti* = 4. See **Four**.

SHUKRANETRA. [S]. Value = 1. The "Eye of Shukra". According to legend, this divinity had one eye destroyed by *Vishnu, thus the symbolism in question. See **One**.

SHÛLA. [S]. Value = 3. "Point". Allusion to the three points of *Shiva's Trident (*trishûlâ*), which symbolise the three aspects of the manifestation (creation, preservation, destruction), as well as the three primordial principles (*triguna*), and the three states of consciousness (*tripurâ*). See **Guna** and **Three**.

SHÛLIN. [S]. Value = 11. This is one of the attributes of *Rudra, who is invoked as "lord of the animals" in the *Shûlagava*, Brahmanic sacrifices of two-year-old calves with the aim of obtaining prosperity. Thus *Shûlin* = Rudra = 11. See **Rudra-Shiva** and **Eleven**.

SHUNGA NUMERALS. Symbols derived from *Brâhmî numerals, arising during the Shunga dynasty (second century BCE). The symbols corresponded to a mathematical system that was not based on place-values and therefore did not possess a zero. See: **Indian written numeral systems (Classification of)**. See Fig. 24.30, 24.52 and 24.61 to 69.

SHÛNYA. Literally "void". This is the principal Sanskrit term for "zero". However, the Sanskrit language (the excellent literary instrument of mathematicians, astronomers and all Indian scholars) has many synonyms for expressing this concept (*abhra, *âkâsha, *ambara, *ananta, *antariksha, *bindu, *gagana, *jaladharapatha, *kha, *nabha, *nabhas, *pûrna, *vishnupada, *vindu, *vyoman, etc.). The words *kha, *gagana, etc., are used for "sky", "firmament", and the words *ambara, *abhra, *nabhas, etc., signify "space", "atmosphere", etc. The word *âkâsha means the fifth "element", "ether", the immensity of space, as well as the essence of all that is uncreated and eternal. There is also the word *bindu, which means "dot" or "point". At least since the beginning of the Common Era, *shûnya* means not only void, space, atmosphere or ether, but also nothing, nothingness, negligible, insignificant, etc. In other words, the Indian concept of zero far surpassed the heterogeneous notions of vacuity, nihilism, nothingness, insignificance, absence and non-being of Greek and Latin philosophies. See **Shûnyatâ, Numerical symbols, Zero, Zero (Graeco-Latin concepts of), Zero (Indian concepts of)** and **Indian atomism**.

SHÛNYA-BINDU. Literally: "void-dot". Name of the graphical representation of zero in the shape of a dot. See **Shûya, Bindu, Numeral 0 (in the shape of a dot)** and **Zero**.

SHÛNYA-CHAKRA. Literally: "void-circle". Name of the graphical representation of zero in the shape of a little circle. See **Shûnya, Numeral 0 (in the shape of a little circle)** and **Zero**.

SHÛNYA-KHA. Literally: "void-space". Name given to the function of zero in numerical representations: it is the *empty space* which marks the absence of units of a given order in positional numeration. See **Kha, Shûnya** and **Zero**.

SHÛNYA-SAMKHYA. Literally: "void-number". Name given to zero as a numerical symbol. It is also the "zero quantity" considered to be a whole number in itself. See *Samkhya, Shûnya* and **Zero.**

SHÛNYATÂ. In Sanskrit, the privileged term for the designation of zero is **shûnya*, which literally means "void". But this word existed long before the discovery of the place-value system. Since Antiquity, this word has constituted the central element of a mystical and religious philosophy, developed as a way of thinking and behaving, namely the philosophy of "vacuity" or *shûnyatâ*. See *Shûnya*. This doctrine is a fundamental concept of Buddhist philosophy and is called the "Middle Way" (*Mâdhyamaka*), which teaches that every thing in the world (*samskrita*) is empty (**shûnya*), impermanent (*anitya*), impersonal (*anâtman*), painful (*dukha*) and without original nature. Thus this vision, which does not distinguish between the reality and non-reality of things, reduces these things to complete insubstantiality.

This philosophy is summed up in the following answer that the Buddha is said to have given to his disciple Shariputra, who wrongly identified the void (**shûnya*) with form (**rûpa*): "That is not right," said the Buddha, "in the *shûnya* there is no form, no sensation, there are no ideas, no volitions, and no consciousness. In the *shûnya*, there are no eyes, no ears, no nose, no tongue, no body, no mind. In the *shûnya*, there is no colour, no noise, no smell, no taste, no contact and no elements. In the *shûnya*, there is no ignorance, no knowledge, or even the end of ignorance. In the *shûnya*, there is no aging or death. In the *shûnya*, there is no knowledge, or even the acquisition of knowledge.

Buddhists did not always use *shûnya* in this sense: in the ancient Buddhism of *Hînayâna* (known as the "Lesser Vehicle"), this notion only applied to the person, whereas in *Mahâyâna* Buddhism (of the schools of the North and known as the "Greater Vehicle"), the idea of vacuity stretched to include all things. To explain the difference between these two concepts, the Buddhists of the schools of the North make the following comparison: in the ancient vision, things were regarded as if they were empty shells, whereas in the *Mahâyâna* the very existence of the empty shells is denied. This concept of the whole of existence being a void should not lead to the conclusion that this is an attitude of nihilism. Far from meaning that things do not exist, this philosophy expresses that things are merely illusions. Through criticising the knowledge of things as being a pure illusion (*mâyâ*), it actually means that absolute truth is independent of the being and the non-being, because it is the *shûnyatâ* or "vacuity". The *shûnyatâ* has a real existence; it is composed of **âkâsha*, or "ether", the last and most subtle of the "five elements" (**pañchabhûta* or "ether") of Hindu and Buddhist philosophies, which is considered to be the essence of all that is uncreated and eternal, and the element which penetrates everything. The **âkâsha* has no substance, yet it is considered to be the condition of all corporeal extension and the receptacle of all matter which manifests itself in the form of one of the other four elements (earth, water, fire, air). According to this philosophy, the *shûnyatâ* is the ether-universe, the only "true universe".

This is why the being and the non-being are considered to be insignificant and even illusory compared to the *shûnyatâ*, which thus excludes any possible mixing with material things, and which, as an unchanging and eternal element, is impossible to describe. In **Mâdhyamika* Buddhism (the followers are still called **Shûnyavâdin* or "vacuitists"), the void has been identified with the absence of self and salvation. Both are meant to achieve redemption, which is only possible in the *shûnyatâ*. In other words, in order to be granted deliverance, vacuity must be achieved; for this, the mind must be purified of all affirmation and all negation at once.

This ontology is inextricably linked to the mysticism of universal vacuity, and represented the great philosophical revolution of Buddhism amongst the schools of the North, implemented by Nâgârjuna, the patriarch of this sect. The *Madhyamakashâstra*, the fundamental text which is traditionally attributed to Nâgârjuna, was translated into Chinese in the year 409 CE, when he had already achieved almost god-like status, and was renowned far beyond the frontiers of India. [See Bareau (1966), pp. 143ff.-; Frédéric (1987); Percheron (1956); Renou and Filliozat (1953)]. This proves that the fundamental concepts of this mysticism were already fully established at the beginning of the Common Era.

These concepts were pushed to such an extent that twenty-five types of *shûnya* were identified. Amongst these figured: the void of non-existence, of non-being, of the unformed, of the unborn, of the non-product, of the uncreated or non-present; the void of the non-substance, of the unthought, of immateriality or insubstantiality; the void of non-value, of the absent, of the insignificant, of little value, of no value, of nothing, etc. This means that in the *shûnyatâvâda*, the philosophical notions of vacuity, nihilism, nullity, non-existence, insignificance and absence were conceived of very early and unified according to a perfect homogeneity under the unique label of *shûnyatâ* ("vacuity"). In this domain at least, India was very advanced in comparison with corresponding Graeco-Latin ideas. See **Zero (Graeco-Latin concepts of)**, **Zero and Zero (Indian concepts of)**.

SHÛNYATÂVADÂ. Name of the Buddhist doctrine of vacuity. See **Shûnya** and **Shûnyatâ**.

SHÛNYAVÂDIN. "Vacuitist". Name given to the followers of the Buddhist philosophy of vacuity. See **Shûnyatâ** and **Mâdhyamika**.

SIDDHA. In Hindu and Jaina philosophies, this is the name given to human beings that become immortals after having obtained liberation.

SIDDHAM NUMERALS. Symbols derived from *Brâhmî numerals and influenced by Shunga, Shaka, Kushâna, Ândhra, Gupta and Nâgarî. Used in Ancient Nepal (sixth to eighth centuries CE). The symbols corresponded to a mathematical system that was not based on place-values and therefore did not possess a zero. Ancestor of Nepali, Tibetan, Mongolian, etc. numerals. Siddham also influenced the shapes of Kutilâ numerals, whence came Bengali, Oriyâ, Kaîthî, Maithilî, Manipurî, etc. numerals. See Fig. 24.41. For systems derived from Siddham, see Fig. 24.52. For graphical development, see Figs. 24.61 to 69. See: **Indian written numeral systems (Classification of)**.

SIDDHÂNTA. [Astronomy]. Generic name of the astronomical texts which describe such things as the calculation for an eclipse of the Moon or the Sun, and the procedures, methods and instruments of observation. Diverse parameters and data are supplied, as well as the procedure for trigonometric operations, etc. See **Indian astronomy (History of)** and **Yuga (Astronomical speculations on)**.

SIDDHÂNTADARPANA. See **Nilakanthasomayâjin**.

SIDDHÂNTASHEKHARA. See **Shrîpati**.

SIDDHÂNTASHIROMANI. See **Bhâkarâchârya**.

SIDDHÂNTATATTVAVIVEKA. See **Kamâlakara**.

SIDDHI. [S]. Value = 8. "Supernatural power". This is an allusion to the *ashtasiddhi*, the eight major *siddhi*, or eight supernatural powers which the *siddha* and the *pûrnayogin* (perfect adepts of the techniques of yoga) are gifted with. See **Eight**.

SIGNS IN THE FORM OF "S" OR "8". See **Numeral 8**, **Serpent (Symbolism of the)** and **Infinity (Indian mythological representation of)**.

SIGNS OF NUMERATION. See Fig. 24.61 to 69. See **Indian numerals**, which gives the complete list of signs of numeration, as well as **Numerical notation**, which gives a list of the main systems of numeration used in India since Antiquity. See also **Indian written numeral systems (Classification of)**, which recapitulates on all the numerical notations of the Indian sub-continent, from both a mathematical and a palaeographic point of view.

SIMHASAMVAT (Calendar). Calendar beginning in the year 1113 CE. Add 1113 to a date in this calendar to find the corresponding year in the Common Era. Formerly used in Gujarat. It was probably abandoned during the thirteenth century. See **Indian calendars**.

SIMPLE YUGA (Non-speculative). See **Yuga (Astronomical speculation on)**.

SINDHI NUMERALS. Symbols derived from *Brâhmî numerals and influenced by Shunga, Shaka, Kushâna, Ândhra, Gupta and Shâradâ. Currently used in the region of Sindh (whose name derives from the river now called the Indus). The symbols correspond to a mathematical system that has place values and a zero (shaped like a small circle). See: **Indian written numeral systems (Classification of)**. See Fig. 24.6, 24.52 and 24.61 to 69.

SINDHU. [S]. Value = 4. "Sea". See **Sâgara**, **Four** and **Ocean**.

SINE (Function). This is referred to as *ardhajyâ*, which literally means: "demi-chord". This is the name used since *Âryabhata (c. 510 CE) by Indian astronomers to denote this function of trigonometry.

SINHALA (SINHALESE) NUMERALS. Symbols derived from *Brâhmî numerals and influenced by Shunga, Shaka, Kushâna, Ândhra, Pallava, Châlukya, Ganga, Valabhî and Bhattiprolu numerals. Currently used mainly in Sri Lanka (Ceylon), in the Maldives and in other islands to the north of the Maldives. (Note that in the north and northwest of Sri Lanka, *Tamil numerals are used by the Tamil inhabitants.). The symbols correspond to a mathematical system that is not based on place values and therefore does not possess a zero. See: **Indian written numeral systems (Classification of)**. See Fig. 24.22, 24.52 and 24.61 to 69.

SIX. Ordinary Sanskrit names for this number: **shad, *shash, *shat.* Here is a list of corresponding numerical symbols: **Anga, Ari, *Darshana, *Dravya, *Guna, Karâka, *Kârttikeyâsya, *Kâya, Khara, *Kumârâsya, *Kumâravadana, Lekhya, Mala, *Mâsârdha, *Râga, *Rasa, Ripu, *Ritu, *Shâdâyatana, *Shaddarshana, *Shâdgunya, *Shanmukha, Shâstra, Tarka.*

These words have the following symbolic meanings or translations: 1. The philosophical points of view *(Darshana)* 2. The six philosophical points of view *(Shaddarshana)*, 3. The bodies *(Kâya)*, 4. The colours *(Râga)*, 5. The musical modes *(Râga)*, 6. The weapons *(Shâstra)*, 7. The limbs *(Anga)*, 8. The **Vedânga (Anga)*, 9. The merits, the qualities, the primordial properties *(Guna)*, 10. The six primordial properties, the six bases, the six categories *(Shadâyatana, Shâdgunya)*, 11. The seasons *(Mâsârdha, Ritu)*, 12. The substances *(Dravya)*, 13. The faces of Kârttikeya-Kumâra *(Kârttikeyâsya, Kumârâsya, Kumâravadana, Shanmukha)*, 14. The sensations, in the sense of "flavours" *(Rasa)*. See **Numerical symbols.**

SIX AESTHETIC RULES OF PAINTING. See *Shâdanga.*

SIXTEEN. Ordinary Sanskrit name: **shaddasha.* Here is a list of corresponding numerical symbols: **Ashti, *Bhûpa, Kalâ* and **Nripa.* These words refer to or are related to the following: 1. A particular element of Indian metric *(Ashti)* 2. The sixteen kings of the legend of the **Mahâbhârata (Bhûpa* and *Nripa)* 3. The "fingers of the Moon" *(Kalâ).* See **Numerical symbols.**

SIXTY. See *Shashti.*

SMALLEST UNIT OF LENGTH. See *Paramânu.*

SMALLEST UNIT OF WEIGHT. See *Paramânu râja.* See also **Indian weights and measures.*

SOGANDHIKA. Name given to the number ten to the power ninety-one. See **Names of numbers** and **High numbers.**

Source: **Vyâkarana* (Pâli grammar) by Kâchchâyana (eleventh century CE).

SOMA. [S]. Value = 1. Name of an intoxicating drink, used in Vedic times for religious ceremonies and sacrifices: "It is a drink made from a climbing plant, with which an offering is made to the gods and which is drunk by Brahmanic priests. This drink plays an important role in the *Rigveda*. It is considered to be capable of conferring supernatural powers and is worshipped as though it were a god. The Hindus also call it the wine of immortality (**amrita*). It is the symbol of the transition from ordinary sensory pleasures to divine happiness *(ânanda)*. K. Friedrichs, etc. "In Indian thought, *Soma* also represents the source of all life and symbolises fertility; thus it is the sperm, the receptacle of the seeds of cyclic rebirth. In this respect, the *soma* is connected to the symbolism of the moon. This is why the *soma* is also the lunar star, a masculine entity compared with a full goblet of the drink of immortality. Thus: *soma* = 1. See *Abja* and **One.**

Source: **Lalitavistara Sûtra* (before 308 CE).

SPECULATIVE YUGA. See *Yuga* (**Astronomical speculation on**).

SQUARE ROOT. [Arithmetic]. See *Varganmûla.* See also **Square roots (How Âryabhata calculated his).**

SQUARE ROOTS (How Âryabhata calculated his). In the chapter of *Ganitapâda* in *Âryabhatîya* devoted to arithmetic and methods of calculation, the astronomer **Âryabhata* (c. 510 CE) described, amongst other operations, the rule for the extraction of square roots [see Arya, *Ganita*, line 4]:

Always divide the even column by twice the square root. Then, after subtracting the square of the even column, put the quotient in the next place. This will give you the square root.

The rule, thus formulated, is a typical example of Âryabhata's extremely concise style, only giving the essential information in his definitions, operations or concepts, any other information being deemed useless for reasons only known by the man himself. Here is the extract again, with the necessary information added for easy understanding:

[After subtracting the largest possible square from the figure found in the last uneven column, then having written the square root of the subtracted number in the line of the square root] *always divide the* [figure in the] *even column* [written on the right] *by twice the square root.*

Then, after subtracting the square [of the quotient] *from the* [figure found in the] *even column* [written on the right], *place the quotient in the next place* [to the right of the figure which is already written down in the line of the square root]. *This will give you the square root* [desired]. [But if there are figures remaining on the right, repeat the process until there are no more of these figures].

[See: Datta and Singh (1938), pp. 169–75; Clark (1930), pp. 23ff; Shukla and Sarma (1976), pp. 36–7; Singh, in BCMS, 18, (1927)]. Here is the reproduction (with no theoretical justification) of the first of these rules, in order to calculate the square root of the number 55, 225, according to the information given notably by Bhâskara (in 629) in his Commentary on the *Âryabhatîya*: First, the number in question is written in the following manner, marking each uneven place with a vertical line and each even place with a horizontal line:

$$
\begin{array}{ccccc}
| & - & | & - & | \\
5 & 5 & 2 & 2 & 5
\end{array}
$$

Then a horizontal line is drawn (to the right of the number in question), in order to write down the successive numbers of the square root:

$$
\begin{array}{ccccc}
| & - & | & - & | \\
5 & 5 & 2 & 2 & 5
\end{array} \quad \underline{\hspace{3cm}}
$$

line of the square root

By beginning the operation with the highest figure of the uneven column, the biggest square it contains is 4, thus the square root is equal to 2. Therefore a 4 is placed in a line underneath and a 2 on the line of square roots:

$$
\begin{array}{ccccc}
| & - & | & - & | \\
5 & 5 & 2 & 2 & 5 \\
4
\end{array} \qquad 2
$$

line of the square root

Then a line is drawn below the 4, which is subtracted from the preceding 5; the result is 1, and this figure is placed under the line in the even position of this first section, without forgetting to return the 4 to the extreme left of this lower line:

$$
\begin{array}{ccccc}
| & - & | & - & | \\
5 & 5 & 2 & 2 & 5 \\
4
\end{array} \qquad 2
$$

line of the square root

4) 1

Next the figure in the even column written immediately to the right (the 5) is considered, and is placed below the bottom line, to the right of the 1:

$$
\begin{array}{ccccc}
| & - & | & - & | \\
5 & 5 & 2 & 2 & 5 \\
4
\end{array} \qquad 2
$$

line of the square root

4) 1   5

Now the number 15 which has been obtained is divided by twice the square root that was previously found (2), in other words by 4; as the quotient found is 3, thus 3 is written on the line of square roots, to the right of the 2 that is already there, without forgetting to record the same figure on the extreme right of the line of the 15:

$$
\begin{array}{ccccc}
| & - & | & - & | \\
5 & 5 & 2 & 2 & 5 \\
4
\end{array} \qquad 2 \quad 3
$$

line of the square root

4) 1   5   (3

The product of the numbers 4 and 3 (placed to the left and right of the line of 15) is 12, and this is placed on the line below 15:

$$
\begin{array}{ccccc}
| & - & | & - & | \\
5 & 5 & 2 & 2 & 5 \\
4
\end{array} \qquad 2 \quad 3
$$

line of the square root

4) 1   5   (3
   1   2

Then 12 is subtracted from the above 15, and the result is placed on the line below, after drawing a line below the number 12:

$$
\begin{array}{ccccc}
| & - & | & - & | \\
5 & 5 & 2 & 2 & 5 \\
4
\end{array} \qquad 2 \quad 3
$$

line of the square root

4) 1   5   (3
   1   2

      3

Then the 2 from the following uneven column is placed next to the 3:

$$
\begin{array}{ccccc}
| & - & | & - & | \\
5 & 5 & 2 & 2 & 5 \\
4
\end{array} \qquad 2 \quad 3
$$

line of the square root

4) 1   5   (3
   1   2

      3   2

And a 9 (the square of the quotient 3 found above, indicated to the right of the line of 15) is placed in the line below the 32:

| – | – | 2 3

5   5   2   2   5

4   line of the square root

---

4)   1   5   (3
     1   2

---

          3   2
              9

A line is drawn and the 9 is subtracted from the 32, then the result is placed below this line:

| – | – | 2 3

5   5   2   2   5

4   line of the square root

---

4)   1   5   (3
     1   2

---

          3   2
              9

---

          2   3

Now the 2 is taken from the even column and placed to the right of the positions of 23:

| – | – | 2 3

5   5   2   2   5

4   line of the square root

---

4)   1   5   (3
     1   2

---

          3   2
              9

---

          2   3   2

Then the number 232 which has been thus obtained is divided by 46, which is double the square root found (23), and as the quotient is 5, the numbers 46 and 5 are written as follows (the divisor 46 on the left and the quotient 5 on the right), by placing a 5 on the line of square roots to the right of the 3:

| – | – | 2 3 5

5   5   2   2   5

4   line of the square root

---

4)   1   5   (3
     1   2

---

          3   2
              9

---

46)   2   3   2   (5

And as the product of 46 times 5 is 230, this number is placed below 232:

| – | – | 2 3 5

5   5   2   2   5

4   line of the square root

---

4)   1   5   (3
     1   2

---

          3   2
              9

---

46)   2   3   2   (5
          2   3   0

Another line is drawn, and the following subtraction is carried out:

| – | – | 2 3 5

5   5   2   2   5

4   line of the square root

---

4)   1   5   (3
     2   2

---

          3   2
              9

---

46)   2   3   2   (5
          2   3   0

                  2

The last figure (5) is lowered and placed to the right of the 2:

| – | – | 2 3 5

5   5   2   2   5

4   line of the square root

---

4)   1   5   (3
     1   2

---

          3   2
              9

---

46)   2   3   2   (5
          2   3   0

---

                  2   5

The last quotient is equal to 5, and the square of this number is taken (25) and subtracted from this last number. As the result is equal to zero, the operation is finished. It is clear that the operation has worked, because the square root of 55 225 is equal to 235.

						2	3	5
	|	–	|	–	|			
5	5	2	2	5				
4					line of the square root			
4)	1	5	(3					
	1	2						
	3	2						
		9						
46)	2	3	2	(5				
	2	3	0					
		2	5					
		2	5					
			0					

Thus it is clear that this procedure is not algebraic (contrary to Kaye's allegations, who gave the unwarranted affirmation that Âryabhata's method was identical to that of Theon of Alexandria), and it is also clear that it is impossible to use Âryabhata's method if the numbers in question are not expressed *in writing* using distinct numerals as a base for the calculations. In other words, the operations described by Âryabhata involve placing the numbers involved in the calculation in two or three blocks of numbers, according to whether it is the square root or the cube root that is being extracted. It can be proved mathematically that these operations could not be carried out using a written numeration that was not based upon the place-value system and did not have a zero.

STHÂNA. Sanskrit term meaning "place". Generally used by Indian scholars to express the place-value system. See *Sthânakramâd, Ankakramena* and *Ankasthâna.*

Source: *Lokavibhâga* (458 CE).

STHÂNAKRAMÂD. Sanskrit term which literally means: "in the order of the position". Often used by Indian scholars in ancient times (fifth – seventh century CE) to indicate that a series of numbers or numerical word-symbols were written according to the place-value system. An example of this is found in the *Jaina cosmological text, the *Lokavibhâga* ("Parts of the universe"), which is the oldest known Indian text to contain an example of the place-value system written in numerical symbols. [See Anon. (1962), chap. IV, line 56, p. 79].

SUBANDHU. Indian poet from the beginning of the seventh century CE. His works notably include a love story entitled *Vâsavadattâ,* where there are precise references to zero written as a dot *(*shûnya-bindu).* See Zero.

SUBSTANCE. [S]. Value = 6. See *Dravya* and Six.

SUBTRACTION. [Arithmetic]. See *Vyavakalita.*

SUDDHA SVARA. These are the seven notes of the *sa-grâmma (Sa, Ri, Ga, Ma, Pa, Dha, Ni),* the first scale in Indian music. The notes are represented by short syllables, each one corresponding to the initial of the Sanskrit name of the note *(Ni = Nishâda; Ga = Gandhâra,* etc.) [see Frédéric, *Dictionnaire,* (1987)].

SUDHÂMSHU. [S]. Value = 1. "Moon". See *Abja* and One.

SUDRISHTI. "That which is seen clearly". Name given to the Pole star, "the Star which never moves". See *Dhruva, Grahâdhâra* and Mount Meru.

SUMERU. One of the names of *Mount Meru. See *Adri, Dvîpa, Pûrna, Pâtâla, Sâgara, Pushkara, Pâvana* and *Vâyu.*

SUN. As a concept associated with the number thousand. See *Sahasrakirana, Sahasrâmshu* and High numbers.

SUN. As a mystical value equal to 7. See Mount Meru.

SUN. [S]. Value = 12. See *Bhânu, Divâkara, Dyumani, Mârtanda, Shasrâmshu, Sûrya* and Twelve.

SUN RAYS. [S]. Value = 12. See *Shasrâmshu* and Twelve.

SUN-MOON (The couple). [S]. Value = 2. See *Ravichandra* and Two.

SUPERNATURAL POWER. [S]. Value = 8. See *Pañchâbhijñâ, Siddhi* and Eight.

SURA. [S]. Value = 33. "Gods". See *Deva* and Thirty-three.

SÛRYA. [S]. Value = 12. Name of the Brahmanic sun god. This symbolism is explained by the "course" of the sun during the twelve months of the year. See *Râshi* and Twelve.

SUTA. [S]. Value = 5. "Son". See *Putra* and Five.

SVARA. [S]. Value = 7. "Note", "syllable". This is probably an allusion to the *suddha svara, the seven notes of the first scale in Indian music. See Seven.

SYLLABLE. [S]. Value = 7. See *Svara* and **Seven.**

SYMBOLISM OF NUMBERS. Here is a list of associations of ideas contained in Indian numerical symbolism, given in arithmetical order (list not exhaustive):

Number One. Concept often directly or symbolically related to: the god *Sûrya; the god *Ganesha; a type of deep concentration (*ekâgratâ); the sacred Syllable of the Hindus (*ekâkshara); a certain monotheist doctrine (*ekântika); the study of the unique reality; the contemplation of Everything (*ekatva); the *moon; the drink of immortality (*soma); the *earth; the *Ancestor; the *Great Ancestor; the *First Father; the *beginning; the *body; the *Self; *Ultimate reality; the superior soul; the *individual soul; the *Brahman; the "*form"; the "*drop"; the "*immense"; "*indestructible"; the *rabbit; the "*Luminous"; the *Pole star; the "Cold rays"; the *eye of Shukra; the *terrestrial world; the "*Bearer"; the *primordial principle; the *rabbit figure; the *Cow; the sour milk, etc. See *Eka* and **One.**

Number Two. Concept often directly or symbolically related to: *duality; the idea of couple, *pair, twins or contrast; the *symmetrical organs; *wings; the *hand; the *arms; the *eyes; *vision; the *ankles; the primordial couple; the couple; *Sun-Moon; the twin gods; the conception of the world; *contemplation; revelation; the *Horsemen; the epithet "Twice born"; the third age of a *mahâyuga (*Yuga); etc. See *Dva* and **Two.**

Number Three. Concept often directly or symbolically related to: the *three "classes" of beings; the *Triple science; the first three *Vedas; the *eyes; the three eyes of *Shiva; the "*three worlds"; the god Shiva; the god Vishnu; the god Krishna; the ritual dress of Buddhist monks (*trichîvara); the "three primary forces"; the "three eras"; the "*three bodies"; the "*three forms"; the "three baskets"; the *three city-fortresses; the *three states of consciousness; the "*three jewels"; the triple town-fortress (*tripura); the town-fortress with the triple rampart (*tripura); a demon with three heads (*trishiras); Shiva's Trident; the principal castes of Brahmanism; the "*three aims"; the "*three letters"; the god *Agni; "fire"; the *god of sacrificial fires; the "three rivers"; the *phenomenal world; the "*aphorism"; *Feminine Energies; the "*merits"; the "*qualities"; the Spirit of the waters; the *Eye; the *points; the *times; the "*three heads"; the *three *Râma; etc. See *Traya, Vajra* and **Three.**

Number Four. Concept often directly or symbolically related to: the "*four oceans"; the "*four stages"; the "*four island-continents"; the four "*great kings"; the "four *months"; the "four *faces"; the "four aims"; the "four *ages"; the "four *ways of rebirth"; water; sea; *ocean; "*horizons"; the *cardinal points; the *arms of Vishnu; the *positions; the *vulva; the *births; the "*Fourth" (as an epithet of Brahma); the conditions of existence; the *Vedas; the *faces of Brahma; the four ages of a *mahâyuga; "*faces"; etc. See *Chatur* and **Four.**

Number Five. This number is considered to be sacred and magic in India and all Indianised civilisations of Southeast Asia. It is often directly or symbolically related to: the *Bow with five flowers; the "*five supernatural powers"; the *five elements of the manifestation; the "five visions of Buddha"; the "five *horizons"; the *gifts of the Cow; the "five *faculties"; the "five *impurities"; the five *heads of Rudra (= Shiva); the "five *mortal sins"; the five *orders of beings; the "five treasures" of Jaina religion (*pañchaparameshtin"); the "sons of Pându"; the *arrows; the characteristics; *Purification; the "*Great Elements"; the *great sacrifices; the *main observances; the *fundamental principles; the *realities; the *truths; the "*true natures"; the *Jewels; the *breaths; the *senses; the *winds; the sense organs; the *faces of Rudra; etc. See *Pañcha;* and **Five.**

Number Six. Concept often directly or symbolically related to: the "six *parts; the "six *bases"; the "six categories"; the six *philosophical points of view; the six aesthetic rules (*shâdanga); the *bodies; the *colours; the *musical modes; the weapons; the limbs; the *merits; the *qualities; the *primordial properties; the *substances; the *seasons; the *Vedânga; the *faces of Kârttikeya (= *Kumâra); the *sensations; the flavours; etc. See *Shad* and **Six.**

Number Seven. Concept often directly or symbolically related to: the seven Buddhas (*saptabuddha); the seven *planets; the "seven paces" (*saptapâdi); the "seven *Jewels" (saptagraha); the "seven *sages"; the *Rishi; "*Purification"; the *horses; the "seven *divine mothers"; the "seven rivers" (*saptasindhava); the seven *horses of Sûrya; the *island-continents; the *seas; the *oceans; the "worlds"; the seven *inferior worlds; the seven *hells; the *mountain; the seven *syllables; the seven *musical notes; the last of the seven *Rishi* (*Atri); the seven *days of the *week; "That which never moves"; *blue lotus flower; the seven *winds; etc. See *Sapta* and **Seven.**

Number Eight. Concept often directly or symbolically related to: the "eight parts (*ashtasansa*)"; the "eight *horizons*"; the "eight *forms*"; the "eight *limbs*" of prostrating oneself (*ashtânga*); the *serpent; the *serpent of the deep; the "eight liberations" (*ashtavimoksha*); the *elephant; the eight "things which augur well"; the "eight *elephants*"; the *guardians of the horizons; the *guardians of the points of the compass; the "*jewel"; the "shapes"; the eight divinities (*Vasu*); the spheres of existence; the *supernatural powers; the acts; the "body" (*tanu*); etc. See *Ashta*, **Serpent (Symbolism of the)** and **Eight.**

Number Nine. Concept often directly or symbolically related to: the nine planets (*navagraha*); the "nine *Jewels*"; the Feast of nine days; the "nine precious stones" (*navaratna*); the *Brahman; the ninth month of the year *chaitradi*; the numeral of the place-value system (*anka*); the "Unborn"; the "*Inaccessible"; the "*Divine Mother"; the divinity *Durgâ; the *holes; the *orifices; the *radiance; the "*Cow"; etc. See *Nava* and **Nine.**

Number Ten. Concept often directly or symbolically related to: the digits; the Feast of the tenth day; the ten powers of a Buddha (*dashabala*); the *descents; the "ten *earths*"; the "ten paradises" (*dashabhûmi*); the "ten stages" of the Buddha (*dashabhûmi*); the *horizons; the *heads of Râvana; the ten *major incarnations of Vishnu (*dashâvatâra*); etc. See *Dasha*, **Ten**, and **Durgâ.**

Number Eleven. Concept often symbolically associated with: the god *Rudra (= *Shiva), who is often referred to by one of his attributes instead of by name ("Supreme Divinity", "*Great God", "*Lord of the universe", "*Lord of tears", "Rumbling", Lord of the animals", "Violent", etc.). See *Ekadasha* and **Eleven.**

Number Twelve. Concept often symbolically associated with: the "brilliant"; the sun; the Sun-god; the "solar fire"; the *sun rays; the "*months"; the *zodiac; the *arms of Kârttikeya; the "*wheel"; the *eyes of Senânî; the sons of Âditî; etc. See *Dvâdasha* and **Twelve.**

Number Thirteen. Concept often symbolically associated with the *god of carnal love and of cosmic desire (*Kâma*) and with the *universe formed by thirteen worlds. See *Trayodasha* and **Thirteen.**

Number Fourteen. Concept often symbolically associated with: the god *Indra, who is often referred to by one of his attributes instead of by name ("*Courage", "Strength", "*Power",

"*Powerful", etc.); the "*human" (in the sense of the progenitor of the human race); the worlds; the fourteen universes (*bhuvana*); the "*Jewels"; etc. See *Chaturdasha* and **Fourteen.**

Number Fifteen. Concept often symbolically associated with: "*wing"; "*day"; etc. See *Pañchadasha* and **Fifteen.**

Number Sixteen. Concept often symbolically associated with: the sixteen *kings of the legend of the *Mahâbhârata*; the "fingers of the moon" (*kalâ*). See **Shaddasha** and **Sixteen.**

Number Twenty. Concept often directly or symbolically associated with: the digits; the *nails; the *arms of Râvana; etc. See *Vimshati* and **Twenty.**

Number Twenty-five. Concept often symbolically associated with: the *fundamental principles; the "*true natures"; the *truths; the *realities; etc. See *Pañchavimshati* and **Twenty-five.**

Number Twenty-seven. Concept often directly related to: the "stars"; "*lunar mansions"; the *constellations; etc. See *Saptavimshati* and **Twenty-seven.**

Number Thirty-two. Concept often directly related to: the teeth. See *Dvatrimshati* and **Thirty-two.**

Number Thirty-three. Concept symbolically associated with: the "*gods"; the "immortals". See *Trâyastrimsha* and **Thirty-three.**

Number Forty-nine. Concept often symbolically associated with the *winds. See *Navachatvârimshati* and **Forty-nine.**

Number thousand. Concept often interpreted in the sense of the multitude or the incalculable, often associated with: the attributes of many Hindu and Brahmanic divinities (the "Thousand arms", the "*Thousand rays" or the "Thousand of the Brilliant" all denote the Sun-god *Sûrya; the "Thousand names" denotes the gods *Vishnu and *Shiva; the "Thousand eyes" refers to the gods Vishnu and Indra; etc.); or mythological figures (such as the demon Arjuna, who is referred to by the name "Thousand arms of Arjuna"). This number is also associated with: the Mouths of the Ganges (*jâhnavîvaktra*); the Arrows of Ravi (= Sûrya); *Ananta (the serpent with a thousand heads); the *lotus with a thousand petals; etc. See *Sahasra* and **Thousand.**

SYMBOLISM OF NUMBERS (Concept of a large quantity). Here is an alphabetical list of English words which have a connection with Indian high numbers, and which can be found as entries in this dictionary: *Arithmetical speculations, *Astronomical speculations, *Billion, *Blue lotus, Conch, *Cosmic cycles,

*Day of Brahma, *Diamond, *Dot, *Earth, *High numbers, *High numbers (Symbolic meaning of), *Hundred billion, *Hundred million, Hundred quadrillion, Hundred quintillion, Hundred thousand, *Hundred thousand million, *Hundred trillion, Incalculable, *Indeterminate, *Infinity, *Kalpa, *Kalpa (Arithmetical-cosmogonical speculations on), *Lotus, *Million, *Moon, *Names of numbers, *Ocean, *Pink lotus, Pink-white lotus, *Powers of ten, *Quadrillion, *Quintillion, *Serpent with a thousand heads, *Ten billion, *Ten million, Ten quadrillion, Ten quintillion, *Ten thousand, *Ten thousand million, *Ten trillion, *Thousand (in the sense of "multitude"), *Thousand, *Thousand million, *Trillion, Unlimited, *White lotus. *Zero. See **High numbers**, which gives a list of the principal corresponding Sanskrit words, as well as all the necessary explanations.

SYMBOLISM OF NUMBERS (Concept of Infinity). Here is an alphabetical list of English words which are connected to the Indian idea of infinity, and which can be found as entries in this dictionary: *Arithmetical speculations, Arithmetical-cosmogonical speculations, *Blue lotus, Conch, *Cosmic cycles, *Cosmogonic speculations, *Day of Brahma, *Diamond, *Dot, *Earth, *Eternity, *High numbers, *High numbers (Symbolic meaning of), Incalculable, Indefinite, *Infinitely big, *Infinity, *Infinity (Indian concepts of), *Kalpa, *Kalpa (Arithmetical-cosmogonical speculations on), *Lotus, *Moon, *Names of numbers, *Ocean, *Pink lotus, Pink-white lotus, *Serpent of infinity and eternity, *Serpent (Symbolism of the), *Serpent with a thousand heads, Sky, *Thousand, Unlimited, *White lotus.

SYMBOLISM OF NUMBERS (Concept of Zero). Here is an alphabetical list of words which are connected to Indian notions of vacuity, the void and zero, and which appear as entries in this dictionary: *Sanskrit terms:* *Abhra, *Âkâsha, *Ambara, *Ananta, *Antariksha, *Bindu, *Gagana, *Jaladharapatha, *Kha, *Khachheda, *Khahâra, *Nabha, *Nabhas, *Pûrna, *Randhra, *Shûnya, *Shûnya-bindu, *Shûnya-chakra, *Shûnya-kha, *Shûnya-samkhya, *Shûnyatâ, *Shûnyavâdin, *Vindu, *Vishnupada, *Vyant, *Vyoman. *English terms:* *Absence, *Atmosphere, *Canopy of heaven, *Dot, *Ether, *Firmament, *Hole, *Indian atomism, *Infinitely small, *Infinity, *Insignificance, *Low numbers, Negligible, *Nihilism, *Non-being, *Non-existence, *Non-present,

*Non-substantiality, *Nothing, *Nothingness, *Numeral 0, Sky, Space, Uncreated, Unformed, Unproduced, *Unthought, *Vacuity, *Void, *Zero, *Zero (Graeco-Latin concepts of), *Zero (Indian concepts of), *Zero and Sanskrit poetry. See also **Durgâ**.

SYMBOLISM OF WORDS WITH A NUMERICAL VALUE. Here is an alphabetical list of English words which correspond to the associations of ideas contained in Sanskrit numerical symbols, which appear as entries in this dictionary (the list is not exhaustive):

*Ablaze (= 3), *Ancestor (= 1), *Ankle (= 2), *Aphorism (= 3), *Arms (= 2), *Arms of Arjuna (= 1,000), *Arms of Kârttikeya (= 12), *Arms of Râvana (= 20), *Arms of Vishnu (= 4), *Arrow (= 5), *Arrows of Ravi (= 1,000), *Atmosphere (= 0). *Bearer (= 1), *Beginning (= 1), *Birth (= 4), *Blind king (= 100), *Body (= 1), *Body (= 6), *Body (= 8), *Brahma (= 1), *Breath (= 5), *Brilliant (= 12).

*Canopy of heaven (= 0), *Cardinal point (= 4), *Characteristic (= 5), *City-fortress (= 3), *Colour (= 6), Condition of existence (= 4), *Constellation (= 27), *Contemplation (= 6), *Courage (= 14), *Cow (= 1), *Cow (= 9).

*Day (= 15), *Day of the week (= 7), *Delectation (= 6), *Demonstration (= 6), *Descent (= 10), Digit (= 10), Digit (= 20), *Divine mother (= 7), *Dot (= 0), *Drop (= 1).

*Earth (= 1), *Earth (= 9), *Element (= 5), *Elephant (= 8), Energy (= 3), *Ether (= 0), *Eye (= 2), *Eye (= 3), *Eye of Shukra (= 1), *Eyes (= 2), *Eyes of Senânî (= 12), *Eyes of Shiva (= 3), *Eyes of Indra (= 1,000).

*Face (= 4), *Faces of Brahma (= 4), *Faces of Kârttikeya (= 6), *Faces of Kumâra (= 6), *Faces of Rudra (= 5), *Faculty (= 5), *Fire (= 3), *Fire (= 12), *Firmament (= 0), *First father (= 1), *Form (= 1), *Form (= 3), *Form (= 8), *Four cardinal points (= 4), *Fourth (= 4), *Fundamental principle (= 5), *Fundamental principle (= 7), *Fundamental principle (= 25).

*Ganges (= 1,000), *Gift of the Cow (= 5), *God of carnal love (= 13), *God of cosmic desire (= 13), *God of sacrificial fires (= 3), *Gods (= 33), *Great Ancestor (= 1), *Great god (= 11), *Great element (= 5), *Great sacrifice (= 5), *Great sin (= 5), *Guardian of the horizons (= 8), *Guardian of the points of the compass (= 8).

*Hand (= 2), He who has three heads (= 3), *Heads of Râvana (= 20), *Hell (= 7), *Hole (= 0), *Horizon (= 4, *Horizon (= 8), *Horizon (= 10), *Horse (= 7), *Horsemen (= 2), *Human (= 14).

*Immense (= 1), *Inaccessible (= 9), *Incarnation (= 10), Indestructible (= 1), *Individual soul (= 1), *Indra (= 14), *Inferior world (= 7), *Infinity (= 0), *Island-continent (= 4), *Island-continent (= 7).

*Jewel (= 8), *Jewel (= 5), *Jewel (= 9), *Jewel (= 14).

*King (= 16). Limb (= 6), *Lord of the universe (= 11), *Luminous (= 1), *Lunar mansion (= 27).

*Main observance (= 5), *Merit (= 6), *Merit (= 3), *Month (= 12), *Moon (= 1), *Mountain (= 7), *Mouths of Jâhnavî (= 1,000), *Musical mode (= 6), *Musical note (= 7).

*Nail (= 20), *Numeral (= 9).

*Ocean (= 4), *Ocean (= 7), *Opinion (= 6), *Orifice (= 9).

*Pair (= 2), *Paradise (= 13), *Paradise (= 14), *Part (= 6), *Passion, *Phenomenal world (= 3), *Philosophical point of view (= 6), *Planet (= 9), *Point (= 3), *Position (= 4), *Power (= 14), *Powerful (= 14), *Precept (= 6), Primordial couple (= 2), *Primordial principle (= 1), *Primordial property (= 3), *Primordial property (= 6), *Progenitor of the human race (= 14), *Purification (= 7).

*Quality (= 3), *Quality (= 6).

*Rabbit (= 1), Rabbit figure (= 1), *Radiance (= 9), *Reality (= 5), *Reality (= 7), *Reality (= 25), *Rudra-Shiva (= 11), Rumbler (= 11). *Sage (= 7), *Season (= 6), *Self (= 1), *Sensation (= 6), *Sense (= 5), *Sense organs (= 5), *Serpent (= 8), *Serpent of the deep (= 8), *Serpent with a thousand heads (= 1,000), Sky (= 0), Son (= 5), Sons of Âditî (= 12), *Sons of Pându (= 5), Sour milk (= 1), Space (= 0), Spirit of the waters (= 3), Star (= 27), State (= 3), State of the manifestation (= 5), Strength (= 14), *Substance (= 6), *Sun (= 12), *Sun (= 1,000), *Sun-god (= 12), *Sun-Moon (= 2), *Sun rays (= 12), *Supernatural power (= 8), Supreme Divinity (= 11), Supreme soul (= 1), *Syllable (= 7), *Symmetrical organs (= 2).

*Taste (= 6), *Terrestrial world (= 1), That which augurs well (= 8), That which must be done (= 5), That which belongs to all humans (= 3), *Thousand (= 12), *Thousand rays (= 12), *Three aims (= 3), *Three bodies (= 3), *Three city-fortresses (= 3), *Three classes of beings (= 3), *Three eyes (= 3), *Three forms (= 3), *Three fundamental properties (= 3), *Three heads (= 3), *Three jewels (= 3), *Three letters (= 3), *Three sacred syllables (= 3),

*Three states (= 3), *Three times (= 3), *Three universes (= 3), *Three worlds (= 3), *Time (= 3), *Tone (= 49), Tooth (= 32), *Triple science (= 3), *True nature (= 7), *True nature (= 25), *Truth (= 5), *Truth (= 7), *Truth (= 25), Twice born (= 2), Twin gods (= 2), Twins, pairs or couples (= 2).

*Ultimate reality (= 1), *Universe (= 13).

*Veda (= 3), *Veda (= 4), *Vedânga (= 6), *Violent (= 11), *Vision (= 6), *Voice (= 3), *Void (= 0), *Vulva (= 4).

Water (= 4), *Week (= 7), *Wheel (= 12), *Wind (= 5), *Wind (= 7), *Wind (= 49), *Wing (= 2), *Wing (= 15), *Word (= 3), *World (= 3), *World (= 7), *World (= 14).

*Yuga (= 2), *Yuga (= 4).

*Zenith (= 0), *Zodiac (= 12).

See **Symbols, Numerical symbols, One, Two, Three, Four, Five, Six, Seven, Eight, Nine, Ten, Eleven, . . . Zero** and **Names of numbers.**

SYMBOLISATION OF THE CONCEPT OF INFINITY. See **Infinity, Infinity (Indian concepts of), Infinity (Mythological representation of)** and **Serpent (Symbolism of the).**

SYMBOLISATION OF THE CONCEPT OF ZERO. See **Zero, Dot** and **Circle.**

SYMBOLS. In the Brahmanic religion, and other religions of the Indian sub-continent, symbols have always been very important. They are either visible and understood by everyone and resume a number of concepts which are difficult to write down (*stûpa*, for example), or they are invisible because they have a sense which the profane cannot see (such as the *bîja*, the *yantra*, the *mudrâ*, etc.).

The symbols are represented by numerous categories of beings (such as animals), objects or even plants. As with Mahâyâna Buddhism, each divinity of Brahmanism possesses a carrier-animal which symbolises the god himself: Garuda for *Vishnu, Nandin for *Shiva, etc.: they also have a *bîja* (a letter-symbol for the corresponding sound to invoke them), *mantras* (or sacred formulas), *yantras* (geometrical diagrams with symbolic meaning) and various "signs" or distinctive marks which allow the faithful to identify the representations of the gods immediately.

The combination of signs is also symbolic, and different from a sole, isolated symbol (like *vajra* and *ganthâ*). Some symbols are raw materials like the *linga* of Shiva or the *shâlagrâma* of Vishnu; others are constructions (such as *stûpas, chaityas,* temples and various sculptures).

As for the plant kingdom, many trees (pipal, banyan, etc.) plants (*tulasî*) and seeds (*rudrâksha*) constitute symbols to Hindus, Buddhists and followers of the Jaina religion. In India, all things are potentially symbolic, not only in philosophy and religion, but also in literature, art and music. The most significant symbols are the attributes of the divinities. The Trident (**trishûlâ*) belongs to Shiva, but like the serpent (**nâga*) or the elephant, it has other meanings. See **Serpent (Symbolism of the)**. The club (*danda, gadâ*) is the sign of the guardians of the gate (*dvârapâla*), but also a symbol of solar energy. The lance (*shakti*) and other weapons: dagger (*kshurikâ*), axe (*parashu*), bow and arrow (*dhanus, *bâna*), shield (*khetaka*), sword (*khadga*), etc., are used to show the power of divinities.

*Lotus flowers are most important to Buddhism, but are also highly symbolic of the pure nature of Hindu divinities.

Other very common symbolic attributes include: musical instruments (the *vînâ* of Sarasvatî, the *damarû* of Shiva-Nâtarâja); the conch (**shankha*); the bell (*ganthâ*); everyday objects (the mirror of Mâyâ, *darpana*); the cord (*pâsha*) that joins the soul to matter; the book (*pushtaka*) which represents all the **Vedas*; etc.

The sun (*chakra*) and the moon (*kulikâ*), the symbols of constellations, all have specific meanings which are either obvious or hidden (esoteric or tantric). There is a lot of symbolism connected to the human body: nudity suggests detachment from contingencies; colour of skin means anger and fury or peace and joy. Hair (in a bun) symbolises Yogin; dishevelled hair represents the mobility of Mâyâ; frizzy, untidy hair means rage.

The number of arms and legs that a divinity possesses is also highly symbolic: the more arms, the more active the god is. When a god only has two arms, this represents "angelic", peaceful qualities. If a god has no attributes whatsoever, this represents neutrality, like the **Brahman. Jewels and ornaments also have precise meanings, which can vary according to era, beliefs and philosophies. [The information in this entry is taken from Frédéric, *Dictionnaire de la civilisation indienne* (1987)].

SYMMETRICAL ORGANS. As symbols for the number two. See *Bâhu, Gulpha, Nayana, Netra, Paksha* and **Two**.

T

TAITTIRÎYA SAMHITÂ. Text derived from the *Yajurveda* "black", which figures amongst the texts of Vedic literature. It is the result of oral transmission since ancient times, and did not appear in its definitive form until the beginning of the Common Era. See **Veda**. Here is a list of the principal names of numbers mentioned in the text: **Eka* (= 1), **Dasha* (= 10), **Sata* (= 10^2), **Sahasra* (= 10^3), **Ayuta* (= 10^4), **Niyuta* (= 10^5), **Prayuta* (= 10^6), **Arbuda* (= 10^7), **Nyarbuda* (= 10^8), **Samudra* (= 10^9), **Madhya* (= 10^{10}), **Anta* (= 10^{11}), **Parârdha* (= 10^{12}). [See **Names of numbers** and **High numbers**. See: **Taittirîya Samhitâ*, IV, 40. 11. 4; VII, 2. 20. 1; Datta and Singh (1938), p. 9; Weber, in: JSO, XV, p. 132-40].

TÂKÂRI NUMERALS. Symbols derived from *Brâhmî numerals and influenced by Shunga, Shaka, Kushâna, Ândhra, Gupta and Shâradâ numerals. Currently used in Kashmir alongside the so-called "Hindi" numerals of eastern Arabs. Also called Tankrî numerals. The symbols correspond to a mathematical system that has place values and a zero (shaped like a small circle). See **Indian written numeral systems (Classification of)**. See Fig. 24.13, 52 and 24.61 to 69.

TAKSHAN. [S]. Value = 8. "Serpent". See *Nâga*, **Eight** and **Serpent (Symbolism of the)**.

TAKSHASA. Name of the king of the **nâga*. See **Serpent (Symbolism)**.

TALLAKSHANA. Name given to the number ten to the power fifty-three. According to the legend of Buddha, this number is the highest in the first of the ten numerations of high numbers defined by the Buddha child during a contest in which he competed against the great mathematician Arjuna. *Tallakshana* contains the word *lakshana*, which literally means "character", "mark", "distinguishing feature". In Buddhism, this word often expresses the "hundred and eight distinctive signs of perfection" which distinguish a Buddha from other human beings (108 being considered a magic and sacred number which symbolises perfection). See **Names of numbers** and **High numbers (Symbolic meaning of)**.

Source: **Lalitavistara Sûtra* (before 308 CE).

TAMIL NUMERALS. Symbols derived from *Brâhmî numerals and influenced by Shunga, Shaka, Kushâna, Ândhra, Pallava, Châlukya,

Ganga, Valabhî, Bhattiprolu and Grantha numerals. Currently in use by the Dravidian population of the state of Tamil nadu (Southeast India). The symbols correspond to a mathematical system that is not based on place values and therefore does not possess a zero. For contemporary symbols, see Fig. 24.17; for historical symbols, see Fig 24, 49. See **Indian written numeral systems (Classification of)**. See also Fig. 24.52 and 24.61 to 69.

TÂNA. [S]. Value = 49. "Tone". In Indian music, this refers to the combinations of seven octaves of seven notes.

TANKRÎ NUMERALS. See Tâkarî Numerals.

TANU. [S]. Value = 1. "Body". This symbolism comes from astrology, where "house I" is that which refers to the person, and by extension the body (*tanu*) of the person, whose horoscope is being prepared. See **One.**

TANU. [S]. Value = 8. "Body". This is an allusion to the *âshtânga*, the "eight limbs" of the human body that are involved in the act of prostrating oneself. See *Ashtânga* and **Eight.**

TAPANA. [S]. Value = 3. "Fire". See *Agni*, **Three** and **Fire.**

TAPANA. [S]. Value = 12. The word means "fire", but here it is taken in the sense of "solar fire" and thus of the Sun-god *Sûrya. See *Sûrya* and **Twelve.**

TASTE. [S]. Value = 6. See *Rasa* and **Six.**

TATTVA. [S]. Value = 5. "Reality, truth, true nature, fundamental principle". Allusion to the five "fundamental principles" identified by Indian philosophers and considered to be the basis for thought. See **Five.**

TATTVA. [S]. Value = 7. "Reality, truth, true nature, fundamental principle". Allusion to the seven "fundamental principles" identified by Jaina philosophy and considered to be the basis of the system for thought. See **Seven.** This symbol is very rarely used to represent this value, except for in the *Ganitasârasamgraha* by the Jaina mathematician *Mahâvîrâchârya [see Datta and Singh (1938), p. 56].

TATTVA. [S]. Value = 25. "Reality, truth, true nature, fundamental principle". Allusion to the twenty-five "fundamental principles" identified by the orthodox philosophy of *Sâmkhya: avyakta* (the "non-manifest"); *buddhi* (intelligence); *ahamkâra* (Ego, the consciousness of the Me); the *tanmâtra* (or "original substances", five subtle elements from which the basic elements are said to derive); the *mahâbhûta* (the five elements of the

manifestation); the *buddhîndriya* (the five "sense organs"); the *karmendriya* (the five organs of activity, namely: the tongue, the hands, the legs, the organs of evacuation, and the reproductive organs); *manas* (the "Ability for reflection"; and *purusha* (the Self, the Absolute, pure consciousness) See **Twenty-five.**

TELINGA NUMERALS. See Telugu numerals.

TELUGU NUMERALS. Symbols derived from *Brâhmî numerals and influenced by Shunga, Shaka, Kushâna, Ândhra, Pallava, Châlukya, Ganga, Valabhî and Bhattiprolu numerals. Currently in use amongst the Dravidian population of Andhra Pradesh (formerly Telingana). Also called *Telinga* numerals. The symbols correspond to a mathematical system that has place values and a zero (shaped like a small circle). For contemporary symbols, see Fig. 24.20; for historical symbols, see Fig. 24, 47. See: **Indian written numeral systems (Classification of).** See. Fig. 24.13, 52 and 24.61 to 69.

TEN. Ordinary name in Sanskrit: •*dasha*. List of corresponding numerical symbols: *Anguli, *Âsha, *Avatâra, *Dish, *Dishâ, *Kakubh, Karman, Lakâra, Pankti, *Râvanshiras.*

These terms translate or designate symbolically: 1. Descendances and incarnations (*Avatâra*); 2. Fingers (*Anguli*); 3. Horizons (*Dish, Dishâ, Âshâ, Kakubh*); 4. The heads of Râvana (*Râvanshiras*). See **Numerical symbols.**

TEN BILLION (= ten to power thirteen; in US expressed as "ten trillion"). See *Ananta, Kankara, Khamba, Makâkharva, Nikharva, Shankha, Shangku.* See also **Names of numbers.**

TEN MILLION (= ten to power seven). See *Arbuda. Koti.* See also **Names of numbers.**

TEN THOUSAND (= ten to power four). See *Ayuta, Dashashasra.* See also **Names of numbers.**

TEN THOUSAND MILLION (= ten to power ten; in US expressed as "ten billion"). See *Arbuda, Kharva, Madhya, Samudra.* See also **Names of numbers.**

TEN TRILLION (in British sense of ten to power nineteen; otherwise called "ten quadrillion"). See *Attata, Mahâshankha Vivaha.* See also **Names of numbers.**

TERRESTRIAL WORLD. [S]. Value = 1. See **One, Prithivî.**

THAI (THAI-KHMER) NUMERALS. Symbols derived from *Brâhmî numerals and influenced by Shunga, Shaka, Kushâna, Ândhra, Pallava, Châlukya, Ganga, Valabhî,

"Pâli" and Vatteluttu numerals. Currently used in Thailand, Laos and Cambodia (Kampuchea). The symbols correspond to a mathematical system that has place values and a zero (shaped like a small circle). See **Indian written numeral systems (Classification of)**. See Fig. 24.24, 52 and 24.61 to 69.

THÂKURÎ (Calendar). Calendar beginning in the year 595 CE. To find the corresponding date in the Common Era, add 595 to a date expressed in the *Thâkurî* calendar. Formerly used in Nepal. See **Indian calendars**.

THIRTEEN. Ordinary Sanskrit name: *trayodasha*. Here is a list of the corresponding numerical symbols: *Aghosha, Atijagatî, *Kâma, Manmatha, *Vishva, *Vishvadeva*.

These words have the following translation or symbolic meaning: 1. The god of carnal love and of cosmic desire (Kâma). 2. The universe comprised of thirteen worlds (*Vishva, Vishvadeva*).

See **Numerical symbols**.

THIRTY. Ordinary Sanskrit name: *trimshat*.

THIRTY-TWO. Ordinary Sanskrit name: *dvatrimshati*. The corresponding numerical symbols are: *Danta and *Rada. These words both mean "teeth". See **Numerical symbols**.

THIRTY-THREE. Ordinary Sanskrit word: *trâyastrimsha*. The corresponding numerical symbols are: *Amara, *Deva, *Sura, Tridasha*. These words have the following meaning: 1. The "gods" (Amara, Deva, Sura) 2. The "immortals", in allusion to the gods (Amara). See **Numerical symbols**.

THOUSAND. Ordinary Sanskrit name: *Sahasra*. Corresponding numerical symbols: *Arjunâkara, *Indradrishti, *Jâhnavîvaktra, *Ravibâna, *Sheshashîrsha*.

These terms name or refer to: 1. The mouth of the Ganges or Jâhnavî (*Jâhnavîvaktra*). 2. The arms of Arjuna (*Arjunâkara*). 3. The arrows of Ravi (*Ravibâna*). 4. The thousand-headed serpent (*Sheshashîrsha*). 5. The eyes of Indra (*Indradrishti*). See **Numerical Symbols**.

THOUSAND. In the sense of "many, a multitude of...". See *Jâhnavîvakta*. See also **High Numbers (Symbolic Meaning of)**.

THOUSAND. In the sense of infinity and eternity. See *Sheshashîrsha*.

THOUSAND. [S]. Value = 12. See *Sahasrâmshu*. **Twelve**.

THOUSAND MILLION. (= ten to power nine, known in US as "one billion"). See *Abja,*

Ayuta, Nahut, Nikharva, Padma, Samudra, Saroja, Shatakoti, Vâdava, Vrindâ. See also **Names of numbers**.

THOUSAND RAYS. [S]. Value = 12. See *Sahasrâmshu*. **Twelve**.

THREE. The ordinary Sanskrit names for this number are: *traya, *trai and *tri. Here is a list of corresponding word-symbols:

*Agni, *Anala, *Âptya, *Bhuvana, *Dahana, Dosha, Gangâmarga, *Guna, *Haranayana, *Haranetra, *Hotri, *Hutâshana, *Îshadrish, *Jagat, *Jvalana, *Kâla, *Krishânu, *Loka, *Mûrti, Nâdî, *Netra, *Pâvaka, *Pinâkanayana, *Purâ, *Râma, *Ratna, Sahodara, *Shakti, *Shankarâkshi, *Shikhin, *Shûla, *Tapana, *Trailokya, *Trayî, Trigata, *Triguna, *Trijagat, *Trikâla, *Trikâya, *Triloka, *Trimûrti, *Trinetra, *Tripura, *Tripurâ, *Triratna, *Trishiras, *Trivarga, *Trivarna, *Tryambaka, *Udarchis, *Vâchana, *Vahni, *Vaishvânara, *Veda, Vishtapa*.

These words have the following translation or symbolic meaning: 1. The god of fire (Agni). 2. "Fire", in allusion to the god of sacrificial fire (Agni, Anala, Dahana, Hotri, Hutâshana, Jvalana, Krishânu, Pâvaka, Shikhin, Tapana, Udarchis, Vahni, Vaishvânara). 3. "That which belongs to all humans" (Vaishvânara). 4. Ablaze (Shikhin). 5. The worlds, the universe (Bhuvana, Loka). 6. The three worlds (Triloka). 7. The phenomenal worlds (Jagat). 8. The three phenomenal world (Trijagat). 9. The "three letters", in allusion to the three sacred syllables (Trivarna). 10. The "aphorism" (Vâchana). 11. Feminine energies (Shakti). 12. The City-Fortresses (Pura). 13. The Three City, Fortresses (Tripura). 14. The "States", in allusion to the States of consciousness (Purâ). 15. The Three states of consciousness (Tripurâ). 16. The "forms" (Mûrti). 17. The three forms (Trimûrti). 18. The Jewels (Ratna). 19. The three Jewels (Triratna). 20. The "qualities", the "primordial properties" (Guna). 21. The "three primordial properties" (Triguna). 22. The Eye, in allusion to the "three eyes" (Netra). 23. The three eyes (Trinetra, Tryambaka). 24. The points (Shûla). 25. Time, in allusion to the "three times" (Kâla). 26. The three times (Trikâla). 27. The triple science (Trayî). 28. The three aims (Trivarga). 29. The three classes of beings (Trailokya). 30. The three bodies (Trikâya). 31. The three states (Tripurâ). 32. The spirit of the waters (Âptya). 33. The eyes of Shiva (Haranetra). 34. The god Shiva (Pinâkanayana). 35. "The one with three heads" (Trishiras). 36. The three Râmas (Râma). See **Numerical symbols**.

THREE AIMS. [S]. Value = 3. See *Trivarga* and **Three**.

THREE BODIES. [S]. Value = 3. See *Trikâya* and **Three**.

THREE CITY-FORTRESSES. [S]. Value = 3. See *Tripura* and **Three**.

THREE CLASSES OF BEINGS. [S]. Value = 3. See *Trailokya* and **Three**.

THREE EYES. [S]. Value = 3. See *Tryambaka* and **Three**.

THREE FORMS. [S]. Value = 3. See *Trimûrti* and **Three**.

THREE HEADS. [S]. Value = 3. See *Trishiras* and **Three**.

THREE JEWELS. [S]. Value = 3. See *Triratna* and **Three**.

THREE LETTERS. [S]. Value = 3. See *Trivarna* and **Three**.

THREE PRIMORDIAL PROPERTIES. [S]. Value = 3. See *Triguna* and **Three**.

THREE SACRED SYLLABLES. [S]. Value = 3. See *Trivarna* and *AUM*.

THREE STATES. [S]. Value = 3. See *Tripurâ* and **Three**.

THREE TIMES. [S]. Value = 3. See *Trikâla* and **Three**.

THREE UNIVERSES [S]. Value = 3. See *Jagat, Loka, Trijagat* and **Three**.

THREE WORLDS. [S]. Value = 3. See *Triloka* and **Three**.

TIBETAN NUMERALS. Symbols derived from *Brâhmî numerals and influenced by Shunga, Shaka, Kushâna, Ândhra, Gupta, Nâgarî and Siddham numerals. Used in areas of Tibet since the eleventh century CE. The symbols correspond to a mathematical system that has place values and a zero (shaped like a small circle). However, it was not always thus: many Tibetan manuscripts show that a structure identical to the archaic Brâhmî system was used in former times. See **Indian written numeral systems (Classification of)**. See Fig. 24.16, 52 and 24.61 to 69.

TILAKA. "Sesame". Name given to the dot that Hindus stick to their foreheads which represents the third eye of *Shiva, the eye of knowledge. See **Poetry, zero and positional numeration**.

TIME. [S]. Value = 3. See *Kâla, Trikâla* and **Three**.

TITHI. Unit of time used in Babylonian tablets which corresponds to a thirtieth of a synodic revolution of the Moon. This length of time is approximately the same as a day or *nychthemer*. See **Indian astronomy (History of)**.

TITHI. [S]. Value = "Day". 15. Allusion to the 15 days of each *paksha* of the month. See *Tithi* and **Fifteen**.

This symbol is notably found in *Varâhamihira:* PnSi, VIII, line 4; Dvivedi and Thibaut (1889); Neugebauer and Pingree (1970–71).

TITILAMBHA. Name given to the number ten to the power twenty-seven. See **Names of numbers** and **High numbers**. Source: *Lalitavistara Sûtra (before 308 CE).

TONE. [S]. Value = 49. See *Tâna*.

TOTAL. [Arithmetic]. See *Sarvadhana*.

TRAI. (TRAYA, TRI). Ordinary Sanskrit terms for the number three which form part of several words which are directly related to the number in question.

Examples: *Trailokya, *Trairâshika, *Trayî, *Triambaka, *Tribhuvana, *Tribhûvaneshvara, *Trichîvara, *Triguna, *Trijagat, *Trikâla, *Trikâlajñâna, *Trikândî, *Trikâya, *Triloka, *Trimûrti, *Trinetra, *Tripitaka, *Tripura, *Tripurâ, *Tripurasundarî, *Triratna, *Trishiras, *Trishûlâ, *Trivamsha, *Trivarga, *Trivarna, *Trivenî, *Trividyâ, *Tryambaka.

For words which are symbolically associated with this number, see **Three** and **Symbolism of numbers**.

TRAILOKYA. [S]. Value = 3. "Three classes". This name denotes the three classes of beings envisaged by Hindu and Buddhist philosophies: the *kâmadhâtu*, beings evolving in desire; the *rûpadhâtu*, those of the world of forms; and the *arûpadhâtu*, those of the world of the formless. See *Trai* and **Three**.

TRAIRÂSHIKA. [Arithmetic]. Sanskrit name for the Rule of Three. See *Trai.*

TRAYA. Ordinary Sanskrit name for the number three. See *Trai.*

TRÂYASTRIMSHA. Ordinary Sanskrit name for the number thirty-three. For words which are symbolically associated with this number, see **Thirty-three**, *Deva* and **Symbolism of numbers**.

TRAYÎ. [S]. Value = 3. "Triple science". Allusion to the *Samhitâ (Rigveda, Yajurveda, Sâmaveda)*, who are the three first *Vedas. See *Trai, Veda* and **Three**.

TRAYODASHA. Ordinary Sanskrit name for the number thirteen. For words which are symbolically associated with this number, see **Thirteen** and **Symbolism of numbers**.

TRETÂYUGA. Name of the second of the four cosmic eras which make up a *mahâyuga. This cycle, which is said to be worth 1,296,000 human years, is regarded as the age during which human beings would live no more than three quarters of their life. See *Mahâyuga* and *Yuga*.

TRI. Ordinary Sanskrit word for the number three. See *Trai*.

TRIAMBAKA. "With three eyes". See *Tryambaka*.

TRIBHUVANA. Name of the "three worlds" of Hindu cosmogony: the skies (*svarga*), the earth (**bhûmi*) and the hells (**pâtâla*). See *Trai*.

TRIBHÛVANESHVARA. "Lord of the three worlds". One of the titles attributed to *Shiva, *Vishnu and *Krishna. See *Trai*.

TRICHÎVARA. "Three garments". Term denoting the ritual costume comprising the loincloth, sash and robe worn by Buddhist monks of the schools of the South (Hînayâna, Theravâda). See *Trai*.

TRIGUNA. [S]. Value = 3. "Three primordial properties", "three primary forces". Symbolism which corresponds to the representation of the group Vishnu-Sattva, Brahma-Rajas and Rudra-Tamas, this group being thus composed of the energies which personify the three main divinities of the Brahmanic pantheon. See *Guna*, **Brahma, Vishnu, Shiva** and **Three**.

TRIJAGAT. [S]. Value = 3. "Three universes". See *Jagat, Triloka* and **Three**.

TRIKÂLA. [S]. Value = 3. "Three times". Allusion to the three divisions of time: the past, the present and the future. See *Kâla* and **Three**.

TRIKÂLAJÑÂNA. From **trikâla*, "three times", "three eras", and from *jñâna*, "knowledge". Name denoting the magic and occult power which is given to the **Siddhi* to enable them to know the past, the present and the future all at once. See *Kâla, Trikâla* and *Trai*.

TRIKÂNDÎ. "Three chapters". This name is sometimes given to the *Vâkyapadîya* of Bhartrihari, famous text of "grammatical philosophy" divided into three *kânda* or "chapters". See *Trai*.

TRIKÂYA. [S]. Value = 3. "Three bodies". Allusion to the three bodies that a Buddha may assume simultaneously: the "body of the Law" (*dharmakâya*), the "body of enjoyment" (*sambhogakâya*) and the "body of magical creation or transformation" (*nirmânakâya*). See **Three**.

TRILLION. See *Akshiti, Antya, Madhya, Mahâpadma, Viskhamba, Vivara* and **Names of numbers**.

TRILOKA. [S]. Value = 3. "Three worlds". In allusion to the worlds of Hindu cosmogony: the Skies (*svarga*), the earth (**bhûmi*) and the hells (**pâtâla*). See **Three**.

TRIMSHAT. Ordinary Sanskrit name for the number thirty.

TRIMÛRTI. [S]. Value = 3. "Three forms". See *Mûrti* and **Three**.

TRINETRA. [S]. Value = 3. "Three eyes". See *Haranetra* and **Three**.

TRIPITAKA. "Three baskets". Term denoting the Buddhist Law written in Sanskrit which constitutes the sacred Scriptures of this religion. The allusion is to the three different baskets into which the three principal compilers placed the three fundamental Buddhist texts: the *vinâyapitaka*, which deals with monastic discipline; the *sûtrapitaka* and the *abhidharmapitaka* which deals with Buddha's teaching [see K. Friedrichs, etc, (1989)]. See *Trai*.

TRIPLE SCIENCE. [S]. Value = 3. See *Trayî* and **Three**.

TRIPURA. [S]. Value = 3. Literally: "Three City-fortress". Name of a triple fortress-town (or triple rampart) which was built by the **Asura* and destroyed by Shiva. See *Pura* and **Three**.

TRIPURÂ. [S]. Value = 3. Literally: "three states". Name which collectively denotes the three states of consciousness of Hinduism. See *Purâ* and **Three**.

TRIPURASUNDARÎ. "Beauty of the three cities". One of the names given to *Pârvatî, the "mountain dweller", daughter of Himâlaya, sister of *Vishnu and wife of *Shiva. See *Trai*.

TRIRATNA. [S]. Value = 3. "Three jewels". See *Ratna* and **Three**.

TRISHATIKÂ. See *Shrîdharâchârya*.

TRISHIRAS. [S]. Value = 3. "He who has three heads". This is the name of the demon with three heads, younger brother of *Râvana, who, according to the legend of **Râmâyana*, was killed by *Râma. See *Râvana* and **Three**.

TRISHÛLA. "Three points". Name of *Shiva's Trident. See *Shûla* and *Trai*.

TRIVAMSHA. Name which collectively denotes the three principal castes of Brahmanism (namely: the Brahmans, the *kshatriya* and the *vaishya*). See *Trai*.

TRIVARGA. [S]. Value = 3. "Three aims". This is an allusion to the three objectives of human existence according to Hindu philosophy, namely: material wealth (*artha*), love with desire (**kâma*) and duty (**dharma*). See **Three**.

TRIVARNA. [S]. Value = 3. "Three letters". This refers to the letters A, U and M of the Indian alphabet, which spell AUM, the sacred Syllable of the Hindus, which means something approximating "I bow". This represents all of the following at once: the divine Word in an audible form; the fullblown *Brahman; the Fire of the Sun; the Unity; the Cosmos; the Immensity of the Universe; the past; the present; the future; as well as all Knowledge. According to Hindu religion, AUM contains the very essence of all the sounds that have been, that are, and that will be made, and within it is reunited the three great powers of the three great divinities of the Brahmanic pantheon [see Frédéric (1987)]. See *AUM, Akshara,* **Mysticism of letters,** *Trai* and **Three.**

TRIVENÎ. "Three rivers". Name sometimes given to the town of Prayâga (now Allâhâbâd) where the following three rivers are said to meet: the Ganges, the Yamunâ and the mythical Sarasvatî. See *Trai.*

TRIVIDYÂ. Name given to the "three axioms" of Buddhist philosophy: *anitya,* the impermanence of all things; *dukha,* pain, suffering; and *anâtmâ,* the non-reality of existence. See *Trai.*

TRIVIMSHATI. Ordinary Sanskrit name for the number twenty-three. For words which are symbolically connected with this number, see **Twenty-three** and **Symbolism of numbers.**

TRUE NATURE. [S]. Value = 5. See *Tattva* and **Five.**

TRUE NATURE. [S]. Value = 7. See *Tattva* and **Seven.**

TRUE NATURE. [S]. Value = 25. See *Tattva* and **Twenty-five.**

TRUTH. [S]. Value = 5. See *Tattva* and **Five.**

TRUTH. [S]. Value = 7. See *Tattva* and **Seven.**

TRUTH. [S]. Value = 25. See *Tattva* and **Twenty-five.**

TRYAKSHAMUKHA. [S]. Value = 5. Synonymous with *Rudrâsya,* "faces of *Rudra".* See **Five.**

TRYAMBAKA. [S]. Value = 3. "With three eyes", "with three sisters". Epithet given to many Hindu divinities, especially Shiva . See *Haranetra, Traya* and **Three.**

TURAGA. [S]. Value = 7. "Horse". See *Ashva* and **Seven.**

TURANGAMA. [S]. Value = 7. "Horse". See *Ashva* and **Seven.**

TURÎYA. [S]. Value = 4. "Fourth". Epithet occasionally given to the Brahman who transcends the three states of consciousness. See *Tripurâ* and **Four.**

TWELVE. Ordinary Sanskrit name: *dvâdasha.*

TWENTY. Ordinary Sanskrit name: *vimshati.* Here is a list of corresponding numerical symbols: *Anguli, *Kriti, *Nakha, *Râvanabhuja. These words express: 1. The arms of Râvana (*Râvanabhuja*). 2. The fingers (*Anguli*). 3. The nails (*Nakha*). 4. An element of Indian metrication (*Kriti*). See **Numerical symbols.**

TWENTY-ONE. Ordinary Sanskrit name: *ekavimshati.* Corresponding numerical symbols: *Prakriti, Svaga* ("heaven"), *Utkriti.*

TWENTY-TWO. Ordinary Sanskrit name: *dva vimshati.* Corresponding numerical symbols: *Akriti, Jâti* ("Caste"), *Kritin.*

TWENTY-THREE. Ordinary Sanskrit name: *trayavimshati* (or *trivimshati*). Corresponding numerical symbol: *Vikriti.*

TWENTY-FOUR. Ordinary Sanskrit name: *chaturvimshati.* Corresponding numerical symbols: *Arhat, *Gâyatrî, Jina, Siddha.*

TWENTY-FIVE. Ordinary Sanskrit name: *panchavimshati.* Corresponding numerical symbol: *Tattva.* This word expresses: 1. The fundamental principles. 2. The "true natures". 3. The realities. 4. The truths.

TWENTY-SIX. Ordinary Sanskrit name: *shadvimshati.* Corresponding numerical symbol: *Utkriti.*

TWENTY-SEVEN. Ordinary Sanskrit name: *saptavimshati.* Corresponding numerical symbols: *Bha, *Uda, *Nakshatra. These words express or symbolise: 1. The "stars" (*Bha, Uda*). 2. The "lunar mansions" (*Nakshatra*). 3. The constellations (*Nakshatra*).

TWO. Ordinary Sanskrit names: *dva, dve, dvi. Corresponding numerical symbols: *Akshi, Ambaka, *Ashvin, *Ashvina, *Ashivinau, Ayana, *Bâhu, *Chakshus, *Dasra, *Drishti, *Dvandva, *Dvaya, *Dvija, Grahana, *Gulpha, Îshana, Janghâ, Jânu, *Kara, Karna, Kucha, Kutumba, *Lochana, Nadikulâ, *Nâsatya, Naya, *Nayana, *Netra, Oththa, *Paksha, Râmananddana, Ravichandra, Vishuvat, *Yama, *Yamala, *Yamau, *Yuga, *Yugala, *Yugma.

These terms symbolically refer to or designate: 1. Twins. pairs or couples (*Ashvin, Ashvina, Ashvinau, Dasra, Dvandva, Dvaya, Dvija, Nâsatya, Ravichandra, Yama, Yamala, Yugala, Yugma*). 2. Symmetrical organs (*Bâhu,*

Gulpha, Kara, Nayana, Netra, Paksha). 3. Wings (*Paksha*). 4. Arms (*Bâhu*). 5. The Horsemen (*Ashvin, Ashvina, Ashvinau*). 6. Ankles (*Gulpha*). 7. The conception of the world, contemplation, revelation, theory (*Drishti*). 8. The primordial couple (*Yama*). 9. The epithet "twice born" (*Dvija*). 10. The twin gods (*Ashvin, Dasra, Nâsatya*). 11. The hand (*Kara*). 12. The pair (*Dvaya*). 13. The Sun-Moon couple (*Ravichandra*). 14. The eye (*Netra, Chakshus*). 15. Eyes (*Lochana*). 16. Vision (*Drishti*). 17. The third age of a *mahâyuga* (*Yuga*). See **Numerical symbols**.

U

UCHCHAISHRAVAS. [S]. Value = 1. This is the name of a wonderful white horse which, according to Brahmani and Hindu mythology, came from the "churning of the sea of milk" and which Indra appropriated. He is considered to be the ancestor of all horses, thus the symbolism in question. See **One**.

UDA. [S]. Value = 27. "Star". This is an allusion to the twenty-seven **nakshatra*. See *Nakshatra* and **Twenty-seven**.

UDADHI. [S]. Value = 4. "Ocean". See *Sâgara*, **Four** and **Ocean**.

UDARCHIS. [S]. Value = 3. "Fire". See *Agni* and **Three**.

UJJAYINÎ. Town situated in the extreme west of what is now the state of Madhya Pradesh. It defines the first meridian of Indian astronomy. See **Indian astronomy (History of)** and **Yuga (Astronomical speculation on)**.

ULTIMATE REALITY. [S]. Value = 1. See *Âtman* and **One**.

UNIQUE REALITY. [S]. Value = 1. See *Âtman* and **One**.

UNIVERSE. [S]. Value = 13. See *Vishva*, *Vishvada* and **Thirteen**.

UPPALA. Pali word which literally means: "(blue) lotus flower (half open)". Name given to the number ten to the power ninety-eight. See **Names of numbers**. For an explanation of this symbolism, see **Lotus** and **High numbers (Symbolic meaning of)**.

Source: **Vyâkarana* (Pali grammar) by Kâchchâyana (eleventh century CE).

URVARÂ. [S]. Value = 1. "Earth". See *Prithivî*.

UTKRITI. [S]. Value = 26. In Sanskrit poetry, this is a metre of four lines of twenty-six syllables per stanza. See **Indian metric**.

UTPALA. Literally: "(blue) lotus flower (half open)". In Hindu and Buddhist philosophies, this lotus (which is never represented in full bloom) notably symbolises the victory of the mind over the body. Name given to the number ten to the power twenty-five. See **Names of numbers**. For an explanation of this symbolism, see **Lotus** and **High numbers (Symbolic meaning of)**.

Source: **Lalitavistara Sûtra* (before 308 CE).

UTSANGA. Name given to the number ten to the power twenty-one (= quintillion). See **Names of numbers** and **High numbers**.

Source: **Lalitavistara Sûtra* (before 308 CE).

V

VÂCHANA. [S]. Value = 3. "Aphorism". From *vâch*, "voice", "speech", "spoken word", and form *annâ*, "nourishment". This is an allusion to the creative and evocative power of sound and acoustic resonance (especially through speech) and to its "indestructible and imperishable" nature, which correspond to the revelation of the **Brahman*, which is said to be resumed in the three letters of the sacred Syllable **AUM*. See *Akshara, Trivarna* and **Three**.

VACUITY. See *Shûnya, Shûnyatâ*, **Zero, Zero (Graeco-Latin concepts of)**, **Zero (Indian concepts of)** and **Zero and Sanskrit poetry**.

VÂDAVA. Name given to the number ten to the power nine. See **Names of numbers** and **High numbers**.

Source: **Kâthaka Samhitâ* (from the start of the Common Era).

VÂDAVA. Name given to the number ten to the power fourteen. See **Names of numbers** and **High numbers**.

Source: **Pañchavimsha Brâhmana* (date uncertain).

VAHNI. [S]. Value = 3. "Fire". See *Agni* and **Three**.

VAIKUNTHA. Celestial home of **Vishnu* and **Krishna*. See *Bhuvana*.

VAISHESHIKA. See *Darshana*.

VAISHVÂNARA. [S]. Value = 3. "that which belongs to all humans". This is one of the Vedic names for **Agni* (= 3), the god of sacrificial fire, who is said to possess the powers of fire, lightning and light. See *Agni* and **Three**.

VÂJASANEYÎ SAMHITÂ. This is a text which forms part of the *Yajurveda* "white", which is one of the oldest Vedic texts. Passed down through oral transmission since ancient times, it only found its definitive form at the beginning of Christianity. See **Veda**. Here is a list of the main names of numbers mentioned in this text:

Eka (= 1), *Dasha* (= 10), *Sata* (= 10^2), *Sahasra* (= 10^3), *Ayuta* (= 10^4), *Niyuta* (= 10^5), *Prayuta* (= 10^6), *Arbuda* (= 10^7), *Nyarbuda* (= 10^8), *Samudra* (= 10^9), *Madhya* (= 10^{10}), *Anta* (= 10^{11}), *Parârdha* (=10^{12}). See **Names of numbers** and **High numbers**.
[See: *Vâjasaneyî Samhitâ*, XVII, 2; Datta and Singh (1938), p. 9; Weber, in: JSO, XV, pp. 132–40; Woepcke (1863).]

VÂJIN. [S]. Value = 7. "Horse". See *Ashva* and **Seven**.

VAJRA. In Hindu and Buddhist philosophies, the *vajra* is the "diamond" that symbolises all that remains when appearances have disappeared. Thus it is the vacuity (*shûnyatâ*) that is as indestructible as a diamond. It is also the missile "with a thousand points", which is said to never miss its target, and made out of bronze by Tvashtri for *Indra. This weapon is a symbol of *linga and divine power. It also indicates a strong, stable and indestructible mind. As a word-symbol, *vajra* has several meanings: the weapon is usually a short bronze baton, which has three, five, seven or nine points at each end. With three points, for example, *vajra* symbolises: the *triratna (or "three jewels" of Buddhism); time in its three tenses (*trikâla); the three aspects of the world (*tribhuvana); etc. [see Frédéric, *Dictionnaire* (1987)]. See *Shûnyatâ* and **Symbols**.

VALABHI NUMERALS. Symbols derived from *Brâhmî numerals and influenced by Shunga, Shaka, Kushâna, Ândhra, Pallava, Châlukya, and Ganga numerals. The system arose at the time of the inscriptions of Valabhi, the capital city of a Hindu-Buddhist kingdom that ruled over present-day Gujarat and Maharastra. The symbols correspond to a mathematical system that is not based on place values and therefore does not possess a zero. See **Indian written numeral systems (Classification of)**. See Fig. 24.44, 52 and 24.61 to 69.

VÂRA. [S]. Value = 7. "Day of the week". This is because of the seven days: *ravivâra* or *âdivâra* (Sunday), *induvâra* or *somavâra* (Monday), *mangalavâra* (Tuesday), *budhavâra* (Wednesday), *brihaspativâra* (Thursday), *shukravâra* (Friday), and *shanivâra* (Saturday). See **Seven**.

VARÂHAMIHÎRA. Indian astronomer and astrologer c. 575 CE. His works notably include *Panchasiddhântikâ* (the "Five *Siddhântas*"), where

there are many examples of the place-value system [see Neugebauer and Pingree (1970–71)]. See **Indian astrology**, **Indian astronomy (History of)** and **Indian mathematics (History of)**.

VARGA. Word used in arithmetic to denote the squaring operation. Synonym: *kriti*. In algebra, the same word is used for the square, in allusion to cubic equations. See *Ghana*, *Varga-Varga* and *Yâvattâvat*.

VARGAMÛLA. Word used in arithmetic to describe the extraction of the square root. See *Pâtîganita*, **Indian methods of calculation** and **Square roots (How Âryabhata calculated his)**.

VARGA-VARGA. Algebraic word for quadratic equations.

VÂRIDHI. [S]. Value = 4. "Sea". See *Sâgara*, **Four** and **Ocean**.

VÂRINIDHI. [S]. Value = 4. "Sea". See *Sâgara*, **Four** and **Ocean**.

VARNA. Literally "letter", in mathematics "symbol". See *AUM*, *Bîja* and *Bîjaganita*.

VARNASAMJÑA. "Syllable system". Name that Haridatta gave to the *katapaya system.

VARNASANKHYÂ. Literally: "letter-numbers". This word denotes any system of representing numbers which uses the vocalised consonants of the Indian alphabet, each one being assigned a numerical value. See **Numeral alphabet**.

VARUNA. Vedic and Hindu divinity of water, the sea and the oceans. See **High numbers (Symbolic meaning of)**.

VASU. [S]. Value = 8. Name in the *Mahâbhârata* which is given to a group of eight divinities, who are meant to correspond, philosophically speaking, to the eight "spheres of existence" of the *Adibhautika*, which in turn represent the visible forms of the laws of the universe. See **Eight**.

VASUDHÂ. [S]. Value = 1. "Earth". See *Prithivî* and **One**.

VÂSUKI. In Brahmanic mythology, this is the name given to the king of the *naga. He is said to have been used by the *deva (the gods) and the *asura (the anti-gods) as a "rope" with which to spin *Mount Meru on its axis in order to churn the sea of milk and thus extract the "nectar of immortality" (*amrita). See **Serpent (Symbolism of the)**.

VASUNDHARÂ. [S]. Value = 1. "Earth". See *Prithivî* and **One**.

VATTELUTTU NUMERALS. Symbols derived from *Brâhmî numerals and influenced by Shunga, Shaka, Kushâna, Ândhra, Pallava, Châlukya, and Ganga, Valabhî, Bhattiprolu and Grantha numerals as well as by Ancient Tamil. Used from the eighth to the sixteenth centuries CE in the Dravidian areas of South India, particularly the Malabar coast. The symbols correspond to a mathematical system that is not based on place values and therefore does not possess a zero. See: **Indian written numeral systems (Classification of)**. See Fig. 24.52 and 24.61 to 69.

VÂYU. "Wind". This is a name for the god of the wind. Other names include: *Marut* ("Immortal"), *Anila* ("Breath of life"), *Vâta* ("Wandering", "He who is in perpetual movement") or *Pâvana* ("Purifier"). According to Brahmanic and Hindu cosmogonies, he is one of the eight *Dikpâla (divinities who guard the horizons and points of the compass), whose task is to guard the northwest "horizon".

VÂYU. [S]. Value = 49. "Wind". This symbolism can be explained by reference to tales of Brahman mythology. One day Vâyu revolted against the *deva, the "gods" who live on the peaks of *Mount Meru. He decided to destroy the mountain, and unleashed a powerful hurricane. However, the mountain was protected by the wings of Garuda, the carrier-bird of *Vishnu, which rendered all the assaults of the wind ineffectual. One day, in Garuda's absence, Vâyu chopped off the peak of *Mount Meru, and threw it into the ocean. That is how the island of Sri Lanka was created. Mount Meru was meant to be the place where the gods lived and met. It was said to be situated at the centre of the universe, above the seven *pâtâla (or "inferior worlds"); it has seven faces, each one facing one of the seven *dvîpa (or "island-continents) and the seven *sâgara ("*oceans"). When Vâyu attacked the mountain, he created seven strong winds, one for each face. Once the summit of the sacred mountain had been rased, the seven winds, thus placed at the centre of the universe and no longer encountering any barrier, each went to one of the seven continents and the seven oceans. Thus: *Vâyu* = 7 × 7 = 49. See other entry entitled *Vâyu*.

VEDA. Name of the oldest sacred texts of India, they are made up of four principal books (namely: the *Rigveda*, the *Yajurveda*, the *Sâmaveda*, and the *Atharvaveda*). These texts and those of derived literature probably date back to ancient times in the history of India. But it is impossible to date them exactly, because they were primarily transmitted orally before being transcribed at a later date. In fact, it is only possible to give them a chronological position in relation to each other. The three *Samhitâ* (the texts of the *Rigveda*, the *Yajurveda* and the *Sâmaveda*) seem to have been composed first. As for the fourth *Veda*, (the *Atharvaveda*), it was followed by the *Brâhmana*, the *Kalpasûtra*, and lastly by the *Âranyaka* and the *Upanishad* [see Frédéric, *Dictionnaire* (1987)].

VEDA. [S]. Value = 3. (Very rarely used as a numerical symbol). The allusion here is probably to the three *Samhitâ* (the *Rigveda*, the *Yajurveda* and the *Sâmaveda*), which constitute the first three texts of the *Veda*. See *Trayî* and **Three**.

VEDA. [S]. Value = 4. (The most frequent value of this word as a numerical symbol.) Here the allusion is to the four principal books of which the *Veda* is composed (the *Rigveda*, the *Yajurveda*, the *Sâmaveda*, and the *Atharvaveda*). See **Four**.

VEDÂNGA. [S]. Value = 6. "Members of the *Veda". Group of six Vedic and Sanskrit texts dealing principally with the Vedic ritual, its conservation and its transmission. See *Darshana*.

VEDIC RELIGION. See **Indian religions and philosophies**.

VIBHUTANGAMÂ. Name given to the number ten to the power fifty-one. See **Names of numbers** and **High numbers**.

Source: *Lalitavistara Sûtra* (before 308 CE).

VIDHU. [S]. Value = 1. "Moon". See *Abja* and **One**.

VIKALPA. Word used in mathematics since the eighth century to designate "permutations" and "combinations".

VIKRAMA. (Calendar). Formerly used in the centre, west and northwest of India. Also called *vikramâdityakâla, vikramasamvat*, or quite simply *samvat*. It began in the year 57 BCE. To find an approximate corresponding date in the Common Era, subtract 57 from a date in the *Vikrama* calendar.

VIKRAMÂDITYAKÂLA (Calendar). See *Vikrama*.

VIKRAMASAMVAT (Calendar). See *Vikrama*.

VIKRITI. [S]. Value = 23. In Sanskrit poetry, this is the metre of four times twenty-three syllables per stanza. See **Indian metric**.

VILÂYATÎ (Calendar). Solar calendar commencing in the year 592 CE. Used in

Bengal and Orissa. To find a date in the Common Era, add 592 to a date expressed in the *Vilâyatî* calendar. See **Indian calendars.**

VIMSHATI. Ordinary Sankrit name for the number twenty. For words which have a symbolic relationship with this number, see **Twenty** and **Symbolism of numbers.**

VINDU. [S]. Value = 0. **Prâkrit* word which has the literal meaning and symbolism of **bindu.* See **Zero.**

VIOLENT. [S]. Value = 11. See **Rudra-Shiva** and **Eleven.**

VÎRASAMVAT (Calendar). Commencing in the year 527 BCE, it is only used in **Jaina texts. To find a corresponding date in the Common Era, subtract 527 from a date expressed in this calendar. See **Indian calendars.**

VISAMJÑAGATI. Name given to the number ten to the power forty-seven. See **Names of numbers** and **High numbers.**

Source: **Lalitavistara Sûtra* (before 308 CE).

VISHAYA. [S]. Value = 5. "Sense, sense organ". See *Shara* and **Five.**

VISHIKHA. [S]. Value = 5. "Arrow". See *Shara* and **Five.**

VISHNU. Name of one of the three major divinities of the Brahmanic and Hindu pantheon (*Brahma, *Vishnu, *Shiva). See *Vishnupada, Pûrna* and **High numbers (Symbolic meaning of).**

VISHNUPADA. [S]. Value = 0. Literally: "foot of Vishnu", and by extension (and characteristically of Indian thought): "zenith", "sky". This is an allusion to the "Supreme step of Vishnu", the zenith, which denotes *Mount Meru, home of the blessed. The symbolism in question also refers to the "Three Steps of Vishnu", symbols of the rising, apogee and setting of the sun, which allowed him to measure the universe. It is also from the "feet of Vishnu" that, according to Hindu mythology, the sacred *Gangâ* (the Ganges) springs and, before it divides into terrestrial rivers, has its source at the summit of Mount Meru (which is situated at the centre of the universe and over which are the heavens or "worlds of Vishnu"). Vishnu rests upon *Ananta, the serpent with a thousand heads who floats on the primordial waters and the "ocean of unconsciousness", during the time that separates two creations of the universe. Thus this symbolism corresponds to the connection in Indian philosophy between

infinity and zero, because Ananta is the serpent of infinity, eternity and of the immensity of space. Space also means sky, which is considered to be the "void" which has no contact with material things. Thus Vishnu is identified with ether (**âkâsha*), an immobile, eternal and indescribable space. In other words, Vishnu is synonymous with vacuity (**shûnyatâ*). See *Abhra, Âkâsha, Kha, Ananta,* **Zero** and **Zero (Indian concepts of).**

VISHVA. [S]. Value = 13. Contraction of the word *Vishvadeva* and a symbol for the number 13. See *Vishvadeva* and **Thirteen.**

VISHVADEVA. [S]. Value = 13. This is an allusion to the universe formed by thirteen paradises or chosen lands (**bhuvana*), and does not include the **vaikuntha.* See *Bhuvana, Vaikuntha* and **Thirteen.**

VISION. [S]. Value = 2. See *Drishti* and **Two.**

VISION. [S]. Value = 6. See *Darshana* and **Six.**

VISKHAMBA. Name given to the number ten to the power fifteen. See **Names of numbers** and **High numbers.**

Source: **Lalitavistara Sûtra* (before 308 CE).

VIVAHA. Name given to the number ten to the power nineteen. See **Names of numbers** and **High numbers.**

Source: **Lalitavistara Sûtra* (before 308 CE).

VIVARA. Name given to the number ten to the power fifteen. See **Names of numbers** and **High numbers.**

Source: **Lalitavistara Sûtra* (before 308 CE).

VOICE. [S]. Value = 3. See *Vâchana* and **Three.**

VOID. [S]. Value = 0. See *Shûnya, Shûnyatâ* and **Zero.**

VRINDÂ. Name given to a plant which is similar to basil, the leaves of which are said to have the power to purify the body and mind. It is believed to be an incarnation of Vishnu: according to the legend, Vrindâ was the wife of a Titan then was seduced by Vishnu. She cursed her husband and transformed him into a *shâlagrâma* stone before killing herself by throwing herself onto a fire of live coals; the plant (still called *tulasî*) was born out of the ashes. See *Ananta, Vishnupada, Samudra,* **Names of numbers** and **High numbers.**

VRINDÂ. Name given to the number ten to the power nine. See **Names of numbers** and **High numbers.** For an explanation of this symbolism, see *Vrindâ* (first entry) and **High**

numbers (Symbolic meaning of).

Source: *Âryabhatîya* (510 CE).

VRINDÂ. Name given to the number ten to the power seventeen. See **Names of numbers** and **High numbers.** For an explanation of this symbolism, see *Vrindâ* (first entry) and **High numbers (symbolic meaning of).**

Source: **Kâmâyana* by Vâlmikî (early centuries CE).

VULVA. [S]. Value = 4. See *Yoni* and **Four.**

VYÂKARANA. See *Kâchchâyana.*

VYAKTAGANITA. Name for arithmetic (literally: "science of calculating the known"), as opposed to algebra, which is called **Avyaktaganita.*

VYANT. [S]. Value = 0. "Sky". The symbolism can be explained by the fact that the sky (or heaven) is the "void" in Indian beliefs. See *Shûnya* and **Canopy of heaven.**

VYARBUDA. Name given to the number ten to the power eight (= hundred million). See **Names of numbers** and **High numbers.**

Source: *Kitab fi tahqiq i ma li'l hind* by al-Biruni (c. 1030 CE).

VYASTATRAIRÂSHIKA. [Arithmetic]. Name of the inverse of the Rule of Three. See *Trairâshika.*

VYAVAHÂRA. Literally: "procedure". Term used in algebra (since the seventh century CE) to denote the solving of equations.

VYAVAKALITA. [Arithmetic]. Sanskrit term for subtraction. Literally: "taken away".

VYAVASTHÂNAPRAJÑAPATI. Name given to the number ten to the power twenty-nine. See **Names of numbers** and **High numbers.**

Source: **Lalitavistara Sûtra* (before 308 CE).

VYOMAN. [S]. Value = 0. Word meaning "sky", "space". See **Zero** and *Shûnya.*

VYUTTKALITA. [Arithmetic]. Sanskrit term for subtraction. See *Vyavakalita.*

W

WAYS OF REBIRTH (The four). See *Chaturyoni* and *Yoni.*

WEEK. [S]. Value = 7. See *Vâra* and **Seven.**

WHEEL. [S]. Value = 12. See *Chakra, Râshi* and **Twelve.**

WHITE LOTUS. As a representation of the numbers ten to power twenty-seven and ten to power 112. See **Pundarika, High Numbers (Symbolic meaning of).**

WIND. [S]. Value = 5. See *Pâvana.*

WIND. [S]. Value = 7. See *Pâvana.*

WIND. [S]. Value = 49. See *Vâyu.*

WING [S]. Value = 15. See *Paksha.* **Fifteen.**

WING. [S]. Value = 2. See *Paksha.* **Two.**

WORLD. [S]. Value = 3. See **Bhuvana.**

WORLD. [S]. Value = 7. See **Loka.**

WORLD. [S]. Value = 14. See **Bhuvana.**

Y

YAMA. [S]. Value = 2. "Primordial couple". Allusion to the couple in Hindu mythology, formed by *Yama* (the first mortal who became god of death) and *Yamî*, his twin sister, wife and his feminine energy (**shakti*). See **Two.**

YAMALA. [S]. Value = 2. Synonym of **Yama.* See **Two.**

YAMAU. [S]. Value = 2. Synonym of **Yama.* See **Two.**

YÂVATTÂVAT. Literally : "as many as". Term used in algebra to denote the "equation" in general.

YONI. [S]. Value = 4. "Vulva". Allusion to the four lips that form the entrance of the vulva. By extension, the word also means "birth". Here, the reference is to the **Chaturyoni* which, according to Hindus and Buddhists, correspond to the "four ways of rebirth". According to this philosophy, there are four different ways to enter the cycle of rebirth (**samsâra*) : either through a viviparous birth (*jarâyuva*), in the form of a human being or mammal; or an oviparous birth (*andaja*), in the form of a bird, insect or reptile; or by being born in water and humidity (*samsvedaja*), in the form of a fish or a worm; or even through metamorphosis (*aupapâduka*), which means there is no "mother" involved, just the force of *Karma.* See **Four.**

YUGA (Definition). "Period". Generic names for the cosmic cycles of Indian speculations which are either based upon Brahmanic cosmogny or the learned astronomy founded by **Âryabhata.* The principal cycle is the **mahâyuga* (or "great period") made up of 4,320,000 human years. This is divided into four successive *yugas.* Thus the words **mahâyuga* and **chaturyuga* (literally : four periods) are treated as synonymous. These four successive ages are named respectively : **kritayuga* (or **satyayuga*), **tretâyuga, *dvaipâyanayuga* (or **dvâparayuga*), and **kaliyuga.* The corresponding lengths can considered to be equal or unequal depending on which system of calculation is used. See other entries entitled *Yuga.*

YUGA (Astronomical speculation on). Since its emergence at the start of the sixth century CE, *learned Indian astronomy* has been marked by its amazing speculation about the cosmic cycles (known as **yugas*), which is very different from the cosmogonical speculations. See *Yuga* (Definition) and *Yuga* (Systems of calculating).

According to this speculation, directly linked to astronomical elements, the **chaturyuga* or cycle of 4,320,000 years is the period at the beginning and end of which the nine elements (namely the sun, the moon, their apsis and node and the planets) are in mean perfect conjunction at the starting point of the longitudes. Thus the durations of the revolutions, previously considered to be the same lengths, are (in this astronomy) subjected to common multiples and general conjunctions. See Indian astronomy (History of) and Indian mathematics (History of). This speculation seems so audacious because it is obviously devoid of any physical meaning.

As for the cycle called **kalpa*, which constitutes an even longer period of time of 4,320,000,000 years, it is delimited, according to **Brahmagupta* (628 CE), by two perfect conjunctions in true longitude of the totality of elements, themselves each matched by a total eclipse of the Sun on the stroke of six in the secular time in **Ujjayinî*. In practice, however, these fictional eras can be reduced to the age of the **kaliyuga*, the present age, which traditionally starts at a theoretical point of departure of the celestial revolutions corresponding to the 18 February 3101 BCE at zero hours. (This moment is fixed itself at the general conjunction in mean longitude at the starting point of the sidereal longitudes of the sun, the moon and the planets, the apogees and node ascending from the moon being respectively at 90° and 180° from these longitudes.) Literally, the word **yuga* signifies "yoke", "link". In ancient Indian astronomy, this term was employed in the very limited sense of the simple "cycle". Thus in the **Jyotishavedânga* ("Astronomic Element of Knowledge"), a *yuga* of five years is used, this being a period at the end of which the sun and the moon are considered to have each completed a whole number of revolutions. On the other hand, in the *Romakasiddhânta* (start of the fourth century CE), the *yuga* is a lunar-solar cycle, the length of which is 2,850 years. These cycles, however, do not constitute an

"astronomical speculation" like the one that began to be developed in Âryabhata's time. No speculative system relating to *yugas* is found in the texts of the **Vedas*. This means that the *yuga* speculations were probably unknown in India during Vedic times and until the early centuries of the Common Era.

Nevertheless, purely arithmetical speculative calculations on these cycles appear in the **Manusmriti* (a significant religious work considered to form the basis of Hindu society), as well as in the much later texts of the *Yâjñavalkyasmriti* and the epic of the **Mahâbhârata*. It is difficult, however, if not impossible, to glean from this a chronology for the history of speculative *yugas*, since a great deal of uncertainty presides over the dates of these documents.

On the other hand, the work of Âryabhata, in which astronomical speculation of *yugas* appears for the first time, is dated in a rather precise manner, to within a few years of 510 CE.

In fact, as far as it is possible to tell, it was Âryabhata who, after the beginning of the sixth century CE, introduced speculative *yugas* into mathematical astronomy and made them generally known in India.

None of the Indian speculative canons (on *yugas*) that are known today is dated before Âryabhata's time. Âryabhata's astronomical speculations on *yugas* use basic numbers, some of which can be seen in the following calculations (**nakshatra* here denoting the twenty-seven lunar mansions divided into equal lengths) :

1 **mahâyuga* = 4,320,000 years = 12,000 (moments) × 360 = 27 (*nakshatra*) × 4 × 4 (phases) × 10,000 = "great period". 1 **yugapâda* = 1,080,000 years = 3,000 (moments) × 360 = 27 (*nakshatra*) × 4 (phases) × 10 000 = one quarter of a "great period". 1 **kaliyuga* = 432,000 years = 1,200 (moments) × 360 = 27 (*nakshatra*) × 4 × 4 (phases) × 1,000 = 1,200 ("divine years") × 360 = one tenth of a "great period".

According to Censorinus, Heraclitus's "great year" was 10,800 years long. On the other hand, the surviving fragments of the work of Babylonian astronomer Berossus (fourth–third century BCE) contain mention of a cosmic period 432,000 years long, which is also called "Great Year" :

1 **Great Year of Heraclitus* = 10,800 years = 30 (moments) × 360. 1 **Great Year of Berossus* = 432,000 years = 1,200 (moments) × 360. In other words, in all the cycles there is the following arithmetical relationship: 1 **yugapada* = 100 times the Great Year of

Heraclitus = 2.5 times the Great Year of Berossus. 1 *kaliyuga* = one Great Year of Berossus = 40 times the Great Year of Heraclitus. 1 *mahâyuga* = 400 times the Great Year of Heraclitus = 10 times the Great Year of Berossus.

From what is known today, it is impossible to establish whether there is any link between Âryabhata's *yugas* and the cosmic periods of the Mediterranean world. What *is* known is that Heraclitus belonged to the time when Persia dominated certain countries of the Greek world as well as part of India, whilst Berossus belonged to the end of the Persian rule and the beginning of the conquests of Alexander the Great . . . So why did Âryabhata develop his remarkable speculation? "As far as Âryabhata was concerned, speculation about *yugas* was just a theory. Convinced of the existence of common multiples of the different revolutions, he had set himself the task of researching the cycles of this astronomy, which was the most advanced of his time, and of which he was fully aware of the value. Whether it was a spontaneous idea, or drawn from a revival of the the *Great Year* 432,000 years long of the Babylonian astronomer Berossus, or even inspired by a wholly verbal, strictly arithmetical speculation, in any case Âryabhata drew out the constants of the mean movements in order to construct these common multiples and general conjunctions, from a single reality in time, that is to say the astronomical reality of 510 CE almost to the year. Of course, the theory was regrettable, but we must not forget the serious and extreme rigour he showed in undertaking such a work." (Billard)

YUGA (Cosmogonical speculations on). According to speculations developed by cosmogonies on what is referred to as the "decline of Proper moral and cosmic Order over the course of time", the *mahâyuga* corresponds to the appearance, evolution and disappearance of a world, and the whole cycle is followed by a new *mahâyuga*, and so on until the destruction of the universe. The four ages of this "great cycle" are considered to be unequal in terms of both length and worth. Qualitatively, this is how things are meant to unfold [see Friedrichs, etc (1989), *Dictionnaire de la sagesse orientale*] :

1. The *kritayuga*, the first of the four *yugas*, is the golden age, during which humans enjoy extremely long lives and everything is perfect. This is the ideal age, where virtue, wisdom and spirituality reign supreme, and there is a total absence of ignorance and vice. Hatred, jealousy, pain, fear and menace are unknown. There is only one god, one sole *Veda*, one law and one religion, each caste fulfilling its duties with the utmost selflessness. This age is said to have lasted 4,800 divine years (*divyavarsha*), which is equal to 1,728,000 human years.

2. The *tretâyuga* is the age during which humans are only believed to live three quarters of their lives. They are now marked by vice, there are the beginnings of laxity in their behaviour and the first ritual sacrifices are carried out. During this age, humans begin to act with intention and in self-interest. Rectitude diminishes by a quarter. The age is said to last 3,600 divine years, or 1,296,000 human years.

3. The *dvâparayuga* is said to be the age during which the forces of Evil equal those of Good, and where honest behaviour, virtue and spirituality are reduced by half. Illnesses have made their appearance and humans now only live half their lives. This age is said to last 2,400 divine years, or 864,000 human years.

4. Finally, the *kaliyuga* or "iron age" is the age we are living in now. "True virtue" is something which has all but disappeared and conflicts, ignorance, irreligion and vice have increased manifold. Illnesses, exhaustion, anger, hunger, fear and despair reign supreme. Living things only live for a quarter of their existence and the forces of evil triumph over good. Only a quarter of the original rectitude displayed by humans remains. This age is meant to have begun in the year 3101 BCE, and is meant to last 1,200 divine years, or 432,000 human years. It is meant to end with a *pralaya* (destruction by fire and water) before a new *chaturyuga* begins.

See *Yuga* (Definition), *Yugas* (Systems of calculating) and any other entry entitled *Yuga*.

YUGA (Systems of calculating). In the traditional system, the four ages of a *chaturyuga* are of unequal length, with the ratios of 4, 3, 2 and 1, from the *kritayuga* to the *kaliyuga* whose length is equal to 1/10 of the *mahâyuga*. In other words, the four successive ages of a *chaturyuga* are divided unequally as follows :

1 *mahâyuga* = 4/10 + 3/10 + 2/10 + 1/10.

Thus the corresponding values in "divine" years: 1 *kritayuga* = 4,800 divine years (= 4/10

Yuga

of *mahâyuga*); 1 *tretâyuga* = 3,600 divine years (= 3/10 of *mahâyuga*); 1 *dvâparayuga* = 2,400 divine years (= 2/10 of *mahâyuga*); 1 *kaliyuga* = 1,200 divine years (= 1/10 of *mahâyuga*); 1 *mahâyuga* = 12,000 divine years.

As one divine year is said to be equal to 360 human years, these cycles have the following durations in human time:

1 *kritayuga* = 1,728,000 human years; 1 *tretâyuga* = 1,296,000 human years; 1 *dvâparayuga* = 864,000 human years; 1 *kaliyuga* = 432,000 human years; 1 *mahâyuga* = 4,320,000 human years.

See *Divyavarsha*. The system for calculating unequal *yugas* was used by a considerable number of Indian astronomers (including *Brahmagupta), as well as by a great many cosmographers, philosophers and authors of religious texts (traditional system).

However, the system used by *Âryabhata consists in dividing the *mahâyuga* in the following manner :

1 *kritayuga* = 1,080,000 human years; 1 *tretâyuga* = 1,080,000 human years; 1 *dvâparayuga* = 1,080,000 human years; 1 *kaliyuga* = 1,080,000 human years; 1 *mahâyuga* = 4,320,000 human years.

In other words, the four cycles of the *chaturyuga* are all considered to be equal here. This is the system of the *yugapâdas or "quarters of a yuga". However, the *mahâyuga* is not the longest unit of cosmic time in these systems of calculation. There is also the cycle called a *kalpa, which corresponds to 12,000,000 divine years:

1 *kalpa* = 12,000,000 divine years = 12,000,000 × 360 = 4,320,000,000 human years. Bearing in mind the length of the *mahâyuga* (= 12,000 divine years), this cycle is thus also defined by :

1 *kalpa* = 1,000 *mahâyuga* = 4,320,000 × 1,000 = 4,320,000,000 human years.

An even longer period exists, the *mahâkalpa, or "great kalpa", which is the length of twenty "ordinary" *kalpas* (20,000 *mahâyugas*) :

1 *mahâkalpa* = 12,000,000 × 20 = 240,000,000 divine years = 240,000,000 × 360 = 86,400,000,000 years.

YUGA. [S]. Value = 2. (This symbol is very rarely used to represent this value.) The allusion here is to the cycle called *Dvâipayanayuga, where, according to Brahmanic cosmogony, men have only lived half of their lives and where the forces of Evil are counteracted by the equal strength of the forces of Good. The duality (*dvaita) between Good and Evil is at the root of this symbolism. See *Yuga* (Definition) and Two.

YUGA. [S]. Value = 4. The allusion here is to the most important of the cosmic cycles of this name : the *mahâyuga* (or "Great Age"), also called *chaturyuga* (or "four ages"). Composed of four successive "ages", in Hindu cosmogony the *mahâyuga* is said to correspond to the appearance, evolution and disappearance of a world. See **Yuga** (Definition) and **Four**.

YUGALA. [S]. Value = 2. Synonym of *Yama. See **Two**.

YUGAPÂDA. "Quarter of a *yuga*". Name given to each of the four cycles of a *chaturyuga*, subdivided into four equal parts, according to the system of calculation used by *Âryabhata. A *yugapâda* thus corresponds to 1,080,000 human years. See *Yuga* (Definition) and **Yuga** (Systems of calculating).

YUGMA. [S]. Value = 2. Synonym of *Yama. See **Two**.

Z

ZENITH. [S]. Value = 0. See *Vishnupada* and **Zero**.

ZERO. Ordinary Sanskrit name for zero : *shûnya. Here is a list of corresponding numerical symbols: *Abhra, *Âkâsha, *Ambara, *Ananta, *Antariksha, *Bindu, *Gagana, *Jaladharapatha, *Kha, *Nabha, *Nabhas, *Pûrna, *Randhra, *Vindu, *Vishnupada, *Vyant, *Vyoman. These words translate or symbolically express :

1. The void (*Shûnya*). 2. Absence (*Shûnya*). 3. Nothingness (*Shûnya*). 4. Nothing (*Shûnya*). 5. The insignificant (*Shûnya*). 6. The negligible quantity (*Shûnya*). 7. Nullity (*Shûnya*). 8. The "dot" (*Bindu, Vindu*). 9. The "hole" (*Randhra*). 10. Ether, or "element which penetrates everything" (*Âkâsha*). 11. The atmosphere (*Abhra, Ambara, Antariksha, Nabha, Nabhas*). 12. Sky (*Nabha, Nabhas, Vyant, Vyoman, Vishnupada*). 13. Space (*Âkâsha, Antariksha, Kha, Vyant, Vyoman*). 14. The firmament (*Gagana*). 15. The canopy of heaven (*Gagana*). 16. The immensity of space (*Ananta*). 17. The "voyage on water" (*Jaladharapatha*). 18. The "foot of Vishnu" (*Vishnupada*). 19. The zenith (*Vishnupada*). 20. The full, the fullness (*Pûrna*). 21. The state of that which is entire, complete or finished (*Pûrna*). 22. Totality (*Pûrna*). 23. Integrity (*Pûrna*). 24. Completion (*Pûrna*). 25. The serpent of eternity (*Ananta*). 26. The infinite (*Ananta, Vishnupada*). See **Numerical symbols**.

ZERO (Graeco-Latin concepts of). Western cultures have obviously had a concept of the void since Antiquity. To express it, the Greeks had the word *oudén* ("void"). As for the Romans, they used the term *vacuus* ("empty"), *vacare* ("to be empty"), and *vacuitas* ("emptiness"); they also had the words *absens*, *absentia*, and even *nihil* (nothing), *nullus* and *nullitas*. But these words actually corresponded to notions that were understood very distinctly from each other. With the help of some appropriate examples, an etymological approach will enable us hereafter to form quite a clear idea of the evolution of the concepts down the ages and to perceive better the essential difference which exists between these diverse notions and the Indian concept of zero. "Presence" (from the Latin *praesens*, present participle of *praesse*, "to be before [*prae*]", "to be facing") is properly speaking the fact of being where one is. But the adjective present also means "what is there in the place of which one is speaking"; this meaning is applicable then both to an object and to a living being. In the figurative sense, applied to people, present means "that which is present in thought to the thing being spoken of " (to be present in thought at a ceremony, despite the physical absence); applied to things, however, it means "that which is there for the speaker, or for what he is aware of ". It is thus a moral or deliberate presence. Another meaning of presence, in opposition this time to the past and the future, is "that which exists or is really happening, either at the moment of speaking or at the moment of which one is speaking". Consequently, this meaning corresponds to the present situation. Figuratively, it is rather a matter of "that which exists for the consciousness at the moment one is speaking", somewhat like a scene one witnessed and which remains present in one's mind.

This preamble allows a better understanding of "absence", since it is a term that is opposed to presence. The word comes from the Latin *absentia*, which derives from *abest*, "is far". Thus it expresses the character of "that which is far from". It is thus by definition the fact of not being present at a place where one is normally or one is expected. And the absent is the person or the thing which is lacking or missing. As for non-presence, it is simply the *void* left by an absence, since it is the space that is not occupied by any being or any thing. If it is an unoccupied place, it is this that is empty, whether it be a seat, an administrative post or even one of the "places" of the place-value system. By dint of thinking solely of the void, some thinkers have arrived at vacuism, a type of physics, according to which there exist spaces where all material reality is void of all existence. It was developed notably by the Epicureans, who conceded the existence of places where all matter, visible or invisible, was absent. Others opted rather for anti-vacuism, like Descartes, who considered an absolute void to be a contradictory notion. Nowadays, it is still sometimes said that nature abhors a vacuum. This aphorism was created by those who held to the physics of the ancient world in order to make sense of the existence of certain phenomena for which they were incapable of providing a satisfactory explanation. It was not until the experiments of the Italian physicist Torricelli on the laws of atmospheric weight, that the lie was given to this idea.

ZERO (Indian concepts of). In Sanskrit, the principal term for zero is *Shûnya*, which literally means "void" or "empty". But this word, which was certainly not invented for this particular circumstance, existed long before the discovery of the place-value system. For, in its meaning as "void", it constituted, from ancient times, the central element of a veritable mystical and religious philosophy, elevated into a way of thinking and existing. The fundamental concept in **shûnyatâvâda*, or philosophy of "vacuity", **shûnyatâ*, this doctrine is in fact that of the "Middle Way" (*Mâdhyamakha*), which teaches in particular that every made thing (*samskrita*), is void (**shûnya*), impermanent (*anitya*), impersonal (*anâtman*), painful (*dukh*) and without original nature. Thus this vision, which does not distinguish between the reality and non-reality of things, reduces the same things to total insubstantiality. See *Shûnya* and *Shûnyatâ*.

This is how the philosophical notions of "vacuity", nihilism, nullity, non-being, insignificance and absence, were conceived early in India (probably from the beginning of the Common Era), following a perfect homogeneity, contrary therefore to the Graeco-Latin peoples (and more generally to all people of Antiquity) who understood them in a disconnected and totally heterogeneous manner.

The concepts of this philosophy have been pushed to such an extreme that it has been possible to distinguish twenty-five types of *shûnya*, expressing thus different nuances, among which figure the void of non-existence, of non-being, of the unformed, of the unborn, of the non-product, of the uncreated or the

Fɪɢ.24ᴅ.10. *The Western concept of nought*

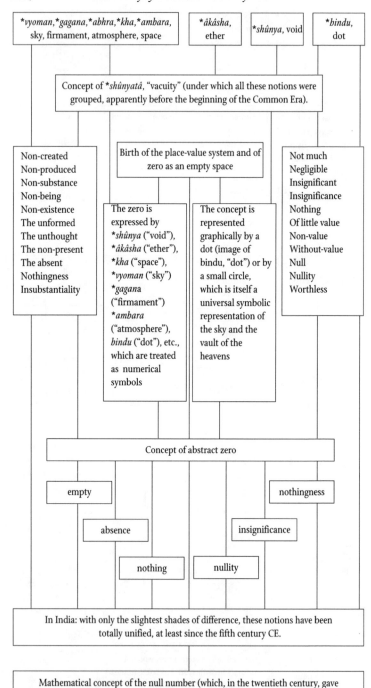

non-present; the void of the non-substance, of the unthought, of immateriality or insubstantiality; the void of non-value, of the absent, of the insignificant, of little value, of no value, of nothing, etc. In brief, the zero could have hardly germinated in a more fertile ground than the Indian mind. Once the place-value system was born, the *shûnya*, as a symbol for the void and its various synonyms (absence, nothing, etc.), naturally came to mark the absence of units in a given order [see Fig. D. 11]. It is important to remember that the Indian place-value system was born out of a simplification of the *Sanskrit place-value system* as a consequence of the suppression of the word-symbols for the various powers of ten. This was a decimal positional numeration which used the nine ordinary names of numbers and the term *shûnya* ("void") as the word that performs the role of zero. Thus the Indian zero has meant from an early time not only the void or absence, but also heaven, space, the firmament, the canopy of heaven, the atmosphere and ether, as well as nothing, a negligible quantity, insignificant elements, the number nil, nullity and nothingness, etc. This means that the Indian concept of zero by far surpassed the heterogenous notions of vacuity, nihilism, nullity, insignificance, absence and non-being of all the contemporary philosophies. See all other entries entitled **Zero**.

The Sanskrit language, which is an incomparably rich literary instrument, possessed more than just one word to express "void". It possessed a whole panoply of words which have more or less the same meaning; these words are related, in a direct or indirect manner, to the universe of symbolism of Indian civilisation. See **Sanskrit**, **Numerical symbols**, **Numeration of numerical symbols**.

Thus words which literally meant the sky, space, the firmament or the canopy of heaven came to mean not only the void but also zero. See *Abhra*, *Âkâsha*, *Ambara*, *Antariksha*, *Gagana*, *Kha*, *Nabha*, *Vyoman* and **Zero**. In Indian thought, space is considered as the void which excludes all mixing with material things, and, as an immutable and eternal element, is impossible to describe. Because of the elusive character and the very different nature of this concept as regards ordinary numbers and numerals, the association of ideas with zero was immediate. Other Indian numerical symbols used to mean

zero were: **pûrna* "fullness", "totality", "wholeness", "completion"; **jaladharapatha*, "voyage on the water"; **vishnupada*, "foot of Vishnu"; etc. To find out more about this symbolism, see the appropriate entries. Such a numerical symbolism has played a role that has been all the more important in the history of the place-value system because it is in fact at the very origin of a representation that we are very familiar with. The ideas of heaven, space, atmosphere, firmament, etc., used to express symbolically, as has just been seen, the concept of zero itself. And as the canopy of heaven is represented by human beings either by a semi-circle or a circular diagram or again by a complete circle, the little circle that we know has thus come, through simple transposition or association of ideas, to symbolise graphically, for the Indians, the idea of zero itself. It has always been true that "The circle is universally regarded as the very face of heaven and the Milky Way, whose activity and cyclical movements it indicates symbolically" [Chevalier and Gheerbrant (1982)]. And so it is that the little circle was put beside the nine basic numerals in the place-value system, to indicate the absence of units in a given order, thereafter acquiring its present function as arithmetical operator (that is to say that if it is added to the end of a numerical representation, the value is thus multiplied by ten). See **Numeral 0** (in the form of a little circle), *Shûnya-chakra*.

The other Sanskrit term for zero is the word **bindu*, which literally means "dot". The dot, it is true, is the most elementary geometrical figure there is, constituting a circle reduced to its centre. For the Hindus, however, the **bindu* (in its supreme form of a **paramabindu*) symbolises the universe in its non-manifest form and consequently constitutes a representation of the universe before its transformation into the world of appearances (*rûpadhâtu*). According to Indian philosophies, this uncreated universe is endowed with a creative energy capable of engendering everything; it is thus in other terms the causal point whose nature is consequently *identical* to that of ""vacuity" (**shûnyatâ*). See **Bindu**, **Paramânu**, **Paramabindu**, *Âkâsha*, **Shûnyatâ** and **Zero**.

Thus this natural association of ideas with this geometrical figure, which is the most basic of them all, yet capable of engendering all possible lines and shapes (*rûpa*). It is the perfect symbol for zero, the most negligible quantity there is, yet also and above all the most basic concept of all abstract mathematics. Thus the dot came to be a representation of zero (particularly in the *Shârâdâ* system of Kashmir and in the notations of Southeast Asia; see Fig. 24. 82) which possesses the same properties as the first symbol, the little circle. See **Numeral 0** (in the form of a dot) and *Shûnya-bindu.*

This is the origin of the eastern Arabic zero in the form of a dot : when the Arabs acquired the Indian place-value system, they evidently acquired zero at the same time. This is why, in Arabic writings, sometimes the sign is given in the form of a dot, sometimes in the form of a small circle. It is the little circle that prevailed in the West, after the Arabs of the Maghreb transmitted it themselves to the Europeans after the beginning of the twelfth century. To return to India, this notion was gradually enriched to engender a highly abstract mathematical concept, which was perfected in *Brahmagupta's time (c. 628 CE); that of the "number zero" or "zero quantity". It is thus that the *shûnya* was classified henceforth in the category of the *samkhya, that is to say the "numbers", so marking the birth of modern algebra. See *Shûnya-samkhya.* So, from abstract zero to infinity was a single step which Indian scholars took early and nimbly. The most surprising thing is that amongst the Sanskrit words used to express zero, there is the term *ananta, which literally means "Infinity". Ananta, according to Indian mythologies and cosmologies, is in fact the immense serpent upon which the god *Vishnu is said to rest between two creations; it represents infinity, eternity and the immensity of space all at once. Sky, space, the atmosphere, the canopy of heaven were, it is true, symbols for zero, and it is impossible not to draw a comparison in these conditions, between the void of the spaces of the cosmos with the multitude represented by the stars of the firmament, the immensity of space and the eternity of the celestial elements. As for the ether (*âkâsha), this is said to be made up of an infinite number of atoms (*anu, *paramânu). This is why, from a mythological, cosmological and metaphysical point of view, the zero and infinity have come to be united, for the Indians, in both time and space. See *Ananta, Shesha, Sheshashîrsha,* **Infinity (Indian mythological representation of)** and **Serpent (Symbolism of the).**

But from a mathematical point of view, however, these two concepts have been very clearly distinguished, Indian scholars having known that one equalled the inverse of the other. See **Infinity, Infinity (Indian concepts of)** and **Indian mathematics (History of).**

To sum up, the Indians, well before and much better than all other peoples, were able to unify the philosophical notions of void, vacuity, nothing, absence, nothingness, nullity, etc. They started by regrouping them (from the beginning of the Common Era) under the single heading *shûnyatâ (vacuity), then (from at least the fifth century CE) under that of the *shûnyakha (the sign zero as empty space left by the absence of units in a given order in the place-value system) before recategorising them (well before the start of the seventh century CE) under the heading of *shûnya-samkhya* (the "zero number") [see Fig. D. 11]. Once again, this indicates the great conceptual advance and the extraordinary powers of abstraction of the scholars and thinkers of Indian civilisation. The contribution of the Indian scholars is not limited to the domain of arithmetic; by opening the way to the generalising idea of number, they enabled the rapid development of algebra and consequently played an essential role in the development of mathematics and all the exact sciences. It is impossible to exaggerate the significance of the Indian discovery of zero. It constituted a natural extension of the notion of *vacuity, and gave the means of filling in the space left by the absence of an order of units. It provided not only a word or a sign, it also and above all became a numeral and a numerical element, a mathematical operator and a whole number in its own right, all at the point of convergence of all numbers, whole or not, fractional or irrational, positive or negative, algebraic or transcendental.

ZERO AND INFINITY. See **Zero, Infinity, Infinity (Indian concepts of), Infinity (Mythological representations of), Serpent (Symbolism of the), Zero (Graeco-Latin concepts of), Zero (Indian concepts of)** and **Indian mathematics (History of).**

ZERO AND SANSKRIT POETRY. In India, the use of zero and the place-value system has been a part of the way of thinking for so long that people have gone as far as to use their principal characteristics in a subtle and very poetic form in a variety of Sanskrit verse. As proof, here is a quotation from the poet Bihârîlâl who, in his *Satsaî,* a famous collection of poems, pays homage to a very beautiful woman in these terms : "The dot [she has] on her forehead Increases her beauty tenfold, Just as a zero dot (*shûnya-bindu*) Increases a number tenfold. " [See Datta and Singh, in: AMM, XXXIII, (1926), pp. 220ff.].

First of all, it should be remembered that the dot that the woman has on her forehead is none other than the *tilaka* (literally: sesame), a mark representing for the Hindus the third eye of *Shiva, that is the eye of knowledge. While young girls put a black spot between their eyebrows by means of a non-indelible colouring matter, married women put a permanent red dot on their foreheads; it would seem then that the homage was being paid to a married woman. It is known that the dot (*bindu*) figures among the numerous numerical symbols with a value equal to zero, and is even used as one of the graphical representations of this concept. See **Zero**, **Shûnya-bindu**, **Numeral 0** (in the form of a dot). This is a very clear allusion to the arithmetical operative property of zero in the place-value system, because if zero is added to the right of the representation of a given number, the value of the number is multiplied by ten (see Fig. 23.26 and 27). Another quotation, taken this time from the *Vâsavadattâ* by the poet *Subandhu (a long love story, written in an extremely elaborate language, swarming with word plays, implications and periphrases):

"And at the moment of the rising of the Moon

With the darkness of the falling night,

It was as if, with folded hands

Like closed blue lotus blossoms,

The stars had begun straightway

To shine in the firmament (*gagana*)...

Like zeros in the form of dots (*shûnya-bindu*),

Because of the emptiness (*shûnyatâ*) of the *samsâra,

Disseminated in space (*kha*),

As if they had been [dispersed]

In the dark blue covering the skin of the Creator [= *Brahma],

Who had calculated their sum total

By means of a piece of Moon in the guise of chalk."

[See *Vâsavadattâ of *Subandhu*, Hall, Calcutta (1859), p. 182; Datta and Singh (1938), p. 81.]

Here too the metaphor used leaves the reader in no doubt; the void (*shûnya*) – which is placed in relation to the emptiness (*shûnyatâ*) of the cycle of rebirths (*samsâra*) – is also implied in its representation in the form of a dot (*shûnya-bindu*), as an operator in the art of written calculation. These concepts really had to have been part of the way of thinking for a long time for the subtleties used in this way to have been understood and appreciated by the wider public of the time.

ZODIAC. [S]. Value = 12. See *Râshi* and **Twelve**.

CHAPTER 25

INDIAN NUMERALS AND CALCULATION IN THE ISLAMIC WORLD

As we saw in the previous chapter, it was indeed the Indians who invented zero and the place-value system, as well as the very foundations of written calculation as we know it today.

These highly significant inventions date back at least as far as the fifth century CE.

However, it was not until five centuries later that the nine basic numerals appeared in Christian Europe.

Another two or three centuries elapsed before zero was first used in Europe, along with the afore-mentioned methods of calculation, and it was later still that these revolutionary new ideas were propagated and fully accepted in the Western world.

Thus the Indian inventions were not transmitted directly to Europe: Arab-Muslim scholars (amongst their numerous fundamental roles) played an essential part as vehicles of Indian science, acting as "intermediaries" between the two worlds.*

Therefore, before we proceed with our history, it is worth knowing a little about the Arabs, in terms of their culture, their way of thinking, their own science and their fundamental contributions to the evolution of science the world over. This will give the reader a clearer idea of the conditions under which this transmission of ideas took place, which led to the internationalisation of Indian science and methods of calculation.

THE SCIENTIFIC CONTRIBUTIONS OF ARAB-ISLAMIC CIVILISATION

In the century following the death of the prophet Mohammed the Islamicised Arabs built up an enormous empire through their conquests. At the beginning of the eighth century CE, the Empire stretched from the Pyrenees to the borders of China, and included Spain, southern Italy, Sicily, North Africa, Tripolitania, Egypt, Palestine, Syria, part of Asia Minor and Caucasia, Mesopotamia, Persia, Afghanistan and the Indus Valley.

---

* Words preceded by an asterisk have entries in the Dictionary (pp. 867-1007).

FIG. 25.1. *Detail of a page from Al bahir fi 'ilm al hisab (The Lucid Book of Arithmetic) by As Samaw'al ibn Yahya al-Maghribi (died c. 1180 in Maragha), a Jewish mathematician, doctor and philosopher from the Maghreb, who converted to Islam [Istanbul, Aya Sofia Library, Ms. ar. 2,718. See Rashed and Ahmed 1972]. This document, which uses "Hindi" numerals to reproduce what is known as "Pascal's triangle", shows that Muslim mathematicians knew about the binomial expansion* (a + b)m, *where "m" is a positive integer, as early as the tenth century. The author admits that this triangle is not his, and attributes it to al-Karaji, who lived near the end of the tenth century [Anbouba; Rashed in DSB].*

Nevertheless, the advance came to a halt when it met with successful resistance: in 718 by the Byzantine army near Constantinople; in 732 by Charles Martel at Poitiers; and in 751 by the Chinese on the border of Sogdiana.

Once the political influence of the "Son of the Arabian Desert" fell into decline, the Empire was controlled for nearly a century by the caliphs of the Omayyad Dynasty (661–750), with Damascus as their capital. Power then went to the Abbasid caliphs (750–1258) who made Baghdad their capital in 772 and reigned over the empire for the next 500 years.

There followed a period of exaltation characteristic of expansion, and this was a highly fertile era of cultural assimilation and scientific development. Arabic culture dominated the world for several centuries, before Mongol invasions, the Crusades, the division of the Empire and the anarchy of internal wars brought it to an end in the thirteenth century.

THE ASSIMILATION OF OTHER CULTURES

When the Arab nomads who had been converted to Islam left the desert to conquer this immense territory, they lived from trading spices, medicines, cosmetics and perfume. Their level of literacy and numeracy was very basic. The little that they knew of science was based on practical applications involving simple formulae, and was often tinged with arithmology, mysticism and all kinds of magical and divinatory practices.

Thus the first Islamicised Arabs initially possessed none of the intellectual means they would need to realise their desire to conquer other lands and to deal with the enormous revenue that taxation and capitation would soon bring to this vast new Empire.

However, through their conquests and trade relations, they found themselves increasingly in contact with people from different cultures: Syrians, Persians, Greek émigrés, Mesopotamians, Jews, Sabaeans, Turks, Andalusians, Berbers, peoples from Central Asia, inhabitants of the shores of the Caspian, Afghans, Indians, Chinese, etc. Thus they discovered cultures, sciences and technologies far superior to their own. They were quick to adapt and to get to grips with the new concepts and knowledge, which scientists, intellectuals and engineers from the conquered lands had accumulated over the ages, and in some cases had developed to quite an advanced level.

THE METROPOLIS OF NEAR EASTERN SCIENCE
BEFORE ISLAM

Long before the Arabic conquest, the philosophy of Aristotle and the sciences of nature, mathematics, astronomy and medicine, according to the doctrines of Hippocrates and Galen, were all taught in Syria and Mesopotamia, notably at the schools of Edesse, Nisibe and Keneshre.

At the same time, Persia constituted an important crossroads and centre of influence for the meeting of Greek, Syrian, Indian, Zoroastrian, Manichaean and Christian cultures.

The Persian King Khosroes Anûshîrwân (531–579) sent a cultural mission to India and brought many Indian scientists to Jundishapur. Nestorian Christians, who had been expelled from the school at Edesse by Byzantine orthodoxy, found refuge in the same town. This is also where the Neo-Platonist philosophers of Athens (such as Simplicius who wrote commentaries on the works of Aristotle and Euclid) were welcomed by King Anûshîrwân when their academy was closed in 529 under the orders of Emperor Justinian (527–565). It was at Edesse, Nisibe, Keneshre and Jundishapur that Greek works were first translated into Syrian or Persian, and that the first works in Sanskrit were discovered. The first translations into Arabic were undertaken at Jundishapur shortly after the constitution of the Islamic Empire [see L. Massignon and R. Arnaldez in HGS; A. P. Youschkevitch (1976)].

BAGHDAD, FIRST ISLAMIC CENTRE OF SCIENTIFIC LEARNING

The importance of these cultural and scientific centres gradually declined during the Abbassid Dynasty, and so the town of Baghdad became the centre of intellectual activity in the Near East, thus playing a vital role in this history.

Founded in 762, then elevated to capital of the Arabic Empire in 772, Baghdad was initially the obvious centre for international trade. Then, owing to both its privileged location, and to the generous action of sovereign patrons, such as caliphs al-Mansur (754–775), Harun al-Rashid (786–809) and al-Ma'mun (813–833), whose subsidies contributed to the development of science and culture in Islam, Baghdad became the most vivacious intellectual centre of the East. This is where Arabic science truly began.

> If we put together the religious and social conditions, we shall under-
> stand the position of Islamic intellectuals and the fillip they gave to
> intellectuals of all creeds and races, by their mobilisation for a common
> labour in the Arabic tongue. For science is one of the Islamic city's insti-
> tutions. Not only do patrons encourage it, but caliphs work to create and
> develop it. It is sufficent to cite Khalid, the "philosopher prince", whose
> actions were "perhaps legendary", or al-Mansur, the founder of
> Baghdad, and al-Ma'mun "who eagerly sent out emissaries to look for
> manuscripts and have them translated" [L. Massignon and R. Arnaldez].

THE GOLDEN AGE OF ARABIC SCIENCE

One of the most outstanding periods in the history of science took place in Islam between the eighth and thirteenth centuries of the Common Era.

This was at a time when Western civilisation was devastated by epidemics, famine and war, and was in no position to relay the cultural heritage of Antiquity. The Arab-Muslim scholars were able to develop not only mathematics, astronomy and philosophy, but also medicine, pharmacy, zoology, botany, chemistry, mineralogy and mechanics.

Through a collective effort, the Muslims and the peoples conquered by Islam collected together all the Greek works that they could find on philosophy, literature, science and technology.

It is sufficient to cite the names of Euclid, Archimedes, Ptolemy, Aristotle, Plato, Galen, Hero of Alexandria, Apollonius, Menelaus, Philo of Byzantium, Plotinus and Diophantus to give an idea of the variety and richness of the works that were translated into Arabic.

These translations and collected works grew in number and circulation, as universities and libraries sprang up all over the Islamic world. Towns such as Damascus, Cairo, Kairouan, Fez, Granada, Cordoba, Bokhara, Chorem, Ghazni, Rey, Merv and Isfahan soon became intellectual and artistic centres which were centuries ahead of the Christian capitals.

"ARABIC" OR "ISLAMIC" SCIENCE?

Arabic science is not necessarily the same thing as Muslim science. The Arabic language was a vehicle for science, which, during that long period of time, became the international language of the scholars of the Muslim world, and consequently the intellectual link between the different races.

Amongst the diverse cultures which were conquered or influenced by Islam was Persia, the birthplace of many brilliant minds, including al-Fazzari, al-Khuwarizmi, al-Razi, Avicenna, al-Biruni, Kushiyar ibn Labban al-Gili, Umar al-Khayyam, Nasir ad din at Tusi and Ghiyat ad din Ghamshid ibn Mas'ud al-Kashi.

During the assimilation of Indian science, the Arabs were helped by many Hindu Brahmans, who were often received at the court of Baghdad by enlightened caliphs.

They were assisted by Persians and Christians from Syria and Mesopotamia, who, being fervent admirers of Indian cultures, had gone so far as to learn Sanskrit.

The Buddhists also greatly contributed, especially those converted to Islam, such as Barmak who was sent to India to study astrology, medicine and pharmacy and who, on his return to Muslim territory, translated many Sanskrit texts into Arabic [see L. Frédéric (1989)].

There were also non-Muslim Arabic scholars, such as the Christian and Jewish intellectuals, who were often referred to as *ahl al kitab*, the "people of books", and whom the caliphs of Baghdad and Cordoba integrated to a certain extent amongst the members of their empires, sometimes allowing them the privileged right to hospitality which they called *dhimma*.

Often mistranslated as "tolerance", *dhimma* really means "right to hospitality", a "protection" that the caliphs sometimes gave to non-Islamic residents. They did also show a certain "tolerance" towards their conquered peoples, sometimes even "allowing" them to profess a different religion. But this tolerance had its limits. The expression of ideas contrary to official dogma, and even more, living by non-orthodox ideas, was vigorously repressed. Non-believers were often considered to be "internal emigrants" and not permitted to rise to the same rank as Muslims. The "Pact of 'Umar" even forced Jews and Christians to wear a "circlet": a round piece of cloth, yellow for the former and blue for the latter. Conversion to Islam offered a number of social, pecuniary and fiscal advantages.

Even the brilliant culture of the Kharezm Province was discriminated out of existence, as al-Biruni (a native of Kharezm) explained: "Thus Qutayba did away with those who knew the script of Kharezm, who understood the country's traditions and taught the knowledge of its inhabitants; he submitted them to tortures so that they were wrapped up in shadows and no one could know (even in Kharezm) what had (preceded) or followed the birth of Islam" [Youschkevitch].

The case of the Maghreb and especially that of Islamic Spain (before the virulence of the Almohads) do still prove that "tolerance" was practised for almost six centuries, in the sense of a greater liberty for Jews and for *Mozarabes* ("Arabic" Christians) who lived peacefully according to their own philosophies, organisations and traditions, with their synagogues, churches and convents [V. Monteil (1977)].

Thus the Christian scholars of the Arabic world often worked as "catalysts" by collecting, translating and commenting on, in Syriac or Arabic, many scientific and philosophical works of Greek or Indian origin. Amongst these men were: the astrologer Theophilus of Edesse, who translated many Greek medical texts into Syriac; the doctor Ibn Bakhtyashu, head of the Jundishapur hospital; the doctor Salmawayh ibn Bunan; the astronomer Yahya ibn Abi Mansur; the doctor Massawayh al-Mardini; the philosopher, doctor, physician, mathematician and translator Qusta ibn Luqa, from Baalbek; and the translators Yahya ibn Batriq, Hunayn ibn Ishaq, Matta ibn Yunus and Yahya ibn 'Adi.

As for Jewish intellectuals, or those issued from Judaism, it is worth mentioning the astronomer Ya'qub ibn Tariq, one of the first scholars of the Empire to study Indian astronomy, arithmetic and mathematics;

astronomers Marshallah and Sahl at Tabari; the astrologer Sahl ibn Bishr; the mathematician converted to Islam As Samaw'al ibn Yahya ibn 'Abbun al-Maghribi, who continued al-Karaji's work on algebra; and the converted doctor and historian Rashid ad din, who compiled a history of China.

There was also the philosopher-rabbi Abu 'Amran Musa ibn Maymun ibn 'Abdallah, better known as Rabbi Moshe Ben Maimon or Maimonides, whose encyclopaedic interests embraced not only philosophy, but also mathematics, astronomy and medicine. Born in Cordoba in 1135, he was initially one of the victims of religious persecution at the hands of the Almohad sovereigns, who forced him to proclaim himself a Muslim for sixteen years. The rabbi remained a Jew, and at the end of this period of time, he went first to Fez, then to Palestine, before settling in Egypt where he became a doctor at the court of the Fatimids in Cairo, until his death in 1204. He wrote many works on medicine (*Aphorisms of Medicine, Tract of Conservation and of the Regime of Health* and *Rules of Morals* being the only ones to have survived). These works were mainly concerned with:

> the treatment of haemorrhoids (a surgical operation which should only be carried out as a last resort), of asthma by a dry climate, nervous depression or "melancholy" through psychotherapy: recovery being seen as a return to equilibrium; and diets, all embraced by a global vision of the human being, always presented in a spirit of compassion and charity [V. Monteil (1977)].

He also wrote the famous *Moreh Nebukhim* (*Guide for the Lost*), in which his Aristotelian philosophy seeks to reconcile faith and reason, and he asserts himself as one of the first intermediaries between Aristotle and the doctors of scholasticism. Another of his fundamental contributions, this time to Judaism, was his *Commentary on Mishna* (1158–1165) and his *Second Law or the Strong Hand* (1170–1180). Before they were even translated into Hebrew or Latin, the medical and philosophical works of Maimonides were generally written in Arabic first. In other words, despite their profound attachment to Judaism, scholars such as Maimonides were authentic Arabic thinkers.

After the Abbasid school of Baghdad (ninth to eleventh century CE), there came the schools of Toledo and Seville, and Jewish scholars such as Yehuda Halevi, Salomon ibn Gabirol (Avicebron) and Abraham ibn Ezra or Abraham bar Hiyya (who would have spoken Hebrew, Arabic, Castilian and even Latin or Greek) acted as intermediaries between the Arabic and Christian worlds.

Of course, Arabic science was also and above all the creation of Muslim scholars. Amongst these men were: al-Fazzari, al-Kindi, al-Razzi, al-Khuwarizmi, Thabit ibn Qurra, al-Battani, Abu Kamil, al-Farabi, al-Mas'udi, Abu'l Wafa, al-Karayi, al-Biruni, Ibn Sina (Avicenna), Ibn al-

Haytham, 'Umar al-Khayyam, Ibn Rushd (Averroes) and Ibn Khaldun (see the Chronology, pp. 519ff. for further information).

Islamic religion played an important role in the scientific discoveries of this civilisation. The Koran preaches humanism in the search for knowledge; one of the necessities imposed by the study of this holy book and of Islamic thought is "the development of scientific research where Revelation, Truth and History are considered in their dialectic relationship as structural terms of human existence" [M. Arkoun (1970)].

The Koran frequently encourages the faithful to look for signs of proof of their faith in the heavens and on Earth: "Search for science from the cradle to the grave, even if you have to travel as far as China . . . Those that follow the path of scientific research will be led by God on the path to Paradise" [L. Massignon and R. Arnaldez in HGS]. (We have not been able to find the source for this advice, which many attribute to Mohammed. But it would seem that it forms an integral part of Islamic culture, at least since the time of Ibn Rushd.)

It is true that the science in question here is knowledge of religious Law (*'ilm*), but in Islam this is not set apart from secular science. Thus we find a whole series of *hadith* about medicine, remedies and the legitimacy of their use. Moreover, scholars and philosophers have not hesitated to quote the texts in order to defend their activities.

Averroes wrote in his *Authoritative Text*: "It is clear in the Koran that the Law invites rational observation of living beings in the search for an understanding of these beings through reason."

This is the opinion of all Muslims who have accepted and cultivated science. It is because the Koran invites the faithful to contemplate the power of Allah in the organisation of the universe that astrology has always been considered the "highest, noblest and most beautiful of sciences" [al-Battani] in the Islamic world.

The patient assimilation of observations and calculations relative to the positions of the planets, the moon's phases, equinoxes, etc., were often directly related to the demands of Islamic religion: the calculation of the exact times of the ritual prayer of the *'asr*, the dates of religious ceremonies, the month of Ramadan, orientation towards Mecca, etc.

This is why, despite the considerations above, the science and culture of the Islamic world should be more accurately termed "Arab-Islamic".

THE SPREAD OF SCIENTIFIC KNOWLEDGE:
ANOTHER ACHIEVEMENT OF ISLAM

Other sciences existed before Islam (in Ancient Greece, Persia, India, China, etc.), but although these were all mainly concerned with the same problems, they all had their own unique way of dealing with them. In other words, before Islam, there was no universal science as we know it today.

In fact, different cultures sought to preserve their knowledge and keep it a secret from the outside world. An example of this is the Neo-Pythagoreans in Greece.

Part of the reason why the Muslims were responsible for the unification of science is their success in conquering other peoples.

International trade played an important role, as did the Arabic geographers, travellers and cosmographers, translators, philologists, lexicographers and writers of commentaries:

By describing different areas of the globe, those unusual men described the marvels of nature, products of the earth, fauna, agriculture and crafts. This was a considerable source of information. Some geographers were also great scholars, experts in all fields, such as the famous al-Biruni [Massignon and Arnaldez].

Another factor was the cultural assimilation by the Muslims of the most diverse of cultures: this began at the time of the caliphs of Damascus, but it was not until the time of the enlightened caliphs of Baghdad and Cordoba that the results were felt.

The "tolerance" of these caliphs towards other cultures, beliefs, customs and traditions for nearly six centuries was also an important factor.

The promotion of study and research in the Koran has already been mentioned in this chapter. This was not only a fundamental condition for the development of Arabic Islamic science, but also one of the main causes for Islam's ready acceptance of the most diverse of cultures. (But it should also be noted that Arab-Islamic science, despite its universal nature, was always oriented towards knowledge of divine Law. It is necessary to wait until the European Renaissance before science gains the non-religious character we now recognise.)

THE DEVELOPMENT OF ARABIC ISLAMIC SCIENCES

The Islamic conquerors were not always in favour of science and culture.

Caliph 'Umar (634–644 CE) ordered the destruction of countless works seized in Persia. His argument was as follows: "If these books contain the key to the truth, Allah has given us a more reliable way to find it. And if these books contain certain falsehoods, they are useless" [see A. P. Youschkevitch (1976)].

There were certainly other similar cases of religious or xenophobic opposition, leading to great cultural losses. In the Islamic world, scholars suffered from the whims of totalitarian leaders. They had to avoid direct confrontation with official dogma if they did not want to lose their state subsidies and risk even greater repression. At the end of the eleventh century, the famous poet, astronomer and mathematician Omar Khayyam reported, in his *Mathematical Treatise*:

> We have witnessed a decline in scholarship, few scholars are left, and those who remain experience vexations. Their troubled times stop them from concentrating on deepening and bettering their knowledge. Most so-called scholars today mask the truth with lies.
>
> In science, they go no further than plagiarism and hypocrisy and use the little knowledge they have for vile material ends. And if they come across others who stand apart for their love of the truth and rejection of falsehood and hypocrisy, they attack them with insults and sarcasm.

But according to Youschkevitch, "this situation could not in the long term stop the triumph of scientific progress. Schools, libraries and observatories were built in the cities. To make a name for themselves, enlightened sovereigns set up academies similar to those founded by European monarchs in the seventeenth and eighteenth centuries. The transmission of knowledge was thus assured; but it was only later that the discovery of printing facilitated it."

However, such opinions were exceptional and not held by caliphs ruling later in Islam. In fact, the role of Islam and of Arabic scholars in the fields of science and culture can never be overstated.

The famous library of Alexandria, the most important one of Ancient Greece, was pillaged and destroyed twice: the first time in the fourth century CE by the Christian Vandals, and the second time (through a perverse paradox of history) by the Muslims in the seventh century. Many original manuscripts of inestimable value disappeared. Many Greek literary and scientific masterpieces would have been lost forever if they had not been collected and translated into Arabic.

It was thanks to the work of the North African philosopher Ibn Rushd (Averroes) that Saint Thomas Aquinas could study and understand the importance of Aristotle's work. Similarly, it is thanks to Avicenna that Albertus Magnus could develop the philosophy of universality. It is largely thanks to the work of Arabic translators that the works by Ancient Greeks are known to us today.

Moreover, the Arabs have never hesitated to underline the importance of Greek science and to express their admiration for it: "The language of the Hellenistic people is Greek; it is the most vast and the most robust of languages" [Sa'id al-Andalusi, *Tabaqat al umam*, in R. Taton (1957), I, p. 432].

Greek culture played a huge part in the development of Arabic science.

But it would be a mistake to believe that the latter was nothing more than the continuation of Greek science. This would be as far-fetched as the opinion that "India, and not Greece, formed the religious ideals of Arabia and inspired its art, literature and architecture".

Of course the framework of Arabic scientific thinking constituted an obvious extension of, and was largely based upon Hellenic science. However, the Arabs used the discoveries of Ancient Greece as a source of inspiration and actually expanded upon them. Moreover, Greece was not the only civilisation to inspire the Arabic scholars. They were also interested in oriental culture, from which they borrowed different elements which they adopted to suit their needs.

Thus their numeral alphabet was forged from a combination of Jewish, Greek and Syriac systems by adopting the corresponding principle to the twenty-eight letters of their own written alphabet.

Through the Christians of Syria and Mesopotamia they discovered the place-value system of the Babylonians, which they used in their tables and their astronomical texts to record sexagesimal fractions. Through trading with Persia and parts of the Indian sub-continent, they also came into contact with Indian civilisation. Thus they discovered Indian arithmetic, algebra and astronomy. Sa'id al-Andalusi (see above) expressed his admiration for Indian culture. He recognised its precedence over Islam and went as far as to call it "a mine of wisdom and the source of law and politics". He also wrote that "the Indian scientists devote themselves to the science of numbers (*'ilm al 'adad*), to the rules of geometry, to astronomy and generally to mathematics . . . they are unrivalled in medicine and the knowledge of treatments". His conclusion, however, is a little subjective. He claims that the intellectual talents and qualities of the Indians are nothing more than the product of "good fortune (*hazz*)", due to "astral influences" [R. Taton (1957), I, p. 432].

The Chinese were another foreign influence. After the battle of Talas in 751, the Arabs learned the secrets of making paper from linen or hemp from their Chinese prisoners. The first factory was built in Baghdad c. 800. It would be another four centuries before paper was used in Europe, through the intermediary of Spain.

At the beginning of the fourteenth century, Rashid ad din, grand vizier of the Mongolian sultan Ghazan Khan a Tabriz, and himself a converted Jew, compiled a library of 60,000 manuscripts, many of which came from Chinese and Indian sources.

In his *Universal History* (*Jami'at tawarikh*), he carefully described how Chinese characters were engraved on wood, and gave their transcription in Arabic. He translated extracts from the best known medical works of China and Mongolia into Arabic and Persian, including *Mejing*, a text on sphygmology (or science of the arterial pulse) by Wang Shuhe (265 –

317), which identifies four standard methods of medical examination, namely observation, auscultation, interrogation and palpation. These would not be studied in Western Europe until the eighteenth century [see V. Monteil (1977)].

However, the Arabs were not content merely to preserve the discoveries of Greece, Babylon, China and India. They wanted to make their own contribution to the world of science.

As they carefully collected, translated and studied works from the past, they added various commentaries, after mixing explanations with original developments, and always maintaining a critical perspective which rejected fixed dogmatism [see A. P. Youschkevitch (1976)].

Thus in mathematics, Greek methods were often mixed with Indian methods, sometimes with Babylonian ones or even, at a later date, with Chinese approaches.

The Arabs combined the strict systematisation of Greek mathematics and philosophy with the practicality of Indian science. This enabled them to make significant progress in the fields of arithmetic, algebra, geometry, trigonometry and astronomy. Through collecting, propagating and teaching the use of Indian numerals and calculation, and by pushing the study of certain remarkable properties of numbers towards the first seeds of a theory of numbers, the Arabs made considerable progress in the field of arithmetic.

Scholars such as al-Khuwarizmi, Abu Kamil, al-Karaji, As Samaw'al al-Maghribi, 'Umar al-Khayyam, al-Kashi and al-Qalasadi led arithmetic towards algebrised operations.

The Arabs (and more generally the Semites) "personalised" the number. Instead of an object which had various properties, it became an active being. They did not see numbers as being static and limited, as the Greeks did. The Arabs were interested in the ordinal, rather than the cardinal numbers: they were not deterred as the Greeks were by the irrational [see L. Massignon and R. Arnaldez].

The assimilation of the classical heritage allowed the mathematicians of Islamic countries to develop algorithms and corresponding problems, and thus achieve a higher level than that reached by Indian or Chinese mathematicians. It also enabled them to find more efficient ways to resolve and generalise these problems than the Chinese and the Indian methods. Where the latter were content to establish a specific rule of calculation, the mathematicians of Islam often managed to develop an entire theory [A. P. Youschkevitch (1976)].

In short, the work of the Arabic scholars involved objectivity, the questioning of the doctrines of the ancient scientists and systematic recourse to analysis, synthesis and experimentation.

The progress of sciences, in terms of knowledge, is dependent on the scientific mind . . . Perhaps some thought that all of science had already been discovered, and that all that remained to do was to assimilate all the information. But this gathering of knowledge was in fact an excellent prelude to methodical research and progress. The need for an inventory led to classifications of the sciences (such as those of al-Farabi or Avicenna) which was enough to cause an evolution of the concept of science. Under the influence of Plato and Aristotle, the Ancient Greeks classified the sciences according to their method, and the degree of intelligibility of their purpose. The Arabs took a more straightforward stance: the sciences exist and they must be ordered so that none is forgotten. The lack of conceptual analysis which characterises Arabic classification of the sciences was in fact an advantage from a purely scientific point of view. Knowledge itself promotes learning and marks out the direction to follow towards the acquisition of further knowledge [L. Massignon and R. Arnaldez].

For the Arabs, then, to know was not to contemplate, but to do; in other words to verify, challenge, experiment, observe, rethink, describe, identify, measure, correct, even complete and generalise. This is the Arab influence on science: it became an operating science following the development of "scientific reason". The Arabs had a great deal of curiosity, love and estimation for knowledge [L. Massignon and R. Arnaldez], which meant that they not only preserved and transmitted the science of Antiquity, but they transformed it and established it along new lines, giving it a new lease of life and originality.

THE ARABIC LANGUAGE: THE AGENT AND VEHICLE OF ISLAMIC SCIENCE

Right from the start of the history of Arab-Islamic science, anything concerning the science had to be written in Arabic if it was to be of any consequence, this language having become the permanent intellectual link between the scholars of various origins during this long period of time.

For many philosophers, mathematicians, physicians, chemists, doctors and astronomers, this language was more than a mere obligation: it was a real passion. It was the preferred language for expressing science and knowledge.

For example, the Persian scientists Avicenna and al-Biruni, rather than writing in Turkish or Persian wrote in Arabic, despite having been born in Kharezm, to the north of Iran in what is now Uzbekistan (formerly in the USSR). Al-Biruni explains his preference for Arabic in his *Kitab as saydana* ("*Treatise on Drugs*" [see V. Monteil (1977), p. 7]:

It is in Arabic that, through translation, the sciences of the world were transmitted [to us] and were embellished and found a place in our hearts. The beauty of the Arabic language has circulated with them in our arteries and veins. Of course, every nation has its own language, the one used for trading and talking to our friends and companions. But personally I feel that if a science found itself eternalised in my own mother-tongue, it would be as surprised as a camel finding itself in a gutter of Mecca or a giraffe in the body of a thoroughbred. When I compare Arabic with Persian (and I am equally competent in both languages) I admit that I prefer invective in Arabic to praise in Persian. You would agree with me if you saw what happens to a scientific work when it is translated into Persian: it loses all its brilliance and has less impact, it becomes muddled and quite useless. Persian is only good for transmitting historical stories about kings or telling tales at evening gatherings.

Of course, al-Biruni's description of Farsi is totally unjustified. Many Muslim scholars from Persia, Afghanistan and the Indus Valley also wrote in Arabic, although Persian is perfectly capable of expressing any concept, as well as the rigour, nuances and foundations of scientific thought. However, al-Biruni's preference for Arabic was not brought about by chance, and was certainly not due to a passing fad.

In terms of structure, Arabic became a much richer language and gradually acquired its scientific character in order to meet the demands of the translation of foreign works and the transposition of scientific texts.

When a scientific text is translated from its original language into an equally well-equipped language, there might be grammatical problems but there are no technical or conceptual difficulties. This was not the case when Greek was first translated into Arabic: vocabulary had to be created, or existing words adjusted to meet the needs of science. There was often an intermediate Syriac word which prepared the way for Arabic. The creation of the scientific Arabic language was not only philological, it also involved two scientific methods: the identification and verification of concepts [L. Massignon and R. Arnaldez].

It was in this scientific spirit that the lexicographers made an inventory of the Arabic language, as scholars had made an inventory of knowledge by attempting to classify different fields of learning through rethinking and evaluating concepts, then organising them in relation to one another. As for those who translated or commented on texts, they looked for Arabic equivalents for foreign terms in lexicons and in nature, and also in the different elements of knowledge, either to introduce new words and concepts, or even to express new ideas using the most ancient vocabulary.

This is how Arabic acquired its unique aptitude for expressing scientific thought, and for developing it in the service of the exact sciences.

This language, which was originally considered to be the language of the Revelation and the fundamental criterion for anyone wishing to belong to the Muslim religion and the Islamic community (*Umma*), became not only the vehicle of international science, but also and above all the essential agent of the Renaissance and the dominant factor in the Arabic scientific revolution.

OTHER ARABIC CONTRIBUTIONS TO THE WORLD OF SCIENCE

The Arabs also contributed significantly in the field of technology, developing upon the knowledge passed down by the Ancient Greeks.

The Greek school (which had turned out such prestigious scholars as Archimedes, Ctesibios, Philon of Byzantium and Hero of Alexandria) had seen the discovery of quite advanced mechanical technology: the endless screw, the hollow screw, pulley blocks, mobile pulleys, levers and weapons; clepsydras (types of clocks activated by water); astrolabes (for observing the positions of the stars and determining their height above the horizon); the construction of automata (devices capable of moving by themselves); the use of the odometer (an instrument designed to measure distances, comprising a series of chains and endless screws, moved by chariot wheels and pulling a needle along a graduated scale which indicated the distance travelled); etc. [see A. Feldman and P. Fold (1979); B. Gille (1980, 1978); C. Singer (1955–7); D. de Solla Price (1975)].

The Greeks of Byzantium carried on the work of the Greeks of Alexandria, and, to a certain extent, were one of the transmission links with mediaeval Europe.

However, the handing-on of the Greek tradition was also and above all the work of the engineers of the Muslim world. Here again, the Arabs gathered all the information, then made improvements and even innovations. Under orders from the caliph Ahmad ibn Mu'tasim, Qusta ibn Luqa al-Ba'albakki translated Hero of Alexandria's work on the traction of heavy bodies into Arabic; others translated or were inspired by the work of Philo of Byzantium [see B. Gille (1978)]. The Arabs also distinguished themselves in the art of clock-making. They even created their own inventions, above all in the field of automata and astronomical clocks, this being not only the legacy of the Greeks but probably also the Chinese, who were likewise experts in this field.

The following were amongst the most famous of the Arabic-Muslim engineers: the Banu Musa ibn Shakir brothers, whose works notably include *Al'alat illati tuzammi bi nafsiha* (*The Instrument Which Plays Itself*, written c. 850), largely inspired by the ideas of Hero of Alexandria; Ibn Mu'adh Abu 'Abdallah al-Jayyani, who wrote *Kitab al asrar fi nata'ij al afkar*, which describes several water clocks (second half of the tenth century); Badi'al-Zaman al-Asturlabi, famous for the automata he built for the Seleucid monarchs (first half of the twelfth century); 'Abu Zakariyya Yahya al-Bayasi, known for his mechanical pipe organs (second half of the twelfth century); and Ridwan of Damascus, made famous by his automata activated by ball-cocks (1203).

The most famous and most productive of the Arabic engineers was Isma'il ibn al-Razzaz al-Jazzari, whose *Kitab fi ma'rifat al hiyat al handasiyya* (*Book of the Knowledge of Ingenious Mechanical Instruments*, 1206) shows a perfect knowledge of Greek traditions and records apparently hitherto unknown innovations. This work not only contains the plans for constructing perpetual flutes, water clocks, mechanical pump systems for fountains, automata activated by ball-cocks and movements transmitted by chains and cords, it also contains descriptions of sequential automata, mainly using camshafts, thus transforming the circular movement of a type of crankshaft into the alternating movement of distribution instruments.

As well as the diverse instruments, there is also the astrolabe which became known in the West (at the same time as the "Arabic" numerals) thanks to Pope Sylvester II (Gerbert of Aurillac), who acquired the astrolabe from the Arabs when he lived as a monk in Spain from 967 to 970 CE.

There was also the compass, that ingenious device which has a magnetic needle and made navigation possible. It was invented by the Chinese at the beginning of the Common Era and was retrieved by the Arabs (in all likelihood in 752 during the battle of Talas), who in turn passed it on to the Europeans during the Renaissance.

The scholars of Islam also made their mark on the science of optics, which led to the invention of the mirror:

> Optics was particularly studied by Ibn al-Haytham. His work included physiological optics and a philosophical discussion of the nature of light, but he is known above all for his research in the field of geometry. He knew about reflection and refraction; he experimented with different mirrors, planes, spheres, parabolas, cylinders, both concave and convex. He wrote a text about the measurement of the paraboloid of revolution. He embarked on actual physical research through his work on the light of the stars, the rainbow, the colours, shadows and darkness. For a scientist of this calibre there was no fixed distinction between mathematical sciences and natural sciences, and Ibn al-Haytham was always shifting between the two [L. Massignon and R. Arnaldez].

Alchemy, too, was a fanciful art, the aim of which was to find the so-called "philosopher's stone" from which could be extracted a miraculous property which would at once cure all illnesses, give eternal life and transform metals: it was a vain science whose basis was denounced by great minds such as al-Kindi, Avicenna and Ibn Khaldun. However, as Diderot pointed out, alchemy, in spite of its frivolous nature, "often led to the discovery of important truths on the great path of the imagination". By stripping it of some of its arithmology and magic, the early Arabian scholars began to prepare the way for the creation and expansion of modern chemistry.

THE FORERUNNERS OF CONTEMPORARY SCIENCE?

Certain Arabic scholars, such as al-Biruni and Averroes, and doctors such as 'Ali Rabban at Tabari and Ibn Massawayh were well ahead of their time.

Perhaps the most significant contribution of the Arabic world, however, came from the historian 'Abd ar Rahman ibn Khaldun (who was born in Tunis in 1332 and died in Cairo in 1406), a visionary of modern science, who was gifted with truly extraordinary insight. One only need read this extract from his *Prolegomena* to appreciate his foresight: "The human world is the next step after the world of apes (*qirada*) where sagacity and wisdom may be found, but not reflection and thought. From this point of view, the first human level comes after the ape world: our observation ends here" [see *Muqaddimah*, p. 190; V. Monteil (1977), p. 101].

This is a surprising opinion for a time when such ideas were practically inconceivable. It would not be until 1859, in Darwin's *Origin of Species*, that these ideas would be presented and even then some time elapsed before they were accepted and developed in the Western world.

Thus, we can see how much Europe owes to this civilisation which is largely unknown or at least unrecognised by the Western public.

SIGNIFICANT DATES IN THE HISTORY OF ARABIC-ISLAMIC CIVILISATION

The following chronology is divided into sections, each representing half a century in the golden age of Arab-Islamic civilisation. Its aim is to give an idea of cultural, literary, scientific and technical activity which ran parallel to military and religious events. The list (which, of course, is not exhaustive) is of scholars and intellectuals, including the most illustrious poets, writers, mathematicians, physicians, astronomers, geographers, engineers, chemists, naturalists and doctors of the Arab world. In some cases, a summary of their fundamental contributions is supplied [see A. A. al-Daffa (1977); M. Arkoun (1970); O. Becker and J. E. Hoffman (1951); E. Dermenghem (1955); EIS; O. Fayzoullaiev (1983); A. Feldman and P. Fold (1979); L. Frédéric (1987 and 1989); L. Gille

(1978); C. Gillespie (1970–80); L. Massignon and R. Arnaldez in HGS; A. Mazaheri (1975); A. Mieli (1938); V. Monteil (1977); R. Rashed (1972); G. Sarton (1927); C. Singer (1955–7); H. Suter (1900–02); G. J. Toomer; K. Vogel (1963); H. J. J. Winter (1953); A. P. Youschkevitch (1976)].

Second half of the sixth century

c. 571 CE. "Year of the Elephant". Supposed birth-date of the prophet Mohammed.

First half of the seventh century

612. Year of the "Revelation", when Mohammed began his prophecy in Mecca.

622. Mohammed and the first followers of the new faith, the "Muslims" (*al muslimin*, from the Arabic word "believers") were expelled from Mecca. They found refuge in Yahtrib, which then became the "Town" of the Prophet or "Medina" (*madinah*). This year marked the beginning of the Muslim calendar, which is called the Hegira (from *hijra*, "expatriation").

624. Battle of Badr. The *qibla* is established, the symbol of the "new people of God". Beginning of the "Muslim institutions".

628. Mecca is seized by Mohammed and his followers.

632. The death of Mohammed.

632–661. Time of the "orthodox" caliphs (Abu Bakr, 'Umar, 'Uthman and 'Ali); capital: Medina.

632–634. Abu Bakr is caliph, the successor of Mohammed.

634. The conquest of Syria by the Arabs, who defeat Heraclius's Byzantine army near Jerusalem.

634–644. 'Umar (Omar) is caliph.

635. The Arabs take Damascus and overturn the Persian Empire.

637. Battle of Kadisiya, defeat of the Persians.

637–638. Founding of the towns Basra and Kufa. The writing of the Koran begins.

637–640. Conquest of Mesopotamia, Khuzistan, Azerbaidjan and Media.

638. Jerusalem is surrendered to Omar.

639. Arabs attack Armenia.

640. The conquest of Palestine.

641. Egypt is conquered by the Arabs.

642. Victory over the Persians.

642–646. The Arabs attack Armenia.

643. The Arabs complete their conquest of Persia and Tripolitania, and arrive in Sind (now Pakistan).

644–656. Rule of 'Uthman (Ottoman).

647. Barka in Tripolitania is taken (now Libya).

649. Cyprus is conquered by the Arabs.

Second half of the seventh century

655. Battle of Lycia, where the Muslim fleet destroys the Byzantine fleet.

656–661. Rule of 'Ali, the son-in-law of the prophet.

657. Battles of Jamal and Siffin, where the followers of 'Ali (then considered to be the first man converted to Islam) fought the followers of Mu'awiyah (rival and hostile descendants of Mohammed's family).

661–750. The Omayyad Dynasty. Capital: Damascus. Rule henceforth becomes hereditary. Effort to centralise Omayyad administration.

665. First attacks in the Maghreb.

670. Successful campaigns in North Africa. Founding of the town of Qairawa (Kairouan, in present-day Tunisia). Appearance and beginning of Shiite and Kharajite movements.

673–678. Siege of Constantinople by the Arabs.

680. Death of Husayn in Kerbala. Martyrdom of the Shiites.

695. First use of coins by the Arabs.

Culture, Science and Technology
Period of:

- the poet Imru' al-Qays.
- the poet Yahya ibn Nawfal al-Yamani.
- Khalil ibn Ahmad (one of the founders of Arabic poetry).

First half of the eighth-century

707. Development of political, "courtly" and urban poetry. First theological-political discussions.

707–718. The Muslims seize the mouths of the River Indus (Sind) and part of the Punjab (India).

709. The Maghreb surrenders to Arabic domination.

711. Musa Ben Nusayir dispatches Tariq ibn Ziyad, who crosses the Straits of Gibraltar (called Jabal Tariq), then successively occupies Seville, Cordoba and Toledo, before continuing north.

712. Arabic conquest of Samarkand (now Uzbekistan).

715. The Arabic Empire extends its confines to China and the Pyrenees.

718. The Arabs meet resistance from the Byzantine army at Constantinople. Thus the Arabic advance comes to a halt at the Taurus mountains.

720. The Arabs cross the Pyrenees and penetrate the kingdom of the Franks. First Arabic colony in Sardinia.

732. The Arabs are defeated at Poitiers by Charles Martel; the end of the Arabic advance in Europe.

Culture, Science and Technology
Period of:

- the Christian doctor Yuhanna ibn Massawayh.
- the poets al-Farazdaq, al-Akhtal and Jarir.
- the mystic thinker Hasan al-Basri.
- the Arabic version of the *Kalila wa Dimna* fables by Ibn al-Muqafa' (ancient Persian tale inspired by the Indian *Pañchatantra*).
- the first paintings of Islamic art.

Second half of the eighth century

750. Abu'l 'Abbas founds the Dynasty of the same name.

750–1258. Abbasid Dynasty. Capital: Baghdad (from 772).

751. Battle of Talas in present-day ex-Soviet Kyrghyzstan, where the Chinese armies are defeated by Arab troops. But Chinese reprisals later stop the Arabic advance at the limits of Sogdania.

754–775. Reign of the Abassid caliph al-Mansur.

756–1031. Omayyad Dynasty in Spain. Capital: Cordoba.

760. Arabic expedition against Kabul (in Afghanistan).

761–911. Rustamid Dynasty in Tiaret.

762. Caliph al-Mansur founds the town of Baghdad.

768. Sind is governed by the Arabs.

786. The Arabs seize Kabul.

786–809. Reign of the Abassid caliph Harun al-Rashid.

786–922. Idrissid Dynasty in the Maghreb. Capital: Fez.

795. Disorder in Egypt.

Culture, Science and Technology
Period of:

- the introduction of Indian science, numerals and calculation to the Islam world.
- the Persian astronomers Abu Ishaq al-Fazzari, and Muhammad al-Fazzari (his son), and of Jewish astronomer Ya'qub ibn Tariq. These are the men who would translate the *Brahmasphutasiddhânta* by Brahmagupta and study, for the first time in Islam, Indian astronomy and mathematics.
- the Persian astrologer al-Nawabakht and his son al-Fadl, chief librarian of caliph Harun al-Rashid.
- the Jewish astronomer Mashallah.
- the Christian Abu Yahya, translator of *Tetrabiblos* by Ptolemy.

- the Persian Christian Ibn Bakhtyashu', first of a large family of doctors, head of the hospital at Jundishapur.
- the Sabaean alchemist Jabir ibn Hayyan (Gebir in mediaeval Latin) who studied chemical reactions and bonds between chemical bodies.
- the alchemist Abu Musa Ja'far al-Sufi who wrote that there are two types of distillation, depending on whether or not fire is used.
- the Christian astrologer Theophilus of Edesse, translator of Greek works.
- the philologist and naturalist al-Asma'i.
- Abu Nuwas, one of the greatest Arabic poets.
- the poet Abu al-'Atahiya.
- the mystic thinker Abu Shu'ayb al-Muqafa.
- and the first production of paper in Islam.

End of the eighth century

At this time, the provinces of Africa, the Maghreb and Spain freed themselves from the links with the caliph of Baghdad.

Ninth – tenth century

This was the time of the development of the Sunni (Hanbali, Maliki, Hanafi, Shafi, Mutazili, Zahiri, etc.) and Shiite (Immami, Zayidi, Ismaeli, etc.) religious movements and of the mystical philosophy of the Sufi; popular Islam prevailed over classical Islam, which was reduced to a few common cultural and religious signs. This time was also marked by the rapid development of Arab-Islamic civilisation in all fields. It was also the era when the *Alf layla wa layla*, the *Thousand and One Nights* was written (anonymous masterpiece of Arabic literature, a collection of tales and legends, such as those of Scheherazade, Ali Baba, Sinbad the sailor, the magic lamp, etc., which have become an integral part of universal mythology).

First half of the ninth century

800. Charlemagne is named Emperor of the West.
800–809. Aghlabite Dynasty in "Ifriqiya" (territory composed of present-day Tunisia and part of Algeria).
813–833. Reign of the Abassid caliph al-Ma'mun who, as a grand patron, would favour cultural and scientific translations.
820–999. Independent indigenous dynasties in eastern Persia: Tahirid (820–873), Saffarid (863–902), Samanid (874–999).
826. Crete is taken by the Arabs.
827–832. Sicily is taken.
846. Sacking of Rome by the Saracens.

Culture, Science and Technology

Period of:

- the founding of the "House of Wisdom" (*Bayt al-Hikma*) in Baghdad, a kind of academy of sciences, where the cultural heritage of Antiquity was welcomed with enthusiasm and where the development of Arab-Islamic science began.
- the Persian astronomer and mathematician al-Khuwarizmi. His work on the Indian place-value system and on algebra with quadratic equations contributed greatly to the knowledge and propagation of Indian numerals, methods of calculation and algebraic procedures, not only in the Muslim world but also in the Christian West. He also wrote an interesting series of problems with examples taken from the methods of merchants and executors which required a great deal of mathematical skill due to the complex structure of the legacies of the Koran.
- the mathematician 'Abd al-Hamid ibn Wasi ibn Turk.
- the Christian translator Yahya ibn Batriq.
- al-Hajjaj ibn Yusuf, translator of Euclid's *Elements*.
- the astronomer and mathematician al-Jauhari, who carried out some of the first work on the parallel postulate.
- the converted Jewish astronomer Sanad ibn 'Ali, who had the observatory built in Baghdad.
- the philosopher al-Nazzam.
- the great philosopher and physician al-Kindi, who was interested in logic and mathematics, and sought to analyse the essence of definition and demonstration; he also wrote about geometrical optics and physiology.
- the philosopher al-Jahiz, author of the famous *Book of Animals*.
- the Persian Christian astronomer Yahya ibn Abi Mansur, who drew up *Al zij al muntahan* (*Established Astronomical Tables*).
- the astronomer Abu Sa'id al-Darir, from the Caspian region, who wrote about the course of the meridian.
- the astronomer al-Abbas, who introduced the tangent function.
- the astronomer Ahmad al-Nahawandi of Jundishapur.
- the astronomer Hasbah al-Hasib, from Merv, who established a table of tangents.
- the astronomer al-Farghani, who wrote an Arabic version of Ptolemy's *Almagestus*.
- the astronomer al-Marwarradhi, from Khurasan.
- the astronomer 'Umar ibn al-Farrukhan, from Tabaristan.
- the Jewish astronomer Sahl at Tabari, from Khurasan.
- the Jewish astronomer Sahl ibn Bishr, from Khurasan.
- the astrologer Abu Ma'shar, from Balkh (Khurasan).
- Ali ibn 'Isa al-Asturlabi, famous maker of astronomical instruments.

- al-Himsi, who made the work of Apollonius known.
- the Banu Musa ibn Shakir brothers, translators, mathematicians and engineers, who wrote a work on automata.
- Ibn Sahda, who translated medical works.
- the Christian doctor Jibril ibn Bakhtyashu'.
- the Christian doctor Salmawayh ibn Bunan.
- the surgeon Abu'l Qasim az Zahrawi (Abulcassis in mediaeval Latin), from Cordoba.
- the Christian pharmacologist Ibn Massawayh, author of *Aphorisms*.
- the writer As Suli.
- the doctor and philosopher 'Ali Rabban at Tabari, author of *Paradise of Wisdom*, inspired by the *Aphorisms* of the Indian Brahman heretic Chanakya of the third century BCE.
- the mystical thinkers Dhu 'an Nun Misri, al-Muhasibi, Ibn Karram, Bistami.
- and the poets Abu Tammam and Buhturi.

Second half of the ninth century

868–905. Tulunid Dynasty in Egypt and Syria.

869. Malta is taken by the Arabs.

875–999. Samanid Dynasty in the north and east of present-day Iran, Tadjikistan and Afghanistan. Capital: Bokhara.

880. Italy is recaptured from the Arabs by Basil I.

Culture, Science and Technology
Period of:

- al-Mahani the geometer and astronomer from the region of Kirman, who studied the problems of the division of the sphere using the cubic equation which bears his name.
- al-Nayrizi (Anaritius in mediaeval Latin), astronomer and mathematician from the Shiraz region, who wrote commentaries on Euclid and Ptolemy.
- the Egyptian mathematician Ahmad ibn Yusuf, who wrote a work dealing with proportions.
- the mathematician Thabit ibn Qurra, who translated Archimedes's treatise on the sphere and the cylinder and who did important work on conic sections; he also produced a brilliantly clear proof of Pythagoras's theorem, the first general rule for obtaining pairs of amicable numbers* and a method for constructing magic squares.

---

* Two numbers are "amicable" if the sum of the distinct divisors of each one (including 1 but excluding the number itself) is equal to the other number. For instance, 220 has divisors 1, 2, 4, 5, 10, 11, 20, 22, 44, 55, 110, which add up to 284; while 284 has divisors 1, 2, 4, 71, 142, which add up to 220. The numbers 220 and 284 form an "amicable pair", and they are the smallest to do so.

- the mathematicians Abu Hanifa Ahmad and al-Kilwadhi.
- al-Battani (Albategnus in mediaeval Latin) the astronomer who accompanied his theory of planets with insights into trigonometry, which were later to be thoroughly investigated by Western astronomers; he determined the inclination of the ecliptic and the precession of the equinoxes with great accuracy using cotangents.
- the astronomer Hamid ibn 'Ali.
- the Persian astrologist Abu Bakr.
- Qusta ibn Luqa al-Ba'albakki, the Christian mathematician and engineer of Greek origin, who in particular translated Hero of Alexandria's *Mechanics* which deals with the traction of heavy objects, as well as works by Autolycus, Theodosius, Hypsicles and Diophantus.
- the Christian doctor Hunayn ibn Ishaq, who translated Greek medical works into Arabic, as well as works by Archimedes, Theodosius and Menelaus.
- the Christian Yahya ibn Sarafyun, who wrote a medical encyclopaedia in Syriac.
- the pharmacologist Sabur ibn Sahl, from Jundishapur, author of a book of antidotes.
- Muhammad Abu Bakr Ben Zakariyya al-Razi (Razhes in mediaeval Latin) the great Persian clinician, alchemist and physician who was thought to be the greatest doctor of his age; he first differentiated between German measles and measles; he described how to equip a chemical "laboratory" and his *Sirr al Asrar* (*The Secret of Secrets*), contains important work on distillation.
- the philosopher Abul Hasan 'Ali ibn Ismail al-Ash'ari, founder of Muslim scholasticism and of the Mutaqallimin school. He expounded a theological system based on an atomism similar to that of Epicurus.
- the geographer al-Ya'qubi.
- the Persian geographer Ibn Khurdadbeh, alias Ibn Hauqal, author of the *Book of Routes and Provinces*.
- the mystical thinker Tirmidhi, known as "the philosopher".
- the poets Mutanabbi and Ibn Sa'ad.

First half of the tenth century

905. End of Tulunid Dynasty in Egypt, power taken by the governors of the caliphs.
909. Beginning of the rule of the Fatimid caliphs in Ifriqiya.
932–1055. The Buyid Dynasty, unifying eastern Persia and Media.
935. Muhammad ibn Tughaj reconquers Alexandria and southern Syria.
943. The Caliphate of Baghdad confers the rule of Egypt to Ibn Tughaj for thirty years.

945. The Buyids enter Baghdad. The Caliphate is now no more than a "legal fiction".

Culture, Science and Technology
Period of:

• Abu Kamil, the great algebraist from Egypt, who continued the work of al-Khuwarizmi, and whose algebraic discoveries were to be used, c. 1206, by the Italian mathematician Leonard of Pisa (or "Fibonacci"); also devised interesting formulas related to the pentagon and decagon.

• the geometer Abu 'Uthman, translator of the tenth book of Euclid's *Elements* and of Pappus's *Commentary*.

• the Christian translators Matta ibn Yunus and Yahya ibn 'Adi.

• Sinan ibn Thabit, mathematician, physician, astronomer and doctor.

• the mathematician Ibrahim ibn Sinan ibn Thabit, who dealt with the problem of constructing conic sections, and who studied the surface of the parabola and conoids.

• the mathematician Abu Nasr Muhammad, who made an interesting discovery with his theorem of sines in plane and spherical trigonometry.

• the mathematician Abu Ja'far al-Khazini, from Khurasan, who worked on algebra and geometry, and who solved al-Mahani's cubic equation by using conic sections.

• the astronomer al-Husayn Ben Muhammad Ben Hamid, known as Ibn al-Adami.

• the astrologist and mathematician al-Imrani, who wrote a commentary on Abu Kamil's algebra.

• the arithmeticians Ali ibn Ahmad and Nazif ibn Yumn al-Qass.

• Bastulus, the famous maker of astronomical instruments.

• the great geographer and mathematician al-Mas'udi.

• the geographer Qudama.

• the geographer Abu Dulaf.

• the geographer Ibn Rusta of Isfahan.

• the Persian geographer Ibn al-Faqih, from Hamadan.

• the geographer Abu Zayd, from Siraf (Arabic-Persian gulf).

• the geographer al-Hamdani, from the Yemen.

• the philosopher al-Farabi (Alpharabius in mediaeval Latin), from Turkestan, who devised a metaphysics based on Aristotle, Plato and Plotinus and who, in his *Ihsa' al 'ulum*, came up with a "Classification of the Sciences" in five branches: linguistics and philology; logic; mathematical sciences, subdivided into arithmetic, geometry, perspective, astronomy, mechanics and gravitation; physics and metaphysics; and finally the political, legal and theological sciences.

- the alchemist and agronomist known as Ibn Wahshiya.
- the mystic thinkers Junayd and Abu Mansur ibn Husayn al-Hallaj.
- the poet Ibn Dawud.
- and the Persian poet Rudaki.

Second half of the tenth century

957. The Byzantines in northern Syria.

961–969. The Byzantines reconquer Crete and Cyprus, as well as Antioch and Aleppo (Syria).

962. A Turkish tribe conquers the Afghan kingdom of Ghazna.

969. The Fatimids of Tunisia occupy Egypt, which puts up no resistance, then settle there.

972–1152. Zirid and Hammadid Dynasties in Ifriqiya.

973. Foundation of *Al Kahira* (Cairo).

998–1030. Reign of Mahmud, or "the Ghaznavid" (because he settled in Ghazna), over what is now Afghanistan, Khurasan and various annexed regions in the north of India.

Culture, Science and Technology
Period of:

- the founding in Cairo of the *Dar al Hikma* (House of Wisdom), a sort of scientific academy similar to that of Baghdad.
- the founding of the al-Azhar university in Cairo.
- the blossoming of the sciences in the Caliphate of Cordoba, thanks to Caliph al-Hakam II, who put together an immense library.
- the mathematician Abu'l Wafa' al-Bujzani, from Quistan, who wrote commentaries on Euclid, Diophantus and al-Khuwarizmi. He introduced the tangent function, and his work on trigonometry led to great improvements in methods of solving spherical triangles where, instead of Ptolemy's formula (derived from Menelaus's theorem) which involved the six great-circle arcs of a quadrilateral, a formula involving the four arcs generated on a transversal of the figure composed of a spherical triangle and the perpendiculars dropped from two of its vertices to opposite sides is used. Thanks to him, the Arabs acquired Diophantus's *Arithmetica* with its studies of algebra and number theory.
- the mathematician al-Uqlidisi (whose name means "the Euclidean") who published an important study of decimal fractions.
- the mathematician Ibn Rustam al-Kuhi, from Tabaristan, who studied geometrical problems posed by Archimedes and Apollonius.
- the Persian mathematician/astronomer Abu'l Fath from Isfahan, who revised the Arabic translation of much of Apollonius's work.

- the mathematician al-Sijzi, from Sigistan, who studied problems of conic intersections and the trisecting of angles.
- the mathematician al-Khujandi, from the Sir Daria region, who established a proof concerning the sine in spherical triangles, and proved that the sum of two cubes cannot equal a cube.
- the mathematicians Sinan ibn al-Fath and Abu Nasr.
- the secret society of the Brothers of Purity (*Ikhwan al-Safa*), whose *Epistles* were a sort of encyclopaedia based on Pythagorean and Neo-Platonic mysticism, and which divided the sciences into four sorts: mathematics; science of physical bodies; science of rational souls; and science of divine laws.
- the Andalusian astronomer and mathematician Maslama ibn Ahmad, based in Cordoba.
- the Persian astronomer and mathematician ʿAbd ar-Rahman al-Sufin, who drew up a catalogue of stars containing the first observation of the Andromeda nebula.
- the great doctor and surgeon Abu'l Qasim from Zakna, near Cordoba, author of the *Kitab al-Tasrif*, which deals with practical surgery, cauterising wounds, tying up arteries, operating on bones, the eyes, etc.
- ʿAli ibn ʿAbbas, a doctor from southern Persia.
- Abu Mansur Muwaffak, a doctor who wrote an important medical treatise in Persian.
- the Andalusian doctor Ibn Juljul.
- the Persian geographer al-Istakhri, from near Persepolis.
- the geographer Buzurg ibn Shahriyyar, from Khuzistan.
- the Palestinian traveller and geographer al-Muqaddasi, from Jerusalem.
- the philosopher Abu Bakr Ahmad ibn ʿAli al-Baqilani.
- the historian Ya ʿqub ibn al-Nadim, author of *Kitab al-Fihrist al ʿulum* (*The Book and Index of Sciences*) containing biographies of contemporary thinkers.
- the mechanical and hydraulic engineer Ibn Muʿadh Abu ʿAbdallah al-Jayyani, author of an important treatise on water clocks.
- and the mystic thinker Tawhidi.

First half of the eleventh century

1000. First clashes in Khurasan between the Ghaznavids and the Seljuks (Turkomans pushed out of Central Asia by the Chinese).

1001–1018. Mahmud the Ghaznavid takes possession of Peshawar, crushes a Hindu coalition and sacks Muttra, one of India's holy cities.

1030. The Ghaznavids crushed by the Seljuks.

1030–1050. The Seljuks occupy various towns in eastern then western Persia (where they clash with the Buyids), before turning away towards Syria and Asia Minor.

Culture, Science and Technology
Period of:

- the mathematician, astronomer, physician and geographer al-Biruni, from Khiva in Kharezm, who travelled widely in India, where he learnt Sanskrit and became acquainted with Indian science; he later took back what he had learnt and wrote numerous works on astronomy, arithmetic and mathematics; he also made a new calculation of trigonometric tables based on Archimedes's premises, an equivalent of Ptolemy's theory.
- the mathematician al-Karaji, who did important work on the arithmetic of fractions; basing himself on the work of Diophantus and Abu Kamil, he devised an algebra in which, alongside the standard forms of second degree equations, he dealt with certain 2n degree equations; his work showed how a rigorous approach can, by using irrational numbers, arrive at forms that are more supple than those of Greek geometric algebra; this was, in fact, the start of a development which would lead to the elimination of geometrical representations in Arabic arithmetic and algebra thanks to the use of symbols.
- the mathematician Kushiyar ibn Labban al-Gili, from the south of the Caspian Sea, who worked on Indian arithmetic and sexagesimal calculations.
- the Persian mathematician An Nisawi, from Khurasan, who continued the work of al-Khuwarizmi in arithmetic and algebra.
- the mathematician Abu'l Ghud Muhammad ibn Layth.
- the mathematician Abu Ja'far Muhammad ibn al-Husayn.
- the astronomer Ibn Yunus, appointed to the Dar al-Hikma observatory in Cairo.
- the mathematician, physician and doctor Ibn al-Haytham (Alhazen in mediaeval Latin) whose *Book of Optics* contains important discoveries about eyesight, the theory of the reflection and refraction of light, the fundamental laws of which he established.
- the Andalusian mathematician al-Kirmani, from Cordoba.
- the Andalusian mathematician and astronomer Ibn al-Samh, from Granada.
- the mathematician and astronomer Ibn Abi'l Rijal (Abenragel in mediaeval Latin), from Cordoba but working in Tunis.
- the Andalusian mathematician and astronomer Ibn al Saffar, from Cordoba.

- the great philosopher Al Husayn ibn Sina (Avicenna), a universal mind as interested in medicine as in mathematics; based on Aristotle's ideas, his philosophy rejected mysticism and theology and dwelt instead on science and nature; his *Canon Medecinae* remained a text-book in Europe until the seventeenth century; in his *Aqsam al 'ulum al 'aqliyya* (or *Division of the Rational Sciences*) he drew up a consistent classification by means of an analytical division to allow a hierarchy of the sciences to be established.
- the Christian philosopher Miskawayh whose rational thought makes him one of Ibn Khaldun's precursors.
- the chemist al-Kathi.
- the Christian doctor Massawayh al-Mardini, settled in Cairo.
- the doctor 'Ali ibn Ridwan, from Cairo.
- the doctor Abu Sa'id 'Ubayd Allah.
- the Andalusian doctor Ibn al-Wafid, from Toledo.
- the Jewish doctor Ibn Janah, author of a treatise on medicinal herbs.
- the doctor Ibn Butlan.
- the oculist 'Ammar.
- the oculist 'Ali ibn 'Isa, author of an important treatise on ophthal-mology.
- the jurist and poet Ibn Hazem.
- the atheist Syrian poet Abu'l 'ala al-Ma'ari.
- and the Persian poet Abu'l Qasim Firduzi, author of the famous *Book of Kings*.

Second half of the eleventh century

1050. Troubled times in Egypt. Order re-established by Badr al-Jamali who then ruled over Egypt until 1121 for the Fatimids.

1055. Tughril Beg, the Seljuk, enters Baghdad as the protector of the Empire of the Caliphs.

1055–1147. Almoravid Dynasty in the Maghreb.

1062. Yusuf Ben Tashfin (the true founder of modern Morocco) founds Marrakech, which becomes the Almoravids' capital.

1069. Yusuf Ben Tashfin takes Fez, an Arab-Islamic centre, then develops its intellectual, artistic and economic activities.

1076. The Seljuks take Damascus and Jerusalem.

1085. The Christians occupy Toledo.

1086. Faced with a dangerous situation in Andalusia, Yusuf Ben Tashfin declares a "Holy War" against Christian Spain, stops the activities of Alphonsus VI of Castile, then annexes the whole of the south of Spain, uniting it with the Maghreb and thus forming the Almoravid Empire.

1090. The Turks arrive between the Danube and the Balkans.

1096. Start of the First (People's) Crusade. The badly organised crusaders are massacred in Asia Minor.

1097–1099. The crusaders take Nicea, defeat the Turks at Dorylaeum then take Jerusalem.

Culture, Science and Technology

Period of:

- the great Persian poet and mathematician 'Umar al-Khayyam (Omar Khayyam), from Nishapur, author of the *Rubaiyat*, the famous collection of poems; he also wrote commentaries on Euclid's *Elements*, worked on the theory of proportions and studied third-degree equations, suggesting geometric solutions for some of them.
- the mathematician Yusuf al-Mu'tamin (the enlightened king of Saragossa).
- the mathematician Muhammad ibn 'Abd al-Baqi.
- the Andalusian astronomer al-Zarqali, from Cordoba, who reworked Ptolemy's *Planisphaerium*.
- the poet, philosopher, mathematician and astronomer al-Hajjami, who played a vital part in reforming the calendar; he also provided an overview of third-degree equations and made an important study of Euclid's postulates.
- the Persian oculist Zarrin Dast.
- the Andalusian geographer and chronicler al-Bakri, from Cordoba.
- the doctors Ibn Jazla and Sa'id ibn Hibat Allah.
- the Andalusian agronomist Abu 'Umar ibn Hajjaj, from Seville.
- the mystic philosopher Abu Hamid al-Ghazali (Algazel in mediaeval Latin), whose teachings stood against Islam's scientific progress.
- the "sociologist" al-Mawardi.
- and the Persian poet Anwari.

First half of the twelfth century

1100. Foundation of the Christian Kingdom of Jerusalem.

1125. Revolt of the Masmudas of the Atlas under Ibn Tumert, the inventor of the Almohad doctrine.

1136. Cordoba, Western Islam's cultural capital, is taken by Ferdinand III, king of Castile and Leon.

1147. 'Abd al-Mu'min, the successor of Ibn Tumert, destroys the power of the Almoravids and proclaims himself Caliph after taking Fez (in 1146) and Marrakech (in 1147). He then extends his conquests to Ifriqiya and reaches Spain.

1147–1269. The Almohad Dynasty in the Maghreb and Andalusia.

1148. The crusaders defeated at Damascus.

Culture, Science and Technology
Period of:

- the Andalusian mathematician Jabir ibn Aflah, from Seville, particularly famous for his work on trigonometry.
- the great Andalusian geographer al-Idrisi, who made important contributions to the development of mathematical cartography.
- the Jewish mathematician from Spain Abraham Ben Meïr ibn 'Ezra (better known as Rabbi Ben Ezra).
- the engineer Badi al-Zaman al-Asturlabi, famous for the automata he made for the Seljuk kings.
- the philosopher Ibn Badja (Avempace in mediaeval Latin and during the Renaissance).
- the philosopher and doctor Abu al-Barakat, author of the *Kitab al mu'tabar* (*Book of Personal Reflection*).
- and the Andalusian philosopher Ibn Zuhr (alias Avenzoar).

Second half of the twelfth century

1150. Allah ud din Husayn, Sultan of Ghur, destroys the Ghazni Empire.

1169–1171. Salah ad din (Saladin), a Muslim of Kurdish origins, succeeds his uncle as Vizier of Egypt then ends the reign of the Fatimids by recognising only the suzerainty of Nur ad din, the unifier of Syria, and the Abbasid Caliph of Baghdad.

1174. Salah ad din succeeds Nur ad din and founds the Ayyubid Dynasty which dominated Egypt and Syria thenceforth. Leaning on the Arab traditionalists, he declares "Holy War" against the Christians of the West, hence reinforcing the links between the eastern peoples and their Arab-Islamic traditions.

1187. Salah ad din takes back Jerusalem. Victory of the Almohads at Gafsa under the Maghribi Sultan Abu Yusuf Ya'qub al-Mansur.

1188. Genghis Khan unifies the Mongols.

1191. Under Muhammad of Ghur, the Islamic Afghan and Turkish tribes of Central Asia try to conquer the north of India, but are pushed back at the very gates of Delhi.

1192. Battle of Tarain: Muhammad of Ghur defeats Prithiviraj and takes Delhi.

1192–1526. Sultanate of Delhi.

1193. The Muslims take Bihar and Bengal.

1195. Victory of the Almohads at Alarcos.

Culture, Science and Technology

Period of:

- the mathematician al-Amuni Saraf ad din al-Meqi.
- the converted Jewish mathematician, philosopher and doctor As Samaw'al ibn Yahya al-Maghribi, from the Maghreb, who continued the work of al-Karaji.
- the Persian encylopaedist and mathematician Fakhar ad din al-Razi.
- the Persian engineer Abu Zakariyya Yahya al-Bayasi, famous for his mechanical pipe organs.
- the great Jewish philosopher Maimonides (Rabbi Moshe Ben Maimon), from Cordoba, whose encyclopaedic interests included astronomy, mathematics and medicine.
- the great Andalusian philosopher Ibn Rushd (Averroes), born in Cordoba and died in Marrakech, the finest flowering of Arab philosophy and a profound influence on the West.
- the Maghrebi philosopher Ibn Tufayl (Abubacer in mediaeval Latin and during the Renaissance).
- the mystical thinker Ruzbehan Baqli.
- the Persian mystical poet Nizami.
- and the Persian poet Khaqani.

First half of the thirteenth century

1202. The Muslims arrive on the banks of the Ganges, at Varanasi (Benares).

1203. Continuation of Muhammad of Ghur's conquest of northern India.

1206–1211. Reign of Qutb ud din (Sultanate of Delhi).

1208. The Albigensian Crusade.

1212. Defeat of the Almohads at Las Navas de Tolosa.

1211–1222. Under Genghis Khan, the Mongols invade China, Transoxiana and Persia, before continuing their migration under Hulagu Khan towards Mesopotamia and Syria.

1211–1227. Reign of Iltumish (Sultanate of Delhi) who obtains recognition of his authority over India from the Caliph of Baghdad. Under this domination, India will remain relatively stable until 1290.

1214–1244. The Banu Marin (Merinids) conquer the north of the Maghreb.

1221. The Mongols press against the borders of the Sultanate of Delhi, but are held back by Iltumish.

1227. Death of Genghis Khan, whose empire stretched from the Pacific to the Caspian Sea.

1248. The Christians take back Seville from the Muslims.

Culture, Science and Technology
Period of:

- the mathematician Muwaffaq al din Abu Muhammad al-Baghdadi.
- the leading court official and patron Abu'l Hasan al-Qifti, author of *Tarikh al huqama* (*Chronology of the Thinkers*).
- Muhammad ibn Abi Bakr, famous maker of astronomic instruments.
- the engineer Ridwan of Damascus, best known for his ball-cock automata.
- the great Persian engineer al-Jazzari, author of the *Book of the Knowledge of Ingenious Mechanical Instruments*, in which he provided plans for perpetual flutes, water clocks, and different sorts of sequential automata using ball-cocks and camshafts.
- Ya'qub ibn 'Abdallah ar Rumi, who produced an important encyclopaedia of Arab geography.
- the esoteric Muslim Ibn 'Arabi.
- and the poets Ibn al-Farid and Shushtari.

Second half of the thirteenth century

1250. The Mamelukes take power in Egypt. The Kingdom of Fez created by the Banu Marin (Merinids).
1254–1517. Reign of the Mamelukes in Egypt.
1258. Hulagu Khan's Mongols retake and sack Baghdad.
1259. Mongol invasion of Syria.
1260. Mongols crushed at the border of Egypt by the Mameluke monarch.
1261. Egypt becomes the centre of the Arab world and also, to a certain degree, of the Islamic world.
1269. In the Maghreb, the Banu Marin take Marrakech and found their own dynasty (the "Merinids").
1291. The Mamelukes take Acre and eliminate the Christians on the Syrian–Palestinian coast.
1297. 'Ala ud din Khalji (Sultanate of Delhi) defeats the Mongols then starts sacking Gujarat and Rajasthan.

Culture, Science and Technology
Period of:

- the mathematician and astronomer Nasir ad din at Tusi, from Tus in Khurasan, who did important work on arithmetic, algebra and geometry; his work undoubtedly marks the high point of Arabic trigonometry, dealing thoroughly with spherical right-angled triangles and successfully broaching the study of spherical triangles in general, even bringing in the polar triangle; in astronomy, he published his famous "Ikhanian" Tables; and, in geometry, he corrected the translations of Greek geometrical works and his discussion of Euclid's propositions was later to inspire the Italian mathematician Saccheri in his initial research in 1773 into non-Euclidean geometry.

- the doctor Ibn al-Nafis of Damascus, who wrote a commentary on Avicenna's *Canon*, with important developments concerning pulmonary circulation.
- the pharmacologist and botanist Ibn al-Baytar.
- the Persian mystical poet and hagiographer Farid ad din 'Attar.
- the Persian poet Sa'adi of Chiraz, author of the *Gulistan*.

First half of the fourteenth century

1306. The Sultans of Delhi repulse the Mongols once more.

1307–1325. The Sultans of Delhi attack the kingdoms of Deccan and reach the south of India, conquering the lands of the Maratha, Kakatiya and Hoysala.

1333. The Moors recapture Gibraltar from the Kingdom of Castile.

Culture, Science and Technology
Period of:

- the great Maghrebi arithmetician Ibn al-Banna al-Marrakushi.
- the converted Jewish doctor and historian Rashid ad din, author of a *Universal History*, in which he reproduced large extracts of the best-known medical works of China and Mongolia.
- the historian al-Umari.
- the moralist Ibn Taymiyya.
- the Andalusian mystic thinker Ibn Abbad of Ronda.
- the great Maghrebi traveller Ibn Battutua who, in thirty years, covered more than 120,000 kilometres in the Islamic world, from Northern Africa to China, via India.
- and the Persian poets Hamdallah al-Mustawfi and Tebrizi.

Second half of the fourteenth century

1356. India is "given" by the Caliph of Baghdad to Firuz Shah Tughluq.

1371. The Ottomans defeat the Serbs at Chirmen.

1389. The Ottomans crush the Serbs at Kosovo Polje.

1390. The Ottomans occupy the remaining territories of the Byzantine Empire in Asia Minor.

1392. The Ottomans arrive in the Balkans.

1398–1399. Timur (Tamerlane) sacks Delhi.

Culture, Science and Technology

Period of:

- the great thinker Ibn Khaldun, from Tunis, remarkable for his rationalism, his feeling for general laws and his extraordinarily acute scientific insights; in many ways a precursor of Auguste Comte.
- the writer Ibn al-Jazzari.
- the writer Taybugha.
- and the Persian poet Hafiz of Chiraz, author of *Bustan*.

First half of the fifteenth century

1400–1401. Incursion of Timur and sacking of Baghdad.

1405. Return to Baghdad of the Jalayrid leaders.

1422. The Ottomans besiege Constantinople.

1400–1468. Constant disputes between Turkomans and Mongols.

1444. The Viceroy of the Baghdad Timurid Dynasty founds his empire in Mesopotamia and Kurdistan.

1447. End of the empire of Timur, independence of Persia and of the Afghan and Indian regions.

Culture, Science and Technology

Period of:

- Ulugh Bek, the enlightened monarch of Samarkand, builder of an observatory equipped with the finest instruments of the age; author of trigonometric tables, among the most precise of the numeric tables produced by Islam's thinkers.
- the Persian mathematician Ghiyat ad din Ghamshid ibn Mas'ud al-Kashi, who did important work on algebra, sexagesimal calculations and arithmetic, especially on the binomial formula, decimal fractions, exponential powers of whole numbers, n roots, the theory of proportions and irrational numbers.
- the historian al-Maqrizi.

Second half of the fifteenth century

1453. Constantinople falls to Sultan Mehmet II. The beginning of the Ottoman Empire, which will later cover Anatolia, Rumelia, Bulgaria, Albania, Greece, the Crimea, Syria, Mesopotamia, Palestine, Egypt, Hejaz, Armenia, Kurdistan and Bessarabia and which, after 1520, will extend its frontiers as far as Hungary, southern Mesopotamia, the Yemen, Georgia, Azerbaidjan, with Tripoli and the whole of Ifriqiya as dependencies (excepting the Maghreb which managed to remain autonomous during this period).

1468. The Turkoman al-Koyunlu establishes his authority in Mesopotamia.

1492. The Catholics Ferdinand and Isabella retake Granada.

1499–1722. Reign of the Safavids in Persia; Shi'ism becomes the official religion.

Culture, Science and Technology
Period of:
- the mathematician al-Qalasadi, who did important work on arithmetic, especially algebra, greatly developing its symbols.
- and the Persian historian Mirkhond.

The sixteenth century

1508. The Safavids push the Turkomans out of Mesopotamia.

1516. Turkish corsairs establish themselves in Algiers.

1517. Ottoman conquest of Syria and Egypt, thus ending the Caliphate of Baghdad and bringing about the fall of the Mamelukes in Egypt.

1524. Babur, a descendant of Timur, invades the Punjab and takes Lahore.

1526. Babur kills the last Sultan of Delhi and takes the throne. The beginning of the Mogul Empire in India and Afghanistan (1526–1707).

1571. Turks defeated by Holy League in naval battle of Lepanto.

1574. The central and eastern regions of North Africa come under Ottoman control.

1578–1603. Beginning of the Sa'adian Dynasty with the reign of al-Mansur (Maghreb).

Culture, Science and Technology
Period of:
- the Turkish arithmetician Tashköprüzada.
- and the Turkish poets Baki and Fuzuli.

The seventeenth century

1672–1727. Beginnings of the 'Alawite Dynasty in the Maghreb with the reign of Mulay Ismail, contemporary of Louis XIV.

Culture, Science and Technology
Period of:
- the mathematician Beha ad din al-Amuli.
- the arithmetician and commentator al-Ansari.
- the writers Hajji Khalifa, 'Abd al-Qadir al-Baghdadi, 'Abd ar Rashid Ben 'Abd al-Ghafur and Ad Damamini.
- the encylopaedist Jamal ad din Husayn Indju.

- the Turkish poets Nefi, Nabî and Karaja Oghlan.
- and the Turkish traveller and writer Evliya Chelebi.

The eighteenth century

1799. Start of the *Nahda* ("Renaissance").

Culture, Science and Technology
Period of:
- the Turkish poet Nedim.
- the Turkish writer and historian Naima.

The nineteenth century

1804. The Wahhabis take Mecca and restore Hanbali Islam.
1805–1849. Reign of Muhammad 'Ali, Pasha of Egypt.
1811–1818. 'Ali defeats the Wahhabis.

Culture, Science and Technology
Period of:
- the Turkish thinkers Namik Kemal, Ziya Pasha, Ahmet Mithat, Chinassi and Avdülhak Hamit.

Beginning of the twentieth century

1918–1922. Reign of Sultan Mehmed IV (whom the Treaty of Sèvres obliged to accept the dismemberment of the Turkish Empire. Turkey was reduced to the landmass of Anatolia).
1922. Mehmed IV overturned by Mustafa Kemal, founder of modern, republican Turkey.
1924. Official end of the Ottoman Empire.

THE ARRIVAL OF INDIAN NUMERALS IN THE ISLAMIC WORLD

How were Indian numerals and calculating methods introduced into Islam?

The Arabs possibly encountered them at the beginning of the eighth century CE, when Hajjaj sent out an army under Muhammad Ben al-Qasim to conquer the Indus Valley and the Punjab.

But it is far more likely that the army had nothing to do with it, and that it was necessary to wait for a delegation of scholars before Indian science was transmitted to the Islamic world.

This is, indeed, Ibn Khaldun's explanation, who says in his *Prolegomena* that the Arabs received science from the Indians, as well as their numerals and calculation methods, when a group of erudite Indian scholars came to the court of the caliph al-Mansur in year 156 of the Hegira (= 776 CE) [see *Muqaddimah*, trans. Slane, III, p. 300].

This is a late source, dating from about 1390. But Ibn Khaldun's version corresponds closely with earlier texts, especially with a tale told by the astronomer Ibn al-Adami in about 900, which is referred to by the court patron Hasan al-Qifti (1172–1288) in his *Chronology of the Scholars*:

> Al-Husayn Ben Muhammad Ben Hamid, known as Ibn al-Adami, tells in his Great Table, entitled *Necklace of Pearls*, that a person from India presented himself before the Caliph al-Mansur in the year 156 [of the Hegira = 776 CE] who was well versed in the *sindhind* method of calculation related to the movement of heavenly bodies, and having ways of calculating equations based on *kardaja* calculated in half-degrees, and what is more various techniques to determine solar and lunar eclipses, co-ascendants of ecliptic signs and other similar things. This is all contained in a work, bearing the name of Fighar, one of the kings of India, from which he claimed to have taken the *kardaja* calculated for one minute. Al-Mansur ordered this book to be translated into Arabic, and a work to be written, based on the translation, to give the Arabs a solid base for calculating the movements of the planets. This task was given to Muhammad Ben Ibrahim al-Fazzari who thus conceived a work known by astronomers as the *Great Sindhind*. In the Indian language *sindhind* means "eternal duration". The scholars of this period worked according to the theories explained in this book until the time of Caliph al-Ma'mun, for whom a summary of it was made by Abu Ja'far Muhammad Ben Musa al-Khuwarizmi, who also used it to compose tables that are now famous throughout the Islamic world [F. Woepcke (1863)].

Much can be learned from this. The repetition of the word *sindhind* is significant; it is the Arabic translation of the Sanskrit **siddhânta*, the general term for Indian astronomic treatises, which contained a complete set of instructions for calculating, for example, lunar or solar eclipses, including the trigonometric formulae for true longitude [see R. Billard in IJHS]. The "*sindhind*" method thus stands for the set of elements contained in such treatises. As for the word *kardaja*, which is also frequently used, it means "sine" and derives from an Arabic deformation of the Sanskrit *ardhajyâ* (literally "semi-chord") which Indian astronomers had used, from the time of **Âryabhata*, for this trigonometric function which is the basis of all calculations in the Indian *siddhânta* system.

This method is presented in the mathematician and astronomer Brahmagupta's (628) *Brahmasphutasiddhânta* and the astrologer *Varâhamihîra's (575) *Pañchasiddhântikâ*. But it was explained long before these treatises in the astronomer *Âryabhata's *Âryabhatîya* (c. 510).

Now, apart from the *Âryabhatîya* (which uses a special form of alphabetic numeration), all Indian astronomers noted their numbers by using Sanskrit numerical symbols: this notation gave them a solid base for noting numeric data and was based on a decimal place-value system using zero. As for their calculations, they used a system quite similar to our own one with their nine numerals plus a tenth sign written as a circle or point and acting as a true zero (see *Zero, etc).

In other words, when the Arabs learnt Indian astronomy, they inevitably came up against Indian numerals and calculation methods, so that the arrival of the two branches of knowledge precisely coincided. This is confirmed by al-Biruni's *Kitab fi tahqiq i ma li'l hind* (c. 1030), which tells of his thirty-year stay in India.

We must now try to date this transmission.

Now, al-Qifti, Ibn al-Adami and other authors agree on the date mentioned in the quotation above; i.e. 156 of the Hegira, or 776 CE. Several facts about Arabic science make this date plausible. According to A. P. Youschkevitch:

> If the arrival of Indian scholars gave the astronomers of Baghdad the possibility of acquainting themselves with the astronomy of the *siddhânta*, there was already much interest in the subject. Three astronomers who worked during the reign of Caliph al-Mansur are known to us, thanks to al-Qifti: Abu Ishaq Ibrahim al-Fazzari (died c. 777) who first made Arabic astrolabes, his son Muhammad (died c. 800), and finally Ya'qub ibn Tariq (died c. 796), who wrote works dealing with spherical geometry and who also compiled various tables.

All we now have to discover is which of the Indian *siddhânta* was adapted by al-Fazzari during the reign of al-Mansur. Now, the Fighar who is mentioned in the text is none other than Vyâgramukha (abbreviated to Vyâgra then deformed into Fighar), an Indian sovereign of the Châpa Dynasty who, according to an inscription, was defeated by Pulakeshin II, king of the Deccan in about 634. His capital was Bhillamâla (now Bhinmal), in the southwest of what is now Rajasthan. And it was precisely under the reign of Vyâgramukha, in the year 550 of the *Shaka* era (i.e. 628 CE), that *Brahmagupta composed his *Brahmasphutasiddhânta* (*Brahma's Revised System*) at the age of thirty.

Thus, one or other of the Indian scholars who arrived in Baghdad in 773 probably gave the caliph a copy of the *Brahmasphutasiddhânta*, along with other Sanskrit works.

It thus seems quite likely that not only Indian astronomy, but mathematics too, were introduced to the Muslims through the work of Brahmagupta.*

What led these Indian scholars to give such a present to al-Mansur? They had been kept for some time in his palace, which gave that enlightened monarch, with his lifelong thirst for knowledge, the opportunity to learn some Indian astronomy and arithmetic. Thus it was that these Brahmans, as worthy representatives of Indian culture, were led to demonstrate to him what seemed to them to be most important, original and ingenious in their science. They then, quite probably, gave the caliph copies of Brahmagupta's *Brahmasphutasiddhânta* and *Khandakhâdyaka*, which contained not only the *siddhânta* method, but also the principle of the decimal place-value system, the zero, calculation methods and the basics of Indian algebra.

It is easy to imagine the enthusiasm of al-Khuwarizmi, Abu Kamil, al-Karaji, al-Biruni, An Nisawi and others, too, who could appreciate the superiority of the Indians' place-value system and methods of calculation.

In his *Chronology of the Scholars*, Abu'l Hasan al-Qifti speaks of their admiration:

> Among those parts of their sciences which came to us, [I must mention] the numerical calculation later developed by Abu Ja'far Muhammad Ben Musa al-Khuwarizmi; it is the swiftest and most complete method of calculation, the easiest to understand and the simplest to learn; it bears witness to the Indians' piercing intellect, fine creativity and their superior understanding and inventive genius [F. Woepcke (1863)].

We must, in passing, admire this author's objectivity and lack of chauvinism, his ability to recognise the superiority of a discovery made by foreigners and his praise for a civilisation which had produced such a superior system to his own culture's.

---

* Even if Brahmagupta made some mistakes (he argued against the rotation of the earth demonstrated by Âryabhata in 520, for example), he was incontestably the greatest mathematician of the seventh century – a reputation he would keep for several centuries among Indian mathematicians and astronomers, and also among many Arabic-Islamic scholars, such as al-Biruni. His work, first presented in his *Brahmasphutasiddhânta* (628) then expanded in his *Khandakhâdyaka* (664), made considerable progress compared to earlier work, including that of Âryabhata and Bhâskara, particularly in algebra, one of his main innovations. Among his fundamental contributions can be cited his own system of a negative or zero arithmetic (with a clear and accurate statement of the rules of algebraic symbols), and his presentation of general solutions to quadratic equations with positive, negative or zero roots.

This quotation also leads us to look at one of the Islamic world's most famous mathematicians: al-Khuwarizmi, who was born in 783 in Khiva (Kharezm) and died in Baghdad in about 850 [see O. Fayzoullaiev (1983); G. J. Toomer in DSB; K. Vogel (1963)]. Little is known about his life, except that he lived at the court of the Abbasid caliph al-Ma'mun, shortly after the time when Charlemagne was made Emperor of the West, and that he was one of the most important of the group of mathematicians and astronomers who worked at the "House of Wisdom" (*Bayt al-Hikma*), Baghdad's scientific academy.

His fame is due to two works which made significant contributions to the popularisation of Indian numerals, calculation methods and algebra in both the Islamic world and the Christian West. One of them, *Al jabr wa'l muqabala* (*Transposition and Reduction*), dealt with the basics of algebra. It has come down to us both in its original Arabic and in Geraldus Cremonensis's mediaeval Latin translation, entitled *Liber Maumeti filii Moysi Alchoarismi de algebra et almuchabala*. This book was extremely famous, to such an extent that we owe to it the term for that fundamental branch of mathematics, "algebra". The first word of its title stands for one of the two basic operations which must be made before solving any algebraic equation. *Al jabr* is the operation of transposing terms in an equation such that both sides become positive; later compressed into *aljabr*, it was translated into Latin as "algebra", giving us the term we know today. As for *Al-muqabala*, it stands for the operation consisting in the reduction of all similar terms in an equation.

According to Ibn al-Nadim's *Fihrist*, al-Khuwarizmi's other work was called *Kitab al jami' wa'l tafriq bi hisab al hind* (*Indian Technique of Addition and Subtraction*). The original has, unfortunately, been lost but several post-twelfth century Latin translations of it survive. It is the first known Arabic book in which the Indian decimal place-value system and calculation methods are explained in detail with numerous examples. Like his other book, it became so famous in Western Europe that the author's name became the general term for the system. Latinised, al-Khuwarizmi first became *Alchoarismi*, then *Algorismi*, *Algorismus*, *Algorisme* and finally *Algorithm*. This term originally stood for the Indian system of a zero with nine digits and their methods of calculation, before acquiring the more general and abstract sense it now has.

Unbeknown to him, al-Khuwarizmi provided the name for a fundamental branch of modern mathematics, and gave his own name to the science of algorithms, the basis for one of the practical and theoretical activities of computing. What more can be said about this great scholar's influence?

FIG. 25.2. *Muhammad Ben Musa al-Khuwarizmi (c. 783–850). Portrait on wood made in 1983 from a Persian illuminated manuscript for the 1200th anniversary of his birth. Museum of the Ulugh Begh Observatory, Urgentsch (Kharezm), Uzbekistan (ex USSR). By calling one of its fundamental practices and theoretical activities the "algorithm", computer science commemorates this great Muslim scholar.*

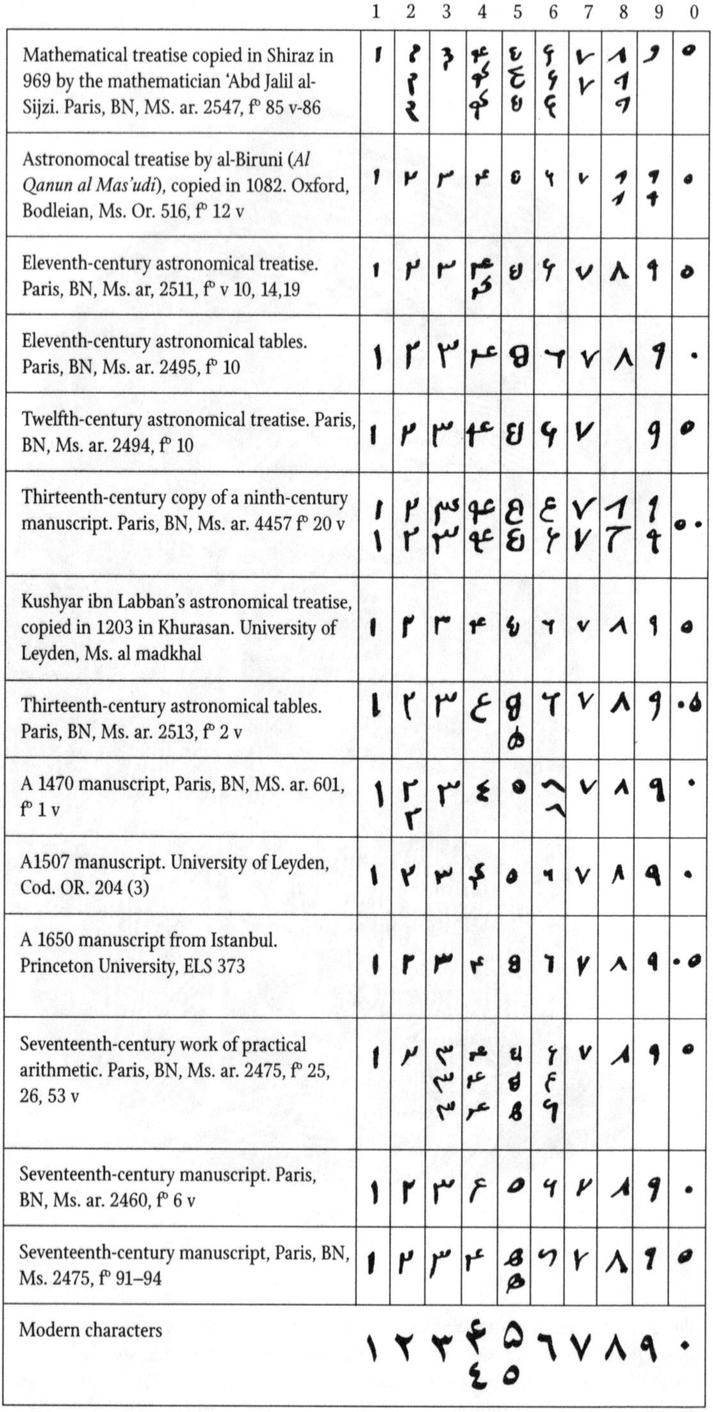

	1	2	3	4	5	6	7	8	9	0
Mathematical treatise copied in Shiraz in 969 by the mathematician 'Abd Jalil al-Sijzi. Paris, BN, MS. ar. 2547, f° 85 v-86										
Astronomocal treatise by al-Biruni (*Al Qanun al Mas'udi*), copied in 1082. Oxford, Bodleian, Ms. Or. 516, f° 12 v										
Eleventh-century astronomical treatise. Paris, BN, Ms. ar, 2511, f° v 10, 14,19										
Eleventh-century astronomical tables. Paris, BN, Ms. ar. 2495, f° 10										
Twelfth-century astronomical treatise. Paris, BN, Ms. ar. 2494, f° 10										
Thirteenth-century copy of a ninth-century manuscript. Paris, BN, Ms. ar. 4457 f° 20 v										
Kushyar ibn Labban's astronomical treatise, copied in 1203 in Khurasan. University of Leyden, Ms. al madkhal										
Thirteenth-century astronomical tables. Paris, BN, Ms. ar. 2513, f° 2 v										
A 1470 manuscript, Paris, BN, MS. ar. 601, f° 1 v										
A1507 manuscript. University of Leyden, Cod. OR. 204 (3)										
A 1650 manuscript from Istanbul. Princeton University, ELS 373										
Seventeenth-century work of practical arithmetic. Paris, BN, Ms. ar. 2475, f° 25, 26, 53 v										
Seventeenth-century manuscript. Paris, BN, Ms. ar. 2460, f° 6 v										
Seventeenth-century manuscript, Paris, BN, Ms. 2475, f° 91–94										
Modern characters	١	٢	٣	٤	٥	٦	٧	٨	٩	٠

FIG. 25.3. *The "Hindi" numerals, used by Eastern Arabs*

THE GRAPHIC EVOLUTION OF INDIAN
NUMERALS IN EASTERN ISLAMIC COUNTRIES

When the Arabs learnt this number-system, they quite simply copied it (Fig. 25.3).

In the middle of the ninth century, the Eastern Arabs' 1 (𝟏), 2 (𝟤), 3 (𝟛), 4 (𝟜), 5 (𝟝), 6 (𝟞) and 9 (𝟫) could easily be confused with their Indian *Nâgarî* prototypes, thus:

𝟏	૨	૩	૪	૬	૬	৭	૬	૭
1	2	3	4	5	6	7	8	9

But Arabic scribes gradually modified them, until they no longer resembled their prototypes (Fig. 25.3).

Such a development was a normal adaptation of the Indian models to the style typical of Arab writing. In other words, as they became integral parts of the writing system and associated with its graphic style, the Indian numerals gradually changed until they looked like a set of original symbols.

But these stylistic changes cannot explain everything. A close examination of Arab manuscripts, dating from the early centuries of Islam, shows that the Indian numerals became inverted.

And thus, in Islamic countries of the Near East:

Indian 1	(૧)	became:	𝟏				
Indian 2	(૨)	became:	૨	then:	۲	and finally:	۲
Indian 3	(૩)	became:	૩	then:	۳	and finally:	۳
Indian 4	(૪)	became:	૪	then:	۴	and finally:	٤
Indian 5	(૬)	became:	૬	then:	۵	and finally:	٥
Indian 6	(૬)	became:	૬	then:	۶	and finally:	٦
Indian 7	(৭)	became:	৩	then:	૭	and finally:	٧
Indian 8	(૬)	became:	૬	then:	૭	and finally:	٨
Indian 9	(૭)	became:	૭	then:	۹	and finally:	٩

This inversion came about for practical, material reasons.

During the early centuries of the Hegira, eastern Arabic scribes used to write the characters of their cursive script from top to bottom, rather than from right to left, in successive lines from left to right. They wrote somewhat as follows:

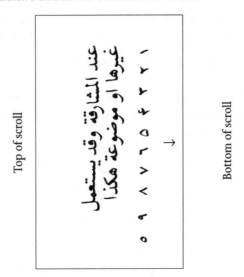

FIG. 25.4A.

Then to read, they turned their manuscript clockwise through 90°, so that the lines could be read from right to left:

Top of scroll

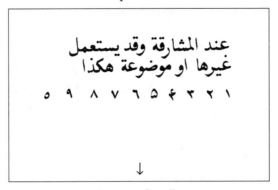

Bottom of scroll

FIG. 25.4B.

This was the old custom of Aramaic scribes of the ancient city of Palmyra, perpetuated then transmitted to the Arabs by Syriac scribes [see M. Cohen (1958)].

It came about for the following reasons, essentially to do with manuscript writing on papyrus, which, until the ninth century, was widely used in the Islamic world.

First of all, stalks were cut into sections, the length of which determined the height of the sheet. The tissue was then cut open with a knife, hammered flat, then the strips thus obtained were laid side by side in two layers at right-angles to each other. They were then struck repeatedly. The finished sheets were glued along the longer sides so that the horizontal fibres were on one side (the facing page) and the vertical ones on the other. Once the horizontal fibres had been placed on the inside and the vertical ones on the outside, the sheet could be rolled up into a scroll [see L. Cottrell (1962)].

In order to write, Arabic scribes (like their Palmyrenean and Syriac predecessors) sat cross-legged, with their robe pulled up as a writing table.

Bearing in mind this position and the fragility of the sheet, it is easy to understand why scribes held their manuscripts lengthways, perpendicular to their bodies, with the head of the scroll to their left, thus writing their cursive script from top to bottom, in successive lines from left to right.

This explains the inversion of most Indian numerals in Arabic manuscripts dating from the early centuries of Islam.

As for zero, it was originally written as a "little circle resembling the letter 'O'," to borrow al-Khuwarizmi's explanation, who was referring to the Arabic letter *ha* (٥), shaped like a small circle [see A. Allard (1957); B. Boncompagni (1857); K. Vogel (1963); A. P. Youschkevitch (1976)].

Several Arabic manuscripts prove that this usage continued in certain places until the seventeenth century.

Here is a pun, typical of twelfth-century Arabic poetry. It occurs in two lines taken from the poem Khaqani composed in praise of Prince Ghiyat ad din Muhammad (c. 1155), to exhort him to free the province of Khurasan from its Oghuzz Turkoman invaders [see A. Mazaheri (1975)]:

Your enemy will be *mutawwaq* ("captured with a metal collar")
Like zero (*al sifr*) on the earthen tablet (*takht al turab*);
At his side will be the units ("of soldiers")
Like a sigh (*aah*) of regret.
It is true; among your subjects, your enemy is nothing.
If we did pay attention to him,
He would merely be a zero to the left of the figures (*arqam*).

The meaning of this fine passage is clearer if we consider that:

- the Arabic word for "sigh" is *aah*, composed of a double *alif* (١) and a single *ha* (٥);
- the first of these two letters looks like the vertical line representing the number 1, while the other resembles zero;
- the phrase "your enemy will be *mutawwaq*" means "your enemy will be captured with a metal collar, as the zero which is shaped like an O" the (hence, by extension: "your enemy will be imprisoned, then hanged").

The poet's metaphor thus plays on the graphic resemblance between the word *aah* (a sigh) and the numerical notation 011 to give the image of the leader of the opposing army being dragged by the neck (0) by the victorious troops (11):

These verses thus mean: "The Turkoman will have a chain round his neck, and be dragged by the troops in front of Sultan Muhammad."

This confirms that the small circle still stood for zero in the twelfth century in certain eastern provinces of the Muslim empire.

This is not surprising, for it is the *Shûnya-chakra* (the "zero-circle"), one of the Indian ways of depicting zero (see **Shûnya; *Shûnya-chakra; *Zero*).

But, in the long term, this circle became so small that it was reduced to a point (Fig. 25.3).

The point is, in fact, the second way the Indians used to depict zero. It appeared at an early period in India and Southeast Asia (see **Shûnya; *Bindu; *Shûnya-bindu; *Zero*). Al-Biruni also speaks of this in his *Kitab fi tahqiq i ma li'l hind*, where he discusses Indian numerals and the Sanskrit numeric symbol system and lists the words symbolising zero: he cites the Sanskrit words **shûnya* ("vacuum", "zero") and **kha* ("space", "zero") before adding *"wa huma 'n naqta"* ("they mean 'point'") [see F. Woepcke (1863)].

To conclude, it was in this stylised and slightly modified form that the nine Indian numerals spread across the eastern provinces of Islam, in a fixed series that was only to be changed in insignificant ways throughout the succeeding centuries, particularly for the numbers 5 and 0 (Fig. 25.3). And these were what Arab authors have always referred to as *arqam al hindi* ("Indian numerals"):

١	٢	٣	۴ or ٤	٥ or ۵	٦ or ۶	٧	٨	٩	·
1	2	3	4	5	6	7	8	9	0

These forms can be found in 'Abd Jalil al-Sijzi (951–1024), al-Biruni (c. 1000), Kushiyar ibn Labban al-Gili (c. 1020) and As Samaw'al al-Maghribi (c. 1160) (Fig. 25.1), and they are still used in all the Gulf countries, from Jordan and Syria to Saudi Arabia, the Yemen, Iraq, Egypt, Iran, Pakistan, Afghanistan, Muslim India, Malaya and Madagascar.

THE WESTERN ARABS' "GHUBAR" NUMERALS

But this was not exactly the origin of our "Arabic" numerals. We inherited them from the Arabs, true enough, but from the Arabs of the West (the inhabitants of North Africa and Spain) and not from the Arabs of the Near East.

Before proceeding further, we should like to quote three revealing passages from manuscripts in the Bibliothèque nationale and translated by Woepcke [F. Woepke (1863), pp. 58–69].

They are three commentaries on mathematical works. In each of them, the commentator's explanations are mixed in with the original text, which is written in red ink to distinguish it from the commentary, which is written in black ink. Thus, in the following extracts, the original text is printed in italics and the commentary in Roman.

First passage

The nine Indian numerals [arqam al hindi] *are as follows*:

1	2	3	4	5	6	7	8	9

Or like this:

1	2	3	4	5	6	7	8	9

. . . which are the "Ghubar" numerals.

Second passage

The author says: *The first order goes from one to nine and is called the order of units.*

These nine symbols, called "ghubar" [= "dust"] numerals, are widely used in the provinces of Andalusia and in the lands of the Maghreb and Ifriqiya. Their origin is said to have occurred when an Indian picked up some fine dust, spread it over a board (*luha*) made of wood, or of some other material, or else over any plane surface, on which he marked the multiplications, divisions or other operations he wanted to carry out. When he had finished his problem, he put it [the board] away in its case until he needed it again.

[In order to memorise their shapes] the following verses have been written about these numerals [in which the shapes of the letters, words and figures mentioned evoke the numerals being referred to]:

	1	2	3	4	5	6	7	8	9	0
Practical arithmetical treatise by Ibn al-Banna al-Marrakushi. Fourteenth century. University of Tunis, Ms. 10 301, fº 25 v. CF. M. Souissi	١	٢	٣	ﭐ	٧	٦	١	٨	٩	
Guide to the Katib (work which gives details of the various number-systems used by scribes, accountants, officials etc.) Manuscript dated to 1571–72 (see Fig. 25.10). Paris, BN, Ms. ar. 4441, fº 22	١	ح	ز	٩ﻉ	ﻉ	٥	٩	٤	٩	
Sharishi, *Kashf al talkhis* ("Commentary on the Arithmetical Treatise..."). Manuscript dated 1611. University of Tunis, M. 2043, fº 16r	١	ح	ﺯ	٩ﻉ	٩	٦	٦	٨	٩	٥
Bashlawi, *Risala fi'l hisab* ("Letter Concerning Arithmetic"). Seventeenth-century manuscript. University of Tunis, Ms. 2043, fº 32 r. Cf. M. Souissi	١	٢	٣	ﭐ	٩	٦	٧	٨	٩	
Anonymous. Arithmetical treatise entitled *Fath al wahhab 'ala nuzhat al husab 'al ghubar* ("Guide to the Art of *Ghubar* Calculations"). Commentary by al-Ansari, written in 1620 and completed by1629. Paris, BN, Ms. ar. 2475, fº 46 r, 152 v and 156 v	١	ز	٦	٩ﻉ	٤	٦	٤	٨	٩	٥
Copy of a treatise of practical arithmetic by Ibn al-Banna (*Talkhis a 'mal al hisab*, "Concise Summary of Arithmetical Operations") Seventeenth century. Paris, BN, Ms. ar. 2464, fº 3v	١	٢	٣	ﭐ	٧	٦	٩	٨	٩	٥
As Sakhawi, *Mukhtasar Fi 'ilm al hisab* ("Summary of Arithmetic"). Eighteenth century. Paris, BN, Ms. ar. 2463, fº 79 v – 80	١	ح	ﺯ	٩ﻉ	٤	٦	٩	٨	٩	

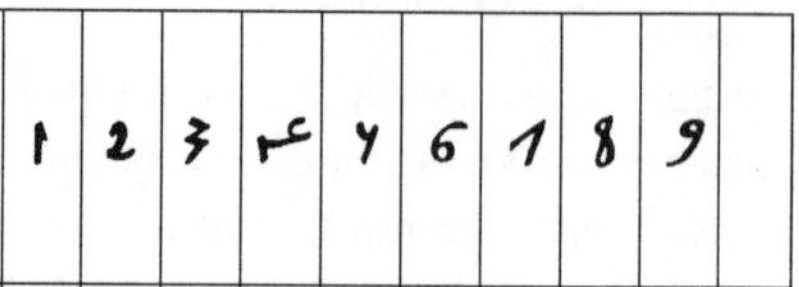

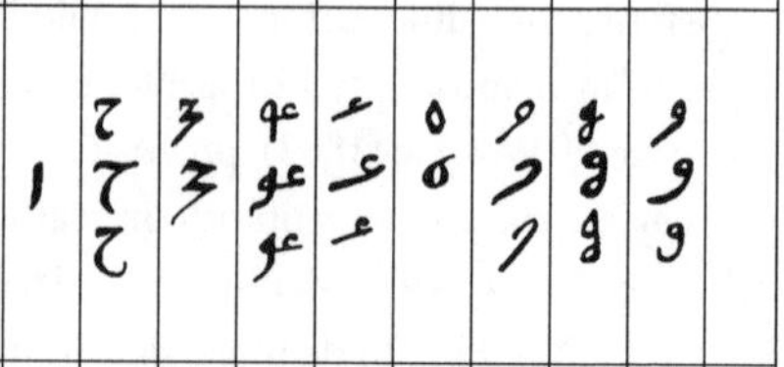

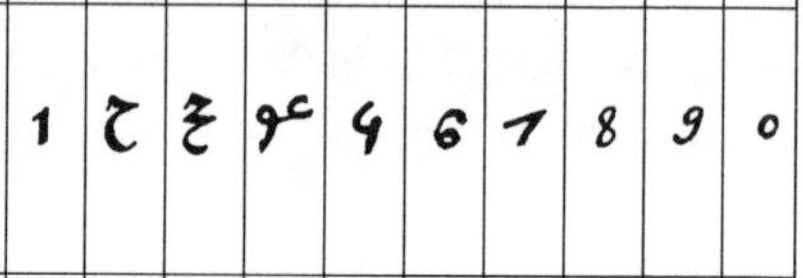

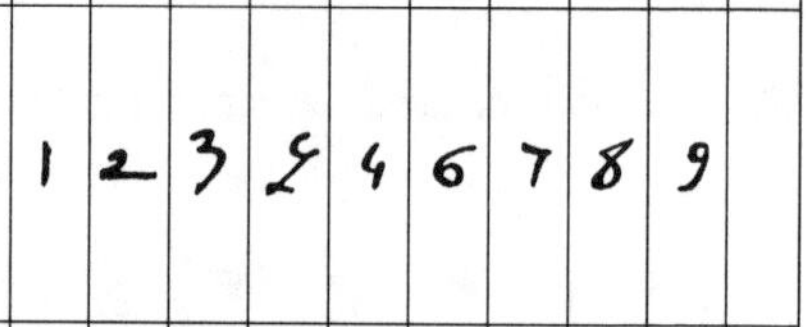

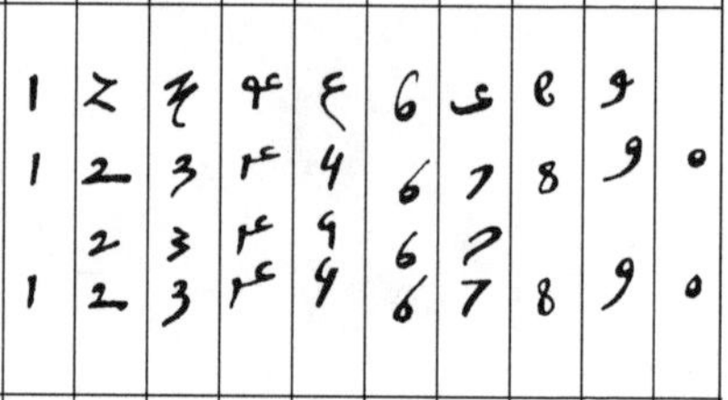

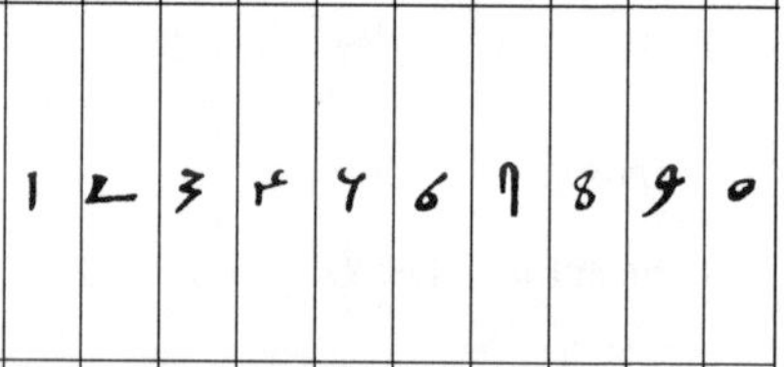

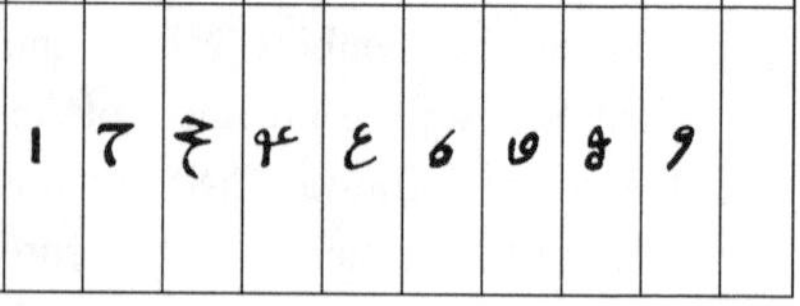

FIG. 25.5. *The Western Arabs' numerals ("Ghubar" script)*

These are an *alif* (|) [for number 1],

And a *ya* (ﺪ) [for 2],

Then the word *hijun* (ﻊ) [for 3].

After that the word *'awun* (ﻉ) [for 4];

And after *'awun*, one traces an *'ayin* (ﻉ) [for 5].

Then a *ha* [final] (ﻩ) [for 6].

And after the *ha*, appears a number [7], which,

When it is written, looks like an iron with a bent head (٦).

The eighth (of these signs is made of) two zeros [*sifran*]

[Connected by] an *alif* (8). And the *waw* (ﻭ) is the

Ninth, which completes the series.

The shape of the *ha* (ﺡ) [sometimes given to number 2] is not pure. Here are the nine signs (which must be written so that) the one appears in the highest place, with the two below it, as follows:

1	2	3	4	5	6	7	8	9

Third passage

The preface deals with the shape of the *Indian signs*, as they were drawn up by the Indian nation, *and these are*, i.e. the Indian signs, *nine figures* which must be formed *as follows*, that is: one, two, three, four, five, six, seven, eight, nine, with the following forms:

1	2	3	4	5	6	7	8	9

which are most often used by us, i.e. the Easterners, but others too are used. *Or*, they must be formed *as follows*:

١	۲	۳	۴	۵	۶	۷	۸	۹
1	2	3	4	5	6	7	8	9

which are not much used by us, while their use is widespread among the Western [Arabs].

Note. The author's meaning is clearly that both series come from India, which is true. The learned al-Shanshuri says in his commentary on the Murshidah: and they are called, i.e. the second way [of forming these signs], Indian, because they were devised by the Indian nation. End of quotation. But they are distinguished by different names, the former are called *Hindi* and the latter *Ghubar*, and they are termed *Ghubari* because people used to spread flour over their board and trace figures in it.

The following verses have been written about these signs . . . [the same as those quoted in the second passage above, with one slight difference which is described in Fig. 25.6].

(But) they have been brought together better in one single verse, as follows:

An *alif* (١) (for number 1),

a *ha* (ح) [for 2],

hizun (ح) (for 3),

'*awun* (ع) [for 4],

an '*ayin* (ع) (for 5),

a *ha* (final) (ﻪ) [for 6],

an inverted *waw* (ﻉ) (for 7),

two zeros [linked by an *alif*] (ﻻ) (for 8),

and a *waw* (ﻭ) [for 9].

Certain points are worthy of note in these passages.

Firstly, we learn that the *Ghubar* numerals were used in the Maghreb (the western region of North Africa, between Constantine and the Atlantic), in Muslim Andalusia and in Ifriqiya (the eastern region of North Africa, between Tunis and Constantine). And it can be observed that they are written in a completely different way from the eastern provinces' *Hindi* numerals.

We have also learnt about a means of calculation: a sort of wooden board sprinkled with dust, the use of which was, as we shall see, linked with *Ghubar* numerals.

We can also see that the tradition of an Indian origin for these numerals had been transmitted by Arab and Maghrebi arithmeticians.

But the most important point concerns the verses written about *Ghubar* numerals, and which ingeniously fix their shapes. The stability of these verses from one manuscript to another is remarkable when one considers that they are not copies of the same source, but two completely independent manuscripts from different periods and locations.

They are an excellent way of memorising the nine numerals, by associating them with certain Arabic letters (or groups of letters), written in the typical style of the old Maghribi and Andalusian script. They were presumably composed to teach pupils how to write the nine Indian numerals in the style of their native province; it is rather as though we gave the shapes of the Roman letters O, I, Z etc. to our children for them to learn the numbers 0, 1, 2 etc.

Figure 25.6 contains further explanations of each line, as it appears in manuscript. The exact forms have been recreated, with reference to local scripts and drawing on parallels with the numerals contained in these manuscripts.

The two oldest known documents which refer to *Ghubar* numerals and calculation date back to 874 and 888 CE [see JASB 3/1907; SC XXIV/1918]. The shapes of the numerals they contain are close to those in Fig. 25.6 and, of course, to those described in the verses quoted above. And, as the most recent manuscript containing these verses comes from the beginning of the nineteenth century, it can be supposed that the forms of the *Ghubar* numerals were fixed centuries ago and passed down from generation to generation in this manner. In other words, an attempt was made to prevent the *Ghubar* numerals from being altered by scribes. These verses can also be found in numerous other arithmetical treatises.

The original forms of these numerals were conserved no doubt because the Maghrebi are attached to traditions coming from the Muslim conquest of Andalusia and North Africa. And that is when these numerals arrived in these regions and were then adapted to the local cursive scripts.

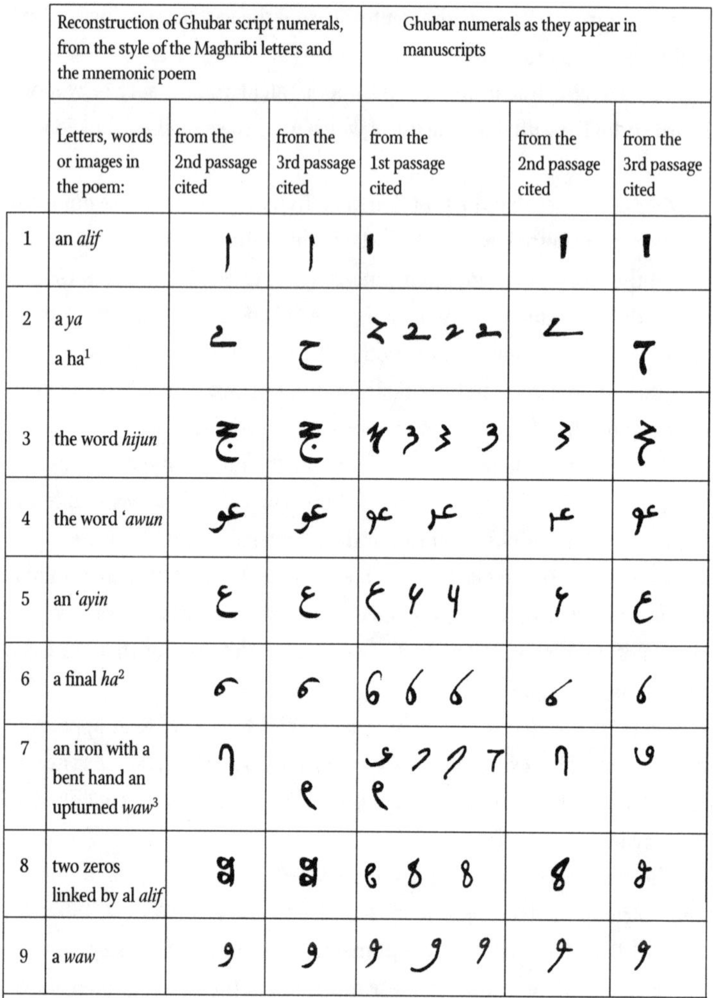

		Reconstruction of Ghubar script numerals, from the style of the Maghribi letters and the mnemonic poem		Ghubar numerals as they appear in manuscripts		
	Letters, words or images in the poem:	from the 2nd passage cited	from the 3rd passage cited	from the 1st passage cited	from the 2nd passage cited	from the 3rd passage cited
1	an *alif*					
2	a *ya* a *ha*[1]					
3	the word *hijun*					
4	the word *'awun*					
5	an *'ayin*					
6	a final *ha*[2]					
7	an iron with a bent hand an upturned *waw*[3]					
8	two zeros linked by al *alif*					
9	a *waw*					

1: The author of the second passage notes that the *ha* "is not pure". This remark, referring here of course to the number 2, seems to mean that the variant similar to this letter (which is also found in manuscripts) was not the original shape of 2 and that it had initially been more like the final form of *ya*, which is often written in this way in the Maghribi script (Fig. 25.8).

2: Such is, in fact, the final form of *ha*, as it occurs in Maghrebi and Andalusian manuscripts (Fig. 25.8A).

3: The existence of this variant of the number 7 (as an upturned *waw*) is confirmed in a marginal note which occurs in the manuscript of the first passage.

FIG. 25.6.

THE TRANSMISSION OF INDIAN NUMERALS TO WESTERN ARABS

The question that now needs answering is how and when Indian arithmetic arrived in North Africa and Spain.

Woepcke provides us with part of the answer:

Even though the unity of the caliphs came to an early end, pilgrimages to Mecca, flourishing trade, individual travels, migrations of entire populations and even wars kept up a constant communication between the various lands inhabited by Muslims. Once Indian arithmetic was known in the East, it inevitably became introduced into the West. A lack of precise information concerning this event in the history of science makes dating it impossible, but we are probably not far from the truth if we say that Indian arithmetic arrived in North Africa and Spain during the ninth century.

It is important to remember the special relationship the Caliphate of Cordoba had with Byzantium, which allowed the circulation of certain ancient texts. It can also be supposed that this facilitated contacts and meetings with representatives of Indian culture in the cosmopolitan world of Byzantium. But we should also bear in mind the contact that the Andalusians and Maghrebi must have had with their eastern cousins, without passing through Byzantium.

The arrival of Indian arithmetic in these regions could easily have come about either through texts written by eastern Arabs, or via more direct contacts with Indian scholars; thus in a similar way to what happened between India and the eastern Arabs.

But we must not overlook the vital role Jewish tradesmen and merchants probably played in this transmission. This is, in fact, suggested by Abu'l Qasim 'Ubadallah, a Persian geographer working in Baghdad. Better known as Ibn Khurdadbeh, he wrote as follows in his work entitled *Book of Routes and Provinces* (c. 850 CE):

Jewish merchants speak Arabic as well as Persian, Greek, Latin and all other European languages. They travel constantly from the Orient to the Occident and from the West to the East, by both land and sea. They take ship from the land of the Latins [*franki*] by the western sea [the Mediterranean] and sail towards Farama; there, they unload their merchandise, place it in caravans and take the overland route to Colzom, on the edge of the eastern sea [the Red

Sea]. From there, they take ship again and sail towards Hejaz [Arabia] and Jidda, before moving on to Sind, India and China. Then they return, bringing with them goods from the east . . . These travels are also made by road. The merchants leave the land of the Latins, go towards Andalusia, cross the patch of sea [the Straits of Gibraltar] and travel across the Maghreb before reaching the African provinces and Egypt. They then travel towards Ramalla, Damascus, Kufa, Baghdad and Basra, before coming to Ahwaz, the Fars, Kerman, the Indus, India and China [quoted in Smith and Karpinski (1911)].

Similar information about these merchants can be found in this extract from the *Gulistân* (*Rose Garden*), written by the Persian poet Sa'adi in the first half of the thirteenth century [see E. Arnold (1899); D. E. Smith and L. C. Karpinski (1911)]:

I met a merchant who had a hundred and forty camels
And fifty porters and slaves . . .
He replied: I want to take Persian sulphur to China,
Which, from what I have heard,
Fetches a high price in that country;
Then procure goods made in China
And take them to Rome (*Rum*);
And from Rome load a boat with brocades for India;
And with that trade for Indian steel (*pûlab*) in Halib;
From Halib, I shall transport glass to the Yemen,
And take back Yemeni painted cloth to Persia.

Unlike Ibn Hauqal, the poet does not specify the origin of this travelling merchant, who may not be Jewish. The Jews have never had a monopoly over international trade. So Jewish traders were merely one of the numerous links in this chain of transmission.

Whether they were or were not Jewish, these tradesmen used numbers as often as they travelled or traded. And, like the various languages they learnt in their business, they must also have become acquainted with the different systems of arithmetic used by the peoples they encountered.

As India was part of their route, they must surely have been obliged to learn Indian numerals and arithmetic, and were thus one form of communication between India and the Maghreb.

FROM HINDI NUMERALS TO GHUBAR NUMERALS – A SIMPLE QUESTION OF STYLE

To return to Arabic numerals, the Indian influence is evident, whether it be on the *Hindi* symbols, or the *Ghubar* (Fig. 25.3 and 6).

Even a rapid comparison between the Indian *Nâgarî* numerals and the *Ghubar* shows of course the presence of the Indian 1, but also 2, 3, 4 (with a slightly different orientation in Arabic), 6, 7, 9 and 0, and even 5 and 8 (Fig. 25.5 and 7).

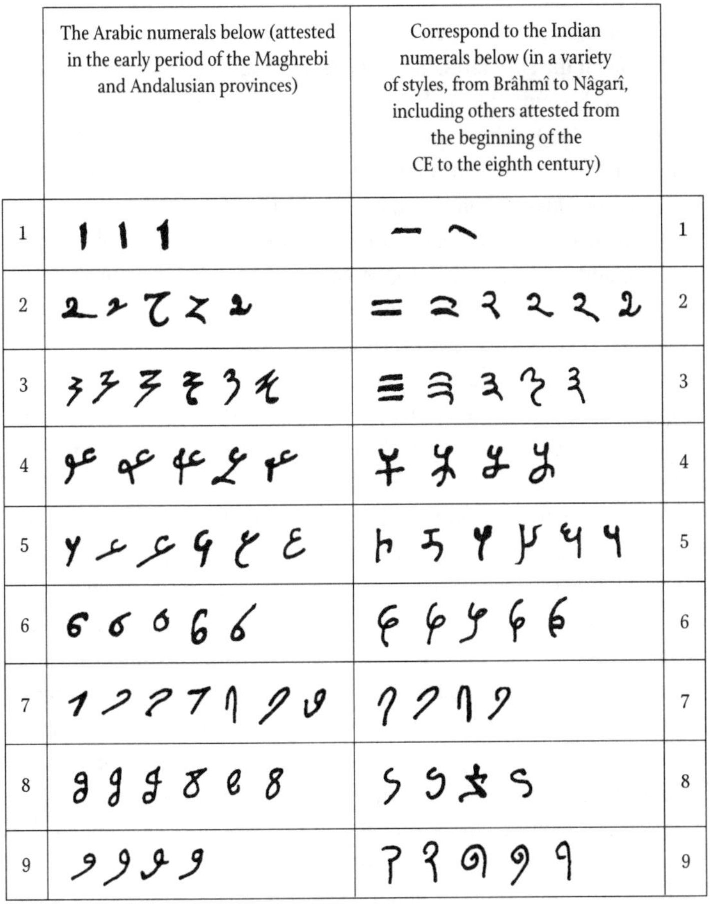

	The Arabic numerals below (attested in the early period of the Maghrebi and Andalusian provinces)	Correspond to the Indian numerals below (in a variety of styles, from Brâhmî to Nâgarî, including others attested from the beginning of the CE to the eighth century)	
1			1
2			2
3			3
4			4
5			5
6			6
7			7
8			8
9			9

Fig. 25.7.

In palaeographic terms, there is thus no difference between the *Hindi* numerals of the Machreq and the *Ghubar* numerals of the Maghreb. Both come from the same source. Any differences between them simply derive from the habits of scribes and copyists in the two regions.

The history of Arabic writing styles helps us to understand these changes more clearly (Fig. 25.8). From the beginning of Islam, two distinct forms of writing evolved: a lapidary cursive style, derived from pre-Islamic inscriptions; and an even more cursive style, from the earliest written Arabic manuscripts, also dating to before the Hegira.

The lapidary cursive style produced the *Kufic* script, for inscriptions and manuscripts, with its characteristic horizontal base line on which the rigid, angular letters are set vertically. According to Ibn al-Nadim's *Fihrist* (987), this script derived from the early habits of the stone-carvers and scribes from Kufa on the Euphrates, hence its name. (Founded in 638 CE, Kufa was a centre of learning under the Omayyad caliphs until the foundation of Baghdad in 762.) This script was also used, during the first centuries of Islam, for legal and religious texts (in particular for the first copies of the Koran, in mosques and on tombstones), which explains its hieratic nature.

It was then gradually replaced by the *naskhi* script, generally used by copyists, and leading to the elegant calligraphy of the "Avicenna" Arabic alphabet which is most commonly used today. Derived from ancient cursive Arabic manuscripts, this style is marked by its smooth rounded forms, broken up into small curved elements. It is also the source of the *nastalik* script, used in Persia, Mesopotamia and Afghanistan, and the *sulus* script of the Turkish Ottoman Empire. With certain exceptions, the form of the letters remained very similar to *Naskhi*.

The difference between the two styles, at least at the beginning, was really due to what they were used for and the material they were written on. While the cursive manuscript style was used for everyday texts on papyrus or parchment, the other one was reserved for inscriptions on stone, wood or metal. The former was traced onto the papyrus, parchment or other smooth surface with a quill or a reed (the famous *qalam*, or "calamus") dipped in thick ink. But the latter was sculpted into stone, carved into wood or engraved into copper. This naturally explains the former's smooth rounded forms, contrasting with the latter's angular rigidity.

If we now return to the numerals and compare the signs contained in Fig. 25.3 and 5, we can see that the cursive *Hindi* numerals are far more rounded than those of the Maghreb, with the base line of the former breaking up into small curves. In other words, the eastern Arabs' numerals follow closely the rules of the *Naskhi* script.

On the other hand, the *Ghubar* numerals, while remaining cursive, are nevertheless obviously more angular, stiff and rigid. A closer look reveals that their curves, down-strokes and angles are absolutely identical to those used in the *Kufic* script. This is, at least, what is revealed in the original of *Kashf al asrar 'an 'ilm al gobar*, by the Andalusian mathematician al-Qalasadi. Its letters and *Ghubar* numerals are written in a way which reflects the pure *Kufic* tradition from the early centuries of Islam. This manuscript dates from the fifteenth century and the Institut des Langues Orientales in Paris possesses a copy of it from a lithograph made in Fez [see A. Mazaheri (1975)].

This is not surprising, for the *Maghribi* (or *African*) script which spread across North Africa, Sudan and Muslim Spain after the ninth century is in fact nothing more than a manuscript *Kufic*.

It should not be forgotten that the Maghrebi and Andalusians were extremely attached to ancient Islamic traditions. This is particularly true of the lapidary cursive style of the first conquerors of the region, the Abbasids of Samara, which gave the Maghribi script its stiffness and rigidity.

By fixing their forms by means of the verses quoted above (Fig. 25.6), they were made to adopt the characteristic shapes of Maghribi letters and thus follow its cursive rules.

To sum up, whatever differences there may be between *Hindi* and *Ghubar* numerals, their common source is demonstrably Indian.

But it was not in their *Hindi* form, but in the *Ghubar* style that Indian numerals migrated from Spain to the Christian peoples of Western Europe, before finally taking the shape they have today.

ARAB RESISTANCE TO INDIAN NUMERALS

It is tempting to think that the Indian system spread through the Islamic world, replacing all other ways of representing numbers and, because of their ingenious simplicity, the corresponding calculation methods were rapidly accepted at all levels of Arab-Islamic society. The author humbly admits that he was wrong in the first edition of the present work in which he subscribed to that idea and neglected the following interesting details. Of course, certain scholars such as al-Khuwarizmi and An Nisawi were sufficiently astute to understand the superiority of this system. But there was an equal number of Muslims who were, sometimes violently, opposed to the use of numerals and even more so to their becoming generalised.

This means that, contrary to what is often believed, the domination of the Indian system was a long, difficult process. Many arithmetical treatises, for example, contain not a single Indian numeral, and sometimes no numerals at all, because the numbers in each line are expressed by their Arabic names. And if Indian numbers are to be found anywhere, then it is most probably, or even one would think inevitably, in arithmetical works.

NAME OF LETTER	NUMERICAL VALUE	SHAPE OF LETTER in Naskhi Arabic	in Maghribi Arabic	in Persian, in the nasta'lik script
alif	1	ا	ا	ا
ba	2	ب	ب	ب
jim	3	ج	ج	ج
dal	4	د	د	د
ha	5	ه	ه	ه
wa	6	و	و	و
zay	7	ز	ز	ر or ر
ḥa	8	ح	ح	ح
ṭa	9	ط	ط	ط
ya	10	ى	ي or ے	ی
kaf	20	ك	ك	ک
lam	30	ل	ل	ل
mīm	40	م	م	م
nūn	50	ن	ن	ن
sin	60	س	س	س
'ayin	70	ع	ع	ع
fa	80	ف	ڢ	ف
sad	90	ص	ص	ص
qaf	100	ق	ڧ	ق
ra	200	ر	ر	ر
shin	300	ش	ش	ش
ta	400	ت	ت	ت
tha	500	ث	ث	ث
kha	600	خ	خ	خ
dhal	700	ذ	ذ	ذ
dad	800	ض	ض	ض
dha	900	ظ	ظ	ظ
ghayin	1,000	غ	غ	غ

FIG. 25.8A. *The Arabic alphabet in the Naskhi and Maghribi scripts*

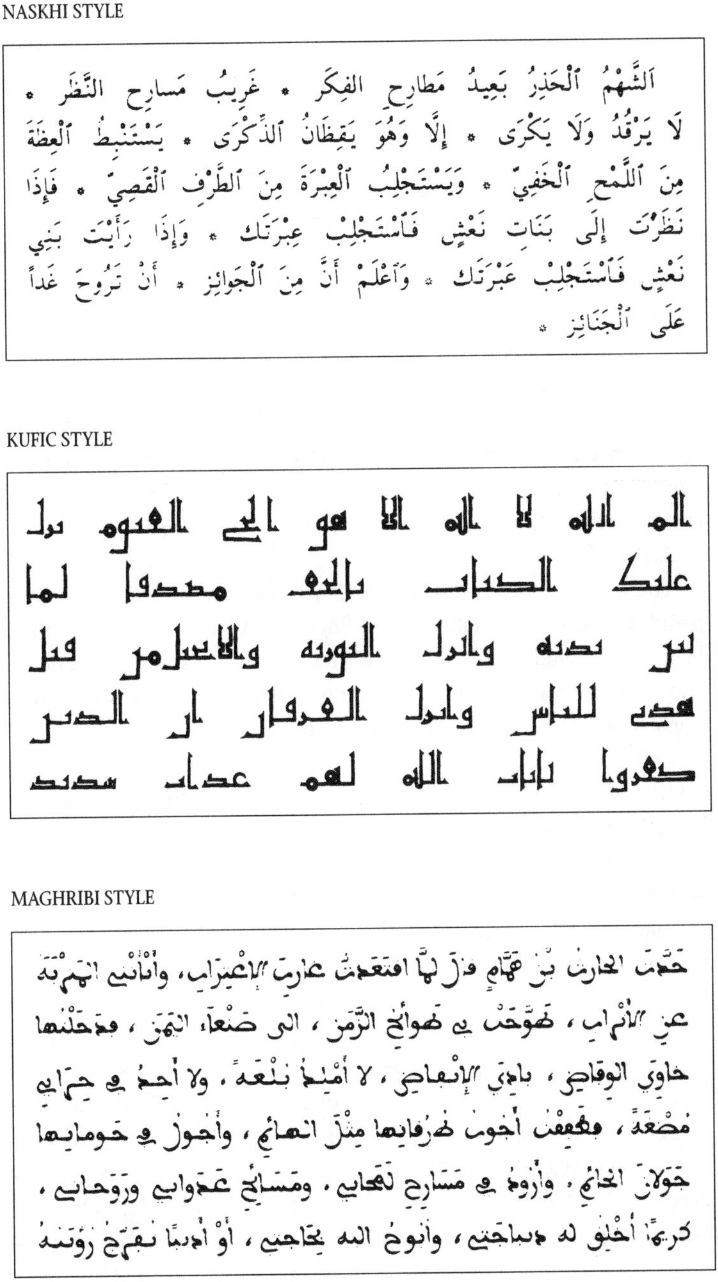

NASKHI STYLE

اَلشَّهْمُ الْحَذِرُ بَعِيدُ مَطَارِحِ الْفِكَرِ • غَرِيبُ مَسَارِحِ النَّظَرِ •
لَا يَرْقُدُ وَلَا يَكْرَى • إِلَّا وَهُوَ يَقِظَانُ الذِّكْرَى • يَسْتَنْبِطُ الْعِظَةَ
مِنَ اللَّمْحِ الْخَفِيِّ • وَيَسْتَجْلِبُ الْعِبْرَةَ مِنَ الطَّرْفِ الْقَصِيِّ • فَإِذَا
نَظَرْتَ إِلَى بَنَاتِ نَعْشٍ فَاسْتَجْلِبْ عِبْرَتَكَ • وَإِذَا رَأَيْتَ بَنِي
نَعْشٍ فَاسْتَجْلِبْ عَبْرَتَكَ • وَاعْلَمْ أَنَّ مِنَ الْجَوَائِزِ • أَنْ تَرُوحَ غَدًا
عَلَى الْجَنَائِزِ •

KUFIC STYLE

MAGHRIBI STYLE

FIG 25.8B. *Different styles of written Arabic [CPIN; see also de Sacy; Sourdel in EIS]*

For Islam, like everywhere else, had its "traditionalists", bookkeepers and accountants who remained deeply attached to previous practices and vigorously opposed to scientific and technological innovations.

THE CONSERVATISM OF ARAB SCRIBES AND OFFICIALS

One of the reasons for this opposition was the conservatism of Arab and Islamic scribes and officials, who long remained attached to their ancestral methods of counting and calculating on their fingers.

Thus, in his *Kitab al mu'allimin* (*Schoolmasters' Book*), al-Jahiz gives this advice, which provides a clear idea of the polemic that must have confronted the users of Indian numerals and the ardent defenders of traditional methods for several generations: "It seems better to teach pupils digital calculation and avoid Indian arithmetic (*hisab al hindi*), geometry and the delicate problems of land measurement." [British Museum Ms. 1129, f° 13r].

This author, who scorned Indian numerals and arithmetic, thus recommended teaching calculation using fingers and joints (*hisab al 'aqd*) as being, to his mind, more useful for the future official scribe of the period. Some accountants even preferred manual calculation to the Arabs' traditional means of calculation, the dust board.

This is, for example, revealed in *Kitab al hisab bila takht bal bi'l yad* ("Treatise on calculation without the board, but with [the fingers of] the hand"), written in 985 by al-Antaki [see A. Mazaheri (1975)].

In his *Adab al kutab*, destined for scribes and accountants, the Persian writer As Suli (died 946) gives the reason for this preference for manual calculation. After mentioning the "nine Indian characters" and "the great simplicity of this system" when expressing "large quantities", he then adds: "Official scribes nevertheless avoid using this system because it requires equipment [i.e. a counting board] and they consider that a system that requires nothing but the members of the body is more secure and more fitting to the dignity of a leader." As Suli then eulogises the official accountants of the Arab-Islamic world, with their supple joints and movements "as fast as the twinkling of an eye". He quotes a certain 'Abdullah ibn Ayub who "compares the jagged lightning fork with the rapidity of the accountant's hand, when he says: 'It seems that its flash [of lightning] in the sky is made up of a scribe's or accountant's two hands!'" Then he concludes: "That is why they content themselves with just the *iqd* [i.e. counting on the fingers] and the system of joints" [see J. G. Lemoine (1932)].

Officials always, of course, claim they are irreplaceable in order to keep their privileged positions. They are thus never happy to see a new simple system becoming generalised, which anyone can use without going through their difficult and mysterious apprenticeship.

This is a universal tendency, which can be witnessed throughout Antiquity, and in Western Europe from mediaeval times up until the French Revolution. If Arab-Islamic scribes and officials violently opposed the introduction of Indian numerals, it was because it could mean an end to their monopoly.

But this traditionalism does not explain everything. We must also consider the multiplicity and diversity of the peoples that made up the Muslim empire. The heterogeneous nature of the cultures and populations of this complex world, along with regional and individual habits, also played a part.

"Culture", as E. Herriot put it, "is what remains when all else has been forgotten." It is the form of knowledge which enables the mind to learn new things. Hence the idea of developing and enriching our various mental faculties by intellectual exercises such as study and research.

But "culture", in any given civilisation, is also the intellectual, scientific, technological and even spiritual inheritance of its people. It is thus the sum of knowledge, which its great minds have assimilated, and which greatly adds to its enrichment.

In this way, Arab-Islamic civilisation was exceptional for its originality, strong culture and deep insights of its thinkers, scholars, poets and artists.

And, to quote P. Foulquié, a culture is also the "collective way people think and feel, the set of customs, institutions and works which, in any given society, are at once the effect and the means of personal culture." Thus (to run Martin du Gard and Mead together), it is the set of virtues, preconceptions, individual habits and works which make up a given nation in its ways of behaving, acquired and transmitted by its members, who are accordingly united by a shared tradition.

Like any other, Arab culture was also composed of varied customs, countless details, endless habits and presumptions, characteristic of its daily existence. Great minds thus coexisted with lesser, more ignorant souls whose unthinking conservatism led them to clutch onto methods that had been useful to their distant ancestors, but which had long since stopped being appropriate to modern times and activities.

TRADITIONAL ACCOUNTANTS VERSUS USERS OF OUTMODED SYSTEMS

When the Arab-Islamic civilisation found itself in contact with the Christian West, some Arab accountants had the curious idea of adopting Latin calculation methods using counters on a board, and thus set about turning the clock back. This was the case with certain Syrian and Egyptian accountants, presumably under the influence of their trading links with the Genoans and Byzantines.

This was severely criticised by the Persian historian Hamdullah who, in his 1339 *Nuzhat al qolub* (work of geography and chronology), says: "In the year 420 [of the Hegira, thus 1032 CE], Ibn Sina invented the 'calculation knots', thus freeing our accountants from the tedium of totting up counters [*mishsara shumari*] on instruments and boards, like the Latin abacus [*takhata yi frenki*] and suchlike" [see A. Mazaheri (1975)].

As an accountant, Hamdullah had certainly been deeply impressed by a calculation method called '*uqud al hisab* ("calculation knots"), recommended for accountancy two centuries before by the famous Ibn Sina (Avicenna), then the finance minister of Persia, under Buyid domination.

To gain a better understanding of this method, we must remember that a "knot" (in Arabic '*aqd* or '*uqda*, the singular of '*uqud* or '*uqad*) had at this time not only its primary meaning, but also signified "class of numbers corresponding to the successive products of the nine units and any power of ten". In other words, the "knot" stood for the decimal system. There was the units *knot*, the tens *knot*, the hundreds *knot* and so on. This same term can be found in al-Maradini [see S. Gandz (1930)] and in Ibn Khaldun's *Prolegomena* [see *Muqaddimah*, trans. Slane, I, pp. 243–4].

By extension, the expression '*uqud al hisab* came to mean "calculation knots", in reference to an ancient way of recording numbers on knotted cords, used by the Arabs in antiquity. The various places of consecutive digits were marked by knots tied in predetermined positions. This system was thus very similar to the South American Incas' *quipus* and the ancient Japanese *ketsujo*, used until recently in the Ryu-Kyu Islands (Fig. 25.9).

The Arabs (presumably before the advent of Islam) had long used these knotted cords as a way of noting numbers for administrative records. The numbers thus tied on the strings recorded accounts and various inventories. This is reminiscent of the tradition, reported by Ibn Sa'ad, according to which Fatima, Mohammed's daughter, counted the ninety-nine attributes of Allah, and the supererogatory eulogies which followed the compulsory prayers, on knotted cords, and not on a rosary. These cords were also used as receipts and contracts. This is shown by the fact that, in Arabic, the word '*aqd* means both "knot" and "contract".

To return to the "calculation knots" which Avicenna is supposed to have invented, it is highly probable that Hamdullah was referring to a means of manual calculation.

The common Arabic expression for "hand counting" is *hisab al yad* (from *hisab*, "counting, calculation", and *yad*, "hand"). It can be found, for example, in al-Antaki and As Suli (*op. cit.*), as well as al-Baghdadi in his *Khizanat al 'adab*.

But in many authors, the word *'aqd* or *'uqda* ("knot") also means the "join" between the finger and the hand, and by extension the "joints" of the finger. For, this *hisab al 'uqud* ("counting with knots") is in fact "counting on the joints of the fingers", by allusion to the "knot" of the joints and the "join" between the fingers and the hand.

There were several ways of counting on fingers in Islam. Although Hamdullah is vague about Avicenna's method, it is possible to work out what it was by elimination. To Hamdullah's mind, the word *'uqud* in the expression *'uqud al hisab* ("calculation knots") could in fact have meant the "order of units" in an enumeration. And, as this concerns a manual method, the "knots" in question could refer to units in a highly evolved decimal system. What comes to mind is that "dactylonomy", similar to deaf and dumb sign language, which was used by the Arabs and Persians for centuries, in which the units and tens were counted on the phalanxes and joints of one hand, while the other one was symmetrically used for the hundreds and thousands (see Chapter 3). This system was famously described in a poem written in rajaz metre, called *Urjuza fi hisab al 'uqud*, composed before 1559 by Ibn al-Harb and dealing with the science of "counting on phalanxes and joints" [see J. G. Lemoine (1932)].

But this cannot be the method referred to by Hamdullah. As Guyard explains: "the word *'uqud*, taken as a noun, stands for the shapes obtained by bending the fingers and, by extension, the numbers thus formed." That is why the units in the manual systems already alluded to were called "knots". But, this same word *'uqud*, taken as an action, means "bending the fingers" [see JA, 6th series, XVIII (1871), p. 109]. And, since he is discussing arithmetic, what Hamdullah is talking about is definitely an action, not a state. It is thus the science of calculating with what may be called "moving knots" which is in question. For the other systems were mere static ways of counting on the fingers and joints of the hand (just simple manual representations of numbers), whereas the technique being envisaged here allows calculations to be made by actively bending the fingers.

By opposing "calculation knots" to the Latin abacus, Hamdullah was thinking of "knots" as an action, bending certain fingers and straightening others, allowing arithmetical operations to be carried out in a much easier way than on the abacus. That is why, according to this admirer of Avicenna, these "moving knots" had freed "our accountants from the tedium of totting up counters" thrown down onto "the Latin abacus and suchlike."

But Hamdullah is guilty of making an historical mistake. The method he attributes to Avicenna had already existed in the Islamic world for a long time.

This is not our accountant historian's only slip. For the method recommended by the famous philosopher was only of use in operations on common numbers. Hence Hamdullah's error of judgment. He had not understood that the Latin abacus, primitive though it was, allowed numbers to be reached that are far higher than can be obtained by any form of manual calculation, no matter how elaborate. For the limits of the human hand set the limits of the method.

Thus it was that, through ignorance of basic practical arithmetic, or perhaps through sheer bloody-mindedness, users of a totally outmoded means of calculation attacked other accountants with methods as primitive as their own. The latter were, of course, to be upbraided for falling for a technique that came from a culture that was quite alien to Islam, and which the former presumably held in disdain.

In this context, it is easy to imagine how both camps violently opposed the introduction of Indian numerals and calculation methods, whose evident superiority over their archaic ways they would never admit.

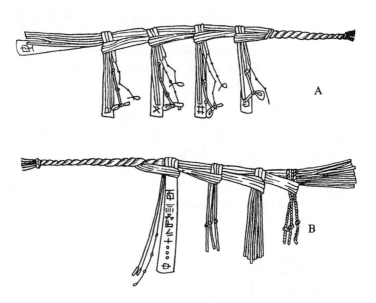

FIG. 25.9. *Japanese* ketsujo

This was a concrete accountancy method, used in ancient Japan and analogous to the quipus of the Incas (Peru, Ecuador and Bolivia). Given the universal nature of this method, this Figure will provide a good idea of how Arabs used knotted cords in the pre-Islamic era and probably also in the early days of Islam (despite lack of evidence).

This ketsujo *stands for(the knots represent sums of money, as used in the Ryu-Kyu Islands, particularly by workmen and tax collectors) [Frédéric 1985, 1986, 1977-1987, 1994]:*

A – cloth account given to the State, or a temple, from left to right:

– Yoshimoto family: 1 jo, 8 shaku, 5 sun and 7 bu;

– 1 jo, 4 shaku, 3 sun and 7 bu;

– Togei family: ibid.

B – Horizontal strand: 20 households.

Others, from right to left: 3 hyo, 1 to, 3 shaku and 2 sai.

THE NUMERICAL NOTATION OF ISLAM'S OFFICIALS

In fact, the Indian system was introduced into the Islamic world in several steps. As operators, and thus as a means of calculation, the numerals were rapidly adopted by mathematicians and astronomers, soon followed by an ever-increasing number of intellectuals, mystics, magi and soothsayers. Meanwhile, others preferred to calculate by using the first nine letters of the Arabic alphabet (from *alef* to *ta*). But as a way of representing numbers (i.e. when noting numerical values and not making calculations), Indian numerals did not completely replace traditional notation until a relatively recent date.

Thus it was that Arab, Persian and Turkish officials continued to favour their own special notations, which had nothing to do with the Indian numerals in public use, for official and diplomatic documents, bills of exchange and administrative circulars until the nineteenth century.

This is shown in the *Guide to the Writer's Art* (1571–1572) [BN, Ms. ar. 4,441], which is a sort of handbook for professional writers. It gives a clear idea of the plurality of the numerical systems used by the scribes, officials and accountants of the Ottoman Empire at the end of the sixteenth century (Fig. 25.10).

Among these varied forms, let us mention the *Dewani* numerals used in Arab administrations, and the *Siyaq* numerals favoured by the accounts offices in the Ottoman Empire's finance ministry and in Persian administrations. These numerals were, originally, simply monograms or abbreviations of the names of the numbers in Arabic, written in an extremely cursive style. Later, they became so stylised and modified that their origins were scarcely recognisable. It is easy to understand how they were used to prevent fraudulent alterations to accounts, while at the same time leaving the general public in the dark as to what amounts were being described [see H. Kazem-Zadeh (1913); A. Chodzko (1852); L. Fekete (1955); A. P. Pihan (1860); C. Stewart (1825)].

We should also like to mention the *Coptic numerals*, used since antiquity by officials in the Arab administration of Egypt in their accounts, which were in fact slightly deformed letter numerals from the ancient alphabet of the Christian Copts of Egypt.

The Dewani numerals

These numerals were used in Arab administrative offices (called *dewan*, hence their name).

FIG. 25.10. *Page from an Arabic work, entitled* Murshida fi sana'at al katib *("Guide to the Writer's Art"). Dated 1571-1572, it is a sort of handbook for professional writers.*

It gives a very clear idea of the numerous different ways Arab-Muslim scribes, accountants and officials wrote down their numbers at the end of the sixteenth century. It contains, counting from the top down: the Ghubar numerals (2nd line) (Fig. 25.5); the Arabic letter numerals (5th line); the Hindi numerals (6th line) (Fig. 25.3); then the Ghubar numerals again (7th line); the Dewani numerals (8th line); the Coptic numerals (9th line); the Arabic letter numerals (10th line); the Hindi numerals (11th line); the Ghubar numerals (12th line); two variants of the Coptic numerals (13th and 14th lines); etc. [BN Paris, Ms. ar. 4441, f° 22]

They are abbreviations of the Arabic numerical nouns. Thus, number 1 is the letter *alif*, standing for *ahad*, "one". Similarly, numbers 5, 10 and 100 correspond to the letters *kha*, *'ayin* and *mim*, standing for *khamsa*, "five", *'ashara*, "ten", and *mi'at*, "hundred".

As for the number 1,000, it is a stylised form of the complete word *alf*, meaning "thousand". Number 10,000 corresponds to a monogram of *'asharat alaf*, "ten thousand" [A. P. Pihan (1860)].

Units

1	ا	4	لعا	7	لعا
2	لا	5	حا	8	رها
3	عـ or للا	6	سا	9	ىعا

Tens

10	عا	40	لعا	70	لعا
20	ىع	50	حا	80	ىـ
30	سا	60	ىا	90	لعا

Hundreds

100	ها	400	لعها	700	ىعها
200	هل	500	حها	800	ىها
300	سها or للها	600	سها	900	ىعها

Thousands

1,000	الڡ or الها	4,000	لعالى	7,000	ىعالى
2,000	الهى	5,000	حالى	8,000	رهالى
3,000	سالى	6,000	سالى	9,000	ىعالى

Ten Thousands

10,000	عالى	40,000	لعلا	70,000	ىعلا
20,000	لرا/ع	50,000	حلا	80,000	ىلا
30,000	سلا	60,000	سلا	90,000	ىعلا

Hundred Thousands

100,000	هالى	400,000	لعوهالى	700,000	ىعا هالى
200,000	لا هالى	500,000	حهالى	800,000	رها هالى
300,000	ىعهالى	600,000	سوالى	900,000	ىعا هالى

Composite numbers

Units are always placed before tens and between the hundreds and tens, as is done in spoken Arabic. Numerals are written from right to left, like the Arabic words they represent in this same order for composite numbers.

11		17		206	
14		21	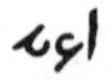	3,478	
15		24	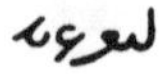	62,789	

The numerals of Egyptian Coptic officials

The Arab administration of Egypt employed Christian Copts, who had their own special accountancy notation. These signs (which can be found in several Arabic manuscripts from this region) are cursive derivatives of the letter numerals of the Coptic alphabet, itself derived from Greek. Numbers up to nine thousand are reached by using the units and underlining them. For the ten thousands, the tens are underlined, as are the hundreds for the hundred thousands. Finally, composite numbers are always topped by a slightly curved line.

Units

1		4		7	
2		5		8	
3		6		9	

Tens

10		40		70	
20		50		80	
30		60		90	

Hundreds

100		400		700	
200		500		800	
300		600		900	

Thousands

1,000	ⅎ	4,000	ⅎ	7,000	ⅎ
2,000	ⅎ	5,000	ⅎ	8,000	ⅎ
3,000	ⅎ	6,000	ⅎ	9,000	ⅎ

Ten Thousands

10,000	ⅎ	40,000	ⅎ	70,000	ⅎ
20,000	ⅎ	50,000	ⅎ	80,000	ⅎ
30,000	ⅎ	60,000	ⅎ	90,000	ⅎ

Hundred Thousands

100,000	ⅎ	400,000	ⅎ	700,000	ⅎ
200,000	ⅎ	500,000	ⅎ	800,000	ⅎ
300,000	ⅎ	600,000	ⅎ	900,000	ⅎ

Composite numbers

16	ⅎ	803	ⅎ	38,491	دصرۃطج ل
45	ⅎ	4,370	ⅎ	752,020	ساییلاج

The Persian Siyaq numerals

These numbers were used in Persian administrations, and were also
favoured by tradesmen and merchants. They are abbreviations of the
words for the numbers in Arabic (and not in Persian). They are written
from right to left, like the Arabic words they represent, as are the composite numbers [A. P. Pihan (1860); see also A. Chodzko (1852); H.
Kazem-Zadeh (1913); C. Stewart (1825)].

Units

1	ⅎ	4	ⅎ	7	ⅎ
2	ⅎ	5	ⅎ	8	ⅎ
3	ⅎ	6	ⅎ	9	ⅎ

Tens

10		40		70	
20		50		80	
30		60		90	

Composite numbers from 11 to 18

For these numbers, the final line of the units is rounded off and rises up towards the top of number ten:

11		14		17	
12		15		18	

Composite numbers from 21 to 99

The units and the other ten digits are linked together in the same way:

21		54		87	
43		76		99	

Hundreds

When written on their own, the hundreds have special signs, sometimes followed by a sort of upturned comma and full stop, which are always omitted in composite numbers. One sign calls for particular attention, because of possible errors (χ). With a line before, it stands for 400, and with no line, 700. The same sign, with an additional curl to the right, stands for 900:

100		400		700	
200		500		800	
300		600		900	

Composite numbers from 101 to 999

101		366		791	
109		377		820	
110		388		896	
111		399		915	
204		472		999	

Thousands

To form the multiples of 1,000, the characteristic patterns of the units are used, with the final stroke lengthened from right to left. In this position, and with a pronounced broadening, it is enough to indicate the presence of the thousand in the combination:

1,000		4,000		7,000	
2,000		5,000		8,000	
3,000		6,000		9,000	

Composite numbers of four digits

The group ·ول stands for the thousand, but only that exact value. For, when followed by hundreds or tens, the group الـ is used (abbreviation of the Arabic word الـف, *alf*, "thousand"):

1,050		1,200		4,377	
1,100		1,250		5,555	
1,150		3,213		9,786	

Ten and Hundred Thousands

After 10,000, the group الـ (abbreviation of the number 1,000) reappears, and the final stroke of the ten thousands is lengthened below the signs, instead of going down vertically:

10,000		99,112	
25,072		110,100	
34,683		245,123	
45,071		300,000	
50,008		456,789	

Other variants of the Persian Siyaq numerals

	Variants noted by Forbes	Variants noted by Stewart		Variants noted by Forbes	Variants noted by Stewart
1			100		
2			200		
3			300	or	
4			400	or	
5			500		
6			600		
7			700	or	
8	or		800	or	
9			900	or	
11			1,000		
22			2,000		
33			3,000		
44			40,000		
55			50,000		
66			100,000		
77			200,000		
88					
99					

Note that the number 100,000 is none other than the Sanskrit word *lakh* ( ), used by the Indians for this amount.

The Siyaq numerals of the Ottoman Empire

These numbers were favoured by the accounts offices in the Ottoman Empire's finance ministry.

They are abbreviations of the words for the numbers in Arabic (and not in Turkish). They are written from right to left, like the Arabic words they represent, as are the composite numbers. Also called *Siyaq*, they are analogous to the Persian numerals of the same name, even though they differ in certain respects.

Note that the point (which stands for 6) normally replaces the other sign (ـل) for the same value in composite numbers. But when this point is placed at the end of a number it is a mere punctuation mark, without any numerical value. Finally, in composite numbers made up of tens and units, the latter always come first, as in Arabic [A. P. Pihan (1860); see also L. Fekete (1955)].

Units

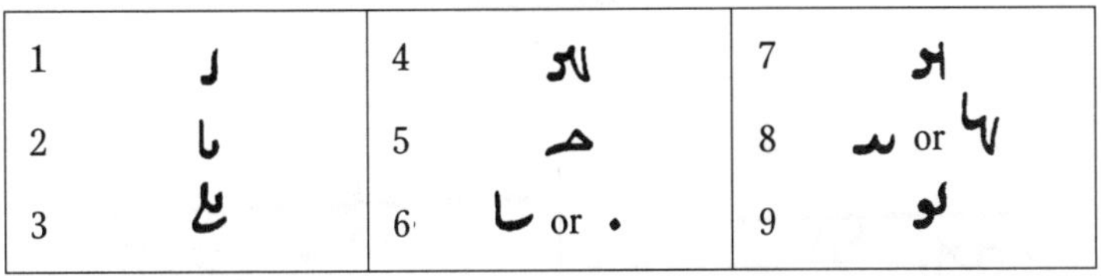
1
2
3
4
5
6
or
7
8
or
9

Tens

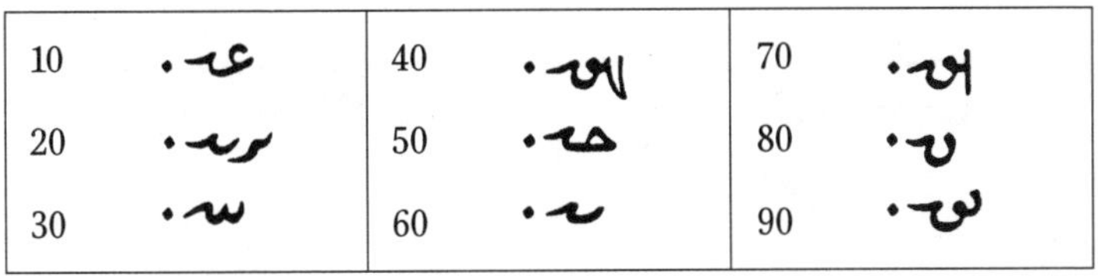
10
20
30
40
50
60
70
80
90

Hundreds

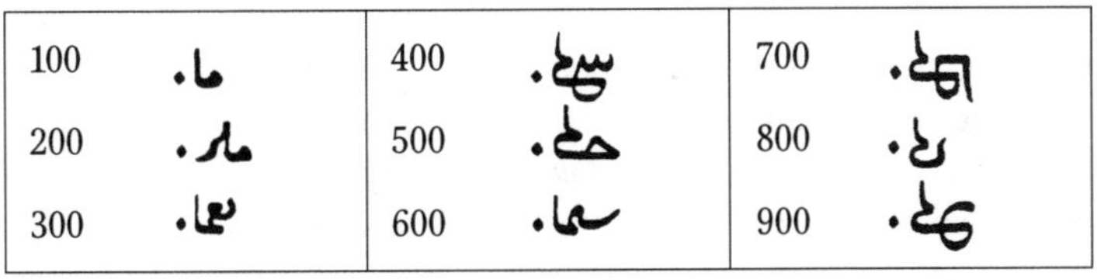
100
200
300
400
500
600
700
800
900

Thousands

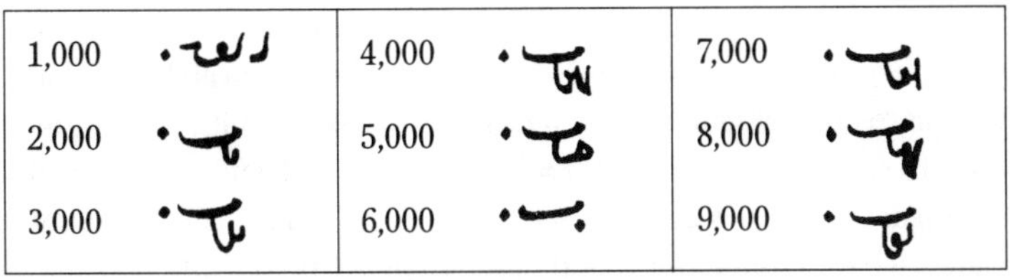
1,000
2,000
3,000
4,000
5,000
6,000
7,000
8,000
9,000

Ten Thousands

10,000	ع	40,000	لم	70,000	لو
20,000	ر	50,000	هـ	80,000	ن
30,000	لس	60,000	س	90,000	و

Composite numbers

Note that for composite numbers containing several digits, the Turks generally used the letter ‎س‎ (*sin*), lengthening its horizontal stroke over the group. This letter stood for the word ‎سيـاق‎ (*siyaq*).

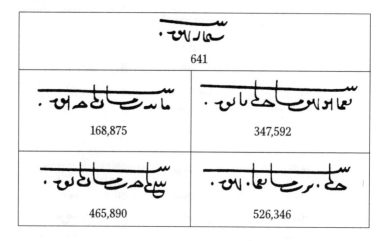

641

168,875	347,592
465,890	526,346

INDIAN NUMERALS' MAIN ARAB RIVAL

Of all rival notations, with which Indian numerals were sometimes mixed in Arabic writings, the most important was certainly Arabic letter numerals. These were known as *huruf al jumal* (literally "letters [for calculating] series") and also as *Abjad* (from its first four letters), because it does not use the letters in the "dictionary" order, or *mu'jama*, (i.e. *alif, ba, ta, tha, jim, ha, kha, dal, dhal*, etc.) but in a special order, called *abajadi* beginning *alif, ba, jim, dal, ha, wa, zay, ḥa, ṭa* etc., attributed as follows: '*a* = 1, *b* = 2, *j* = 3, *d* = 4, *h* = 5, *w* = 6, *z* = 7, *ḥ* = 8, *ṭ* = 9 etc. This is not, of course, a simple number-series (like one going from 1 to 26 by means of the Roman alphabet), but a true place-value system, the first nine letters being the units, the next nine the tens (*y* = 10, *k* = 20, *l* = 30, *m* = 40, *n* = 50, etc.), the following nine the hundreds (*q* = 100, *r* = 200, *sh* = 300, *ta* = 400, etc.) and, finally, the twenty-eighth letter standing for one thousand (*gh* = 1,000). Note that the *al abajadi* order is very close to Hebrew, Greek and Syriac letter numerals and is obviously the older order because it derives directly from the original Phoenician alphabet (see Fig. 25.8A). There were some differences between East and West. In the former *sin, sad, shin, dad, dha* and *ghayin* stood for 60, 90, 300, 800, 900 and 1,000, but in the latter 300, 60, 1,000, 90, 800 and 900 respectively.

Islamic scholars and writers often preferred to use this system. One example is the *Kitab fi ma yahtaju ilahyi al kuttab min 'ilm al hisab (Book of Arithmetic Needed by Scribes and Merchants)*, written by the geometer and astronomer Abu'l Wafa al-Bujzani between 961 and 976.

The first two parts deal with calculating with whole numbers and fractions, the third with surfaces of plain figures, the volumes of solid bodies and the measurement of distances. The last four parts deal with various arithmetical problems, such as in business transactions, taxation, units of measurement, exchanges of currency, cereals and gold, paying and maintaining an army, constructing buildings, dams etc. [A. P. Youschkevitch (1976)]. Now, in this book, which was especially conceived for practical use, the Indian decimal place-value system is never used. All numbers are expressed by Arabic letter numerals.

A further significant example: the *Kitab al kafi fi'l hisab (Summary of the Science of Arithmetic)*, written by the mathematician al-Karaji towards the end of the tenth century. It is rather similar to Abu'l Wafa's work and, like many later books, contains no mention of Indian numerals.

True, these works were especially for scribes, accountants and merchants, and we know that this form of arithmetic was favoured not only by scribes but also by officials and tradesmen. That is why this system stood up for so long against the new Indian way, which was supported by al-Khuwarizmi, An Nisawi and many others [A. P. Youschkevitch (1976)].

More surprisingly, the same phenomenon can be found in many Arabic works dealing with algebra, geometry and geography, which also contain only the letter system.

Works on astronomy

For astronomic treatises and tables, this was for a long time the only system the Arabs used.

It may be useful to remind ourselves of certain points concerning the sexagesimal system which the Arabs had inherited from the Babylonians, via Greek astronomers.

Babylonian scholars used a place-value system with base 60 and, from around the fourth century CE, they had a zero. These cuneiform numerals were the vertical wedge for units and the slanting wedge for tens (see Fig. 13.41). As for zero, it was represented either by a double oblique vertical, or by two superimposed slanting wedges.

This system was then adopted by Greek astronomers (at least from the second century BCE), but only to express the sexagesimal fractions of units (negative powers of 60). Otherwise, instead of using cuneiform signs, the Greeks had their own letter numerals, from α to θ for the first nine numbers, the next five (ι to ν) for the first five tens, with all the intermediate numbers expressed as simple combinations of these letters. Influenced by the Babylonians, they introduced a zero expressed either as sign written in various different ways (presumably the result of adapting old Mesopotamian cuneiform into a cursive script), or as a small circle topped by a horizontal stroke (probably the letter *omicron* (*o*) the initial letter of *ouden*, "nothing", and topped with a stroke to avoid confusion with the letter o which stood for 70); or else as an upturned 2 (probably a cursive variant of the above) (see Fig. 13.74A and *Zero).

Arab astronomers also took over the Greek sexagesimal system, adapting it to their own alphabet. Note that to express zero in their sexagesimal calculations, the Arabs hardly ever used the Indian signs (the circle and the point). Instead they used a sign written in a variety of different ways (including the upturned 2 referred to above) which they had also inherited from the Greeks. Woepcke has this to say about the Arabs' sexagesimal system:

> Rather than [Indian] numerals, the Arabs preferred an alphabetic notation for their astronomic tables. They apparently found it more convenient. This use is confirmed in Arabic manuscripts containing astronomic tables, in which Indian numerals are rarely met with. The Arabs sometimes used them to express very large numbers, for example degrees over the circumference [see JA April–May (1860)]. However, this exception was unnecessary. Sexagesimal calculation, just as it had divided the degree into minutes, seconds, thirds etc., also had higher values, superior to the degree so that it was unnecessary to go higher than 59 in this notation. This is revealing about the relationship between sexagesimal calculation and alphabetic notation: it is after the number 60, useless in a rigorous sexagesimal system [i.e. based on place-value], that the divergence between the African and Asian alphabetic notations began [F. Woepcke (1857), p. 282].

The Arabs wrote an expression such as 0° 20' 35'' as follows (reading from right to left):

HL	K	0
35	20	0

O being zero, K the letter *kaf* = 20 and the group LH, or *lam-ha*, the juxta-position of *lam* (= 30) and *ha* (= 5).

To sum up, in their sexagesimal calculations and tables, Arab astronomers generally used their alphabet in the way described above (see Fig. 13.76). An exception to this rule was Abu'l Hasan Kushiyar ibn Labban al-Gili (971–1029), who wrote the *Maqalatan fi osu'l hisab al hind* (*Two Books Dealing with Calculations Using Indian Numerals*), the second book of which is concerned with base 60. The "tables of sixty" (*jadwal al sittini*) are expressed in the traditional Arabic letter numerals, but the operations are made using Indian numerals [Aya Sofia Library, Istanbul, Ms. 4857, f° 274 r and following; see A. Mazaheri (1975), pp. 96–141]. But, so far as I know, this is the only author to break the rule stated above.

Books of magic and divination

The underlying reason for this preference is suggested in a work dealing with Arab astrology and white magic, dating from 1631 CE [BN Paris, Ms ar 2595, f° 1–308]. The author, a certain al-Gili (not to be confused with the mathematician cited above), uses a number of magic alphabets to name the spirits and the seven planets and shows how to make talismans "using Indian numerals", according to "the secret virtues of Arabic numerals". When drawing up "judgements of nativities" (i.e. horoscopes) the writer speculates about the numerals' "magical properties" in what he calls *hisab al jumal* (or "calculation of series") in which the letters of the Arabic alpha-bet are used, each with a number attached to it. He then draws up two lists of numbers, one called the *jumal al kabir* ("large series"), the other *jumal as saghir* ("little series") [see P. Casanova (1922); A. Winkler (1930)]. The author then explains how remarkable it is that "Arabic calculation (*bi hisab al 'arabi*) is always used for the little series, and Indian calculation (*bi hisab al hindi*) for the large series". In other words, the "large series" is always expressed in Indian numerals, the "little series" in Arabic letter numerals.

Why this difference? Large series were designed to give numerical values, true arithmetical numbers, while the little series was compared with it in order to give a name to each numerical value and determine its alleged secret virtues. For the author is of course referring to letter numerals when he mentions "the secret virtues of Arabic numerals". For him, the Indian numerals had no hidden powers.

Thus, the *Abjad* system (also called *Huruf al jumal*) was considered by the Arabs as "more their own than any other" [F. Woepcke (1857)]. They even gave their own name to it: *hisab al 'arabi* ("Arabic Calculation").

Arab magi and soothsayers presumably wanted to make a clear distinc-tion between a system which they considered to be typically Arabic and part of Muslim traditions and practices, and another, the arithmetical superiority of which they were willing to recognise, but which remained in their eyes foreign and "not sacred".

A strange "machine" for thinking out events

To gain a clearer idea of the Arabs' magical and divinatory practices, let us listen to Ibn Khaldun who, in his *Prolegomena*, describes that strange "machine" for thinking out events which is known as the *za'irja*. It inspired Ramon Lull (died 1315) in his famous *Ars Magna*, and, even at the end of the seventeenth century, Leibnitz was still one of its admirers.

It is claimed that by using an artificial system, we can know about the contents of the invisible world. This is the *za'irjat al 'alam* ["circular chart of the universe"] supposedly invented by Abu'l 'Abbas as Sibti, from Ceuta, one of the most distinguished of the Maghribi *Sufi*. Near the end of the sixth century [of the Hegira = twelfth century CE], As Sibti was in the Maghreb while Ya'qub al-Mansur, the Almohad monarch, was on the throne.

The construction of the *za'irja* ["circular chart"] is a wondrous piece of work. Many highly placed persons like to consult it to obtain useful knowledge from the invisible world. They try to use enigmatic procedures and sound out its mysteries in the hope of reaching their goals. What they use is a large circle, containing other concentric circles, some of which refer to the celestial spheres, and others to the elements, the sublunary world, spirits, all sorts of events and various forms of knowledge. The divisions of each circle are the same as the sphere they represent; and the signs of the zodiac, plus the four elements [air, earth, water and fire] are found within them. The lines which trace each division continue as far as the centre of the circle and are called "radii".

On each radius appears a series of letters, each with a numerical value, some of which belong to the writing of records, that is to say to signs which Maghribi accountants and other officials still use for writing numbers. [The author is of course referring to the monograms and abbreviations of the Arabic names of the numbers, called *Dewani* numerals].

There are also some *gobar* numerals [Fig. 25.5].

Inside the *za'irja*, between the concentric circles, can be found the names of the sciences and various sorts of place name. On the other side of the chart of circles, there is a figure containing a large number of squares, separated by vertical and horizontal lines. This chart is fifty-five squares high, by one hundred and thirty-one squares across.

[The author does not say that many of these squares are empty.] Some of these contain numbers [written in Indian numerals], and others letters. The rule which determines how the characters are placed in the squares is unknown to us, as is the principle that determines which squares are to be filled and which remain empty. Around the *za'irja* are found some lines of verse, written in the *tawil* metre, rhymed on the syllable *la*. This poem explains how to use the chart to obtain the answer to a question. But its lack of precision and vagueness mean that it is a veritable enigma.

On one side of the chart is a line of verse written by Abu 'Abdallah Malik ibn Wuhaib [fl. 1122 CE], one of the West's most distinguished soothsayers. He lived under the Almoravid Dynasty and belonged to the *uleima* of Seville. This line is always used when consulting the *za'irja* in this way, or in any other way to obtain an answer to a question. To have an answer, the question must be written down, but with all the words split up into separate letters. Then, the sign of the zodiac is located [in the astronomical tables] and the degree of that sign as it rises above the horizon [i.e. its ascendant] coinciding with the moment of the operation. Then, on the *za'irja*, the radius is located which forms the initial boundary of the sign of the ascendant. This radius is followed to the centre of the circle, and thence to the circumference, opposite the place where the sign of the ascendant is indicated, and all the letters found on this radius, from beginning to end, are copied out.

Also noted are the numerical signs [Indian numerals] written between the letters, which are then transformed into letters according to the *hisab al jumal* system [the "series calculation", used when replacing Indian numerals by letters and vice versa]. Sometimes units must be converted into tens, tens into hundreds, and vice versa, but always under the rules drawn up for the *za'ijra*. The result is placed next to the letters which make up the question. Then the radius which marks the third sign from the ascendant is examined. All the letters and numbers on this radius are written down, from its beginning to the centre, without going to the circumference. The numbers are then replaced by letters, according to the procedure already described, the letters being placed one beside the other.

Then, the verse written by Malik ibn Wuhaib, the key for all operations is taken and, once it has been split into separate letters, it is put to one side. After that, the number of the degree of the ascendant is multiplied by what is called the sign's *'asas* [literally "base" or "foundation" an algebraic term for the index of a power, but here standing for the number of degrees between the end of the last sign of the zodiac and the sign which is the ascendant at the time of the operation, the distance being taken in the opposite direction from the normal order of the signs].

To obtain this *'asas*, we count backwards, from the end of the series of signs; this is the opposite of the system used for ordinary calculations which starts at the beginning of the series. The product thus obtained is multiplied by a factor called the great *'asas* and the fundamental *dur* ["circuit", or "period", in astronomy used for the time it takes a point to make a complete orbit of the earth. A planet's *dur* is thus either its orbit, or the time taken to return to any given point in the heavens. But in the *za'irja*, *dur* also stands for certain numbers used for selecting the letters which will give the required answer].

The results are then applied to the squares on the chart, according to the rules governing the operation, and after using a certain number of *dur*. In this way, several letters are extracted [from the chart], some of which are eliminated, while the rest are placed opposite Ibn Wuhaib's verse.

Some of these letters are also placed among the letters forming the words of the question, which have already had others added to them. Letters of this series are eliminated when they occupy places indicated by the *dur* numbers. [Thus:] As many letters are counted as there are digits in the *dur*, when the last *dur* figure is arrived at, the corresponding letter is rejected; this operation is repeated until the series of letters is exhausted.

It is then repeated using other *dur*. The isolated letters remaining are put together and produce [the answer to the question asked by] a certain number of words forming a verse, in the same metre and rhyme as the key verse, composed by Malik ibn Wulaib. Many highly placed persons have become absorbed in this pursuit and eagerly use it in the hope of learning the secrets of the invisible world. They believed that the relevance of the answers showed that they were accurate. This belief is absolutely unfounded.

The reader will already have understood that the secrets of the invisible world cannot be discovered by such artificial means.

It is true that there is some connection between the questions and answers, in that the answers are intelligible and relevant, as in a conversation.

It is also true that the answers are obtained as follows: a selection is made between the letters in the question and on the radii of the chart.

The products of certain factors are applied to the squares on the chart, whence some letters are extracted; certain letters are eliminated by several selections using the *dur*, and the rest are then placed opposite the letters making up the verse [of Malik ibn Wahaib].

Any intelligent person who examines the connections between the various steps in this operation will discover its secret. For these mutual connections give the mind the impression that it is in communication with the unknown, and also provide the way of going there. The faculty of noticing the connections between things is most often found in people used to spiritual exercises, and practice increases the power of reasoning and adds new strength to the faculty of reflection. This effect has already been explained on numerous occasions.

This idea has resulted in the fact that almost everybody has attributed the invention of the *za'irja* to people [the *Sufi*, Muslim esoterics], who had purified their souls by spiritual exercise.

Thus, the *za'irja* I have described is attributed to As Sibti [a *Sufi*]. I have seen another one, invented, it is said, by Sahl ibn 'Abdallah and must admit that it is an astounding work, a remarkable production of a profound spiritual application.

To explain why As Sibti's *za'irja* gives a versified answer, I tend to think that the use of Ibn Wuhaib's verse as a starting point influences the answer and gives it the same metre and rhyme.

To support this view, I have seen an operation made without this verse as a starting point, and the answer was not versified. We shall speak further of this later. Many people refuse to accept that this operation is serious and that it can answer one's questions. They deny that it is real and look on it as something suggested by fancy and imagination. If they are to be believed, people who use the *za'irja* take letters from a verse they have composed as they see fit and insert them among the letters making up the question and those from the radii. They then work by chance and without any rules; finally they produce the verse, pretending that it has been obtained by following a fixed procedure.

Such an operation would only be an ill-conceived game. No one using it would be capable of grasping the connections between beings and knowledge, or of seeing how different the operations of perception are from those of the intelligence. The observers would also be led to deny anything they do not perceive.

To answer those who call the *za'irja* a piece of juggling, suffice it to say that we have seen operations performed on it respecting the rules and, according to our considered opinion, they are always carried out in the same way and follow a genuine system of rules. Anybody possessed of a certain degree of penetration and attention would agree with this, once one of these operations has been witnessed.

Arithmetic, a science producing absolutely clear results, contains many problems which the intelligence cannot understand at once, because they include connections which are hard to grasp and elude observation.

How much more so, then, for the art of the *za'irja*, which is so extraordinary and whose connections with its subject are so obscure?

We shall cite one rather difficult problem here, to illustrate this point. Take several *dirhams* [silver coins] and, beside each coin, place three *fulus* [copper coins]. With the sum of the *fulus* you buy one bird, and with that of the *dirhams* several more at the same price. How many birds have you bought? The answer is nine. We know that there are twenty-four *fulus* to a *dirham*; so three fulus are the eighth of a *dirham*. Now, since each unit is made up of eight eighths, we can suppose that when making this purchase we have brought together the eighth of each *dirham* with the eighths of the other *dirhams*, and that each of these sums is the price of one bird. With the *dirhams* we have then bought just eight birds; the number of eighths in a unit; add to that the bird purchased with the *fulus* and we have nine birds in all, since the price in *dirhams* is the same as that in *fulus*.

This example shows us how the answer is hidden implicitly in the question and is arrived at by knowing the hidden connections between the quantities given in the problem.

The first time we encounter a question of this sort, we imagine that it belongs to a category that can be solved only by applying to the invisible world. But mutual connections allow us to extract the unknown from what is known. This is especially true of things in the sentient world and the sciences.

As for future events, they are secrets that cannot be known precisely because we are ignorant of their causes and have no certain knowledge of them.

From what we have explained, it can be seen how a procedure which, by using the *za'irja*, extracts an answer from the words of the question is a matter of making certain combinations of letters, which had initially been ordered to ask the question, appear in a different form.

For anyone who can see the connection between the letters of the question, and those of the answer, the mystery is now clear.

People capable of seeing these connections and using the rules we have explained can thus easily arrive at the solution they require.

Each of the *za'irja's* answers, seen under a different light, is like any other answer, according to the position and combination of its words; that is, it can either be negative or positive.

To return to the first point of view, the answer has another characteristic: its indications are in the class of predictions and their accordance with events [in other words, as Slane emphasises in modern terms, these indications are part of the category of agreements between discourse and the extrinsic].

But we shall never know [about future events] if we use procedures such as the one just described.

What is more, mankind is forbidden to use it for these ends. God communicates knowledge to whomsoever he wants; [for, as the Koran says (sura 2, verse 216) *God knows, but you know not.*] [See *Muqaddimah*, pp. 213–19; cf. Slane's translation, pp. 245–53.]

We must salute, in passing, Ibn Khaldun's eminently modern rationality, categorically rejecting the rather strange practices of Arab astrologers and soothsayers, which were in fact outlawed by Islam.

To this can be added the strange "revelation calculation" [*hisab 'an nim*], which soothsayers used in time of war to predict which of the two sovereigns would conquer or be conquered. Here is how Ibn Khaldun describes it in his *Prolegomena:*

The numerical value of the letters in each sovereign's name was added up. Then each sum was reduced until it was under nine. The two remainders were compared. If one was higher than the other, and if both were odd or both even, then the king whose name had provided the lower figure would win. If one was even and the other odd, the king whose name had provided the higher figure would win. If both remainders were equal and even, then the king who had been attacked would vanquish. But if both were equal and odd, then the attacking king would be victorious [*Muqaddimah*, cf. Slane's translation, I, pp. 241–2].

The underlying reasons for the preponderance of Arabic letter numerals

Thus, the system of Arabic letter numerals was favoured as a way of writing numbers not only by scholars, mathematicians, astronomers, physicians and geographers, but also by authors of religious works, mystics, alchemists, magicians, astrologers, soothsayers, scribes, officials and tradesmen, among both Arabs and Muslims.

The system was so common in the Islamic world, that Arab poets even invented a particular form of literary composition which used the letter numerals. These *ramz* were versified according to the arithmetic equalities or progressions of the numerical values of the letters in each line.

Even historians, and the lapidaries of North Africa, Spain, Turkey and Persia, were (at least in later periods) fond of a technique called *tarikh*, i.e. "chronograms", which consists in grouping a set of letters, the numerical value of which when added together produces the date of some past or future event, into one meaningful or significant word, or else into a short phrase.

This shows how the representation of numbers was of vital importance in the history of Islam. It was, of course, directly linked with both the meanings and the characters of Arabic writing, since the "numerals" were simply letters of the alphabet. This numerical notation was always written from right to left, like words, and, as for ordinary letters, the characters were generally joined up and slightly modified depending on whether they were isolated, initial, medial or final.

Thus, for poets enamoured of the *ramz*, these "letter numerals" or numerical letters were an integral part of their artistic expression, mirroring the beauty of the language. For artists, they also harmonised with the art of calligraphy, reflecting both their individual perspectives and the emotional state in which the work was created. And for those with a mystical bent, these same "numerals" allowed them to produce graphic or versified symbolic expressions, at once literal and numerical, of their quest for Allah.

Meanwhile, the scribes, who adhered to their characteristic embellishments, were able to give these numerical letters the same grace, balance and rhythm as the ordinary letters in their miniatures and illuminations.

All of which confirms the perfect continuity between this system and the purest of Arab and Islamic traditions, and the fundamental practices of Muslim mysticism.

It must not be forgotten that the Arabic script is considered to embody a Revelation and the spreading of the word of the Prophet; it is thus the basic criterion for belonging to the Islamic community (the *Umma*). It is this close connection between the Muslim religion and the Arabic script which gives the Arabic alphabet its privileged, almost fundamentally sacred, position. Tradition even has it that the reed pen, the famous *qalam* ("calamus"), was the first of all Allah's creations.

For the *Hurufi* ("Letterers"), sects based on beliefs attached to the symbolic meaning of Arabic letters, a name was the essence of the thing named.

And, as all names are supposed to be contained in the letters of the discourse, the entire universe was the product of Arabic letters. In other words, from these letters proceeded the universe. Hence the association between the "science of letters" (*'ilm al huruf*), the "science of words" (*'ilm al simiya*) and the "science of the universe" (*'ilm al 'alam*).

The mystic al-Buni was one follower of this belief. He established correspondences between the Arabic letters and what he thought were the elements of the visible world: the four elements (water, earth, air and fire), the celestial spheres, the planets and the signs of the zodiac. And, as there are twenty-eight letters, he associated them with the twenty-eight lunisolar mansions [see E. Doutté (1909)].

God is a force translated by the Word; he acts through his voice and so, by inference, through the very letters of the Arabic alphabet.

Thus the "sciences" of letters and words, once mastered, would reveal the attributes of Allah as they are manifested in nature through the Arabic letters. According to these doctrines:

> The Arabic letter symbolises the mystery of being, through its fundamental unity derived from the divine Word and its countless diversity resulting in virtually infinite combinations; it is the image of the multitude of creation, and even the very substance of the beings it names. Together, they are regarded as manifestations of the *Word Itself*, inseparable attributes of the Divine Essence, as indestructible as the Supreme Truth. Like the divine being, they are immanent in all things. They are merciful, noble and eternal. Each of them is invisible (hidden) in the Divine Essence [J. Chevalier and A. Gheerbrant (1982)].

This is why, according to the precepts of Islam, the Koran, as the Revelation of the Prophet Mohammed, cannot be read in another language than Arabic, nor can it be transcribed into a different script. For this Book, seen as one of the expressions of the Word of Allah, is identified with the Divine Essence. To quote Doutté:

> This conception takes us back to ancient times, when the Romans, by the word *litterae*, and the Nordic peoples, by the word "rune", meant the entirety of human knowledge. Nearer to the Arabs, in the Semitic world, the Talmud teaches that letters are the essence of things. God created the world by using two letters; Moses on going up to heaven met God who was weaving crowns with letters. Ibn Khaldun has much to say about these doctrines and gives a theory of written talismans [see *Muqaddimah*, trans. Slane, II, pp. 188–95]: as the letters composing them were formed from the elements which make up each being, they could act upon them.

> Such is the basis of *'ilm al huruf* and *'ilm as simiya*, Islam's mystical "sciences" of letters and words.

One category of letters, whose magical powers have a religious origin and are thus characteristic of Arab magic, are those at the beginning of certain suras of the Koran, and whose meaning is totally unknown (or else jealously guarded by Muslim mystics). For example, sura II begins with *alif, lam, mim*; sura III with *alif, lam, mim, sad* etc. Orthodox Muslims call these letters *mustabih*, and say that their meaning is impenetrable for the human mind; thus, unsurprisingly, they have been adopted by magicians.

Al-Buni calls them *al huruf an nuraniya*; there are fourteen of them, exactly half the number (28) of lunar mansions, from which he draws further speculations. Each of them, he points out, is the initial of one of the names of God. Two of these groups, which contain five letters, have particularly attracted magicians. They are supposed to have extraordinary virtues and many *herz* ("talismans") have been made using them.

If letters have magical powers, then these powers are increased when they are written separately. In the Arabic script, individual letters are more perfectly formed than when they are joined up.

But the letters' most singular properties come from their numerical values. Two different words can have the same numerical total. The mysticism of letters then says that they are equivalent. In the Cabbala, this is the principle of "gematria". It is also a favourite of Muslim magic. Not only are words linked together by the numbers expressed by their letters, but these very letters can reveal their magical virtues through a numerical evaluation of the letters and words   [E. Doutté (1909)].

In other words, Arabic words have a numerical value. A reciprocal logic even had it that numbers were charged with the semantic meaning of the word or words they corresponded to. Hence, as with the Cabbala, ciphered messages, "secret languages" and all sorts of speculations were cooked up by mystics, numerologists, alchemists, magi and soothsayers. Their aim was to stop laymen understanding and harmonising with these esoteric meanings, which supposedly held a hidden truth, or else to compose cryptographic texts wrapped in apparently indecipherable allegories and puzzles, or to use them for a variety of interpretations, conclusions, practices and predictions (see Chapter 20).

It can thus be seen how a numerical value was added to the letters' symbolic, magical and mystical powers, thus giving them the broadest and most effective range of meaning.

Words have always fascinated us, but numbers even more so. Since time immemorial, numbers have been the mystic's ideal tool. They do not express only arithmetical values but, inside what was considered to be their visible exteriors, numbers also contained magical and occult forces which ran on an unseen current, rather like an underground stream. Such ideas could be either for good or evil, depending on their inherent nature.

The magical and mystical character of numbers is a common human belief. Their importance in Mesopotamia, ancient Egypt, pre-Columbian America, China and Japan is beyond our scope. As are the theories and doctrines of the Pythagoreans and Neo-Platonists who, struck by the importance of numbers and their remarkable properties, made them into one of the bases of their metaphysics, believing that numbers were the principle, the source and the root of all beings and things. But what should be emphasised is the direct link between a belief in the magic of numbers and the fear of enumeration, present among the Hebrews (see for example Exodus, 30: 12 and II Samuel 24: 10), the Chinese and Japanese (who are particularly superstitious about the number four), and also among several African, Oceanian and American peoples, who find numbers repellent.

It should be said in passing, that the ancient fear of enumeration reveals the difficulties humans have always had in assimilating the concept of number, which they see, and rightly so, as highly abstract.

It is this very link between magic and the ancient fear of numbers which forbids, for example, North African Muslims from pronouncing numbers connected with people dear to them or personal possessions. For, according to this belief, giving the number of an entity allows it to be circumscribed. If you provide the number of your brothers, wives or children, your oxen, ewes or hens, the sum of your belongings, or even your age, you are giving Satan, who is ever on the lookout, the possibility to use the hidden power of these numbers. You thus allow him to act upon you and do evil to the people or things you so imprudently enumerated.

A sort of superstitious reciprocity led to the making of *herz* in the form of magic squares: talismans with alleged beneficent powers, such as curing female sterility, bringing happiness to a home or attracting material riches.

As a passing remark, Islamic religion and traditions see the number five as a good omen in, for example, the five *takbir* of the Muslim profession of faith, *Allah huwa akbar* ("God is Great"); the five daily prayers; the five days dedicated to 'Arafat; the five fundamental elements of the pilgrimage to Mecca, the five witnesses of the pact of the *Mubahala*; and the five keys to the mystery in the Koran (6: 59; 31: 34). There is also Thursday, called in Arabic *al khamis* ("the fifth"), which is a particularly sacred day. Then there are: the five goods given as a tithe; the five motives for ablution; the five sorts of fasting; the pentagram of the five senses and of marriage; the five generations that mark the end of tribal vengeance; and so on. Naturally, there are the five fingers of the hand, placed under special protection in memory of the five fingers of the "hand of Fatima", the daughter of Mohammed and Khadija, and wife of 'Ali, the Prophet's cousin [see J. Chevalier and A. Gheerbrant (1982); E. Doutté (1909); EIS; T. P. Hugues (1896)].

Even today if you foolishly ask Tunisians, Algerians or Moroccans how old they are, or how much money they have, they will cast off the evil eye by vaguely replying "a few" if they are polite, or else curtly say "five", or even brusquely slap the five fingers of their hand over your own "evil eye".

To sum up, each of the twenty-eight Arabic letters, as an ordinary letter, was supposed to have its own symbolic meaning, magical power and creative force. But as a numeral or written in cipher, each was linked with a number and, as such, was directly in touch with the supposed idea, power and force contained in that number. A name is the outward sign of the Word, considered to be one of the main magical and mystical forces. As it is made up of letters, and thus of the corresponding numbers too, it is easy to see why the Arabs' alphabetic numbering (as a particular case among their multiple ways of evaluating their letters and words) was for mystics, magi and soothsayers a product of sound, sign and number, and hence had powers that transcended the ordinary alphabet.

We can now see how important this system was at all levels of Islamic society. And we can also see why the Indian place-value system was considered by most authors to be something absolutely alien to their culture and traditions.

The direction it was written in added to its relative unpopularity. It ran from left to right (one hundred and twenty-seven, for example, being written as 127), the opposite way to Arabic script. And as the numerical letters were written from right to left, from the highest digit to the lowest, and obeyed the rules of the Arabic cursive script, they were favoured above any other system.

The direction of Indian numerals had been highly practical for Indian mathematicians and accountants, whose script went from left to right. But this fact (which caused obvious problems for people accustomed to writing from right to left) raised difficulties for Arab-Muslim scholars.

They would certainly have solved this problem if they had inverted the original order of the Indian decimal system, by writing something like this:

$$8 \quad 7 \quad 6 \quad 3 \quad 2$$
$$\longleftarrow$$

when an Indian would have written:

$$23{,}678 \; (= \; 2 \times 10^4 + 3 \times 10^3 + 6 \times 10^2 + 7 \times 10 + 8).$$

They would thus have completely adapted the Indian system to their own script. But this idea apparently never occurred to the Arabs, or else they refused to break with the Indian tradition.

Another reason, the last we shall give here, for this opposition was as follows.

During their relations with India, the Arabs were in contact with the Hindî, but also with the Punjabî, the Sindhî, the Mahârâshtrî, the Manipurî, the Orissî, the Bihârî, the Multanî, the Bengalî, the Sirmaurî and even the Nepalî. A glance at Chapter 24 will confirm how much the writing of numerals in India varied, not only from one period to another, but also in different regions, and even with different scribes (Fig. 24.3 to 52). What was a 2 for some became something like a 3, 7, or 9 for others, for palaeographic reasons. In other words, a lack of standardisation meant that the written form of Indian numerals remained unstable. But for mathematicians and astronomers numbers had to remain the same and be absolutely consistent. How could one transmit the fundamental data of a work of astronomy, for example, if the numerical value of observations and results could be variously interpreted, depending on the time, place and habits of the user? What is more, if a scribe or copyist made a mistake, it might never be noticed. These numerals were therefore not sufficiently rigorous for works dealing with mathematics, geography or astronomy in which value was of prime importance. Hence the preference for numerical letters, which did not present such a problem.

Need we add that, if the so-called "Arabic" numerals had really been invented by the Arabs, then they would have been used more widely and adopted by Muslims much more rapidly? There is also a good chance that these numerals would have been written from right to left, like the Arabic script.

These important facts add to the indisputable evidence that our present number-system comes from India.

Among other imperishable merits, the Arab-Islamic civilisation did certainly transmit our modern numerals and methods of calculation to mediaeval Europe, which was at the time at a much lower scientific and cultural level. In gratitude for this basic contribution, Europe then named these numerals after the people who had provided them. But to say that Islam was the cradle of these numerals would be to fall into the trap laid by an erroneous term, which even Arab and Muslim scholars never used in their writings or vocabulary.

DUST-BOARD CALCULATION

The time has come to discuss Indian calculation methods, which not only played an important role in the transmission of Indian numerals throughout the Islamic world, but also profoundly influenced how techniques evolved.

Many good reasons lead us to suppose that, from earliest times, Arab-Muslim arithmeticians in the East and the West made their calculations by sketching out the nine Indian numerals in loose soil, with a pointer, stick or just with a finger. This was known as *hisab al ghubar* ("calculating on dust") or *hisab 'ala at turab* ("calculating on sand").

But they did not always write on the ground; they also had other methods. Their most common tool seems to have been the counting board, what is called in the East *takht al turab* or *takht al ghubar* (from *takht*, "tablet" or "board", *turab*, "sand", and *ghubar*, "dust"), which was also known in the Maghreb and Andalusia as the *luhat al ghubar* (*luhat* being a synonym of *takht*).

Several Persian poets refer to it, at least from the twelfth century on, such as Khaqani, in his eulogy for Prince Ala al-Dawla Atsuz (1127–1157) [A. Mazaheri (1975)]:

> The seven climates tremble with quartan fever;
> And dust will cover the vaulted sky,
> Like the accountant's board (*takht*).

Or the mystical poet Nizami (died 1203) [Nizami (1313), cited by A. Mazaheri (1975)]:

> From the system of nine heavens
> [Marked] with nine figures,
> [God] cast the Indian numerals
> Onto the earth board.

This counting board was favoured not only by professional Arab accountants, mathematicians and astronomers, but also by magi, soothsayers and astrologers.

In about 1155, Nizami told this story, which features the philosopher al-Kindi (ninth century) [Nizami as above]: "Al Kindi asked for the dust board, got up and [with his astrolabe] read the height of the sun, the hour and traced the horoscope on the sand board (*takht al turab*) . . . "

It consisted of a board of wood, or of any other material, on which was scattered dust or fine sand, so that the Indian numerals could be traced out in it and calculations made. Powder, or sometimes even flour were also used, as our sources indicate. The word *ghubar* in fact means "powder" or "any powdery substance" as well as "dust".

This counting board was not unique to Arab arithmeticians. It was also used long before Islam by the Indians (see **Pâtîganita*).

TRACES OF THE OLD PERSIAN ABACUS FROM THE TIME OF DARIUS

Old abacuses from time of Darius and Alexander were also used, at least in Persia during the first centuries of the Hegira. Calculations were made by throwing down pebbles or counters, and certain Persian accountants kept up this method (see Fig. 16.72 and 73).

The following is, of course, just a hypothesis, but it is supported by much of the evidence. The Persian verb "to count", "to calculate" is *endakhten*, which also means "to throw". At this time, arithmetical operations were carried out on tables or rugs, divided by horizontal and vertical lines, on which the counters were placed, their value changing as they moved from one column to another.

It is also interesting to note that the action which corresponds to the verb *endakhten* ("count", "calculate") is *endaza*, which means three things: "throwing", "counting" and "calculating". This is shown in this brief quotation from *Kalila wa Dimna*, a famous Persian fable, here in a twelfth-century version by Abu al Ma'ali [see A. Mazaheri (1975)]: "Having carefully listened to his mother's words / The lion threw them backwards (*baz endakht*) with his memory." This is so subtle that a commentary is necessary.

Even for a lion, "throwing words backwards" is meaningless. But if we take the verb to mean "to calculate" or, by extension, "to measure" we can then see that the king of the jungle had thought over, or "weighed", his mother's words.

But let us not take etymology too far in order to explain something which had already almost vanished from the old country of the Sassanids, for these words had lost their numerical meaning by the thirteenth century. And the instrument itself, rightly considered as cumbersome and impractical, had been rejected by the region's professional accountants at an early date. (Note also that they rejected the Chinese abacus, introduced by Mongol invaders during the thirteenth century; but the unpopularity of this excellent apparatus was due to the Persians' hatred of Genghis Khan and his successors.)

THE BOARD AS A COLUMN ABACUS

To return to calculations made on the ground, or else in dust scattered over a flat surface or board, there were of course different ways of working. Here is the most rudimentary.

The arithmeticians began by tracing several parallel lines on the surface to be used, thus marking out a series of columns which corresponded to the places of the decimal system. Then they drew the nine numerals inside each one. In this way, they immediately acquired a place value.

The Arabs, like the ancient Indian arithmeticians, would write a number such as 4,769 by tracing the number 9 in the units column, the number 6 in the tens column, the number 7 in the hundreds column and the number 4 in the thousands column.

Ten thousands	Thousands	Hundreds	Tens	Units
	4	7	6	9

So there was no need for zero. It was sufficient just to leave the column empty, as in our next example which represents 57,040:

Ten thousands	Thousands	Hundreds	Tens	Units
5	7		4	

As for the calculations, they were carried out in the dust, then erased.

There is a clear trace of this in the etymology of the Sanskrit words *gunara, hanana, vadha, kshayam,* etc., used by the Indians to mean "multiplication". Literally, they mean "to destroy" or "to kill", in allusion to the successive wiping out of intermediary products, as our example will now show.

Let us suppose that an accountant wants to multiply 325 by 28.

The first thing to do is trace out the four columns required. Then, inside them, we place 325 and 28 as follows, with the highest place of the multiplicand in the same column as the lowest place of the multiplier.

	3	2	5
2	8		

We then multiply the upper 3 by the lower 2. As this equals 6, we place this figure to the left of the upper 3:

$$\downarrow$$

6	3	2	5
2	8		

Then we multiply the upper 3 by the lower 8. As this equals 24, we wipe out the 3 and replace it with 4 (the unit column of 24, the partial product):

$$\downarrow$$

6	4	2	5
2	8		

And, to the 6 we add 2 (the tens digit of 24):

6	4	2	5
2	8		

Then, after erasing, we have:

8	4	2	5
2	8		

The first step has now been carried out, both columns of the multiplier 28 having acted on the hundreds column of the multiplicand (the upper 3 of the initial layout).

We then proceed to the second step by moving all the numbers of the multiplier one place to the right:

8	4	2	5
	2	8	

Then, by using the tens digit of the multiplicand (the upper 2 of the initial layout), we multiply 2 by 2. As this equals 4, we then add 4 to the 4 which lies immediately to the left of upper 2:

8	4	2	5
	2	8	

Then, after erasing, we have:

8	8	2	5
	2	8	

We then multiply the same upper 2 by the lower 8. This makes 16, so we replace, after erasing, the upper 2 with 6 (the units digit of the result):

8	8	6	5
	2	8	

We then add 1 (the tens digit of 16, as above) to the 8 just to the left of the new 6:

8	8	6	5
	2	8	

Then, after erasure, we have:

8	9	6	5
	2	8	

We have now finished the second step, since both digits of the multiplier 28 have operated on the tens digit of the multiplicand (the upper 2 of the initial layout).

We then begin the next step by moving the numbers of the multiplier one column to the right again:

8	9	6	5
		2	8

This time we multiply the units digit of the multiplicand (the upper 5 of the initial layout) by the lower 2. This comes to 10, so we leave untouched the upper 6 (there being no unit digit in the number 10), but add 1 (the tens digit of 10) to the 9 immediately to the left of the 6:

8	9	6	5
		2	8

But as this makes 10 again, we wipe out the 9, leave the space empty (because of zero units in 10) and add 1 to the 8 just to the left of this blank column:

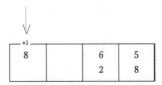

Then, after erasure, we have:

9		6	5
		2	8

We then multiply the upper 5 by the lower 8. As this makes 40, we wipe out the upper 5, but leave the space empty because there is no unit in the product found:

9		6	
		2	8

But we then add the 4 of the product to the upper 6:

9		6	
		2	8

As this again makes 10, we wipe out the 6, leave the space blank and add 1 to the number (zero) in the empty space immediately to the left:

9			
		2	8

Then, after erasure, we have:

9	1		
		2	8

And, as the lowest place of the multiplier is now in the lowest place of the multiplicand (here, the units column of the abacus), we know that the multiplication of 325 by 28 has been completed.

All we have to do know is to read the number on the upper line, nine thousand, one hundred, no tens, no units; so the result is 9,100:

9	1		

This method thus consists in carrying out a number of steps corresponding to the number of places in the multiplicand, each being subdivided into a series of products of one of the digits of the multiplicand successively operated on by all the digits of the multiplier.

In this case, the procedure (now called the operation's "algorithm") has three main phases, each subdivided into two simple steps consisting of calculating a partial product; hence six simple steps in all:

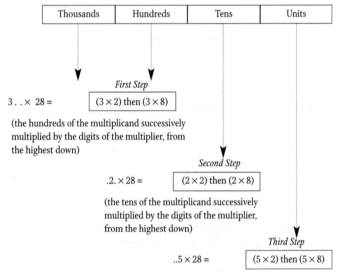

Thousands	Hundreds	Tens	Units

First Step

$3 . . \times 28 =$ (3×2) then (3×8)

(the hundreds of the multiplicand successively
multiplied by the digits of the multiplier, from
the highest down)

Second Step

$.2. \times 28 =$ (2×2) then (2×8)

(the tens of the multiplicand successively
multiplied by the digits of the multiplier,
from the highest down)

Third Step

$..5 \times 28 =$ (5×2) then (5×8)

(the units of the multiplicand successively
multiplied by the digits of the multiplier,
from the highest down)

In other words, this "algorithm" works according to the following formula:

$$325 \times 28 = (3 \times 100 + 2 \times 10 + 5) \times (2 \times 10 + 8)$$
$$= (3 \times 2) \times 1,000 + (3 \times 8) \times 100$$
(first step)
$$+ (2 + 2) \times 100 + (2 \times 8) \times 10$$
(second step)
$$+ (5 \times 2) \times 10 + 5 \times 8$$
(third step)

This counting board thus allows us to carry out calculations without using zero, which explains why certain Arab manuscripts dealing with Indian numerals and methods of calculation make no mention of it.

In certain parts of North Africa, this method continued to be used until the end of the seventeenth century, which explains why the *Ghubar* numerals of the Maghreb generally come down to us in incomplete series, with the zero missing (Fig. 25.5).

But in the East, it gradually disappeared after the tenth or eleventh century and was replaced by more highly developed methods. It is true that this system is long, tiresome and requires considerable concentration and practice.

In fact, very little distinguishes it from methods used in Antiquity. The reason for this has less to do with the numerals themselves than with the method used. It makes no difference whether we trace out the nine Indian numerals, the first nine letters of the Greek or Arabic alphabet, or even the first nine Roman numerals. The principle would still remain virtually the same.

The Indians, as we have seen, certainly used such a system early in their history. But they abandoned it as soon as they had developed their own place-value system and their arithmetic allowed simpler rules to be found.

To carry out arithmetical operations, the early Indians used whatever was to hand. Like everybody else, they presumably began by using pebbles, or similar objects. Then, or perhaps at the same time, they carried out operations on their fingers. But during the next stage, when they developed their first written numerals, they conceived of the idea of drawing several parallel columns, putting the units in the first one, the tens in the second, the hundreds in the third, and so on. They thus invented the column abacus, as others did before and after them. But instead of using pebbles, counters or reeds, they preferred their own nine numerals, which they traced out in dust with a pointer inside the appropriate columns. This was the birth of their dust abacus, which they later improved by working on a table or board covered with sand or dust, instead of the ground.

But this system could not evolve further, so long as it continued to be a column abacus; this concept in fact trapped the human mind for centuries, preventing us from thinking out simpler and more practical rules.

This once again highlights the importance of the discovery of the place-value system. This principle had, of course, long been present in the way calculations were made, but without anybody noticing it. The creative genius of the Indians then brought together all the necessary ideas for discovering the perfect number-system. They had to:

- get rid of stones, reeds, knotted cords, manual techniques or, more generally, any concrete method;
- eliminate any notions of ideogrammatic representation (writing numerals as numbers of lines, points etc.), which certainly came later than the previous system, but was just as primitive;
- eliminate any notation of numbers higher than or equal to the base of the calculation system;

- keep only the nine numerals, in a decimal system, and apply place-value to them;
- replace all existing systems by this group of nine numerals, independent one from the other, and which visually represented only what they were supposed to represent;
- get rid of the abacus and its now useless columns, and apply the new principle to the numerals which were freed from any direct visual intuition;
- fill the gap now created by this method when a place was not filled by a numeral;
- think of replacing this gap by a written sign, acting as zero in the strict arithmetical and mathematical sense of the term.

To sum up, it was by rejecting the abacus that Indian scholars discovered the place-value system.

This raises a question concerning the arithmeticians of the Maghreb and Andalusia, who continued to use the dust abacus and its associated methods for several centuries: did the Western Arabs not know about zero and the place-value system? The answer is no, because these arithmeticians knew the *Hindi* numerals which, as we know, were based on the place-value system and included zero.

In other words, they were aware that the numerals they used could also be manipulated with zero and its associated rules. This is shown in certain Maghrebi manuscripts, in which zero is drawn as a circle (Fig. 25.5).

Why, then, did they not use them for "written calculation" instead of using a dust board? The answer seems to lie in the attachment the Maghrebi and Andalusians always felt for traditions coming from the time of the conquest of North Africa and Spain. Thus, the use of the dust board/abacus has the same traditionalist explanation as their cursive script, derived directly from the *Kufic*.

In fact, the Arabs inherited various arithmetical methods from the Indians, ranging from the most primitive to the most highly developed. In their thirst for knowledge, they presumably took from the Indians everything they could find in terms of calculation methods, without realising that certain things could well be left alone. We should not forget that India is a veritable sub-continent, cut up into regions, peoples, practices, customs and traditions, and it has always been difficult, if not impossible, to see it as a whole.

It is because they came into contact with people who used methods already abandoned by the scholars, and decided to uphold this tradition, that certain Arab-Muslim arithmeticians remained stuck in such a rudimentary rut for several generations.

THE COLUMNLESS BOARD

But this was not, of course, the case for all the Arabs. Others were lucky or bright enough to take up the dust board freed of its columns.

Among them was al-Khuwarizmi. In his *Kitab al jama' wa'l tafriq bi hisab al hind* (*Book of Addition and Subtraction According to Indian Calculations*) he had not only explained the decimal place-value system when applied to Indian numerals, but also recommended "writing the zeros so as not to mix up the positions" [A. P. Youschkevitch (1976), p. 17]. There was also Abu'l Hasan 'Ali ibn Ahmad an Nisawi (died c. 1030), whose *Al muqni' fi'l hisab al hind* (*Complete Guide to Indian Arithmetic*) followed the same sources and methods as the previous work.

Abu'l Hasan Kushiyar ibn Labban al-Gili (971–1029) also deserves a mention. The first chapter of Book I of his *Maqalatan fi osu'l hisab al hind* (*Two Books Dealing with Calculations Using Indian Numerals*) begins as follows:

The aim of any calculation is to find an unknown quantity. To do this, at least [one of these] three operations is necessary: multiplication [*al madrub*], division [*al qisma*] and [extraction of] the square root [*al jadr*] . . . There is also a fourth operation, less often used, which is the extraction of the side of a cube.

But before learning how to carry out these operations, we must familiarise ourselves with each of the nine numerals [*huruf*], the position [*rutba*] of each in relation to the others in the [place-value] system [*al wad'*] . . .

Here are the nine numerals [written in the *Hindi style*, but here updated]:

9 8 7 6 5 4 3 2 1.

[Thus positioned], they represent a number and each stands in a position [*martaba*].

The first is the image of one, the second of two, the third of three . . . and the last of nine. What is more, the first is in the position of the units, the second in the tens, the third in the hundreds, the fourth in the thousands . . .

As for the number formed by these numerals, it must be read: nine hundred and eighty-seven million six hundred and fifty-four thousand three hundred and twenty-one.

[When writing] a number [containing several place-values] we must put a zero [*sifr*, literally "void"] in each place where there is no numeral. For example, to write ten, we put a zero in the place of the units; to write a hundred, we put two zeros, one in the place of the missing tens and one in the place of the units.

Here are these two figures:

Ten: 10

Hundred: 100

There are no exceptions to this rule.

For any of the nine numerals under consideration, the one immediately to its left stands for tens, the next one to the left for hundreds, and the next one to the left for thousands.

In the same way, any of the nine numerals under consideration stands for the tens of the numeral immediately to its right, for the hundreds of the next numeral to its right, for the thousands for the following one, and so on [f° 267v and 268r; A. Mazaheri (1975), pp. 75–76].

These scholars had thus understood that the place-value system and zero removed the need for columns on a counting board.

So, like the Indians, they entered into the era of modern "written calculation".

But they now had to know off by heart the tables giving the results of the four basic operations on these numerals. This is what the Persian mathematician Ghiyat ad din Ghamshid ibn Mas 'ud al-Kashi explains in his *Miftah al hisab* (*Key to Calculation*), in which he reproduces one of these tables: "Here is the table for multiplying numbers inferior to ten. The arithmetician should learn it by heart and know it perfectly, for it can also be used for the multiplication of numbers superior to ten " [see A. Mazaheri (1975)].

Calculating on a columnless board by erasing intermediate results

Our first example of this method comes from the work of Kushiyar ibn Labban al-Gili, cited above [f° 269v to 270v]:

We want to multiply three hundred and twenty-five by two hundred and forty-three.

We put them on the board as follows:

$$3 \quad 2 \quad 5$$
$$2 \quad 4 \quad 3$$

the first numeral [on the right] of the bottom number being always under the last numeral [on the left] of the top number.

We then multiply the upper three by the lower two; this makes six, which we place above the lower two, to the left of the upper three, thus:

6   3   2   5

2   4   3

If the six had contained tens, these would have been placed to its left.

Then we multiply the upper three again by the lower four; this makes twelve, of which we place the two above the four and add the one [which represents the tens] to the six of sixty, obtaining seventy, thus:

7   2   3   2   5

2   4   3

Then we multiply the upper three by the lower three; this makes nine, which replaces the upper three:

7   2   9   2   5

2   4   3

We then advance the bottom number one place towards the right, thus:

7   2   9   2   5

2   4   3

And we multiply the two above the lower three by the lower two; this makes four which, added to the two above the lower two, makes six:

7   6   9   2   5

2   4   3

Then we multiply the upper two again by the lower four; this makes eight, which we add to the nine above the four:

7   7   7   2   5

2   4   3

Then we multiply the upper two again by the lower three; this makes six, which replaces the upper two above the lower three:

7   7   7   6   5

2   4   3

We then advance the bottom number one place [towards the right], thus:

7   7   7   6   5

2   4   3

Finally, we multiply the upper five by the lower two; this makes ten, which we thus add to the tens position above the lower two:

$$
\begin{array}{ccccc}
7 & 8 & 7 & 6 & 5 \\
 & & 2 & 4 & 3
\end{array}
$$

Then we multiply the five again by the lower four; this makes two [tens]. Added to the tens [in the position above] the four, [these two numbers] together make nine:

$$
\begin{array}{ccccc}
7 & 8 & 9 & 6 & 5 \\
 & & 2 & 4 & 3
\end{array}
$$

Finally, we multiply the five by the lower three; this makes fifteen, thus leaving the five alone, we just add one [the tens digit] to the tens, thus:

$$
\begin{array}{ccccc}
7 & 8 & 9 & 7 & 5 \\
 & & 2 & 4 & 3
\end{array}
$$

The [upper] number is the one we wanted to calculate.

This method thus consists in applying the same number of steps as there are places in the multiplicand, each being subdivided into as many products of one of its numbers and the successive digits of the multiplier.

The same method, with some variants, can be found in, for example al-Khuwarizmi and An Nisawi, as well as numerous Indian mathematicians such as Shrîdharâchârya (date uncertain), Nârâyana (1356), Bhâskarâchârya (1150), Shrîpati (1039), Mahâvîrâchârya (850), etc. [B. Misra (1932) XIII, 2; H. R. Kapadia (1935), 15; B. Datta and A. N. Singh (1938), pp. 137–43].

THE DUST BOARD SMEARED WITH A TABLET OF MALLEABLE MATTER

Despite being freed of columns, this approach remained primitive. It was merely a written imitation of older methods and could hardly develop further because of the limitations imposed by the medium.

The dust board was certainly very practical for calculation methods with or without the abacus columns, and especially for the technique of wiping out intermediate results, as this passage from *Psephophoria kata Indos* shows (by Maximus Planudes (1260–1310), a Byzantine monk):

It would perhaps not be superfluous to show another multiplication method. But it is extremely inconvenient when done with ink and paper, while it is suited for use on a board covered with sand. For it is necessary to wipe out certain numbers, then replace them with others; when using ink, this leads to much inextricable confusion, but with sand it is easy to wipe out a number with one's finger and replace it with others. This method of writing numbers in sand is especially useful, not only for multiplication, but for other operations as well . . . [BN Paris. Ancien Fonds grec, Ms 2381, f° 5v, ll. 30–35; Ms 2382, f° 9r, ll. 13–25; Ms 2509, f° 105v, ll. 2–10] [see A. Allard (1981); H. Waeschke (1878); F. Woepcke (1857), p. 240].

But the dust board became increasingly impractical as the numerals began to resemble one another more and more.

Just take a wooden board, sprinkle it with dust or flour, then draw numbers on it in the usual way. Then try to carry out one of the operations we have seen, following the same method. You will immediately see how hard it is to replace one number with another. If you sprinkle the number to be removed, or use a flat instrument to wipe it out, the very nature of the powdery matter means you risk wiping out all the adjacent numbers as well.

Attempts were made to get round this problem by leaving a large space between the different numbers. But there are limits to the size of the board, and this means that longer, more complicated calculations would require a larger space. What is more, by wiping out intermediate results, this method limits the contribution of the human memory and makes spotting intermediate mistakes extremely difficult. Hence an obvious block on finding out simpler and more practical methods.

It is possible to guess what replaced sand calculation and the use of the dust board in certain Islamic countries.

As we have seen, in Persian and Mesopotamian provinces, the preceding method of calculation was also called *takht al turab* (or in Persian *takhta-yi khak*), literally "board of sand". This expression is found, for example in the *Jami' al hisab bi't takht wa't turab*, by the mathematician and astronomer Nasir ad din at Tusi (1201–1274). This work's title can be translated literally as "Collection of arithmetic using a board and dust" [A. P. Youschkevitch (1976), p. 181, n. 71].

But the Arabic word *turab*, and its Persian equivalent *khak*, means not only "sand" or "dust", but also "earth", "clay" and "cement". Hence the difficulty in precisely translating this author's ideas: for Persian and Mesopotamian arithmeticians, did this word mean only "sand" and "dust", or did it also cover a wad of clay? We can, in fact, suppose that for reasons linked to climate and the nature of the soil in different regions, these arithmeticians were led to use clay for carrying out their calculations, rather than a board scattered with sand. This hypothesis is reasonable, given the limited number of material solutions. It becomes even more probable when we remember that, in these regions, clay tablets had been used for writing for thousands of years. It is sufficient to remember the Sumerians, the Elamites, the Babylonians, the Assyrians and the Acheminid Persians, the distant precursors of these Persian and Mesopotamian arithmeticians, to support the idea that, even under Islam, these peoples had not forgotten their ancient writing materials.

According to this hypothesis, these arithmeticians would then have smeared soft clay over their boards and traced numbers on them with a stylus, pointed at one end and flattened at the other. This is why the Arabic expression *takht al turab*, and its Persian equivalent *takhta-yi khak*, as in At Tusi's book cited above, could be translated by "board smeared with clay".

This hypothesis can be applied to the regions of Persia, Mesopotamia and Syria, but less so to other Muslim provinces.

If we return to the "board", the Arabic word *luha*, used by the Maghrebi and Andalusians for this article had, and always has had, as broad a range of meaning as its Eastern equivalent *takht* (which comes from the Persian *takhta*, itself derived from the Sassanid *takhtag*). Both words mean "table", but also "board", "plank", "tablet" and "plate" or "plaque", be it of wood, leather, metal, earth or even clay.

At a certain time, it is not impossible that wax came to replace the dust or flour used on the board in the Maghreb, and elsewhere. In other words, it can be supposed that the Maghrebi and other Islamic peoples calculated on tablets covered with wax, like those of the ancient Romans, using a stylus with a flattened tip for rubbing out.

All of these techniques perhaps coexisted, each being favoured at different times, in different regions and according to local customs. It is extremely unlikely that people living in such a vast and varied world as Islam would have all used the same method.

CALCULATING WITHOUT INTERMEDIATE ERASURES

What is certain is that the Arab arithmeticians' next step was to "calculate without erasures, by crossing out and writing above their intermediate results".

This method is found, for example, in the *Kitab al fusul fi'l hisab al hind* (*Treatise on Indian Arithmetic*), written in Damascus in 952 (or 958) by Abu'l Hasan Ahmad ibn Ibrahim al-Uqlidisi. It can also be found in works by An Nisawi (1052), al-Hassar (c. 1175), al-Qalasadi (c. 1475), etc., in which it is described as the *a'mal al hindi* ("method of the Indians") or else as *tarik al hindi* (literally "way of the Indians") [see A. Allard (1976), pp. 87–100; A. Saidan (1966); H. Suter BMA, II, 3, pp. 16–17; F. Woepcke (1857), p. 407].

Here are the rules, applied to the product of 325 and 243:

As before, we begin by placing the multiplicand above the multiplier, thus:

```
    3   2   5        ← Multiplicand
2   4   3            ← Multiplier
```

We then multiply the upper 3 by the lower 2; this makes 6, which we place on the line above the multiplicand, in the same column as the 2 of the multiplier:

```
6
    3   2   5        ← Multiplicand
2   4   3            ← Multiplier
```

And we cross out the 2 of the multiplier:

```
6
    3   2   5        ← Multiplicand
2̸   4   3            ← Multiplier
```

Then we multiply the upper 3 by the lower 4; this makes 12, we carry forward the 1 and place the 2 on the same line as the 6, above the 4:

```
6   2
    3   2   5        ← Multiplicand
2̸   4   3            ← Multiplier
```

Then we add the carried-forward number to the 6; so we cross out 6 and write 7 on the line above, just over the crossed-out number:

```
7
6̸   2
    3   2   5        ← Multiplicand
2̸   4   3            ← Multiplier
```

And we cross out the 4 of the multiplier:

```
7
6̸   2
    3   2   5        ← Multiplicand
2̸   4̸   3            ← Multiplier
```

We then multiply the upper 3 by the lower 3; this makes 9, which we write in the same column as the 3 of the multiplier, but on the line above the multiplicand:

```
       7
       6́  2   9
              3   2   5        ← Multiplicand
       2́  4́  3              ← Multiplier
```

And we cross out the 3 of the multiplier:

```
       7
       6́  2   9
              3   2   5        ← Multiplicand
       2́  4́  3́             ← Multiplier
```

The first step of the operation has now been completed, so we write the multiplier 243 again on the line below, but moving one column to the right, after having crossed out the 3 of the multiplicand:

```
       7
       6́  2   9
              3́  2   5        ← Multiplicand
       2́  4́  3́
          2   4   3          ← Multiplier
```

Then we multiply the 2 of the multiplicand by the 2 of the multiplier; hence 4, which we add to the 2 to the right of the already crossed-out 6 on the line above the multiplicand; we thus cross out this 2, and write 6 on the line above, in the same column:

```
       7   6
       6́  2́  9́
              3́  2   5        ← Multiplicand
       2́  4́  3́
          2   4   3          ← Multiplier
```

And we cross out the 2 of the multiplier:

```
       7   6
       6́  2́  9́
              3́  2   5        ← Multiplicand
       2́  4́  3́
          2́  4   3          ← Multiplier
```

We then multiply the 2 of the multiplicand by the 4 of the multiplier; this makes 8, which we add to the 9 in the same column in the line above the multiplicand; this makes 17, we carry forward 1 and place 7 on the line above (just over the 9), after crossing out the 9:

```
7   6   7
6́   2́   9́
        3́   2   5        ← Multiplicand
2́   4́   3́
    2́   4   3            ← Multiplier
```

Then we add the carried-forward 1 to the 6 on the top line; we thus cross out this 6 and write a 7 on the line above, in the same column:

```
        7
7   6́   7
6́   2́   9́
        3́   2   5        ← Multiplicand
2́   4́   3́
    2́   4   3            ← Multiplier
```

And we cross out the 4 of the multiplier:

```
        7
7   6́   7
6́   2́   9́
        3́   2   5        ← Multiplicand
2́   4́   3́
    2́   4́   3            ← Multiplier
```

Then we multiply the 2 of the multiplicand by the 3 of the multiplier; this makes 6, so we write 6 in the same column as the 2 in the line just above:

```
        7
7   6́   7
6́   2́   9́   6
        3́   2   5        ← Multiplicand
2́   4́   3́
    2́   4́   3            ← Multiplier
```

And we cross out the 3 of the multiplier:

```
        7
7   6́   7
6́   2́   9́   6
        3́   2   5        ← Multiplicand
2́   4́   3́
    2́   4́   3́           ← Multiplier
```

The second step has now been completed, so we write the multiplier 243 once again on the line below, moving one column to the right, after having crossed out the 2 of the multiplicand:

```
          7
    7   6́   7
    6́   2́   9́   6
              3́   2́   5        ← Multiplicand
    2́   4́   3́
         2   4́   3́
         2   4   3        ← Multiplier
```

Then we multiply the 5 of the multiplicand by the 2 of the multiplier; this makes 10, we carry forward 1, but add nothing to the 7 in the same column as the 2 on the second line above the multiplicand. We then add the carried-forward 1 to the 7 on the top line; we cross out this 7 and write 8 on the line above:

```
          8
          7́
    7   6́   7
    6́   2́   9́   6
              3́   2́   5        ← Multiplicand
    2́   4́   3́
         2́   4́   3́
         2   4   3        ← Multiplier
```

And we cross out the 2 of the multiplier:

```
          8
          7́
    7   6́   7
    6́   2́   9́   6
              3́   2́   5        ← Multiplicand
    2́   4́   3́
         2́   4́   3́
         2́   4   3        ← Multiplier
```

Then we multiply the 5 of the multiplicand by the 4 of the multiplier; this makes 20, we carry forward the 2, but add nothing to the 6 in the same column as the 2 on the line just above the multiplicand. Then we add the carried-forward 2 to the 7 in the column just to the left; we cross out this 7 and write 9 on the line above:

```
      8
      7̷ 9
  7  6̷ 7̷
  6̷ 2̷ 9̷ 6
        3̷ 2̷ 5        ← Multiplicand
  2̷ 4̷ 3̷
     2̷ 4̷ 3̷
        2̷ 4  3        ← Multiplier
```

And we cross out the 4 of the multiplier:

```
      8
      7̷ 9
  7  6̷ 7
  6̷ 2̷ 9̷ 6
        3̷ 2̷ 5        ← Multiplicand
  2̷ 4̷ 3
     2̷ 4̷ 3̷
        2̷ 4̷ 3        ← Multiplier
```

Finally, we multiply the 5 of the multiplicand by the 3 of the multiplier; this makes 15, so we write a 5 above the 5 of the multiplicand:

```
      8
      7̷ 9
  7  6̷ 7
  6̷ 2̷ 9̷ 6  5
        3̷ 2̷ 5        ← Multiplicand
  2̷ 4̷ 3
     2̷ 4̷ 3̷
        2̷ 4̷ 3        ← Multiplier
```

Then we add the carried-forward 1 to the 6 immediately to the left on the same line; we cross out this 6 and write 7 on the line above:

```
      8
      7̷ 9
  7  6̷ 7̷ 7
  6̷ 2̷ 9̷ 6̷ 5
        3̷ 2̷ 5        ← Multiplicand
  2̷ 4̷ 3
     2̷ 4̷ 3̷
        2̷ 4̷ 3        ← Multiplier
```

And we cross out the 3 of the multiplier and the 5 of the multiplicand:

```
              8
              7̷  9
       7  6̷  7̷  7
       6̷  2̷  9̷  6̷  5
              3̷  2̷  5̷      ← Multiplicand
       2̷  4̷  3̷
          2̷  4̷  3̷
             2̷  4̷  3̷       ← Multiplier
```

As the operation has now been completed, all we have to do is read the uncrossed-out numerals, from left to right, to obtain the result:

```
              8
              7̷  9
       7  6̷  7̷  7
       6̷  2̷  9̷  6̷  5
              3̷  2̷  5̷
       2̷  4̷  3̷
          2̷  4̷  3̷
             2̷  4̷  3̷
          ↓   ↓   ↓   ↓   ↓
          ∨   ∨   ∨   ∨   ∨
325 × 243  7   8   9   7   5
```

The advantage of this method over the preceding one is the possibility to check the operation and so spot any errors. This is why it was used by many Muslim arithmeticians for some time; and that is also why it survived in Europe until the end of the eighteenth century.

The disadvantage was to make the writing of calculations extremely crowded and their progression difficult to follow.

This can be seen in the following example of division "à la française", as explained in F. Le Gendre's *Arithmétique*:

```
                        1  2
                        9  3
                     0  1  2
                  1  8  2  9
                  9  9  3  0
               0  0  1  8  2
            1  8  8  2  9  9
            9  9  9  3  0  0
         0  0  0  1  8  8  2
      1  8  8  8  2  9  9  9
      9  9  9  9  1  8  8  8  2
      0  0  0  0  2  9  9  9  9
   1  9  9  9  9  1  0  0  0  0  7
```

| 199993 | | | | | | | | | | | | | | | 0 |
| (quotient) | 9 9 9 9 9 9 9 9 9 9 | | | | | | | | | | (remainder) |

```
         9  9  9  9  9  9  9  9
            9  9  9  9  9  9
               9  9  9  9
                  9  9
```

We will not weary the reader by explaining this extraordinarily complex system. Suffice it to say that this represents 19,999,100,007 divided by 99,999 [Le Gendre, *Arithmétique en sa perfection* (Paris 1771), p. 54].

It can thus be understood that division, even when written down, long remained beyond the scope of the average person.

It is also true that this work was not meant for a large public. As the author himself makes clear, the "perfection" of this arithmetic was based on "the usage of financiers, experienced people, bankers and merchants".

FROM THE WOODEN BOARD TO PAPER OR BLACKBOARD

However, long before Le Gendre, several Arab and Indian arithmeticians embarked on a far better way, omitting intermediate results and thus dropping the technique of constant erasure. But this method, and its consequent change of writing medium, called for a greater application of memory.

The changes in methods and the changes in materials affected each other, long before they resulted in our present day techniques. (See *Pâtî, *Pâtîganita.)

It can be supposed that, as in India, Muslim arithmeticians at some time started using a blackboard, or else a wooden board painted black, with chalk to write and cross out their numbers, and a cloth for erasing them. (See *Pâtîganita*.)

By using chalk and keeping or, even better, rubbing out intermediate results, certain Indian and Arab arithmeticians were able to free their imaginations and work towards the methods we now use.

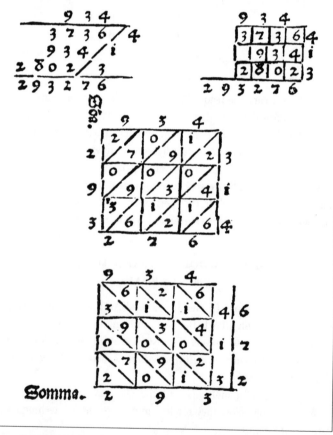

FIG. 25.11. *Page of an anonymous Italian arithmetical treatise, published in Treviso in 1478. It contains various forms of "jealousy" multiplication (per gelosia). [Document in the Palais de la Découverte]*

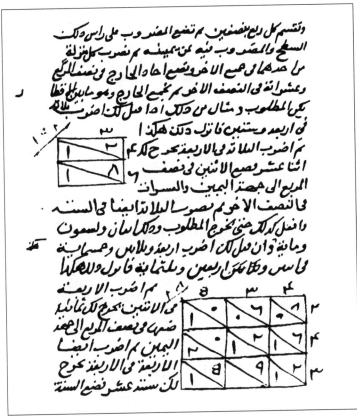

FIG. 25.12. *Page of an Arabic treatise dealing with written calculation using Indian numerals, explaining a method of multiplication "by a table" (known in the West as* per gelosia*). To the left we have the product of 3 and 64 and bottom right the product of 534 and 342. Sixteenth-century copy of* Kashf al mahjub min 'ilm al ghubar *(see below). Paris, BN, Ms. ar. 2473, f° 9)*

"JEALOUSY" MULTIPLICATION

Here follows an example of a highly developed technique, which the Arabs must have invented in around the thirteenth century. At the end of the Middle Ages it was transmitted to Europe, where it was known as multiplication *per gelosia* ("by jealousy"), an allusion to the grid used in the operation which is reminiscent of the wooden or metal lattices through which jealous wives, and especially husbands, could see without being seen. It is described in an anonymous work published in Treviso in 1478 (Fig 25.11) and in the *Summa de arithmetica, geometria, proporzioni di proporzionalita* by Luca Pacioli, an Italian mathematician (Venice, 1494). The Arabs called this system "multiplication on a grid" (*al darb bi'l jadwal*) and it was described in about 1470 by Abu'l Hasan 'Ali ibn Muhammad al-Qalasadi, in his *Kashf al mahjub min 'ilm al ghubar* (*Revelation of the Secrets of the Science of Arithmetic*), the word *ghubar* here being used for "written arithmetic" in general and not in the original sense of "dust" (Fig. 25.12). But there is a much earlier version, from about 1299, by Abu'l 'Abbas Ahmad ibn Muhammad ibn al-Banna al-Marrakushi in his *Talkhis a'mal al hisab* (*Brief Summary of Arithmetical Operations*) [see A. Marre (1865); H. Suter (1900–02), p. 162]. But in India there is no trace of it before the middle of the seventeenth century. It was described there for the first time in 1658 by Ganesha in his *Ganitamañjari* [see B. Datta and A. N. Singh (1938), pp. 144–5].

The layout is quite simple, and the final result is arrived at, rather as in our present-day system, by adding together the products of various numbers contained in the multiplier and multiplicand.

Let us multiply 325 by 243. There are three digits in the multiplier and three in the multiplicand, we thus draw a square grid with three columns and three lines.

Above the grid we write the numbers 3, 2 and 5 of the multiplicand from left to right; then we write the 2, 4 and 3 of the multiplier up the right-hand side of the grid:

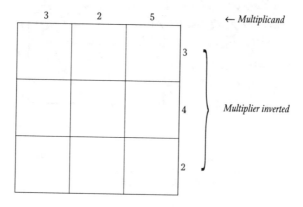

We then divide each square of the grid in half by drawing a diagonal from the top left-hand corner to the bottom right-hand corner. Then, in each square we write the product of the number on the same line to the right and the number in the same column at the top. This product must, of course, be inferior to 100.

We then write the tens digit in the left-hand triangle of the square and the units digit in the right-hand triangle. If either of these digits is missing, then we must write zero.

In the first upper right-hand square we thus write the product of 5 and 3, i.e. 15, by placing the 1 in the left-hand triangle and the 5 in the right-hand triangle.

And so on, thus:

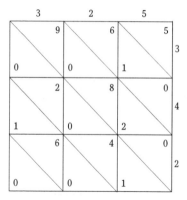

Outside the grid, we then add up the numbers contained in each oblique strip, beginning with the 5 in the top right-hand corner. We then proceed from right to left and from the top to the bottom. When necessary we carry forward any tens digits and add them onto the next strip and we thus obtain all the digits of the final result outside the grid, thus:

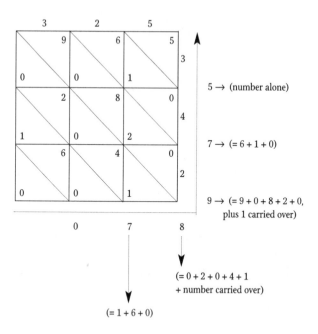

The result is then obviously read from left to right, following the arrow, therefore 78,975:

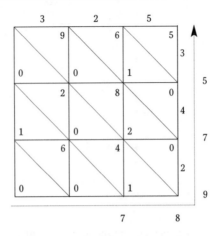

Note that the Arabs often wrote the resulting digits along an oblique segment, perpendicular to the main diagonal, to the left of the grid; the result, of course, still reads from left to right:

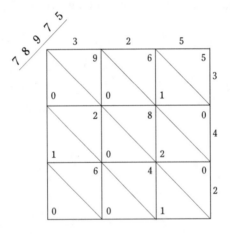

This method may seem long in comparison with the one we now use. But its advantage is that the final result is grouped together at the end whereas in our modern system it is produced gradually during the intermediate steps.

Other ways of proceeding

Instead of following a falling diagonal, we follow a rising one; and instead of writing the multiplier backwards, we write it the correct way round.

Hence the following arrangement, which is used in the same way, except that the additions appear to the left of the grid:

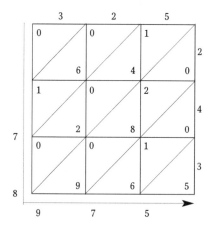

Other possibilities also exist, of course, by placing the digits of the multiplier to the left rather than to the right of the grid.

Simplified techniques

In the anonymous work cited above, we also find another layout alongside the preceding ones. Instead of noting down all the details of the operation, we simply give the results, which certainly requires a greater effort of memory, especially when it comes to carrying numbers forward during the intermediary steps (Fig. 25.11).

As both the multiplicand and multiplier have three digits, we draw a rectangular grid with four columns and three lines, the extra column being used for noting partial results with more digits than in the multiplicand.

Then we place the digits of the multiplicand and multiplier thus:

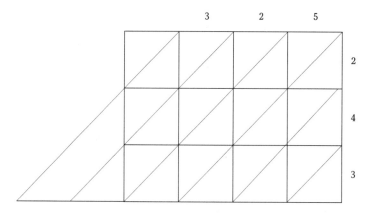

We then calculate the products at the intersections of the columns and lines. But, as there are here no diagonals, only one digit must be written in each square, with the tens digit being added onto the following square on the left.

On the first line, to the right, the first square thus gives 10; we note the 0 and carry forward the 1 to the next square on the left. Its own result is 4, to which the 1 is added, making 5; and so on:

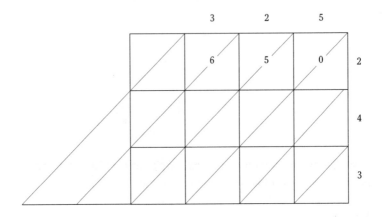

On the second line we thus obtain:

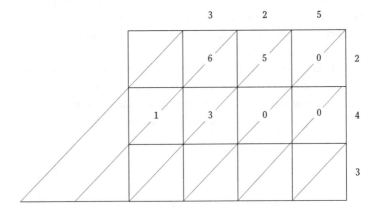

And, finally, on the third:

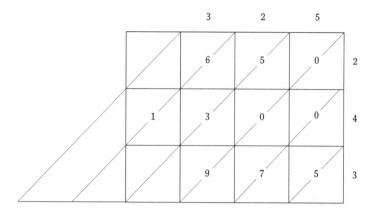

The result is then obtained by adding together the numbers along each line parallel to the rising diagonal from right to left; it is then read from left to right:

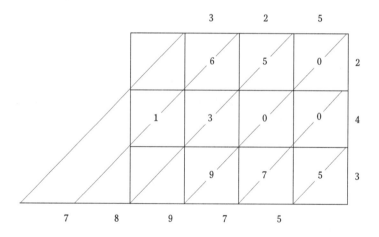

Note that we can work the other way round, by writing the digits of the multiplier backwards on the left. But we must then follow the falling diagonal from left to right to obtain the result:

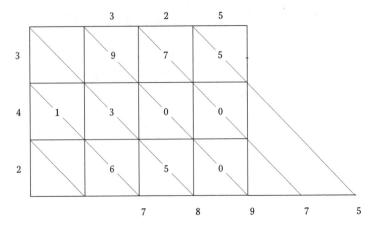

An even simpler technique

At the end of the fifteenth century, the following more highly developed variant could be found in Europe.

We draw a line and write the digits of the multiplicand above it then, below to the right, the digits of the multiplier obliquely rising from left to right:

$$\begin{array}{ccc} \underline{3 \quad 2 \quad 5} \\[2pt] 3 \\[2pt] 4 \\[2pt] \underline{2} \end{array}$$

To the left of the 3 of the multiplier, we then write its products with the digits of the multiplicand:

$$\begin{array}{ccc} \underline{3 \quad 2 \quad 5} \\[2pt] 9 \quad 7 \quad 5 \quad\quad 3 \\[2pt] 4 \\[2pt] \underline{2} \end{array}$$

Then we do the same with the 4:

$$\begin{array}{c} \underline{3 \quad 2 \quad 5} \\[2pt] 9 \quad 7 \quad 5 \quad\quad 3 \\[2pt] 1 \quad 3 \quad 0 \quad 0 \quad\quad 4 \\[2pt] \underline{2} \end{array}$$

Then with the 2:

	3	2	5	
	9	7	5	3
1	3	0	0	4
6	5	0		2

We then add up the products, which gives us:

	3	2	5	
	9	7	5	3
1	3	0	0	4
6	5	0		2
7	8	9	7	5

This method is thus as highly developed as our own (which we append here to facilitate comparison), the only difference being the position of the multiplier:

	3	2	5
	2	4	3
	9	7	5
1	3	0	0
6	5	0	
7	8	9	7	5

NASIR AD DIN AT TUSI'S METHOD

Here now is a multiplication technique, already existing in the thirteenth century, particularly in the *Miftah al Hisab* (*Key to Calculation*) by Ghiyat ad din Ghamshid ibn Mas'ud al-Kashi, from the Persian town of Kashan; this work was completed in 1427 [see A. P. Youschkevitsch (1976), p. 181, n. 67].

But it was known and used two centuries earlier. It can be found, with a slight variation, in Nasir ad din at Tusi's *Jami' al hisab bi't takht wa't turab* (*Collection of Arithmetic Using a Board and Dust*), which dates back to 1265 and was copied by his disciple Hasan ibn Muhammad an Nayshaburi in 1283 [translated by S. A. Akhmedov and B. A. Rosenfeld; see A. P. Youschkevitsch 1976, p. 181, n. 71].

Let us multiply 325 by 243. The multiplicand and multiplier are placed as follows:

	3	2	5
	2	4	3

We multiply the 5 of the multiplicand by the 3 of the multiplier and place the result beneath the line, being careful to respect the place values:

```
        3  2  5
        2  4  3
        ─────────
              1  5
```

Then we multiply the 3 of the multiplicand (not the 2, which is for the moment ignored) by the 3 of the multiplier and we place the product beneath the same line, on the left of the previous one:

```
        3  2  5
        2  4  3
        ─────────
        9  1  5
```

We now return to the 2 of the multiplicand, which we multiply by 3 and this time place the result on the line below, one step to the left:

```
        3  2  5
        2  4  3
        ─────────
        9  1  5
           6
```

We draw another horizontal below these results and then multiply the 5 of the multiplicand by the 4 of the multiplier, placing the result one step to the left:

```
        3  2  5
        2  4  3
        ─────────
        9  1  5
           6
        ─────────
        2  0
```

Then we multiply the 3 of the multiplicand by the 4 of the multiplier and write the product on the same line on the left of the preceding result:

```
        3  2  5
        2  4  3
        ─────────
        9  1  5
           6
        ─────────
     1  2  2  0
```

Then we return to the 2 of the multiplicand, which we multiply by the 4 of the multiplier and place the result on the line below the preceding ones, one step to the left:

```
        3  2  5
        2  4  3
     ─────────────
        9  1  5
           6
     ─────────────
  1  2  2  0
           8
```

Then we draw another line and carry out the preceding operations with the
2 of the multiplier, placing the first product one step towards the left. With
the 5 of the multiplicand we obtain:

```
        3  2  5
        2  4  3
     ─────────────
        9  1  5
           6
     ─────────────
  1  2  2  0
        8
     ─────────────
     1  0
```

Omitting the 2, with the 3 we obtain (on the same line):

```
        3  2  5
        2  4  3
     ─────────────
        9  1  5
           6
     ─────────────
  1  2  2  0
        8
     ─────────────
  6  1  0
```

And with the 2 of the multiplicand, on the line below and one step to the
left, we have:

```
        3  2  5
        2  4  3
     ─────────────
        9  1  5
           6
     ─────────────
  1  2  2  0
        8
     ─────────────
  6  1  0
     4
```

The intermediate steps are thus over. All we have to do now is draw another
line and add up the partial results, position after position, from the right to
the left, and so easily obtain our result:

```
            3  2  5
            2  4  3
         _____
            9  1  5
               6
         _____
      1  2  2  0
               8
         _____
      6  1  0
      4
         _____
   7  8  9  7  5
```

THE INDIAN MATHEMATICIAN
BHÂSKARÂCHÂRYA'S METHODS

In his *Lîlâvatî*, Bhâskharâchârya (c. 1150) often uses a more highly developed method than the preceding one, which he called, in Sanskrit, *sthânakhanda* (literally "separation of positions"). There are several variants, the main ones being as follows [see J. Taylor (1816), pp. 8–9; B. Datta and A. N. Singh 1938, p. 147]:

To multiply 325 by 243, we begin by setting out the operation like this, separating the three digits of the multiplier and copying the digits of the multiplicand three times:

```
   2  4  3      2  4  3      2  4  3
      3            2            5
   _____   _____   _____
```

We begin multiplying with the 5. First we take the product of 5 and 3 and write the full result below the line, without carrying anything forward; we then move to the product of 5 and 2 (skipping the product of 5 and 4) and write the result on the same line, again without carrying forward, just to the left of the first result:

```
   2  4  3      2  4  3      2  4  3
      3            2            5
                          _____
                          1  0  1  5
```

Then we take the product of 5 and the 4 we had omitted and write the result below the others, one step to the left:

```
   2  4  3      2  4  3      2  4  3
      3            2            5
                          _____
                          1  0  1  5
                       2  0
```

We draw a line below these results and add them up:

```
2  4  3      2  4  3      2  4  3
   3            2            5
_____    _____    _____
                        1  0  1  5
                           2  0
            _____
            1  2  1  5
```

We then multiply using the 2, in the same way, placing the sum of the partial results one step to the left from the first one:

```
2  4  3      2  4  3        2  4  3
   3            2              5
_____    _____    _____
               4     6   1  0  1  5
               8            2  0
            _____
            1  2  1  5
            4  8  6
```

Then we multiply by 3 in the same way, placing the sum of the partial results one step to the left from the previous one:

```
2  4  3      2  4  3        2  4  3
   3            2              5
_____    _____    _____
6     9      4     6     1  0  1  5
1  2            8            2  0
            _____
            1  2  1  5
            4  8  6
         7  2  9
```

We draw a final line, add up the totals and obtain the result:

```
2  4  3      2  4  3        2  4  3
   3            2              5
_____    _____    _____
6     9      4     6     1  0  1  5
1  2            8            2  0
            _____
            1  2  1  5
            4  8  6
         7  2  9
      _____
      7  8  9  7  5
```

Another method

One variant of Bhâskarâchârya's method uses a layout like this:

```
    3  2  5
  2  4  3
  _____
```

We then multiply the 3 of the multiplicand (the highest place) by each of the numbers in the multiplier (this time from the lowest first):

```
    3  2  5
  2  4  3
  _____
  7  2  9
```

Then we multiply the 2 of the multiplicand by each of the numbers of the multiplier, placing the result on the line below, one step to the right from the previous result:

```
     3  2  5
   2  4  3
   _____
   7  2  9
      4  8  6
```

Finally, we carry out the same procedure with the 5 of the multiplicand and move one step more to the right to note the result:

```
     3  2  5
   2  4  3
   _____
   7  2  9
      4  8  6
      1  2  1  5
```

We draw a line and add up the partial results to obtain the final answer:

```
     3  2  5
   2  4  3
   _____
   7  2  9
      4  8  6
      1  2  1  5
   _____
   7  8  9  7  5
```

THE INDIAN MATHEMATICIAN BRAHMAGUPTA'S METHODS

Long before Bhâskarâchârya, Brahmagupta, in his *Brahmasphutasiddhânta* (628 CE) described four even more highly developed methods, which he called *gomûtrikâ, khanda, bheda* and *isa* [S. Dvivedi (1902), p. 209; H. T. Colebrooke (1817); B. Datta and A. N. Singh (1938), p. 148].

Here, as an example, is the method called *gomūtrika* (which, in Sanskrit, literally means "like the trajectory of a cow's urine", an allusion to the zigzagging of the arithmetician's eyes as he carries out the operation).

To multiply 325 by 243, we begin with the following layout, copying the multiplicand onto three successive lines, moving one step to the right as we go down. We place the digits of the multiplier vertically from top to bottom starting on the top line:

```
2      3 2 5
4        3 2 5
3          3 2 5
          _____
```

On the first line we then mentally multiply the 2 of the multiplier by the 5 (the lowest digit) of the multiplicand; this makes 10, we write 0 on a lower line in the same column as this 5 and carry forward the 1 which will be added to the next product:

```
2      3 2 5
4        3 2 5
3          3 2 5
          _____
               0
```

Then we multiply the same 2 by the 2 of the multiplicand, which makes 4, and which is added to the carried-forward 1. The result is placed under the line, to the left of the 0:

```
2      3 2 5
4        3 2 5
3          3 2 5
          _____
             5 0
```

Then we multiply the same 2 by the 3 of the multiplicand; this makes 6, which we place under the line to the left of the 5:

```
2      3 2 5
4        3 2 5
3          3 2 5
          _____
           6 5 0
```

We then move to the line with the multiplier 4 and carry out the same steps, this time placing the results on a line below the 650, one step to the right, thus:

– with the product of 4 and 5:

```
2          3  2  5
4             3  2  5
3                3  2  5
          _____
           6  5  0
                      0
```

– then with 4 and 2 (adding the 2 carried forward):

```
2          3  2  5
4             3  2  5
3                3  2  5
          _____
           6  5  0
                   0  0
```

– and with 4 and 3 (adding the 1 carried forward):

```
2          3  2  5
4             3  2  5
3                3  2  5
          _____
           6  5  0
           1  3  0  0
```

We then go down to the line with the multiplier 3, carrying out the same steps, this time with the partial results on a line below the 1,300, one step to the right, thus:

– with the product of 3 and 5:

```
2          3  2  5
4             3  2  5
3                3  2  5
          _____
           6  5  0
           1  3  0  0
                      5
```

– then with 3 and 2 (adding the 1 carried over):

```
2          3  2  5
4             3  2  5
3                3  2  5
          _____
           6  5  0
           1  3  0  0
                   7  5
```

– and finally with 3 and 3:

```
2          3  2  5
4             3  2  5
3                3  2  5
           ─────────────
           6  5  0
           1  3  0  0
                 9  7  5
```

All we have to do now is add up the partial results to obtain the final answer:

```
2          3  2  5
4             3  2  5
3                3  2  5
           ─────────────
           6  5  0
           1  3  0  0
                 9  7  5
           ─────────────
        7  8  9  7  5
```

Other variants of this method

Another layout Brahmagupta used was as follows, with the multiplicand copied three times on three successive lines, each moving one step to the left in comparison with the line above, and with the multiplier placed on the right, from the bottom to the top:

```
          3  2  5      3
       3  2  5         4
    3  2  5            2
   ─────────────
          9  7  5
    1  3  0  0
    6  5  0
   ─────────────
    7  8  9  7  5
```

Brahmagupta's method was, thus, highly developed. There was just one more small step to be taken for it to become as efficient as our present-day technique. This was, in fact, what happened as is shown in Brahmagupta's works, which contain the following extremely interesting variant.

Instead of copying the multiplicand three times, Brahmagupta wrote it just once in the layout below, in which the multiplier is written as in the preceding example, i.e. from the bottom to the top and below the initial line, each partial result being noted opposite and to the left of the number that produces it:

$$
\begin{array}{cccc}
 & 3 & 2 & 5 \\
\hline
 & 9 & 7 & 5 & \quad 3 \\
1 & 3 & 0 & 0 & \quad 4 \\
6 & 5 & 0 & & \quad 2 \\
\hline
7 & 8 & 9 & 7 & 5
\end{array}
$$

This is exactly the same as the method which Italian mathematicians in the second half of the fifteenth century (Luca Pacioli, etc.) had deduced from simplifying the *per gelosia*, and laid out as follows (Fig. 25.11):

$$
\begin{array}{ccccc}
 & & 3 & 2 & 5 \\
\hline
 & & 9 & 7 & 5 & \qquad 3 \\
 & 1 & 3 & 0 & 0 & \quad 4 \\
 & 6 & 5 & 0 & \quad 2 \\
\hline
 & 7 & 8 & 9 & 7 & 5
\end{array}
$$

In other words, from as early as the beginning of the seventh century, Indian mathematicians had a way of multiplying that was far simpler than the "jealousy" method; a procedure which, with a mere change in the layout of the numbers, was to give rise to our present-day technique.

It can now be seen just how advanced the Indians and their Arab successors were in this field.

CHAPTER 26

THE SLOW PROGRESS OF INDO-ARABIC NUMERALS IN WESTERN EUROPE

All that is now left to tell is the story of how India's discoveries reached the Christian West through Arabic intermediaries. As is well known, this transmission did not happen in a day. Quite the contrary!

When they first encountered numeral systems and computational methods of Indian origin, Europeans proved so attached to their archaic customs, so extremely reluctant to engage in novel ideas, that many centuries passed before written arithmetic scored its decisive and total victory in the West.

RENAISSANCE ARITHMETIC: AN OBSCURE AND COMPLEX ART

I was borne and brought up in the Countrie, and amidst husbandry: I have since my predecessours quit me the place and possession of the goods I enjoy, both businesse and husbandry in hand. I cannot yet cast account either with penne or Counters [Montaigne, *Essays*, Vol. II (1588), p. 379].

These words were written by one of the most learned men of his day: Michel de Montaigne, born 1533, was educated by famous teachers at the College de Guyenne, in Bordeaux, travelled widely thereafter, and came to own a sumptuous library. He was a member of the *parlement* of Bordeaux and then mayor of that city, as well as a friend of the French kings François II and Charles IX. And he admits without the slightest embarrassment, that he cannot "cast account" – or, in modern language, do arithmetic!

Could he have been aware of the fabulous discoveries of Indian scholars, already over a thousand years old? Almost certainly not. Cultural contacts between Eastern and Western civilisations had been very limited ever since the collapse of the Roman Empire. Montaigne might have known of two ways, at most, of doing sums: with "Counters" on a ruled table or abacus; and using written Arabic numerals ("with penne"). The first operating method stands in the highly complicated tradition of Greece and Rome; the second, which Montaigne would no doubt have ascribed to the Arabs, was in fact the invention of Indian scholars. But no one had thought of teaching it to him; Montaigne, like most of his contemporaries, no doubt viewed it with mistrust and suspicion.

The following anecdote gives a good picture of the arithmetical state of Europe in the fifteenth and sixteenth centuries. A wealthy German merchant, seeking to provide his son with a good business education, consulted a learned man as to which European institution offered the best training. "If you only want him to be able to cope with addition and subtraction," the expert replied, "then any French or German university will do. But if you are intent on your son going on to multiplication and division – assuming that he has sufficient gifts – then you will have to send him to Italy."

It has to be said that arithmetical operations were not in everyone's grasp: they constituted an obscure and complex art, the specialist preserve of a privileged caste, whose members had been through a long and rigorous training which had allowed them to master the mysterious and infinitely complicated use of the classical (Roman) counter-abacus.

A student of those days needed several years of hard work as well as a long voyage to master the intricacies of multiplication and division – something not far short of a PhD curriculum, in today's terms.

The great respect in which such scholars were held provides a measure of the difficulty of the operational techniques. Specialists would take several hours of painstaking work to perform a multiplication which a child could now do in a few minutes. And tradesmen who wanted to know the total of the week's or the month's takings were obliged to employ the services of such counting specialists (Fig. 26.1).

FIG. 26.1. *Arithmetician performing a calculation on a counter-abacus. From a fifteenth-century European engraving, reproduced from Beauclair, 1968*

This situation did not alter in the conservative bureaucracies of the European nations throughout the seventeenth and eighteenth centuries. Samuel Pepys, for example, became a civil servant after taking a degree at Cambridge, and after a time in the Navy, became a clerk to the Admiralty. From 1662, he was in charge of naval procurement. Though thoroughly well-educated by the standards of the day, Samuel Pepys was nonetheless quite unable to make the necessary calculations for checking the purchases of timber made by the Admiralty. So he resolved to educate himself afresh:

> Up at 5 a-clock... By and by comes Mr Cooper, Mate of the *Royall Charles*, of whom I entend to learn Mathematiques; and so begin with him today... After an hour's being with him at Arithmetique, my first attempt being to learn the Multiplication table, then we parted till tomorrow; and so to my business at my office again... [Pepys, *Diary*, (1985), p. 212].

He eventually mastered the techniques, and was so proud of himself that he sought to teach his wife addition, subtraction and multiplication. But he didn't dare launch her into the subtleties of long division.

It is now perhaps easier to understand why skilled abacists were long regarded in Europe as magicians enjoying supernatural powers.

THE EARLIEST INTRODUCTION OF "ARABIC" NUMERALS IN EUROPE

All the same, even before the Crusades, Westerners could have made full and profitable use of the Indian computational methods which the Arabs had brought to the threshold of Europe from the ninth century CE. A channel of transmission existed, and it was by no means a paltry one.

A French monk with a thirst for knowledge, named Gerbert of Aurillac, could indeed have played the same role in the West as had the learned Persian al-Khuwarizmi, in the Arab-Islamic world. In the closing stages of the tenth century CE, Gerbert – who was to become Pope in the year 1000 – could have broadcast in the West the discoveries of India which had reached North Africa and the Islamic province of Andalusia (Spain) some two centuries earlier. But he found no followers in this respect.

In order to understand the circumstances attendant on the first arrival of Indian numerals in Western Europe, we have to remember the long-drawn-out sequels of the collapse of the Roman Empire and the ensuing Barbarian invasions.

From the end of the Roman Empire in the fifth century until the end of the first millennium, Western Europe was continually laid waste by epidemics, by famine, and by warfare, and suffered centuries of political instability, economic recession, and profound obscurantism. The so-called "Carolingian renaissance" in the Benedictine monasteries of the ninth century may have revitalised the idea and structure of education in the era of Charlemagne and also laid the bases of mediaeval philosophy, but it actually brought only minor and temporary relief to the general situation.

Scientific knowledge available at that time was very elementary, if not entirely deficient. The few privileged men who received any "education" learned first to read and to write. They went on to grammar, dialectics and rhetoric, and sometimes also to the theory of music. Finally, they received basic instruction in astronomy, geometry and arithmetic.

"Theoretical" arithmetic in the High Middle Ages was drawn from a work attributed to the Latin mathematician Boethius (fifth century CE) who had himself drawn handsomely on a second-rate work by the Greek Nicomacchus of Gerasa (second century CE). As for "practical" arithmetic, it consisted mainly in the use of Roman numerals, and in operations with counters on the old abacus of the Romans; it also included the techniques of finger-counting transmitted by Isidore of Seville (died 636 CE) and by Bede (died 735 CE).

In these almost completely "dark ages", even the memory of human arts and sciences was almost lost. But a sudden reawakening occurred in the eleventh and twelfth centuries:

> A massive demographic explosion brought many consequences in its wake – the development of virgin lands, the growth of towns and of the monastic orders, the crusades, and the construction of ever larger churches. Prices rose, the circulation of money accelerated, and, as sovereign states began to control feudal anarchy, trade also began to prosper. An increase in international contacts created a favourable environment for the introduction of Arabic science in the West [G. Beaujouan (1947)].

Gerbert of Aurillac was certainly one of the most prominent scientific personalities of this whole period. Born in southwest France c. 945 CE, he took holy orders at the monastery of Saint-Géraud at Aurillac, where his sharp mind and passion for learning soon marked him out. He learned mathematics and astronomy from Atton, the Bishop of Vich, and then, probably as a result of a visit to Islamic Spain from 967 to 970, he absorbed the lessons of the Arabic school. He learned to use an astrolabe, he learned the Arabic numeral system, as well as the basic arithmetical operations in the Indian manner. From 972 to 987, Gerbert was in charge of the Diocesan school of Reims, and then, after a period as abbot of Bobbio (Italy), he became an adviser to Pope Gregory V and became successively Archbishop of Reims, Archbishop of Ravenna, and finally Pope Sylvester II, from 2 April 999 until his death on 12 May 1003.

Legend has it that Gerbert went as far as Seville, Fez and Cordoba to learn Indo-Arabic arithmetic and that he disguised himself as a Muslim pilgrim in order to gain entrance to Arab universities. Though that is not impossible, it is more likely that he remained in Christian Spain at the monastery of Santa Maria de Ripoll, a striking example, according to Beaujouan, of the hybridisation of Arabic and Isidorian traditions. The little town of Ripoll (near Barcelona, in Catalonia) had indeed long served as a meeting point for the Islamic and Christian worlds.

One thing is nonetheless quite certain: Gerbert brought back to France all the techniques necessary for modern arithmetic to exist. His teaching at the diocesan school at Reims was highly influential and did much to reawaken interest in mathematics in the West. And it was Gerbert who first introduced so-called Arabic numerals into Europe. Arabic numerals, indeed – but alas, only the first nine! He did not bring back the zero from his Spanish sojourn, nor did he include Indian arithmetical operations in his pack.

So what happened? Gerbert's initiative actually met fierce resistance: his Christian fellows clung with conservative fervour to the number-system and arithmetical techniques of the Roman past. Most clerics of the period, it has to be said, thought of themselves as the heirs of the "great tradition" of classical Rome, and could not easily countenance the superiority of any other system. The time was simply not ripe for a great revolution of the mind.

A Victorian howler

The mediaeval forms of the Arabic numerals are found in a manuscript entitled *Geometria Euclidis* (*Euclidian Geometry*) which for a long time was attributed to Boethius, a Roman mathematician of the fifth century CE. The text itself claims that the nine numeral symbols shown and their use in a place-value system had been invented by Pythagoreans and derived from the use of the table-abacus in Ancient Greece. For this reason, the shapes of the nine Indo-Arabic numerals used in the Middle Ages came to be called "the *apices* of Boethius", even though, as we have seen, there is no possibility whatsoever that a Roman of the fifth century, let alone Greek Pythagoreans, could have known or understood Indo-Arabic arithmetic.

The solution of this conundrum is very simple. As modern analyses have shown, *Geometria Euclidis* was put together by an anonymous compiler in the eleventh century, and its attribution to Boethius is entirely apocryphal.

Early forms of Arabic numerals in the West

The earliest actual appearances of Arabic numerals in the West are to be found in the *Codex Vigilanus* (copied by a monk named Vigila at the Monastery of Albelda, Spain, in the year 976) and the *Codex Aemilianensis*, copied directly from the *Vigilanus* in the year 992 at San Millan de la Cogolla, also in northern Spain (see Fig. 26.2).

These nine figures are clearly integrated in the cursive script of "full Visigothic of the Northern Spanish type" (in the terms of R. L. Burnam, 1912–25), but their Indian origin is nonetheless quite manifest. Both manuscripts give the numerals shapes that are very close to the *Ghubar* figures of the Western Arabs.

From the early eleventh century, the nine figures appear in a whole variety of shapes and sizes in a great number of mss copied in more or less every corner of the European continent. The variations in shape and style are the result of palaeographic modifications occurring in different periods and places, as can be seen in Figure 26.4.

However, contrary to first appearances, "Arabic" numerals did not first spread through the West by manuscript transmission, but through a piece of calculating technology called *Gerbert's abacus*. In other words, the numerals were disseminated not by writing but by the oral transmission of the knowledge necessary to learn how to operate the entirely new kind of abacus that Gerbert of Aurillac had promoted from around 1000 CE, and thereafter by his numerous disciples in Cologne, Chartres, Reims and Paris.

Let us recall that for many centuries the Christianised populations of Western Europe had expressed number almost exclusively through the medium of Roman numerals, a very rudimentary system of notation whose operational inefficacy lay at the root of all the difficulties experienced in

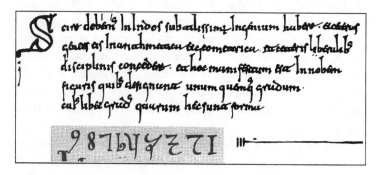

Fig. 26.2. *Detail from the* Codex Vigilanus *(976 CE, Northern Spain). The first known occurrence of the nine Indo-Arabic numerals in Western Europe. Escurial Library, Madrid, Ms. lat. d.I.2, f° 9v. See Burnam (1920), II, plate XXIII*

calculation throughout that long period of the "dark ages". First-millennium mediaeval arithmeticians made their calculations just like their Roman predecessors, through a complicated game of counters placed on tables marked out with rows and columns delimiting the different decimal orders.

On a Roman abacus, you place as many unit counters in a column for a specific decimal order as there are units in that order. But just before 1000 CE, it occurred to Gerbert of Aurillac, who had seen Arabic counting methods during his time in Andalusia, to reduce the number of counters used and so to simplify the material complexity of computation on an abacus.

Gerbert's system involved jettisoning multiple unit counters and replacing them with single counters in each decimal column, the new horn "singles" being marked with one of the nine numerals he had brought back from the Arabs. These number-tokens were called *apices* (*apex* in the singular) and were each dubbed with a specific name that has nothing to do with the number shown (though a few of them seem to hark back to Arabic and Hebrew number-names).

The *apex* for 1 was called *Igin*
for 2 was called *Andras*
for 3 was called *Ormis*
for 4 was called *Arbas*
for 5 was called *Quimas*
for 6 was called *Caltis*
for 7 was called *Zenis*
for 8 was called *Temenias*
for 9 was called *Celentis*

Fig. 26.3.

So the one, two, three, four, five, six, seven, eight or nine unit-counters in each column of the Roman abacus were replaced by a single *apex* bearing on it the corresponding numeral in "Arabic" script.

When a decimal order was "empty", the abacist simply put no *apex* in the corresponding column. So to represent the number 9,078, you put the apex for 8 in the unit column, the *apex* for 7 in the tens column, and the *apex* for 9 in the thousands column, leaving all the other columns empty.

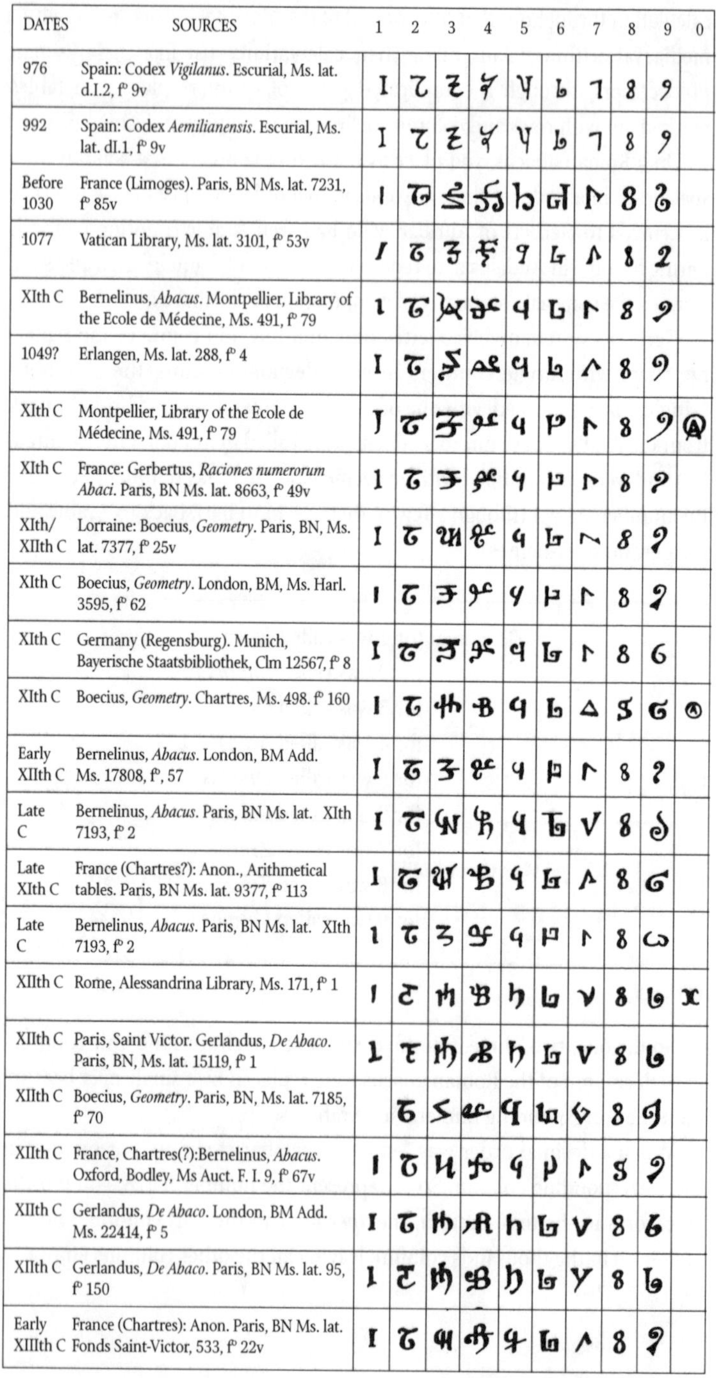

DATES	SOURCES	1	2	3	4	5	6	7	8	9	0
976	Spain: Codex *Vigilanus*. Escurial, Ms. lat. d.I.2, fº 9v										
992	Spain: Codex *Aemilianensis*. Escurial, Ms. lat. dI.1, fº 9v										
Before 1030	France (Limoges). Paris, BN Ms. lat. 7231, fº 85v										
1077	Vatican Library, Ms. lat. 3101, fº 53v										
XIth C	Bernelinus, *Abacus*. Montpellier, Library of the Ecole de Médecine, Ms. 491, fº 79										
1049?	Erlangen, Ms. lat. 288, fº 4										
XIth C	Montpellier, Library of the Ecole de Médecine, Ms. 491, fº 79										
XIth C	France: Gerbertus, *Raciones numerorum Abaci*. Paris, BN Ms. lat. 8663, fº 49v										
XIth/ XIIth C	Lorraine: Boecius, *Geometry*. Paris, BN, Ms. lat. 7377, fº 25v										
XIth C	Boecius, *Geometry*. London, BM, Ms. Harl. 3595, fº 62										
XIth C	Germany (Regensburg). Munich, Bayerische Staatsbibliothek, Clm 12567, fº 8										
XIth C	Boecius, *Geometry*. Chartres, Ms. 498. fº 160										
Early XIIth C	Bernelinus, *Abacus*. London, BM Add. Ms. 17808, fº, 57										
Late XIth C	Bernelinus, *Abacus*. Paris, BN Ms. lat. 7193, fº 2										
Late XIth C	France (Chartres?): Anon., Arithmetical tables. Paris, BN Ms. lat. 9377, fº 113										
Late XIth C	Bernelinus, *Abacus*. Paris, BN Ms. lat. 7193, fº 2										
XIIth C	Rome, Alessandrina Library, Ms. 171, fº 1										
XIIth C	Paris, Saint Victor. Gerlandus, *De Abaco*. Paris, BN, Ms. lat. 15119, fº 1										
XIIth C	Boecius, *Geometry*. Paris, BN, Ms. lat. 7185, fº 70										
XIIth C	France, Chartres(?):Bernelinus, *Abacus*. Oxford, Bodley, Ms Auct. F. I. 9, fº 67v										
XIIth C	Gerlandus, *De Abaco*. London, BM Add. Ms. 22414, fº 5										
XIIth C	Gerlandus, *De Abaco*. Paris, BN Ms. lat. 95, fº 150										
Early XIIIth C	France (Chartres): Anon. Paris, BN Ms. lat. Fonds Saint-Victor, 533, fº 22v										

FIG. 26.4. *Mediaeval apices. Sources: BSMF X (1877), p. 596; Burnam (1920), II, plates XXIII & XXIV; Folkerts (1970) Friedlein (1867) p. 397; Hill (1915) Smith and Karpinski (1911) p. 88*

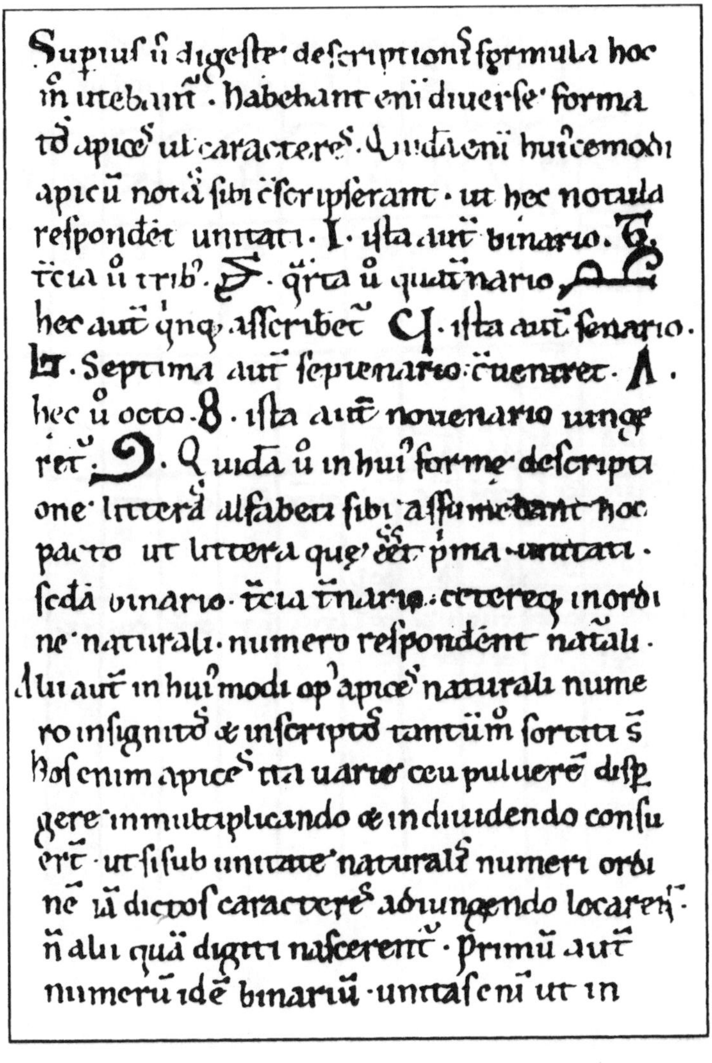

Supius ū digeste descriptioñ formula hoc
in utebant̄ . habebant eñi diuerse forma
tō apicē ut caracteres. Quidā eñi huicemodi
apicū notā sibi c̄scripserant · ut hec notula
respondēt unitati. I. ista aut̄ binario. Ƶ.
tc̄ia ū trib̄. Ƽ. q̄ta ū quatinario, ㄴㄷ
hec aut̄ q̄nq̄, asscribēt ㄷ . ista aut̄ senario.
Ŀ . Septima aut̄ septenario c̄uenaret Λ .
hec ū octo 8 . ista aut̄ nouenario utngq̄
rēt ℈ . Quidā ū in huiʾ forme descripti
one litterā alfabeti sibi assumebant hoc
pacto ut littera que̅ d̄t pma vnitati .
secā binario. tc̄ia t̄naria. ceterecq̄ inordi
ne naturali. numero respondēt natuli .
Alii aut̄ in huiʾmodi opʾapicēʾ naturali nume
ro insignitō & inscriptō tantūm sortiti s̄
hos enim apicēʾ ita uario ceu puluere disp̄
gere inmultiplicando & indiuidendo c̄nsu
ert̄ · ut sisub unitate naturali numeri ordi
ne ū dictos caracteres ad iungendo locareñ
ñ alii quā digiti nascerent̄ · primū aut̄
numerū idē binariū · unitas eni ut in

FIG. 26.5. Apices *in an eleventh-century Latin manuscript. Berlin, Ms. lat. 8° 162 (n), f° 74. From Folkerts (1970)*

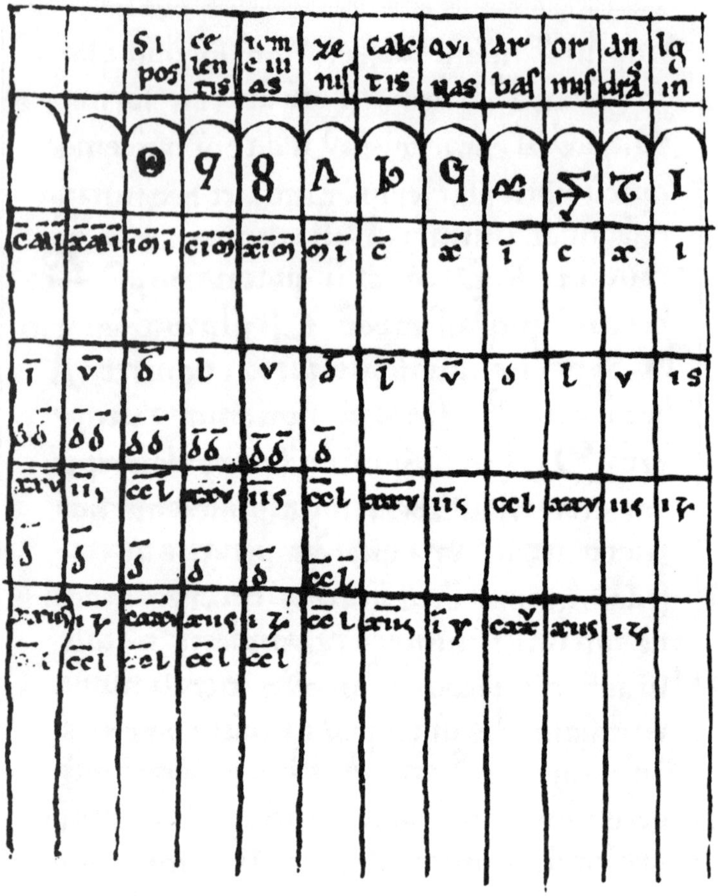

FIG. 26.6. Apices *and the columns of Gerbert's abacus in an eleventh-century Latin manuscript. Berlin, Ms. lat. 8° 162 (n), f° 73v. From Folkerts (1970)*

So at this early stage the Arabic numerals introduced by Gerbert served only to simplify the use of an abacus identical in structure to that of classical Rome. Indeed, some mediaeval arithmeticians continued to use Roman numerals – or even the letters of the Greek numeral alphabet ($\alpha = 1$, $\beta = 2$, ... $\theta = 9$) – on their *apices*.

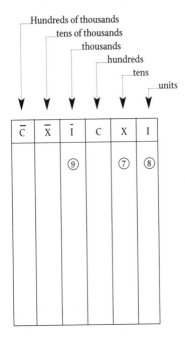

FIG. 26.7.

ARITHMETICAL OPERATIONS ON GERBERT'S ABACUS

The following example shows how sums can be done on Gerbert's abacus without zero. The fact that it is possible to complete these operations correctly explains why mediaeval manuscripts of the eleventh and twelfth centuries contain no symbols for zero, nor ever even mention the concept. The nine symbols of Indian origin spread around Europe, but only in very restricted circles, since the whole business of counting was in the hands of a tiny elite of arithmeticians, appropriately called *abacists*.

To multiply 325 by 28

1. Place the *apices* for multiplicand (325) on the bottom row of the abacus, putting the "5" counter in the units column, the "2" counter in the tens column, and the "3" counter in the hundreds column:

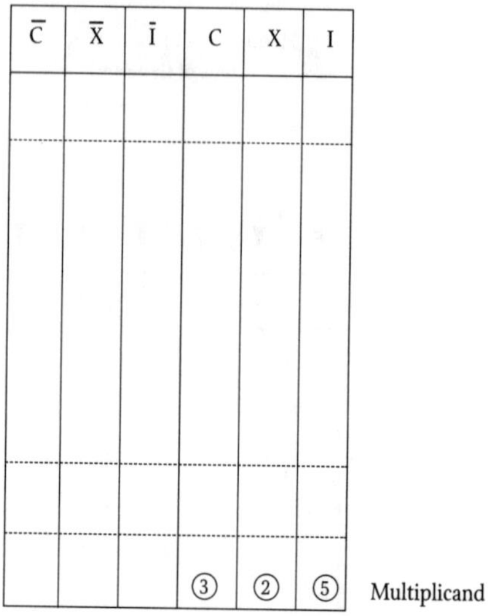

$\overline{C}$	$\overline{X}$	$\overline{I}$	C	X	I
			③	②	⑤

FIG. 26.8A.   Multiplicand

2. Then place the *apices* for the multiplier (28) on the top row of the abacus, putting the "2" and "8" counters in the tens and units columns respectively:

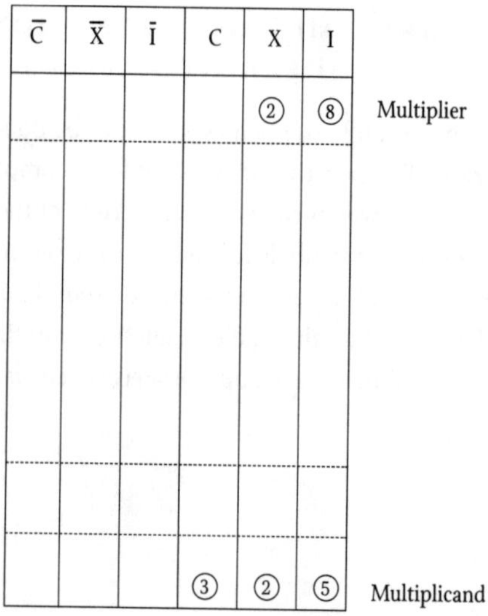

$\overline{C}$	$\overline{X}$	$\overline{I}$	C	X	I
				②	⑧
			③	②	⑤

FIG. 26.8B.   Multiplicand

3. Now find the product of the 8 × 5 in the units column. Since the product is 40, place a "4" counter in the upper part of the central rows of the abacus, in the tens column, leaving the units column empty, thus:

$\bar{C}$	$\bar{X}$	$\bar{I}$	C	X	I	
				②	⑧	Multiplier
				④		
			③	②	⑤	Multiplicand

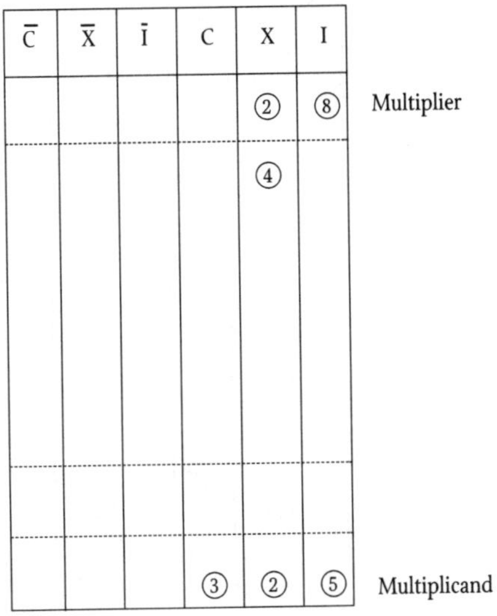

Fig. 26.8c.

4. Now find the product of the counter in the units column by the one in the tens columns, in other words 2 × 8. The product being 16, place a "1" counter in the hundreds column and a "6" counter in the tens column, still leaving the units column empty, thus:

$\bar{C}$	$\bar{X}$	$\bar{I}$	C	X	I	
				②	⑧	Multiplier
				④		
			①	⑥		
			③	②	⑤	Multiplicand

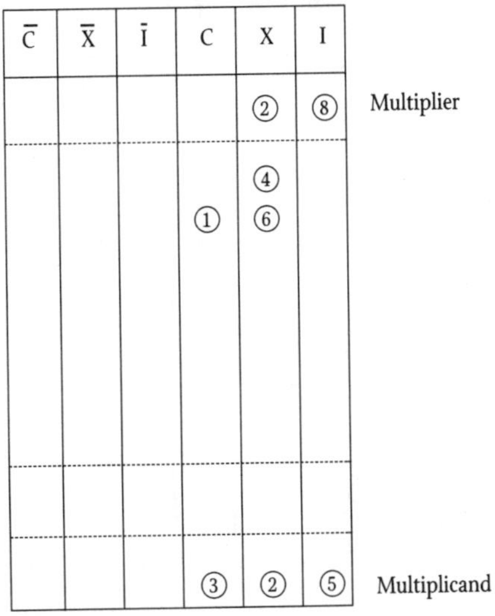

Fig. 26.8d.

5. Now multiply the same unit 8 by the *apex* in the hundreds column, in other words 3 × 8. The product being 24, place a "2" counter in the thousands column and a "4" counter in the hundreds column:

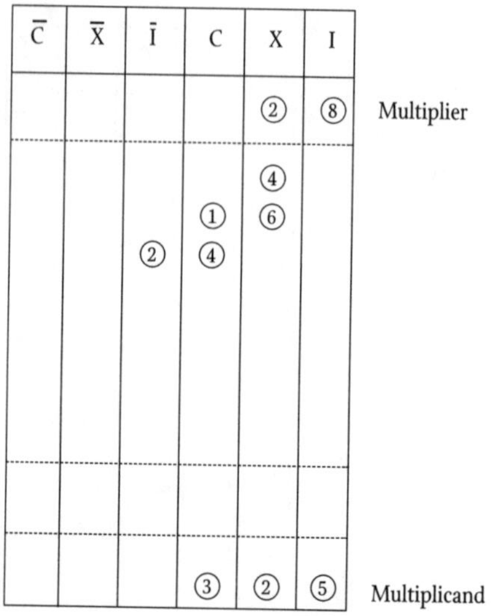

C̄	X̄	Ī	C	X	I	
				②	⑧	Multiplier
				④		
			①	⑥		
		②	④			
			③	②	⑤	Multiplicand

FIG. 26.8E.

6. Since the multiplying of the "8" is now complete, remove the "8" token from the abacus before turning attention to the "2" in the multiplier:

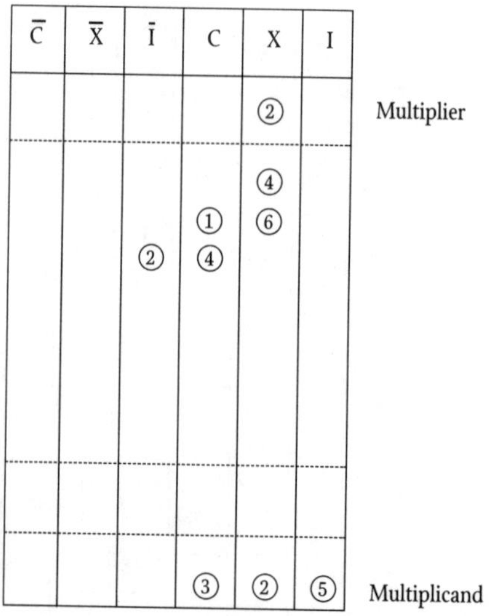

C̄	X̄	Ī	C	X	I	
				②		Multiplier
				④		
			①	⑥		
		②	④			
			③	②	⑤	Multiplicand

FIG. 26.8F.

7. Now multiply the 2 by the 5 in the multiplicand. Since the 2 is in the tens column, the product (10) requires us to place a "1" counter in the hundreds column, thus:

$\overline{\text{C}}$	$\overline{\text{X}}$	$\overline{\text{I}}$	C	X	I	
				②		Multiplier
				④		
			①	⑥		
		②	④			
			①			
			③	②	⑤	Multiplicand

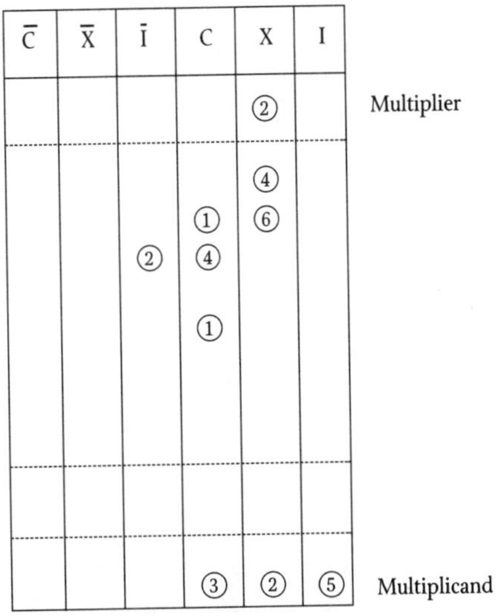

FIG. 26.8G.

8. Now multiply the 2 of the multiplier by the 2 in the multiplicand, giving the answer 4. Since both factors are in the tens columns, the result (four tens of tens) is registered by placing a "4" in the hundreds column, thus:

$\overline{\text{C}}$	$\overline{\text{X}}$	$\overline{\text{I}}$	C	X	I	
				②		Multiplier
				④		
			①	⑥		
		②	④			
			①			
			④			
			③	②	⑤	Multiplicand

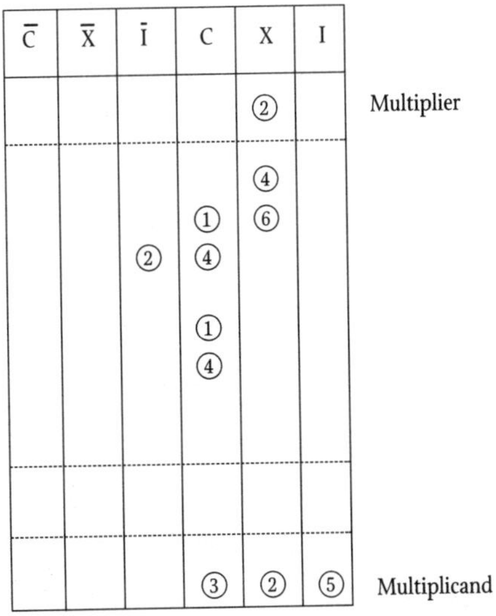

FIG. 26.8H.

9 Now multiply the same 2 by the 3 in the hundreds column of the multiplicand. The product, 6, means six tens of hundreds, so we place a "6" counter in the thousands column, thus:

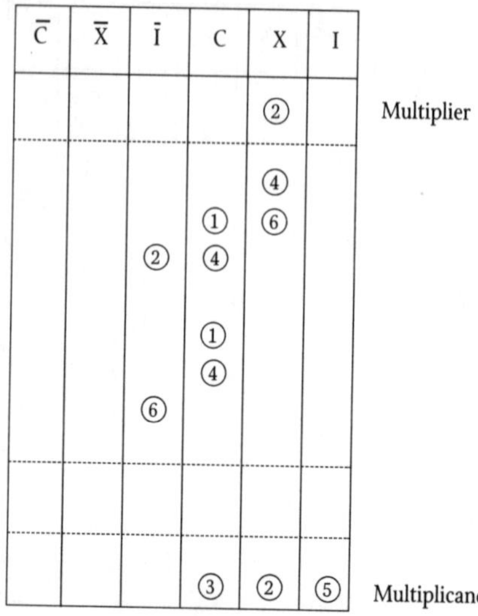

C̄	X̄	Ī	C	X	I	
				②		Multiplier
				④		
			①	⑥		
		②	④			
			①			
			④			
		⑥				
			③	②	⑤	Multiplicand

FIG. 26.8I.

10. Since the multiplying of the 2 is now complete, remove the "2" token from the multiplier line of the abacus:

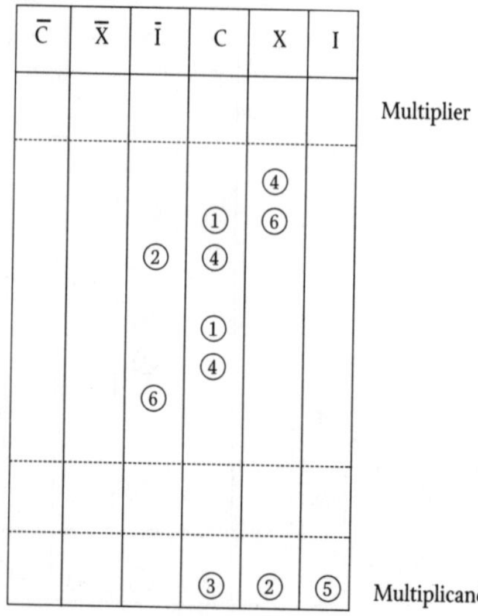

C̄	X̄	Ī	C	X	I	
						Multiplier
				④		
			①	⑥		
		②	④			
			①			
			④			
		⑥				
			③	②	⑤	Multiplicand

FIG. 26.8J.

11. As all the multiplications of the highest number in the multiplier are now also complete, all that remains is to sum the partial products on the board, replacing counters whose total is more than 10 by a unit counter in the next-leftmost column. Since the "4" and "6" of the tens column total 10, they are taken off the board and replaced by a "1" counter in the hundreds column, thus:

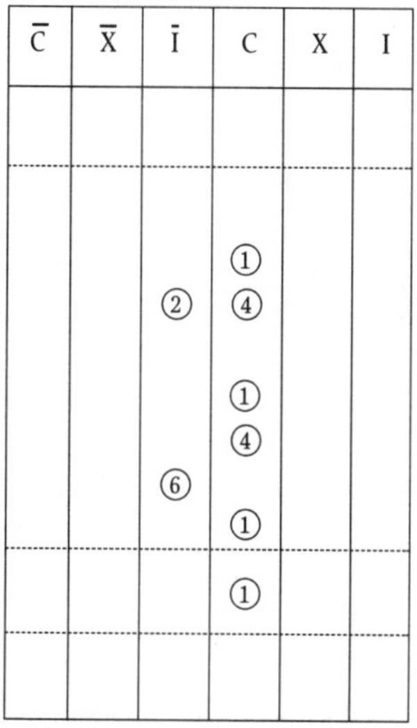

$\bar{C}$	$\bar{X}$	$\bar{I}$	C	X	I	
						Multiplier
			①			
		②	④			
			①			
			④			
		⑥				
			①			Remainder
			①			Product
						Multiplicand

Fig. 26.8K.

12. Now sum the tokens in the hundreds column. As they total 11, remove all counters bar the "1", and place a "1" in the thousands column, thus:

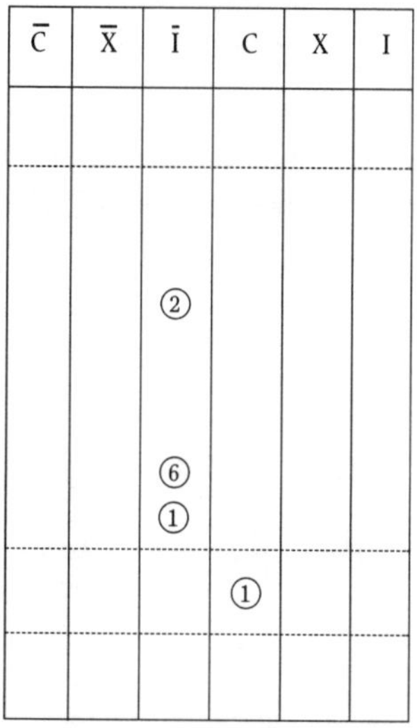

$\bar{C}$	$\bar{X}$	$\bar{I}$	C	X	I	
						Multiplier
		②				
		⑥				
		①				Remainder
			①			Product
						Multiplicand

Fig. 26.8L.

13. Finally, sum the tokens in the thousands column, which gives 9, so remove all the tokens and replace them by a "9" in the thousands column, thus:

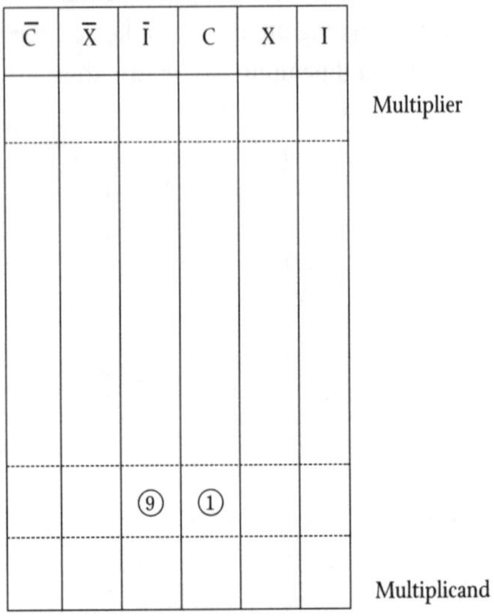

C̄
X̄
Ī
C
X
I
Multiplier
Product
9
1
Multiplicand

FIG. 26.8M.

14. The result of the operation is therefore 9,100 (since the tens and units columns are empty). This example shows how Gerbert's abacus made arithmetical operations long and complicated; its use presupposed lengthy training and a high degree of intelligence.

FROM "ARABIC" NUMERALS TO EUROPEAN APICES

The shapes of the Arabic numerals brought back from Spain by Gerbert of Aurillac were represented with the most fantastical variations on European horn *apices*. Consider the following versions of "4" found over the first two hundred years of the second millennium CE:

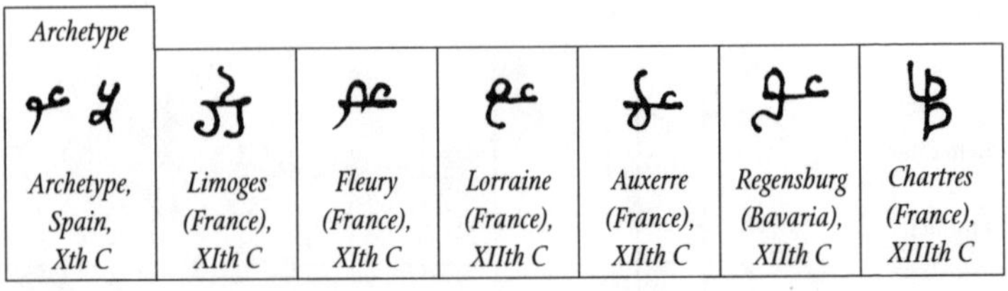

Archetype
Archetype, Spain, Xth C
Limoges (France), XIth C
Fleury (France), XIth C
Lorraine (France), XIIth C
Auxerre (France), XIIth C
Regensburg (Bavaria), XIIth C
Chartres (France), XIIIth C

FIG. 26.9.

Styles obviously varied from one region to another, from one school to another, even from one engraver to another, in a period that had no concept of standardisation. Indeed, what we can see happening in these examples is the adaptation of the *Ghubar* forms of the Arabic numerals to the very different styles of writing practised in different parts of Europe. So in Italy we see numerals assimilated to the round shapes and wide openings of Italic script, in England to the narrower and more angular shapes of English script, in Germany to the thicker and squarer writing style of German script, and in France and Spain we see them being shaped in harmony with the dominant styles of Carolingian script.

A similar phenomenon has already been observed in India and in the Indic civilisations of Southeast Asia. Scribes and stone-carvers adapted the basic nine symbols to their own indigenous writing styles and applied their own aesthetic sense to the shapes, so that there quickly resulted widely differing sets of numerals that at first glance seem quite unrelated.

Similar diversity has been seen in the Arabic world too, where scribes and copyists adapted the same basic figures to the different scripts used in different areas of the Arabic-speaking world.

So there is no reason for the Western world not to have also generated a range of distinctive variations on the numeral set. However, as Beaujouan has pointed out, there was a supplementary factor in the West. All the different shapes found, he insists, are virtually superimposable on each other provided they are rotated by some degree. That is particularly noticeable for the 3, 4, 5, 6, 7 and 9 (see Fig. 26.4).

The reason is that the *apices* were often placed on the abacus without any particular regard for the original orientation of the shape. In some schools, for example, the *apices* were placed upside-down, so the 5 was sometimes found with its "tail" at the bottom. The 9 was sometimes placed on its right side, sometimes on its left side, and sometimes placed upside down so that it looked like today's 6.

Some scribes and stone-carvers simply replaced the original shape of the numeral with the shape that they had grown accustomed to, or which seemed more "logical" in their eyes. Confusion became generalised, and even mathematical course-books often taught the numbers upside-down and back-to-front.

The obvious solution would have been to mark the top or bottom of each horn *apex* with a dot, but people were content merely to distinguish the two figures that could most easily be confused by writing the 6 with sharp, angular lines and the 9 with curved and flowing lines.

However, mediaeval *apices* did not actually give rise directly to our current numerals. After the Crusades, these early forms of the numerals were simply abandoned, and shapes closer to the original Arabic forms were re-introduced – and it is these later arrivals, which eventually stabilised into standard forms, that ultimately gave rise to modern "Arabic" numerals.

THE SECOND INTRODUCTION OF ARABIC NUMERALS IN EUROPE

We might have expected Pope Sylvester II to have opened the millennium onto a new era of progress in the West, thanks to the numerals and operational techniques he had brought back from the Arabic-Islamic world. But such expectations would be vain: the ignorance and conservatism of the Christian world blocked the way.

Although modern numerals and number-techniques were in fact available from the late tenth century, they were used only in the most rudimentary ways for over two hundred years. They served solely to simplify archaic counting methods and to give rise to rules of procedure which, according to William of Malmesbury, "perspiring abacists barely comprehended themselves".

Some arithmeticians even put up a solid resistance to the new-fangled figures from the East by inscribing their *apices* with the Greek letter-numerals from $\alpha = 1$ to $\theta = 9$, or the Roman figures I to IX. Anything was better than having recourse to the "diabolical signs" of the "satanic accomplices" that the Arabs were supposed to be!

Gerbert of Aurillac also suffered at the hands of the rearguard. It was rumoured that he was an alchemist and a sorcerer, and that he must have sold his soul to Lucifer when he went to taste of the knowledge of the Saracens. The accusation continued to circulate for centuries until finally, in 1648, papal authorities reopened the tomb of Sylvester II to make sure that it was not still infested by the devil!

The dawn of the modern age did not really occur until Richard Lionheart reached the walls of Jerusalem. From 1095 to 1270, Christian knights and princes tried to impose their religion and traditions on the Infidels of the Middle East. But what they actually achieved was to bring back to Europe the cultural riches they encountered in the Holy Land. It was these campaigns – or rather, their secondary consequences – that finally allowed the breakthrough which Gerbert of Aurillac, for all his knowledge and energy, had failed to achieve at the end of the tenth century. For the wars implied a whole range of contacts with the Islamic world, and a number of clerks travelling with the armies learned the written numerals and arithmetical methods of the Indo-Arabic school.

Gerbert's abacus thus slowly fell into disuse. Gradually, numerals written on sand or dust, instead of being engraved on horn-tipped *apices*, led to the disappearance of the columns on the abacus. This allowed much simpler, much faster and more elegant operations, which now came to be called *algorisms*, after al-Khuwarizmi, the first Islamic scholar who had generalised their application.

DATES	SOURCES	1	2	3	4	5	6	7	8	9	0
XIIth C	Toledo (Spain): Astronomical Tables. Munich, Bayerische Staatsbibliothek, Clm 18927, fº 1r, 1v										
XIIth C	Algorism. Munich, Bayerische Staatsbibliothek, Clm 13021, fº 27r.										
XIIth C	Algorism. Paris, BN, Ms. lat. 15461, fº 1										
XIIth C	Algorism. Paris, BN, Ms. lat. 16208, fº 3										
XIIth C	Algorism. Paris, BN, Ms. lat. 16208, fº 4										
XIIth C	Algorism. Paris, BN, Ms. lat. 16208, fº 67										
XIIth C	Algorism. Paris, BN, Ms. lat. 16208, fº 68										
XIIth C	Algorism. Vienna Nat. Library, Cod. Vin. 275, fº 33										
Late XIIth C	France: Astronomical Tables. Berlin, Cod. lat. Fol. 307, ff. 6, 9, 10, 28.										
XIIIth C	London, BM Ms. Arund 292, fº 107v										
After 1264	England: Algorism. London, BM Ms. Add. 27589, fº 28										
1256	Paris, BN, Ms. lat. 16334										
1260–1270	London, BM Ms. Royal 12 E IV										
Late XIIIth C	Paris, BN, Ms. lat. 7359, fº 50v										
XIIIth C	Paris, BN, Ms. lat. 15461, fº 50v										
Around 1300	London, BM Ms. Add. 35179										
Mid-XIVth C	London, BM Ms. Harl. 2316, ff. 2v-11v										
Mid-XIVth C	London, BM Ms. Harl. 80, fº 46r										
Around 1429	London, BM Ms. Add. 7096, fº 71										
XVth C	England: Algorism. London, BM Ms. Add. 24059, fº 22r										
XVth C	Italian manuscript. London, BM Ms. Add. 8784, fº 50r-51										
Around	*Quodlibetarius*. Erlangen, Ms. nº 1463										

FIG. 26.10. *The second form of European numerals (algorisms). For more details, see Hill, 1915*

FIG. 26.11. *Numerals including zero in a thirteenth-century Latin manuscript. Paris, BN, Ms. lat. 7413, part II. Facsimile in the Ecole des Chartes, AF 1113*

So the first European "algorists" were born at the gates of Jerusalem. But unlike the "abacists", the new European counting experts were obliged to adopt the zero, to signify missing orders of magnitude, otherwise computations written in sand would lead to confusing representations of number and mistaken operations. At last, then, true "Arabic" numerals including zero, and the arithmetical tradition that had been born long before in India, were able to make their way into Europe.

There were of course other contacts with the Islamic world on the other side of the Mediterranean, by way of Sicily and most especially through Spain and North Africa. It was in Spain that a huge wave of translations began in the twelfth century, bringing into Latin works written in Arabic, and even more importantly Greek and Sanskrit texts already translated into Arabic. Thanks to translators like Adelard of Bath and to centres of scholarship at Cordoba and Toledo, the resources available for acquiring knowledge of arithmetic, mathematics, astronomy, natural sciences and philosophy swelled almost by the day; and it was by means of translation from the Arabic that the   West eventually became familiar with the works of Euclid, Archimedes, Ptolemy, Aristotle, al-Khuwarizmi, al-Biruni, Ibn Sina, and many others.

Between them, the Crusaders at Jerusalem and the scholars of Toledo were ensuring the more or less rapid death of the abacus and of abacism.

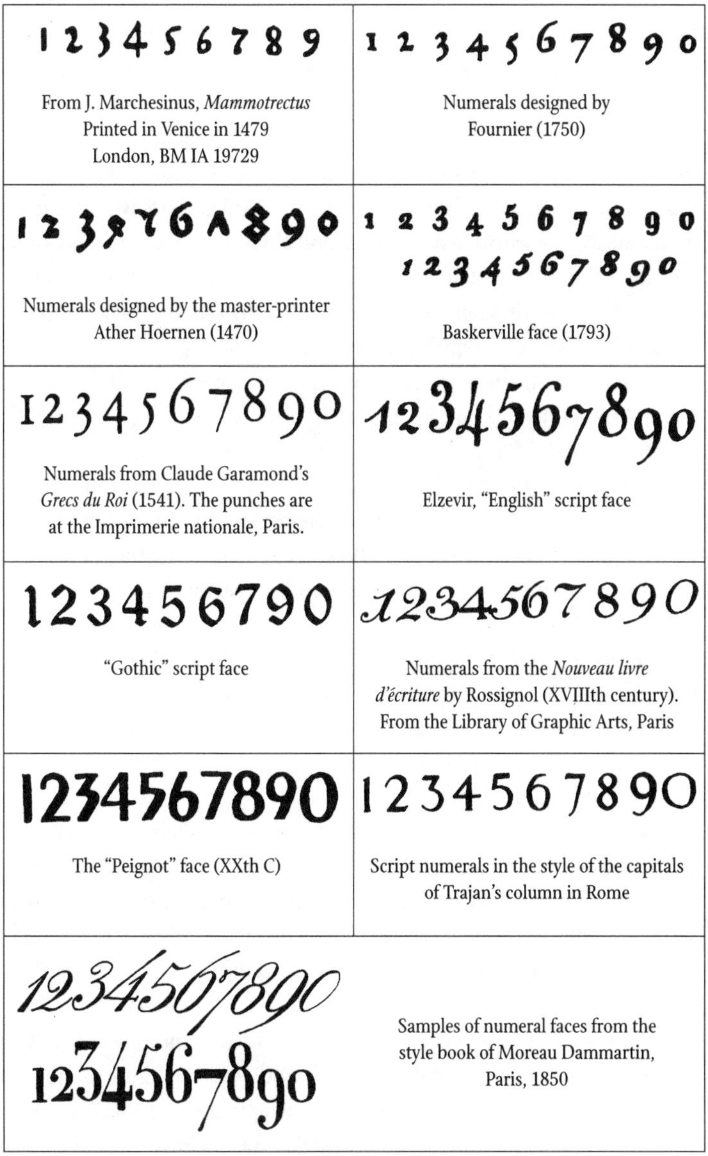

From J. Marchesinus, *Mammotrectus*
Printed in Venice in 1479
London, BM IA 19729

Numerals designed by
Fournier (1750)

Numerals designed by the master-printer
Ather Hoernen (1470)

Baskerville face (1793)

Numerals from Claude Garamond's
Grecs du Roi (1541). The punches are
at the Imprimerie nationale, Paris.

Elzevir, "English" script face

"Gothic" script face

Numerals from the *Nouveau livre
d'écriture* by Rossignol (XVIIIth century).
From the Library of Graphic Arts, Paris

The "Peignot" face (XXth C)

Script numerals in the style of the capitals
of Trajan's column in Rome

Samples of numeral faces from the
style book of Moreau Dammartin,
Paris, 1850

FIG. 26.12. *The development of printed numerals since the fifteenth century*

The spread of "algorism" was given renewed impetus from the start of the thirteenth century by a great Italian mathematician, Leonard of Pisa (c. 1170–1250), better known by the name of Fibonacci. He visited Islamic North Africa and also travelled to the Middle East. He met Arabic arithmeticians and learned from them their numeral system, the operational techniques, the rules of algebra and the fundamentals of geometry. This education was what underlay the treatise that he wrote in 1202 and which was to become the algorists' bible, the *Liber abaci* (*The Book of the Abacus*). Despite its title, Fibonacci's treatise (which assisted greatly the spread of Arabic numerals and the development of algebra in Western Europe) has no connection with Gerbert's abacus or the arithmetical course-books of that tradition – for it lays out the rules of written computation using both the zero and the rule of position. Presumably Fibonacci used "abacus" in his title in order to ward off attacks from the practical abacists who effectively monopolised the world of accounting and clung very much to their counters and ruled tables. At all events, from 1202 the trend began to swing in favour of the algorists, and we can thus mark the year as the beginning of the democratisation of number in Europe.

Resistance to the new methods was not easily overcome, however, and many conservative counting-masters continued to defend the archaic counter-abacus and its rudimentary arithmetical operations.

Professional arithmeticians, who practised their art on the abacus, constituted a powerful caste, enjoying the protection of the Church. They were inclined to keep the secrets of their art to themselves; they necessarily saw algorism, which brought arithmetic within everyone's grasp, as a threat to their livelihood.

Knowledge, though it may now seem rudimentary, brought power and privilege when it represented the state of the art, and the prospect of seeing it shared seemed fearful, perhaps even sacrilegious, for its practitioners. But there was another, more properly ideological reason for European resistance to Indo-Arabic numerals.

Even whilst learning was reborn in the West, the Church maintained a climate of dogmatism, of mysticism, and of submission to the holy scriptures, through doctrines of sin, hell and the salvation of the soul. Science and philosophy were under ecclesiastical control, were obliged to remain in accordance with religious dogma, and to support, not to contradict, theological teachings.

The control of knowledge served not to liberate the intellect, but to restrict its scope for several centuries, and was the cause of several tragedies. Some ecclesiastical authorities thus put it about that arithmetic in the Arabic manner, precisely because it was so easy and ingenious, reeked of magic and of the diabolical: it must have come from Satan himself! It was only a short step from there to sending over-keen algorists to the stake, along with witches and heretics. And many did indeed suffer that fate at the hands of the Inquisition.

The very etymology of the words "cypher" and "zero" provides evidence of this.

FIG. 26.13. *Written arithmetic using "Arabic" numerals. European engraving, sixteenth century. Paris, Palais de la Découverte*

When the Arabs adopted Indian numerals and the zero, they called the latter *sifr*, meaning "empty", a plain translation of the Sanskrit *shûnya*. *Sifr* is found in all Arabic manuscripts dealing with arithmetic and mathematics, and it refers unambiguously to the null figure in place-value numbering. (See for example the manuscripts in the Bibliothèque nationale, Paris, shelf-marks Ms. ar. 2457, f° 85v–86; Ms. ar. 2463, f° 79v–80; Ms. ar. 2464, f° 3v; Ms. ar. 2473, f° 9; Ms. ar. 2475, f° 45v–46r; and University of Tunis Ms. 10301, f° 25v; Ms. 2043, f°16v and 32v.) Etymologically, *sifr* means "empty" and also "emptiness" (the latter can also be expressed by *khalâ* or *farâgh*).

The stem SFR can also be found in words meaning "to empty" (*asfara*), "to be empty" (*safir*) and "have-nothing" (*safr al yadyn*, literally "empty hands", that is to say, "he who has nothing in his hands".

When the concept of zero arrived in Europe, the Arabic word was assimilated to a near-homophone in Latin, *zephyrus*, meaning "the west wind" and, by rather convenient extension, a mere breath of wind, a light breeze, or – almost – nothing. In his *Liber Abaci*, Fibonacci (Leonard of Pisa) used the term *zephirum*, and the term remained in use in that form until the fifteenth century:

> The nine Indian figures [*figurae Indorum*] are the following: 9, 8, 7, 6, 5, 4, 3, 2, 1. This is why with these nine figures and the sign 0, called *zephirum* in Arabic, all the numbers you may wish can be written [Fibonacci, as reproduced by B. Boncompagni (1857)].

However, in his *Sefer ha mispar* (*Number Book*), Rabbi Ben Ezra (1092–1167) used the term *sifra* [see M. Silberberg (1895) p. 2; D. E. Smith and Y. Ginsburg (1918)]. In various spellings, the Arabic term *sifra* (*cifra, cyfra, cyphra, zyphra, tzyphra...*) continued to be used to mean "zero" by some mathematicians for many centuries: we find it in the *Psephophoria kata Indos* (*Methods of Reckoning of the Indians*) by the Byzantine monk Maximus Planudes (1260–1310) [A. L. Allard, (1981)], in the *Institutiones mathematicae* of Laurembergus, published in 1636, and even as late as 1801 in Karl Friedrich Gauss's *Disquisitiones arithmeticae* (Gauss must have been one of the very last scholars to write in Latin).

In popular language, words derived from *sifr* soon came to be associated not with figures in general but with "nothing" in particular: in thirteenth-century Paris, a "worthless fellow" was called a *cyfre d'angorisme* or a *cifre en algorisme*, i.e. "an arithmetical nothing".

However, it was Fibonacci's term, *zephirum*, which gave rise to the modern name of zero, by way of the Italian *zefiro* (*zero* is just a contraction of *zefiro*, in Venetian dialect). The first known occurrence of the modern form of the word occurs in *De arithmetica opusculum* by Philippi Calandri and which, despite its Latin title, was written in Italian, and published in Florence in 1491. There is absolutely no doubt that *zero* owes its spread to French (*zéro*) and Spanish (*cero*) (and later on to English and other languages) to the enormous prestige that Italian scholarship acquired in the sixteenth century.

Meanwhile, Arabic *sifr* had also developed into the French word *chiffre*, the English *cipher*, German *Ziffer*, Spanish *cifra*. To begin with, the Latin terms *figuris* and *numero* were used to refer to the set of number-symbols (in English they still are called *figures* or *numerals*, more or less interchangeably); but from about 1486 in French, we find *chiffre* being used not to mean zero, but to mean a figure or numeral; and a similar development can be found in sixteenth- and seventeenth-century mathematical texts written in Latin, such as those by Willichius (1540), Conrad Rauhfuss Dasypodius of Strasbourg (*Institutionum Mathematicarum*, 1593), and the *Chronicle of Theophanes* (1655).

FIG. 26.14. *The Quarrel of the Abacists (to the left) and the Algorists (to the right). Adapted from an illustration in Robert Recorde (1510-1558),* The Grounde of Artes *(1558)*

Why did the original name of zero come to be used for the whole set of Indo-Arabic numerals? The answer lies in the attitude of the Catholic authorities to the counting systems borrowed from the Islamic world. The Church effectively issued a veto, for it did not favour a democratisation of arithmetical calculation that would loosen its hold on education and thus weaken its power and influence; the corporation of accountants raised its own drawbridges against the "foreign" invasion; and in any case the Church preferred the abacists – who were most often clerics as well – to keep their monopoly on arithmetic. "Arabic" numerals and written calculation were thus for a long while almost underground activities. Algorists plied their skills in hiding, as if they were using a secret code.

All the same, written calculation (on sand or by pen and ink) spread amongst the people, who were keenly aware of the central role played by zero, then called *cifra*, or *chifre*, or *chiffre*, or *tziphra*, etc. By a very common form of linguistic development, known as synecdoche, the name of the part (in this case, zero) came to be used for the whole, as in a kind of shorthand, so that words derived from *sifr* came to mean the entire set of numerals or any one of them. Simultaneously, it also came to mean "a secret", or a secret code – a cipher.

So the history of words for zero also tell the history of our culture: each time we use the word "cipher", we are also reviving a linguistic memory of the time when a *zero* was a dangerous *secret* that could have got you burned at the stake.

It is now easier to understand why in the mid-sixteenth century Montaigne could not "cast account" either "with penne or with Counter". For even with the introduction of written arithmetic, multiplication and division long remained outside the grasp of ordinary mortals, given the complicated operating techniques that were used. It was not until the end of the eighteenth century that simpler techniques were generalised and brought basic arithmetical operations even to those with little taste for sums.

The quarrel between the *abacists* (the defenders of Roman numerals and of calculations done on ruled boards with counters) and the algorists, who supported the written calculation methods originally invented in India, actually lasted several centuries. And even after the latter's victory, the use of the abacus was still so firmly entrenched in people's habits that all written sums were double-checked on the old abacus, just to make sure.

Until relatively recently, the British Treasury still used the abacus to calculate taxes due. And because the reckoning-board was called an *exchequer* (related to the words for *chess* and *chess-board* in various European languages), the Finance Minister of the United Kingdom is still called the Chancellor of the Exchequer.

Even long after written arithmetic with Arabic numerals had become the sole tool of scientists and scholars, European businessmen, financiers, bankers and civil servants – all of whom turned out to be more conservative than men of learning – found it hard to abandon entirely the archaic methods of the bead and counter-abacus.*

Only the French Revolution had the strength to cut through the muddle and to implement what many could see quite clearly, that written arithmetic was to counting-tokens as walking on a well-paved road was to wading through a muddy stream. The use of the abacus was banned in schools and government offices from then on.

Calculation and science could thenceforth develop without hindrance. Their stubborn and fierce old enemy had finally been put to rest.

---

* Translator's note: my father was trained as an accountant in the City of London in the late 1920s. Although he had of course learned modern arithmetic at school, he was required to learn how to tally sums on a bead abacus before being allowed to draw a wage. (DB)

FIG. 26.15. *Wood-block engraving from Gregorius Reisch,* Margarita Philosophica *(Freiburg, 1503). Lady Arithmetic (standing in the centre) gives her judgment by smiling on the arithmetician (to our left, her right) working with Arabic numerals and the zero (the numerals also adorn her dress). The quarrel of the abacists and algorists is over, and the latter have won.*

CHAPTER 27

BEYOND PERFECTION

That then was how numerical notation was brought to its full completion, democratised, and universalised: after a long history of twists and turns, with leaps forward and steps backward, ideas lost and found again, and with the friction between different systems used in conjunction ultimately generating the flash of genius on which it is all based: the decimal place-value system.

Is the story really at an end? After such a long and eventful history, could there not be more adventures to come? No, there could not. This really is the end. Our positional number-system is perfect and complete, because it is as economical in symbols as can be and can represent any number, however large. Also, as we have seen, it is the most efficacious in that it allows everyone to do arithmetic.

True, the development of computers and of electronic calculators with liquid crystal displays in the last half century has brought some changes in the graphical representation of the "Arabic" numerals. They have taken on more schematic shapes that would no doubt have horrified the scribes and calligraphers of yesteryear. In reality, however, these changes have had no effect whatever on the structure of the number-system itself. The numerals have been redesigned to meet the physical constraints of the display media, while also meeting the requirement to be readable both by machines and by the human eye.

Of course, as we have seen many times, a different base could have been used for our number-system. The base 12, for example, is in many ways more convenient than our decimal base; and the base 2 is well adapted to electronic computers which usually can recognise only two different states, symbolised by 0 and 1, of a physical system (perforation of a tape, or direction of magnetisation or of a current, etc.). But a change of base would change nothing in the structure of the number-system: this would continue to be a positional system and would continue to possess a zero, and its fundamental rules would be identical to those which we know already for our decimal system.

In short, the invention of our current number-system is the final stage in the development of numerical notation: once it was achieved, no further discoveries remained to be made in this domain.

The difficulties encountered on the road to a fully finished number-representation bear witness, on a limited front though one rich in possibilities, to true progress in human affairs.

From the beginning, human beings have shown the unique characteristic of harnessing the forces of nature to their development, their survival and their domination over other species, through discovering the laws of nature by means of observing the effects of their actions on their environment. Instead of following immutably programmed instinct, they act, seek to understand the "why" of things, ponder, and create.

In his novel *Les Animaux dénaturés*, Vercors recounts a telling story. A tribe of "primitive" people share a valley with a colony of beavers. The valley is swept by a flood. The beavers, driven by their hereditary instincts, build a dam and thus protect their dens. The humans, on the other hand, guided by their grand wizard, climb the sacred hill and meditate, begging mercy from their gods; this, however, does not prevent their village from being devastated by the flood.

At first sight, the behaviour of the humans seems stupid. But on reflection we see something really profound in it, for it is the germ of all future civilisation. They were certainly wrong to attribute the disaster to supernatural forces but, despite appearances, their reaction leapt beyond the mere instinct of the beavers, since they sought to understand the true cause of their misfortune. Humanity has surely passed through such phases: we know how far our tribulations have brought us.

This is not the place to retrace the evolution of the human race since the time of the first hominids. We must rather recall that human beings are characterised above all, not by what is innate and does not need to be learned, but by the predominance of what they can adjoin to their nature from learning, experience and education.

In other words, humankind is universally an intelligent social animal, and is differentiated from other higher animals by, above all, the predominance of what is acquired over what is inborn.

That fundamental truth has not always been, nor indeed yet is, obvious to everyone. For reasons ranging from the political to the criminal, this question has been subjected to systematic mystification in order that irrelevant criteria, such as the colour of the skin or the shape of the face, may be used to demonstrate the supposed superiority of one race over others.

The principal motivations and the basic ideas of racist and segregationist philosophies are directed towards maintaining great confusion between the notion of race and the ideas of a people, of a tribe, of an ethnicity and of a linguistic group, and towards cultivating a belief that there are so-called superior races who have a kind of natural right to exploit or even to suppress so-called inferior ones.

These indefensible racist mystifications, which the Nazis elevated to political ideology during Germany's Third Reich and which throughout the Second World War gave rise to the greatest barbarity of all time and led millions of innocents to slaughter, reflected an appalling eugenic mentality whose spirit still haunts the world decades after Nazism was crushed. All those who may have forgotten it, or who would wish that it should be forgotten, need to be reminded that "one man is not the same as another" but at the same time "one race is not unequal to another, still less is one people unequal to another" (J. Rostand, *Hérédité et Racisme*, p. 63).

As to the colour of the skin, this in fact (according to François Jacob) depends on the intensity of sunlight or, as the Arab philosopher Ibn Khaldun expressed it around 1390: "The climate gives the skin its colour. Black skin is the result of the greater heat of the South" [*Muqaddimah, Prolegomena*, p. 170; see V. Monteil (1977), p. 169].

The concept of race, in fact, is strictly biological, while that of people is historical. We talk, therefore, not of the French race but of the French people, which is made up of a mixture of several races. Nor is there a Breton race, but there is a Breton people; no Jewish race, but the Jewish people; no Arab race but Arab culture; no Latin race but Latin civilisation; and neither Semitic nor Aryan races, but Semitic and Aryan languages.

According to R. Hartweg (GLE Vol. 8, p. 976) the concept of race is "one of the categories of zoological classification. It denotes a relatively broad grouping within a species, a kind of sub-species, a collection of individuals of common origin which share a number of sufficiently meaningful biological characters." It therefore "rests on genetic, anatomical, physiological and pathological criteria. The difficulty with attempting to apply a racial classification to humankind therefore arises from: 1. the choice of criteria; 2. the fact that there are at present very few races which might be considered relatively 'pure', because of inter-breeding; 3. the transitory nature of the definition of any given race since races, like humanity itself, undergo continual evolution." D. L. Julia (1964) has the following view of this question:

> From the biological point of view, the notion of race as applied to humans is very imprecise. Features such as skin colour or facial structure are definite morphological characters, but they are biologically vague. Even if we suppose that different races exist, criteria such as physical strength, or intelligence (as measured by IQ tests), show no systematic variation. Though the people of industrialised nations may have weaker constitutions than those of African nations, for example, and although culture and education may seem less prevalent among the latter than among Western peoples, nonetheless this has no bearing on the physical potential of the former nor on the intellectual potential of the latter. On the other hand, differences of character – whereby we traditionally

contrast the intellectual strictness of the "whites" with the intuitive mind and generous spirit of the "blacks", or the openness of both of these with the feline suppleness and deep capacity for dissimulation of the "yellow" peoples – bear no relationship to a scale of values. Differences of character should not be a source of conflict, but an occasion for learning and therefore of enrichment: in coming to understand other people, any persons of any race will come to better understand themselves as individuals, and learn wisdom for the conduct of their own lives.

In short, "racist theories are gratuitous constructs, based on tendentious and immature anthropological ideas" (J. Rostand, *Hérédité et Racisme*, p. 57). "The truth is, that there is no such thing as a pure race, and to base politics on ethnographic analysis is to base it on a chimera" (E. Renan, *Discours et Conférences*, pp. 93–4).

In the domain of the history of numbers, at least, we have seen that human intelligence is universal and that the progress has been achieved in the mental, cultural and collective endowment of the whole of humankind. From the Cro-Magnon to the modern period, no fundamental change in the human brain has in fact occurred: only cultural enrichment of mental furnishings. This means that all human beings, whether white, red, black or yellow, whether living in the town, the country or the bush, have without exception equal intellectual potential. Individuals will develop the possibilities of their intelligence, or not, according to their needs, their environments, their social circumstances, their cultural heritage and their diverse individual aptitudes. These strictly personal individual differences are what determine whether one mind will be more or less enlightened, more or less inventive, than another.

As was stated in the Preface, number and simple arithmetic nowadays seem so obvious that they often seem to us to be inborn aptitudes of the human brain.

This was no doubt why the great German mathematician Kronecker said "God created the integers, the rest is the work of Man", whereas in fact the whole is an invention, the pure creation of the human mind; as the German philosopher Lichtenberg said: "Mankind started from the principle that every magnitude is equal to itself, and has ended up able to weigh the sun and the stars."

And the invention is of purely human origin: no god, no Prometheus, no extra-terrestrial instructor, has given it to the human race.

The actual history of numbers serves also, incidentally, to refute all those popular stories of extra-terrestrials who came to Earth to civilise the human race. Had we been visited by a scientifically and technologically advanced civilisation from outer space, we would not first have learned from it mysterious methods of erecting megaliths, but a number-system based on the principle of position and endowed with a zero. There is abundant documentation which proves that these were of late appearance, and that historically there was a great variety of number-systems in use. Quite sufficient to disprove any extra-terrestrial source for arithmetic – and therefore for everything else.

This profoundly human invention is also the most universal of inventions. In more than one sense, it binds humanity together. There is no Tower of Babel for numbers: once grasped, they are everywhere understood in the same way. There are more than four thousand languages, of which several hundred are widespread; there are several dozen alphabets and writing systems to represent them; today, however, there is but one single system for writing numbers. The symbols of this system are a kind of visual Esperanto: Europeans, Asiatics, Africans, Americans or Oceanians, incapable of communicating by the spoken word, understand each other perfectly when they write numbers using the figures 0, 1, 2, 3, 4... , and this is one of the most notable features of our present number-system. In short, numbers are today the one true universal language. Anyone who thinks that number is inhuman would do well to reflect on this fact.

The invention and democratisation of our positional number-system has had immeasurable consequences for human society, since it facilitated the explosion of science, of mathematics and of technology.

This in its turn gave rise to the mechanisation of arithmetical and mathematical calculations.

Yet all the elements needed to construct a true calculating engine had already been in existence, known and utilised since ancient times by scholars and engineers such as Archimedes, Ctesibius or Hero of Alexandria – such devices as levers, the endless screw, gears, toothed wheels, etc. But when we look at the numerical notations which they used at the time we can see that it would have been out of the question for them to conceive of, let alone to construct, such machines.

Nor did the technology of the time permit their actual construction: not until the start of the seventeenth century, when clockwork mechanisms underwent enormous development, would the first implementations of such devices be seen. Without a positional number-system with a zero, Schickard and Pascal would have been unable to imagine the components of their calculating machines. Pascal, for example, would not have thought of the transferrer (a counter-balanced pawl which, when one counting-wheel advanced from "9" to "0" after completing a revolution, advanced the next wheel through one step), nor of the totalisator (a device which, for each power of ten, had a cylinder bearing two enumerations from "0" to "9", in opposite directions, one used for additions and the other for subtractions).

To sum up: if the positional number-system with a zero had not existed, the problem of mechanising the process of calculation would never have found a solution; still less would it have been conceivable to automate the process. This, however, is another story, the story of automatic calculation, which begins with the classical calculating or analytical engines, passes on to machines for sorting and classifying data, and culminates in the emergence of the computer.

These powerful developments would never have seen the light of day, had the Indian discovery of positional notation not influenced the art of calculation itself. Since, however, it evidently did, we are led to look far beyond the domain of mere figures into the universe of number itself.

Note first that, unlike almost all earlier systems, our modern number-system allows us to write out straightforwardly any number whatever, no matter how large it may be. But modern mathematicians have introduced a simplification in the representation of very large numbers by means of so-called "scientific" notation which makes use of the powers of ten. For example, 1,000 may be written as 10^3, a million as 10^6 and a billion as 10^9, the small number in the exponent denoting the number of zeros in the standard representation of the number. For a billion, for example, we write down three figures instead of ten.

As it stands this is no more than an abbreviated notation, which effects no change in the number-system being used. Nonetheless, it is more than mere shorthand, since it lends itself to the procedure known as exponentiation ("raising to the power") which stands to multiplication as multiplication stands to addition, since we can write:

$$a^m \times a^n = a^{m+n}; \; a^m/a^n = a^{m-n}; \; (a^m)^n = a^{mn}$$

Using this notation, a very large number such as

72,965,000,000,000,000,000,000,000,000,000 (27 zeros)

can be written more economically as

$$72,965 \times 10^{27}$$

which simply indicates that by adjoining 27 zeros to 72,965 the complete representation of the number is obtained. We can also use "floating-point" notation, and express the first number as a decimal fraction followed by the appropriate power of ten, as in

$$7.2965 \times 10^{31}$$

which indicates that the decimal point is to be moved 31 places to the right in order to obtain the complete representation.

Most pocket calculators and electronic computers have a facility of this kind which allows them to show numerical results which exceed the decimal capacity of the display (or at least to show their approximate values).

The positional number-system gave rise to great advances in arithmetic, because it showed the properties of numbers themselves more clearly. It similarly enabled mathematicians of recent times to unify apparently distinct concepts, and to create theories which had previously been unthinkable.

Fractions, for example, had been known since ancient times, but owing to the lack of a good notation they were for long ages written using notations which were only loosely established, which were not uniform, and which were ill adapted to practical calculation.

Originally, remember, fractions were not considered to be numbers. They were conceived as relations between whole numbers. But, as methods of calculation and arithmetic developed, it was observed that fractions obeyed the same laws as integers, so that they could be considered as numbers (an integer, therefore, being a fraction whose denominator was unity). As a result, where numbers had previously served merely for counting, they now became "scales" which could be put to several uses. Thereafter, two magnitudes would no longer be compared "by eye"; they could be conceived as subdivided into parts equal to a magnitude of the same kind which served as a unit of reference. Despite this advance, however, the ancients, with their inadequate notations, were unable to unify the notion of fraction and failed to construct a coherent system for their units of measurement.

Using their positional notation with base 60, the Babylonians were the first to devise a rational notation for fractions. They expressed them as sexagesimal fractions (in which the denominator is a power of sixty) and wrote them much as we now write fractions of an hour in minutes and seconds:

$$33m\ 45s\ (= 33/60h + 45/3600h).$$

They did not, however, think of using a device such as the "decimal point" to distinguish between integers and sexagesimal fractions of unity, so that the combination [33; 45] could as well mean 33h 45m as 0h 33m 45s. They had, so to speak, a "floating notation" whose ambiguities could only be resolved by context.

The Greeks then tried to make a general notation for vulgar fractions, but their alphabetic numerals were ill-adapted for the purpose and so they abandoned the attempt. Instead, they adopted the Babylonian sexagesimal notation.

Our modern notation for vulgar fractions is due to the Indians who, using their decimal positional number-system, wrote a fraction such as 34/1,265 much as we do now:

34 (numerator)
1,265 (denominator).

This notation was adopted by the Arabs, who brought it into its modern form by introducing the horizontal bar between numerator and denominator.

Then, following the discovery of "decimal" fractions (in which the denominator is a power of ten), people gradually became aware of the importance of extending the positional system in the other direction, i.e. of representing numbers "after the decimal point", and this is what finally allowed all fractions to be written without difficulty, and which showed the integers to be a special kind of fraction, in which no figures appear after the decimal point.

The first European to make the decisive step towards our modern notation was the Belgian Simon Stevin. Where we would write 679.567, he wrote:

679(0) 5(1) 6(2) 7(3)

which stood for 679 integer units, 5 decimal units of the first order (tenths), 6 of the second order (hundredths) and 7 of the third (thousandths).

Later on, the Swiss Jost Bürgi simplified this notation by omitting the superfluous indication of decimal order, and by marking the digit representing the units with the sign°:

$679\overset{\circ}{5}67$

At the same time, the Italian Magini replaced the ring sign by a point placed between the units digit and the tenths digit, creating the decimal-point notation which is still the standard usage in English-speaking countries:

679.567

In continental Europe, a comma is commonly used instead of the point, and this was introduced at the start of the seventeenth century by the Dutchman Snellius:

679,567

This rationalisation of the concept and of the notation of fractions had immeasurable consequences in every domain. It led to the invention of the "metric system", built entirely on the base 10 and completely consistent: in 1792, the French Revolution offered "to all ages and to all peoples for their greater good" this system which replaced the old systems of arbitrary, inconsistent and variable units. We know full well the fantastic progress that this brought in every practical domain, by virtue of the enormous simplification of every kind of calculation.

Once established, positional decimal notation opened up the infinite complexity of the universe of number, and led to prodigious advances in mathematics.

In the sixth century BCE the Greek mathematicians, following Pythagoras, discovered that the diagonal of a square "has no common measure" with the side of the square. It can be observed by measurement, and deduced by reason, that the diagonal of a square whose side is one metre long has a length which is not a whole number of metres, nor of centimetres, nor millimetres... . In other words, the number $\sqrt{2}$ (which is its mathematical magnitude) is an "incommensurable" number. This was the moment of discovery of what we now call "irrational" numbers, which are neither integers nor fractions.

This discovery greatly perturbed the Pythagoreans, who believed that number ruled the Universe, by which they understood the integers and their simpler combinations, namely fractions. The new numbers were called "unmentionable", and the existence of these "monsters" was not to be divulged to the profane. According to the Pythagorean conception of the world, this inexplicable error on the part of the Supreme Architect must be kept secret, lest one incur the divine wrath.

But the secret soon became well known to right-thinking people who were prepared to mention the unmentionable, to name the unnameable, and who delivered it up to the profane world. That perfect harmony between arithmetic and geometry, which had been one of the fundamental tenets of the Pythagorean doctrine, was seen to be a vain mystification.

Once we are free of these mystical constraints, we can accept that there are numbers which are neither integers nor fractions. These are the "irrational" numbers, of which examples are $\sqrt{2}$, $\sqrt{3}$, the cube root of 7 and of course the famous π.

Nevertheless, this class of numbers remained ill defined for many centuries, because the defective number-systems of earlier times did not allow such numbers to be represented in a consistent manner. They were in fact designated by words, or by approximate values which had no apparent relation to each other. Lacking the means to define them correctly, people were obliged to admit their existence but were unable to incorporate them into a general system.

Modern European mathematicians, with the benefits of effective numerical notation and continual advances in their science, finally succeeded where their predecessors had failed. They discovered that these irrational numbers could be identified as decimal numbers where the series of digits after the decimal point does not terminate, and does not eventually become a series of repetitions of the same sequence of digits. For example: $\sqrt{2} = 1.41421356237...$ This was a fundamental discovery: this property characterises the irrational numbers.

Of course, a fraction such as 8/7 also possesses a non-terminating decimal representation:.

$$8/7 = 1.142857142857142857...$$

but its representation is periodic: the sequence "142857" is indefinitely repeated, with nothing else intervening: we can therefore, for instance, easily determine that the 100th decimal digit will be "8" , since 16 repetitions will take us to the 96th place, and four more digits will give the digit "8".

On the other hand, the irrational numbers do not follow such a pattern. Their decimal expansion is not periodic, and there is no rule which allows us to determine easily what digit will be in any particular place. This is precisely the respect in which the vulgar fractions (what we today call "rational numbers") differ from the irrational numbers.

However, nowadays this is not how irrational numbers are defined. Instead, an algebraic criterion is used, according to which an irrational number is not the solution of any equation of the first degree with integer coefficients. The number 2, for example, is the solution of $x - 2 = 0$, and the fraction 2/3 is the solution of $3x - 2 = 0$. On the other hand, it can be proved that the number $\sqrt{2}$ cannot be the solution of any equation of this kind, and so it is irrational.

Nonetheless, the concept of such numbers would not have been fully understood without the introduction of a further extension of the notion of number: the " algebraic" numbers. This concept was discovered in the nineteenth century by the mathematicians Niels Henrik Abel of Norway, and Évariste Galois of France. An algebraic number is a solution of an algebraic equation with integer coefficients. Clearly this holds for any integer or fractional number, but it also holds for any irrational number which can be expressed by radicals. For example, $\sqrt{2}$ is a solution of the equation $x^2 - 2 = 0$, and the cube root of 5 is a solution of the equation $x^3 - 5 = 0$. The set of algebraic numbers, therefore, includes both the set of rational numbers (which itself includes the integers) and the set of all numbers that can be expressed by the use of radicals.

However, even this is inadequate to contain all numbers. After the discoveries of Liouville, Hermite, Lindemann and many others, we know that there are additional "real numbers", which are not integers, or fractions, or even algebraic irrational numbers. These are the so-called "transcendental" numbers, which cannot occur as a solution of any algebraic equation with integer or fractional coefficients. They are, of course, irrational; but they cannot be expressed by the use of radicals. There are infinitely many of them; examples include the number "π" (the area, and also half the circumference, of a circle with unit radius), the number "e" (the base of the system of natural logarithms invented in 1617 by the Scottish mathematician John Napier), the number "log 2" (the decimal logarithm of 2) and the number "cos 25°" (cosine of the angle whose measure is 25 degrees). However, we cannot here let ourselves be carried away into the further reaches of the theory of numbers.

Now, if it is possible as we have seen to write any number whatever in a simple and rational way, no matter how large or strange it may be, then we

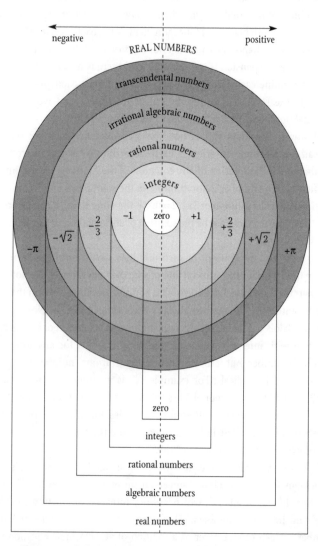

FIG. 27.1. *The successive algebraic extensions of the concept of number*

may well ask if there is a last number, greater than all the others. We can directly see from the positional notation that this cannot be so, since if we write down the decimal representation of an integer then all we have to do is to add a zero at the right-hand end, to multiply this number by 10. Proceeding indefinitely in this way, we readily see that the sequence of integers has no limit. All the more so for the fractions and the irrational numbers, for which we can demonstrate that there exist "several infinities" between any two consecutive integers.

From the dawn of history, people came up against the dilemma of the infinite (see the article *Infinity, in the Dictionary). Since then, however, the concept of infinity has been made perfectly precise and objective, and presents no fundamental obscurity – at least, not such as the common mind attributes to it. Infinity has its own symbol: ∞, like a figure 8 on its side, called "lemniscate" by some and introduced quite recently into mathematical notation by the English mathematician John Wallis who first employed it in 1655. But we can hardly prove the existence of infinity – the impossibility of counting all numbers – since infinity, nowadays, is taken as an axiom, a mathematical hypothesis, on which the whole of contemporary mathematics is based.

It is but one step from infinity to zero, and it is a step which leads us on to algebra, since the null is the opposite of the unlimited.

For thousands of years, people stumbled along with inadequate and useless systems which lacked a symbol for "empty" or "nothing" . Similarly there was no way of conceiving of "negative" numbers (–1, –2, –3, etc.), such as we nowadays use routinely to express, for example, sub-zero temperatures or bank accounts in deficit. Therefore a subtraction such as "3 – 5" was for a long time considered to be impossible. We have seen how the discovery of zero swept away this obstacle so that ordinary ("natural") numbers were extended to include their "mirror images" with respect to zero.

That inspired and difficult invention, zero, gave rise to modern algebra and to all the branches of mathematics which have come about since the Renaissance (see the article *Zero in the Dictionary).

Algebra would not however have blossomed as it did if, as well as the zero, there had not also been another, equally important discovery made by Franciscus Vieta in 1591 and brought to perfection by René Descartes in 1637: this is the use of letters as mathematical symbols, which inaugurated a completely new era in the history of mathematics.

Algebra, in fact, is a generalisation of arithmetic. An x or a y, or any other letter, is a new sort of "number": it stands for any number, whose value is unknown. One might say that it is a sign in wait for a number, holding the place for one or more figures yet to come, just as the zero sign itself filled the place of a digit corresponding to a missing decimal order of magnitude.

But this is no merely formal artifice. Using a letter to stand for a parameter or an unknown value finally freed algebra from enslavement to words, leading to the creation of a kind of "international language" which is understood unambiguously by mathematicians the world over.

In its turn, literal notation underwent a further liberation from certain restrictions acquired in its everyday usage. The symbol x or y did not simply represent a number: it could be considered in itself, independently of what kind or size of thing it represented. Thus the symbol itself transcends what it represents and becomes a mathematical object in its own right, obeying the laws of calculation. Mathematical arguments and calculations could therefore be abbreviated and systematised, and abstraction became directly accessible. Leibnitz wrote that "This method spares the work of the mind and the imagination, in which we must economise above all. It enables us to reason with small cost in effort, by using letters in place of things in order to lighten the load on the imagination." In turn, the spread of algebra throughout Europe brought about great scientific progress, and led to substantial refinement of operational symbolism in its widest sense.

Taking a very rapid overview of the history of mathematics, this science arose in Ancient Greece when her philosophers and mathematicians brought a decisive advance into human thought: that combination of abstraction, generalisation, synthesis and logical reasoning which had previously lain hid in shadow. The Greeks, however, were enamoured of what is beautiful and simple and, consequently, of what is divine. They thereby cut themselves off from the world of reality and therefore from applied mathematics. The epic Græco-Latin era was succeeded in the West by the long dark night of the Middle Ages, feebly lit up from time to time by a few individuals of no great stature.

It was the Arabs who took over. They were well placed to assimilate the whole of the Ancient Greek legacy, together with Indian science, saving the essentials from oblivion, and they developed and propagated it according to "scientific reasoning".

In due course, the first great European universities were founded and the pursuit of knowledge was resumed: the Western world once again awoke and initiated the study of nature based on independence of thought. This great dawning derives above all from the work of Fibonacci, *Liber abaci* (1202) which, over the next three centuries, was to prove a rich source of inspiration for the development of arithmetic and algebra in the West. But the West also established numerous contacts with Arabic and Islamic culture from the eleventh century onwards, whereby European mathematicians came to know not only the works of Archimedes, Euclid, Plato, Ptolemy, Aristotle and Diophantus, but also became acquainted with the work of Arab, Persian and Indian thinkers and learned the methods of calculation which had been invented in India.

The true renaissance – or rather the true awakening – of mathematics in Europe would not take place until the seventeenth century, first of all in the work of René Descartes who made full use of the new knowledge in his invention of algebraic and analytic geometry. Pascal later opened new questions in considering the problems of mathematical infinity, followed in this by Newton who also, with Leibnitz, ushered in the era of the infinitesimal calculus.

During the eighteenth century the spirits of Greek and of Cartesian mathematics were sustained together, leading on to a synthesis which, continuing into the nineteenth century, gave rise to the invention of determinants and matrices and the development of vector calculus.

In the nineteenth century, Gauss, Cauchy and Picard completed the Græco-Cartesian edifice. Lobachevsky questioned the foundations of Euclidean geometry and invented non-Euclidean geometry. On the last night of his all too short and dramatic life, the young Évariste Galois, a political revolutionary, left for the world his creation of the first abstract algebraic structures. George Boole laid the foundations of mathematical logic and Georg Cantor worked out the fundamentals of the theory of sets and of modern topology. The century closed with Hilbert's publication of his axiomatisation of geometry, which became the model for the modern axiomatic study of mathematics.

Since then, the explosion of modern mathematics has been characterised by an ever more pronounced algebraic approach: unlike the ancient mathematics which was based on very specific concepts of line and of magnitude, its basis is the universal and very abstract concept of a set. This recent unification in terms of logic and the theory of sets has made mathematics, for the first time in its history, an undivided subject.

And, finally, this unity in abstraction of modern mathematics laid the foundation of the computer science which is being developed today.

Therefore we must pay tribute to all the mathematicians, be they English, French, American, Italian, Russian, German, Japanese or any other, who have brought mathematics to its present extraordinary flowering, for which the words of Arthur Cayley in 1883 are still a beautifully apt description: "It is difficult to give an idea of the vast extent of modern mathematics. The word 'extent' is not the right one: I mean extent crowded with beautiful detail – not an extent of mere uniformity such as an objectless plain, but a tract of beautiful country to be rambled through and studied in every detail of hillside and valley, stream, rock, wood, and flower. But, as for everything else, so for mathematical beauty – beauty can be perceived but not explained." We may not, however, omit from this roll of honour to the glory of Western mathematics the Indian civilisation which invented the modern number-system in which the later great discoveries are rooted. Nor should we omit the Arabic and Islamic civilisation which carried the flame whilst the West slept.

There is a last great question. Could modern mathematics, in all its rigour, and in all its principles, with its theoretical extensions and practical applications which have revolutionised the way we live – could mathematics have possibly occurred in the absence of a positional numerical notation so perfect as the one we have? It seems incredibly unlikely. Modern science and technology may have their roots in antiquity, but they could only flourish as they have in the context of the modern era and in the framework of a number-system as revolutionary and efficient as our positional decimal system, which originated in India. To move mountains, the mind requires the simplest of tools.

And so our history of numbers is now completed. However, it is itself but a chapter in another history, the history of the representations of the world, and that history, beyond doubt, will never be completed.

LIST OF ABBREVIATIONS

Where appropriate, cross-references to fuller information
in the Bibliography are given in the form: "see: AUTHOR*"*

AA	*The American Anthropologist*	Menasha, Wisconsin
AAN	*American Antiquity*	
AANL	*Atti dell'Accademia Pontificia de' Nuovi Lincei*	Rome
AAR	*Acta Archaeologica*	Copenhagen
AAS	*Annales archéologiques syriennes*	Damascus
AASOR	*Annual of the American School of Oriental Research*	Cambridge, MA
AAT	*Aegypten und altes Testament*	
ABSA	*Annual of the British School in Athens*	London
ACII	*Appendice al Corpus Inscriptionum Italicorum*	see: GAMURRINI
ACLHU	*Annals of the Computation Laboratory of Harvard University*	Cambridge, MA
ACT	*Astronomical Cuneiform Texts*	See: NEUGEBAUER
ACOR	*Acta Orientalia*	Batavia
ADAW	*Abhandlungen der deutschen Akademie der Wissenschaften zu Berlin*	
ADFU	*Ausgrabungen der deutschen Forschungsgemeinschaft in Uruk-Warka*	Berlin
ADO	*Annals of the Dudley Observatory*	Albany, NY
ADOGA	*Ausgrabungen der deutschen Orient-Gesellschaft in Abusir*	
ADP	*Archives de Psychologie*	Geneva
ADSM	*Album of Dated Syriac Manuscripts*	see: HATCH
AEG	*Aegyptus. Rivista italiana di egittologia e papirologia*	Milan
AESC	*Annales. Economies, Sociétés, Civilisations*	Paris
AFD	*Annales d'une famille de Dilbat*	see: GAUTIER
AFO	*Archiv für Orientforschung*	Graz
AGA	*Aegyptologische Abhandlungen*	Wiesbaden
AGM	*Abhandlungen zur Geschichte der Mathematik*	Leipzig

AGMNT	*Archiv für Geschichte der Mathematik, der Naturwissenschaften und der Technik*	
AGW	*Abhandlungen der Gesellschaft der Wissenschaften*	Göttingen
AHC	*Annals of the History of Computing*	IEEE, New York
AHES	*Archive for the History of the Exact Sciences*	
AI	*Arad Inscriptions*	see: AHARONI
AIEE	American Institute of Electrical Engineers	New York, NY
AIHS	*Archives internationales d'histoire des sciences*	Paris
AIM	*Artificial Intelligence Magazine*	
AJ	*Accountants Journal*	
AJA	*American Journal of Archaeology*	New York, NY
AJPH	*American Journal of Philology*	New York, NY
AJPS	*American Journal of Psychology*	New York, NY
AJS	*American Journal of Science*	New York, NY
AJSL	*American Journal of Semitic Languages and Literature*	Chicago, IL
AKK	*Akkadika*	Brussels
AKRG	*Arbeiten der Kaiserlichen Russischen Gesandschaft zu Peking*	Berlin
AM	*American Machinist*	
AMA	*Asia Major*	Leipzig & London
AMI	*Archaeologische Mitteilungen aus Iran*	Berlin
AMM	*American Mathematical Monthly*	
AMP	*Archiv der Mathematik und Physik*	
ANS	*Anatolian Studies*	London
ANTH	*Anthropos*	Göteborg
ANTHR	*Anthropologie*	Paris
AOAT	*Alter Orient und Altes Testament*	Neukirchen-Vluyn
AOR	*Analecta Orientalia*	Rome
AOS	*American Oriental Series*	New Haven, CT
APEL	*Arabic Papyri in the Egyptian Library*	see: GROHMANN
ARAB	*Arabica. Revue d'études arabes*	Leyden
ARBE	*Annual Report of the American Bureau of Ethnology*	Washington, DC

ARBS	*Annual Report of the Bureau of the Smithsonian Institution*	Washington, DC
ARC	*Archeion*	Rome
ARCH	*Archeologia*	Rome
ARCHL	*Archaeologia*	London
ARCHN	*Archaeology*	New York, NY
ARM	*Armenia*	
ARMA	*Archives royales de Mari*	Paris
AROR	*Archiv Orientálni*	Prague
ARYA		see: SHUKLA & SARMA
AS	*Automata Studies*	Princeton, NJ, 1956
ASAE	*Annales du service de l'antiquité de l'Egypte*	Cairo
ASB	*Assyriologische Bibliothek*	Leipzig
ASE	*Archaeological Survey of Egypt*	London
ASI	*Archaeological Survey of India*	New Delhi
ASMF	*Annali di scienze matematiche e fisiche*	Rome
ASNA	*Annuaire de la société française de numismatique et d'archéologie*	Paris
ASOR	*American School of Archaeological Research*	Ann Arbor, MI
ASPN	*Annales des sciences physiques et naturelles*	Lyon
ASR	*Abhandlungen zum schweizerischen Recht*	Bern
ASS	*Assyriological Studies*	Chicago, IL
ASTP	*Archives suisses des traditions populaires*	
ASTRI	*L'Astronomie indienne*	see: BILLARD
AT	*Annales des Télécommunications*	Paris
ATU	*Archaische Texte aus Uruk*	see: FALKENSTEIN
ATU2	*Zeichenliste der Archaischen Texte aus Uruk*	see: GREEN & NISSEN
AUT	*Automatisme*	Paris
BAB	*Bulletin de l'Académie de Belgique*	Brussels
BAE	*Bibliotheca Aegyptica*	Brussels
BAMNH	*Bulletin of the American Museum of Natural History*	New York, NY
BAMS	*Bulletin of the American Mathematical Society*	

BAPS	*Bulletin de l'Académie polonaise des Sciences*	Warsaw
BARSB	*Bulletin de l'Académie royale des sciences et belles-lettres de Bruxelles*	Brussels
BASOR	*Bulletin of the American School of Oriental Research*	Ann Arbor, MI
BCFM	*Bulletin du Club français de la médaille*	Paris
BCMS	*Bulletin of the Calcutta Mathematical Society*	Calcutta
BDSM	*Bulletin des sciences mathématiques*	Paris
BEFEO	*Bulletin de l'Ecole française d'Extrême-Orient*	Paris & Hanoi
BEPH	*Beiträge zur englischen Philologie*	Leipzig
BFT	*Blätter für Technikgeschichte*	Vienna
BGHD	*Bulletin de géographie historique et descriptive*	Paris
BHI	*Bulletin hispanique*	Bordeaux
BHR	*Bibliothèque d'humanisme et de Renaissance*	Geneva
BIFAO	*Bulletin de l'Institut français d'antiquités orientales*	Cairo
BIMA	*Bulletin of the Institute of Mathematical Applications*	
BJRL	*Bulletin of the John Rylands Library*	Manchester, UK
BLPM	*Bulletin de liaison des professeurs de mathématiques*	Paris
BLR	*Bell Laboratories Record*	Murray Hill, NJ
BMA	*Biblioteca Mathematica*	
BMB	*Bulletin for Mathematics and Biophysics*	
BMET	*Bulletin du Musée d'ethnologie du Trocadéro*	Paris
BMFRS	*Biographical Memoirs of Fellows of the Royal Society*	London
BMGM	*Bulletin of the Madras Government Museum*	Madras
BRAH	*Boletín de la Real Academia de la Historia*	Madrid
BSA	*Bulletin de la Société d'anthropologie*	Paris
BSC	*Bulletin scientifique*	Paris
BSEIN	*Bulletin de la Société d'encouragement pour l'industrie nationale*	Paris
BSFE	*Bulletin de la Société française d'Egyptologie*	Paris
BSFP	*Bulletin de la Société française de philosophie*	Paris
BSPF	*Bulletin de la Société préhistorique française*	Paris

BSI	*Biblioteca Sinica*	Paris
BSM	*Bulletin de la Société mathématique de France*	Paris
BSMA	*Bulletin des sciences mathématiques et astronomiques*	Paris
BSMF	*Bollettino di bibliografia e di storia delle scienze matematiche e fisiche*	Rome
BSMM	*Bulletin de la Société de médecine mentale*	Paris
BSNAF	*Bulletin de la Société nationale des antiquaires de France*	Paris
BSOAS	*Bulletin of the School of Oriental and African Studies*	London
BST	*Bell System Technology*	Murray Hill, NJ
CAA	*Contributions to American Archaeology*	Washington, DC
CAAH	*Contributions to American Anthropology and History*	Washington, DC
CAH	*The Cambridge Ancient History*	Cambridge, UK, 1963
CAPIB	*Corpus of Arabic and Persian Inscriptions of Bihar*	Patna
CAW	Carnegie Institution	Washington, DC
CDE	Chronique d'Egypte, in: *Bulletin périodique de la Fondation égyptienne de la reine Elisabeth*	Brussels
CENT	*Centaurus*	Copenhagen
CETS	*Comparative Ethnographical Studies*	
CGC	*A Catalogue of Greek Coins in the British Museum*	see: POOLE
CHR	*China Review*	
CIC	*Corpus des Inscriptions du Cambodge*	see: COEDÈS
CIE	*Corpus inscriptionum etruscarum*	1970
CIG	*Corpus inscriptionum graecarum*	see: BOECKH, FRANZ, CURTIUS & KIRKHOFF
CII	*Corpus inscriptionum Iudaicorum*	see: FREY
CIIN	*Corpus inscriptionum Indicarum*	London, Benares & Calcutta, 1888–1929

CIL	*Corpus inscriptionum latinarum*	Leipzig & Berlin, 1861–1943
CIS	*Corpus inscriptionum semiticarum*	Paris, 1889–1932
CJW	*Coins of the Jewish War*	see: KADMAN
CNAE	*Contributions to North American Ethnology*	Washington, DC
CNP	*Corpus Nummorum Palaestiniensium*	Jerusalem
COWA	*Relative Chronologies in Old World Archaeology*	see: ERICHSEN
CPH	*Classical Philology*	Chicago, IL
CPIN	*Le Cabinet des poinçons de l'Imprimerie national*	Paris, 1963
CR	*Classical Review*	
CRAI	*Comptes-rendus des séances de l'Académie des Inscriptions et Belles-Lettres*	Paris
CRAS	*Comptes-rendus des séances de l'Académie des Sciences*	Paris
CRCIM	*Comptes-rendus du Deuxième Congrès international de Mathématiques de Paris*	see: DUPORCK
CRGL	*Comptes-rendus du Groupe linguistique d'études hamito-sémitiques*	Paris
CRSP	*Comptes-rendus de la Société impériale orthodoxe de Palestine*	
CSKBM	*Catalogue of Sanskrit Buddhist Manuscripts in the British Museum*	London
CSMBM	*Catalogue of Syriac Manuscripts in the British Museum*	see: WRIGHT
CTBM	*Cuneiform Texts from Babylonian Tablets in the British Museum*	London, 1896
D	*Le Temple de Dendara*	see: CHASSINAT
DAA	*Denkmäler aus Aegypten und Aethiopien*	see: LEPSIUS
DAB	*Dictionnaire archéologique de la Bible*	Paris: Hazan, 1970
DAC	*Dictionnaire de l'Académie française*	
DAE	*Deutsche Aksoum-Expedition*	Berlin
DAFI	*Cahiers de la Délégation archéologique française en Iran*	Paris

DAGR	*Dictionnaire des antiquités grecques et romaines*	see: DAREMBERG & SAGLIO
DAR	*Denkmäler des Alten Reiches im Museum von Kairo*	see: BORCHARDT
DAT	*Dictionnaire archéologique des techniques*	Paris: L'Accueil, 1963
DCI	*Dictionnaire de la civilisation indienne*	see: FRÉDÉRIC
DCR	*Dictionnaire de la civilisation romaine*	see: FREDOUILLE
DG	*Demotisches Glossar*	see: ERICHSEN
DgRa	*De Gestis regum Anglorum libri*	see: MALMESBURY
DI	*Der Islam*	
DJD	*Discoveries in the Judaean Desert of Jordan*	Clarendon Press, Oxford
DMG	*Documents in Mycenean Greek*	see: VENTRIS & CHADWICK
DR	*Divination et rationalité*	Paris: Le Seuil, 1974
DS	*Der Schweiz*	
DSB	*Dictionary of Scientific Biography*	see: GILLESPIE
DTV	*Dictionnaire de Trévoux*	Paris, 1771
E	*Le Temple d'Edfou*	see: CHASSINAT
EA	*Etudes asiatiques*	Zürich
EBR	*Encyclopaedia Britannica*	London
EBOR	*Encyclopédie Bordas*	Paris
EC	*Etudes crétoises*	Paris
EE	*Epigrafia etrusca*	see: BUONAMICI
EEG	*Elementa epigraphica graecae*	see: FRANZ
EENG	*Electrical Engineering*	
EG	*Epigraphia greca*	see: GUARDUCCI
EI	*Epigraphia Indica*	Calcutta
EIS	*Encyclopédie de l'Islam*	Leyden, 1908–1938
EJ	*Encyclopaedia Judaica*	Jerusalem
EMDDR	*Entwicklung der Mathematik in der DDR*	Berlin, 1974
EMW	*Enquêtes du Musée de la vie wallone*	
ENG	*Engineering*	Paris

EP	*Encyclopédie de la Pléiade*	Paris
EPP	*L'Ecriture et la psychologie des peuples*	Paris: A. Colin, 1963
ERE	*Encyclopaedia of Religions and Ethics*	Edinburgh & New York, 1908–1921
ESIP	*Ecritures. Systèmes idéographiques et pratiques expressives*	see: CHRISTIN
ESL	*L'Espace et la lettre*	Paris: UGE, 1977
ESM	*Encyclopédie des sciences mathématiques*	Paris, 1909
EST	*Encyclopédie internationale des sciences et des techniques*	Paris, 1972
EUR	*Europe*	
EXP	*Expedition*	Philadelphia, PA
FAP	*Fontes atque Pontes. Eine Festgabe für Helmut Brunner*	see: AAT 5 (1983)
FEHP	*Facsimile of an Egyptian Hieratic Papyrus*	see: BIRCH
FIH	*Das Mathematiker-Verzeichnis im Fihrist des Ibn Abi Jakub an Nadim*	see: SUTER
FMAM	*Field Museum of Natural History*	Chicago, IL
FMS	*Frühmitelalterliche Studien*	Berlin
GIES	Glasgow Institute of Engineers and Shipbuilders in Scotland	
GKS	*Das Grabdenkmal des Königs S'ahu-Re*	see: BORCHARDT
GLA	*De sex arithmeticae practicae specibus Henrici Glareani epitome*	Paris, 1554
GLO	*Globus*	
GORILA	*Recueil des inscriptions en linéaire A*	see: GODART & OLIVIER
GT	*Ganitatilaka, by Shrîpati*	see: KAPADIA
GTSS	*Ganitasârasamgraha by Mahâvîra*	see: RANGACARYA
HAN	*Hindu-Arabic Numerals*	see: SMITH & KARPINSKI
HESP	*Hesperis. Archives berbères et Bulletin de l'Institut des Hautes Etudes marocaines*	
HF	*Historical Fragments*	see: LEGRAIN

HG	*"Hommages à H. G. Güterboch" in Anatolian Studies*	Istanbul, 1974
HGE	*Handbuch der griechischen Epigraphik*	see: LARFELD
HGS	*Histoire générale des sciences*	see: TATON
HLCT	*Haverford Library Collection of Cuneiform Tablets*	New Haven, CT
HMA	*Historia mathematica*	
HMAI	*Handbook of Middle American Indians*	Austin, TX
HNE	*Handbuch der Nordsemitischen Epigraphik*	see: LIDZBARSKI
HOR	*Handbuch der Orientalistik*	Leyden & Cologne
HP	*Hieratische Paläographie*	see: MÖLLER
HPMBS	*The History and Palaeography of Mauryan Brahmi Script*	see: UPASAK
HUCA	*Hebrew Union College Annual*, ed. S. H. Blank	
IA	*Indian Antiquary*	Bombay
IDERIC	Institut d'études et de recherches interethniques et interculturelles	Nice
IEJ	*Israel Exploration Journal*	Jerusalem
IESIS	*Indian Epigraphy and South Indian Scripts*	see: SIVARAMAMURTI
IHE	*Las Inscripciónes hebraïcas de España*	see: CANTERA & MILLAS
IHQ	*Indian Historical Quarterly*	Calcutta
IJES	*International Journal of Environmental Studies*	
IJHS	*Indian Journal of History of Science*	
IMCC	*Listes générales des Inscriptions et Monuments du Champa et du Cambodge*	see: COEDÈS & PARMENTIER
INEP	*Indian Epigraphy*	see: SIRCAR
INM	*Indian Notes and Monographs*	
INSA	*Die Inschriften Asarhaddons, König von Assyrien*	see: BORGER
IOS	*Israel Oriental Studies*	
IP	*Indische Palaeographie*	see: BUHLER
IR	*Inscription Reveal, Documents from the Time*	

	of the Bible, the Mishna and the Talmud	Jerusalem, 1973
ISCC	*Inscriptions sanskrites du Champa et du Cambodge*	see: BARTH & BERGAIGNE
IS	*Isis, revue d'histoire des sciences*	
JA	*Journal asiatique*	Paris
JAI	*Journal of the Anthropological Institute of Great Britain*	
JAOS	*Journal of the American Oriental Society*	Baltimore, MD
JAP	*Journal of Applied Psychology*	
JASA	*Journal of the American Statistical Association*	
JASB	*Journal of the Asiatic Society of Bengal*	Calcutta
JB	*Jinwen Bián*	see: RONG REN
JBRAS	*Journal of the Bombay branch of the Royal Asiatic Society*	Bombay
JCS	*Journal of Cuneiform Studies*	New Haven, CT
JEA	*Journal of Egyptian Archaeology*	London
JFI	*Journal of the Franklin Institute*	
JFM	*Jahrbuch über die Fortschritte der Mathematik*	
JHS	*Journal of Hellenic Studies*	London
JIA	*Journal of the Institute of Actuaries*	
JJS	*Journal of Jewish Studies*	London
JNES	*Journal of Near Eastern Studies*	Chicago, IL
JPAS	*Journal and Proceedings of the Asiatic Society of Bengal*	Calcutta
JRAS	*Journal of the Royal Asiatic Society*	London
JRASB	*Journal of the Royal Asiatic Society of Bengal*	
JRASI	*Journal of the Royal Asiatic Society of Great Britain and Ireland*	London
JRSA	*Journal of the Royal Society of the Arts*	London
JRSS	*Journal of the Royal Statistical Society*	
JSA	*Journal de la société des américanistes*	Paris
JSI	*Journal of Scientific Instruments*	London
JSO	*Journal de la société orientale d'Allemagne*	

KAI	*Kanaanaïsche und Aramaïsche Inschriften*	see: DONNER & RÖLLIG
KAV	*The Kashmirian Atharva-Veda*	Baltimore, MD, 1901
KR	*The Brooklyn Museum Aramaic Papyri*	see: KRAELING
KS	*Keilschriften Sargons, König von Assyrien*	see: LYON
LAA	*Annals of Archaeology and Anthropology*	Liverpool
LAL	*Lalitavistara Sûtra*	see: LAL LITRA
LAT	*Latomus*	Brussels
LAUR	*Petri Laurembergi Rostochiensis Institutiones arithmeticae*	Hamburg, 1636
LBAT	*Late Babylonian Astronomical and Related Texts*	see: PINCHES & STRASMAIER
LBDL	*Late Old Babylonian Documents and Letters*	see: FINKELSTEIN
LEV	*Levant*	
LIL	*Lîlâvatî by Bhâskata*	see: DVIVEDI
LOE	*The Legacy of Egypt*	see: HARRIS
LOK	*Lokavibhâga*	see: ANONYMOUS
MA	*Mathematische Annalen*	
MAA	*Les Mathématiques arabes*	see: YOUSHKETVITCH
MACH	*Machriq*	Baghdad
MAF	*Mémorial de l'artillerie française*	Paris
MAGW	*Mitteilungen der Anthropologischen Gesellschaft in Wien*	Vienna
MAPS	*Memoirs of the American Philosophical Society*	Philadelphia, PA
MAR	*Die Mathematiker und Astronomen der Araber und ihre Werke*	see: SUTER
MARB	*Mémoires de l'Académie royale de Bruxelles*	Brussels
MARI	*Mari. Annales de recherches interdisciplinaires*	Paris
MAS	*Memoirs of the Astronomical Society*	
MCM	*Memoirs of the Carnegie Museum*	Washington, DC
MCT	*Mathematical Cuneiform Texts*	see: NEUGEBAUER & SACHS

MDP *Mémoires de la délégation archéologique en Susiane* (vols. 1–5), continued as*: Mémoires de la Délégation en Perse* (vols. 6–13), *Mémoires de la mission archéologique en Perse* (vols.14–30), *Mémoires de la mission archéologique en Iran* (vols.31–40), *Mémoires de la délégation archéologique en Iran* (vols.41–)

MDT *Mémoires de Trévoux*

MFO *Mélanges de la Faculté orientale*   Beirut

MG Morgenländische Gesellschaft

MGA *Mathematical Gazette*

MIOG *Mitteilungen des Instituts für österreichische Geschichtsforschung*   Innsbruck

MM *Mitteilungen für Münzsammler*   Frankfurt/Main

MMA *Memoirs of the Museum of Anthropology*   Ann Arbor, MI

MMO *Museum Monographs*   Philadelphia, PA

MNRAS *Monthly Notes of the Royal Astronomical Society*

MP *Michigan Papyri*   Ann Arbor, MI

MPB *Mathematisch-physikalische Bibliothek*   Leipzig

MPCI *Mémoire sur la propagation des chiffres indiens*   SEE: WOEPKE

MSA *Mémoires de la société d'anthropologie*   Paris

MSPR *Mitteilungen aus der Sammlung der Papyrus Rainer*

MT *Mathematics Teacher*

MTI *Mathematik Tijdschrift*

MUS *Mélanges de l'université Saint-Joseph*   Beirut

N *Le Nabatéen*   SEE: CANTINEAU

NA *Nature*   London

NADG *Neues Archiv der Gesellschaft für ältere deutsche Geschichtskunde*   Hanover

NAM *Nouvelles Annales de Mathématiques*   Paris

NAT *La Nature*   Paris

NAW *Nieuw Archief voor Wiskunde*

NAWG	*Nachrichten der Akademie der Wissenschaften zu Göttingen*	Göttingen
NC	*Numismatic Chronicle*	London
NCEAM	*Notices sur les caractères étrangers anciens et modernes*	see: FOSSEY
NEM	*Notices et Extraits des Manuscrits de la Bibliothèque nationale*	Paris
NMM	*National Mathematics Magazine*	
NNM	*Numismatical Notes and Monographs*	New York
Nott	*Christophori Nottnagelii Professoris Wittenbergensis Institutionum mathematicarum*	Wittenberg, 1645
NS	*New Scientist*	London
NYT	*New York Times*	
NZ	*Numismatische Zeitschrift*	Vienna
OED	*Oxford English Dictionary*	
OIP	*Oriental Institute Publications*	Chicago, IL
OR	*Orientalia*	Rome
PA	*Popular Astronomy*	
PEQ	*Palestine Exploration Quarterly*	London
PFT	*Persepolis Fortification Tablets*	see: HALLOCK
PGIFAO	*Papyrus grecs de l'Institut français d'Archéologie orientale*	Cairo
PGP	*Paläographie der griechischen Papyri*	see: SEIDER
PHYS	*Physis*	Buenos Aires
PI	*The Paleography of India*	see: OJHA
PIB	*Paleographia Iberica*	see: BURNAM
PLMS	*Proceedings of the London Mathematical Society*	London
PLO	*Porta Linguarum Orientalum*	Berlin
PM	*The Palace of Minos*	see: EVANS
PMA	*Periodico matematico*	
PMAE	*Papers of the Peabody Museum*	Cambridge, MA
PPS	*Proceedings of the Prehistoric Society*	

PR	*Physical Review*	
PRMS	*Topographical Bibliography*	see: PORTER and MOSS
PRS	*Proceedings of the Royal Society*	London
PRU	*Le Palais royal d'Ugarit*	see: SCHAEFFER
PSBA	*Proceedings of the Society of Biblical Archaeology*	London
PSREP	*Publications de la société royale égyptienne de papyrologie*	Cairo
PTRSL	*Philosophical Transactions of the Royal Society*	London
PUMC	*Papyri in the University of Michigan Collection*	see: GARETT-WINTER
QSG	*Quellen und Studien zur Geschichte der Mathematik, Astronomie und Physik*	Berlin
RA	*Revue d'Assyriologie et d'Archéologie orientale*	Paris
RACE	Real Academia de Ciencias Exactas, Físicas y Naturales	Madrid
RAR	*Revue archéologique*	Paris
RARA	*Rara Arithmetica*	see: D. E. SMITH
RB	*Revue biblique*	Saint-Etienne
RBAAS	*Report of the British Society for the Advancement of Science*	London
RCAE	*Report of the Cambridge Anthropological Expedition to the Torres Straits*	Cambridge, 1907
RdSO	*Revista degli Studi Orientali*	Rome
RE	*Revue d'Egyptologie*	Paris
REC	*Revue des Etudes Celtiques*	Paris
REG	*Revue des Etudes Grecques*	Paris
REI	*Revue des Etudes Islamiques*	Paris
RES	*Répertoire d'épigraphie sémitique*	Paris
RFCB	*Reproduccion fac similar*	see: SELER
RFE	*Recueil de facsimilés*	see: PROU
RH	*Revue historique*	Paris

RHA	*Revue de Haute-Auvergne*	Aurillac
RHR	*Revue de l'Histoire des Religions*	Paris
RHS	*Revue d'Histoire des Sciences*	Paris
RHSA	*Revue d'Histoire des Sciences et de leurs applications*	Paris
RMM	*Revue du Monde musulman*	Paris
RN	*Revue numismatique*	Paris
RRAL	*Rendiconti della Reale Accademia dei Lincei*	Rome
RSS	*Rivista di Storia della Scienza*	Florence
RTM	*The Rock Tombs of Meir*	see: BLACKMAN
S	*Aramäische Papyrus und Ostraka*	see: SACHAU
SAOC	*Studies in Ancient Oriental Civilizations*	Chicago, IL
SC	*Scientia*	
SCAM	*Scientific American*	New York
SE	*Studi etruschi*	Florence
SEM	*Semitica*	Paris
SGKIO	*Studien zur Geschichte und Kultur des islamischen Orients*	Berlin
SHAW	*Sitzungsberichte der Heidelberger Akademie der Wissenschaften*	Heidelberg
SHM	*Sefer ha Mispar*	see: SILBERBERG
SIB	*Scripta Pontificii Instituti Biblici*	Rome
SIP	*Elements of South Indian Paleography*	see: BURNELL
SJ	*Science Journal*	
SKAW	*Sitzungsberichte der kaiserlichen Akademie der Wissenschaften*	Vienna
SM1	*Scripta Minoa, 1*	see: EVANS
SM2	*Scripta Minoa, 2*	see: EVANS & MYRES
SMA	*Scripta Mathematica*	
SME	*Studi medievali*	Turin
SMS	*Syrio-Mesopotamian Studies*	Los Angeles, CA
SPA	*La scrittura proto-elamica*	see: MERIGGI
SPRDS	*Scientific Proceedings of the Royal Dublin Society*	Dublin

SS	*Schlern Schriften*	Innsbruck
STM	*Studia Mediterranea*	Pavia
SUM	*Sumer*	Baghdad
SVSN	*Mémoires de la société vaudoise des sciences naturelles*	Lausanne
SWG	*Schriften der Wissenschaftlichen Gesellschaft in Strassburg*	Strasburg
TA	*Tablettes Albertini*	see: COURTOIS, LESCHI, PERRAT & SAUMAGNE
TAD	*Türk Arkeoloji Dergisi*	
TAPS	*Transactions of the American Philosophical Society*	
TASJ	*Transactions of the Asiatic Society of Japan*	Yokohama
TCAS	*Transactions published by the Connecticut Academy of Arts and Sciences*	New Haven, CT
TDR	*Tablettes de Drehem*	see: GENOUILLAC
TEB	*Tablettes de l'époque babylonienne ancienne*	see: BIROT
TH	*Theophanis Chronographia*	Paris, 1655
TIA	*Thesaurus Inscriptionum Aegypticum*	see: BRUGSCH
TLE	*Testimonia Linguae Etruscae*	1968
TLSM	*Transactions of the Literary Society of Madras*	Madras
TMB	*Textes mathématiques de Babylone*	see: THUREAU-DANGIN
TMIE	*Travaux et mémoires de l'Institut d'Ethnologie de Paris*	Paris
TMS	*Textes mathématiques de Suse*	see: BRUINS & RUTTEN
TRAR	*Trattati d'Aritmetica*	see: BONCOMPAGNI
TRIA	*Transactions of the Royal Irish Academy*	Dublin
TSA	*Tablettes sumériennes archaïques*	see: GENOUILLAC
TSM	*Taylor's Scientific Memoirs*	London
TTKY	*Türk Tarih Kurumu Yayinlarindan*	Ankara
TUTA	*Tablettes d'Uruk*	see: THUREAU-DANGIN
TZG	*Trierer Zeitschrift zur Geschichte und Kunst des Trierer Landes*	Trier

UAA	*Urkunden des Aegyptischen Altertums*	see: STEINDORFF
UCAE	*University of California Publication of American Archaeology and Ethnology*	Berkeley, CA
UMN	*Unterrichtsblätter für Mathematik und Naturwissenschaften*	
URK	*Hieroglyphischen Urkunden der griechischen-römischen Zeit*	see: SETHE
URK.I	*Urkunden des Alten Reichs*	see: SETHE
URK.IV	*Urkunden der 18.ten Dynastie*	see: SETHE & HELCK
UVB	*Vorläufiger Bericht über die Ausgrabungen in Uruk-Warka*	Berlin
VIAT	*Viator. Medieval and Renaissance Studies*	Berkeley, CA
WKP	*Wochenschrift für klassische Philologie*	
WM	*World of Mathematics*	
YI	*Xiao dun yin xu wenzi: yi bián*	see: DONG ZUOBIN
YOS	*Yale Oriental Series*	New Haven, CT
ZA	*Zeitschrift für Assyriologie*	Berlin
ZAS	*Zeitschrift für Aegyptische Sprache und Altertumskunde*	Berlin
ZDMG	*Zeitschrift der Deutschen Morgenländischen Gesellschaft*	Wiesbaden
ZDP	*Zeitschrift des Deutschen Palästina-Vereins*	Leipzig & Wiesbaden
ZE	*Zeitschrift für Ethnologie*	Braunschweig
ZKM	*Zeitschrift für die Kunde des Morgenlandes*	Göttingen
ZMP	*Zeitschrift für Mathematik und Physik*	
ZNZ	*Zbornik za Narodni Zivot i Obicaje juznih Slavena*	Zagreb
ZOV	*Zeitschrift für Osterreichische Volkskunde*	Vienna
ZRP	*Zeitschrift für Romanische Philologie*	Tübingen

BIBLIOGRAPHY

ABBREVIATIONS USED ARE LISTED ON PP. VIII–XVI ABOVE

*(Dates in brackets are the first edition when a more
recent edition has been published)*

1. WORKS AVAILABLE IN ENGLISH

N. ABBOT, "(No title)". *JRAS* : pp.277–80. London, 1938.

A. ABOUL-SOUF, "Tell-es-Sawwan. Excavations ... Spring 1967". *SUM* 24: pp.3ff. Baghdad, 1968.

Y. AHARONI, "Hebrew ostraca from Tel Arad". *IEJ* 16: pp.1–10. Jerusalem, 1966.

Y. AHARONI, "The Use of Hieratic Numerals in Hebrew Ostraca and the Shekel". *BASOR* 184: pp.13–19. Ann Arbor, 1966.

Y. AHARONI, *Arad Inscriptions*. Jerusalem: Bialik Institute/Israel Exploration Society, 1975.

Q. AHMAD, "A Note on the Art of Composing Chronograms". *CAPIB* 10: pp.367–74. Patna, 1973.

AL-BIRUNI, *Chronology of Ancient Nations*. Transl. E. C. Sachau. London, 1879.

AL-BIRUNI (1888), *Alberuni's India, an account of the Religion, Philosophy, Literature, Geography, Chronology, Astronomy, Customs, Laws and Astrology of India about AD 1030*. Transl. E. C. Sachau. London: Paul, Trench & Trübner, 1910.

A. A. AL-DAFFA, *The Muslim Contribution to Mathematics*. Atlantic Highlands, NJ: Humanities Press, 1977.

E. ALFÖLDI-ROSENBAUM, "The Finger Calculus in Antiquity and the Middle Ages". *FMS* 5: pp.1–9. Berlin, 1971.

AL-JAZZARI (Ibu al-Razzaz), *The Book of Knowledge of Ingenious Mechanical Devices*. Transl. D.R.Hill. Dordrecht, 1974.

J. DE ALWIS, *A Grammar of the Singhalese Language*. Colombo, 1852.

A. ANBOUBA, "Al-Sammaw'al". In: C. Gillespie, *DSB* 12: pp.91–4. New York: Scribners, 1970.

ANONYMOUS, *Sûryasiddhânta*. Bombay, 1955.

ANONYMOUS, *Lokavibhaga*. Sholapur: P. Balchandra Siddhanta-Shastri, 1962.

F. ANTON, *The Art of the Maya*. New York: Putnam, 1970.

E. ARNOLD, *Gulistân*. New York, 1899.

N. AVIGAD, "A Bulla of Jonathan the High Priest". *IEJ* 25: pp.8–12. Jerusalem, 1975.

L. M. BAGGE, "The Early Numerals". *CR* 20: pp.259–67.

G. H. BAILLIE, *Clocks and Watches. An Historical Bibliography*. Vol. I. London: Holland Press, 1978.

R. BAKER, *The Science Reference Library ... Inventors and Inventions that have Changed the Modern World*. London: British Library, 1976.

W. BALASINSKI and W. MROWKA, "On Algorithms of Arithmetical Operations". *BAPS* 5: pp.803–4. Warsaw, 1957.

JAMES DYER BALL, *Things Chinese*. London: S. Low & Marston, 1900.

W. W. R. BALL, *A Short Account of the History of Mathematics*. London: Macmillan, 1888.

V. H. BARKLEY, *Historia Numorum. A Manual of Greek Numismatics*. Oxford: Clarendon Press, 1911.

FRANCIS PIERREPONT BARNARD, *The Casting Counter and the Counting Board: a chapter in the history of numismatics and early arithmetic*. Oxford: Clarendon Press, 1916.

R. D. BARNETT, "The Hieroglyphic Writing of Urartu". In: *Hommages à H. G. Güterbock*: pp.43ff. Istanbul: Anatolian Studies, 1974.

G. A. BARTON, "Documents from the Temple Archives of Telloh". *HLCT*, Pt. I. New Haven, 1918.

E. C. BAYLEY, "On the Genealogy of Modern Numerals". *JRASI* 14: p.335. London, 1847.

E.A. BECHTEL, "Finger-Counting among the Romans in the Fourth Century". *CPH* 4: pp.25–31. Chicago, 1909.

ERIC TEMPLE BELL (1937), *Men of Mathematics*. New York: Dover, 1965.

ERIC TEMPLE BELL, *The Development of Mathematics*. New York: McGraw Hill, 1940.

C. BENDALL, "Table of Numerals". *CSKBM*. London, 1902.

P. K. BENEDICT, "Chinese and Thai Numeratives". *JAOS* 65. Baltimore, 1945.

S. R. BENEDICT, *A Comparative Study of the Early Treatises Introducing the Hindu Art of Reckoning into Europe*. PhD thesis, University of Michigan at Ann Arbor, 1914.

ARIEL BENSION, *The Zohar in Moslem and Christian Spain*. London: Routledge, 1922. New York: Hermon, 1974.

K. T. BHARAT, *Vedic Mathematics*. Delhi, 1970.

ROGER BILLARD, "Āryabhata and Indian Astronomy". *IJHS* 12/2.

S. BIRCH, *Facsimile of an Egyptian Hieratic Papyrus of the Reign of Ramses II*. London: British Museum, 1876.

A. N. BLACKMAN, "The Rock Tombs of Meir". *ASE* 25. London, 1924.

J. BONNYCASTLE, *Introduction to Arithmetic*. London, 1810.

J. BORDAZ, "The Suberde Excavations in Southwest Turkey. An Interim Report". *TAD* 17: pp.43–61. 1969.

J. BOTTERO, E. CASSIN and J. VERCOUTTER, *The Near East: The Early Civilizations*. London: Weidenfeld & Nicolson, 1967.

J.G. BOURKE, "Medicine Men of the Apache". *ARBE* 9: pp.555ff. Washington, DC, 1892.

CHARLES P. BOWDITCH, *The Numeration, Calendar Systems and Astronomical Knowledge of the Maya*. Cambridge: Cambridge University Press, 1910.

C. B. BOYER, "Fundamental Steps in the development of numeration". *IS* 35: pp.153–69. 1944.

C. B. BOYER (1968), *A History of Mathematics*. New York: Wiley. Revised by Uta C. Merzbach, new ed., 1991.

R. J. BRAIDWOOD and B. HOWE, "Prehistoric Investigations in Iraqi Kurdistan". *SAOC* 3. Chicago, 1960.

P. N. BRETON, *Illustrated History of Coins and Tokens relating to Canada*. Montreal: Breton, 1894.

W. C. BRICE, "The Writing System of the Proto-Elamite Account Tablets". *BJRL* 45: pp.15–39. Manchester, 1962.

J. A. BRINKMAN, "Mesopotamian chronology of the historical period". *AM* : pp.335–7.

SIR JAMES BROOKE, *Ten Years in Sarawak*. Vol. I. pp. 139–40. [The author (1803–1868) was Rajah of Sarawak.]

BURGESS and WHITNEY, "Translation of the Sûrya-Siddhânta". *JAOS* 6: pp.141–498. Baltimore, 1860.

R. BURNAM, "A Group of Spanish Manuscripts". *BHI* 22: pp.229–33. Bordeaux, 1920.

A. C. BURNELL, *Elements of South Indian Paleography from the Fourth to the Seventeenth Centuries AD*. London, 1878.

C. CAHEN and R. S. SERJEANT, "A fiscal survey of medieval Yemen". *ARAB* 4: pp.23–32. Leyden, 1957.

FLORIAN CAJORI (1897), *A History of Mathematics*. New York, 1980.

FLORIAN CAJORI, "A Notable Case of Finger-Counting in America". *IS* 8: pp.325–7. 1926.

C. A. CAMPBELL, "A first season of excavations at the Urartian citadel of Kayalidere". *ANS* 16: pp.89ff. London, 1966.

A. D. CAMPBELL, *A Grammar of the Teloogoo Language*. Madras, 1820.

F. CAREY, *A Grammar of the Burman Language*. Serampore, 1814.

F. H. CHALFANT, "Chinese Numerals". *MCM* 4. Washington, DC, 1906.

J. CHALMERS, "Maipua and Namaua Numerals". *JAI* 27: pp.141ff. 1898.

B. H. CHAMBERLAIN, "A Quinary System of Notation Employed in Luchu". *JAI* 27: pp.338–95. 1898.

J. CHARTER, *A Grammar of the Cingalese Language*. Colombo, 1815.

A. B. CHASE, *The Rhind Mathematical Papyrus*. Oberlin, 1927.

D. C. CHENG, "The Use of Computing Rods in China". *AMM* 32: pp.492–9. 1925.

EDWARD CHIERA, *They Wrote on Clay*. Chicago: University of Chicago Press, 1938.

J. E. CIRLOT, *A Dictionary of Symbols*. Transl. Jack Sage. New York: Philosophical Library, 1962.

W. E. CLARCK, *The Âryabhatiya of Âryabhata*. Chicago, 1930.

CODRINGTON, *Melanesian Languages*. Pp.211–12. London, (no date).

MICHAEL D. COE, *The Maya Scribe and his World*. New York: Grolier, 1973.

MICHAEL D. COE, *Breaking the Maya Code*. New York: Thames and Hudson, 1992.

I. B. COHEN, *Album of Science from Leonardo to Lavoisier, 1450–1800*. New York: Scribner, 1980.

H. T. COLEBROOKE, *Algebra with Arithmetic and Mensuration from the Sanscrit*. London, 1817.

L.L.CONANT, *The Number Concept, its Origin and Development*. New York: Simon & Schuster, 1923.

L. L. CONANT, "Counting". *WM* 1: pp.432–41. 1956.

H. DE CONTENSON, "New Correlations between Ras Shamra and Al-'Amuq". *BASOR* 172: pp.35–40. Ann Arbor, 1963.

H. DE CONTENSON, "Tell Ramad". *ARCHN* 24: pp.278-85. New York, 1971.

C. S. COON, "Cave Exploration in Iran". *MMO* : 75. Philadelphia, 1951.

A. C. CROMBIE, *History of Science from St Augustine to Galileo*. London, 1961.

CUNNINGHAM, "Book of Indian Eras". *JRAS* : pp.627ff. London, 1913.

F. H. CUSHING, "Manual Concepts: A Study of the Influence of Hand-Usage on Culture-Growth". *AA* 5: pp.289–317. Menasha, Wis., 1892.

F. H. CUSHING, "Zuñi Breadstuffs". *INM* 8: pp.77ff. 1920.

TOBIAS DANTZIG (1930), *Number: The Language of Science*. New York: Macmillan, 1967.

B. DATTA, "The Scope and Development of Hindu Ganita". *IHQ* 5: pp.497–512. Calcutta, 1929.

B. DATTA and A.N. SINGH (1938), *History of Hindu Mathematics*. Bombay: Asia Publishing, 1962.

M. DAUMAS, Ed. (1969), *A History of Technology and Invention*. New York: Crown,1979.

P. DÉDRON and J. ITARD, *Mathematics and Mathematicians*. Transl. J. Field. Milton Keynes: Open University Press, 1974.

CHARLES DICKENS, Speech to the Administrative Reform Association. In: K. J. Fielding, *The Speeches of Charles Dickens*: pp.204–5. Hemel Hempstead: Harvester Wheatsheaf, 1855.

DAVID DIRINGER (1948), *The Alphabet. A Key to the History of Mankind*. London: Hutchinson, 1968. New York: Funk & Wagnalls, 1968.

R. B. DIXON, "The Northern Maidu". *BAMNH* 17: pp.228 & 271. New York, 1905.

B. DODGE, *The Fihrist of Al Nadim*. New York, 1970.

R. DRUMMOND, *Illustrations of the Grammatical Parts of the Guzerattee, Mahratta and English Languages*. Bombay, 1808.

S. DVIVEDI, *Trishâtika by Shrîdharâchârya*. Benares, 1899.

S. DVIVEDI, "Brâhmasphutasiddhânta, by Brahmagupta". *The Pandit, Vol.* 23–24. Benares, 1902.

S. DVIVEDI, *Lîlavâti by Bhâskarâ, II*. Benares, 1910.

S. DVIVEDI, *Siddhântatattvaviveka, by Kamâlakara*. Benares, 1935.

S. DVIVEDI and G. THIBAUT (1889), *The Pañchasiddhântikâ, the Astronomical Work of Varâha Mihira*. Lahore: Motilal Banarsi Dass, 1930.

W. C. EELS, "Number Systems of North American Indians". *AMM* 20: pp.293–8. 1913.

R. W. ERICHSEN, "Papyrus Harris I". *BAE* 5. Brussels, 1933.

R. W. ERICHSEN, Ed., *Relative Chronologies in Old World Archaeology*. Chicago, 1954.

SIR ARTHUR EVANS, *Scripta Minoa*. Vol. 1. Oxford: Clarendon Press, 1909.

SIR ARTHUR EVANS, *The Palace of Minos*. 5 vols. London: Macmillan, 1921–1936.

SIR ARTHUR EVANS and SIR JOHN MYRES, *Scripta Minoa*. Vol. 2. Oxford: Clarendon Press, 1952.

HOWARD EVES, *An Introduction to the History of Mathematics*. New York: Harcourt Brace, 1990.

H. W. FAIRMAN, "An introduction to the study of Ptolemaic signs and their values". *BIFAO* 43. Cairo (no date).

JOHN FAUVEL and JEREMY GRAY (1987), *The History of Mathematics. A Reader*. Contains many of the texts quoted by Ifrah, from Aristotle to Nicomacchus of Gerasa and Bacon. Milton Keynes: Open University, 1987.

P. L. FAYE, "Notes on the Southern Maidu". *UCAE* 20: pp.44ff. Berkeley, 1923.

A. FELDMAN and P. FOLD, *Scientists and Inventors. The People who Made Technology, from the earliest times to the present*. London, 1979.

L. N. G. FILON, "The Beginnings of Arithmetic". *MGA* 12: p.177. 1925.

J. J. FINKELSTEIN, "Late Old Babylonian Documents and Letters". *YOS* 13. New Haven, 1972.

J. C. FLEET, "The Last Words of Asoka". *JRAS* : pp.981–1016. London, 1909.

W. FORBES, *A Grammar of the Goojratee Language*. Bombay, 1829.

E. FÖRSTEMANN, "Commentary on the Maya manuscript in the Royal Library of Dresden". *PMAE* 6. Cambridge, 1906.

L. FRÉDÉRIC, *Encyclopaedia of Asian Civilizations*. Paris, 1977–1987.

D. H. FRENCH, "Excavations at Can Hasan". *ANS* 20: p.27. London, 1970.

R. FUJISAWA, "Note on the Mathematics of the Old Japanese School". In: Duporck, *op. cit.*: p. 384. 1902.

C. J. GADD, "Omens expressed in numbers". *JCS* 21: pp.52ff. New Haven, 1967.

T.V. GAMKRELIDZE and V. V. IVANOV, *Indo-European and the Indo-Europeans*. In Russian. Tbilisi, 1984.

SOLOMON GANDZ, "The Knots in Hebrew Literature". *IS* 14: pp.189ff. 1930.

SOLOMON GANDZ, "The Origin of the Ghubar Numerals". *IS* 16: pp.393–424. 1931.

SOLOMON GANDZ, *Studies in Hebrew Astronomy and Mathematics*. New York: Ktav, 1970.

N. R. GANOR, "The Lachich Letters". *PEQ* 99: pp.74–7. London, 1967.

A. GARDINER, *Egyptian Grammar*. Oxford: Oxford University Press, 1950.

J. GARETT-WINTER, *Papyri in the University of Michigan Collection*. (University of Michigan Studies, Vol. 3). Ann Arbor, 1936.

W. GATES, *Codex of Dresden*. Baltimore, 1932.

I. J. GELB, *The Study of Writing. The Foundations of Grammatology*. Chicago: University of Chicago Press, 1963.

L.T. GERATY, "The Khirbet El-Kôm Bilingual Ostracon". *BASOR* 220: pp.55–61. Ann Arbor, 1975.

J. C. L. GIBSON, *Textbook of Syrian Semitic Inscriptions*. Vol. 1. Oxford: Clarendon Press, 1971.

C. GILLESPIE, Ed., *Dictionary of Scientific Biography*. New York: Scribners, 1970–1980.

RICHARD J. GILLINGS (1972), *Mathematics in the Time of the Pharaohs*. New York: Dover Press, 1982.

S. R. K. GLANVILLE, "The Mathematical Leather Roll in the British Museum". *JEA* 13: pp.232–9. London, 1927.

A. GLASER, *A History of Binary Numeration*. Philadelphia: Tomash, 1971.

B. R. GOLDSTEIN, "The Astronomical Tables of Levi Ben Gerson". *TCAS* 45. New Haven, 1974.

R. GOPALAN, *History of the Pallava of Kâñchî*. Madras, 1928.

C. H. GORDON, "Ugaritic Textbook". *AOR* 38/7: pp.42–52. Rome, 1965.

J. GOUDA, *Reflections on the Numerals "one" and "two" in Ancient Indo-European Language*. Utrecht, 1953.

J. GOW, *A Short History of Greek Mathematics*. New York: Chelsea, 1968.

F. L. GRIFFITH, "The Rhind Mathematical Papyrus". *PSBA* 13: pp.328ff. See also Vol. 14, pp. 26ff.; Vol. 16, pp.164ff. London, 1891.

A. GROHMANN, *Arabic Papyri in the Egyptian Library*. Vol. 4. Cairo, 1962.

B. GUNN, "Finger Numbering in the Pyramid Texts". *ZAS* 57: pp.283ff. Berlin, 1922.

B. GUNN, "The Rhind Mathematical Papyrus". *JEA* 12: p.123. London, 1926.

L. V. GURJAR, *Ancient Indian Mathematics and Vedha*. Poona, 1947.

A. C. HADDON, "The Ethnography of the Western Tribes of the Torres Straits". *JAI* 19: pp.305ff. 1890.

J. B. S. HALDANE, *Science and Indian Culture*. Calcutta, 1966.

F. HALL, *The Sûrya-Siddhânta, or an Ancient System of Hindu Astronomy*. Amsterdam, 1975.

H. HALL, *The Antiquities of the Exchequer*. London: Elliott Stock, 1898.

R. T. HALLOCK (1969). *Persepolis Fortification Tablets*. Oriental Institute
Publications, Vol. 42. Chicago: University of Chicago Oriental Institute, 1969.

G. B. HALSTED, *On the Foundation and Technique of Arithmetic*. Chicago, 1912.

J. R. HARRIS, Ed., *The Legacy of Egypt*. Oxford: Oxford University Press, 1971.

A. P. HATCH, *An Album of Dated Syriac Manuscripts*. Boston, 1946.

E. C. HAWTREY, "The Lengua Indians of the Paraguayan Chaco". *JAI* 31:
pp.296ff. 1902.

T. L. HEATH, *A History of Greek Mathematics*. Oxford: Clarendon Press, 1921.

B. W. HENDERSON, *Selected Historical Documents of the Middle Ages*, pp.20ff.
London, 1892.

T. HIDEOMI, *Historical Development of Science and Technology in Japan*. Tokyo, 1968.

D. R. HILL, *The Book of Knowledge of Ingenious Mechanical Devices*. Transl. D. R. Hill.
Dordrecht: Reidel, 1974.

G. F. HILL, "On the early use of Arabic numerals in Europe". *ARCHL* 62:
pp.137–90. London, 1910.

G. F. HILL, *The Development of Arabic Numerals in Europe*. Oxford:
Clarendon Press, 1915.

E. HINCKS, "On the Assyrian Mythology". *TRIA* 12/2: pp.405ff. 1855.

E. W. HOPKINS, *The Religions of India*. Boston, 1898.

A. W. HOWITT, "Australian Message Sticks and Messengers". *JAI* 18:
pp.317–19. 1889.

T. P. HUGUES, *Dictionary of Islam*. London, 1896.

G. HUNT, "Murray Island, Torres Straits". *JAI* 28: pp.13ff. 1899.

GEORGES IFRAH, *From One to Zero*. New York: Viking, 1985.

B. INDRAJI, "On Ancient Nagari Numeration". *JBRAS* 12: pp.404ff. Bombay, 1876.

B. INDRAJI, "The Western Kshatrapas". *JBRAS* 22: pp.639–62. Bombay, 1890.

R. A. K. IRANI, "Arabic Numeral Forms". *CENT* 4: pp.1–13. Copenhagen, 1955.

R. V. IYER, "The Hindu Abacus". *SMA* 20: pp.58–63. 1954.

J. JANSSEN, *Commodity Prices from the Rasmessid Period*. Leyden: E. J. Brill, 1975.

H. JENKINSON, "Exchequer Tallies". *ARCHL* 72: pp.367–80. London, 1911.

H. JENKINSON, "Medieval Tallies, public and private". *ARCHL* 74: pp.289–351.
London, 1913.

L. KADMAN, "The Coins of the Jewish War of 66-73 CE". *CNP* 2/3.
Jerusalem, 1960.

H. KALMUS, "Animals as Mathematicians". *NA* 202: pp.1156–60. London, 1964.

H. J. KANTOR, "The relative chronology of Egypt". In: Erichsen, *Relative
Chronologies*: pp.10ff. Chicago, 1954.

H. R. KAPADIA, *Ganitatilaka by Shrîpati*. (Gaikward Sanskrit Series). Baroda:
Gaikward, 1935.

L. C. KARPINSKI, "Hindu numerals in the Fihrist". *BMA* 11: pp.121-4. 1911.

L. C. KARPINSKI, *Robert of Chester's Algoritmi de numero Indorum*. New York, 1915.

L. C. KARPINSKI, *The History of Arithmetic*. New York: Russell & Russell, 1965.

J. T. KAUFMAN, "New Evidence for Hieratic Numerals". *BASOR* 188: pp.67–9. Ann
Arbor, 1967.

G. R. KAYE, "Notes on Indian Mathematics". *JPAS* 8: pp.475–508. Calcutta, 1907.

G. R. KAYE, "The Use of the Abacus in Ancient India". *JPAS* 4: pp.293–7.
Calcutta, 1908.

G. R. KAYE, *Indian Mathematics*. Calcutta & Simla, 1915.

G. R. KAYE, *Bakhshâlî Manuscript. A Study in Mediaeval Mathematics.* Lahore, 1924.

G. R. KAYE, *Hindu Astronomy.* Calcutta, 1924.

R. G. KENT, "Old Persian Grammar, texts, lexicon". *AOS* . New Haven, 1953.

V. J. KERKHOF, "An Inscribed Stone Weight from Schechem". *BASOR* 184: pp.20ff. Ann Arbor, 1966.

E. K. KINGSBOROUGH, *Antiquities of Mexico.* 3 vols. London, 1831–1848.

G. KLEINWÄCHTER, "The Origin of the Arabic Numerals". *CHR* 11: pp.379–81. Cont. in Vol. 12: pp.28–30. 1882.

M. KLINE, *Mathematical Thought from Ancient to Modern Times.* Oxford: Oxford University Press, 1972.

Y. V. KNOROZOV, "The Problem of the Study of Maya Hieroglyphic Writing". *AA.* Menasha, Wisc., 1946.

C. G. KNOTT, "The Abacus in its Historic and Scientific Aspects". *TASJ* 14. Yokohama, 1886.

O. KOEHLER, "The Ability of Birds to count". *WM* 1: pp.489ff. 1956.

E. KRAELING, *The Brooklyn Museum Aramaic Papyri.* New Haven, 1953.

R. LABAT, "Elam and Western Persia, c. 1200–1000 BC". In: *CAH*: Cambridge, 1963

R. LABAT, "Elam, c. 1600–1200 BC". In: *CAH*: Cambridge, 1963

A. TERRIEN DE LACOUPERIE, "The Old Numerals, the Counting Rods and the Swan Pan in China". *NC* 3. London, 1888.

S. LANGDON, *JRAS* : pp.169–73. London, 1925.

A. LANGSDORF and D. F. McCOWN, "Tall-i-Bakun". *OIP* 59. Chicago, 1942.

T. LATTER, *A Grammar of the Language of Burma.* Calcutta, 1845.

H. P. LATTIN, "The Origins of our Present System of Notation according to the Theories of Nicolas Bubnov". *IS* 19. 1933.

J. D. LEECHMAN and M. R. HARRINGTON, "String Records of the Northwest". *INM* 9. 1921.

L. LEGRAIN, *Historical Fragments.* (Publications of the Babylonian Section, Vol. 13). Philadelphia: University Museum, 1922.

R. LEMAY, "The Hispanic Origin of our Present Numeral Forms". *VIAT* 8: pp.435–59. University of California Press, 1977.

J. LESLIE, *The Philosophy of Arithmetic.* Edinburgh: Constable, 1817.

H. LEVEY and PETRUCK, *Principles of Hindu Reckoning.* Madison, 1965.

C. LEVIAS, Ed., *Jewish Encyclopedia.* New York, 1905.

LUCIEN LÉVY-BRÜHL (1922), *How Natives Think.* Transl. Lilian A. Clare. New York: Washington Square Press, 1966.

SAUL LIEBERMANN, *Hellenism in Jewish Palestine.* New York: Jewish Theological Seminary, 1950.

S. J. LIEBERMANN, "Of Clay Pebbles, Hollow Clay Balls, and Writing". *AJA* 84: pp.339–58. New York (no date).

S. J. LIEBERMANN, "A Mesopotamian Background for the so-called Aggadic 'Measures' of Biblical Hermeneutics?" *HUCA* 58: pp.157–223. 1987

A. LIVINGSTON, *Mystical and Mythological Explanatory Works of Babylonian and Assyrian Scholars.* Oxford: Clarendon Press, 1986.

L. LELAND LOCKE, *The Ancient Quipo or the Peruvian Knot Record.* New York: American Museum of Natural History, 1923.

W. K. LOFTUS, *Travels and Researches in Chaldaea and Susiana.* London, 1857.

W. J. MacGee, "Primitive Numbers". *ARBE* 19: pp.821–51. Washington, DC, 1897.

E. J. H. MacKay, *Further Excavations at Mohenjo Daro*. Delhi, 1937–38.

G. Mallery, "Picture Writing of the American Indians". *ARBE* 10: pp.223ff. Washington, DC, 1893.

M. E. L. Mallowan, *The Development of Cities from Al Ubaid to the end of Uruk V.* Vol. 1. Cambridge: Cambridge University Press, 1967.

J. Marshall, *Mohenjo Daro and the Indus Civilisation*. London, 1931.

D. D. Mehta, *Positive Science in the Vedas*. New Delhi, 1974.

Karl W. Menninger (1957), *Number Words and Number Symbols: A cultural history of numbers*. Transl. P. Broneer. Boston: MIT Press, 1969.

B. D. Meritt, H. T. Wade-Gery and M. F. McGregor, *The Athenian Tribute Lists*. Cambridge: Harvard University Press, 1939.

H. Midonick, *The Treasury of Mathematics*. New York: Penguin, 1965.

R. A. Miller, *The Japanese Language*. Chicago, 1967.

Mingana, "Arabic Numerals". *JRAS* : pp.139–48. London, 1937.

B. Misra, *Siddhânta-Shiromani by Bhâskâra II*. Vol. 1. Calcutta, 1932.

R. Lal Mitra, *Lalitavistara Sûtra*. Calcutta, 1877.

P. Moon, *The Abacus*. New York: Gordon & Breach, 1971.

A. M. T. Moore, "The Excavations at Tell Abu Hureyra". *PPS* 41: pp.50–77. 1975.

A. de Morgan, *Arithmetical Books from the Invention of Printing up to the Present Time*. London: Taylor & Walton, 1847.

Sylvanus G. Morley, *An Introduction to the Study of Maya Hieroglyphs*. Washington, DC: Bureau of American Ethnology, 57. 1915.

Sylvanus G. Morley and F. R. Morley, "The Age and Provenance of the Leyden Plate". *CAAH* 5/509. Washington, DC, 1939.

P. Mortenson, "Excavations at Tepe Guran". *AAR* 34: pp.110–21. Copenhagen, 1964.

S. Moscati, Ed., *An Introduction to the Comparative Grammar of the Semitic Languages*. Wiesbaden, 1969.

J. V. Murra, *The Economic Organisation of the Inca State*. Chicago, 1956.

R. Van Name. "On the Abacus of China and Japan". *JAOS* X: p.cx. Baltimore (no date).

J. Naveh, "Dated Coins of Alexander Janneus". *IEJ* 18: pp.20–5. Jerusalem, 1968.

J. Naveh, "The North-Mesopotamian Aramaic Script-type". *IEJ* 2: pp.293–304. Jerusalem, 1972.

Joseph Needham, "Mathematics and the Science of the Heavens and the Earth". In: J. Needham, *Science and Civilization in China*, 3. Cambridge: Cambridge University Press, 1959.

Oscar Neugebauer, "The Rhind Mathematical Papyrus". *MTI* : pp.66ff. 1925.

Oscar Neugebauer (1952), *The Exact Sciences in Antiquity*. Princeton: Princeton University Press, 1969.

Oscar Neugebauer, *Astronomical Cuneiform Texts*. London, 1955.

Oscar Neugebauer, "Studies in Byzantine Astronomical Terminology". *TAPS* 50/2: p.5. 1960.

Oscar Neugebauer, *A History of Ancient Mathematical Astronomy*. Berlin & New York, 1975.

O. Neugebauer and D. Pingree, *The Pañcasiddhantika of Varahamihira*. Copenhagen: Munksgård, 1970–1971.

OSCAR NEUGEBAUER and A. SACHS, "Mathematical Cuneiform Texts". *AOS* 29. New Haven, 1945.

E. NORDENSKJÖLD, "Calculation with years and months in the Peruvian Quipus". *CETS* 6/2. 1925.

E. NORDENSKJÖLD, "The Secret of the Peruvian Quipus". *CETS* 6/1. 1925.

M. D' OCAGNE (1893), *Simplified Calculation*. Transl. M. A. Williams & J. Howlett. Cambridge: MIT Press, 1986.

G. H. OJHA, *The Paleography of India*. Delhi, 1959.

A. L. OPPENHEIM, "An Operational Device in Mesopotamian Bureaucracy". *JNES* 18: pp.121–8. Chicago, 1959.

A. L. OPPENHEIM, *Ancient Mesopotamia. A Portrait of a Dead Civilization*. Chicago: University of Chicago Press, 1964.

O. ORE, *Number Theory and its History*. New York: McGraw-Hill, 1948.

T. OZGÜÇ, "Kültepe and its vicinity in the Iron Age". *TTKY* 5/29. Ankara, 1971.

R. A. PARKER, *Demotic Mathematical Papyri*. London, 1972.

A. PARPOLA, S. KOSKIENNEMI, S. PARPOLA and P. AALTO, *Decipherment of the Proto-Dravidian Inscriptions of the Indus Civilisation*. Copenhagen, 1969.

J. PEET, *A Grammar of the the Malayalim Language*. Cottayam, 1841.

T. E. PEET, *The Rhind Mathematical Papyrus*. London: British Museum, 1923.

SAMUEL PEPYS, *The Shorter Pepys*. Ed. Robert Latham. Berkeley: University of California Press, 1985.

FREDERICK PETERSON, *Ancient Mexico*. New York: Capricorn, 1962.

T. G. PINCHES and J. N. STRASMAIER, *Late Babylonian Astronomical and Related Texts*. Providence, 1965.

R. S. POOLE, *A Catalogue of Greek Coins in the British Museum*. Bologna, 1963.

B. PORTER and R. L. B. MOSS, *Topographical Bibliography of Ancient Egyptian Hieroglyphs*. Oxford: Clarendon Press, 1927–1951.

M. A. POWELL, "Sumerian Area Measures and the Alleged Decimal Substratum". *ZA* 62/2: pp.165–221. Berlin, 1972.

M. A. POWELL, "The Antecedents of Old Babylonian Place Notation". *HMA* 3: pp.417–39. 1976.

M. A. POWELL, "Three Problems in the History of Cuneiform Writing". *VL* 15/4: pp.419–40. 1981.

S. POWERS, "Tribes in California". *CNAE* 3: pp.352ff. Washington, DC, 1877.

J. PRINSEP, "On the Inscriptions of Piyadasi or Asoka". *JPAS*. Calcutta, 1838.

T. PROUSKOURIAKOFF, *An Album of Maya Architecture*. Vol. 558. Washington: Carnegie Institute, 1946.

T. PROUSKOURIAKOFF, "Sculpture and the Major Arts of the Maya Lowland". *HMAI* 2. Austin, 1965.

J. M. PULLAM, *The History of the Abacus*. London: Hutchinson, 1968.

J. E. QUIBBEL, *Hierakonpolis*. London: Quaritch, 1900.

A. F. RAINEY, "The Samaria Ostraca in the light of fresh evidence". *PEQ* 99: pp.32–41. London, 1967.

M. RANGACARYA, *Ganitâsarasamgraha, by Mahâv"ra*. Madras, 1912.

R. RASHED, "Al-Karaji". In: C. Gillespie, *DSB*, 7: pp.240–6. New York, 1970.

C. RAWLINSON, "Notes on the Early History of Babylonia". *JRAS* 15: pp.215–59. London, 1855.

ROBERT RECORDE, *The Grounde of Artes*. London, 1558.

G. A. REISNER, *A History of the Giza Necropolis*. Vol. 1. London & Oxford, 1942.

C. T. E. RHENIUS, *A Grammar of the Tamil Language*. Madras, 1846.

J. L. RICHARDSON, "Digital Reckoning Used among the Ancients". *AMM* 23: pp.7–13. 1916.

J. T. ROGERS, *The Story of Mathematics*. London: Hodder & Stoughton, 1966.

BERTRAND RUSSELL (1928), *Introduction to the Philosophy of Mathematics*. London: Allen & Unwin, 1967.

A. SAIDAN, "The Earliest Extant Arabic Arithmetic". *IS* 57. 1966.

S. SAMBURSKI, "On the Origin and Significance of the term Gematria". *JJS* 29: pp.35–8. London, 1978.

K. V. SARMA, *Grahachâranibandhana, by Haridatta*. Madras, 1954.

K. V. SARMA, *Drgganita, by Parameshvara*. Hoshiarpur, 1963.

K. V. SARMA, *Siddhântadarpana, by Nîlakanthasomayâjin*. Madras: Adyar Library and Research Centre

K. V. SARMA and K. S. SASTRI, *Vâkyapañchâdhyayi*. Madras, 1962.

G. SARTON, *Introduction to the History of Science*. Baltimore: Johns Hopkins, 1927.

G. SARTON, "The First Explanation of Decimal Fractions in the West". *IS* 23: pp.153–245. 1935.

B. D. SASTRI, *Siddhântashiromani, by Bhâskarâchârya*. Benares, 1929.

K. S. SASTRI, *The Karanapadhati by Putumanasomayâjin*. Trivandrum, 1937.

L. SATTERTHWAITE, *Concepts and Structures of Maya Calendrical Arithmetic*. Philadelphia, 1947.

F. C. SCESNEY, *The Chinese Abacus*. New York, 1944.

DENISE SCHMANDT-BESSERAT, "The earliest precursor of writing". *SCAM (June)* : pp.38–47. New York, 1978.

DENISE SCHMANDT-BESSERAT, "Reckoning before writing". *ARCHN* 32/3: pp.22–31. New York, 1979.

DENISE SCHMANDT-BESSERAT, "The envelopes that bear the first writing". *TC* 21/3: pp.357–85. Austin, 1980.

DENISE SCHMANDT-BESSERAT, "From Tokens to Tablets". *VL* 15/4: 321–44. 1981.

J. F. SCOTT, *A History of Mathematics*. London: Taylor & Francis, 1960.

R. B. Y. SCOTT, "The Scale-Weights from Ophel". *PEQ* 97: pp.128–39. London, 1965.

J. B. SEGAL, "Some Syriac Inscriptions of the 2nd–3rd century AD". *BSOAS* 16: pp.24–8. London, 1954.

A. SEIDENBERG, "The Ritual Origin of Counting". *AHES* 2: pp.1–40. 1962.

E. SENART, "The Inscriptions in the cave at Karle". *EI* 7: pp.47–74. Calcutta, 1902.

E. SENART, "The Inscriptions in the cave at Nasik". *EI* 8: pp.59–96. Calcutta, 1905.

R. SHAFER, "Lycian Numerals". *AROR* 18: pp.250–61. Prague, 1950.

E. M. SHOOK, "Tikal Stela 29". *Expedition* 2: pp.29–35. Philadelphia, 1960

K. S. SHUKLA and K. V. SARMA, "Âryabhatiya of Âryabhata". *Indian National Science Academy* 77. Delhi, 1976.

L. G. SIMONS, "Two Reckoning Tables". *SMA* 1: pp.305–8. 1932.

C. SINGER, *A History of Technology*. Oxford: Clarendon Press, 1955–1957.

D. C. SIRCAR, *Indian Epigraphy*. Delhi, 1965.

C. SIVARAMAMURTI, "Indian Epigraphy and South Indian Scripts". *BMGM* 3/4. Madras, 1952.

A. SKAIST, "A Note on the Bilingual Ostracon from Khirbet-el-Kom". *IEJ* 28: pp.106–8. Jerusalem, 1978.

D. SMELTZER, *Man and Number*. London: A. & C. Black, 1958.

DAVID EUGENE SMITH, "An Ancient English Algorism". In: *Festschrift Moritz Cantor*. Leipzig, 1909.

D. E. SMITH, *Rara Arithmetica*. Boston, 1909.

D. E. SMITH, "Computing jetons". *NNM* 9. New York, 1921.

D. E. SMITH, Ed., *A Source Book in Mathematics*. New York: McGraw-Hill, 1929.

D. E. SMITH, *History of Mathematics*. New York: Dover, 1958.

D. E. SMITH and JEKUTHIEL GINSBURG, "Rabbi Ben Ezra and the Hindu-Arabic Problem". *AMM* 25: pp.99–108. 1918.

D. E. SMITH AND L. C. KARPINSKI, *The Hindu-Arabic Numerals*. Boston: Ginn & Co, 1911.

D. E. SMITH AND Y. MIKAMI, *History of Japanese Mathematics*. Chicago: Open Court, 1914.

D. E. SMITH and M. SALIH, "The Dust Numerals amongst the Ancient Arabs". *AMM* 34: pp.258–60. 1927.

V. A. SMITH, "Asoka Notes". *IA* 37: pp.24ff. Also Vol. 38 (1909): pp.151–9. Bombay, 1908.

D. DE SOLLA-PRICE, "Portable Sundials in Antiquity". *CENT* 14: pp.242–66. Copenhagen, 1969

D. DE SOLLA-PRICE, "Gears from the Greeks. The Antikythera Mechanism". *TAPS*. November 1974

D. DE SOLLA-PRICE, *Gears from the Greeks*. New York: Science History Publications, 1975.

E. A. SPEISER, *Excavations at Tepe Gawra*. Vol. 1. Philadelphia: University of Pennsylvania Press, 1935.

H. J. SPINDEN, *Maya Art and Civilization*. Indian Hills, CO, 1959.

G. STACK, *A Grammar of the Sindhi Language*. Bombay, 1849.

C. STEWART, *Original Persian Letters and Other Documents*. London, 1825.

DIRK J. STRUIK, Gerbert d'Aurillac. In: C. Gillespie, *DSB*, 5: pp.364–6. New York, 1970.

DIRK J. STRUIK (1948), *A Concise History of Mathematics*. Mineola: Dover Books, 1987.

A. SUTTON, *An Introductory Grammar of the Oriya Language*. Calcutta, 1831.

SZEMERENYI, *Studies in the Indo-European System of Numerals*. Heidelberg, 1960.

C. TAYLOR, *The Alphabet*. London, 1883.

J. TAYLOR, *Lilawati*. Bombay, 1816.

J. E. TEEPLE, "Maya Astronomy". *CAA* 403/1: pp.29–116. Washington, DC, 1931.

J. ERIC THOMPSON, "Maya Chronology: the Correlation Question". *Publications of the Carnegie Institute of Washington* 456. Washington, DC, 1935.

J. ERIC THOMPSON, "Maya Arithmetic". *CAAH* 528/7–36. Washington, DC, 1942.

J. ERIC THOMPSON, *Maya Hieroglyphic Writing*. Washington, DC: Carnegie Institute, 1950

J. ERIC THOMPSON, *The Rise and Fall of Maya Civilization*. Norman: University of Oklahoma Press, 1954.

J. ERIC THOMPSON, "A Commentary on the Dresden Codex". *MAPS* 93. Philadelphia, 1972.

R. C. THOMPSON and M. E. L. MALLOWAN, "The British Museum Excavations at Nineveh". *LAA* 20: pp.71–186. Liverpool, 1933.

A. J. TOBLER, *Excavations at Tepe Gawra*. Philadelphia: University of Pennsylvania Press, 1950.

N. M. TOD, "The Greek Numeral Notation". *ABSA* 18: pp.98–132. London, 1911. Also Vol. 28: pp.141ff. 1926.

N. M. TOD, "Three Greek Numeral Systems". *JHS* 33: pp.27–34. London, 1913. Also Vol. 24: pp.54–67. 1919.

N. M. TOD (1918), "The Macedonian Era". *ABSA* 23: pp.206–17. London, 1918. Also Vol. 24 (1919): pp.54–67.

N. M. TOD, "The Greek Acrophonic Numerals". *ABSA* 37: pp.236–58. London, 1936. Also Vol. 24: pp.54–67. 1919.

N. M. TOD, "The Alphabetic Numeral System in Attica". *ABSA* 45: pp.126–39. London, 1950.

G. J. TOOMER, "Al Khowarizmi". In: C. Gillespie, *DSB*, 7: 358–65. New York, 1970.

E. B. TYLOR, *Primitive Culture*. London, 1871.

C. S. UPASAK, *The History and Paleography of Mauryan Brahmi Script*. Patna: Nalanda, 1960.

G. VAILLANT, *The Aztecs of Mexico*. London: Pelican, 1950.

MICHAEL VENTRIS and J. CHADWICK, *Documents in Mycenean Greek*. Cambridge: Cambridge University Press, 1956.

J. WALLIS, *A Treatise on Algebra, both Historical and Practical*. London, 1685.

A. N. WHITEHEAD and BERTRAND RUSSELL, *Principia Mathematica*. Cambridge, 1925–1927.

J. G. WILKINSON, *The Manner and Customs of the Ancient Egyptians*. London, 1837.

F. H. WILLIAMS, *How to Operate the Abacus*. London, 1941.

H. J. J. WINTER, "Formative Influences in Islamic Sciences". *AIHS* 6: pp.171–92. Paris, 1953.

C. E. WOODRUFF, "The Evolution of Modern Numerals from Tally-Marks". *AMM*. August 1909.

C. L. WOOLEY, *Ur Excavations: The Royal Cemetery*. London: Oxford University Press, 1934.

W. WRIGHT, *Catalogue of Syriac Manuscripts in the British Museum*. London: British Museum, 1870.

F. A. YELDHAM, *The Story of Reckoning in the Middle Ages*. London: Harrap, 1926.

Y. YOSHINO (1937), *The Japanese Abacus Explained*. Dover Books: New York, 1963.

C. ZASLAVSKY, "Black African Traditional Mathematics". *MT* 63: pp.345–52. 1970.

C. ZASLAVSKY, *Africa Counts*. Boston: Prindle Webster & Schmidt, 1973.

2. WORKS IN OTHER LANGUAGES

H. Adler, J. Bormann, W. Kammerer, I. O. Kerner and N. J. Lehmann. "Mathematische Maschinen". In: *EMDDR*. Berlin: VEB, 1974.

B. Aggoula, "Remarques sur les inscriptions hatréennes". *MUS* 47. Beirut, 1972.

E. Aghayan, "Mesrop Maschtots et la création de l'alphabet arménien". *ARM:* pp.10–13. January 1984.

A. Al-karmali, "Al 'uqad". *MACH* 3: pp.119–69. Baghdad, 1900.

A. Allard, *Les Plus Anciennes Versions latines du XIIe siècle issues de l'arithmétique d'Al Khwarizmi*. Louvain, 1975.

A. Allard, "Ouverture et résistance au calcul indien". In: *Colloque d'histoire des sciences: Résistance et ouverture aux découvertes*. Louvain, 1976.

A. Allard, *Le "Grand Calcul selon les Indiens" de Maxime Planudes: sa source anonyme de 1252*. Louvain, 1981.

A. Allard, *Le calcul indien (algorismus)*. Paris: Blanchard, 1992.

V. Alleton, *L'Ecriture chinoise*. Paris: PUF, 1970.

A. Amiet, *La Glyptique mésopotamienne archaïque*. Paris: Editions du CNRS, 1961.

P. Amiet, "Il y a cinq mille ans les Elamites inventaient l'écriture". *ARCH* 12: pp.20–2. Paris, 1966.

A. Amiet, *Elam*. Auvers-sur-Oise: Archée, 1966.

A. Amiet, "Glyptique susienne des origines à l'époque des Perses achéménides". *MDP* 43/2. 1967.

A. Amiet, *Bas-reliefs imaginaires de l'Ancien Orient d'après les cachets et les sceaux-cylindres*. Paris, 1973.

A. Amiet, *Les Civilisations antiques du Proche-Orient*. Paris: PUF, 1975.

A. Amiet, "Alternance et dualité. Essai d'interprétation de l'histoire Elamite". *AKK* 15: pp.2–22. Brussels, 1979.

A. Amiet, "Comptabilité et écriture à Suse". In: A. M. Christin, *ESIP*: pp.39–44. Paris: Le Sycomore, 1982

A. Apokin and L. E. Maistrov (1974). *Rasvitie Vyichistltelynih Masin*. Moscow: Nauka, 1974.

A. J. Arberry, *Le Soufisme*. Paris, 1952.

M. Arkoun, "Chronologie de l'Islam". In: Kasimirski, *Le Koran*. Paris: Garnier-Flammarion, 1970.

Arnauld and Nicole (1662). *Logique de Port-Royal, ou l'art de penser*. Paris: Hachette, 1854.

J. Auboyer, "La Monnaie en Inde". In: *DAT* 2: p.712. . Paris, Editions de l'Accueil, 1963.

A. Aymard and J. Auboyer, "L'Orient et la Grèce antique". In: A. Aymard and J. Auboyer, *Histoire générale des civilisations*, I. Paris: PUF, 1961.

E. Aymonier, "Quelques notions sur les inscriptions en vieux khmer". *JA*: pp.441–505. Paris, April 1883.

E. Babelon, "Moneta". In: C. Daremberg and E. Saglio, *DAGR*: pp.1963ff. Paris, 1873.

J. Babelon, *Mayas d'hier et d'aujourd'hui*. Paris, 1967.

J. Babelon, *ARCH* 22: p.21. Paris, 1968.

M. Baillet, J. T. Milik and R. De Vaux, "Les Petites Grottes de Qumran". *DJD* 3. Oxford, 1961.

L. Bakhtiar, *Le Soufisme, expression de la quête mystique*. Paris: Le Seuil, 1977.

A. Bareau, "Bouddhisme". In: *Les Religions de l'Inde*, 3. Paris, 1966.

P. Barguet, "Les Dimensions du temple d'Edfou et leurs significations". *BSFE* 72. Paris, March 1975.

E. Barnavi, *Histoire universelle des Juifs, de la Genèse à la fin du XXe siècle*. Paris: Hachette, 1992.

A. Barth and A. Bergaigne, "Inscriptions sanskrites du Champa et du Cambodge". *NEM* 27. Paris, 1885.

L. Baudin, "L'Empire socialiste des Inkas". *TMIE* 5. Paris, 1928.

M. Baudouin, *La Préhistoire par les étoiles*. Paris, 1926.

Bayer, *Historia regni Graecorum Bactriani*. Saint Petersburg, 1738.

W. de Beauclair, *Rechnen mit Maschinen. Eine Bildgeschichte der Rechnentechnik*. Braunschweig: Vieweg u. S., 1968.

G. Beaujouan, "Etude paléographique sur la rotation des chiffres et l'emploi des apices du Xe au XIIe siècle". *RHS* 1: pp.301–13. 1947.

G. Beaujouan, *Recherches sur l'histoire de l'arithmétique au Moyen Age*. Paris: Ecole des Chartes (unpublished thesis abstract), 1947.

G. Beaujouan, "Les soi-disant chiffres grecs ou chaldéens". *RHS* 3: pp.170–4. 1950.

G. Beaujouan, "La Science dans l'Occident médiéval chrétien". In: R. Taton, *HGS*, 1: pp.517–34. Paris: PUF, 1957.

G. Beaujouan, "L'enseignement du Quadrivium". In: *Settimane di studio del centro italiano di studi sull'alto medioevo*: pp.650ff. Spoleto, 1972.

O. Costa de Beauregard, *Le Second Principe de la Science du temps*. Paris: Le Seuil, 1963.

O. Becker and J. E. Hoffmann, *Geschichte der Mathematik*. Bonn: Athenäum Verlag, 1951.

Bede, The Venerable, "De Temporum ratione". In: Migne, *Patrologia cursus completus. Patres latini*: 90. Paris, 1850.

L. Benoist, *Signes, Symboles et Mythes*. Paris: PUF, 1977.

E. Benvéniste, *Problèmes de linguistique générale*. Paris: Gallimard, 1966, 1974.

D. Van Berchem, "Notes d'archéologie arabe. Monuments et inscriptions fatimides". *JA*: pp.80 & 392. Paris, 1891–1892.

D. Van Berchem, "Tessères ou calculi? Essai d'interprétation des jetons romains en plomb". *RN* 39: pp.297–315. Paris, 1936.

H. Berlin, "El glifo 'emblema' de las inscripciones mayas". *JSA* 47: pp.111–19. Paris, 1958.

J. J. G. Berthevin, *Eléments d'arithmétique complémentaire*. Paris, 1823.

M. Betini, *Apiaria universae philosophicae, mathematicae*. Vol. 2, 1642.

P. Beziau, *Origine des chiffres*. Angers, 1939.

O. Biermann, *Vorlesungen über mathematische Näherungsmethoden*. Braunschweig, 1905.

R. Billard, *L'Astronomie indienne. Investigation des textes sanskrits et des données numériques*. Paris: Ecole française d'Extrême-Orient, 1971.

E. Biot, "Notes sur la connaissance que les Chinois ont eue de la valeur de position des chiffres". *JA* December: pp.497–502. Paris, 1839.

M. Birot, *Tablettes d'époque babylonienne ancienne*. Paris, 1970.

K. Bittel, *Les Hittites*. Paris: Gallimard, 1976.

A. Blanchet and A. Dieudonné, *Manuel de numismatique*. Paris: Picard, 1930.

MARC BLOCH, *Apologie pour l'histoire, ou le métier d'historien*. Paris: Armand Colin, 1949.

R. BLOCH, "Etrusques et Romains: problèmes et histoire de l'écriture". In: *EPP*: pp.183–298. 1963.

R. BLOCH, "La Monnaie à Rome". In: *DAT*, 2: pp.718ff. Paris, 1964.

BOBYNIN, *Enseignement mathématique*, p.362. Paris, 1904.

A. BOECKH, J. FRANZ, E. CURTIUS and A. KIRCKHOFF, *Corpus Inscriptionum Graecarum*. Berlin, 1828–1877.

M. BOLL, *Histoire des mathématiques*. Paris: PUF, 1963.

B. BONCOMPAGNI, *Trattati d'aritmetica*. Rome, 1857.

B. BONCOMPAGNI, Ed., *Liber abaci. Scritti di Leonardo Pisano*. Contains the works of Fibonacci (Leonard of Pisa). Rome, 1857.

LUDWIG BORCHARDT, "Das Grabdenkmal des Königs S'ahu-Re'". *ADOGA* 7. 1913.

LUDWIG BORCHARDT, *Denkmäler des Alten Reiches im Museum von Kairo*. (General Catalogue of Egyptian Antiquities in the Cairo Museum). *DAR* Vol. 57. Berlin, 1937.

R. BORGER, "Die Inschriften Asarhaddons, König von Assyrien". *AFO* 9. Graz, 1956.

J. BOTTERO, *La Religion babylonienne*, pp. 59–60. Paris: PUF, 1952.

J. BOTTERO, "Textes économiques et administratifs". *ARMA* 7: pp.326–41. Paris, 1957.

J. BOTTERO, "Symptômes, signes, écritures en Mésopotamie ancienne". In: *Divination et rationalité*: pp.159ff. Paris: Le Seuil, 1974.

J. BOTTERO, "De l'aide-mémoire à l'écriture". In: A.-M. Christin, *Ecritures. Systèmes idéographiques et pratiques expressives*: pp.13–37. Paris: Le Sycomore, 1982.

A. BOUCHÉ-LECLERCQ (1879), *Histoire de la divination dans l'antiquité*. Paris, 1963.

B. BOURDON, *La Perception visuelle de l'espace*. Paris: Costes, 1902.

P. BOUTROUX, *L'Idéal scientifique des mathématiciens*. Paris: Alcan, 1922.

J. BOUVERESSE, J. ITARD and E. SALLÉ, *Histoire des mathématiques*. Paris: Larousse, 1977.

J.-P. BOUYOU-MORENO, *Le Boulier chinois*. Paris: Self-published.

F. BRAEUNIG, *Calcul mental*. Paris, 1882.

H. BREUIL, "La formation de la science préhistorique". In: C. Zervos, *L'Art de l'époque du renne en France*: pp.13–20. Paris, 1959.

M. BRÉZILLON, *Dictionnaire de la préhistoire*. Paris: Larousse, 1969.

H. BRUGSCH, *Thesaurus inscriptionum Aegyptiacarum*. Leipzig, 1883.

E. M. BRUINS and M. RUTTEN, "Textes mathématiques de Suse". *MDP* 34. 1961.

H. BRUNS, *Grundlinien des wissenschaftlichen Rechnens*. Leipzig, 1903.

L. BRUNSCHVICG, *Les Etapes de la philosophie mathématique*. Paris: Alcan, 1912.

N. M. BUBNOV, *Origin and History of Our Numbers*. In Russian. Kiev, 1908.

N. M. BUBNOV, *Arithmetische Selbständigkeit der europäischen Kultur*. Berlin: Friedländer, 1914.

N. M. BUBNOV (1899). *Gerberti opera mathematica*. Hildesheim, 1963.

G. BÜHLER, *Indische Paleographie*. Strasbourg, 1896.

G. BUONAMICI, *Epigrafia etrusca*. Florence, 1932.

A. BURLOUD, *La Pensée conceptuelle*. Paris: Alcan, 1927.

R. BURNAM, *Paleographica Iberica. Facsimiles de manuscrits portugais et espagnols du IXe au XVIe siècle*. Paris, 1912–1925.

R. Cagnat, *Cours d'épigraphie latine*. Paris, 1890.

R. Cagnat, "Revue des publications épigraphiques relatives à l'antiquité romaine". *RAR* 3/34: pp.313–20. Paris, 1899.

P. Calandri, *De Arithmetica opusculum*. Florence, 1491.

A. Dom Calmet, "Recherches sur l'origine des chiffres de l'arithmétique". *MDT* pp.1620–35. September 1707.

C. D. du Cange (1678). *Glossarium ad scriptores mediae et infimae latinitatis*. Paris, 1937–1938.

F. Cantera and J. M. Millas, *Las Inscripciones hebraïcas de España*. Madrid: Consejo superior de Investigaciones cientificas, 1956.

G. Cantineau, *Le Nabatéen*. Vol. 1. Paris: Leroux, 1930.

M. Cantor, *Kulturleben der Völker*, pp.132–6. Halle, 1863.

M. Cantor, *Vorlesungen über die Geschichte der Mathematik*. Leipzig, 1880. See also third edition, 1: p.133. Leipzig, 1907.

P. Casanova, "Alphabets magiques arabes". *JA* 17/18: pp.37–55; 19/20: pp.250-67. Paris, 1921–1922.

G. Casaril, *Rabbi Simeon Bar Yochaï et la Cabbale*. Paris: Le Seuil, 1961.

Casiri, *Bibliotheca Arabo-Hispanica Escurialensis*. Madrid, 1760–1770.

A. Caso, *Calendario y escritura de las antiguas culturas de Monte-Albán*. Mexico, 1946.

A. Caso, *Las calendarios prehispánicos*. Mexico: Universidad Nacional Autónoma, 1967.

E. Cassin, "La monnaie en Asie occidentale". In: *DAT*, 2: pp.712–15. Paris: A. Colin, 1963.

Cataneo, *Le pratiche delle due prime mathematiche*. Venice, 1546.

J. M. A. Chabouillet, *Catalogue général et raisonné des camées et pierres gravées de la Bibliothèque impériale*. Paris, 1858.

C. Chadefaud, "De l'expression picturale à l'écriture égyptienne". In: *ESIP*: pp.81–99. Paris: Le Sycomore, 1982.

E. Chassinat, *Le Temple d'Edfou*. Cairo, 1897–1960.

E. Chassinat, *Le Temple de Dendara*. Cairo, 1934.

E. De Chavannes, *Les Documents chinois découverts par Aurel Stein*. Oxford, 1913.

Chen Tsu-Lung, "La Monnaie en Chine". *DAT* 2: pp.711–12. Paris: L'Accueil, 1963–1964.

J. Chevalier and A. Gheerbrant, *Dictionnaire des Symboles*. Paris: Laffont, 1982.

A. Chodzko, *Grammaire persane*. Paris: Imprimerie impériale, 1852.

A.-M. Christin, *Ecritures. Systèmes idéographiques et pratiques expressives*. Paris: Le Sycomore, 1982.

N. Chuquet (1484), *Triparty en la science des Nombres*. Ms, BN Paris, partly published by A. Marre, Rome, 1880–1881.

E. Claparède, In: *L'Invention*. Paris, Alcan/Centre international de synthèse, 1937.

G. Coedès and H. Parmentier, *Listes générales des inscriptions et des monuments du Champa et du Cambodge*. Hanoi: Ecole française d'Extrême-Orient, 1923.

Georges Coedès, *BEFEO* 24: pp.3–4. Paris, 1924

Georges Coedès, *Corpus des Inscriptions du Cambodge*. Paris: Paul Geuthner, 1926.

Georges Coedès, "Les Inscriptions malaises deçrivijaya". *BEFEO* 30: pp.29–80. Paris, 1930–1931.

Georges Coedès, "A propos de l'origine des chiffres arabes". *BSOAS* 6/2: pp.323–8. London, 1931.

Georges Coedès, *Les Etats hindouisés d'Indonésie et d'Indochine*. Paris: De Boccard, 1964.

Georges Coedès, *Inscriptions du Cambodge*. Paris: Ecole française d'Extrême-Orient, 1966.

D. Cohen, *Dictionnaire des racines sémitiques ou attestées dans les langues sémitiques*. The Hague: Mouton, 1970.

M. Cohen, "Traité de langue amharique". *TMIE* 24: pp.26ff. Paris, 1970.

M Cohen, *La Grande Invention de l'écriture et son évolution*. Paris: Klincksieck, 1958.

G. S. Colin, "Une nouvelle inscription arabe de Tanger". *HESP* 5: pp.93–9. 1924.

G. S. Colin, "De l'origine des chiffres de Fez". *JA* XIII/1: pp.193–5. Paris, 1933.

G. S. Colin, "Hisab al-Djummal". In: *EIS*: p.484. Paris, 1954.

G. S. Colin, "Abdjad". In: *EIS*: p.100. Paris, 1954.

J. -P. Collette, *Histoire des mathématiques*. Montreal: Renouveau pédagogique, 1973.

Combe, Sauvaget and Wiet, *Répertoire chronologique d'épigraphie arabe*. Cairo, 1931–1956.

A. Conrady, *De chinesischen Handschrift und sonstigen Kleinfunde Sven Hedins in Lou-Lan*. Stockholm, 1920.

G. Conteneau, "Notes d'iconographie religieuse assyrienne". *RA* 37: pp.154–65. Paris, 1940.

H. de Contenson, "Nouvelles données sur le néolithique précéramique ... à Ghoraifé". *BSPF* 73/3: pp.80–2. Paris, 1976.

A. Cordoliani, "Contribution à la littérature ecclésiastique au Moyen-Age". *SME* 3/1: pp.107–37. Turin, 1960. See also Vol. 3/2: pp.169–208. Turin, 1961.

L. Costaz, *Grammaire syriaque*. Beirut, 1964.

Leonard Cottrell, Ed., *Dictionnaire encyclopédique d'archéologie*. Paris: SEDES, 1962.

P. Couderc, *Le Calendrier*. Paris: PUF, 1970.

C. Courtois, L. Leschi, C. Perrat and C. Saumagne, *Tablettes Albertini. Actes privés de l'époque vandale*. Paris: Gouvernement général de l'Algérie, 1952.

D. A. Crusius, *Anweisung zur Rechnenkunst*. Halle, 1746.

M. Curtze, *Petri Philomeni de Dacia in Algorismus Vulgarem Johannis de Sacrobosco commentarius, una cum Algorismo ipso*. Copenhagen, 1897.

A. Cuvillier, *Cours de philosophie*. Paris: Armand Colin, 1954.

A. Dahan-Dalmedico and J. Pfeiffer, *Histoire des mathématiques. Routes et dédales*. Montreal: Etudes vivantes, 1982.

L.C. Damais, "Etudes balinaises". *BEFEO* 50/1: pp.144–52. Paris, 1960.

P. Damerov and R. K. Englund, "Die Zahlzeichensysteme der archaischen Texte aus Uruk". *ATU* 2/5. Berlin, September 1985.

M. Danloux-Dumesnils, *Esprit et bon usage du système métrique*. Paris: Blanchard, 1965.

C. Daremberg and E. Saglio, *Dictionnaire des antiquités grecques et romaines*. Paris: Hachette, 1873–1919.

Dasypodius, *Institutionum Mathematicarum*. Strasbourg, 1593.

F. Daumas, *Les Dieux de l'Egypte*. Paris: PUF, 1970.

S. Debarbat and A. Ten, *Le Mètre et le système métrique*. Paris & Valence, 1993.

M. J. A. DECOURDEMANCHE, "Notes sur quatre systèmes turcs de notation numérique secrète". *JA* 9/14: pp.258ff. Paris, 1899.

J.J. DEHOLLANDER, *Handleiding bij de beoefening der Javaansche Taal en Letterkunde.* Breda, 1848.

A. DEIMEL, "Sumerische Grammatik der archaïschen Texte". *OR* 9/pp.43–44, 182–98. Rome, 1924.

A. DEIMEL, *Sumerisches Lexikon.* Scripta Pontificii Instituti Biblici. Rome, 1947.

H. DELACROIX , *Le Langage et la pensée.* Paris: Alcan, 1922.

H. DELACROIX, *Psychologie de l'art.* Paris: Alcan, 1927.

X. DELAMARRE, *Le Vocabulaire indo-européen.* Paris: Maisonneuve, 1984.

A. DÉLÉDICQ, "Numérations et langues africaines". *BLPM* 27: pp.3–9. Paris, 1981.

M. DELPHIN, "L'astronomie au Maroc". *JA* 17. Paris, 1891.

P. DEMIÉVILLE, *Matériaux pour l'enseignement du chinois.* Paris, 1953.

J. G. DENNLER, "Los nombres indigenas en guarani". *PHYS* 16. Buenos Aires, 1939.

E. DERMENGHEM, *Mahomet et la tradition islamique.* Paris: Le Seuil, 1955.

M. DESTOMBES, "Les chiffres coufiques des instruments astronomiques arabes". *RSS* 2/3: pp.197–210. Florence, 1960.

M. DESTOMBES, "Un astrolabe carolingien et l'origine de nos chiffres arabes". *AIHS* 58/59: pp.3–45. Paris, 1962.

J. DHOMBRES, *Nombre, mesure et continu: épistémologie et histoire.* Paris: Nathan, 1978.

J. DHOMBRES, "La fin du siècle des lumières: un grand élan scientifique". In: G. Ifrah, *Deux Siècles de France,* 111: pp.5–12. Paris: Total Information, 1989.

J. DHOMBRES, *Une Ecole révolutionnaire en l'an III. Leçons de mathématiques.* Paris: Dunod, 1992.

J. DHOMBRES, "Résistance et adaptation du monde paysan au système métrique". In: A. Croix and J. Quéniart, *La Culture programme,* pp.128–42. Roanne, 1993.

DENIS DIDEROT, "Arithmétique". In: D. Diderot and J. d'Alembert, *L'Encyclopédie,* I: pp.680–4. Paris, 1751.

E. DOBLHOFFER, *Le Déchiffrement des écritures.* Paris: Arthaud, 1959.

M. DOBRIZHOFFER, *Auskunft über die Abiponische Sprache.* Leipzig: Platzmann, 1902.

P. J DOLS, "La vie chinoise dans la province de Kan-Su". *ANTH* 12-13: pp.964ff. Göteborg, 1917–1918.

E. DOMLUVIL, "Die Kerbstöcke der Schafhirten in der mährischen Walachei". *ZOV* 10: pp.206–10. Vienna, 1904.

DONG-ZUOBIN, *Xiao dun yin xu wenzi: yi bián.* Taipei, 1949. In Chinese.

H. DONNER and W. RÖLLIG, *Kanaanaische und Aramäische Inschriften.* Wiesbaden, 1962.

B. VON DORN, "Über ein drittes in Russland befindliches Astrolabium mit morgenländischen Inschriften". *BSC* 9/5: pp.60–73. Paris, 1841.

F. DORNSEIFF(1925), *Das Alphabet in Mystik und Magie.* Leipzig/Berlin, 1977.

G. DOSSIN, "Correspondance de Iasmah-Addu". *ARMA* 5: Letter 20, pp. 36–7. Paris, 1952.

E. DOUTTE, *Magie et Religion dans l'Afrique du Nord.* Algiers, 1909.

A. DRAGONI, *Sul metodo aritmetico degli antichi Romani.* Cremona, 1811.

G. DUMESNIL, "Note sur la forme des chiffres usuels". *RAR* 16: pp.342–48. Paris, 1890.

P. Dupont-Sommier, "La Science hébraïque ancienne". In: R. Taton, *HGS*, 1: pp.141–51. Paris, 1957.

E. Duporck, Ed., *Comptes rendus du deuxième Congrès international de mathématiques de Paris*. Paris, 1902.

J.-M. Durand, "Espace et écriture en cunéiforme". In: Christin, *ESIP*: pp.51–62. Paris, 1982.

J.-M. Durand, "A propos du nom du nombre 10 000 à Mari". *MARI* 3: pp.278–9. Paris, 1984.

J.-M. Durand, "Questions de chiffres". *MARI* 5: pp.605–22. Paris, 1987.

R. Duval (1881), *Traité de grammaire syriaque*. Amsterdam, 1969.

W. Dyck, *Katalog matematische usw. Instrumente*, Munich: Wolf und Sohn, 1892.

A. Eisenlohr, *Ein mathematisches Handbuch der Alten Aegypter*. Leipzig, 1977.

Mircea Eliade, *Traité d'histoire des religions*. Paris, 1949.

R. W. Erichsen, *Demotisches Glossar*. Vol. 5. Copenhagen, 1954.

Erpenius, *Grammatica Arabica*. Leyden, 1613.

J. Essig, *Douze, notre dix futur*. Paris: Dunod, 1955.

J. Euting, *Nabatäische Inschriften aus Arabien*. Berlin, 1885.

P. Ewald, *Neues Archiv der Gesellschaft für ältere deutsche Geschichtskunde* 8: pp.258–357. 1883.

A. Falkenstein, "Archäische Texte aus Uruk". *ADFU*. Berlin, 1936.

A. Falkenstein, *Das Sumerische. Handbuch der Orientalistik*. Leyden: Brill, 1959.

F.-G. Faraut, *Astronomie cambodgienne*. Phnom Penh: F. H. Schneider, 1910.

O. Fayzoullaiev, *Mohammed al-Khorazmii*. Tashkent, 1983.

L. Fekete, *Die Siyaqat-Schrift in der türkischen Finanzverwaltung*. Vol. I. Budapest, 1955.

J. Feller, Ed., *Dictionnaire de la psychologie moderne*. Paris: Marabout, 1967.

E. Fettweis, "Wie man einstens rechnete". *MPB* 49. Leipzig, 1923.

J.-G. Février, *Histoire de l'écriture*. Paris: Payot, 1959.

J. Filliozat, "La Science indienne antique". In: R. Taton, *HGS*: pp.159ff. Paris, 1957.

P. S. Filliozat, *Le Sanskrit*. Paris: PUF, 1992.

O. Fine, *De arithmetica practica*, fo. 11–12. Paris, 1544.

L. Finot, "Note d'épigraphie: les inscriptions de Jaya Parameçvaravatman Ier, roi du Champa". *BEFEO* 15: pp.39–52. Paris, 1915.

H. Fleisch, *Traité de philologie arabe*. Beirut, 1961.

M. Folkerts, *Boethius' Geometrie, II. Ein Mathemathisches Lehrbuch des Mittelalters*. Wiesbaden: Franz Steiner, 1970.

Formaleoni, *Dei fonti degli errori nella cosmografia e geografia degli Antichi*. Venice, 1788.

C. Fossey, *Notices sur les caractères étrangers anciens et modernes*. Paris: Imprimerie nationale, 1948.

M. Foucaux, *Grammaire tibétaine*. Paris: Imprimerie nationale, 1858.

P. Foulquié, *Dictionnaire de la langue philosophique*. Paris: PUF, 1982.

J. Franz, *Elementa epigraphica graecae*. Berlin, 1840.

L. Frédéric, *Japon, l'Empire éternel*. Paris: Le Félin, 1985.

L. Frédéric, *Japon intime*. Paris: Le Félin, 1986.

L. Frédéric, *Dictionnaire de la civilisation indienne*. Paris: Laffont, 1987.

L. Fréréric, *Le Lotus*. Paris: Le Félin, 1987.

L. Frédéric, *Dictionnaire de Corée*. Paris: Le Félin, 1988.

L. Frédéric, *L'Inde de l'Islam*. Paris: Arthaud, 1989.

L. Frédéric, *Les Dieux du Bouddhisme*. Paris: Flammarion, 1992.

L. Frédéric, *L'Inde mystique et légendaire*. Paris: Rocher, 1994.

L. Frédéric, *Dictionnaire de la civilisation japonaise*. Paris: Laffont, 1994.

J.-C. Fredouille, *Dictionnaire de la civilisation romaine*. Paris: Larousse, 1968.

J. B. Frey, *Corpus Inscriptionum Iudaicorum*. Rome, 1936 – 1952.

G. Friedlein, *Boethii de Institutione arithmeticae*, pp. 373–428. Leipzig: Teubner, 1867.

G. Friedlein, *Die Zahlzeichen und das elementare Rechnen des Griechen und Römer und des Christlichen Abendlandes vom 7. bis 13. Jahrhundert*. Erlangen, 1869.

J. Friedrich, *Geschichte der Schrift*. Heidelberg, 1966.

K. Friedrichs, I. Fischer-Schreiber, F. K. Erhard and M. S. Diener, *Dictionnaire de la sagesse orientale*. Paris: Laffont, 1989.

W. Froehner, "Le Comput digital". *ASNA* 8: pp.232–8. Paris, 1884.

A. de La Fuye, J. T. Belaiew, R. de Mecquenem and J. M. Unvala, "Archéologie, Métrologie et Numismatique susiennes". *MDP* 25: pp.193ff. 1934.

A. M. von Gabain (1950). *Alttürkische Grammatik*. Leipzig: Harassowitz, 1950.

C. Gallenkamp, *Les Mayas. Découverte d'une civilisation perdue*. Paris: Payot, 1979.

G. F. Gamurrini, *Appendice al Corpus Inscriptionum Italicorum*. Florence, 1880.

Gaudefroy-Demombynes, *Grammaire de l'arabe classique*. Paris: Blachère, 1952.

M. J. E. Gautier, *Archives d'une famille de Dilbat*. Cairo, 1908.

R. Gemma-Frisius, *Arithmetica practicae methodus facilis*. Paris: Peletier, 1563.

Le Gendre, *Arithmétique en sa perfection*. Paris, 1771.

P. Gendrop, *Arte prehispánico en Mesoamerica*. Mexico City: Trillas, 1970.

H. de Genouillac, *Tablettes sumériennes archaiques*. Paris: Paul Geuthner, 1909.

H. de Genouillac, *Tablettes de Drehem*. Paris, 1911.

J. Gernet, "La Chine, aspects et fonctions psychologiques de l'écriture". In: *EPP*. Paris, 1963.

L. Gerschel, "Al Longue crôye. Autour des comptes à crédit". *EMW* 8: pp.256–92. 1959.

L. Gerschel, "Comment comptaient les anciens Romains". In *Latomus* 44: pp.386–97. 1960.

L. Gerschel, "La conquête du nombre". *AESC* 17: pp.691–714. Paris, 1962.

L. Gerschel, "L'Ogam et le nombre". *REC* 10/1: pp.127–66; 10/2: pp.516–77. Paris, 1962.

R. Ghirshman, *L'Iran, des origines à l'Islam*. Paris: Payot, 1951.

O. Gillain, *La Science égyptienne. L'Arithmétique au Moyen Empire*. Brussels: Editions de la Fondation Egyptologique, 1927.

B. Gille, *Les Ingénieurs de la Renaissance*. Paris: Le Seuil, 1978.

B. Gille, *Les Mécaniciens grecs*. Paris: Le Seuil, 1980.

R. Girard, *Les Popol-Vuh. Histoire culturelle des Mayas-Quichés*. Paris: Payot, 1972.

M. Gmür, "Schweizerische Bauermarken und Holzurkunden". *ASR*. Berne, 1917.

L. Godart and J.-P. Olivier, "Recueil des inscriptions en Linéaire A". *EC* 21. Paris, 1976.

G. Godron, "Deux notes d'épigraphie thinite". *RE* 8: pp.91ff. Paris, 1951.

I. Godziher, "Le Rosaire dans l'Islam". *RHR* 21: pp.295–300. Paris, 1890.

V. Goldschmidt, *Die Entstehung unseren Ziffern*. Heidelberg, 1932.

M. Gomez-Moreno, "Documentacion Goda en Pizarra". *BRAH.* Madrid, 1966.

J. Goschkewitsch, "Über das Chinesische Rechensbrett". *AKRG* 1: p.293. Berlin, 1858.

M. Granet, *La Pensée chinoise.* Paris: Albin Michel, 1988.

M. W. Green and H. J. Nissen, *Zeichenliste der archaischen Texte aus Uruk.* Vol. 2. Berlin: Max-Planck-Institut, 1985.

A. Grohmann, *Einführung und Chrestomathie zur arabischen Papyruskunde.* Prague, 1954.

H. Grotefend, *Taschenbuch der Zeitrechnung des deutschen Mittelalters.* Hanover/Leipzig, 1891.

M Guarducci, *Epigraphia greca.* Rome, 1967.

O. Guérard and P. Jouguet, *Un livre d'écolier du IIIe siècle avant J.-C.* Cairo: Société royale égyptienne de papyrologie, 1938.

E. Guillaume, "Abacus". In: Daremberg and Saglio, *Dictionnaire*: pp.1–5. Paris, 1873.

G. Guitel, "Comparaison entre les numérations aztèque et égyptienne". *AESC* 13: pp.687–705. Paris, 1958.

G. Guitel, "Signification mathématique d'une tablette sumérienne". *RA* 57: pp.145–50. Paris, 1963.

G. Guitel, "Classification hierarchisée des numérations écrites". *AESC* 21: pp.959–81. Paris, 1966.

G Guitel, *Histoire comparée des numérations écrites.* Paris: Flammarion, 1975.

G Gundermann, *Die Zahlzeichen.* Giessen, 1899.

S. Günther, "Die quadratischen Irrationalitäten der Alten". *AGM* 4. Leipzig, 1882.

G. Hager, *Memoria sulle cifre arabiche.* Milan, 1813.

L. Hambis, *Grammaire de la langue mongole écrite.* Paris: Maisonneuve, 1945.

L. Hambis, "La Monnaie en Asie centrale et en Haute Asie". In: *DAT*, 2: p.711. Paris, 1963.

E. T. Hamy, "Le Chimpu". *NAT* 21. Paris, 1892.

J. Harmand, "Le Laos et les populations sauvages de l'Indochine". *TDM* 38/2: pp.2–48. Also Vol. 39/1: pp.241–314. Paris, 1879.

J. Harmand, "Les races indochinoises". *MSA* 2/2: pp.338–9. Paris, 1882.

G. P. Harsdörffer, *Delitae Mathematicae et Physicae.* Nürnberg, 1651.

Charles Higounet, *L'Ecriture.* Paris: PUF, 1969.

M Höfner, *Altsüdarabische Grammatik.* Leipzig, 1943.

K. Homeyer, *Die Haus- und Hofmarken.* Berlin, 1890.

E. Hoppe, "Das Sexagesimalsystem und die Kreisteilung". *AMP* 3/15.4: pp.304–13. 1910.

E. Hoppe, "Die Entstehung des Sexagesimalsystems und der Kreisteilung". *ARCH* 8: pp.448–58. Paris, 1927.

B. Hrozny, "L'inscription hittite hiéroglyphe 'Messerschmidt' ". *AROR* 11: pp.1–16. Prague, 1939.

E Hübner, *Exempla scripturae epigraphicae Latinae e Caesaris dictatoris morte ad aetatem Justiniani.* Berlin, 1885.

Huet, *Demonstratio Evangelica ad serenissimum Delphinum.* Paris, 1690.

H. Hunger, "Kryptographische Astrologische Omina". *AOAT* 1: pp.133ff. 1969.

H. Hunger, "Spätbabylonische Texte aus Uruk". *ADFU* 1/9. Berlin, 1976.

M. Hyades, "Ethnographie des Fuégiens". *BSA* 3/10: pp.327ff. Paris, 1887.

C. Idoux, "Inscriptions lapidaires". *BCFM* 19: pp.24–35. Paris, 1959.

C. Idoux (1959). "Quand l'écriture naît sur l'argile". *BCFM* 19: pp.210ff. Paris, 1959.

P. Ivanoff, *Maya*. Paris: Nathan, 1975.

François Jacob, *Logique du vivant*. Paris: Gallimard, 1970.

H. von Jacobs (1896). *Das Volk der Siebener-Zähler*. Berlin, 1896.

E. Jacquet, "Mode d'expression symbolique des nombres employés par les Indiens, les Tibétains et les Javanais". *JA* 16: pp.118–21. Paris, 1835.

Jenner, in *JA* CCLXII/1–2: pp.176–91. Paris, 1974.

H. Jensen, *Die Schrift in der Vergangeneit und Gegenwart*. Berlin, 1969.

R. Jestin, *Tablettes sumériennes de Shuruppak*. Paris, 1937.

P. Joüon, "Sur les noms de nombre en sémitique". *MFO* 6: pp.133–9. Beirut, 1913.

D. Julia, *Dictionnaire de la philosophie*. Paris: Larousse, 1964.

P. Kaplony, "Die Inschriften der ägyptischen Frühzeit". *ÄGA* 8. Wiesbaden, 1963.

H. Kazem-Zadeh, "Les Chiffres Siyâk et la comptabilité persane". *RMM* 30: pp.1–51. 1913.

L. Keimer, "Bemerkungen zum Schiefertafel von Hierakonpolis". *AEG* 7. Milan, 1926.

H. Kern, *Verspreide Geschriften*. The Hague, 1913–1929.

G. Kewitsch, "Zweifel an der astronomischen und geometrischen Grundlage des 60-er Systems". *ZA* 18: pp.73–95. Berlin, 1904.

G. Kewitsch, "Zur Entstehung des 60-er Systems". *AMP* 3/18–2: pp.165ff. 1911.

G. Kewitsch, "Zur Entstehung des 60-er Systems". *ZA* 24: pp.265–83. Berlin, 1915.

Athanasius Kircher, *Oedipi Aegyptiaci*. Rome, 1653.

Köbel, *Ain new geordnet Rechenbiechlin*. Augsburg, 1514.

A. König, "Wesen und Verwendung der Rechenpfennige". *MM* 4. Frankfurt/Main, 1927.

M. Korostovtsev (1947). "L'hiéroglyphe pour 10 000". *BIFAO* 45: 81–88. 1947.

F. Kretzschmer and E.Heinsius, "Über eine Darstellung altrömischer Rechnenbretter". *TZG* 20: pp.96ff. 1951.

W. Kubitschek, "Die Salaminische Rechentafel". *NZ* 31: pp.393–8. Vienna, 1900.

A. Kuckuck, *Die Rechenkunst im sechzenten Jahrhundert*. Berlin, 1974.

R. Labat, "L'écriture cunéiforme et la civilisation mésopotamienne". In: *EPP*. Paris, 1963

R. Labat, "Jeux numériques de l'idéographie susienne". *ASS* 16: pp.257–60. Chicago, 1965.

R. Labat and F. Malbran-Labat, *Manuel d'épigraphie akkadienne*. Paris: Paul Geuthner, 1976.

G. Lafaye (1873). "Micatio". In: Daremberg and Saglio, *Dictionnaire*. Paris, 1889-1890.

Maurice Lambert, "La période présargonique. La vie économique à Shuruppak". *SUM* 9: pp.198–213. Baghdad, 1953.

Maurice Lambert, "La naissance de la bureaucratie". *RH* 224: pp.1–26. Paris, 1960.

Maurice Lambert, "Pourquoi l'écriture est née en Mésopotamie". *ARCH* 12: pp.24–31. Paris, 1966.

Maurice Lambert, *La Naissance de l' écriture en pays de Sumer*. Paris: Société des Antiquités nationales, 1976.

Mayer Lambert, *Traité de grammaire hébraïque*. Paris: E. Leroux, 1931.

W. Larfeld, *Handbuch der griechischen Epigraphik*. Leipzig, 1902–1907.

E. Laroche, *Les Hiéroglyphes hittites*. Vol. 1. Paris: CNRS, 1960.

Petrus Laurembergus, *Pet. Laurembergi Rostochiensis Instituiones arithmeticae*. Hamburg, 1636.

A. Lebrun, "Recherches stratigraphiques à l'acropole de Suse". *DAFI* 1: pp.163–214. Paris, 1971

A. Lebrun and F. Vallat, "L'origine de l'écriture à Suse". *DAFI* 8: pp.11–59. Paris, 1978.

G. Lefebvre, *Grammaire de l'égyptien classique*. Cairo: Institut français d'archéologie orientale, 1956.

J. Leflon, *Gerbert*. Saint-Wandrille, 1946.

J. C. Houzeau de Lehaie, "Fragments sur le calcul numérique". *BAB* 2/39: pp. 487–548. Continued in Vol. 2/40:pp.74–139 & 455–524. Brussels, 1875.

H. Lehmann, *Les Civilisations précolombiennes*. Paris: PUF, 1973.

A. Lemaire, *Inscriptions hébraïques*. Paris: Le Cerf, 1977.

A. Lemaire, "Les Ostraca paléo-hébreux des fouilles de l'Ophel". *LEV* 10: pp.156–61. 1978.

A. Lemaire and P. Vernus, "Les Ostraca paléo-hébreux de Qadesh Barnéa". *OR* 49: pp.341–5. Rome, 1980.

A. Lemaire and P. Vernus, L'Ostracon paléo-hébreu No. 6 de Tell Qudeirat. In: *FAP*: pp.302–27. 1983.

J. G. Lemoine, "Les Anciens procédés de calcul sur les doigts en Orient et en Occident". *REI* 6: pp.1–58. Paris,1932.

K. Lepsius, *Denkmäler aus Aegypten und Aethiopien*. Berlin, 1845.

André Leroi-Gourhan, *Le Geste et la Parole*. Paris: Albin Michel, 1964.

C. Leroy (1963). "La monnaie en Grèce". In: *DAT*, 2: pp.716–18. 1963.

J. Leupold, *Theatrum Arithmetico-Geometricum*. Leipzig, 1727.

M. Lidzbarski (1898) *Handbuch der Nordsemitischen Epigraphik*. Hildesheim, 1962.

U. Lindgren, *Gerbert von Aurillac und das Quadrivium*. Wiesbaden: Sudhoffs Archiv, 1976.

François Le Lionnais, *Les Grands Courants de la Pensée mathématique*. Paris: Blanchard, 1962.

E. Littmann, "Sabäische, griechische und altabessinische Inschriften". *DAE* 4. Berlin, 1913.

A. Loehr, "Méthodes de calcul du XVIe siècle". *SS* 9: p.8. Innsbruck, 1925.

E. Löffler, "Die arithmetischen Kentnisse der Babyloner und das Sexagesimalsystem". *AMP* 17/2: pp.135–44. 1910.

D. Lombard, *La Chine impériale*. Paris: PUF, 1967.

M. G. Loria, *Guida allo studio della storia delle Matematiche*. Milan: Hoepli, 1916.

J. Lüroth, *Vorlesungen über numerisches Rechnen*. Leipzig, 1900.

D. G. Lyon, "Keilschriften Sargons König von Assyrien". *ASB* 5. Leipzig, 1883.

E. J. H. MacKay, *La Civilisation de l'Indus. Fouilles de Mohenjo-Dar et de Harappa*. Paris, 1936.

L. Malassis, E. Lemaire and R. Grellet, "Bibliographie relative à l'arithmétique, au calcul simplifié et aux instruments à calculer". *BSEIN* 132: pp.739–57. Paris, 1920.

Bertil Malmberg, *Le Langage, signe de l'humain*. Paris: Picard, 1979.

WILLIAM OF MALMESBURY, *De gestis regum Anglorum libri*. London, 1596.

J. MARCUS, "L'écriture zapotèque". *PLS* 30: pp.48–63. Paris, 1980.

J. MARQUES-RIVIÈRE, *Amulettes, talismans et pentacles*. Paris: Payot, 1972.

A. MARRE, *Le Khulusat al hisab de Beha ad din al 'Amuli*. Rome, 1864.

A. MARRE, "Le Talkhys d'Ibn Albanna". *AANL* 17. Rome, 1865.

ANDRÉ MARTINET, *Des Steppes aux Océans: L'Indo-européen et les Indo-Européens*. Paris: Payot, 1986.

G. MASPÉRO, *Etudes de mythologie et d'archéologie égyptienne*. Paris, 1893–1916.

G. MASPÉRO, *Histoire ancienne des peuples de l'Orient classique*. Vol. 1: pp. 323–6. Paris, 1896.

H. MASPÉRO, *Les Documents chinois découverts par Aurel Stein*. Oxford, 1951.

H. MASPÉRO, *La Chine antique*. Paris, 1965.

L. MASSIGNON and R. ARNALDEZ, "La science arabe". In: R. Taton, *Histoire générale des sciences*: pp.431–71. Paris: PUF, 1957–1964.

O. MASSON, "La civilisation égéenne. Les écritures crétoises et mycéniennes". In: *EPP*: pp.93–9. Paris: A. Colin,1963.

C. LE MAUR, *Elementos de Mathematica pura*. Madrid, 1778.

A. MAZAHERI, "Les origines persanes de l'arithmétique". In *IDERIC* Vol. 8. Nice, 1975.

R. DE MECQUENEM, "Epigraphie proto-élamite". *MDP* 31. 1949.

ANTOINE MEILLET and MARCEL COHEN, *Les Langues du Monde*. Paris: CNRS, 1952.

G. MENENDEZ-PIDAL, "Los llamados numerales arabes en Occidente". *BRAH* 145: pp.179–208. Madrid, 1959.

P. MERIGGI, *La Scrittura proto-elamica*. Rome: Academia nazionale dei Lincei, 1971–1974.

A. MÉTRAUX, *Les Incas*. Paris: Le Seuil, 1961.

C. MEYER, *Die Historische Entwicklung der Handelsmarken in der Schweiz*. Bern, 1905.

C. B. MICHAELIS, *Grammatica syriaca*. Rome, 1829.

P. H. MICHEL, *Traité de l'astrolabe*. Paris, 1947.

A. MIELI, *La Science arabe et son rôle dans l'évolution scientifique mondiale*. Leyden, 1938.

A. MIELI, *Panorama general de la historia de la Ciencia*. Buenos Aires, 1946–1951.

J. T. MILIK, *Dédicaces faites par les dieux*. Vol. 1. Paris, 1972.

J.T. MILIK, "Numérotation des feuilles des rouleaux dans le Scriptorium de Qumran". *SEM* 27: pp.75–81. Paris, 1977.

G. MÖLLER, *Hieratische Paläographie*. Leipzig, 1911–1922.

T. MOMMSEN, *Die Unteritalischen Dialekte*. Leipzig, 1840.

V. MONTEIL, *La Pensée arabe*. Paris: Seghers, 1977.

J.-F. MONTUCLA (1798). *Histoire des mathématiques*. Paris: A. Blanchard, 1968.

GEORGES MOUNIN, *Histoire de la linguistique*. Paris: PUF, 1967.

C. MUGLER (1963). "Abaque". In: *DAT*, 1: 19–20.

A. MULLER (1971). *Les Ecritures secrètes*. Paris: PUF, 1971.

A. NAGL, "Der arithmetische Traktat von Radulph von Laon". *AGM* 5: pp.98–133. Leipzig, 1890.

A. NAGL, "Die Rechentafel der Alten". *SKAW* 177. Vienna, 1914.

A. NATUCCI, *Sviluppo storico dell'aritmetica generale e dell'algebra*. Naples, 1955.

F. NAU, "Notes d'astronomie indienne". *JA* 10/16: pp.209ff. Paris, 1910.

G. H. F. NESSELMANN, *Die Algebra der Griechen*. Berlin, 1842.

OSCAR NEUGEBAUER, "Zur Entstehung des Sexagesimalsystems". *AGW* 13/1. Göttingen, 1927.

OSCAR NEUGEBAUER, "Zur Aegyptischen Buchrechnung". *ZÄS* 64: pp.44–8. Berlin, 1929.

OSCAR NEUGEBAUER, *Vorgriechische Mathematik*. Berlin, 1934.

K. NIEBUHR, *Beschreibung von Arabien*. Copenhagen, 1772. French translation pub. Paris, 1779.

A. P. NINNI, *Sui segni prealfabeticci usati anche ora nella numerazione scritta dai pescatori Clodieusi*. Venice, 1889.

E. NORDENSKJÖLD, "Le Quipu péruvien du musée du Trocadéro". *BMET*. Paris, January 1931.

M. NOTH, "Exkurs über die Zahlzeichen auf den Ostraka". *ZDP* 50: pp.250ff. Leipzig, 1927.

NOTTNAGELUS, *Christophori Nottnagelii Professoris Wittenbergensis Institutionum mathematicarum*. Wittenberg, 1645.

J. NOUGAYROL, "Textes hépatoscopiques d'époque ancienne". *RA* 60: pp.65ff. Paris, 1945.

A. OLLERIS, *Les Oeuvres de Gerbert*. Paris, 1867.

L. PACIOLI, *Summa de arithmetica*. Venice, 1494.

G. A. PALENCIA, *Los Mozarabes de Toledo en los siglos XII e XIII*. Madrid, 1930.

A. PARROT, *Assur*. Paris: Gallimard, 1961.

H. PEDERSEN, *Vergleichende Grammatik der keltischen Sprachen*. Göttingen, 1909.

J. PEIGNOT and G. ADAMOFF, *Le Chiffre*. Paris, 1969.

PELETARIUS, *Commentaire sur l'Arithmétique de Gemma Frisius*. Lyon, 1563.

M. PERCHERON, *Le Bouddha et le bouddhisme*. Paris: Le Seuil, 1956.

P. PERDRIZET, "Isopséphie". *REG* 17: pp.350–60. Paris, 1904.

P. PERNY, *Grammaire de la langue chinoise*. Paris: Maisonneuve & Leroux, 1873.

C. PERRAT, "Paléographie médiévale". *HM* 11: pp.585–615. Paris, 1961.

J. PERROT, "Le gisement natoufien de Mallaha". *ANTHR* 70: pp.437–84. Paris, 1966.

J. PERROT, Ed., *Les Langues dans le monde ancien et moderne*. Paris: CNRS, 1981.

J. PIAGET, *Le Langage et la pensée chez l'enfant*. Neuchâtel: Delachaux & Niestlé, 1923.

J. PIAGET, *La Naissance de l'intelligence chez l'enfant*. Neuchâtel: Delachaux & Niestlé, 1936.

J. PIAGET, *La Construction du réel chez l'enfant*. Neuchâtel: Delachaux & Niestlé, 1937.

J. PIAGET, *La Représentation du monde chez l'enfant*. Paris: Alcan, 1938.

J. PIAGET and SZEMINSKA, *Le Genèse du nombre chez l'enfant*. Neuchâtel: Delachaux & Niestlé, 1941.

J. PIAGET, *La Formation du symbole chez l'enfant*. Neuchâtel: Delachaux & Niestlé, 1945.

J. PIAGET, *Psychologie de l'intelligence*. Paris: A. Colin, 1947.

J. PIAGET, *La Représentation de l'espace chez l'enfant*. Paris: PUF, 1948.

J. PIAGET and B. INHELDER, *La Psychologie de l'enfant*. Paris: PUF, 1966.

J. PIAGET AND B. INHELDER, *L'Image mentale chez l'enfant*. Paris: PUF, 1966.

PICCARD, "Mémoire sur la forme et la provenance des chiffres". *SVSN*: pp.176 and 184. 1859.

H. Piéron, *Vocabulaire de la psychologie*. Paris: PUF, 1979.

A. P. Pihan, *Notice sur les divers genres d'écritures des Arabes, des Persans et des Turcs*. Paris, 1856.

A. P. Pihan, *Exposé des signes de numération usités chez les peuples orientaux anciens et modernes*. Paris, 1860.

B. Piotrovsky, *Ourartou*. Geneva & Paris, 1969.

A. Poebels, *Grundzüge der sumerischen Grammatik*. Rostock, 1923.

Henri Poincaré, *La Science et l'hypothèse*. Paris: Flammarion, 1902.

Henri Poincaré, *Science et méthode*. Paris: Flammarion, 1909.

G. Poma-de-Ayala, *Codex péruvien*. Paris: Institut d'ethnologie, 1963.

E. Porada, *Iran Ancien*. Paris, 1963.

F. A. Pott, *Die quinäre und vigesimale Zählenmethode bei Völkern aller Weltteile*. Halle, 1847.

J. Prätorius (1599), "Compendiosa multiplicatio". *ZMP* 40: p.7. 1895.

M. Prou, *Recueil de fac-similés d'écritures du Ve au XVIIe siècle*. Paris, 1904.

Guy Rachet, *Dictionnaire de l'archéologie*. Paris: Laffont, 1983.

Guy Rachet, *Civilisations et archéologie de la Grèce pré-hellénique*. Paris: Le Rocher, 1993.

Ramus, *Arithmeticae libri duo, geometriae septem et viginti*. Basel, 1569.

Ramus, *Scholarum mathematicarum libri unus et triginta*. Basel, 1569.

H. Ranke, "Das altägyptische Schlangenspiel". *SHAW* 4: pp.9–14. Heidelberg, 1920.

R. Rashed and S. Ahmed, *Al-Bahir en algèbre d'As-Samaw'al*. Damascus: University of Damascus, 1972.

R. Rashed, *Essai d'histoire des mathématiques par Jean Itard*. Paris: A. Blanchard, 1984.

P. Reichlen, "Abaque, Calcul". In: *DAT*, 1: pp.18 and 188. 1963.

W. J. Reichmann, *La Fascination des nombres*. Paris: Payot, 1959.

S. Reinach, *Traité d'épigraphie grecque*. Paris: E. Leroux, 1885.

Reinaud, *Mémoire sur l'Inde*. Paris, 1894.

Ernest Renan. "Discours et conférences".

L. Renou, *Grammaire sanscrite*. Paris, 1930.

L. Renou and J. Filliozat, *L'Inde classique. Manuel des études indiennes*. Hanoi, 1953.

L. Renou, L. Nitti and N. Stchoupak, *Dictionnaire Sanscrit-Français*. Paris, 1959.

L. Reti, *Léonard de Vinci*. Paris: Laffont, 1974.

J. B. Reveillaud, *Essai sur les chiffres arabes*. Paris, 1883.

A. Riegl, "Die Holzkalender des Mittelalters und der Renaissance". *MIOG* 9: pp.82–103. Innsbruck, 1888.

M. E. de Rivero and J. D. de Tschudi, *Antiquités péruviennes*, p.95. Paris, 1859.

P. Rivet, *Les Cités Mayas*. Paris: Payot, 1954.

J. Rivoire, *Histoire de la monnaie*. Paris: PUF, 1985.

L. Rodet. "Sur la véritable signification de la notation numérique inventée par Âryabhata". *JA* 16/7: pp.440–85. Paris (no date).

L. Rodet, "Les Souan-pan des Chinois et la banque des argentiers". *BSM* 8. Paris, 1880.

A. Rödiger, "Über die im Orient gebräuchliche Fingersprache für den Ausdruck der Zahlen". *ZDMG* : pp.112–29. Wiesbaden, 1845.

M. Rodinson, "Les Sémites et l'alphabet". In: *EPP*. Paris: Armand Colin, 1963.

A. Rohrberg, "Das Rechnen auf dem chinesischen Rechenbrett". *UMN* 42: pp, 34–7. 1936.

Rong-Gen, *Jinwen-bián*. In Chinese. Beijing, 1959.

M. Rösler , "Das Vigesimalsystem im Romanischen". *ZRP* 26. Tübingen, 1910.

L. De Rosny, "Quelques observations sur la langue siamoise et son écriture". *JA*. Paris, 1855.

C. P. Ruchonnet, *Elements de calcul approximatif*. Lausanne, 1874.

J. Ruska, "Zur ältesten arabischen Algebra und Rechenkunst". *SHAW* 2. Heidelberg, 1917.

J. Ruska, "Arabische Texte über das Fingerrechnen". *DI* 10: pp.87–119. 1920.

J. Ruska, "Zahl und Null bei Jabir ibn Hayyan". *AGMNT* IX. 1928.

M. Rutten, *La Science des Chaldéens*. Paris: PUF, 1970.

Eduard Sachau, *Aramäische Papyrus und Ostraka aus Elephantine*. Berlin: Königliches Museum, 1911.

J. A. Sanchez-Perez, *La Arithmetica en Roma, en India, y en Arabia*. Madrid, 1949.

S. Sauneron, *L'Egyptologie*. Paris: PUF, 1968.

A. Schaeffer, *Le Palais royal d'Ugarit*. (Mission de Ras Shamra). Paris: Imprimerie nationale, 1955–1957.

V. Scheil, "Documents archaïques en écriture proto-élamite". *MDP* 6. 1905.

V. Scheil, "Notules". *RA* 12: pp.158–60. Paris, 1915

V. Scheil, "Textes de comptabilité proto-élamites". *MDP* 17–26. 1923–1935.

M. Schmidl, "Zahl und Zählen in Afrika". *MAGW* 45: pp.165–209. Vienna, 1915.

E. Schnippel, "Die Englischen Kalenderstäbe". *BEPH*, 5. Leipzig, 1926.

A. de Schodt, *Le jeton considéré comme instrument de calcul*. Brussels, 1873.

Gershon Scholem, *Les Grands Courants de la mystique juive*. Paris, 1950.

Gershon Scholem, *La Kabbale et sa symbolique*. Paris, 1966.

S. Schott, "Eine ägyptische Schreibpalette als Rechenbrett". *NAWG* 1/5: pp.91–113. Göttingen, 1967.

L. von Schroeder, *Pythagoras und die Inder*. Leipzig, 1884.

H. Schubert, *Mathematische Mussestunden*. Leipzig, 1900.

W. von Schulenburg, "Die Knotenzeichen der Müller". *ZE* 29: pp.491–4. Brunswick, 1897.

E. von Schuler, "Urartäische Inschriften aus Bastam". *AMI* 3: pp.93ff. (Also Vol. 5 [1972], pp.117ff.) Berlin, 1970.

F. Secret, *Les Kabbalistes chrétiens de la Renaissance*. Paris, 1964.

Sédillot, *Matériaux pour servir à l'histoire comparée des sciences mathématiques chez les Grecs et les Orientaux*. Paris, 1845–1849.

R. Seider, *Paläographie der griechischen Papyri*. Vol. 1. Stuttgart, 1967.

E. Seler, *Reproduccion fac similar del codice Borgia*. Mexico & Buenos Aires, 1963.

E. Senart, *Les Inscriptions de Piyadasi*. Paris, 1887.

H. Serouya, *La Kabbale, sa psychologie mystique, sa métaphysique*. Paris: Grasset, 1957.

H. Serouya, *Maimonide*. Paris: PUF, 1963.

H. Serouya, *La Kabbale*. Paris: PUF, 1972.

J. A. Serret, *Traité d'arithmétique*. Paris, 1887.

K. Sethe, *Hieroglyphische Urkunden der griechisch-römischen Zeit*. Leipzig, 1904–1916.

K. Sethe and W. Helck (1905–1908), *Urkunden der 18.ten Dynastie*. Leipzig, 1930.

K. Sethe, "Von Zahlen und Zahlworten bei den alten Aegyptern". *SWG* 25. 1916.

K. Sethe, "Eine altägyptische Fingerzahlreim". *ZAS* 54: pp.16–39. Berlin, 1916.

K. Sethe, *Urkunden des alten Reichs*. Leipzig, 1932–1933.

M. Silberberg, *Sefer Ha Mispar: Das Buch der Zahl, ein hebräisch-arithmetisches Werk des Rabbi Abraham Ibn Ezra*. Frankfurt/Main, 1895.

N. Sillamy, *Dictionnaire de la psychologie*. Paris: Larousse, 1967.

E. Simon, "Über Knotenschriften und ähnliche Kennzeichen der Riukiu-Inseln". *AMA* 1: pp.657–67. Leipzig/London, 1924.

M. Simoni-Abbat, *Les Aztèques*. Paris, 1976.

F. Škarpa, "Rabos u Dalmaciji". *ZNZ* 29/2: pp.169–83. Zagreb, 1934.

W. D. Smirnoff, "Sur une inscription du couvent de Saint-Georges de Khoziba". *CRSP* 12: pp.26–30. 1902.

A. Sogliano, "Isopsepha Pompeiana". *RRAL* 10: pp.7ff. Rome, 1901.

H. Sottas and E. Driotton, *Introduction à l'étude des hiéroglyphes*. Paris, 1922.

D. Soubeyran, "Textes mathématiques de Mari". *RA* 78: pp.19–48. Paris, 1984.

Jacques Soustelle, *La Pensée cosmologique des anciens Mexicains*. Paris, 1940.

Jacques Soustelle, *La Vie quotidienne des Aztèques*. Paris: Hachette, 1955.

Jacques Soustelle, *Les Quatre Soleils. Souvenirs et réflexions d'un ethnologue au Mexique*. Paris: Plon, 1967.

Jacques Soustelle, *Les Olmèques*. Paris, 1979.

F. G. Stebler, "Die Tesseln im Oberwallis". *DS* 1: pp.461ff. 1897.

F. G. Stebler, "Die Hauszeichen und Tesseln der Schweiz". *ASTP* 11: pp.165–205. 1907.

G. Steindorff, *Urkunden des Aegyptischen Altertums*. Leipzig, 1904–1916.

M. Steinschneider, "Die Mathematik bei den Juden". *BMA* : pp.69ff. 1893.

M. Sterner, *Geschichte der Rechenkunst*. Munich & Leipzig, 1891.

J. Stiennon, *Paléographie du Moyen Age*. Paris, 1973.

H. Stoy, *Zur Geschichte des Rechnenunterrichts*. Jena, 1876.

G. Stresser-Péan (1957). "La science dans l'Amérique précolombienne". In: R. Taton, *HGS*, 1: pp. 419–29.

P. Subtil, "Numération égyptienne". *BLPM* 27: pp.11–23. Paris, 1981.

P. Subtil, "Numération babylonienne". *BLPM* 27: pp.25–38. Paris, 1981.

P. Subtil, "Numération chinoise". *BLPM* 27: p.50. Paris, 1981.

H. Suter. "Das Rechenbuch des Abû Zakarija el-Hassar". *BMA* 2/3: p.15.

H. Suter, "Das mathematiker-Verzeichnis im Fihrist des Ibn Abî Ja'kûb an Nadim". *ZMP* 37/Supp.: pp.1–88. 1892.

H. Suter, *Die Mathematiker und Astronomen der Araber und ihre Werke*. Leipzig, 1900.

M. Sznycer, "L'origine de l'alphabet sémitique". In: *L'Espace et la lettre*: pp.79–119. Paris: UGE, 1977.

P. Tannery, "Sur l'étymologie du mot 'chiffre'". *RAR* . Paris, 1892.

P. Tannery, *Leçons d'arithmétique théorique et pratique*. Paris, 1900.

P. Tannery, "Prétendues notations pythagoriciennes sur l'origine de nos chiffres". *Mémoires scientifiques* 5: p.8. 1922.

N. Tartaglia, *General trattato di numeri e misuri*. Venice, 1556.

R. Taton, Ed. *Histoire générale des sciences*. Paris: PUF, 1957–1964

R. Taton, *Histoire du calcul*. Paris: PUF, 1969.

J.-B. TAVERNIER (1679), *Voyages en Turquie, en Perse et aux Indes*. Part II, pp.326–7. Paris, 1712.

THEOPHANES, *Theophanis Chronographia*. Paris, 1655.

F. THUREAU-DANGIN, *Une relation de la huitième campagne de Sargon*. Paris: Paul Geuthner, 1912.

F. THUREAU-DANGIN, "L'exaltation d'Ishtar". *RA* 11: pp.141–58. Paris, 1914.

F. THUREAU-DANGIN, *Tablettes d'Uruk*. Paris, 1922.

F. THUREAU-DANGIN, "L'origine du système sexagésimal". *RA* 26: p.43. Paris, 1929.

F. THUREAU-DANGIN, *Esquisse d'une histoire du système sexagésimal*. Paris: Paul Geuthner, 1932.

F. THUREAU-DANGIN, "Une nouvelle tablette mathématique de Warka". *RA* 31: pp.61–9. Paris, 1934.

F. THUREAU-DANGIN, "L'équation du deuxième degré dans la mathématique babylonienne". *RA* 33: pp.27–48. Paris, 1936.

F. THUREAU-DANGIN (1938). *Textes mathématiques de Babylone*. Leyden, 1938.

W. C. TILL, *Koptische Grammatik*. Leipzig, 1955.

L. F. C. TISCHENDORFF, *Notila editionis Codicis Bibliorum Sinaitici*. Leipzig, 1860.

E. TISSERANT, *Specimena Codicum Orientalum*. Rome, 1914.

DOM TOUSTAINT and DOM TASSIN, "Chiffres". In: *Nouveau Traité de diplomatique*, 3: pp.508–33. Paris, 1757.

J. TRENCHANT, *L'arithmétique ... avec l'art de calculer aux jetons*. See preface. Lyon: Michel, 1561.

F. VALLAT, "Les tablettes proto-élamites de l'acropole de Suse". *DAFI* 3: pp.93–104. Paris, 1973.

J. M. VALLICROSA, *Assaig d'historia de les idees fisiques i matematiques a la Catalunya medieval*. Vol. 1. Barcelona, 1931.

J. M. VALLICROSA, *Estudios sobre historia de la Ciencia española*. Barcelona, 1949.

J. M. VALLICROSA, *Nuevos estudios sobre historia de la Ciencia española*. Barcelona, 1960.

B. CARRA DE VAUX, "Sur l'origine des chiffres". *SC* 21: pp.273–82. 1917.

B. CARRA DE VAUX, "Tar'ikh". In: H. A. R. Gibb, *Encyclopédie de l'Islam*: pp.705–6. Paris: Maisonneuve, 1970–1992.

R. DE VAUX, "Titres et fonctionnaires égyptiens à la cour de David et de Salomon". *RB* 48: pp.394–405. Saint-Etienne, 1939.

VERCORS, *Les Animaux dénaturés*. Paris: Albin Michel, 1952.

J. VERCOUTTER, "La Monnaie en Egypte". In: *Dictionnaire archéologique des techniques* 2: pp.715–16. Paris, 1963.

J. VERCOUTTER, *L'Égypte ancienne*. Paris: PUF, 1973.

J. VÉZIN, "Un nouveau manuscrit autographe d'Adhémar de Chabannes". *BSNAF* : pp.50–51. Paris, 1965.

G. LEVI DELLA VIDA, *DI* 10: p.243. 1920.

G. LEVI DELLA VIDA, "Appunti e quesiti di storia letteraria: numerali greci in documenti arabo-spagnoli". *RdSO* 14: pp.281–3. 1933.

A. VIEYRA, *Les Assyriens*. Paris, 1961.

G. VILLE, Ed., *Dictionnaire d'archéologie*. Paris: Larousse. 1968.

A. VISSIÈRE, "Recherches sur l'origine de l'abaque chinois". *BGHD* : pp.54–80. Paris, 1892.

K. Vogel, *Die Grundlagen der ägyptischen Arithmetik*. Munich, 1929.

K. Vogel, *Mohammed ibn Mussa Alchwarizmi's Algorismus*. Aalen, 1963.

H. Vogt, "Haben die alten Inder den Pythagoreischen Lehrsatz und das Irrationale gekannt?" *BMA* 7/3: pp.6–20.

P. Voizot, "Les chiffres arabes et leur origine". *NAT* 27: p.222. Paris, 1899.

I. Vossius (1658). *Pomponi Meloe libri tres de situ orbis*. Frankfurt, 1700.

I. Vossius, *De Universae mathesos Natura et constitutione*, pp.39–40. Amsterdam: J. Blaue, 1660.

H. Waeschke, *Rechnen nach der indischen Methode. Das Rechenbuch des Maximus Planudes*. Halle, 1878.

J. Wallis, *Opera Mathematica*. Oxford, 1695.

J. F. Weidler, *De characteribus numerorum vulgaribus*. Wittenberg, 1727.

J. F. Weidler, *Spicilegium observationum ad historiam notarum numeralium pertinentium*. Wittenberg, 1755.

H. Weissenborn, *Zur Geschichte der Einführung der jetzigen Ziffern in Europa durch Gerbert*. Berlin, 1892.

J. E. Wessely, "Die Zahl neunundneunzig". *MSPR* 1: pp.113–16. 1887.

L. Weyl-Kailey, *Victoire sur les maths*. Paris: Laffont, 1985.

Willichius, *Arithmeticae libri tres*. Strasbourg, 1540.

A. Winkler, "Siegel und Charakter in der muhammedanischen Zauberei". *SGKIO* 7. Berlin,1930

C. de Wit, "A propos des noms de nombre dans les textes d'Edfou". *CDE* 37: pp.272–90. Brussels, 1962.

F. Woepke, "Sur le mot *kardaga* et sur une méthode indienne pour calculer le sinus". *NAM* 13: pp.392ff. Paris, 1854.

F. Woepke, "Sur une donnée historique relative à l'emploi des chiffres indiens par les Arabes". *ASMF*. Rome, 1855.

F. Woepke, *Sur l'Introduction des de l'Arithmétique indienne en Occident*. Rome, 1857.

F. Woepke, "Mémoire sur la propagation des chiffres indiens". *JA* 6/1: pp.27–79, 234–90, 442–59. Paris, 1863.

Woisin, *De Graecorum notis numeralibus*. Leipzig, 1886.

H. Wuttke, *Geschichte der Schrift und des Schrifttums*. Vol. 1: pp.62ff. Leipzig, 1872.

A.P. Youschkevitch, *Geschichte der Mathematik im Mittelalter*. Leipzig, 1964.

A.P. Youschkevitch, *Les Mathématiques arabes*. Paris: Vrin, 1976.

K. Zangenmeister, *Die Entstehung der römischen Zahlzeichen*. Berlin, 1887.

Charles Zervos, *L'Art de l'époque du renne en France*. Paris, 1959.

INDEX

Aba, tomb of 99

abacus 242–259, 405–413, 654–655, 720, 1099–1110; abacists and algorists 1141, 1165–1167; Akkadian 271–274; Assyro-Babylonian 273; Chinese 243, 413, 555–576, 1099; French 569; Gerbert's 1144–1145, 1148, 1149–1156, 1162; Greek 392–396, 407; Inca 603; Indian 855, 1105; Latin 1071–1072; *Liber Abaci* 710–711, 1162, 1164; Mesopotamian tablet 303–304; multiplication 407–410, 1100–1105; Persian 1098–1099, 1111–1112; Roman 366, 397–406, 410–413, 1140, 1142, 1145, 1149; Russian 569; suan pan 565–576; Sumerian sexagesimal 244–259, 272; Table of Salamis 393, 394–397; wax or sand 405–413, 1112

ibn Abbad 1041

al-Abbas 1029

ibn 'Abbas, 'Ali 1034

'Abbas, caliph Abu'l 1027

Abbasid caliphs 1010, 1014, 1027

ibn 'Abdallah, Abu 'Amran Musa ibn Maymun *see* Maimonides

ibn 'Abdallah, Ahmed ibn 'Ali 493

ibn 'Abdallah, 'Ali 493

ibn 'Abdallah, Sahl 1089

Abel, Niels Henrik 1177

Abenragel 700, 714, 1035

Abjad numerals (ABC) 477–478, 485–486, 511–513, 1082–1097

aboriginals, Australian 7, 32

Abraham 141, 496, 504, 716

Abrasax 507

absolute quantity 38

abstraction 27; counting 15, 34–35, 146 numbers 5, 41 *see also* calculation; model collections; place-value system

Abulcassis 1030

Abusir 768

Abyssinia 186, 482, 762

Academy of Sciences (France) 80

accounting 196–233, 366, 1068–1072; balance sheet 210–214; Cretan 347–349;

Elamite 198–208; Japan 565, 566, 567; Jews 461; Mayan 594–598; Mesopotamian 256; pocket calculator 410–413; Roman 366, 410–413 Sumerian 237–241, 255 *see also* bullae; calculi; quipus; tablet

Acor 799

acrophonic number-systems 361, 419, 762

Adab 156

Adad 315

Adam (first man) 498

al-Adami, Ibn 1032, 1045, 1046

addition: abacus 247, 399, 558–559; calculi, Sumerian use 237; Egypt, Ancient 340; suan pan 571–573

additive principle 451, 637–646, 654–660, 682–689; Americas 599, 603; India 781, 856; Roman numerals 366; Sheban 364

Adelard of Bath 406, 712, 1160

ibn 'Adi, Yahya 1013, 1032

Afghanistan 741, 760, 1030, 1033, 1042; and Arabs 1010, 1027, 1033; counting 182, 569; numerals 446, 723, 1054; writing 740, 1064

Aflah, Jabir ibn 1038

Africa: Arabian provinces 1028; base five in 67; Central 6, 39; counting 15, 89, 186, 243; East 139; Maghribi script 1065; number mysticism 180–181, 1095; South 6; West 34, 44–45, 135, 143

Africa, North: and Arab-Islamic world 1026, 1043, 1160; calculation 1104; counting 418–419; Goths 442; Morra 97; number mysticism 485, 490, 513, 1092; numerals 473, 478, 699–700, 1055, 1058–1059, 1061; writing 485, 1065

Agade *see* Akkadian Empire

ages of the world 839

Aggoula, B. 658–659

Ah Puch, god of death 611

Aharoni, Y. 462

Ahmad, Abu Hanifa 1031

Ahmad, Ali ibn 1032

ibn Ahmad, Khalil 111, 1026

ibn Ahmad, Maslama 1034

Ahmed 713, 1009
Ainus 67, 598
Akhiram 416
al-Akhtal 1027
Akkad 262
Akkadian Empire 156, 261–284; bullae
    193–195; counting 271–274; Mari 142,
    156, 261, 277–284, 659–660; number
    mysticism 180, 311–313; numbers 174,
    260–261, 264–271, 277–284 writing
    253–259, 262–264, 311–313 *see also*
    Assyrian; Babylonian civilisation
Aksharapallî numerals 763
Aksum 482, 762
Albania 61–63, 65, 1042
Albategnus 1031
Albright, W. F. 277
alchemy 1023–1024, 1094
Alexander the Great 263, 501, 760, 802
Alexandria 1017, 1031
Algazel 1037
algebra 1162, 1179–1180; Arabs and 1029,
    1035, 1040, 1042, 1043, 1048;
    Brahmagupta 825, 865, 1047
algebraic numbers 1177–1179
Algeria 486, 1028, 1043; counting 93, 127,
    1096
algorists 1160–1161, 1166
algorithms 1104, 1159; al-Khuwarizmi 1048,
    1159
Alhazen 1035
'Ali, Abu'l Hassan 112
'Ali, caliph 1025, 1026, 1096
ibn 'Ali, Hamid 1031
ibn 'Ali, Sanad 716, 1029
Ali (language) 39
alien intervention theory 1171–1172
Allah 90, 113, 418, 419, 1015, 1016, 1092,
    1093; attributes of 17, 96–97, 136,
    511–513, 1070, 1093
Allah, Abu Sa'id 'Ubayd 1036
Allah, Sa'id ibn Hibat 1037
Allard, A. L. 717, 855, 1053, 1111, 1112,
    1164
Alleton, V. G. 521, 523, 524, 532
almanacs 382–383
alphabet 414–418; Greek 372; Hebrew
    420–427; palaeo-Hebrew 414, 456;
    Samaritans 414
alphabetic numerals 305–307, 414–513,

645–646, 953–954; Arabic 308,
    472–481, 1018, 1082–1097; Aryabhata's
    852; Ethiopian 482–484; Greek
    427–436, 443, 454–455, 466, 468, 645,
    653, 708; Hebrew 308, 443, 456–461,
    466, 467–468, 680, 711; Indian 765;
    Syriac 469–473, 645; Varnasankhyâ's
    763
Alpharabius 1032
Alphonsus VI of Castile 1036
Americans, native: counting 15, 123, 135,
    139, 243, 384; number mysticism
    180–181, 1095 use base five 67
    see also Maya
amicable numbers 1031
Amiet, P. 153, 195
'Ammar 1036
Ammonites 414
Amon 321
Amorites 73, 263
amp 82
al-'Amuli, Beha ad din 713, 1043
Anaritius 1030
Anatolia 145, 187, 190 *see also* Hittite;
    Ottoman Empire; Turkey
Anbouba 1009
al-Andalusi, Sa'id 1017, 1018
Andalusia 1036, 1038; abacus 1098, 1106,
    1112; and Arabs 1010; numerals 1055,
    1061–1062, 1065
Andhra numerals 781–784
Andromeda nebula 1034
Anglo-French, word for money 139
Anglo-Saxon, number names 61–66
animals, counting abilities 1–3
anka (numerals) 724, 818, 820
Annam 532–534 *see also* Vietnam
al-Ansari 1043
al-Antaki 1068, 1070
Antichrist 509
Anu 180, 314
Anûshîrwân, King Khosroes 1011
Anuyogadvâra Sûtra 838
Anwari 111, 1037
Api language 67
apices 1143–1158; of Boethius 1143
Apollonius of Perga 433–434, 709, 1012,
    1033
apostles, New Testament 504
Apuleus 106–107

Aquinas, St Thomas 1017

Arab-Islamic civilisation 99–100, 157, 307, 308, 362–365, 766, 1008–1138; language 262, 263–267, 414–416, 1012, 1020–1022; number-systems 111, 307, 445, 685, 864, 1073–1082, 1175; alphabetic numerals 469–481, 1018, 1091–1097; counting 73–75, 89, 93–94, 127, 134–135, 185–186, 843–845; Indian numerals 723–724, 1008–1069 science 1010, 1015–1016, 1027 *see also* Baghdad; Muslims

ibn 'Arabi 1040

Arabic numerals 45, 108, 772–780, 1055–1065, 1168; in Europe 1139–1167; origins 699–705, 758

Arad 416, 462

Aramaean Indian writing *see* Kharoshthî writing

Aramaeans 261, 461, 740; number-system 73, 267, 443–451, 649–651, 658, 690

Aramaic script 414–416, 461, 469–470, 740–741, 761–762, 767; cryptography 485–486; Jews adopt 456, 467, 468

Aranda people 6, 138

Arawak, base five in 67

archaic numerals 163–164, 167–174, 178–179; accounting tablets 208, 232; bullae 201–202; calculi 242; Sumerian 147–152, 160–162, 178–179, 192–193, 208, 226

Archimedes 405, 435, 709, 1022, 1030, 1031, 1035, 1172, 1180; Arabic translation 1012, 1161; high numbers 653

Ardha-Mâgadhî 754

are (unit of measurement) 80

Argos 429

Aristarch of Samos 433

Aristophanes 89

Aristotle 35, 1010–1011, 1012, 1017, 1020, 1161, 1180

arithmetic 5, 15–16, 185, 401, 406, 1042, 1043; during Renaissance 1139–1141; early 146, 185–187; systems 362, 430–435, 486, 871–872

Arithmetic, Lady 402, 1167

Arithmetical Introduction 82–83

Arjabhad 842

Arjuna 833–835

Arkoun, M. 1015, 1024

Armenia 270, 569, 1025, 1042; alphabet 414; numerals 61–65, 439–440, 645, 646

Arnaldez, R. 1011, 1015, 1016, 1019

Arnold, Edwin 830, 1062

Artaxerxes, King 106

Aryabhata 763, 825, 828, 842, 852, 880–888, 1046

Aryans 758–759

as (Roman unit) 178, 412–413

Asankhyeya 888

al-Ash'ari, Abul Hasan 'Ali ibn Ismail 1031

Ashtâdhyâyî 765

Asia 155, 792–801, 1010; counting 67, 89, 91–94 number-systems 181, 812–814 *see also* China; India

Asia Minor 146, 155, 351–352, 429, 1035

Asianics 261

al-Asma'i 1028

Asoka, Emperor 738–741, 760, 761–762, 828, 854, 857; edicts 768–770, 781

Assemblée constituante 80

Assurbanipal, library of 313

Assyrian Empire 261, 262–263, 352; counting 74, 192; language 262; number-systems 178, 271, 274–275, 451

Assyro-Babylonian civilisation 263, 272–273; number-system 14, 267, 274–275, 649–671, 690

astrology 311, 1084, 1093, 1098; "Greek" 827; Indian 821–822, 913; and Koran 1015

Astronomie Indienne 874

astronomy 177, 1031, 1034, 1035, 1045, 1088; Arabic 476, 1046–1047, 1083–1085; Babylonian 298, 299, 305–308, 802; Chinese 542–545; Greek 305–307, 803, 1084; *Ikhanian Tables* 1041; Indian 805–809, 819–822, 849–852, 871, 874, 913–914, 1012–1013; Mayan 581–584, 604–614, 618–619, 629–632; sexagesimal system 176–177, 183–184, 273, 307, 1083–1085; tables 284, 306–311, 388, 1029 trigonometrical 827 *see also* lunar cycle

al-Asturlabi, Ali ibn 'Isa 1029

al-Asturlabi, Badi al-Zaman 1023, 1038

al-Atahiya, Abu 1028
Athens 355–357, 429, 455
'Attar, Farid ad din 1041
Attica 358, 429
Atton, Bishop of Vich 1142
Auboyer, J. 766
Augustine 504
aureus (Roman money) 411
Aurillac, Gerbert of 712, 1023, 1142–1143, 1149, 1156
Australia 6–7, 32, 138, 180–181
Austria 128
Autolycus 1031
Avempace 1038
Avenzoar 1038
Averroes 1015, 1017, 1024, 1039
Avestan 59, 60, 61–65
Avicebron 1014
Avicenna 714, 1012, 1014, 1017, 1020, 1024, 1036, 1041, 1070–1071; "Avicenna" Arabic alphabet 1064
Avigad 457
Awan 156
Axayacatl 590
Ayala, Guaman Poma de 132, 604
Aymard, A. 766
Aymonier, E. 794
Azerbaidjan 1025, 1042
Aztecs 590–593, 618; monetary system 139–140, 592, 593, 599; number-system 68, 84, 89, 598–603, 639, 641, 684–685; writing 592–593, 598–602
Aztlan 590

al-Ba'albakki, Qusta ibn Luqa 1022, 1031
Babel, tower of 310
Babur 1043
Babylonian civilisation 156, 261–315, 352–353; arithmetic 75, 192, 271, 298–305; cryptograms 308–313; number-system 178, 180, 271–300, 451, 662–671, 678, 694–696, 802–803; writing 261, 269, 298, 308
Bachelard, Gaston 873
ibn Badja (Avempace) 1038
Baghdad 1011, 1027, 1028, 1031–1032, 1036, 1040, 1042; House of Wisdom 1029, 1046–1047, 1048
al-Baghdadi, 'Abd al-Qadir 1043

al-Baghdadi, Muwaffaq al din Abu Muhammad 713, 1040, 1070
Bahrain 93
bakers, counting methods 125–127, 135
ibn Bakhtyashu', Jibril 1013, 1028, 1030
Baki 1043
Bakr, Abu, caliph 1025, 1031
Bakr, Muhammad ibn Abi 1040
al-Bakri 1037
balance sheet 210–214
Bâlbodh writing 748
Bali 798–799, 800, 801, 829; numerals 737, 738, 754, 758
Balkans 1042
Balmés, R. 38
Baltic 61–62
Bamouns, decimal counting 74
Banda 68–69, 84
Banka Island 795
Banu Musa ibn Shakir brothers 1022, 1030
banzai 538
Baoule 74
al-Baqi, Muhammad ibn 'Abd 1037
al-Baqilani, Abu Bakr Ahmad ibn 'Ali 1034
Baqli, Ruzbehan 1039
al-Barakat, Abu 1038
barayta 497
Barguet, P. 344
Barmak 1012
Barnabas 504
barter 138–145
Barton, G. 170
base numbers 41–88, 186; auxilliary 840–846; eleven 76; five 67, 84–88, 120, 376–378; 'm' 698; six 276 *see also* binary; decimal; duodecimal; sexagesimal system
Basil I 1030
Basilides the Gnostic 507
Baskerville face numerals 1161
Basques 71–73
Basra 1025
al-Basri, Hasan 1027
Bastulus 1032
Batak numerals 754, 758
Bath *see* Adelard
ibn Batriq, Yahya 1013, 1029
al-Battani (Albategnus) 1014, 1031
ibn Battutua 1041
Bavaria 1156

al-Bayasi, 'Abu Zakariyya Yahya 1023, 1039

Bayer 705

Bayley, E. C. 117, 760–762

al-Baytar, ibn 1041

Beast, number of 509–511

Beauclair, W. 1140

Beaujouan, G. 699, 1142

Becker, O. 176, 1024

Bede, Venerable 94–95, 96, 100–105, 108, 391, 437, 1142

Beg, Tughril 1036

Bek, Ulugh 1042

Belgium 58, 126

Belhari 790

Belize 586

Bengal 93, 96, 767, 1038

Bengâlî numerals 727–728, 749, 755, 756, 829, 864

Beni Hassan 98–99

Benin, Yedo 598

Béquignon sisters 48

Berbers 74, 1010

Bereshit Rabbati 496

Bergamo, Gnosticism 510

Bernelinus 1146

Beschreibung von Arabien 92

Bessarabia 1042

Bete 67

Bettini, Mario 700

Bhadravarman, King 801

Bhagavad Gîtâ 832

Bhâskara 825, 890, 1047; *Aryabhatîya, Commentary on* 816–818, 828, 865

Bhâskarâchârya (Bhâskara I) 815–818

Bhâskarâchârya (Bhâskara II) 823, 850, 890, 1110; multiplication method 1132–1134

Bhattiprolu writing 742, 753, 757–758

Bhoja 815

Bible;

Old Testament 261, 496–498; Daniel 266; Deuteronomy 497; Esther, Book of 266; Exodus 141; Ezekiel 467; Ezra 266; Genesis 261, 496, 498; Leviticus 497; Micah 501; Nehemiah 266; Pentateuch 266; Prophets, Books of the 266; Psalms 416, 424; Samuel 141 Zechariah 266 *see also Torah*

New Testament 503; Gospels 475, 503, 504; Matthew 503; Revelation (Apocalypse) 502–503, 509

Bihar 492, 1038

bijection 16

Billard, R. 800, 802, 815–822, 824, 850–851, 852, 873, 1045

billion 841, 842–843

binary principle 7, 14, 171, 270, 324

binary system 75–77, 114, 1168

binomial formula 1042

Biot, E. 553, 660

Birman numerals 864

Birot, M. 260, 269

birth-date 614

al-Biruni, Muhammad ibn Ahmed Abu'l Rayhan 1012, 1014, 1016, 1024, 1035, 1161; Indian numerals 823, 864; *Kitab fi tahqiq i ma li'l hind* 714–715, 723–724, 806, 840–845, 1046, 1047, 1054; *Tarikh ul Hind* 492

Bisaya writing 754, 758

ibn Bishr, Sahl 1014, 1029

Bistami 1030

biunivocal correspondence 16

Black Stone, The 285

blackboard 1119–1121

black-letter Hebrew (modern) 414, 420, 456

Bloch, O. 718, 842

Bloch, R. 372

boards: checkerboard 555–565; columnless 1107–1112; dust 1097–1112; wax 405–410, 1112 wooden; 124, 1055, 1059 *see also* abacus; tally sticks

Bodhisattva *see* Buddha

body counting systems 5, 20–34, 41, 418–419 and base 83–88 *see also* finger counting

Boecius 1146

Boethius 91, 1142, 1143

Bokhara 1012, 1030

Bolivia 133–134, 1072

Boncompagni, B. 406, 712, 717, 1053

bones 119–122, 526 *see also* tally sticks

Book of Animals 1029

Book of Kings 1036

Boole, George 1181

Boorstin, D. J. 176

Borchardt 105

Borda 80

Borneo 737, 754

Botocoudos 6, 138
Bottero, J. 153, 154–155, 313
Bouché-Leclerq, A. 707
Bourdin, P. 38
Boursault 403
boustrophedon writing 363, 429
Brahma 740, 823, 826, 831–832, 841–842, 870
Brahmagupta 815, 825, 865, 892;
 Brâhmasphutasiddhânta 828, 865, 1027, 1046–1047, 1134; multiplication method 1134–1138
Brâhmî numerals 743–745, 751, 755–780, 791, 827, 854–860, 892
Brâhmî writing 738–743, 781–785
Brasseur de Bourbourg 588
Brazil *see* Botocoudos
de Brébeuf, Georges 403
Breton 61–65, 71, 72
Brice, W. C. 210
bride, price of 138
Brieux 402
Britain, Great 177, 419; Treasury 1166
British Honduras 586
British Museum 284, 333, 334, 344
British New Guinea 21–22, 23
Brooke 32–34
Brothers of Purity (Ikhwan al-safa) 1034
Bruce Hannah, H. 48
Le Brun, Alain 196, 210
Bubnov, N. 703, 788
Buddha 136, 827, 829–830, 844
Buddhism 17, 136, 801, 803–804, 803–805, 824, 873, 1012
Bugis 24, 754, 758
Bühler 765–766, 864
Buhturi 1030
al-Bujzani, Abu'l Wafa' 1033, 1083
Bulgaria 1042
bullae 187, 192–204, 236, 457
ibn Bunan, Salmawayh 1013, 1030
Bungus, Petrus 389, 510
al-Buni 1093, 1094
Buonamici, G. 385
Bureau des Longitudes (Paris) 81
Burgess 809
Bürgi, Jost 1175
Burma, numerals 736–737, 755, 756, 758, 763; writing 751, 754
Burnam, R. L. 713, 1144, 1146

Burnell, A. C. 765–766, 864
Burnham 109, 391
Bushmen 6, 138
ibn Butlan 1036
Byzantine Empire 435, 469, 707, 1022, 1033, 1061; arithmetic 655

Cabbala 424, 1094
Cadmos 428
Caesar, Julius 9
Cagnat 389
Cai Jiu Feng 546
La Caille 81
Cairo 1012, 1033, 1035
Cajori, F. 699, 701–702, 855
Cakchiquels, Annals of the 589
calamus reed 1064, 1092
Calandri, Philippi 1164
calculation 73, 256–257, 340–344, 655, 1068, 1112–1119; abacus 1069, 1071; Babylonian 298–305; Egyptian 74, 340–344, 655; Mayan 594–598, 604, 629–632; North Africa 1104 tables 246–252, 284, 397–400, 403–404, 555–565, 1097–1112 *see also* abacus; body counting; calculi; notched bones; string; tally sticks
calculator, pocket; first 410–413
calculi 185, 192–204, 229–230, 242, 244–245, 271, 328; Elamites 199–200, 274; Roman abacus 398–401; Sumerian 234–241, 254, 255
calculus 1181
calendar 31, 95, 467, 1037; ciphers 382–383; Hebrew 420, 424–425; lunar 34, 582, 801; Mayan 68, 582–583, 604, 610–632; Roman 9; Shaka 801
Callisthenes 501
Calmet, Dom 706
Cambodia 801, 825; inscriptions 795–798, 849 numerals 737, 793, 814, 864 *see also* Khmer
Cambridge Expeditions 19–23
Campeche 586, 594
Canaan 446, 467
candela 82
Canossa, Darius vase 392, 393
Cantera 423
Canton 532
Cantor, Georg 1181

Cantor, Moritz 175
caoshu- writing 522–525
Capella, Martianus 406
cardinal numeration 35–39, 43–48, 377;
    Attic system 356–357; reckoning
    devices 26–34, 186; Yoruba 70
Carib 67
Carolinas Islands 135
Carolingian script 1157
Carra de Vaux, B. 703, 716, 788
cartography 1038
Casanova, P. 1085
Catalonia 437
Cataneo 709–710
Catherwood, Frederick 588
Cato 379
cattle 139
Cauchy 1181
Cayley, Arthur 1181
Ce Yuan Hai Jing 553
Celebes Islands 737
Celtic numbers 71, 72, 73
censuses 131
centesimal-decimal system 279–283
Central America 588–592, 603; counting
    15, 594–596; numbers 316, 614
    trading methods 139–140 *see*
    also Maya
Ceylon 651 *see also* Vedda
cha lum numerals 736
Chalcidean alphabet 372
Chalfant 527
chalk 1120
chalkos 356, 392, 394–397
Chalmers, J. 24
Chamealî numerals 749, 755
Champa 795–801, 824; inscriptions 828,
    829, 849; numerals 754, 757, 814, 829
Chanakya 1030
Chandra, Hema 838–839
Changal, Stela of 813
Chapultepec 590
Charlemagne 1028
Charles III, king of Spain 486
Charpin, Dominique 170
Chassinat 344
de Chavannes 522
Chelebi, Evliya 1044
Cheremiss, tally sticks 127–128
chess 633–636

Chevalier, J. 862, 873, 1093, 1096
chevrons 289–291
Chhedi 894
Chiapas province 586, 594
Chichén Itzá 587
Chilám Balám, Books of 589
children 3–5, 16, 419
chimpu 70 *see also* string, knotted
China 97, 514–534, 541–580, 750; abacus
    555–576, 1099; counting 74, 93, 93–4,
    95, 117, 128, 135, 673–674, 843–844;
    high numbers 541–545, 653, 843–844;
    monetary system 140, 145; number
    mysticism 1095; number-system 316,
    328, 514–580, 652, 660–674, 693,
    694–696, 737; outside influences
    803–805, 1010, 1018, 1027, 1039
Chinassi 1044
Chinese Turkestan writing 751, 757, 827
Chodzko, A. 1073, 1077
Chogha Mish 196
Chorem 1012
chóu 243, 555–565
Christ 492
Christianity 1012; Arabic 469–470, 1013;
    Central America 588–589; demonised
    Arabic numerals 1162–1163 isopsephy
    259–261 *see also* Crusades
chronograms 490–494, 1092
    see also codes and ciphers
Chuquet, Nicolas 842–843
Chuvash 127–128
Cicero 89, 98, 379, 398
circle 178
City of God, The 504
Claparède, E. 717
classification of sciences 1020, 1032, 1036
clay objects: accounting 149, 153, 210–215
    tokens 185, 191–192 *see also* calculi;
    tablet
clock-making 1022, 1023
Coatepec 592
Coatlan 592   ʼ
Code Napoléon 128
codes and ciphers 158–161, 248–262,
    553–554 *see also* mysticism; numerology
Codex Aemilianensis 1144
Codex Mendoza 68, 591, 593, 599
Codex Morley 584
Codex Selden 584

Codex Telleriano Remensis 602
Codex Tro-Cortesianus 584, 590, 611
Codex Vigilanus 712, 1146
Codices, Hebrew 424
Codrington, M. 7, 34
Coe, M. D. 585, 628
Coedès, G. 794–795, 801, 814
Cohen, M. 362, 473, 476, 478, 740, 759, 1052
Cohen, R. 465
coins 145–146, 358, 371, 1026
Colin, G. S. 478, 490, 493–494
columnless board 1107–1112
Comte, Auguste 1042
Conant, L. L. 34, 86
concrete numeration 37, 41–42, 327–328
La Condamine 79
Congo, early money 141
conic sections 1031, 1032
Conrady, A. 128
Constantinople 1026, 1042
Contenau, G. 310
contracts 127–128, 135
Coomaraswamy 826
Copán 582, 614, 628
Coptic 329, 438
Copts 105
Cordoba 1012, 1033, 1037, 1160
Cordovero, Moses 495
Corinth 429
correspondence 38–39; biunivocal 16; one-for-one 16–19, 27–30, 34, 186, 374, 379
Cortez 592
Cos 358
Cottrell, L. 1053
de Coulanges, F. 720
Coulomb 81
counting 15–16, 34–40, 146; cuneiform ideogram meaning 254; methods 119–122, 130–137, 191–192; rhymes 419; systems *see under* body counting; correspondence; mapping; *see also under* specific race/country
cowrie shells as currency 139
Crafte of Nombrynge, The 710
Creation 425, 492, 715; Mayan Long Count 619, 625, 627
Cremonensis, Geraldus, *Liber Maumeti filii Moysi Alchoarismi de algebra et almuchabala* 1048

Crete 347–351, 1028, 1033; Linear A and Linear B 447, 639, 640; number-system 14, 347–351, 639, 640, 684
Crimea 442, 1042
Cro-Magnon man 119–122
Crusades 1037, 1038, 1039, 1158–1162
cryptography 308–315, 485–490, 508–511, 1094–1095 *see also* mysticism; numerology
Ctesibius 1022, 1172
cubes 714, 1034; roots 558, 575, 1177
cubit 275
cufik *see* kufic
cuneiform notation 167–169, 263–268, 277, 352–353; codes and ciphers 308–315; decimal 267, 270–272; numerals 160–161, 172–174, 193–194, 242, 283; script 208, 234, 288–291; tablets 253–261
Cunningham 760–762
Curr 7
currency 78, 138–146, 356–360, 603
Curtze 710
curviform notation 242, 253
Cushing, F. H. 25, 384
Cuvillier, A. 720
Cyclades 429
cylinders 1030
cypher, etymology of 1163–1166
Cyprian 504
Cyprus 1033
Cyril of Alexandria, Saint 107
Cyrillic alphabet 414
Cyrus of Persia 263
Czech, number names 61–62, 65

dà zhuàn writing style 549
Dadda III 825
al-Daffa, A. A. 1024
DAFI (French Archeological Delegation in Iran) 196–198, 199, 203–206, 210, 273
DAGR 435
ad Dahabi, Ahmad 493
Dahomey 34
Daishi, Kôbô 580
Dalmatia 380–381
Damais 798, 799
Damamini, Ad 1043
Damascus 1012, 1025, 1026, 1036, 1038

Damerov 178, 179
Dammartin, Moreau 1161
Dan 67, 496
Dantzig, T. 7, 39, 41, 67, 88, 655
Daremberg, C. 433, 843
al-Darir, Abu Sa'id 1029
Darius, King 134, 394
Darwin, Charles 1024
Das, S. C. 48
Dast, Zarrin 1037
Dasypodius, Conrad 704–705, 1165
Datta, B. 1132
Datta, B. 699, 715, 716, 763, 785, 787, 815, 825, 832, 855, 864, 1110, 1122
d'Auxerre, Rémy 406
da'wa 511–513
Dayak 32–34
De computo vel loquela digitorum 100, 108
De pascha computus 504
De ratione temporum 94, 100
Dead Sea Scrolls 416, 457
decimal system 44–67, 73–83, 696–698, 1070; Ben Ezra 680; Chinese 514, 545–554; counting 131–132, 186, 271–277, 377, 407–410; Cretan 347–348, 351; Egyptian 74, 316; fractions 553, 1042, 1174–1175; hieroglyphs 327; Mesopotamia 270–284; metric system 79–82; proto-Elamite 232; Semitic 264; Sumerian 179, 183
Decourdemanche, M. J. A. 485, 488
Dedron, P. 91, 433, 434, 843
Deimel, A. 158, 162, 172, 234, 254, 256–257
Delambre 81
Delhi 1038, 1041, 1043
Demetrius II 457
demotic writing 335
denarius 411–412
Dendara, temple of 344
Dene-Dindjie Indians 87–88
denier, French unit 178
Denmark 61, 71, 73
Dermenghem, E. 1024
Descartes, René 80, 390, 1179, 1181
Destombes, M. 707
Devambez, P. 358
Devanagari numerals *see* Nâgarî numerals
Dewani numerals 1073–1076, 1086

Dharmarâksha 838
Dharmashâstra 826
Dhombres, Jean 82
Dibon-Gad 415
Dickens, Charles 125
dictionary of Indian numerical symbols 867–1007
Diderot 1024
Diener, M. S. 873
digital 114
Dilbat 268–269
Ding zhu suan fa 673–674
Diocletian 510
Diophantus of Alexandria 433, 1012, 1031, 1033, 1180
Diringer 528
disability, spatio-temporal 5
divination 159, 269, 549–556 *see also* mysticism
Divine Tetragram 427, 498 *see also* Yahweh
division: à la française 1118–1119; abacus 247–253, 403–404, 558, 563; calculi 234–241; Egypt; Ancient 340, 343
Dobrizhoffer, M. 7
Dodge 715
Dogon 139
Dogrî numerals 728–729, 749, 755, 756, 829
Dols, P. J. 93
Donner 447
Dornseiff, F. 501
Doutté, E. 1093–1094, 1096
dozen 77, 177, 183
drachma 356–358, 394–397
Dravidian numerals 734–735, 753–754
Dresden Codex 589–590, 604, 607–608
duality 60
Duclaux, J. 722
Duke of York's Island 7
Dulaf, Abu 1032
Dumesnil, Georges 700
Dumoutier 532, 533
dung (counting device) 19
duodecimal system 77–79, 81, 177–184
duplications, abacus 404
Dupuy, Louis 80
Durand, J-M. 283, 659
dust-board calculation 1097–1112
Dutch, number names 61–65
ibn Duwad 1033

Dvivedi, S. 815, 818, 865

dyadic principle *see* binary principle

e, number 1178

Ea 314

Easter, determining 95

Ebla tablets 262, 283

eclipse 1045

Ecuador 133, 1072

Edesse 1011

Edfu, temple of 344

Edomites 414

Egine 358

Egypt, Ancient 141–145, 316, 317, 324–325, 462, 508, 765–766; and Arabs 1025, 1030, 1031, 1033, 1036, 1038, 1040, 1042, 1043, 1076; sign language 105

    calculation 74, 340–344, 655; abacus 1069–1070; fingers 97–99, 118, 182

    number-system 14, 175, 316–346, 637–646, 671–688, 767; alphabetic numerals 454–455, 465–466, 476; Arabic numerals 699–700; Indian numerals 723–724, 1054; number mysticism 1094–1095

    writing 316–317, 414, 771; hieratic 332–339, 462–468; secret 485–486

Egyptian Mathematical Leather Roll (EMLR) 344

eight 64, 779; Chinese 527; Egypt 345; Hebrew 421; Indian 807; Japanese 535

eight hundred, Hebrew 423, 424

eight thousand, Aztec 598; Mayan 603

eighteen, Egypt 346

eighty, cryptographic 486; Hebrew 421, 459

eighty-eight, Japanese 578

Eisenstein 186

El Obeid 134–135

El Salvador 586, 594

Elam 197–198, 261, 263; accounting 192, 198–208, 214–233; calculi 274; cryptograms 312; numerals 14, 74, 187, 192–233, 284; proto-Elamite script 207–233

Elema's body counting system 22, 23

Elephantine 416, 444, 447–451, 453, 459, 767, 768

eleven, base 78

Eliezer of Damascus 496–497

Elogium of Duilius 370

Elzevir script face 1161

EMLR (Egyptian Mathematical Leather Roll) 344

end of the world 839

engineering 1022–1023

England: clog almanacs 382, 384; number names 57, 61–66, 844; score (twenty) 70–71; tally sticks 125

English, Old 139

English script 1157

Englund 178, 179

Enlíl 314, 315

Ephesus 389

Ephron the Hittite 141

Epicurus 1031

Epidaurus 362

epigraphy 786–801, 792–801, 826

Epistle of Barnabas 504

Equador 604

equations 554, 563, 1009, 1037, 1047

equivalence between sets 2–3

Erhard, F. K. 873

Erichsen, R. W. 671

Erpenius 705

Erse, Old 61–65

Eskimos 67

Essene sect 458–459

Essig 78–79

estranghelo 469

Ethiopia, number-system 267, 466, 484–484, 693

Ethiopian numerals 762

Etruscans: abacus 243, 397; alphabet 414, 416; number-system 14, 74, 370–372, 641, 685; Roman numerals 384–385

Eubeus 429

Euboea 358

Euclid 1011, 1012, 1030, 1041, 1161, 1180; *Elements* 1029, 1032, 1037

Euclidian geometry 1180

Europe 79, 1024, 1027, 1128–1129; Arabic numerals 1139–1167, 1158–1167

European numerals 772–780

Evans, Sir Arthur 347–350

Eve 498

evolution theory 1024

Ewald 713

Exaltation of Ishtar 310

exponential powers 1042, 1172–1173

Ezra, Rabbi Abraham Ben Meïr ibn 680–681, 711, 1014, 1038, 1164

al-Fadl 1027
Fairman, H. W. 1344
Falkenstein 156, 158
al-Faqih, ibn 1032
Far East 135, 532–541; 545–554, 577–580 *see also individual countries*
Fara 168, 196 *see also* Suruppak
al-Farabi 1014, 1020, 1032
Faraut, F. G. 801
al-Farazdaq 112, 1027
al-Farghani 1029
Farhangi Djihangiri 100
al-Farid, ibn 1040
al-Farrukhan, 'Umar ibn 1029
Farsi language 1022
Fath, Abu'l 1033
al-Fath, Sinan ibn 1034
Fatima (Mohammed's daughter) 136, 1070, 1096
Fayzoullaiev, O. 1024, 1048
al-Fazzari, Abu Ishaq Ibrahim 1012, 1014, 1027, 1046
al-Fazzari, Muhammad Ben Ibrahim 1027, 1045, 1046
feet and inches 178
Fekete, L. 1073, 1081
Feldman, A. 1022, 1024
Fénelon 403
Ferdinand II 1043
Février, J. G. 124, 127, 135, 151, 362, 416, 428, 740, 751, 762
Fez 493, 1012, 1027, 1036, 1037, 1040, 1065
Fibonacci 718, 1032, 1162, 1180; *Liber Abaci* 710–711, 1162, 1164
fifteen 315, 346, 427
fifty 359, 360, 363, 364, 421; cryptogram 180, 314, 315, 486; Roman numerals 367, 368, 376
fifty thousand 359, 386, 387
fifty-three 589
fifty-two 618–619
Fihrist al alum, Al Kitab al 715, 1048, 1064
Fijians 34
Filliozat, J. 658, 864, 873
finger counting 39–40, 51, 89–118, 328, 1142 and base 44, 93–95 *see also* Bede; body counting

Finkelstein 261
Finot, L. 800
Firduzi, Abu'l Qasim 111, 1036
Fischer-Schreiber, I. 873
five 63, 345, 380, 776, 871, 1095–1096; Attic 355; base 67, 84–88, 120, 376–377; Chinese 527; Greek 359, 360; Hebrew 421; Indian 808; Mayan 604; Minaean 363, 364; Roman 367, 368, 375, 376; rule of 14; Sheban 363, 364
five hundred 359, 360, 367, 368, 423–424
five thousand 359, 360, 386, 387
Fleet 864
floating-point notation 1173
Fold, P. 1022, 1024
Folge 716
Folkerts, M. 1146, 1147
Formaleoni 175
Formosa 140
fortune-telling *see* mysticism
forty 180, 314, 421, 486
forty-nine 871
forty-two 540
Fossey 533
Foulquié, P. 717, 1069
four 62, 345, 421, 775, 807; base 181; Chinese 527, 531; Japanese 535; limit of 9–14, 34, 40, 769; mysticism 181, 540
four hundred 421, 598, 603
four thousand 540
Fournier 1161
fourteen 315, 346
fractions 1083, 1084, 1173, 1177; Babylonian 294, 299, 803; decimal 553, 1042, 1174–1175; Egyptian 329–332; Indian 836–837; Maya 584
France 79, 80–82, 97, 139, 1140, 1157; counting 60, 71, 125–127, 569; French Revolution 79, 404, 1166, 1175–1176; metric 42–43 number-system 58, 61–66, 178, 842–843, 1156 *see also* DAFI
Franz, J. 355
Frédéric, Louis 514, 535, 580, 722, 735, 740, 765, 804, 821, 838, 868, 873, 1012, 1024, 1072
Freigius 387, 390
French National Archives 81
Friedrichs, K. 873
Frieldlein 1146

Frohner 106
Fuegians 6
Fulah 67
Fuzuli 1043

von Gabain, A. M. 50
ibn Gabirol, Salomon (Avicebron) 1014
Galba, Emperor 391
Galen 501, 1011, 1012
Gallenkamp, C. 581, 610, 612, 615
Galois, Évariste 1177, 1181
games 577–580
Gamkrelidze, T. V. 758
gán mà zí writing 524, 525
Gandhara 446
Gandz, S. 1070
Ganesha 1122
Gani, Jinabhadra 786, 825
Ganitasârasamgraha 786, 829
Garamond, Claude 1161
du Gard, Martin 1069
Garett Winter, J. 307
Gauss, Karl-Friedrich 1164, 1181
Gautama Siddhânta *see* Buddha
Gautier, M. J. E. 268
Gebir 1028
gematria 494–502, 1094
Le Gendre, F. 1118–1119
Gendrop 582, 584
Genjun, Nakane 566
de Genouillac 166, 170
geography 1083, 1097
Geometria Euclidis 1143
geometry 177, 1068, 1083, 1162, 1181;
    base 176–177, 184; base 12 78;
    Non-Euclidian 1041
Georgia 414, 441, 1042
Geraty 460
Gerbert of Aurillac 712, 1023, 1142–1143,
    1149, 1158
Germany: counting 125, 135, 401, 402, 404;
    language 52, 61–66, 139, 1157; number-
    system 57, 58, 61–66
Gernet, J. 526–527, 528
Gerschel, L. 7, 124, 128–129, 379–380
Gerson, Levi Ben 309
gestures, number 14–19, 58–59 *see also*
    body counting systems
Gettysburg Address 70
al-Ghafur, 'Abd ar Rashid Ben 'Abd 1043

al-Ghazali, Abu Hamid 1037
Ghaznavid 111, 1033, 1034
Ghazni 1012
Gheerbrant, J. 862, 873, 1093, 1096
ghubar numerals 758, 1055–1065,
    1086–1087, 1098, 1104, 1144, 1157
Gibil 315
Gibraltar 1026, 1041
Gideon 504
Giles 525, 544
Gilgamesh 155
al-Gili, Abu'l Hasan Kushiyar ibn Labban
    714, 1012, 1035, 1054, 1085, 1107–1110
Gill, Wyatt 19, 23
Gille, B. 1022
Gille, L. 1024–1025
Gillespie, C. 1025
Gillings, Richard J. 343–344
Ginsburg, Y. 711, 1164
Girard 611
Glareanus 706
glyphs *see* Maya
glyptics 155, 162
Gmür 124, 128–129
gnomon 584
Gnosticism 505–508
Goar, Father 706
gobar numerals *see* ghubar numerals
gods: God (Judaeo-Christian) 506–507,
    1091; Mayan 587, 610–616; names and
    numbers 313, 314–315, 506–507 and
    spirits 528 *see also* Allah
Godziher, I. 97
Goldstein, B. R. 309
Golius 705
Gondisalvo, Domingo 712
Goths 61–66, 414, 442; Gothic script face
    1161
Gourmanches 74
Govindasvâmin 815, 824
Goyon, J. C. 344
Granada 1012
Grantha writing 754, 757
Greece, Ancient 355–365, 372–374,
    501–502; and Arabs 1010–1011, 1017,
    1022, 1042; astronomy 157, 305–307,
    803, 1084; currency 145–146, 357–358;
    Greek Myth 707–708, 719, 790; isopse-
    phy 494, 501–504, 708; science 1017,
    1020, 1022

Greeks, Ancient; abacus 392–396, 407;
  counting 74, 185–186, 243, 430–431,
  842, 843
  number-system 14, 61–65, 306, 307,
  641, 678–679, 684, 686–688;
  acrophonic 355–365, 394–397, 419;
  alphabetic numerals 372–374,
  427–436, 454–455, 466, 468, 645;
  Arabic numerals 700, 703, 705,
  706–709; fractions 1174; high
  numbers 653, 843–844
  writing 59, 317, 349–350; alphabet
  414–416, 429; papyri 306, 307
Green 178
Greenland 67, 598
Gregory V, Pope 1142
Griaule, M. 139
Grohmann, A. 476
gross 77, 177
gua-n zí writing 522, 525
Guarani 67
Guarducci 355
Guatemala 614, 624, 625 *see also* Maya
  civilisation
Guéraud, O. 431
Guide to the Writer's Art 1073, 1074
Guitel, G. 355, 522, 541, 699, 787, 794,
  843
Guitel, Geneviève 419, 674, 682, 804, 861
Guitel, R. L. 851
Gujarâtî numerals 726–727, 749, 756, 829,
  864
Gundermann, G. 355
Gupta dynasty 826
Gupta numerals 745, 750–751, 775, 776,
  781–784, 829, 906
Gupta writing 742, 755, 756, 826–827
Gurkhalî numerals *see* Nepali numerals
Gurûmukhî numerals 726, 749, 755, 756,
  829
Gwalior 748, 776, 780, 787–790, 824, 829

Haab, Mayan solar calendar 611–614, 618
Habuba Kabira 196, 200
Haddon, A. C. 7, 23
Hadiths 90
Hafiz of Chiraz 1042
Haggai 266
Haghia Triada 348, 351
Haguenauer, C. 535, 536–537, 539

Hajjaj, Abu 'Umar ibn 1037, 1044
al-Hajjami 1037
al-Hakam II, caliph 1033
Halevi, Yehuda 1014
Halhed, N. 95
al-Hallaj, Abu Mansur ibn Husayn 1033
Hambis, L. 49, 139
al-Hamdani 1032
Hamdullah 1070–1071
Hamid, al-Husayn Ben Muhammad Ben
  1032
Hamit, Avdülhak 1044
Hammurabi 156, 263, 277, 283; Code of 166
al Hanbali, Mawsili 105, 112
hand, counting with 47–61, 68 *see also* body
  counting; finger counting
hangu-l alphabet (Korea) 538
Hanoi 798
Harappâ 738, 758
al-Harb, *Urjuza fi hasab al 'uqud* 1071
Haridatta 763, 815, 824, 852
Harmand, J. 123, 127
al Harran, Sinan ibn al Fath min ahl 715
Harris Papyrus, The 333, 768
Harsdörffer, Georg Philip 700
Hartweg, R. 1170
haruspicy *see* mysticism
Hasan, Ali ibn Abi'l Rijal abu'l (Abenragel)
  714
al-Hasib, Hasbah 1029
al-Hassar 1112
Hassenfrantz 81
Hattusa 352, 353
ibn Hauqal 1031, 1062
Haüy 81
Havasupai 243
Hawaii 135, 243
Hawtrey, E. C. 25
al-Haytham, Abu Ali al Hasan ibn al Hasan
  ibn 714, 1014–1015, 1023, 1035
ibn Hayyan, Jabir 1028
ibn Hazem 1036
Hebrew number-system 74, 264–267, 283,
  419–427, 678–679; accounting
  461–466; alphabetic numerals 308,
  420–427, 443, 456–461, 466, 467–468,
  470, 645, 680; and Arabic numerals 706;
  Ben Ezra 680, 711; mysticism 468, 490,
  494–501, 1095
Hebrews: calendar 420, 424–425 language

139, 266, 267, 414–417, 420–427, 461
  see also Israel; Jews
Hejaz 1042
Helen of Troy 98
Heliastes, tablets of 419
Henan 526
heqat (Egyptian unit) 330–332
Heraclius 1025
Heraklion 347
herdsmen *see* shepherds
Hermite 1177
Hero of Alexandria 1012, 1022, 1031, 1172
Herodotus 134, 428
Herriot, E. 1069
Hierakonpolis 320–324
hierarchy relation 35, 43
hieratic script 332–339, 462–468
hieroglyphs: Cretan 347–351; Egyptian
  317–346; Hittite 351–354, 639, 640;
  Mayan 584–590, 610–616, 620–632
high numbers 583–584, 653, 843–845,
  1173; China and Japan 541–545; India
  829–846, 856, 867, 907–912; Roman
  385–392
Higounet, C. 147, 166, 427
Hilbert 1181
Hill 1146, 1159
Himalayas 767
al-Himsi 1030
Hindi language 747
Hindi numerals 723, 1009, 1050, 1051, 1058,
  1063–1065, 1107
Hinduism 740, 801, 826, 873; calendar 95
Hippocrates 1011
Hippolytus 505
Hiragana 534
Hisabal Jumal 494
Hittite number-system 61–64, 74, 351–354,
  639, 640, 684
Hiyya, Abraham bar 1014
Hoernan, Ather 1161
Hoffmann, J. E. 176, 1024
Höfner, M. 362
Homer 138–139
Honduras 586, 594, 614
Hôpital des Quinze-Vingts 71
Hoppe, E. 176, 177
Horace 406
horoscope 1085
Horus 331, 346

Houailou 67
Hrozny 353
Huang ji 546
Hübner, E. 366
Huet, P. D. 706
Hugues, T. P. 1096
Huitzilopchtli 590
Hunan 532
hundred 45–46, 350, 380; Chinese 514–515,
  519, 527; Greek 355, 356, 359, 360;
  Hebrew 421; hieroglyphic 322, 328,
  348, 354, 637–638; Japanese 536, 537;
  Mesopotamian 267–268, 270, 276,
  279–281, 363, 447–451; Roman
  numerals 368, 376
hundred and eight 578
hundred and seven 346
hundred thousand 272, 322, 328, 386–387,
  637–638
Hungary 1042
Hunger, H. 300, 311
Hunt, G. 7
al-Husayn, Abu Ja'far Muhammad ibn 1026
Husayn, Allah ud din 1035, 1038
Huygens, Christian 79
hybrid systems 647, 651–652, 656–658,
  678; Aramaean-Indian 759–760;
  classification 689–694; Tamil
  731–732
Hyde, Thomas 308
Hypsicles 1031

I Ching 135
Icelandic, Old 61–65
ideographic representation 151–155, 190,
  207, 264, 283, 317–321; Akkadian
  311–313; Chinese 519, 531, 534; Linear
  A script 348 *see also* hieratic script;
  hieroglyphs
al-Idrisi 1038
Ifriqiya 1028, 1031, 1033, 1037, 1042–1043;
  ghubari numerals 1055, 1058–1059
Ikhanian Tables 1041
Iliad 138–139, 419
Iltumish 1039
Imperial measurements 178
al-Imrani 1032
Inca civilisation 74, 130–133, 243, 603
incalculable, Indian 831–837
inch 178

India 699–866, 1010, 1011, 1027, 1033, 1038, 1039, 1041, 1042; astrology 821–822, 913–914; astronomy 805–809, 819–822, 849–852, 1012–1013; writing 414, 849–852

Indian number-system 652, 670, 679–681, 709–866, 1054; calculation 679, 857–862, 1122; chronograms 492; counting 74, 93, 94, 95, 96, 182, 1105; dictionary of numerical symbols 867–1007; fractions 836–837, 1174; high numbers 829–846, 856, 867; Indian numerals 722–754, 765–785; in Islamic world 1008–1138; place-value system 655–658, 694, 786–805, 819–829

Indju, Jamal ad din Husayn 1043

Indo-Aramaic 446

Indochina 94, 126–127, 801

Indo-Europeans 40, 53–60

Indonesia 723, 801

Indrâji, B. 763–764

Indus civilisation 74, 316, 758

infants *see* children

infinity 711, 825, 829–832, 839, 867, 927–930, 1179

Intaille 397

integers 1179; aspects of 38–39

International Standards system (IS) 82

Inuit 67, 84, 598

invoices 149, 210, 213–214

Iran 155, 263, 1030; accounting 187, 189, 192, 196–198; counting 182, 569 number-system 368, 534 *see also* DAFI

Iraq 99; accounting 196, 234–235; counting 93, 182 number-system 251, 368, 534 *see also* Sumer

Irish 61–65, 71, 72

irrational numbers 1042, 1176–1179

'Isa, 'Ali ibn 1036

Isabella I 1043

Isaiah 505

Isfahan 1012

ibn Ishaq, Hunayn 1013, 1031

Ishtar 315

Isidore of Seville 107, 1142

Isis 331, 506

Iskhi-Addu, King 142

Islam *see* Muslims

Islamic world *see under* Arab-Islamic civilisation

Ismail, Mulay 1043

Isma'il, Sultan 494

Isme-Dagan, King 142

isopsephy 494, 501–511

Israel 97–98, 212, 239 *see also* Hebrews; Jews

al-Istakhri 1034

Italic script 414, 417, 1157

Italy 97, 1030; number-system 58, 466; writing 422, 429, 1157

Itard, J. 91, 433, 434, 843

Itzcoatl 590

Ivan IV Vassilievich 186

Ivanoff, P. 584, 586

Ivanov, V. V. 758

Iyer 855

Jacob, François 1170

Jacob, Simon 404

Jacobites 469

Jacques 873

Jacquet 864

Jaggayyapeta 743

Jaguar Priests 589

al-Jahiz 716, 1029, 1068

Jainas 838–839, 868; *Lokavibhâga* 819–823, 825, 827, 847

Jalalabad writing 740

al-Jamali, Badr 1036

Jami'at tawarikh (Universal History) 1018

ibn Janah 1036

Janus (god) 89

Japan 598, 750; games 577–580; mysticism 1095; number-system 67, 534–554, 567–568, 763, 1070, 1072

Jarir 1027

al-Jauhari 1029

Jaunsarî numerals 749, 755

Java 800–801, 824, 828; Kawi writing 754, 757, 795; Sanskrit 813

Javanese numerals 738, 772–773, 777, 864

al-Jayyani, ibn Mu'adh Abu 'Abdallah 1022–1023, 1034

ibn Jazla 1037

al-Jazzari, Isma'il ibn al-Razzaz 1023, 1040, 1042

jealousy multiplication 1120–1129, 1138

Jefferson 79

Jelinek 119
Jemdet 196
Jemdet Nasr 155, 213
Jensen 428
Jerome, Saint 107
Jerusalem 233–456, 462, 1036, 1037, 1038, 1161
Jestin 172, 235
Jesus 503–505
Jews 261–262, 467–468, 502, 1010, 1013, 1061–1062; mysticism 490, 494–501 number-system 71, 157–158, 238 *see also* Hebrews; Israel
Jiangxian, Old Man of 548–549
Jinkoki 545
Jiu zhang suan shu 563
John, St 502, 509
John of Halifax 710
John of Seville 711
Jonglet, René 126
Jordan 446, 723, 1054
Jouguet, P. 431
Judaea 462, 467–468
Julia, D. L. 1170–1171
Julian calendar 95
Juljul, ibn 1034
Junayd 1033
Jundishâpûr 1011, 1013
Justinian, Emperor 1011
Justus of Ghent 91
Juvenal 106, 397

Kabul 1027
Kabyles 127
Kadman 456
Kai yuan zhan jing 804, 824
Kairouan 1012
kaishu- writing 522, 524, 525
Kaîthî numerals 727, 749, 755, 756, 829
Kalaman, King 478
Kâlidâsa 826
Kalila wa Dimna 633, 826, 1027, 1099
kalpa 933–934
al Kalwadzani, Abu Nasr Muhammad Ben Abdullah 715
Kamalâkara 815
Kamil, 'Abu 1014, 1019, 1032, 1035, 1047
Kamilarai people 6
Kampuchea 737
Kandahar 740

Kangshi, Emperor 673
Kanheri 790
kanji ideograms 534
Kanjô Otogi Zoshi 566
Kannara numerals 735, 753–754, 757, 864
Kapadia, H. R. 1110
al Karabisi, Ahmad ben 'Umar 715
al-Karaji 1009, 1014, 1019, 1035, 1039, 1047, 1083
Karlgren, B. 532
Karnata numerals 735
Karoshthî numerals 759–762
Karpinski, L. C. 406, 699, 710, 712, 715, 716, 749, 750, 785, 787, 1062, 1146
ibn Karram 1030
Karystos 527
al-Kashi 1019
al-Kashi, Ghiyat ad din Ghamshid ibn Mas'ud 1012, 1042, 1108, 1129
Kashmir 724, 728–729, 749, 827, 829, 863
Katakana 534
Katapayâdi numerals 763
al-Kathi 1036
Kawi writing 754, 757, 795, 829
Kaye, G. R. 703, 788–792, 801, 855
Kazem-Zadeh, H. 1073, 1077
kelvin 82
Kemal, Mustafa 1044
Kemal, Namik 1044
Keneshre 719, 1011
Kenriyû, Miyake 557
Kerameus, Father Theophanus 502–503
Kern, H. 709, 813, 824
ketsujo 1070, 1072
Kewitsch, G. 176–177
ibn Khaldun, 'Abd ar Rahman 511, 713, 718, 1015, 1024, 1036, 1042, 1093; *Prolegomena* 1044–1045, 1070, 1086, 1091, 1170
Khalid 1011
Khalifa, Hajji 1043
Khaliji, 'Ala ud din 1040
Khan, Genghis 751, 1038, 1039, 1099
Khan, Haluga 1039, 1040
Khaqani 111, 113, 1039, 1098
Kharezm Province 1013
Kharoshthî writing 740, 759–762
Khas Boloven 126–127
KhaSeKhem, King 323
Khatra 446

Khayyam, Omar 1012, 1015, 1017, 1019, 1037
al-Khazini, Abu Ja'far 1032
Khirbet el Kom 460
Khirbet Qumran 457
Khmer 754, 795–801, 828; number-system 67, 763, 793–796
Khmer numerals 737, 758, 829, 864
Khorsabad 275, 310
khoutsouri 441
Khoziba 509
Khudawadî numerals 726, 749, 755, 756
al-Khujani 1034
Khurasan 1033, 1035
ibn Khurdadbeh 1031, 1061–1062
al-Khuwarizmi, Abu Ja'far Muhammad Ben Musa 716–717, 1012, 1014, 1019, 1029, 1032, 1033, 1045, 1048–1049, 1053, 1065, 1083, 1107, 1110, 1161; algorithms 1048, 1159
Khuzistan 1025
kilogram 80, 81, 82
al-Kilwadhi 1031
al-Kindi 716, 1014, 1024, 1029, 1098
king, ideogram for 311–312
Kircher, A. 411, 442
al-Kirmani 1035
Kis 156, 196, 261
Kitab al arqam (Book of Figures) 714
Kitab fi tahqiq i ma li'l hind 714–715, 723–724, 840–845
Knossos 347–351
knot, meaning decimal system 1070
Köbel 401, 703
Kochî numerals 749, 755, 756
Kokhba, Simon Bar 456
Koran 1015, 1025, 1029, 1093–1094
Korea 538, 545–554
Kota Kapur 795
al-Koyunlu 1043
Kronecker 1171
Kshatrapa numerals 781–784
Kufa, founded 1025
Kufic script 476, 1064–1065
al-Kuhi, ibn Rustam 1033
Kulango 67
Kuluî numerals 749, 755, 756
Kululu 353
Kumi 372
Kurdistan 1042

Kushana numerals 781–784
Kutilâ numerals 749, 755, 756
Kyosuke, K. C. 598

Labat, R. 162, 167, 192
Lafaye, G. 98
Lagas 156, 180
Lagrange 80, 81
Lakhish 416, 462
Lalitavistara Sûtra 827, 829–837
Lalla 815
Lalou, M. 48
Lambert, Meyer 174, 266
Lampong writing 754
Landa, Diego de 588–589, 615
Landa numerals 749, 755, 756
Landsberger, D. 253
Langdon, S. 211, 225, 226, 758–759
Lao Tse 135
de Laon, Radulph 406
Laos 737, 754
Laplace, P. S. 80, 81, 709
Larfeld 355
Laroche, E. 352–353
Larsa 284
laser 82
Latin 100, 139, 185, 379; alphabet 414; number names 9, 58, 61–66
Laurembergus 705, 1164
Lavoisier 80
Law of 10 Frimaire, Year VIII 81
Laws of 18 Germinal, Year III 79, 81
ibn Layth, Abu'l Ghud Muhammad 1035
LCM (lowest common multiple) 179
Lebanon 723
Leclant 99
Lehmann-Haupt 175
Leibnitz 1086, 1180
Lemoine, J. G. 93, 95, 107–108, 1068, 1071
Lengua people 25
Lenoir 81
Leonard of Pisa *see* Fibonacci
Leonidas of Alexandria 501
Lepsius 144
letter numerals *see* Abjad numerals; alphabetic numerals
Lettres of Malherbe 98
Leupold, Jacob 110, 701
Levey 855
Levias, C. 706

Lévy-Bruhl, L. 6, 34, 86–88
Leydon Plate 623–625
Li, J. M. 538
Li Ye 553, 554
Liber de Computo 107
Liber etymologiarum 107
Libya 723, 1026
Lichtenberg 1071
Lidzbarski, M. 415, 768
Liebermann, S. J. 190, 192, 253, 272
ligatures 333–335, 447, 482, 770, 855
Light of Asia, The 830
Lîlâvatî 850
Limbu numerals 750
Lincoln, Abraham 70
L'Inde Classique 873
Lindemann 1177
Linear A script 348–351
Linear B script 348–351
lines, grouping of 854–855
Liouville 1177
lìshu- writing 521–522, 524
Lithuania 61–65
Lives of Famous Men 106
Lobachevsky 1181
Locke 131
Löfler 175
logarithms, natural 1178
logic 1180
Lokavibhâga 819–823, 825, 827, 847
Lombard, D. 518
Lombok 738
London, Royal Society of 79
Long Count 619, 625
Lot of Sodom 496
Louis IX France 71
Louis XIV 1043
Louvre 284
Lucania 369
Lull, Ramon 1086–1087
de Luna, Juan 712
lunar cycle 30–32, 95, 424; calendars 34,
    35, 582, 801; eclipse 1045; mansions
    1093; and numerology 180
ibn Luqa, Qusta 1013
Luther, Martin 510–511
Lutsu 124
Lycians 14, 74
Lydian civilisation 14
Lyon 275

al Ma'ali, Abu 1099
al-Ma'ari, Abu'l 'ala 1036
Macassar writing 754
Maccabeus, Simon 457
MacGregor, Sir William 23
Machtots, Mesrop 439
Madagascar 243–244, 723, 1054
Madura 738
Magadha 754
Maghreb 477, 494, 1013, 1027, 1028, 1036,
    1043; calculation 1098, 1106, 1112;
    numerals 699–700, 758, 1055–1065,
    1104; writing 1064–1067
al-Maghribi, As Samaw'al ibn Yahya
    105–106, 713, 1009, 1014, 1019, 1039,
    1054
Maghribi script 1064–1067
magic 485–513, 583, 592, 1085–1086
    talismans 262, 522, 554 *see also*
    mysticism
Magini 1175
Magnus, Albertus 1017
Mahâbhârata 826
Mahâjanî writing 749, 755, 756
al-Mahani 1030, 1032
Mahârâshtrî writing 748, 755, 756
Mahâvîrâchâryâ 785, 815, 824, 829, 1110
Mahmud 1033, 1034
Mahommed *see* Mohammed
Maidu 243
Maimon, Rabbi Moshe Ben *see*
    Maimonides
Maimonides 1039
Maithilî numerals 728, 749, 755, 756, 829
Majami 105
al Maklati, Muhammad Ben Ahmed 493
Maknez, chronograms 493
Malagasy 74
Malay, Old 754, 757, 795, 800
Malaya 1054
Malayâlam numerals 652, 656, 657, 672,
    693, 694, 733–734, 754, 757
Malaysia 74, 723, 799–801, 824, 828
Maldives 735
Malherbe, M. 67, 98, 534
Mali 139
al Malik, 'Abd 493
Malinke 68–69, 84
Mallia 348
Mallon 438

Malta 1030

al-Ma'mun, caliph 1011, 1028, 1045, 1048

Manaeans 14

Manchuria 532

Manchus 74

Mandealî numerals 749, 755

Manipurî numerals 749, 755, 829

Mann 70

al-Mansur, caliph 1011, 1027, 1045, 1046, 1047

al-Mansur, Sultan Abu Yusuf Ya'qub 1038, 1043, 1086

Mansur, Yahya ibn Abi 1013, 1029

many, concept of 6–7, 59, 181

mapping 16–19, 27–30, 38, 41

al-Maqrizi 1042

al-Maradini 1070

Marâthî numerals 725, 748, 755, 756, 829, 864

Marchesinus, J. 1161

al-Mardini, Massawayh 1013, 1036

Marduk 285–286, 310, 315

Mari 142, 156, 261, 277–284, 659–660, 692

Maronites 469–470

Marrakech 1036, 1037, 1040

al-Marrakushi, Abu'l Abbas Ahmad ibn al-Banna 713, 1041, 1122

Marre, A. 713, 1122

Martel, Charles 1010, 1027

Martial 397

Martinet, A. 758

Mârwarî numerals 749, 755, 756, 829

al-Marwarradhi 1029

Mashallah 706, 1027

Ma'shar, Abu 1029

Mashio, C. 598

Maspéro, H. 143, 522, 526

ibn Massawayh, Yuhanna 1024, 1030

Massignon, L. 1011, 1015, 1016, 1019, 1021, 1023, 1025

Masson, O. 349

al-Mas'udi 1014, 1032

Materialen zum Sumerischen Lexikon 253

Mathematical Treatise 1017

Mathematics in the Time of the Pharaohs 343–344

Mathews 525, 544

Mathura numera782-784 3755

Matlazinca 590

Matzusaki, Kiyoshi 567

Maudslay, Alfred 588

Le Maur, Carlos 701

al-Mawardi 1037

Maximus, Claudius 107

Maya civilisation 139, 581–632; astronomy 618–619, 629–632; calculation 594–598, 629–632; calendars 68, 589, 610–632; mysticism 610–616, 619–632, 3587; writing 584–590, 598, 612–616, 620–632

Mayan number-system 14, 68, 84, 316, 594–611, 678; positional 631–632, 661–667, 694, 696, 847; zero 628–632, 669–671, 847

Mazaheri, A. 714, 1025, 1053, 1065, 1068, 1070, 1085, 1099, 1108

Mead 1069

measurement 157, 176, 298, 308

Mebaragesi 156

Mecca 1025, 1044, 1061, 1095

Méchain 81

de Mecquenem, R. 211, 225–226

Media 1025, 1031

mediating objects *see* model collections

Medina 1025

Mediterranean 414, 435

Mehmed IV, Sultan 1044

Mehmet II, an 1042

Mei Wen Ding 548–549, 557

Mejing 1018–1019

Melanesian Languages 7

Melos 429

Mendoza, Don Antonio de 593

Menelaus 1012, 1031, 1033

Menna, Prince 117

Menninger, K. 371, 541, 554, 660, 673, 699, 843

al-Meqi, al-Amuni Saraf ad din 714, 1039

Merida, bishop of 588

meridian expedition 80–81

Merv 1012

Mesha, King 415

Mesopotamia 181, 316, 467; Arabs 1010, 1025, 1039, 1042, 1043; Babylonian era 261–315; counting; abacus 253–259, 1111–1112; bullae 192, 196; calculi 187–189; clay tablets 162–171 India 740, 1012; Mari 277–284; mysticism 180, 181, 1095;

number-system 157, 185, 187–209, 261–315, 637–646; Aramaic numerals 446; decimals 270–284; letter-numerals 475; zero 299–300, 669

writing 212, 539 *see also* Akkadian Empire; Elam; Semites; sexagesimal system; Sumerians

Mesopotamia, counting, bullae 196

Messiah (Jewish) 495–496

La Mesure de la Terre 79

metal, as currency 140–145

Metaphysics 35

Metonic cycle 382

metric system 1175; history 79–82

metrology 356

Mexico 68, 586, 587–588, 592–594, 598, 602

Mexico City 590, 598

Mî-s'on 800, 801, 814

Middle East: calculi 187–190; language 100, 414, 435, 485–490; Semites 261–262

Midrash 495

Mieli, A. 1025

Mikami 566

Miletus 429

Milik, J. T. 457

Millas 423

Miller, J. 79, 535, 537

milliard 843

million 272, 322, 329, 637–638, 841–844

Minaeans 362–364

Minoan civilisation 74, 347–351

Minos, King 347

minus (mathematical concept) 172

Mirkhond 1043

Miskawayh 1036

Misra, B. 1110

al Misri, Abu Kamil Shuja' ibn Aslam ibn Muhammad al Hasib 716

Misri, Dhu 'an Nun 1030

Mithat, Ahmet 1044

Mithras 508

Mitsuyoshi, Yoshida 545

Miwok 243

Mixtecs 68, 598, 602–603

mkhedrouli 441

mnemonics 852, 1060

Moab 414, 415

model collections 16, 19, 32, 41

modern numerals 636–637, 674–681, 699–722, 723, 758, 839–865, 1168–1182

Modî numerals 748, 755, 756

Mogul Empire 1043

Mohammed 90, 96–97, 111, 113, 1015, 1025, 1093

Mohenjo-daro 738, 758

Mohini 95

mole 82

Molière 71

Möller, G. 671, 768

Mommsen, T. 366

Môn writing 754

money 78, 138–146, 356–360, 603

Monge 80, 81

Mongolia 40, 49–50, 74, 93, 97, 1099

Mongolian Empire 751, 1038, 1039, 1040

Mongolian numerals 751, 757, 777, 779, 780

monks 135–136; Zen 579

Montaigne, Michel de 402, 403, 1139, 1166

Monteil, V. 1013, 1014, 1019, 1024, 1025, 1170

Montezuma 590, 593

Montucla, J. F. 708, 709

moon 424, 467, 809 *see also* lunar cycle

Moor 866

Morazé, Charles 678, 682

Moreh Nebukhim (Guide for the Lost) 1014

Morley, S. G. 584, 619, 627

Morocco 97, 493, 1096

Morra 97–100

Moses 498, 1093

Moss 144

Mota 34

Motecuhzoma I 301, 590, 593

Mouton, Abbé Gabriel 80

Moya, Juan Perez de 106

Mozarabes (Arabic Christians) 1013

al Mu'aliwi, 'Ali Ben Ahmad Abu'l Qasim al Mujitabi al Antaki 716

Mu'awiyah 1026

Mudara, Muhammedal 493

Muhammad, Abu Nasr 1032

Muhammad of Ghur 1038, 1039

al-Muhasibi 1030

al Mulk, Nizam 111

Multânî numerals 749, 755, 756, 829

multiplication methods: abacus 247, 399–404, 407–410, 558, 559–562,

573–574, 1100–1105, 1149–1156; calculi 237; fingers 114–117; tables 301–305, 431, 1108–1110, 1141; written 301–305, 340–344, 1121–1138

multiplicative principle 447, 451, 482, 514–515, 529, 647–653

al-Mu'min, 'Abd 1037

al-Muqaddasi 1034

Muqaddimah 511, 713, 1024, 1045, 1070, 1091, 1093, 1170

al-Muqafa', ibn 1027

al-Muqafa, Abu Shu'ayb 1028

Murabba'at 462

Murray Islanders 6–7, 23

Muslims: finger gestures 90, 96, 99–100, 111–113; Hisabal Jumal 490, 494; magic talismans 513, 1030, 1094; prayer 9, 50, 71 *see also* Arab-Islamic civilisation

al Mustadi 494

al-Mustawfi, Hamdallah 1041

al-Mu'tamin 1037

Mutanabbi 1031

ibn Mu'tasin, Ahmad 1022

Muwaffak, Abu Mansur 1034

Mycenae 349

Myres 350

myriad 47, 433–435

Mysticae numerorum significationes opus 389

mysticism, number: Arabs 1010, 1092–1097; China 528; fear of numbers 418, 539–540; India 849, 1073; Mayan 629–631; sacred symbols 180–181, 317, 468; soothsayers 511–513, 526, 1089, 1091–1097, 1098 *see also* codes and ciphers; magic

'n' roots 1042

Nabataean numerals 414, 443–446, 767

Nabî 1044

al-Nadim, Ya'qub ibn 715, 1034; *Fihrist* 715, 1048, 1064

al-Nafis ibn 1041

Nâgarî numerals 715, 724, 725, 755, 756, 787–788, 829, 864, 948, 1050, 1063

Nâgarî writing 715, 742, 747–749, 764, 827

al-Nahawandi, Ahmad 1029

Naima 1044

Nakshatra 822

names of numbers 24–26, 34, 41, 61–65, 265–267; games 310; gods' names used

184; Indian 948–950; Mayan 594–596; prayer words 418–419

Nânâ Ghât 743, 761, 763, 785, 828, 857–859

Napier, John 1178

Nârâyana 1110

Narmer, King 320–323

Nâsik 744, 761, 763, 782–784, 857

Naskhi script 1064, 1066–1067

Nasr 196

Nasr, Abu 1034

nastalik script 1064, 1066

Nathan, Ferdinand 570

Natural History 89, 387, 392, 842

Nau, F. 719

Naveh 453, 454, 457

al-Nawabakht 1027

al-Nayrizi (Anaritius) 1030

Nayshaburi, Hasan ibn Muhammad an 1129

al-Nazzam 1029

Nebuchadnezzar II 263, 462

Nedim 1044

Needham, Joseph 97, 517, 525, 527, 544, 547, 556, 575, 804

Nefi 1044

negative numbers 545, 554, 563, 1179

Negev, A. 141

Nemaea 361

Nepal 742, 755, 756, 763, 767, 827

Nepali numerals 730, 750, 755, 756, 772–780, 782–784, 864

Nergal 315

Nero 502, 510

Nesselmann, G. H. F. 426

Nestor, King 106

Nestorian sect 469

Neugebauer, O. 176, 177, 292, 298, 306, 815, 818

New Guinea 21–22, 23, 598

New Hebrides 67

New Mexico 384

Newberry 99

Newton, Isaac 80

Nichomachus of Gerasa 82–83

Nicobar Islands 737

Nicomacchus of Gerasa 1142

Niebuhr, Karsten 91–92

Nigeria 135

Nîlakanthaso-mayâjin 815

Nile 508

nine 65, 780; Chinese 527; Egyptian 346; Hebrew 421; Indian 808; Japanese 540, 541

nine hundred 423–424

nineteen 346

ninety 421, 459, 486, 578

ninety-nine 578

ninety-three 111

Nineveh 196, 200, 263, 284

Ninni, A. P. 385

Ninurta 315

Nippur 156, 253, 467

Nisawi, Abu'l Hasan 'Ali ibn Ahmad an 714, 1035, 1047, 1065, 1083, 1107, 1110, 1112

Nisibe 1011

Nissen 178

Nizami 1039, 1098

Nommo the Seventh 139

non-equivalence between sets 2–3

non-Euclidian geometry 1181

notched bones 11 *see also* tally sticks

Nottnagelus 706

Nougayrol, J. 285

Nougayrot, J. 163

nought *see* zero

Nubians 74

Numa, King 89

number-systems: alphabetic numerals 305–307, 414–513, 953–954; Arab-Islamic 1008–1138; Chinese 514–580; Cretan 347–351; Dictionary 867–1007; Egyptian 316, 317–346; Europe 1141–1149, 1158–1167; Greek 355–365, 427–436, 454–455; Hebrew 419–427, 456–461; historical classification 682–698; Hittite 351–354; Indian 722–866; Mayan 581–632; Mesopotamian 185, 187–209, 261–315; modern 636–637, 674–681, 699–722, 1168–1182 Roman 365–392 *see also* abacus; accounting; base numbers; body counting; calculation; decimal; mysticism; position, rule of; Sumerian; zero

numerology 93–94, 161, 360, 554 *see also* codes and ciphers; mysticism

Nur ad din 1038

Nusayir, Musa Ben 1026

Nusku 315

Nuwas, Abu 1028

Nuzi, Palace of 193–195

Oaxaca Valley 590, 598

Oaxahunticu *see* Maya calendars

obols 356–357, 394–397

Oceania 15, 19–23, 67, 84, 1095

Odyssey 419

Oedipi Aegyptiaci 442

Oghlan, Karaja 1044

Okinawa 135

Omayyad dynasty 1010, 1027

Omri, King 462

one 61, 380, 772; Aztec 598; Chinese 527; Greek 350, 355, 359, 360, 363; Hebrew 421; hieroglyphic 322, 328, 344, 348, 354, 637–638; Indian 806, 807; Maya 604; Roman numerals 367, 368, 375, 376; Sumerian 161, 288–289

one hundred and eight 136

one-for-one correspondence 16–19, 27–30, 34, 186, 374, 379

Opera mathematica 175

Ophel, accounting 462

Oppenheim, A. L. 194–195, 255

Ora 422

oral numeration 46–48, 519–520, 594

Orchomenos 358

order relation 35

ordinal numeration 35–39, 44, 356, 377–378

Ore 541, 843

Oriental Research Institute (Baghdad) 193

Origin of Species 1024

Orissî numerals 728

Oriyâ numerals 728, 749, 755, 756, 829, 864

Orontes 106

Oscan alphabet 414

Osiris 331, 508

Ostraca 416, 462, 465

Otman, Khalif 112

Ottoman Empire 1041–1043, 1044, 1073; secret writing 485–490; Siyaq numerals 1080–1082

oudjat 331–332

ounce 178

ownership, mark of 127

oxen 138

Özgüç 353

Pacific Islands 138, 243

Pacioli, Luca 110, 1121, 1138

pairing 7, 38

Pakistan 182, 760, 1026; numerals 723, 749, 1054; phalanx-counting 182

palaeography 770, 790, 796–800, 826, 1064, 1144

palaeo-Hebraic alphabet 414–416, 456, 462, 466

Palamedes 428

Palenque 582, 620, 621, 628

Palestine 134, 461, 467, 1025, 1042; numerals 446, 461–465, 470, 482

Pâlî writing 736, 742, 753–754, 757–758

palindromes, numerical 786

Pallava dynasty 743; numerals 781–784

Palmyra 414, 443–446, 485, 1053

Palmyrenean numerals 767

Pañchasiddhântikâ 815, 818–819, 865

Pañchatantra 633, 826, 1027

Pânini 764–765

Pâninîyam 765

Pantagruel 98

paper making 1018, 1028

Papias 406

Pappus of Alexandria 433–434, 1032

Papuans 21, 23

papyrus 1052–1053

Paraguay 25

Parameshvara 815

parchment, Maya 589–590

Pardes Rimonim 495

Paris, B. N. 710, 711

Paris Codex 590

Paris (of Athens) 98

parity, concept of 7–8

Parmentier, H. 814, 824, 825

Parrot, André 277

Pascal, Blaise 552, 1172, 1181

Pascal's triangle 552, 1009

Pasha, Ziya 1044

pebbles 19, 26, 185–187, 242–244; counting 244

Péguy, Charles 717

Peignot script face 1161

Peking 532

Peletarius 704

Pellat, C. 105

Peloponnese 145, 358

pendulums 79

Pepys, Samuel 1141

perception, limits of 8–15

Perdrizet, P. 502, 505, 508

Pergamon 501

Perny, P. 97, 525, 531

Persia 134, 508, 740, 1011, 1012; abacus 1098–1099, 1111–1112; number-system 74, 111–112, 490, 492, 1077–1082, 1092; writing 469, 485, 1064

Persian Gulf *see* Sumerians

Peru 133–134, 604, 1072

Peruvian Codex 604

Peten, Lake 586, 588

Peter, Simon 504

Peterson, F. A. 612, 615, 621, 622, 628

Petitot 87

Petra 446

Petruck 855

Petrus of Dacia 710

Phaestos 348

phalanx-counting 182–184

Pheidon, King 145

Philippines 754

Phillipe, André 125–126

Philo of Byzantium 1012, 1022

philology 826

philosopher's stone 1023–1024

Philosophica Fragmenta 398

Phoenicians 706; alphabet 414–417, 428–429, 467; number-system 14, 74, 267, 443–446, 690; writing 362, 454, 462, 767

phonograms 80, 136, 265 *see also* hieroglyphs

Phrygia 466

pi 1176, 1178

Piaget 3–5

Picard 1181

Picard, Abbé Jean 79

pictograms 149–155, 163–164, 187–191, 207–209, 599 *see also* hieroglyphs

Piéron, H. 718

Pihan, A. P. 525, 531, 699, 751, 1073, 1074, 1077, 1081

Pingree, D. 815, 818–819

Pinyin system 518

Pisa, Leonard of *see* Fibonacci

Pizarro 130

place-value system 635–636, 1105–1106, 1162; abacus 563–565, 855–862, 1108 discovery 563–565, 662–665, 786–801,

819–829 *see also* position, rule of; positional systems

Planudes, Maximus 710, 718, 1110–1111, 1164

plates, lead 353

Plato 1011, 1012, 1020, 1032, 1180

Plaut 535, 537

Plautus 379

Pliny the Elder 89, 387, 842

Plotinus 1012, 1032

plurality 60

Plutarch 89, 106

Pô Nagar 795, 798, 799, 828

Poincaré, H. 721

Polish 61–65

Polybius 392

Polynesia 7, 138

polynomials 554

Pompeii 501

Popilius Laenas, C. 369

Popol Vuh 589

Porter 144

Portugal 61, 65, 97

Posener, G. 99

position, rule of 43–44, 278, 282–303, 655–667, 678–680 *see also* place-value system

positional systems: Arabic 363–364; Babylonian 282–300; Chinese 545–554; historical classification 694–698; India 810–829; Mayan 604–611, 631–632

Pott, F. A. 70

Powell, M. A. 158, 235

powers: abacus 558; cubed 714; exponential 1042, 1172–1173; negative 305, 545; squared 633–636, 714; ten 545, 840–846, 867, 1172

Práh Kuhâ Lûhon 796–797

Prasat Roban Romas 814

prayer-beads 11, 96–97, 191

Pre-Sargonic era 155, 167, 172–173

Prescott, W. H. 133

priests, Mayan 610–611

primitive societies, barter in 138–146; counting 6, 15, 19–34, 87–88

Prinsep, J. 762–763

Prithiviraj 1038

Prolegomena 511, 713, 1024, 1044–1045, 1070, 1086, 1091, 1170

Prophet, the *see* Mohammed

Proto-Elamite number-system 639

Psammetichus, King 99

Psammites, The 653

Pseudo-Callisthenes 501

Ptolemy 1012, 1027, 1029, 1030, 1035, 1161, 1180

Ptolemy I 501

Ptolemy II 454

Ptolemy V 327

Pudentilla, Aemilia 106–107

Puebla region 590

Pulisha Siddhanta 842

Punjab 446

Punjabî numerals 725, 749, 755, 756, 829, 864

Putumanasomayâjin 815

puzzles, number 344–346

Pygmies 6, 138

Pylos 349

Pythagoras 502, 1016, 1176

Pythagoras' theorem 295, 1030

al-Qalasadi 1065

al-Qalasadi, Abu'l Hasan 'Ali ibn Muhammad 713, 1019, 1043, 1112, 1121–1122

Qasim, Abu'l 1034

al-Qasim, Muhammad Ben 1044

al-Qass, Nazif ibn Yumn 1032

al-Qays, Imru' 1026

al-Qifti, Abu'l Hasan 1040, 1045, 1046, 1047

quadrillion 842–843

Quahuacan 68

Quauhnahuac 68

Qubbat al Bukhari 493

Qudama 1032

Quetzalcoatl 587

quinary systems 14, 84–88, 182–184

Quintana Roo 586, 594

Quintilian 89

quintillion 843

quipucamayoc 132–133, 604

quipus 124, 131–133, 604, 1070, 1072 *see also* string, knotted

Quiriguà 583, 620, 621, 626–627, 630

ibn Qurra, Thabit 1014

Qutan Xida *see* Buddha

Qutayba 1013

Qutb ud din 1039

Rabban, 'Ali 1030
Rabelais 98
Rachet, Guy 155, 262
Raimundo of Toledo 712
Râjasthani numerals 749, 755, 756
Ramus 704
Ramz 490
Rangacarya, M. 815, 825
rank-ordering 27
Ras Shamra 266, 418
Rashed, R. 713, 1009, 1025
Rashi 496
Rashid ad din 1014, 1018–1019, 1041
al-Rashid, Harun 1011, 1027
rational numbers 1177–1178
Razhes 1031
al-Razi, Fakhar ad din 713, 1012, 1039
al-Razi, Muhammad Abu Bakr Ben
    Zakariyya (Razhes) 1031
ready-made mappings 19, 30, 34
real numbers 1179
Rebecca, wife of Isaac 503
rebus 592–593, 600–602
receipts 131, 134
Recorde, Robert 704, 1165
recurrence 35
Red Sea 93
Redjang writing 754
Reinach 355
Reinaud 716
Reisch, Gregorius 1167
Relación de las Cosas de Yucatán 588
Renaissance 1044
Renou, L. 658, 849, 864, 873
Rey 1012
Rhangabes 394
Rhind Mathematical Papyrus (RMD) 334
Richard Lionheart 1158
Richer 80
Ridwan, 'Ali ibn 1036
Ridwan of Damascus 1023, 1040
Riegl 383
Rif 494
right-angled triangles 295, 1030
Rijal, ibn Abi'l 1035
Rivero, Diego 90
de Rivero, M. E. 133
RMD (Rhind Mathematical Papyrus) 334
Robert of Chester 712
Robin, C. 364, 365

Rodinson, M. 362
Röllig 447
rômaji 534
Roman Empire 9, 98, 178, 1028, 1139, 1141;
    abacus 243, 397–406, 410–413, 1141,
    1144–1145, 1149; calculation 74, 134,
    185, 654–655, 842; currency 106, 146
Roman numerals 14, 365–392, 641–643,
    686; used in Europe 407, 1142–1143
Romance languages 58–60
Romanian 61–65
Rong Gen 527
roots, square and cube 305, 558, 575, 825,
    1107, 1177–1178
rosaries 135–136
Rosenfeld, B. A. 1129
Rostand, J. 1171
Rudaki 1033
Ruelle, C. E. 435
Rumelia 1042
Rumi, Ya'qub ibn 'Abdallah ar 1040
ibn Rushd (Averroes) 1015, 1017, 1039
Russia 61–65, 127–128, 139, 414, 569
ibn Rusta of Isfahen 1032
Rutten, M. 180
Ryu-Kyu islands 135, 1070

ibn Sa'ad 112, 136, 1031, 1070
Sa'adi of Chiraz 1041, 1062
Saanen 381
Sabaeans 14, 1010
Saccheri 1041
Sachau, E. 444, 459
sacred symbols 93–94, 162, 239 *see also*
    mysticism
de Sacrobosco, Jean 704, 710
Sa'ddiyat 658
Saffar, ibn al 1035
Saffarid dynasty 1028
Saglio, E. 433, 843
ibn Sahda 1030
Sahdad 196
ibn Sahl, Sabur 1031
Saidan, A. 716, 1112
Sakhalin, Ainu of 598
Saladin 1038
Salamis, Table of 393–397
Samanid dynasty 1028, 1030
Samaria 416, 462
Samaritans 414, 456

Samarkand 1026, 1042
al-Samh, ibn 1035
Sanayî, Abu'l Majîd 111
sangi 545–554
Sankheda 791
Sankhyâyana Shrauta Sûtra 832
Sanskrit 139, 853; high numbers 841,
843–846, 855–856; number names
53–54, 59, 60, 61–65, 795–800,
809–828, 1046; oral counting 839–849;
Pânini 765 Shiddhamatrikâ 750 see also
Brâhmî; Nâgarî writing
ibn Sarafyun, Yahya 1031
Sarapis 501
Sarasvati (goddess) 866
Sardinia 1026
Sargon I The Elder 262
Sargon II 270, 275, 310
Sari 454
Sarma, K. V. 815, 817, 825
Sarton, G. 1205
Sarvanandin 820–823
Sastri, B. D. 815
Sastri, K. S. 815
Satan 1095, 1162–1163
Satires 406
Satraps 743, 802
Saudi Arabia 723, 1054
Saul 141
Saxon, Old 61–65
Scandinavia 125, 383
Scheil, J. 197, 211, 224–225
Scheil, V. 223
Schickard 1172
Schmandt-Besserat, Denise 187–191, 194
Schnippel, E. 382, 384
Scholem, Gershon 424, 501
Schopenhauer 35
Schott 99
Schrimpf, R. 545
science, classification 1020, 1032, 1036;
Koran 1015
scientific notation 1173
Scots Gaelic 61–62
Scythians 741
seals, cylinder 199–201, 206–207
Sebokt, Severus 719–720, 801, 825
secret writing 248–250 see also mysticism
Sefer ha mispar (Number Book) 711
Semites 156, 261–264; alphabet 414–416,

742 number-system 40, 264–284,
443–454, 690 see also Arab-Islamic;
Assyrian; Babylonian; Hebrews;
Phoenician cultures; see also particularly
Akkadian
Seneca 89, 391
Senegal 598
Sennacherib 285
separation sign 290–291
septillion 843
Serere 67
Sessa, legend of 633–636
sestertius 411
Seth 331
sets, theory of 1181
seven 64, 778, 871; Chinese 527; Egyptian
345; Hebrew 421; Indian 808; Japanese
541
seven hundred 423–424
seventeen 346
seventy 421
seventy-seven 577
de Sévigné 403
Seville 1014, 1040
sexagesimal system 157–161, 175–184, 245,
270, 307; Akkadian 260–261, 268–269,
467; astronomy 176–177, 184, 273,
307–308, 1083–1085; Babylonian
260–315; calculation 245–259, 273,
1042; proto-Elamite 232
sextillion 843
Sezhong, King 538
Shah Nameh 111
Shahadah, prayer of 90
ibn Shahriyyar, Buzurg 1034
Shaka calendar 801, 974
Shamash 315
al Shamishi system 486
Shan 74
shàng deng number-system 542–545
sháng fa-ng dà zhuàn writing 524
Shankarâchârya 823
Shankaranârâyana 763, 815, 824, 852
al-Shanshuri 1058
Shâradâ numerals 729–730, 749–750, 755,
756, 829, 863, 864, 975
Shâradâ writing 729–730, 742, 827
Shaturanja (early chess) 633–636
Sheba 362–365, 641
shekel 141

shells 44–45, 70

Shem, son of Noah 261, 498

Shen Nong 135

shepherds: and base 10 44–45; bullae 195, 200; counting methods 90, 374–379, 418–419; pebbles 19; quipus 133; tally sticks 18, 124

Sher of Bihar, King 492

Shiite Islam 1028

Shiraz 476

Shiva 875

Shivaism 801

Shojutsu Sangaka Zue 557

Shook 626

Shrîdhârâchârya 815, 1110

Shrîpati 815, 1110

Shukla, K. S. 815, 817, 825

shûnya (zero) 811, 977

Shuri 135

Shushtari 1040

Siamese numerals 737, 763, 794

Siamese writing 754

Siberia 139

Sibti, Abu'l 'Abbas as 1086, 1089

Sicily 372, 429, 1028, 1160

Siddham numerals 755, 756

Siddham writing 742, 750, 827

Siddhamatrikâ writing 750

siddhânta *see* India, astronomy

Siddim, Valley of 496

sign language 100–114

al-Sijzi, 'Abd Jalil 1034, 1054

Silberberg, M. 680, 711, 1164

silent numbers 418–419

Sillamy, N. 3, 717

Simiand, F. 720

Simonides of Ceos 428

Simplicius 1011

Sin 315

ibn Sina, Al Husayn *see* Avicenna

Sinan, Ibrahim ibn 476

Sindhî numerals 726, 749, 755, 756, 829, 864

Singapore 532

Singer, C. 1022, 1025

Singh, A. N. 699, 715, 716, 785, 787, 815, 825, 832, 855, 864, 1110, 1122, 1132

Singhalese numerals 672, 691, 735–736, 754, 757, 763

singularity 59

Sino-Annamite writing 532–534

Sino-Japanese numerals 534–541, 545

Sino-Korean number-system 538

Sircar 864

Sirmaurî numerals 749, 755, 756

six 63, 315, 777; Chinese 527; Egyptian 345; Hebrew 421; Indian 808

six hundred 161, 423–424

six hundred and sixty-six 509–511

sixteen 427

sixty 176, 179, 180; base 75, 157; Hebrew 421; Mesopotamian 161, 274–277, 288–289, 314

Siyaq numerals 485, 1077–1082

Skaist 460

Skandravarman, King 746

Skarpa, F. 380–381

Slane, I. 1070, 1091, 1093

Slavonic Church 61–64

Smirnoff, W. D. 509

Smith, D. E. 390, 406, 557, 699, 710, 711, 715, 716, 749, 750, 785, 787, 1062, 1146, 1164

Snellius 1075

sol, French unit 178

solar cycle 94, 95; calendar; Maya 582; eclipse 1045

de Solla Price, D. 1022

Solomon Islanders 34

Solomon's ring, legend of 701

Solon 392, 403

Sommerfelt, A. 6

soothsayers 261–262, 269, 551–556

see also mysticism

soroban 565, 566, 576

Soubeyran, D. 281–282, 283, 660

Sourdel 1067

Soustelle, Jacques 68, 139–140, 467–468

South America, counting 6, 15, 67, 243; Inca civilisation 130–133, 603

South Borneo 32–34

Spain 97, 490, 1036, 1037, 1092; and Arabs 486, 1013, 1027, 1028, 1043, 1160; Central America 588–593, 604; number-system 58, 61–66, 705, 1055, 1061–1062, 1065, 1156; Spanish Inquisition 1163; writing 423, 1065, 1157

Spanish Inquisition 1163

spatio-temporal disabilities 5

spheres 1030

Spinoza 390

spirits, malign 539–540

square alphabet 414, 420, 456

square roots 305, 558, 575, 825, 1107, 1177–1178

squares (power of two) 633–636, 714

Sri Lanka 6, 731, 735

Stars and Stripes 568

Steinschneider, M. 680, 711

Stele of the Vultures 166

Stephen, E. 34

Stephens, John Lloyd 588, 589

sterling currency 78

Stevin, Simon 1175

Stewart, C. 1073, 1077

sticks as counting devices 15–16, 125
  see also tally sticks

stone, as medium 317

string, knotted 124, 130–137

Su Yuan Yu Zhian 551

Suan Fa Tong Zong 117, 556, 575

suan pan 288-294 *see* abacus, Chinese

suan zí notation 545–554, 564–565, 803

Subandhu 824

subha (prayer) 96

subtraction 340; abacus 247, 399, 558–559, 573

subtractive principle 171–172, 643

succession 38–39

Sudan 187, 189, 1065

Suetonius 502

Sufi 1028, 1086, 1089

al-Sufi, Abu Musa Ja'far 715, 1028

al-Sufin, 'Abd ar-Rahman 1034

Sulawezi 754

Suli, As 1030, 1068, 1070

sulus script 1064

Sumatar Harabesi 453

Sumatra 124, 754, 795

Sumer 196–198, 262–263

Sumerians 147–176; bullae 192, 200–202, 210–214; calculation 157–160, 182, 234–259, 273–277; number-system 14, 147–184, 192–193, 210–233, 236, 270–271, 276–277, 287–289, 638–639, 686–687; Sumerian-Akkadian synthesis 267–269, 277, 288 writing 77–81, 86–90, 107 *see also* Mesopotamia

Sumerisches Lexikon 234

Sunda 738

Sunni Islam 1028

superstition *see* mysticism

Suruppak 168, 169–170, 234, 237

Sûrya Siddhânta 809

Susa 196–208, 216–223, 229, 230–233, 273, 290, 302, 308–313

Susinak-sar-Ilâni, King 311

Suter, H. 713–714, 1025, 1112, 1122

Swedish 61–65

Switzerland 58, 125, 128–129, 381, 402

Sylvester II *see* d'Aurillac, Gerbert

symbolism 149–150, 984–988

synonyms 807–829, 848–852, 863

Syria 99, 283, 1038, 1039; Arabs 1010, 1025, 1030, 1031, 1033, 1035, 1038, 1039, 1040, 1043; Hittites 352; India 740, 1012 writing 414, 415, 453, 486 *see also* Ugaritic people

Syrian number-system 444, 446, 482; alphabetic numerals 465–466, 645; calculation 93, 182, 190, 196, 1069–1070, 1112; Indian numerals 718–723, 1054; Mari 277–284

Sznycer, M. 416

al-Tabari, Marshallah 1014

at-Tabari, Sahl 1014, 1029

at Tabari, 'Ali Rabban 1024

Tabasco 586, 594

tables: astronomical 284, 306–311, 388, 1029, 1041; mathematical 246–252, 284, 398–400, 403–404, 555–565, 1097–1112; multiplication 301–305, 431, 1108–1110, 1141

tablet: accounting 149, 152–153, 195–236, 261; Babylonian 261, 268–269, 272, 287–288, 311–312; calculation 237–242, 287–288, 295, 1111–1112; clay 147–154, 162–171, 178, 179, 190–242, 256–259, 260–262; Cretan 347–350; Ebla tablets 262, 283; Heliastes 419; Hittite 353; proto-Elamite 197, 204; Sumerian 147–149, 190, 195–197, 207–208, 237–242, 260, 272, 1111–1112; Tablet of Fate 285; wooden (abacus) 256–259

Tabriz, Ghazan Khan a 1018

Tadjikistan 1030

Tadmor 486

al Tadmuri system 486
Tagala writing 754
Tagalog numerals 758
ibn Tahir, Mutahar 715
Tahirid dynasty 1028
Tâkarî numerals 728–729, 749, 755, 756, 829
talent (money) 356, 357, 392, 394–397
talismans 513, 1030, 1094
Tall-i-Malyan 196
Talleyrand 80
tally sticks 18, 19, 27–32, 119–129, 374–385
Talmud 495
Tamanas 67, 84, 598
Tamil numerals 652, 656, 672, 693, 731–733, 735–736, 754, 757
ibn Tamin, Abu Sahl 716
Tammam, Abu 1030
Tangier 493–494
Tankrî numerals 728–729
Tao Te Ching 135
Taoism 873
al Tarabulusi, Ahmad al Barbir 112–113
Tarasques 590
Tarikh 490
Tarikh ul Hind 492
ibn Tariq, Ya'qub 1013, 1027, 1046
Tartaglia, N. 704
Tashfin, Yusuf Ben 1036
Tashköprüzada 1043
Taton, R. 1017, 1018
Tavernier, J. B. 95
Tawhidi 1034
tax collection 124–125, 131, 134, 592, 593, 599
Tayasal 588
Taybugha 1042
Taylor, C. 760–762
Taylor, J. 1132
ibn Taymiyya 1041
Tchen Yon-Sun 541
Tebrizi 1041
Tel-Hariri excavation 277
Telinga numerals 734
Tell Qudeirat 462, 465
Tello 196
Telugu numerals 734, 753, 757, 829, 864
ten 65; Chinese 514–515, 527; decimal system 44–60, 73–83; Greek 355, 359,

360, 363; Hebrew 421; hieroglyphic 322, 328, 346, 348, 350, 354, 637–638; mysticism 82–83, 315; powers of 840–846, 867, 1172; Roman numerals 368, 375; sexagesimal system 157–161, 180–184; tally sticks 380
ten thousand: Aramaic 450; Babylonian 272, 282–283; Chinese 514–515, 519; Greek 355, 359, 360, 432–435; Hebrew 266; hieroglyphic 322, 328, 350, 351, 637–638; Japanese 537, 538; Roman numerals 386–387
Ten Years in Sarawak 34
Tenochtitlán 590, 592, 593, 599, 600
Tepe Yahya 196, 198
ternary principle 171, 270, 324, 444
Tertullian 89
Tetrabiblos 1027
Tetuan 493
Texcoco, Lake 590
ibn Thabit, Ibrahim ibn Sinan 1032
Thai numerals 737, 754, 758, 772–773, 864
Thebes 99, 428
Théodoret 106
Theodosius 1031
Theon of Alexandria 175
Theophanes 706, 1165
Theophilus of Edesse 1013, 1028
Thera 429
Thespiae 358
Thibaut, G. 818
Thibaut of Langres 504
thirty 180, 315, 421
thirty-six thousand 161
Thompson, J. E. 611, 619, 621, 628, 630
Thot 331–332, 345
thousand 46; Aramaic 450; Chinese 328, 514–515, 519, 527; Greek 355, 359, 360, 363; hieroglyphic 322, 328, 348, 350, 354, 637–638; Japanese 536, 537; Mesopotamian 267–268, 276, 282–283, 452; Roman numerals 368, 369, 376, 385–387
Thousand and One Nights 1028
three 62, 774; base 75; Chinese 527; Egyptian 345; Hebrew 421; Indian 807; many as 4, 60, 181; ternary principle 14, 171
three hundred 421, 504
three hundred and sixty five 504

three thousand, six hundred 288–289
three thousand, six hundred 161, 180, 275, 276
Thureau-Dangin, F. 158, 176, 177, 271, 296, 310, 802
Thutmosis 325
Tiberius, Emperor 391, 504
Tibet: counting 74, 136; number-system 48–49, 730, 733, 763, 832; writing 742, 751, 827
Tibetan numerals 730, 757, 772–774, 777, 779
Tijdschrift 800
Tikal 582, 624, 625–626
time 30–34, 52, 94–95, 130–131, 583–584, 610; and base 78, 157, 308
Timur 1042
Tiriqan, King 156
Tirmidhi 1031
al Tirmidhî, Abù Dawud 97
Tizapan 590
Tlatelolco 592
Tod, N. M. 355, 358–359, 361–362, 455
tokens *see* abacus; calculi; currency; tally sticks
Tokharian language 60, 61–65
Tokyo 536, 540, 568
Toledo 491, 1014, 1036, 1160
Toltecs 587
Toluca 68
Toomer, G. J. 1025, 1048
topology 1181
Torah 215, 218, 239, 253–254, 256
    see also Bible
Torkhede 790
Torres Straits 7, 8, 19–23, 23
Trajan's column 1161
transcendental numbers 1176–1179
triangles, spherical 1040–1041
trigonometry 827, 1032, 1038, 1040–1041
trillion 842–843
Tripoli 1042
Truffaut, François 3
Tschudi, J. D. 133
ibn Tufayl (Abubacer) 1039
ibn Tughaj, Muhammad 1031
Tughluq, Firaz Shah 1041
Tula 590
Tulu numerals 754, 757

Tumert, ibn 1037
Tunisia 1026, 1028, 1033, 1096
ibn Turk, Abu al-Hamid ibn Wasi 1029
Turkestan, Chinese; writing 751, 757, 827
Turkey 1010, 1044; mysticism 485–492, 1092; Russian abacus 569 writing 180, 248–250 *see also* Ottoman Empire
Turkish, Ancient 50–53
at Tusi, Nasir ad din 1012, 1040–1041, 1111, 1112, 1129–1132
twelve, base *see* duodecimal system
twenty 84; base *see* vigesimal system; Egyptian 346; Japanese 536, 537; mysticism 180, 315, 486; Semitic 421, 446–447
twenty six 497
two 61, 773; base *see* binary system; Chinese 527; Egyptian 345; Hebrew 421; Indian 806, 807
two hundred 421
Tyal tribe 140
Tylor, E. B. 6
Tyrol 381–382
Tzolkin, Mayan calendar 611, 618

Uaxactún 628
Uayeb 615
Ugaritic people 74, 266, 283, 418, 478
Ulrichen 381
al umam, Tabaqat 1017
'Umar, caliph 1016, 1025
al-'Umari 1041
Umbrian alphabet 414
Umna 156
unciae (Roman ounce) 412
United Kingdom, Chancellor of the Exchequer 1166
United Kingdom *see also* England; Scots Gaelic; Welsh
United States 177, 844
Universal History (Jami'at tawarikh) 1018
Untash Gal 198
al-Uqlidisi, Abu'l Hasan Ahmad ibn Ibrahim 716, 1033, 1112
Ur 156, 168, 174, 262–263
Urartu 14, 74, 270
Urmia, Lake 469
Uruk 156, 165, 205, 212; clay tablets 147–149, 190, 196, 293, 296; number-system 179, 196, 310

'Uthman, Abu 1032
'Uthman, caliph 1025, 1026
Utu-Hegal, King 156
Uyghurs 52
Vâjasaneyi Sâmhita 837–838
Vâkyapañchâdhyâyi 815
Valabhi numerals 782–784
Vallat, F. 210
value, concept of 138–146
Vandel, A. 722
Varâhamihîra 815, 818–819, 825, 865, 995, 1046
Varnasankhyâ numerals 763
Varnasankhyâ's *see also* mysticism
Vatteluttu numerals 754, 757
Veda, Mannen 577
Vedas 53, 838
Vedda people 6, 138
Venezuela 67, 598
Ventris, Michael 350
Venus 581, 609, 618–619
Vera Cruz 592
Vercors 1169
Vercoutter, Jacques 316
verse 849–852, 860–861
Vervaeck, L. 717
Vida, Levi della 112, 706
Vieta, Franciscus 1179
Vietnam 532–534, 754, 801 *see also* Champa
vigesimal system 68–73, 84, 594–619; Aztec 599–603; Mayan 594–595, 599–609, 614, 620; Sumer 158
Vigila 712, 1144
Vikrama 996
de Ville-Dieu, Alexandre 710
Vishmvâmitra 830
Vishnu 95
Visperterminen 381–382
Vissière, A. 546
Vitruvius 379
Vocabularium 406
Vogel, K. 717, 1025, 1048, 1053
Voizot, P. 700
von Wartburg, W. 842
Vossius, I. 705
vulgar fractions 1175
Vyâgramukha (Fighar) 1045, 1046

Waeschke, H. 855, 1111
Wafa, Abu'l 1014
al-Wafid, ibn 1036
al Wahab Adaraq, Abd 493
ibn Wahshiya 1033
Walapai 243
Wallis, John 175, 709, 712, 1179
Wang Shuhe 1018–1019
Waqqas, Muhammedal 493
Warka 200
Warka, Lady of 156
wax calculating board 405–410, 1112
Weber 837–838
wedge 288–291, 313
Weidler, J. F. 706
weights and measures 357, 358, 467; International Bureau of 81–82
Welsh 61–65, 71, 72
Wessely, J. E. 508
West Bank 187, 189
Weyl-Kailey, L. 5
Whitney 809
Wiedler 700
Wieger 527
Wilkinson 99
William of Malmesbury 712, 1158
Willichius 710, 1165
wind, evokes numbers 871
Winkler, A. 1085
Winter, H. J. J. 1025
de Wit, C. 344
Woepcke, F. 473, 699, 712, 713, 714, 723, 806, 830, 833, 843, 845, 853, 864, 1045, 1047, 1055, 1061, 1084, 1111, 1112
Wolof 67
Woods, Thomas Nathan 567
Wright, W. 471, 472
writing 59, 155, 208–209, 532–534, 538, 740; styles 171; 186, 539 *see also under specific race/country; mysticism*
writing materials 163–164, 767–770, 847, 855; chalk 1120; papyrus 589, 1052–1053; reeds 163, 167, 1064, 1092
ibn Wuhaib, Abu 'Abdallah Malik 1087, 1088, 1089
Wulfila 442

xià deng number-system 542–545
Xiao dun 526–529
xíngshu- writing 522–523, 525

Yaeyama 135
Yahweh 137, 415, 427, 496, 498
Yahya, Abu 1027
Yamamoto, Masahiro 545, 577
al Yaman, Hudaifa ibn 112
al-Yamani, Yahya ibn Nawfal 111, 1026
Yang Hui 559
Yang Sun 555
al-Ya'qubi 1031
Yaxchilán 620
year, days of 176
Yebu 68–69, 135
Yedo 598
Yehimilk 416
Yemen 723, 1042, 1054 *see also* Sheba
Yishakhi, Rabbenu Shelomoh 496
Yong-le da dian 517
Yoruba 68–69, 598
Youschkevitch, A. P. 475, 717, 1011, 1013,
    1016, 1019, 1025, 1046, 1053, 1083, 1107,
    1111, 1129
Yoyotte, J. 99
Yucatan 586, 588–589, 594, 612
yuga (cosmic cycle) 809–810, 827,
    999–1001
Yum Kax 611
ibn Yunus, Matta 1013, 1032, 1035
ibn Yusuf, Ahmad 1030
ibn Yusuf, al-Hajjaj 1030

Zahrawi, Abu'l Qasim az (Abulcassis) 1030
za'irja 1086–1091
Zapotec 67, 316, 598, 602–603
al-Zarqali 1037
Zaslavsky, C. 69, 84, 598
Zayd, Abu 1032
Zen 579, 873
Zencirli, Aramaic numerals 447
zero 45, 635, 668–680, 695–697, 702, 717,
    819, 1001–1007, 1160; abacus 720,
    856–862, 1104; absence 282, 291–294,
    673, 720, 731, 733, 736, 1104;
    Babylonian 297–300; Chinese 520,
    549–550, 803–804; Europe 1163–1164;
    Greek 306; imperfect 670–671; India
    729, 786, 808, 811, 817–819, 828,
    853–854, 862–865; Islamic world 1053,
    1054; Mayan 604–610, 628–632
zhong deng system 542–545
Zhu Shi Jie 551
Zimri-Lim 277
ibn Ziyad, Tariq 1026
zodiac 178, 1085, 1086–1089, 1093
ibn Zuhr (Avenzoar) 1038
Zulus 6
Zumpango 590
Zuñi 25, 384